SMOOTH MANIFOLDS AND FIBRE BUNDLES WITH APPLICATIONS TO THEORETICAL PHYSICS

SMOOTH MANIFOLDS AND FIBRE BUNDLES WITH APPLICATIONS TO THEORETICAL PHYSICS

Steinar Johannesen

Oslo and Akershus University College of Applied Sciences
Norway

CRC Press
Taylor & Francis Group
Boca Raton London New York

CRC Press is an imprint of the
Taylor & Francis Group, an **informa** business
A CHAPMAN & HALL BOOK

CRC Press
Taylor & Francis Group
6000 Broken Sound Parkway NW, Suite 300
Boca Raton, FL 33487-2742

First issued in paperback 2020

ISBN-13: 978-1-4987-9671-2 (hbk)
ISBN-13: 978-0-367-65825-0 (pbk)

Library of Congress Cataloging-in-Publication Data

Names: Johannesen, Steinar.
Title: Smooth manifolds and fibre bundles with applications to theoretical physics / Steinar Johannesen.
Description: Boca Raton : CRC Press, [2016] | Includes bibliographical references and index.
Identifiers: LCCN 2016030311| ISBN 9781498796712 (hardback) | ISBN 9781498796729 (e-book)
Subjects: LCSH: Manifolds (Mathematics) | Fiber bundles (Mathematics) | Differential equations. | Lie groups.
Classification: LCC QA613.2 .J64 2016 | DDC 514/.72--dc23
LC record available at https://lccn.loc.gov/2016030311

Visit the Taylor & Francis Web site at
http://www.taylorandfrancis.com

and the CRC Press Web site at
http://www.crcpress.com

Contents

Preface

The intention of this book is to provide the necessary mathematical background from the theory of smooth manifolds, fibre bundles, Lie groups and differential geometry for students interested in classical mechanics and the general theory of relativity, and for students interested in the wide range of applications in diverse areas of mathematics. Although these mathematical subjects become more and more important in the recent developments of theoretical physics, it is hard to find books in physics with a self-contained presentation of the mathematical foundation needed in the physical theory. The purpose of this book is to fill this gap. This book can as well be used as a textbook in a pure mathematics course in differential geometry, giving a self-contained and systematic approach based on principal fibre bundles and jet bundles, and also containing the necessary background theory of smooth manifolds and Lie groups. It is assumed that the reader has a good knowledge of basic analysis, linear algebra and point set topology (see [13], [14], [20] and [22]).

The first chapter is an introduction to differential geometry for curves and surfaces in the 3-dimensional Euclidean space $\mathbf{R}^3$, in order to obtain a good intuition of important concepts in differential geometry such as curvature and torsion. A leading goal in the next chapters will be to generalize these concepts to spaces of higher dimension, with 4-dimensional spacetimes used in the theory of relativity as important examples.

In Chapters 2 to 6 we start by presenting the fundamental theory of smooth manifolds, including the tangent bundle, global flows for vector fields, tensor bundles, differential forms, Frobenius' theorem for distributions, integration on manifolds and Stokes' theorem.

Chapter 7 is devoted to metrical structures in vector bundles and pseudo-Riemannian manifolds. Symplectic manifolds are also introduced, and they are used to describe Hamiltonian and Lagrangian systems, including the Hamilton-Jacobi theorem for time-dependent systems.

Chapters 8 and 9 contain the theory of Lie groups and group actions, which is needed for the last part of the book. They include the relationship between a Lie group and its Lie algebra, the exponential map, the closed subgroup theorem, the Lorentz group, the adjoint representation, semidirect products and affine spaces. Finally, some applications to Hamiltonian systems with symmetry, including Noether's theorem, are given.

Chapters 10, 11 and 12 are a crucial part of the book, describing fibre bundles and differential geometry which is the foundation for the general theory of relativity. The main topic is the theory of Ehresmann connections in principal fibre bundles, which is specialized to linear connections and metrical connections in pseudo-Riemannian manifolds. The Schwarzschild metric describing black holes and the Robertson-Walker metric for homogeneous and isotropic universe models are included as important examples. In Chapter 11 we use Ehresmann connections to describe isometric immersions and the second fundamental form, and in Chapter 12 we introduce jet bundles and give an alternative definition of connections described as jet fields.

Finally, Chapter 13 is an appendix which contains some necessary topics which are used elsewhere in the book. A main subject is the theory of covering spaces and homotopy theory which is used in the theory of Lie groups in Chapter 8.

Steinar Johannesen

Acknowledgment In May 2013 Steinar Johannesen had a stroke and he was not able to take care of the final publication of this book on which he spent many years writing. His legal guardian, Erik Andersen, will heartly thank two of Steinar Johannesen's colleagues, professor Jan O. Kleppe and professor Øyvind Grøn, for assisting in the process of having this book published. He also thanks Dr. Johannes Kleppe for having contributed so much in preparing the final manuscript of the book.

Erik Andersen
Ole Moes vei 7 E
N-1165 Oslo

Chapter 1

INTRODUCTION

In this chapter we will investigate some important results in the differential geometry of curves and surfaces in the 3-dimensional Euclidean space $\mathbf{R}^3$, in order to obtain a good intuition of geometric concepts such as curvature and torsion. These concepts and results will be generalized to higher dimensional manifolds of various types in the next chapters.

SPACE CURVES

1.1 Arc length Let $c : I \to \mathbf{R}^3$ be a smooth curve in $\mathbf{R}^3$ defined on an interval I. If I is not an open interval, this means that c is the restriction of a smooth curve defined on an open interval containing I. We say that the curve c is *regular* if $c'(t) \neq 0$ for every $t \in I$. If $\phi : J \to I$ is a diffeomorphism, then $c \circ \phi : J \to \mathbf{R}^3$ is also a regular curve called a *reparametrization* of c.

Given a regular curve $\gamma : I \to \mathbf{R}^3$, we define its *arc length function* $s : I \to \mathbf{R}$ *with initial point* $t_0 \in I$ by

$$s(t) = \int_{t_0}^{t} \|\gamma'(u)\| \, du$$

for $t \in I$. The curve $\alpha = \gamma \circ s^{-1} : J \to \mathbf{R}^3$ defined on the interval $J = s(I)$ is a reparametrization of γ, called a *parametrization by arc length*, satisfying $\|\alpha'(t)\| = 1$ for $t \in J$.

1.2 The Frenet frame Let $c : I \to \mathbf{R}^3$ be a smooth curve parameterized by arc length. The vector-valued function $\mathbf{t} = c' : I \to \mathbf{R}^3$ is called the *unit tangent vector field* of c. Since

$$\mathbf{t}(s) \cdot \mathbf{t}(s) = 1$$

for $s \in I$, it follows that

$$\mathbf{t}'(s) \cdot \mathbf{t}(s) = 0$$

for $s \in I$. The vector-valued function $\mathbf{t}' : I \to \mathbf{R}^3$ is called the *curvature vector field* of c, and the function $\kappa : I \to \mathbf{R}$ defined by

$$\kappa(s) = \|\mathbf{t}'(s)\|$$

for $s \in I$ which measures how fast the unit tangent vector is changing, is called the *curvature function* of c. Assuming that $\kappa > 0$, we now have a vector-valued function $\mathbf{n} : I \to \mathbf{R}^3$, called the *principal normal vector field* of c, defined by

$$\mathbf{n}(s) = \mathbf{t}'(s) / \|\mathbf{t}'(s)\|$$

for $s \in I$ so that

$$\mathbf{t}' = \kappa \mathbf{n}. \tag{1}$$

We also have a vector-valued function $\mathbf{b} : I \to \mathbf{R}^3$ defined by

$$\mathbf{b}(s) = \mathbf{t}(s) \times \mathbf{n}(s)$$

for $s \in I$, called the *binormal vector field* of c. The vector fields **t**, **n** and **b** form an orthonormal frame field along the curve c, called the *Frenet frame field* of c. The plane through $c(s)$ spanned by $\mathbf{n}(s)$ and $\mathbf{b}(s)$ is called the *normal plane* at $c(s)$ for $s \in I$. The planes through $c(s)$ spanned by $\mathbf{t}(s)$ and $\mathbf{n}(s)$, and by $\mathbf{t}(s)$ and $\mathbf{b}(s)$, are called the *osculating plane* and the *rectifying plane* at $c(s)$, respectively.

Since

$$\mathbf{n}(s) \cdot \mathbf{t}(s) = 0 \quad \text{and} \quad \mathbf{n}(s) \cdot \mathbf{n}(s) = 1$$

for $s \in I$, it follows that

$$\mathbf{n}'(s) \cdot \mathbf{t}(s) = -\mathbf{n}(s) \cdot \mathbf{t}'(s) = -\kappa \quad \text{and} \quad \mathbf{n}'(s) \cdot \mathbf{n}(s) = 0$$

for $s \in I$. Introducing the *torsion function* $\tau : I \to \mathbf{R}$ of c defined by

$$\tau(s) = \mathbf{n}'(s) \cdot \mathbf{b}(s)$$

for $s \in I$, we therefore have that

$$\mathbf{n}' = -\kappa \mathbf{t} + \tau \mathbf{b}. \tag{2}$$

In the same way we see that

$$\mathbf{b}'(s) \cdot \mathbf{t}(s) = -\mathbf{b}(s) \cdot \mathbf{t}'(s) = 0 \quad , \quad \mathbf{b}'(s) \cdot \mathbf{n}(s) = -\tau(s) \quad \text{and} \quad \mathbf{b}'(s) \cdot \mathbf{b}(s) = 0$$

for $s \in I$, which implies that

$$\mathbf{b}' = -\tau \mathbf{n}. \tag{3}$$

Since **b** is perpendicular to the osculating plane, we see that the torsion τ measures the rate at which the curve deviates from being a plane curve.

Equations (1), (2) and (3) are called the *Frenet formulae*. They express the rate of change of the Frenet frame in terms of the frame itself. Since the Frenet frame field is adapted to the curve, with the vector field **t** tangent to the curve and the vector fields **n** and **b** orthogonal to the curve, it gives information about the shape of the curve in terms of the curvature κ and the torsion τ.

CURVES ON SURFACES

1.3 The Gauss' formulae Consider a surface in $\mathbf{R}^3$ parameterized by a smooth map $\mathbf{r} : U \to \mathbf{R}^3$ defined on an open subset U of $\mathbf{R}^2$. Using the coordinates $x = (x^1, x^2)$ in U, the surface and the map $\mathbf{r}$ are said to be *regular* if the vector fields

$$\mathbf{e}_1 = \frac{\partial \mathbf{r}}{\partial x^1} \quad \text{and} \quad \mathbf{e}_2 = \frac{\partial \mathbf{r}}{\partial x^2}$$

are everywhere linearly independent so that they span a *tangent plane* at each point on the surface with a *unit normal vector*

$$\mathbf{u} = \frac{\mathbf{e}_1 \times \mathbf{e}_2}{\|\mathbf{e}_1 \times \mathbf{e}_2\|} .$$

The derivative of $\mathbf{e}_j$ with respect to $\mathbf{e}_k$, which is given by

$$\mathbf{e}_{jk} = \frac{\partial \mathbf{e}_j}{\partial x^k} = \frac{\partial^2 \mathbf{r}}{\partial x^k \partial x^j} ,$$

can be decomposed in a component in the tangent plane and a component along the unit normal $\mathbf{u}$ by the *Gauss' formula*

$$\mathbf{e}_{jk} = \sum_{i=1}^{2} \Gamma^i_{jk} \mathbf{e}_i + a_{jk} \mathbf{u} \tag{1}$$

where $a_{jk} = \mathbf{e}_{jk} \cdot \mathbf{u}$ for $1 \leq j,k \leq 2$. The coefficients Γ^i_{jk} are called the *Christoffel symbols* of the second kind and will be determined in 1.5. Since $\mathbf{e}_{jk} = \mathbf{e}_{kj}$, we see that Γ^i_{jk} and a_{jk} are symmetric in the indices j and k.

1.4 The metric on a surface Let $\gamma : I \to U$ be a regular curve in U, and consider the corresponding curve $\alpha = \mathbf{r} \circ \gamma : I \to \mathbf{R}^3$ on the surface. Since

$$\frac{d\alpha}{dt} = \sum_{i=1}^{2} \frac{d\gamma^i}{dt} \mathbf{e}_i$$

it follows that

$$\left\| \frac{d\alpha}{dt} \right\|^2 = \sum_{j,k=1}^{2} g_{jk} \frac{d\gamma^j}{dt} \frac{d\gamma^k}{dt} ,$$

where

$$g_{jk} = \mathbf{e}_j \cdot \mathbf{e}_k \tag{1}$$

for $1 \leq j,k \leq 2$. The dot product defines a *metric* in $\mathbf{R}^3$, and g_{jk} are the components of the induced metric in each tangent plane with respect to the basis consisting of $\mathbf{e}_1$

and $\mathbf{e}_2$. The induced metric g is a symmetric bilinear function in each tangent plane, given by

$$g(\mathbf{e}_j, \mathbf{e}_k) = g_{jk}$$

for $1 \le j,k \le 2$. Using the coefficients a_{jk} in Gauss' formulae we also define the *shape tensor* II on the surface, which is a symmetric bilinear function in each tangent plane given by

$$\mathrm{II}(\mathbf{e}_j, \mathbf{e}_k) = a_{jk}$$

for $1 \le j,k \le 2$. Classically g was denoted by I, and I and II were called the *first* and *second fundamental form* on the surface. We will follow the modern use of the notion "form" which is reserved for tensors which are skew symmetric multilinear functions in each tangent plane.

1.5 The Christoffel symbols By taking the derivative of Equation (1) in 1.4 with respect to $\mathbf{e}_l$ and using the Gauss' formulae we obtain

$$\frac{\partial g_{jk}}{\partial x^l} = \mathbf{e}_{jl} \cdot \mathbf{e}_k + \mathbf{e}_j \cdot \mathbf{e}_{kl} = \sum_{i=1}^{2} (g_{ik} \Gamma^i_{jl} + g_{ij} \Gamma^i_{kl}) = [jl,k] + [kl,j], \tag{1}$$

where

$$[jk,l] = \sum_{i=1}^{2} g_{il} \Gamma^i_{jk}$$

for $1 \le j,k,l \le 2$ is called *Christoffel symbols* of the first kind. Permuting the indices in Equation (1) and using that $[jk,l]$ is symmetric in the indices j and k, we have that

$$[jk,l] = \tfrac{1}{2} \{([lj,k] + [kj,l]) + ([lk,j] + [jk,l])$$
$$- ([jl,k] + [kl,j])\} = \frac{1}{2} \left\{ \frac{\partial g_{lk}}{\partial x^j} + \frac{\partial g_{lj}}{\partial x^k} - \frac{\partial g_{jk}}{\partial x^l} \right\},$$

and

$$\Gamma^i_{jk} = \sum_{l=1}^{2} g^{il} \, [jk,l]$$

where (g^{ij}) is the inverse of the matrix (g_{ij}).

1.6 The Weingarten formulae The curvature of the surface is determined by the rate of change of the unit normal vector $\mathbf{u}$. Since $\mathbf{u} \cdot \mathbf{u} = 1$, we have that $\mathbf{u}_k \cdot \mathbf{u} = 0$ for $k = 1,2$, where

$$\mathbf{u}_k = \frac{\partial \mathbf{u}}{\partial x^k}$$

is the derivative of $\mathbf{u}$ with respect to $\mathbf{e}_k$. Hence $\mathbf{u}_k$ lies in the tangent plane and is given by the *Weingarten formula*

$$\mathbf{u}_k = -\sum_{i=1}^{2} a^i_k \mathbf{e}_i \tag{1}$$

for $k=1,2$. From Gauss' formulae and the equations $\mathbf{u}\cdot\mathbf{e}_j=0$ it follows that

$$a_{jk}=\mathbf{e}_{jk}\cdot\mathbf{u}=-\,\mathbf{e}_j\cdot\mathbf{u}_k=\sum_{i=1}^{2} g_{ij}\,a^i_k$$

for $1\le j,k\le 2$, so that

$$a^i_k=\sum_{j=1}^{2} g^{ij}\,a_{jk} \tag{2}$$

for $1\le i,k\le 2$. The coefficients a^i_k in the Weingarten formulae are said to be obtained from the coefficients a_{jk} in Gauss' formulae by *raising* the first index.

We now define the *shape operator* S on the surface, which is a linear operator in each tangent plane given by

$$S(\mathbf{e}_k)=\sum_{i=1}^{2} a^i_k\,\mathbf{e}_i \tag{3}$$

for $k=1,2$. Given arbitrary tangent vector fields

$$\mathbf{t}_i=\sum_{k=1}^{2} v^k_i\,\mathbf{e}_k$$

for $i=1,2$, we have that

$$\mathrm{II}\,(\mathbf{t}_1,\mathbf{t}_2)=\sum_{j,k=1}^{2} a_{jk}\,v^k_1\,v^j_2=\sum_{j,k=1}^{2}\left(\sum_{i=1}^{2} g_{ij}\,a^i_k\right)v^k_1\,v^j_2$$

$$=\sum_{i,j=1}^{2} g_{ij}\left(\sum_{k=1}^{2} a^i_k\,v^k_1\right)v^j_2=g\,(S(\mathbf{t}_1),\mathbf{t}_2)\,.$$

As the shape tensor II is symmetric, it follows that the shape operator S is self-adjoint in each tangent plane.

1.7 The Darboux frame Let $s:I\to\mathbf{R}$ be the arc length function of the curve α defined in 1.4 with initial point $t_0\in I$, and let $\beta=\gamma\circ s^{-1}$ be a reparametrization of γ so that $c=\alpha\circ s^{-1}=\mathbf{r}\circ\beta$ is a parametrization of α by arc length. From the relation

$$\mathbf{t}=\frac{dc}{dt}=\sum_{i=1}^{2}\frac{d\beta^i}{dt}\,\mathbf{e}_i \tag{1}$$

and Gauss' formulae it follows that

$$\mathbf{t}'=\kappa\,\mathbf{n}=\sum_{i=1}^{2}\frac{d^2\beta^i}{dt^2}\,\mathbf{e}_i+\sum_{j,k=1}^{2}\frac{d\beta^j}{dt}\,\frac{d\beta^k}{dt}\,\mathbf{e}_{jk}$$

$$=\sum_{i=1}^{2}\left\{\frac{d^2\beta^i}{dt^2}+\sum_{j,k=1}^{2}\Gamma^i_{jk}\,\frac{d\beta^j}{dt}\,\frac{d\beta^k}{dt}\right\}\mathbf{e}_i+\sum_{j,k=1}^{2} a_{jk}\,\frac{d\beta^j}{dt}\,\frac{d\beta^k}{dt}\,\mathbf{u}\,.$$

As the first term after the last equality is orthogonal to both $\mathbf{u}$ and $\mathbf{t}$, it must be parallel to

$$\mathbf{e} = \mathbf{u} \times \mathbf{t}\,.$$

The vector fields $\mathbf{t}$, $\mathbf{e}$ and $\mathbf{u}$ form an orthonormal frame field along the curve c, called the *Darboux frame field* of c, which is adapted to the surface as well as to the curve. Now we have that

$$\mathbf{t}' = \kappa\,\mathbf{n} = \kappa_g\,\mathbf{e} + \kappa_n\,\mathbf{u}, \tag{2}$$

where $\kappa_g = \kappa\,\mathbf{n}\cdot\mathbf{e} = \kappa\,\mathbf{u}\cdot\mathbf{b}$ and $\kappa_n = \kappa\,\mathbf{n}\cdot\mathbf{u}$ are called the *geodesic* and *normal* curvatures of the curve c. They are given by

$$\kappa_g\,\mathbf{e} = \sum_{i=1}^{2}\left\{\frac{d^2\beta^i}{dt^2} + \sum_{j,k=1}^{2}\Gamma^i_{jk}\,\frac{d\beta^j}{dt}\,\frac{d\beta^k}{dt}\right\}\mathbf{e}_i \tag{3}$$

and

$$\kappa_n = \sum_{j,k=1}^{2} a_{jk}\,\frac{d\beta^j}{dt}\,\frac{d\beta^k}{dt} = \mathrm{II}\,(\mathbf{t},\mathbf{t}). \tag{4}$$

The curves c with vanishing geodesic curvature satisfy the differential equations

$$\frac{d^2\beta^i}{dt^2} + \sum_{j,k=1}^{2}\Gamma^i_{jk}\,\frac{d\beta^j}{dt}\,\frac{d\beta^k}{dt} = 0$$

for $i = 1,2$ and are called *geodesics*. These are the curves on the surface which correspond to straight lines in a plane. Introducing the *geodesic torsion* $\tau_g = \mathbf{e}'\cdot\mathbf{u}$ of the curve c, we obtain the formulae

$$\mathbf{e}' = -\kappa_g\,\mathbf{t} + \tau_g\,\mathbf{u} \quad \text{and} \quad \mathbf{u}' = -\kappa_n\,\mathbf{t} - \tau_g\,\mathbf{e}$$

which together with formula (2) express the rate of change of the Darboux frame in terms of the frame itself.

1.8 Gaussian curvature From Equation (4) in 1.7 we see that the normal curvature κ_n of a curve c parameterized by arc length through a point p on the surface only depends on its unit tangent vector $\mathbf{t}$ at this point which is given by Equation (1) in 1.7. The value of

$$\kappa_n(\mathbf{t}) = \mathrm{II}\,(\mathbf{t},\mathbf{t}) = g\,(S(\mathbf{t})\,,\mathbf{t})$$

for all unit vectors $\mathbf{t}$ in the tangent plane at the point p gives information about the curvature of the surface at p. Since the set of unit tangent vectors is compact, there is a maximal and a minimal value of the normal curvature. They are called the *principal curvatures*, and the corresponding unit tangent vectors define the *principal directions*. If the maximal and minimal values are equal, all directions are called principal.

As the shape operator S is self-adjoint, there is an orthonormal basis $\{\mathbf{t}_1,\mathbf{t}_2\}$ in the tangent plane at p consisting of eigenvectors for S with eigenvalues λ_1 and λ_2, respectively. Since

$$\kappa_n(a\mathbf{t}_1 + b\mathbf{t}_2) = a^2\lambda_1 + b^2\lambda_2 = \lambda_1 + b^2(\lambda_2 - \lambda_1)$$

for every $(a,b) \in \mathbf{R}^2$ with $\|a\mathbf{t}_1 + b\mathbf{t}_2\|^2 = a^2 + b^2 = 1$, these eigenvalues are the principal curvatures. Their product

$$K = \lambda_1\lambda_2$$

is called the *Gaussian curvature* of the surface at p. By (2) and (3) in 1.6 we see that the matrix of S with respect to the basis $\{\mathbf{e}_1, \mathbf{e}_2\}$ is $B^{-1}A$, where $A = (a_{ij})$ and $B = (g_{ij})$. We therefore have that

$$K = \det(B^{-1}A) = \frac{\det A}{\det B}\,.$$

Chapter 2

SMOOTH MANIFOLDS AND VECTOR BUNDLES

The purpose of this chapter is to generalize the differential calculus in Euclidean spaces to a kind of topological spaces called manifolds which are only locally Euclidean. The idea is to use local coordinate systems around each point which overlap smoothly. We will investigate how tangent vectors at a point transform under coordinate changes. Using this we will make equivalence classes forming a tangent space at each point of the manifold. Together they constitute the tangent bundle which is our first example of a vector bundle.

SMOOTH MANIFOLDS

2.1 Definition A topological space M is said to be *locally Euclidean* if it is Hausdorff and if each point p in M has an open neighbourhood U homeomorphic to an open set in $\mathbf{R}^n$ for some integer $n \geq 0$. A map $x : U \to \mathbf{R}^n$ sending U homeomorphically onto an open set $x(U) \subset \mathbf{R}^n$, is called a *coordinate map* at p. Its component functions $x^i : U \to \mathbf{R}$, $i = 1, ..., n$, are called *coordinate functions*, and U is called a *coordinate neighbourhood* of p. The pair (x, U) is called a *local coordinate system* or a *local chart*.

If (x, U) and (y, V) are two local charts on M, the open sets $x(U \cap V)$ and $y(U \cap V)$ are mapped homeomorphically onto each other by the maps $y \circ x^{-1}$ and $x \circ y^{-1}$, which are called the *coordinate transformations*. We say that the local charts (x, U) and (y, V) are C^∞*-related* if the coordinate transformations $y \circ x^{-1}$ and $x \circ y^{-1}$ are C^∞.

An *atlas* on M is a family of C^∞-related local charts $\mathscr{A} = \{(x_\alpha, U_\alpha) \mid \alpha \in A\}$ where $\{U_\alpha | \alpha \in A\}$ is an open cover of M.

A *smooth structure* on M is an atlas $\mathscr{D}$ which is maximal in the sense that if (x, U) is a local chart on M which is C^∞-related to every chart in $\mathscr{D}$, then $(x, U) \in \mathscr{D}$.

A locally Euclidean space M which is second countable, is called a *topological manifold*. If it also has a smooth structure, it is called a *smooth manifold*.

2.2 Proposition If $\mathscr{A}$ is an atlas on a locally Euclidean space M, then there is a unique smooth structure $\mathscr{D}$ on M containing $\mathscr{A}$.

PROOF: Let $\mathscr{D}$ be the set of all charts on M which are C^∞-related to every chart in $\mathscr{A}$. In order to prove that $\mathscr{D}$ is an atlas, we must prove that $y \circ x^{-1}$ is C^∞ for any pair of local charts (x,U) and (y,V) in $\mathscr{D}$. Let a be a point in the domain $x(U \cap V)$ of $y \circ x^{-1}$, and let (x_α, U_α) be a local chart in $\mathscr{A}$ around $x^{-1}(a)$. Then $y \circ x^{-1}$ is C^∞ at a, since $y \circ x^{-1}|_{x(U \cap V \cap U_\alpha)}$ is the composition of the C^∞ coordinate transformations $x_\alpha \circ x^{-1}$ and $y \circ x_\alpha^{-1}$. Hence $y \circ x^{-1}$ is C^∞ at every point in $x(U \cap V)$, and it is clear that $\mathscr{D}$ is the unique maximal atlas containing $\mathscr{A}$. □

2.3 Definition A topological space X is called a *Lindelöf space* if every open cover of X contains a countable subcover.

2.4 Proposition If a topological space X is second countable, then it is a Lindelöf space.

PROOF: Let $\mathscr{O}$ be an open cover of X, and let $\mathscr{B}$ be a countable basis for its topology. Then this topology also has a countable basis $\{O_i \mid i \in \mathbf{N}\}$ consisting of those sets in $\mathscr{B}$ which are contained in at least one open set in $\mathscr{O}$. If for each positive integer i we choose an open set U_i in $\mathscr{O}$ such that $O_i \subset U_i$, then $\{U_i \mid i \in \mathbf{N}\}$ is a countable subcover of $\mathscr{O}$. □

2.5 Corollary In a second countable space X there cannot exist an uncountable family of disjoint nonempty open sets.

PROOF: Suppose there is such a family $\{U_\alpha | \alpha \in A\}$ in X. Then $U = \bigcup_{\alpha \in A} U_\alpha$ is a second countable subspace of X which is not a Lindelöf space, since $\{U_\alpha | \alpha \in A\}$ is an open cover of U with no countable subcover. We thus obtain a contradiction to Proposition 2.4. □

2.6 Proposition A locally Euclidean space M with an atlas $\mathscr{A}$ is second countable if and only if $\mathscr{A}$ contains a countable atlas.

PROOF: Let $\{(x_i, U_i) \mid i \in \mathbf{N}\}$ be a countable atlas contained in $\mathscr{A}$, and let $\{O_{ij} \mid j \in \mathbf{N}\}$ be a countable basis for the topology on $x_i(U_i)$ for each i. Then the family $\{x_i^{-1}(O_{ij}) \mid (i,j) \in \mathbf{N} \times \mathbf{N}\}$ is a countable basis for the topology on M which is therefore second countable.

The converse statement follows from Proposition 2.4. □

2.7 Remarks The dimension n of the Euclidean space $\mathbf{R}^n$ having an open subset homeomorphic with a coordinate neighbourhood of a point p in a smooth manifold M, is uniquely determined by p. For if $x : U \to \mathbf{R}^n$ and $y : V \to \mathbf{R}^m$ are two coordinate maps at p, then the derivative $D(y \circ x^{-1})(x(p)) : \mathbf{R}^n \to \mathbf{R}^m$ of the coordinate transformation $y \circ x^{-1}$ is an isomorphism with inverse $D(x \circ y^{-1})(y(p))$, and hence $n = m$.

This result is also true more generally for topological manifolds, where the coordinate transformation $y \circ x^{-1}$ is a homeomorphism between the nonempty open sets $x(U \cap V)$ and $y(U \cap V)$ in $\mathbf{R}^n$ and $\mathbf{R}^m$, respectively. The proof then depends on a deep result from algebraic topology by Brouwer saying that two nonempty, open subsets of $\mathbf{R}^n$ and $\mathbf{R}^m$ can only be homeomorphic when $n = m$ (topological invariance of dimension, cf. [23]).

n is called the dimension of M at p and is denoted $\dim_p(M)$. We say that the manifold M is *n-dimensional* and write $\dim(M) = n$ or simply M^n if it has dimension n at all points p.

A manifold M being locally Euclidean, is also locally connected, i.e., each point has a basis of connected neighbourhoods. Hence the connected components are open subsets of M and are therefore also manifolds. A manifold M can therefore be written as a disjoint union of connected manifolds which are both open and closed in M.

2.8 Proposition A connected manifold has the same dimension at all points.

PROOF: Suppose M is connected, and let $f : M \to \mathbf{Z}$ be the function defined by $f(p) = \dim_p(M)$. Since f is continuous, $f(M)$ is also connected and hence consists of only one number n since $\mathbf{Z}$ has the discrete topology. □

2.9 Examples

(a) The Euclidean space $\mathbf{R}^n$ is a smooth manifold with atlas $\{(id, \mathbf{R}^n)\}$, where $id : \mathbf{R}^n \to \mathbf{R}^n$ is the identity map.

(b) An n-dimensional vector space V is a smooth manifold with the topology defined in Proposition 13.117 in the appendix, and with atlas $\{(x, V)\}$ where $x : V \to \mathbf{R}^n$ is any linear isomorphism.

(c) Let M be an open subset of a smooth manifold N with smooth structure $\mathscr{D}$. Then M is also a smooth manifold with smooth structure $\mathscr{D}|_M = \{(x, U) \in \mathscr{D} | U \subset M\}$.

(d) An open interval I is a smooth manifold with atlas $\{(r, I)\}$, where $r : I \to \mathbf{R}$ is the inclusion map. We call (r, I) the standard local chart on I.

(e) The n-sphere is the set $\mathbf{S}^n = \{x \in \mathbf{R}^{n+1} \mid \|x\| = 1\}$. $\mathbf{S}^n$ is a smooth manifold with an atlas consisting of $2n+2$ local charts (f_i, U_i) and (g_i, V_i) for $i = 1, \ldots, n+1$, where $U_i = \{x \in \mathbf{S}^n | x^i > 0\}$ and $V_i = \{x \in \mathbf{S}^n | x^i < 0\}$ are open half spheres, and $f_i : U_i \to \mathbf{R}^n$ and $g_i : V_i \to \mathbf{R}^n$ are homeomorphisms onto the open ball $\mathbf{D}^n = \{y \in \mathbf{R}^n \mid \| y \| < 1\}$ given by

$$f_i(x) = g_i(x) = (x^1, \ldots, x^{i-1}, x^{i+1}, \ldots, x^{n+1}).$$

Their inverses are given by

$$f_i^{-1}(y) = (y^1, \ldots, y^{i-1}, (1 - \| y \|^2)^{\frac{1}{2}}, y^i, \ldots, y^n) \quad \text{and}$$
$$g_i^{-1}(y) = (y^1, \ldots, y^{i-1}, -(1 - \| y \|^2)^{\frac{1}{2}}, y^i, \ldots, y^n)$$

from which we see that the coordinate transformations $f_j \circ f_i^{-1}$, $g_j \circ g_i^{-1}$, $g_j \circ f_i^{-1}$ and $f_j \circ g_i^{-1}$ are C^∞.

The smooth structure on $\mathbf{S}^n$ can also be defined by an atlas consisting of only two local charts $(\phi_1, \mathbf{S}^n \setminus \{p\})$ and $(\phi_2, \mathbf{S}^n \setminus \{a\})$, where $p = (0,0,...,1)$ and $a = (0,0,...,-1)$ are the north and south pole, and where $\phi_1 : \mathbf{S}^n \setminus \{p\} \to \mathbf{R}^n$ and $\phi_2 : \mathbf{S}^n \setminus \{a\} \to \mathbf{R}^n$ are the stereographic projections given by

$$\phi_1(x) = (1 - x^{n+1})^{-1}(x^1, ..., x^n) \quad \text{and}$$
$$\phi_2(x) = (1 + x^{n+1})^{-1}(x^1, ..., x^n).$$

To find the inverse of ϕ_1, we must solve the equation $y = \phi_1(x)$ with respect to x. Using $\| x \| = 1$, we see that

$$\| y \|^2 = \frac{1 - (x^{n+1})^2}{(1 - x^{n+1})^2} = \frac{1 + x^{n+1}}{1 - x^{n+1}}, \text{ so that}$$

$$1 + \| y \|^2 = \frac{2}{1 - x^{n+1}} \quad \text{and} \quad 1 - x^{n+1} = \frac{2}{1 + \| y \|^2}.$$

Hence the inverse of ϕ_1 is given by

$$\phi_1^{-1}(y) = \left(\frac{2y^1}{1 + \|y\|^2}, ..., \frac{2y^n}{1 + \|y\|^2}, \frac{\|y\|^2 - 1}{\|y\|^2 + 1} \right),$$

and similarly

$$\phi_2^{-1}(y) = \left(\frac{2y^1}{1 + \|y\|^2}, ..., \frac{2y^n}{1 + \|y\|^2}, \frac{1 - \|y\|^2}{1 + \|y\|^2} \right).$$

From this we see that

$$\phi_2 \circ \phi_1^{-1}(y) = \phi_1 \circ \phi_2^{-1}(y) = \frac{y}{\|y\|^2}$$

which proves that the coordinate transformations $\phi_2 \circ \phi_1^{-1}$ and $\phi_1 \circ \phi_2^{-1}$ are C^∞.

We also see that the two atlases defined above are C^∞-related and hence give the same differeniable structure on $\mathbf{S}^n$.

(f) The product $M \times N$ of two smooth manifolds M and N is again a smooth manifold. If $\{(x_\alpha, U_\alpha) \mid \alpha \in A\}$ and $\{(y_\beta, V_\beta) \mid \beta \in B\}$ are atlases for M and N, respectively, then $\{(x_\alpha \times y_\beta, U_\alpha \times V_\beta) \mid (\alpha, \beta) \in A \times B\}$ is an atlas for $M \times N$. In the same way we can form the product manifold $M_1 \times \cdots \times M_n$ with n factors.

(g) The n-dimensional torus $\mathbf{T}^n = \mathbf{S}^1 \times \cdots \times \mathbf{S}^1$ is the product manifold of n circles.

SMOOTH MAPS

2.10 Definition Let $f : M \to N$ be a continuous map between two smooth manifolds M and N. We say that f is *differentiable of class* C^k at a point p in M, for $0 \le k \le \infty$, if $y \circ f \circ x^{-1}$ is C^k in an open neighbourhood of $x(p)$ for local charts (x,U) and (y,V) around p and $f(p)$, respectively. If this is true for one pair of local charts, it is true for every other pair since the local charts around a point in a smooth manifold are C^∞-related.

If f is differentiable of class C^k at every point in an open subset O of M, we say that f is differentiable of class C^k on O, and if $O = M$ we just say that f is differentiable of class C^k. This is equivalent to saying that $y \circ f \circ x^{-1}$ is C^k for all local charts (x,U) and (y,V) on M and N, respectively. We will be interested almost exclusively in the C^∞-case, and by a *smooth map* we mean a map which is differentiable of class C^∞.

A smooth map f is called a *diffeomorphism* if it is bijective and has a smooth inverse. We say that f is a *local diffeomorphism* at a point p in M if it is a diffeomorphism from an open neighbourhood of p onto an open neighbourhood of $f(p)$.

2.11 Remark For a function $f : M \to \mathbf{R}$ we use the standard atlas $\{(id, \mathbf{R})\}$ on $\mathbf{R}$. Hence f is smooth at a point p in M if $f \circ x^{-1} : x(U) \to \mathbf{R}$ is C^∞ in an open neighbourhood of $x(p)$ for a local chart (x,U) around p. f is smooth if $f \circ x^{-1}$ is C^∞ for all local charts (x,U) on M.

2.12 Definition Let V be an open neighbourhood of a point p in M^n, and let $f : V \to \mathbf{R}$ be a continuous function which is smooth at p. If (x,U) is a local chart around p, we define the i-th *partial derivative* of f at p with respect to (x,U) to be

$$\frac{\partial f}{\partial x^i}(p) = D_i(f \circ x^{-1})(x(p)) ,$$

i.e., the i-th partial derivative of $f \circ x^{-1}$ at $x(p)$. In the same way we define the *m-th order partial derivative* of f at p with respect to (x,U) to be

$$\frac{\partial^m f}{(\partial x^{i_1})^{k_1} \cdots (\partial x^{i_r})^{k_r}}(p) = D_{i_1}^{k_1} \cdots D_{i_r}^{k_r}(f \circ x^{-1})(x(p))$$

for $(i_1, ..., i_r) \in I_n^r$ and non-negative integers $k_1, ..., k_r$ with $m = \sum_{i=1}^r k_i$.

An n-tuple $k = (k_1, \ldots, k_n)$ of non-negative integers is called a *multi-index*. Its *length* and *factorial* are defined by

$$|k| = \sum_{i=1}^n k_i \quad \text{and} \quad k! = \prod_{i=1}^n k_i! ,$$

and we set

$$\frac{\partial^{|k|} f}{\partial x^k}(p) = \frac{\partial^{|k|} f}{(\partial x^1)^{k_1} \cdots (\partial x^n)^{k_n}}(p) .$$

We let 1_i denote the multi-index where all the components are equal to zero except the i-th component which is 1.

2.13 Remark If $M = \mathbf{R}^n$ and the local chart is $(id, \mathbf{R}^n)$, we have

$$\frac{\partial f}{\partial x^i}(p) = D_i f(p) \quad \text{and} \quad \frac{\partial^m f}{(\partial x^{i_1})^{k_1} \cdots (\partial x^{i_r})^{k_r}}(p) = D_{i_1}^{k_1} \cdots D_{i_r}^{k_r} f(p)$$

in accordance with the usual notation for partial derivatives of functions defined on an open subset of $\mathbf{R}^n$.

2.14 Proposition Let f and g be functions which are smooth at a point p in M, and let (x, U) be a local chart around p. Let $a \in \mathbf{R}$. Then

(1) $\frac{\partial (f+g)}{\partial x^i}(p) = \frac{\partial f}{\partial x^i}(p) + \frac{\partial g}{\partial x^i}(p)$

(2) $\frac{\partial (af)}{\partial x^i}(p) = a \frac{\partial f}{\partial x^i}(p)$

(3) $\frac{\partial (fg)}{\partial x^i}(p) = \frac{\partial f}{\partial x^i}(p)\, g(p) + f(p) \frac{\partial g}{\partial x^i}(p)$

PROOF : Follows from the definition and the corresponding rules for partial derivatives of functions defined on an open subset of a Euclidean space. □

2.15 Definition Let p be a point in the smooth manifold M, and let $\mathscr{F}_p$ be the algebra of all smooth functions $f : V \to \mathbf{R}$, each defined on some open neighbourhood V of p. Then a linear functional l on $\mathscr{F}_p$ is called a *local derivation* at p if it satisfies the relation $l(fg) = l(f)\, g(p) + f(p)\, l(g)$ for every $f, g \in \mathscr{F}_p$.

2.16 Example If (x, U) is a local chart around a point p in the smooth manifold M, then it follows from Proposition 2.14 that the operator $l = \left.\frac{\partial}{\partial x^i}\right|_p$ on $\mathscr{F}_p$ taking f to $\frac{\partial f}{\partial x^i}(p)$ is a local derivation at p. It is called the i-th *partial derivation* at p with respect to (x, U).

PARTITIONS OF UNITY

2.17 Definition A family $\{U_\alpha | \alpha \in A\}$ of sets in a topological space X is said to be *locally finite* if each point p in X has a neighbourhood V such that $V \cap U_\alpha \neq \emptyset$ for only a finite number of indices $\alpha \in A$.

If $\{U_\alpha | \alpha \in A\}$ and $\{V_\beta | \beta \in B\}$ are two covers of X, we say that $\{U_\alpha | \alpha \in A\}$ is a *refinement* of $\{V_\beta | \beta \in B\}$ if for every $\alpha \in A$, there is a $\beta \in B$ such that $U_\alpha \subset V_\beta$.

2.18 Definition By a *partition of unity* on a smooth manifold M we mean a family $\{\phi_\alpha | \alpha \in A\}$ of nonnegative smooth functions on M such that their supports form a locally finite cover of M and we have that

$$\sum_{\alpha \in A} \phi_\alpha(p) = 1$$

for every $p \in M$.

We say that a partition of unity $\{\phi_\alpha | \alpha \in A\}$ is *subordinate* to a cover $\{V_\alpha | \alpha \in A\}$ of M if $\operatorname{supp}(\phi_\alpha) \subset V_\alpha$ for every $\alpha \in A$.

2.19 Lemma If $\{F_\alpha | \alpha \in A\}$ is a locally finite family of closed sets in a topological space X, then $F = \bigcup_{\alpha \in A} F_\alpha$ is closed.

PROOF: If p is a point in F^c, there is an open neighbourhood V of p and a finite subset B of the index set A such that $V \cap F_\alpha = \emptyset$ for every $\alpha \in A - B$. This implies that $V \cap \bigcap_{\alpha \in B} F_\alpha^c$ is an open neighbourhood of p contained in F^c, showing that F^c is open and hence that F is closed in X. □

2.20 Lemma If a topological space X is locally compact and second countable, then it is covered by a sequence $\{G_i\}_{i=1}^\infty$ of open sets with compact closure such that $\overline{G_i} \subset G_{i+1}$ for all i.

PROOF: Since X is locally compact, we know that each point in X has an open neighbourhood with compact closure, and these neighbourhoods form an open cover of X which by Proposition 2.4 has a countable subcover $\{U_i\}_{i=1}^\infty$. Let $G_1 = U_1$, and suppose that

$$G_k = U_1 \cup \cdots \cup U_{i_k}\,.$$

Then there is a smallest integer i_{k+1} greater than i_k such that

$$\overline{G_k} \subset U_1 \cup \cdots \cup U_{i_{k+1}}\,,$$

and we define

$$G_{k+1} = U_1 \cup \cdots \cup U_{i_{k+1}}\,.$$

Since $i_k \geq k$, we have that $G_k \supset U_k$ which shows that $\{G_i\}_{i=1}^\infty$ covers X, and the other properties of $\{G_i\}_{i=1}^\infty$ in the proposition are clearly satisfied. □

2.21 Lemma Let $\mathscr{O}$ be an open cover of a smooth manifold M. Then there is a countable atlas $\{(x_i, U_i) | i \in \mathbf{N}\}$ on M such that $\{U_i | i \in \mathbf{N}\}$ is a locally finite refinement of $\mathscr{O}$ with $x_i(U_i) = B_2(0)$ for each $i \in \mathbf{N}$, and the family $\{x_i^{-1}(B_1(0)) | i \in \mathbf{N}\}$ is an open cover of M.

PROOF : Let $\{G_j\}_{j=1}^{\infty}$ be a sequence of open sets in M as given by Lemma 2.20, and let $G_{-1} = G_0 = \emptyset$. For each integer $j \geq 0$ and each point $p \in G_{j+1} - G_j$ there is an open set V_p in $\mathscr{O}$ containing p and a local chart (x_p, U_p) around p such that

$$U_p \subset (G_{j+1} - \overline{G}_{j-1}) \cap V_p .$$

We may as well assume that $x_p(U_p) = B_2(0)$, if necessary by shrinking U_p so that $x_p(U_p)$ becomes an open ball centered at $x_p(p)$, and then combining x_p with a translation and a scaling mapping this open ball onto $B_2(0)$. We let $W_p = x_p{}^{-1}(B_1(0))$.

Then $\{W_p \,|\, p \in \overline{G}_{j+1} - G_j\}$ is an open cover of $\overline{G}_{j+1} - G_j$ which contains a finite subcover since $\overline{G}_{j+1} - G_j$ is compact. We thus obtain a sequence $\{p_i\}_{i=1}^{\infty}$ in M such that $\{W_{p_i} \,|\, i \in \mathbf{N}\}$ is an open cover of M, and the family $\{U_{p_i} \,|\, i \in \mathbf{N}\}$ is locally finite since $G_j \cap U_p = \emptyset$ for $p \in \overline{G}_{k+1} - G_k$ when $k \geq j+1$. If we set $x_i = x_{p_i}$ and $U_i = U_{p_i}$, the family $\{(x_i, U_i) \,|\, i \in \mathbf{N}\}$ is hence an atlas on M with the desired properties. □

2.22 Lemma The non-negative real function $f : \mathbf{R} \to \mathbf{R}$ defined by

$$f(x) = \begin{cases} e^{-1/x} & \text{for} \quad x > 0 \\ 0 & \text{for} \quad x \leq 0 \end{cases}$$

is smooth on $\mathbf{R}$, and $f(x) > 0$ for $x > 0$.

PROOF : We must show that f is smooth at 0. We see by induction that

$$f^{(n)}(x) = e^{-1/x} x^{-2n} P_n(x)$$

for $x > 0$, where $P_n(x)$ is a polynomial of degree $\leq n$. Indeed, this is clearly true for $n = 0$, and assuming that it is true for n we have that

$$f^{(n+1)}(x) = e^{-1/x} x^{-2(n+1)} P_{n+1}(x)$$

where

$$P_{n+1}(x) = (1 - 2nx) P_n(x) + x^2 P_n'(x).$$

From this it follows by induction that $f^{(n)}(0) = 0$ which is clearly true for $n = 0$. Assuming that it is true for n we have that

$$\lim_{x \to 0^+} \frac{f^{(n)}(x) - f^{(n)}(0)}{x} = \lim_{x \to 0^+} e^{-1/x} x^{-2n-1} P_n(x) = 0$$

since, by using the change of variable $x = 1/t$, we have that

$$\lim_{x \to 0^+} e^{-1/x} x^{-m} = \lim_{t \to \infty} \frac{t^m}{e^t} = 0$$

for every non-negative integer m.

The last equality follows from the fact that the exponential function e^t grows faster that any power of t as $t \to \infty$. This can be seen from the series expansion

$$e^t = \sum_{n=0}^{\infty} \frac{t^n}{n!}$$

which implies that

$$e^t \geq \frac{t^{m+1}}{(m+1)!}$$

so that

$$0 \leq \frac{t^m}{e^t} \leq \frac{(m+1)!}{t}$$

for $t > 0$. □

2.23 Theorem Let $\mathscr{O}$ be an open cover of a smooth manifold M. Then there is a partition of unity $\{\phi_\alpha | \alpha \in A\}$ on M subordinate to $\mathscr{O}$.

PROOF: Let $\{(x_i, U_i) \mid i \in \mathbf{N}\}$ be a countable atlas on M satisfying the properties given in Lemma 2.21, and for each $i \in \mathbf{N}$, let $h_i : M \to \mathbf{R}$ be the non-negative function defined by

$$h_i(p) = \begin{cases} f(1 - \| x_i(p) \|^2) & \text{for} \quad p \in U_i \\ 0 & \text{for} \quad p \in M - U_i \end{cases}$$

where f is the function defined in Lemma 2.22. We see that each h_i is smooth on M with $\operatorname{supp}(h_i) \subset U_i$, and that $h_i(p) > 0$ if and only if $p \in {x_i}^{-1}(B_1(0))$. Since the family $\{{x_i}^{-1}(B_1(0)) | i \in \mathbf{N}\}$ is a locally finite cover of M, we have a well-defined smooth function

$$h = \sum_{i=1}^{\infty} h_i$$

on M with $h(p) > 0$ for $p \in M$. The family $\{\psi_i | i \in \mathbf{N}\}$ where $\psi_i = h_i/h$ is hence a partition of unity on M such that $\{\operatorname{supp}(\psi_i) | i \in \mathbf{N}\}$ is a locally finite refinement of $\mathscr{O}$.

If $\mathscr{O} = \{V_\alpha | \alpha \in A\}$, we have a map $\mu : \mathbf{N} \to A$ so that $\operatorname{supp}(\psi_i) \subset V_{\mu(i)}$ for all $i \in \mathbf{N}$. If we define $\phi_\alpha = \sum_{\mu(i)=\alpha} \psi_i$ for $\alpha \in A$, it follows from Lemma 2.19 that $\{\phi_\alpha | \alpha \in A\}$ is a partition of unity on M subordinate to $\mathscr{O}$. □

2.24 Corollary If F is a closed and O an open subset of a smooth manifold M with $F \subset O$, then there exists a smooth function $f : M \to \mathbf{R}$ with $0 \leq f(p) \leq 1$ for all $p \in M$ such that $f(p) = 1$ if $p \in F$ and $\operatorname{supp}(f) \subset O$.

PROOF: By Theorem 2.23 there is a partition of unity $\{f, g\}$ on M subordinate to the open cover $\{O, M - F\}$. Then $\operatorname{supp}(f) \subset O$, and since $\operatorname{supp}(g) \subset M - F$ and $f + g = 1$ it follows that $f(p) = 1$ if $p \in F$. Hence f has the desired properties. □

2.25 Corollary Let M be a smooth manifold and W a finite dimensional vector space. If $g: O \to W$ is a smooth map defined on an open subset O of M, and if F is a closed subset of M with $F \subset O$, then there exists a smooth map $f: M \to W$ defined on M which coincides with g on F.

PROOF: By Corollary 2.24 there exists a smooth function $h: M \to \mathbf{R}$ with $h(q) = 1$ for $q \in F$ and supp$(h) \subset O$, and we let

$$f(q) = \begin{cases} h(q)g(q) & \text{for} \quad q \in O \\ 0 & \text{for} \quad q \notin O \end{cases} . \qquad \square$$

THE RANK OF A MAP

2.26 Definition Let $f: M^n \to N^m$ be a map which is smooth at a point p in M, and let (x, U) and (y, V) be local charts around p and $f(p)$, respectively. We define the *rank* of f at p to be the rank of the $m \times n$-matrix $\left(\frac{\partial(y^i \circ f)}{\partial x^j}(p)\right)$. This does not depend on the local charts (x, U) or (y, V). For let (x', U') and (y', V') be another pair of local charts around p and $f(p)$, respectively. Then we have that $y' \circ f \circ x'^{-1} = (y' \circ y^{-1}) \circ (y \circ f \circ x^{-1}) \circ (x \circ x'^{-1})$, and the derivatives $D(x \circ x'^{-1})(x'(p))$ and $D(y' \circ y^{-1})(y(f(p)))$ of the coordinate transformations are isomorphisms.

A point p in M is called a *critical point* of $f: M^n \to N^m$ if the rank of f at p is smaller than $\min(n, m)$. Otherwise p is called a *regular point* of f.

If p is a critical point of f, the value $f(p)$ is called a *critical value* of f. Other points in N are called *regular values* of f.

2.27 Remark If f has rank k at p, then f has rank $\geq k$ in some neighbourhood of p, for some $k \times k$ submatrix of $\left(\frac{\partial(y^i \circ f)}{\partial x^j}\right)$ has non-zero determinant at p, and hence by continuity in a neighbourhood of p.

In particular, if $f: M^n \to N^m$ has rank $k = m$ or $k = n$ at p, then f has rank k in some neighbourhood of p.

2.28 The Rank Theorem

(1) If $f: M^n \to N^m$ has rank k at p, then there are local charts (x, U) and (y, V) around p and $f(p)$, respectively, and a C^∞ map $h: x(U) \to \mathbf{R}^{m-k}$ such that

$$y \circ f \circ x^{-1}(a^1, \ldots, a^n) = (a^1, \ldots, a^k, h^1(a), \ldots, h^{m-k}(a)) \quad \text{for } a \in x(U).$$

(2) If $f : M^n \to N^m$ has rank k in a neighbourhood of p, then there are local charts (x,U) and (y,V) around p and $f(p)$, respectively, such that

$$y \circ f \circ x^{-1}(a^1,\dots,a^n) = (a^1,\dots,a^k,0,\dots,0) \text{ for } a \in x(U).$$

PROOF OF (1): Let (u,W) and (y,V) be local charts around p and $f(p)$, respectively. By a permutation of the coordinate functions u^i and y^j we may assume that the $k \times k$-submatrix $A = \left(\frac{\partial(y^\alpha \circ f)}{\partial u^\beta}(p) \right)$, where $1 \le \alpha, \beta \le k$, has non-zero determinant.

Let $f_1 : u(W) \to \mathbf{R}^k$ and $f_2 : u(W) \to \mathbf{R}^{m-k}$ be the component maps of $y \circ f \circ u^{-1} : u(W) \to \mathbf{R}^k \times \mathbf{R}^{m-k}$, i.e. $y \circ f \circ u^{-1}(a,b) = (f_1(a,b), f_2(a,b))$ for $(a,b) \in u(W) \subset \mathbf{R}^k \times \mathbf{R}^{n-k}$. Then

$$D(y \circ f \circ u^{-1})(u(p)) = \begin{pmatrix} D_1 f_1(u(p)) & D_2 f_1(u(p)) \\ D_1 f_2(u(p)) & D_2 f_2(u(p)) \end{pmatrix}$$

where $D_1 f_1(u(p)) = A$ is non-singular.

Let $\phi : u(W) \to \mathbf{R}^k \times \mathbf{R}^{n-k}$ be the C^∞-map defined by $\phi(a,b) = (f_1(a,b), b)$ for $(a,b) \in u(W) \subset \mathbf{R}^k \times \mathbf{R}^{n-k}$. Then

$$D\phi(u(p)) = \begin{pmatrix} D_1 f_1(u(p)) & D_2 f_1(u(p)) \\ 0 & I_{n-k} \end{pmatrix}$$

is non-singular, and by the inverse function theorem, ϕ is a local diffeomorphism on an open neighbourhood O of u(p).

We let (x,U) be the local chart around p with $U = u^{-1}(O)$ and $x = \phi \circ u|_U$. Then if $\phi(a,b) = (c,d) \in x(U)$, we have $y \circ f \circ x^{-1}(c,d) = y \circ f \circ u^{-1}(a,b) = (f_1(a,b), f_2(a,b)) = (c, h(c,d))$ where $h = f_2 \circ \phi^{-1} : x(U) \to \mathbf{R}^{m-k}$ is C^∞.

PROOF OF (2): Let (x,U) and (v,V) be local charts around p and $f(p)$, respectively, so that $v \circ f \circ x^{-1} : x(U) \to \mathbf{R}^k \times \mathbf{R}^{m-k}$ has the form in (1), i.e., $v \circ f \circ x^{-1}(a,b) = (a, h(a,b))$ for $(a,b) \in x(U) \subset \mathbf{R}^k \times \mathbf{R}^{n-k}$.

By choosing smaller coordinate neighbourhoods U and V, we may assume that f has rank k in U, and that $x(U) = O_1 \times O_2$ and $v(V) = O_3 \times O_4$ for open sets $O_1 \subset O_3 \subset \mathbf{R}^k$, $O_2 \subset \mathbf{R}^{n-k}$ and $O_4 \subset \mathbf{R}^{m-k}$. From (1) we see that O_3 may be replaced by O_1. Since

$$D(v \circ f \circ x^{-1}) = \begin{pmatrix} I_k & 0 \\ D_1 h & D_2 h \end{pmatrix},$$

we have that $D_2 h = 0$ in $x(U)$, and hence there is a C^∞ map $\overline{h} : O_1 \to \mathbf{R}^{m-k}$ so that $h(a,b) = \overline{h}(a)$ for $(a,b) \in O_1 \times O_2$.

Let $\phi : O_1 \times \mathbf{R}^{m-k} \to O_1 \times \mathbf{R}^{m-k}$ be the C^∞-map defined by $\phi(a,b) = (a, b - \overline{h}(a))$. ϕ is a diffeomorphism with inverse $\psi : O_1 \times \mathbf{R}^{m-k} \to O_1 \times \mathbf{R}^{m-k}$ given by $\psi(c,d) = (c, d + \overline{h}(c))$.

We let (y,V) be the local chart around $f(p)$ with $y = i \circ \phi \circ v$, where $i : O_1 \times \mathbf{R}^{m-k} \to \mathbf{R}^k \times \mathbf{R}^{m-k}$ is the inclusion map. Then $y \circ f \circ x^{-1}(a,b) = \phi \circ v \circ f \circ x^{-1}(a,b) = \phi(a, \overline{h}(a)) = (a,0)$ for $(a,b) \in x(U)$. □

2.29 Proposition Let $f : M \to N$ be a smooth map from a connected manifold M to a manifold N, and suppose that f has rank 0 on M. Then f is a constant map.

PROOF : If p is a point in $f(M)$, then $f^{-1}(p)$ is a nonempty closed subset of M by the continuity of f. We want to show that it is also open. Since M is connected, this implies that $f^{-1}(p) = M$ so that f is constant on M.

Let n and m be the dimensions of M and N, respectively, and let q be an arbitrary point in $f^{-1}(p)$. By the second part of the rank theorem, there are local charts (x,U) and (y,V) around q and p, respectively, such that

$$y \circ f \circ x^{-1}(a^1,...,a^n) = (0,...,0)$$

for $a \in x(U)$. This shows that $U \subset f^{-1}(p)$ and completes the proof that $f^{-1}(p)$ is open in M. □

SUBMANIFOLDS

2.30 Definition A smooth map $f : M^n \to N^m$ is called an *immersion* at a point p in M if it has rank n at p, and it is called a *submersion* at p if it has rank m at p. f is called an *immersion* (*submersion*) if it is an immersion (submersion) at all points p in M. An immersion $f : M \to N$ which is a homeomorphism onto its image $f(M)$ endowed with the subspace topology, is called an *embedding*.

Let M and N be smooth manifolds of dimensions n and m, respectively, with $M \subset N$. M is called an *immersed submanifold* of N if the inclusion map $i : M \to N$ is an immersion. It is called a *submanifold* if the inclusion map is an embedding. The number $m-n$ is called the *codimension* of M in N, and M is called a *hypersurface* of N if the codimension is 1.

2.31 Remark Let $f : M \to A$ be a bijection from a smooth manifold M to a set A. Then A can be made into a smooth manifold in a natural way so that f is a diffeomorphism. First define a topology on A so that f is a homeomorphism by defining a subset O of A to be open iff $f^{-1}(O)$ is open in M. If $\mathscr{D}$ denotes the smooth structure on M, we define the smooth structure on A to be $\{(x \circ f^{-1}, f(U)) \mid (x,U) \in \mathscr{D}\}$.

If $f : M \to N$ is a one-to-one immersion, the image $f(M)$ can be made into a smooth manifold in the manner just described. $f(M)$ is then an immersed submanifold of N. If $f : M \to N$ is an embedding, $f(M)$ will have the subspace topology inherited from N and hence is a submanifold of N.

2.32 Examples

(a) The map $f : \mathbf{R} \to \mathbf{R}^2$ defined by $f(t) = (2\sin(t), \sin(2t))$ is an immersion which is not one-to-one. Its image is the *Geronos lemniscate* L given by the equation $x^4 - 4x^2 + 4y^2 = 0$.

(b) The maps $f_1 = f|_{<-\pi,\pi>}$ and $f_2 = f|_{<0,2\pi>}$ are one-to-one immersions which are not embeddings. They make the image L into an immersed submanifold of $\mathbf{R}^2$ in two different ways.

(c) The map $f_3 = f|_{<-\frac{\pi}{2},\frac{\pi}{2}>}$ is an embedding which makes an open subset of L into a submanifold of $\mathbf{R}^2$.

(d) An open subset M of a smooth manifold N, with the smooth structure defined in Example 2.9 (c), is a submanifold of N.

2.33 Remarks (b) of the above example shows that given a subset A of a smooth manifold M, there is not in general a unique manifold structure (i.e., topology and smooth structure) on A which makes it an immersed submani- fold of M. However, we will see in Corollary 2.40 that there is at most one such smooth structure for a given topology on A.

(c) is an example of the next proposition which says that an immersion is locally an embedding.

2.34 Proposition A smooth map $f : M^n \to N^m$ is an immersion if and only if one of the following equivalent assertions are satisfied :

(1) Every point p in M has is an open neighbourhood U such that $f|_U$ is an embedding.

(2) For each point p in M, there exists local charts (x,U) around p in M and (y,V) around $f(p)$ in N such that $f(U) = \{q \in V \,|\, y^{n+1}(q) = \ldots = y^m(q) = 0\}$ and $x^1(q) = y^1 \circ f(q)\, , \ldots, x^n(q) = y^n \circ f(q)$ for $q \in U$.

PROOF: Assume that $f : M^n \to N^m$ is an immersion, and let p be a point in M. By the rank theorem there are local charts (x,U) around p in M and (v,W) around $f(p)$ in N such that the map $v \circ f \circ x^{-1} : x(U) \to \mathbf{R}^n \times \mathbf{R}^{m-n}$ is given by $v \circ f \circ x^{-1}(a) = (a,0)$ for $a \in x(U)$, in particular $f(U) \subset W$. Then $f|_U$ is an embedding since $f|_U = i \circ v^{-1} \circ (v \circ f \circ x^{-1}) \circ x$, where $v \circ f \circ x^{-1}$ is an embedding, x and v^{-1} are diffeomorphisms and $i : W \to N$ is the inclusion map. This proves (1).

In order to prove (2), we let (y,V) be the local chart around $f(p)$ in N with $V = v^{-1}(v(W) \cap (x(U) \times \mathbf{R}^{m-n}))$ and $y = v|_V$. Then $f(U) \subset V$, and we have that $y \circ f(U) = y \circ f \circ x^{-1}(x(U)) = y(V) \cap (\mathbf{R}^n \times \{0\})$ and $y \circ f(q) = y \circ f \circ x^{-1}(x(q)) = (x(q),0)$ for $q \in U$.

Conversely, assume that (2) is satisfied. Then $y \circ f \circ x^{-1}(a) = (a,0)$ for $a \in x(U)$, and hence f is an immersion at p. The same obviously holds if (1) is satisfied. Since this is true for all points p in M, we have that f is an immersion. □

2.35 Corollary Let M and N be smooth manifolds of dimensions n and m, respectively, with $M \subset N$. Then M is an immersed submanifold of N if and only if one of the following equivalent assertions are satisfied :

(1) Every point p in M has is an open neighbourhood U which is a submanifold of N.

(2) For each point p in M, there exists local charts (x,U) and (y,V) around p in M and N, respectively, such that $U = \{q \in V \mid y^{n+1}(q) = \ldots = y^m(q) = 0\}$ and $x^1(q) = y^1(q), \ldots, x^n(q) = y^n(q)$ for $q \in U$.

PROOF : Follows from Proposition 2.34 applied to the inclusion map $i : M \to N$. □

2.36 Proposition Let M be a subset of a smooth manifold N^m. Then M is a submanifold of N of dimension n if and only if the following assertion is satisfied:

For each point p in M, there exists a local chart (y,V) around p in N having the *submanifold property* $y(V \cap M) = y(V) \cap (\mathbf{R}^n \times \{0\})$.

PROOF : Assume that M is a submanifold of N of dimension n, and let p be a point in M. Let (x,U) and (y,V) be local charts around p in M and N, respectively, satisfying assertion (2) of Corollary 2.35. Since M is a subspace of N, we have that $U = M \cap O$ for some open set O in N. Replacing V by $V \cap O$, we see that the assertion above is satisfied.

Conversely, assume that M is a subset of N satisfying the assertion above. We give M the subspace topology, and we define a smooth structure on M as follows. Let $i : \mathbf{R}^n \to \mathbf{R}^n \times \mathbf{R}^{m-n}$ and $p : \mathbf{R}^n \times \mathbf{R}^{m-n} \to \mathbf{R}^n$ be the natural injection and projection defined by $i(a) = (a,0)$ and $p(a,b) = a$. A local chart (y,V) in N having the submanifold property $y(V \cap M) = y(V) \cap (\mathbf{R}^n \times \{0\})$ gives rise to a local chart (x,U) on M with coordinate neighbourhood $U = V \cap M$ and coordinate map $x = p \circ y|_U$. If (x',U') is another local chart on M obtained in the same way from (y',V'), then (x,U) and (x',U') are C^∞-related since $x' \circ x^{-1} = p \circ y' \circ y^{-1} \circ i$. We thus obtain a smooth structure on M which by Corollary 2.35 makes it a submanifold of N. □

2.37 Proposition Let M be an immersed submanifold of N, and let $f : P \to N$ be a smooth map with $f(P) \subset M$. If f is continuous considered as a map into M, then it is also smooth as a map into M.

PROOF : Let p be a point in P, and let n and m be the dimensions of M and N. By Corollary 2.35 there are local charts (x,U) and (y,V) around $f(p)$ in M and N, respectively, such that $U = \{q \in V \mid y^{n+1}(q) = \ldots = y^m(q) = 0\}$ and $x^1(q) = y^1(q), \ldots, x^n(q) = y^n(q)$ for $q \in U$.

Since f is continuous as a map into M, $f^{-1}(U)$ is an open subset of P containing p. Hence there is a local chart (z,W) around p in P with $W \subset f^{-1}(U)$. We have that $x^j \circ f \circ z^{-1} = y^j \circ f \circ z^{-1}$ for $j = 1, \ldots, n$, and since $f : P \to N$ is smooth, these functions are C^∞ at $z(p)$. This proves that f is smooth at p as a map into M, and since p was arbitrary, this completes the proof. □

2.38 Corollary Let M be a submanifold of N, and let $f : P \to N$ be a smooth map with $f(P) \subset M$. Then f is also smooth as a map into M.

PROOF: Since M has the subspace topology, f is continuous as a map into M. The result then follows from the proposition. □

2.39 Remark If M is only an immersed submanifold of N, the assumption that f is continuous as a map into M is necessary. Consider for instance the immersed submanifold L of $\mathbf{R}^2$ defined in Example 2.32 with manifold structure given by the immersion f_2. Then $f_3 : <-\frac{\pi}{2}, \frac{\pi}{2}> \to \mathbf{R}^2$ is a smooth map which is discontinuous considered as a map into L.

2.40 Corollary Let A be a subset of a smooth manifold N, and fix a topology on A. Then there is at most one smooth structure on A which makes it an immersed submanifold of N.

PROOF: Let P and M be equal as topological spaces and let each have a smooth structure which makes it an immersed submanifold of N. Then $id : P \to M$ is a homeomorphism, and hence by Proposition 2.37 a diffeomorphism proving that P and M must have the same smooth structure. □

2.41 Proposition Let $f : M^n \to N^m$ be smooth, and let q be a point in N. If f has constant rank k in a neighbourhood of $f^{-1}(q)$, then $f^{-1}(q)$ is a closed submanifold of M of dimension $n-k$.

PROOF: Let p be a point in $f^{-1}(q)$. By the rank theorem there are local charts (u,U) around p in M and (y,V) around q in N such that the map $y \circ f \circ u^{-1} : u(U) \to \mathbf{R}^k \times \mathbf{R}^{m-k}$ is given by $y \circ f \circ u^{-1}(a,b) = (a,0)$ for $(a,b) \in u(U) \subset \mathbf{R}^k \times \mathbf{R}^{n-k}$. If $u(p) = (a_0, b_0)$, then $u(U \cap f^{-1}(q)) = u(U) \cap (\{a_0\} \times \mathbf{R}^{n-k})$.

Let $r : \mathbf{R}^k \times \mathbf{R}^{n-k} \to \mathbf{R}^{n-k} \times \mathbf{R}^k$ be the map defined by $r(a,b) = (b, a-a_0)$. Then the local chart (x,U) where $x = r \circ u$ has the submanifold property $x(U \cap f^{-1}(q)) = x(U) \cap (\mathbf{R}^{n-k} \times \{0\})$. Since p was an arbitrary point in $f^{-1}(q)$, this proves that $f^{-1}(q)$ is a closed submanifold of M of dimension $n-k$. □

2.42 Example The smooth map $f : \mathbf{R}^{n+1} \to \mathbf{R}$ defined by $f(x) = \|x\|^2$ has rank n in $\mathbf{R}^{n+1} \setminus \{0\}$. Hence the n-sphere $\mathbf{S}^n = f^{-1}(1)$ is a closed submanifold of $\mathbf{R}^{n+1}$ of dimension n.

This manifold structure coincides with the one defined in Example 2.9 (e). To see this, we note that $id \circ f_i^{-1}$ and $id \circ g_i^{-1}$ have rank n on D. Hence the inclusion map $i : \mathbf{S}^n \to \mathbf{R}^{n+1}$ is an embedding when we use the atlas consisting of the local charts (f_i, U_i) and (g_i, V_i) on $\mathbf{S}^n$, and the standard atlas consisting of the local chart $(id, \mathbf{R}^{n+1})$ on $\mathbf{R}^{n+1}$. The manifold structure making $\mathbf{S}^n$ a submanifold of $\mathbf{R}^{n+1}$ is uniquely determined by Corollary 2.40.

2.43 Proposition Let $f : M^n \to N^m$ be a submersion. Then f is an open mapping.

PROOF : Let W be an open subset of M, and suppose that $q \in f(W)$. Choose a point $p \in W$ with $f(p) = q$. By the rank theorem there are local charts (x,U) and (y,V) around p and q, respectively, with $U \subset W$ such that the map $y \circ f \circ x^{-1} : x(U) \to \mathbf{R}^m$ is given by $y \circ f \circ x^{-1}(a,b) = a$ for $(a,b) \in x(U) \subset \mathbf{R}^m \times \mathbf{R}^{n-m}$. As the projection $pr_1 : \mathbf{R}^m \times \mathbf{R}^{n-m} \to \mathbf{R}^m$ on the first factor is an open mapping, it follows that $f(U) = y^{-1} \circ pr_1 \circ x(U)$ is an open neighbourhood of q contained in $f(W)$. Hence $f(W)$ is open in N for every open subset W of M, which shows that f is an open mapping. □

VECTOR BUNDLES

2.44 Definition Let M be a smooth manifold. The product manifold $M \times \mathbf{R}^n$ is called an n-dimensional *product bundle* over M. The projection $pr_1 : M \times \mathbf{R}^n \to M$ on the first factor is a surjective smooth map called the *projection* of the bundle. For a point p in M, the inverse image $pr_1{}^{-1}(p) = \{p\} \times \mathbf{R}^n$ is called the *fibre* over p. It has a natural vector space structure such that the projection $pr_2 : \{p\} \times \mathbf{R}^n \to \mathbf{R}^n$ on the second factor is a linear isomorphism, i.e., with operations $(p,u)+(p,v) = (p,u+v)$ and $k\,(p,u) = (p,ku)$ for vectors u and v in $\mathbf{R}^n$ and real numbers k.

The product bundle is an example of a more general bundle called a vector bundle which we define next. A vector bundle is a smooth manifold which looks locally like a product bundle.

2.45 Definition Let M be a smooth manifold. A smooth manifold E is called an n-dimensional *vector bundle* over M if the following three conditions are satisfied :

(i) There is a surjective smooth map $\pi : E \to M$ which is called the *projection* of the *total space* E onto the *base space* M.

(ii) For each point $p \in M$, $\pi^{-1}(p)$ is an n-dimensional vector space called the *fibre* over p.

(iii) For each point $p \in M$ there is an open neighbourhood U around p and a diffeomorphism $t : \pi^{-1}(U) \to U \times \mathbf{R}^n$ such that the diagram

$$\begin{array}{ccc} \pi^{-1}(U) & \xrightarrow{\;t\;} & U \times \mathbf{R}^n \\ & {\scriptstyle \pi_U} \searrow \quad \swarrow {\scriptstyle pr_1} & \\ & U & \end{array}$$

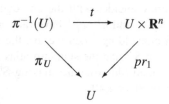

is commutative, and t is a linear isomorphism on each fibre. The pair $(t,\pi^{-1}(U))$ is called a *local trivialization* around p.

We often use the projection $\pi : E \to M$ instead of the total space E to denote the bundle in order to indicate more of the bundle structure in the notation. The fibre $\pi^{-1}(p)$ is also denoted by E_p.

A family of local trivializations $\{(t_\alpha,\pi^{-1}(U_\alpha))|\alpha \in A\}$, where $\{U_\alpha|\alpha \in A\}$ is an open cover of M, is called a *trivializing cover* of E.

2.46 Remark If $(t,\pi^{-1}(U))$ is a local trivialization, we let the map $t_p : E_p \to \mathbf{R}^n$ be the linear isomorphism defined by $t_p = pr_2 \circ t|_{E_p}$ for each $p \in U$. We have that $t(v) = (p,t_p(v))$ when $v \in E_p$.

If $\mathscr{E} = \{e_1,...,e_n\}$ is the standard basis for $\mathbf{R}^n$, then $\mathscr{B}_p = \{t_p^{-1}(e_1),...,t_p^{-1}(e_n)\}$ is a basis for the fibre $\pi^{-1}(p)$.

2.47 Proposition If $\pi : E \to M$ is an n-dimensional vector bundle and p is a point in M, then there is a local trivialization $(t,\pi^{-1}(U))$ around p where U is the coordinate neighbourhood of a local chart (x,U) on M.

PROOF : If $(t_1,\pi^{-1}(U_1))$ is a local trivialization and (x_2,U_2) a local chart around p, let $U = U_1 \cap U_2$ and $x = x_2|_U$, and let $t : \pi^{-1}(U) \to U \times \mathbf{R}^n$ be the diffeomorphism induced by t_1. □

2.48 Proposition Let $\pi : E \to M$ be a map from a set E to a smooth manifold M, and let $\{U_\alpha|\alpha \in A\}$ be an open cover of M. Suppose that for each $\alpha \in A$, there is a bijection $t_\alpha : \pi^{-1}(U_\alpha) \to U_\alpha \times \mathbf{R}^n$ such that the diagram

$$\begin{array}{ccc} \pi^{-1}(U_\alpha) & \xrightarrow{\ t_\alpha\ } & U_\alpha \times \mathbf{R}^n \\ & {\scriptstyle \pi_{U_\alpha}} \searrow \quad \swarrow {\scriptstyle pr_1} & \\ & U_\alpha & \end{array}$$

is commutative. Moreover, suppose that for each pair $\alpha,\beta \in A$, the map $t_\beta \circ t_\alpha^{-1}$ is a diffeomorphism from $(U_\alpha \cap U_\beta) \times \mathbf{R}^n$ onto itself which is a linear isomorphism on each fibre $\{p\} \times \mathbf{R}^n$ for $p \in U_\alpha \cap U_\beta$.

Then there is a unique topology and smooth structure on E and a unique vector space structure on each fibre $\pi^{-1}(p)$ for $p \in M$ such that $\pi : E \to M$ is an n-dimensional vector bundle and $(t_\alpha,\pi^{-1}(U_\alpha))$ is a local trivialization for each $\alpha \in A$.

PROOF : We first define a topology τ on E such that the projection π is continuous and the bijections $t_\alpha : \pi^{-1}(U_\alpha) \to U_\alpha \times \mathbf{R}^n$ are homeomorphisms for every $\alpha \in A$. The continuity of π implies that each $\pi^{-1}(U_\alpha)$ must be open in τ, and hence a set O

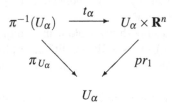

in E must belong to τ if and only if $O \cap \pi^{-1}(U_\alpha) \in \tau$ for every $\alpha \in A$. The requirement that the bijections t_α are homeomorphisms now defines τ uniquely as the collection of sets O in E such that $t_\alpha(O \cap \pi^{-1}(U_\alpha))$ is open in $U_\alpha \times \mathbf{R}^n$ for every $\alpha \in A$. This collection τ is clearly a topology on E, and we must prove that it satisfies the above requirements.

We first show that $\pi^{-1}(U_\alpha)$ is open in τ for each $\alpha \in A$. For every $\beta \in A$ we have that $t_\beta(\pi^{-1}(U_\alpha) \cap \pi^{-1}(U_\beta)) = (U_\alpha \cap U_\beta) \times \mathbf{R}^n$ which is open in $U_\beta \times \mathbf{R}^n$, thereby proving the assertion.

We next show that t_α is a homeomorphism for every $\alpha \in A$. If $O \subset \pi^{-1}(U_\alpha)$ belongs to τ, we have that $t_\alpha(O \cap \pi^{-1}(U_\alpha))$ is open in $U_\alpha \times \mathbf{R}^n$, which proves that t_α is an open map. To show that it is continuous, let O be an open set in $U_\alpha \times \mathbf{R}^n$. For every $\beta \in A$ we have that $t_\beta(t_\alpha^{-1}(O) \cap \pi^{-1}(U_\beta)) = (t_\beta \circ t_\alpha^{-1})(O \cap (U_\alpha \cap U_\beta) \times \mathbf{R}^n)$ which is open in $U_\beta \times \mathbf{R}^n$, and hence $t_\alpha^{-1}(O) \in \tau$ showing that t_α is continuous.

From this it also follows that the projection π is continuous since $\pi_{U_\alpha} = pr_1 \circ t_\alpha$ is the composition of the continuous maps t_α and pr_1 in the above commutative diagram for each $\alpha \in A$.

Now let $\{(x_\beta, V_\beta) \mid \beta \in B\}$ be an atlas on M such that $\{V_\beta \mid \beta \in B\}$ is a refinement of the open cover $\{U_\alpha \mid \alpha \in A\}$. We then have a map $\phi : B \to A$ such that $V_\beta \subset U_{\phi(\beta)}$ for every $\beta \in B$. By the assumptions in the proposition we now see that the family $\mathscr{B} = \{((x_\beta \times id) \circ t_{\phi(\beta)}, \pi^{-1}(V_\beta)) \mid \beta \in B\}$ is an atlas on E. To show that E is a smooth manifold, we must show that it is Hausdorff and second countable. By Proposition 2.6 we know that M has an atlas of the above form where B is countable, thus showing that E is second countable.

To see that the E is Hausdorff, let p_1 and p_2 be two different points in E. If p_1 and p_2 lie in different fibres, they can be separated by the neighbourhoods $\pi^{-1}(W_1)$ and $\pi^{-1}(W_2)$, where W_1 and W_2 are disjoint neighbourhoods of $\pi(p_1)$ and $\pi(p_2)$ in M. If p_1 and p_2 lie in the same fibre $\pi^{-1}(q)$ with $q \in U_\alpha$, they can be separated by the neighbourhoods $t_\alpha^{-1}(O_1)$ and $t_\alpha^{-1}(O_2)$, where O_1 and O_2 are disjoint neighbourhoods of $t_\alpha(p_1)$ and $t_\alpha(p_2)$ in $U_\alpha \times \mathbf{R}^n$.

This shows that E is a smooth manifold with a unique smooth structure such that the bijections $t_\alpha : \pi^{-1}(U_\alpha) \to U_\alpha \times \mathbf{R}^n$ are diffeomorphisms for every $\alpha \in A$. We see that the projection π is smooth since $\pi_{U_\alpha} = pr_1 \circ t_\alpha$ is the composition of the smooth maps t_α and pr_1.

We finally show that there is a unique vector space structure on each fibre $\pi^{-1}(p)$ for $p \in M$ such that $(t_\alpha, \pi^{-1}(U_\alpha))$ is a local trivialization for each $\alpha \in A$. If t_α is to be a linear isomorphism on the fibre $\pi^{-1}(p)$ where $p \in U_\alpha$, we must define addition and scalar multiplication on this fibre by

$$u + v = t_\alpha^{-1}(t_\alpha(u) + t_\alpha(v)) \quad \text{and} \quad ku = t_\alpha^{-1}(k\, t_\alpha(u))$$

for elements u and v in $\pi^{-1}(p)$ and real numbers k. If $p \in U_\beta$ for another index $\beta \in A$, we have that $t_\beta = (t_\beta \circ t_\alpha^{-1}) \circ t_\alpha$ is also a linear isomorphism since it is the composition of the linear isomorphisms t_α and $t_\beta \circ t_\alpha^{-1}$. Hence each fibre has a well-defined vector space structure such that $(t_\alpha, \pi^{-1}(U_\alpha))$ is a local trivialization for each $\alpha \in A$, and this completes the construction of the vector bundle $\pi : E \to M$. □

2.49 Definition Let $\pi_1 : E_1 \to M_1$ and $\pi_2 : E_2 \to M_2$ be two vector bundles. A *bundle map* from π_1 to π_2 is a pair $(\tilde{f}, f)$ of smooth maps $\tilde{f} : E_1 \to E_2$ and $f : M_1 \to M_2$ such that the diagram

$$\begin{array}{ccc} E_1 & \xrightarrow{\tilde{f}} & E_2 \\ {\scriptstyle \pi_1}\downarrow & & \downarrow{\scriptstyle \pi_2} \\ M_1 & \xrightarrow{f} & M_2 \end{array}$$

is commutative, and $\tilde{f}$ is linear on each fibre. We let $\tilde{f}_p$ denote the induced linear map $\tilde{f}_p : \pi_1^{-1}(p) \to \pi_2^{-1}(f(p))$ between the fibres for each $p \in M_1$. If $\tilde{f}$ and f are diffeomorphisms, $(\tilde{f}, f)$ is called an *equivalence*.

2.50 Definition Let $\pi_1 : E_1 \to M$ and $\pi_2 : E_2 \to M$ be two vector bundles over the same base space M. A *bundle map* from π_1 to π_2 *over* M is a smooth map $f : E_1 \to E_2$ such that the diagram

$$\begin{array}{ccc} E_1 & \xrightarrow{f} & E_2 \\ & {\scriptstyle \pi_1}\searrow \quad \swarrow{\scriptstyle \pi_2} & \\ & M & \end{array}$$

is commutative, and f is linear on each fibre. We let f_p denote the induced linear map $f_p : \pi_1^{-1}(p) \to \pi_2^{-1}(p)$ between the fibres for each $p \in M$. If f is a diffeomorphism, it is called an *equivalence over M* .

2.51 Definition A vector bundle $\pi : E \to M$ is said to be *trivial* if it is equivalent over M with a product bundle.

2.52 Definition Let $\pi : E \to M$ be an n-dimensional vector bundle, and let V be an open subset of M and $i : V \to M$ be the inclusion mapping. Then a smooth map $s : V \to E$ with $\pi \circ s = i$ is called a *section* of π on V, or a *lifting* of i to E. The set of all sections of π on V is denoted by $\Gamma(V; E)$.

If $(t, \pi^{-1}(U))$ is a local trivialization of π with $U \subset V$ and $s : V \to E$ is a map with $\pi \circ s = i$, then it follows from the commutative diagram in Definition 2.45 that $pr_1 \circ t \circ s = id_U$. Hence $s|_U$ is completely determined by the map $h : U \to \mathbf{R}^n$ given by $(t \circ s)(p) = (p, h(p))$ for $p \in U$. h is called the *local representation* of s on U. We see that s is smooth on U if and only if the local representation h is smooth.

2.53 Proposition If $\pi : E \to M$ is a vector bundle and V is an open subset of

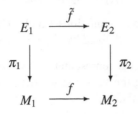

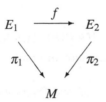

M, then the set $\Gamma(V;E)$ of sections of π on V is a module over the ring $\mathscr{F}(V)$ of C^∞-functions on V with operations defined by

$$(s_1+s_2)(p) = s_1(p)+s_2(p) \quad \text{and} \quad (fs)(p) = f(p)s(p)$$

for sections s_1, s_2 and s and C^∞-functions f on V. The zero element in $\Gamma(V;E)$ is the *zero section* $\zeta : V \to E$ on V defined by $\zeta(p) = 0$ for $p \in V$.

PROOF : By considering local representations we see that s_1+s_2, fs and ζ are sections of π on V, and using the vector space structure on each fibre we see that the axioms for a module are satisfied. □

2.54 Proposition Let $\pi : E \to M$ be an n-dimensional vector bundle and $s : M \to E$ be a map with $\pi \circ s = id_M$, and let $\mathscr{E} = \{e_1, ..., e_n\}$ be the standard basis for $\mathbf{R}^n$. Then s is a section of π on M if and only if the map $a : U \to \mathbf{R}^n$ defined by

$$s(p) = \sum_{i=1}^{n} a_i(p)\, t_p^{-1}(e_i)$$

for $p \in U$ is smooth for every local trivialization $(t, \pi^{-1}(U))$ in a trivializing cover of E .

PROOF : Since

$$(t \circ s)(p) = (p, a(p))$$

for $p \in U$, a is the local representation of s on U as defined in Definition 2.52. □

2.55 Remark Definition 2.52 and Proposition 2.54 can be generalized in the following way. Let $\pi : E \to M$ be an n-dimensional vector bundle, and let $f : N \to M$ be a smooth map from a smooth manifold N. Then a smooth map $F : N \to E$ with $\pi \circ F = f$ is called a *section* of π *along* f, or a *lifting* of f to E. The set of all liftings of f to E is denoted by $\Gamma(f;E)$.

If $(t, \pi^{-1}(U))$ is a local trivialization of π and $F : N \to E$ is a map with $\pi \circ F = f$, then it follows from the commutative diagram in Definition 2.45 that $pr_1 \circ t \circ F = f_U$. Hence $F|_{f^{-1}(U)}$ is completely determined by the map $a : f^{-1}(U) \to \mathbf{R}^n$, called the *local representation* of F on U, given by

$$(t \circ F)(p) = (f(p), a(p))$$

for $p \in f^{-1}(U)$. If $\mathscr{E} = \{e_1, ..., e_n\}$ is the standard basis for $\mathbf{R}^n$, this relation is also equivalent to

$$F(p) = \sum_{i=1}^{n} a_i(p)\, t_{f(p)}^{-1}(e_i)$$

for $p \in f^{-1}(U)$. We see that F is smooth on $f^{-1}(U)$ if and only if the local representation a is smooth.

2.56 Proposition Let $\pi : E \to M$ be an n-dimensional vector bundle over a smooth manifold M.

(1) If $s_1 : V \to E$ is a section of π on an open subset V of M, and if W is a closed subset of M with $W \subset V$, then there exists a section $s : M \to E$ of π on M which coincides with s_1 on W.

(2) If $v \in E_p$ for a point $p \in M$, then there exists a section $s : M \to E$ of π on M with $s(p) = v$.

PROOF OF (1): By Corollary 2.24 there exists a smooth function $f : M \to \mathbf{R}$ with $f(q) = 1$ for $q \in W$ and $\operatorname{supp}(f) \subset V$, and we let

$$s(q) = \begin{cases} f(q)\, s_1(q) & \text{for } q \in V \\ 0 & \text{for } q \notin V \end{cases} .$$

PROOF OF (2): Let $(t, \pi^{-1}(U))$ be a local trivialization around p, and let $\mathscr{E} = \{e_1, \ldots, e_n\}$ be the standard basis for $\mathbf{R}^n$. If

$$v = \sum_{i=1}^{n} a_i t_p^{-1}(e_i) ,$$

we have a section $s_1 : U \to E$ of π on U given by

$$s_1(q) = \sum_{i=1}^{n} a_i t_q^{-1}(e_i)$$

for $q \in U$. Using part (1) there therefore exists a section $s : M \to E$ of π on M which coincides with s_1 on the closed subset $\{p\}$ of M with $\{p\} \subset U$ so that $s(p) = v$. □

2.57 Proposition Let $\pi_1 : E_1 \to M_1$ and $\pi_2 : E_2 \to M_2$ be two vector bundles of dimension n, and let $(\tilde{f}, f)$ be a bundle map from π_1 to π_2 where $\tilde{f}$ induces a linear isomorphism between the fibres $\pi_1^{-1}(p)$ and $\pi_2^{-1}(f(p))$ for every $p \in M_1$. If s_2 is a section of π_2 on an open subset V_2 of M_2 and $V_1 = f^{-1}(V_2)$, then the map $s_1 : V_1 \to E_1$ defined by

$$s_1(p) = \tilde{f}_p^{-1} \circ s_2 \circ f(p)$$

for $p \in V_1$ is a section of π_1 on V_1.

s_1 is called the *pull-back* of s_2 by $\tilde{f}$ and is denoted by $\tilde{f}^*(s_2)$.

PROOF: Let p be a point in V_1, and let $(t_1, \pi_1^{-1}(U_1))$ and $(t_2, \pi_2^{-1}(U_2))$ be local trivializations of π_1 and π_2 around p and $f(p)$, respectively, with $U_2 \subset V_2$ and $U_1 \subset f^{-1}(U_2)$. Then the map $t_2 \circ \tilde{f} \circ t_1^{-1} : U_1 \times \mathbf{R}^n \to U_2 \times \mathbf{R}^n$ is smooth and induces a linear isomorphism $t_{2,f(q)} \circ \tilde{f}_q \circ t_{1,q}^{-1}$ between the fibres $\{q\} \times \mathbf{R}^n$ and $\{f(q)\} \times \mathbf{R}^n$ for every $q \in U_1$.

Now $t_i \circ s_i(q) = (q, h_i(q))$ for $i = 1, 2$, where the local representations $h_i : U_i \to \mathbf{R}^n$ of s_i are given by

$$h_1(q) = t_{1,q} \circ \tilde{f}_q^{-1} \circ s_2(f(q))$$

and

$$h_2(q) = t_{2,q} \circ s_2(q).$$

Combining this it follows that

$$h_1(q) = (t_{2,f(q)} \circ \tilde{f}_q \circ t_{1,q}^{-1})^{-1} \circ h_2(f(q))$$

for $q \in U_1$. If $\mathscr{E}$ is the standard basis for $\mathbf{R}^n$, we have that

$$m_{\mathscr{E}}^{\mathscr{E}}\{(t_{2,f(q)} \circ \tilde{f}_q \circ t_{1,q}^{-1})^{-1}\} = m_{\mathscr{E}}^{\mathscr{E}}(t_{2,f(q)} \circ \tilde{f}_q \circ t_{1,q}^{-1})^{-1}.$$

which shows that s_1 is smooth in a neighbourhood of every point p in V_1 and hence is a section of π_1 on V_1. □

2.58 Proposition An n-dimensional vector bundle $\pi : E \to M$ is trivial if and only if there are n sections $s_1, ..., s_n$ of π on M which are everywhere linearly independent, i.e., $s_1(p), ..., s_n(p) \in \pi^{-1}(p)$ are linearly independent for every $p \in M$.

PROOF: Suppose first that E is trivial, and let $f : E \to M \times \mathbf{R}^n$ be an equivalence over M. Then the sections $s_i : M \to E$ defined by $s_i(p) = f^{-1}(p, e_i)$ for $p \in M$ and $i = 1, ..., n$, where e_i is the i-th standard basis vector in $\mathbf{R}^n$, clearly are everywhere linearly independent.

Conversely, suppose there are n sections $s_1, ..., s_n$ of π on M which are everywhere linearly independent, and let $f : E \to M \times \mathbf{R}^n$ be the map defined by

$$f(\sum_{i=1}^{n} a_i s_i(p)) = (p, a)$$

for $p \in M$ and $a \in \mathbf{R}^n$. To show that f is an equivalence over M, let $(t, \pi^{-1}(U))$ be a local trivialization in E, and let $h_i : U \to \mathbf{R}^n$ be the local representation of s_i on U for $i = 1, ..., n$ so that $(t \circ s_i)(p) = (p, h_i(p))$ for $p \in U$. Writing the vectors in $\mathbf{R}^n$ as column vectors, we then have a smooth map $h : U \to \mathrm{Gl}(n, \mathbf{R})$, where $h(p)$ is the matrix with $h_1(p), ..., h_n(p)$ as column vectors for $p \in U$. Combining this, we have that

$$t \circ f^{-1}(p, a) = t(\sum_{i=1}^{n} a_i s_i(p)) = (p, \sum_{i=1}^{n} a_i h_i(p)) = (p, h(p)a)$$

and hence also

$$f \circ t^{-1}(p, b) = (p, h(p)^{-1} b)$$

for every $p \in U$ and $a, b \in \mathbf{R}^n$. Since this holds for every local trivialization $(t, \pi^{-1}(U))$ in E, it follows that f is an equivalence over M. □

2.59 Corollary Let $\pi : E \to M$ be an n-dimensional vector bundle, and let $s_1, ..., s_n$ be n sections of π on an open subset U of M which are everywhere linearly independent, i.e., $s_1(p), ..., s_n(p) \in \pi^{-1}(p)$ are linearly independent for every $p \in U$. Then we have a unique local trivialization $(t, \pi^{-1}(U))$ in E associated with

$s_1,...,s_n$ such that $t_p(s_i(p)) = e_i$ for $p \in U$ and $i = 1,...,n$, where e_i is the i-th standard basis vector in $\mathbf{R}^n$.

A map $s : U \to E$, with $\pi(s(p)) = p$ for every $p \in U$, is a section of π on U if and only if the map $a : U \to \mathbf{R}^n$ defined by

$$s(p) = \sum_{i=1}^{n} a_i(p)\, s_i(p)$$

for $p \in U$ is smooth.

PROOF : The existence and uniqueness of the local trivialization $(t, \pi^{-1}(U))$ follows from the last part of the proof of Proposition 2.58, replacing π with the bundle $\pi|_U$.

The last part of the corollary follows from the relation

$$(t \circ s)(p) = (p, a(p))$$

which holds for $p \in U$, and which shows that a is the local representation of s on U as defined in Definition 2.52. □

2.60 Definition Let $\pi : E \to M$ and $\pi' : E' \to M$ be two vector bundles over M of dimensions n and k, respectively. Then π' is called a *subbundle* of π if E' is a submanifold of E and the inclusion map $i : E' \to E$ is a bundle map over M.

2.61 Remark The inclusion map i being a bundle map is equivalent to the assertion that $\pi' = \pi|_{E'}$ and that each fibre $\pi'^{-1}(p)$ in E' is a k-dimensional subspace of the fibre $\pi^{-1}(p)$ in E for all $p \in M$.

2.62 Proposition Suppose that $\pi : E \to M$ is an n-dimensional vector bundle, and that E' is a subset of E such that $\pi^{-1}(p) \cap E'$ is a k-dimensional subspace of the fibre $\pi^{-1}(p)$ for every $p \in M$. Then E' is a k-dimensional subbundle of E if and only if one of the following equivalent assertions are satisfied :

(1) For each point p in M there is an open neighbourhood U around p and k sections $s_1,...,s_k$ of π on U such that $s_1(q),...,s_k(q)$ is a basis for $\pi^{-1}(q) \cap E'$ for all $q \in U$.

(2) For each point p in M, there exists a local trivialization $(t, \pi^{-1}(U))$ around p in E having the *subbundle property* $t(\pi^{-1}(U) \cap E') = U \times \mathbf{R}^k \times \{0\} \subset U \times \mathbf{R}^k \times \mathbf{R}^{n-k}$.

PROOF : Suppose first that E' is a k-dimensional subbundle of E with projection $\pi' = \pi|_{E'}$, and let $(t', \pi'^{-1}(U))$ be a local trivialization around p in E'. Then the sections $s_j : U \to E$ defined by $s_j(q) = i \circ t'^{-1}(q, e_j)$ for $j = 1,...,k$, where e_j is the j-th standard basis vector in $\mathbf{R}^k$ and $i : E' \to E$ is the inclusion map, clearly satisfy assertion (1).

We next prove that assertion (1) implies (2). If p is a point in M, let $s_j : U_1 \to E$ for

$j = 1, ..., k$ be the sections given in (1) defined on an open neighbourhood U_1 around p. Let $(t_2, \pi^{-1}(U_2))$ be a local trivialization around p in E, and put $U_0 = U_1 \cap U_2$. If $h_j : U_0 \to \mathbf{R}^n$ is the local representation of s_j as defined in Definition 2.52, we have a smooth map $f : U_0 \to L(\mathbf{R}^k, \mathbf{R}^n)$ given by

$$f(q)(a) = \sum_{j=1}^{k} a_j h_j(q)$$

for $q \in U_0$ and $a \in \mathbf{R}^k$, where $f(p)$ is of rank k. If $\pi_1 : \mathbf{R}^k \times \mathbf{R}^{n-k} \to \mathbf{R}^k$ and $\pi_2 : \mathbf{R}^k \times \mathbf{R}^{n-k} \to \mathbf{R}^{n-k}$ are the projections, there is a permutation σ in S_n such that the standard matrix of $\pi_1 \circ \pi_\sigma \circ f(p)$ is non-singular, where $\pi_\sigma : \mathbf{R}^n \to \mathbf{R}^n$ is the linear isomorphism obtained by permuting the coordinates in $\mathbf{R}^n$ with σ as defined in 13.8 in the appendix.

Now let $A_r(q)$ be the standard matrix of $\pi_r \circ \pi_\sigma \circ f(q)$ for $q \in U_0$ and $r = 1, 2$. Then A_1 is non-singular in an open neighbourhood U of p contained in U_0. If $g(q) : \mathbf{R}^n \to \mathbf{R}^n$ is the linear isomorphism with standard matrix

$$\begin{pmatrix} A_1(q) & 0 \\ A_2(q) & I_{n-k} \end{pmatrix}^{-1} = \begin{pmatrix} A_1(q)^{-1} & 0 \\ -A_2(q)\,A_1(q)^{-1} & I_{n-k} \end{pmatrix}$$

for $q \in U$, and $i_1 : \mathbf{R}^k \to \mathbf{R}^k \times \mathbf{R}^{n-k}$ is the linear map defined by $i_1(a) = (a, 0)$ for $a \in \mathbf{R}^k$, then the map $g : U \to L(\mathbf{R}^n, \mathbf{R}^n)$ is smooth, and we have that

$$\pi_\sigma \circ f(q) = g(q)^{-1} \circ i_1$$

so that

$$g(q) \circ \pi_\sigma \circ f(q) = i_1 \ .$$

Hence the local trivialization $(t, \pi^{-1}(U))$ around p given by $t(v) = (id_U \times g(\pi(v)) \circ \pi_\sigma) \circ t_2(v)$ for $v \in \pi^{-1}(U)$ satisfies the subbundle property in (2). Indeed we have that

$$t_2(\pi^{-1}(q) \cap E') = \{q\} \times f(q)(\mathbf{R}^k)$$

which implies that

$$t(\pi^{-1}(q) \cap E') = \{q\} \times g(q) \circ \pi_\sigma \circ f(q)(\mathbf{R}^k) = \{q\} \times \mathbf{R}^k \times \{0\}$$

for $q \in U$.

Finally we assume that assertion (2) is satisfied and show that E' is a subbundle of E. It only remains to show that E' is a k-dimensional submanifold of E. Let v be a point in E', and let $(t_1, \pi^{-1}(U_1))$ be a local trivialization in E having the subbundle property and (x_2, U_2) a local chart on M around $\pi(v)$. Then $((x_2 \times id) \circ t_1, \pi^{-1}(U_1 \cap U_2))$ is a local chart around v on E having the submanifold property, and the result hence follows from Proposition 2.36. □

2.63 Remark The sections $s_1, ..., s_k$ in assertion (1) of the proposition are said to form a *local basis* for E' on U.

A section s of $\pi : E \to M$ on an open subset U of M is said to *belong to* or *lie in* the subbundle E' of E, and we write $s \in E'$, if $s(q) \in E'$ for all $q \in U$.

If s' is a section of $\pi' : E' \to M$ on U, then $s = i \circ s'$ is a section of $\pi : E \to M$ on U which belongs to E'. Conversely, each section $s \in E'$ is obtained in this way from a unique section s'.

2.64 Proposition Let $\pi_1 : E_1 \to M$ and $\pi_2 : E_2 \to M$ be subbundles of the vector bundle $\pi : E \to M$ so that

$$E_p = E_{1p} \oplus E_{2p}$$

for every $p \in M$, and let $f_1 : E \to E$ and $f_2 : E \to E$ be maps which are projections on E_{1p} and E_{2p} in each fibre E_p, i.e.,

$$v = f_1(v) + f_2(v) \quad \text{where} \quad f_1(v) \in E_{1p} \quad \text{and} \quad f_2(v) \in E_{2p}$$

for $v \in E_p$. Then f_1 and f_2 are bundle maps over M.

PROOF: Let $\{s_1, ..., s_n\}$ and $\{s_{n+1}, ..., s_{n+m}\}$ be local bases for E_1 and E_2, respectively, on an open neighbourhood U of a point $p_0 \in M$. Then $\{s_1, ..., s_{n+m}\}$ is a local basis for E on U, and by Corollary 2.59 there is a unique local trivialization $(t, \pi^{-1}(U))$ in E such that $t_p(s_i(p)) = e_i$ for $p \in U$ and $i = 1, ..., n+m$, where e_i is the i-th standard basis vector in $\mathbf{R}^{n+m}$. Now we have that

$$t \circ f_1 \circ t^{-1}(p,a,b) = (p,a,0) \quad \text{and} \quad t \circ f_2 \circ t^{-1}(p,a,b) = (p,0,b)$$

for $(p,a,b) \in U \times \mathbf{R}^n \times \mathbf{R}^m$, showing that f_1 and f_2 are bundle maps over M. □

2.65 Proposition Let $\pi : E \to M$ and $\pi' : E' \to M$ be two vector bundles over a smooth manifold M of dimensions n and m, respectively, and let $f : E \to E'$ be a bundle map over M so that $\dim \ker(f_p) = k$ for every $p \in M$. Then

$$\ker(f) = \bigcup_{p \in M} \ker(f_p) \quad \text{and} \quad \operatorname{im}(f) = \bigcup_{p \in M} \operatorname{im}(f_p)$$

are subbundles of E and E', respectively, of dimensions k and $n-k$, with projections $\pi'' : \ker(f) \to M$ and $\pi''' : \operatorname{im}(f) \to M$ sending the sets $\ker(f_p)$ and $\operatorname{im}(f_p)$ to p for each $p \in M$.

PROOF: We first show that $\ker(f)$ is a k-dimensional subbundle of E. Given a point p on M, we choose local trivializations $(t, \pi^{-1}(U))$ and $(t', \pi'^{-1}(U'))$ around p in E and E', respectively. Let $i_1 : \mathbf{R}^k \to \mathbf{R}^k \times \mathbf{R}^{n-k}$ and $i_2 : \mathbf{R}^{n-k} \to \mathbf{R}^k \times \mathbf{R}^{n-k}$ be the linear maps defined by $i_1(a) = (a,0)$ and $i_2(b) = (0,b)$ for $a \in \mathbf{R}^k$ and $b \in \mathbf{R}^{n-k}$, and let $\pi_2 : \mathbf{R}^k \times \mathbf{R}^{n-k} \to \mathbf{R}^{n-k}$ and $\pi_2' : \mathbf{R}^{m-n+k} \times \mathbf{R}^{n-k} \to \mathbf{R}^{n-k}$ be the projections on the second factor. Then there are permutations σ and σ' in S_n and S_m, respectively, such that the standard matrix of $\pi_2' \circ \pi_{\sigma'} \circ t_p' \circ f_p \circ t_p^{-1} \circ \pi_\sigma^{-1} \circ i_2$ is non-singular, where $\pi_\sigma : \mathbf{R}^n \to \mathbf{R}^n$ and $\pi_{\sigma'} : \mathbf{R}^m \to \mathbf{R}^m$ are the linear isomorphisms obtained by permuting the coordinates in $\mathbf{R}^n$ and $\mathbf{R}^m$ with σ and σ' as defined in 13.8 in the appendix.

Now let $A_r(q)$ be the standard matrix of $\pi_2' \circ \pi_{\sigma'} \circ t_q' \circ f_q \circ t_q^{-1} \circ \pi_\sigma^{-1} \circ i_r$ for $q \in U \cap U'$ and $r = 1,2$. Then A_2 is non-singular in an open neighbourhood V of p contained in $U \cap U'$. If $g(q) : \mathbf{R}^n \to \mathbf{R}^n$ is the linear isomorphism with standard matrix

$$\begin{pmatrix} I_k & 0 \\ A_1(q) & A_2(q) \end{pmatrix}$$

for $q \in V$, then the map $g : V \to L(\mathbf{R}^n, \mathbf{R}^n)$ is smooth, and we have that

$$\pi_2' \circ \pi_{\sigma'} \circ t_q' \circ f_q \circ t_q^{-1} \circ \pi_\sigma^{-1} = \pi_2 \circ g(q)$$

so that

$$\pi_2' \circ \pi_{\sigma'} \circ t_q' \circ f_q = \pi_2 \circ g(q) \circ \pi_\sigma \circ t_q \,.$$

This shows that

$$\begin{aligned} \ker(f_q) &= \ker(\pi_{\sigma'} \circ t_q' \circ f_q) \subset \ker(\pi_2' \circ \pi_{\sigma'} \circ t_q' \circ f_q) \\ &= (g(q) \circ \pi_\sigma \circ t_q)^{-1}(\mathbf{R}^k \times \{0\}) \end{aligned}$$

for $q \in V$. As $\dim\ker(f_q) = k$, the above inclusion is actually an equality. Thus we have a local trivialization $(t'', \pi^{-1}(V))$ around p given by $t''(v) = (id_V \times g(\pi(v)) \circ \pi_\sigma) \circ t(v)$ for $v \in \pi^{-1}(V)$ which satisfies

$$t''(\pi^{-1}(q) \cap \ker(f)) = \{q\} \times \mathbf{R}^k \times \{0\}$$

for every $q \in V$, completing the proof that $\ker(f)$ is a k-dimensional subbundle of E.

To see that $\operatorname{im}(f)$ is a subbundle of E', let $s_1, \ldots, s_n$ be a local basis for E on an open neighbourdood U of a point p on M. Then the sections $f \circ s_1, \ldots, f \circ s_n$ in E' span $\operatorname{im}(f)$ on U. Hence $n-k$ of these sections form a local basis for $\operatorname{im}(f)$ on an open neighbourhood V of p contained in U, showing that $\operatorname{im}(f)$ is an $n-k$-dimensional subbundle of E'. □

THE TANGENT BUNDLE

2.66 Let p be a point on a smooth manifold M of dimension n, and let $c : I \to M$ be a smooth curve on M passing through p, where I is an open interval containing t_0, and $c(t_0) = p$. By the tangent vector to the curve c at p we mean an object which is represented by the vector $(x \circ c)'(t_0)$ in a local chart (x, U) around p. If (y, V) is another local chart around p, we have that

$$(y \circ c)'(t_0) = D(y \circ x^{-1})(x(p))\,(x \circ c)'(t_0)\,.$$

Note that an arbitrary vector $v \in \mathbf{R}^n$ is the representation in the local chart (x, U) of a curve c defined by $c(t) = x^{-1}(x(p) + (t - t_0)\,v)$. In order to obtain the tangent vectors to M at p, we therefore consider the set of all pairs (x,v) where (x,U) is a local chart around p and $v\in\mathbf{R}^n$, and we say that two pairs (x,v) and (y,w) are equivalent, and write $(x,v)\sim_p(y,w)$, if

$$w=D(y\circ x^{-1})(x(p))\,v\,.$$

By the chain rule, this is an equivalence relation, and the equivalence class of (x,v) is denoted by $[x,v]_p$ and is called a *tangent vector* to M at p. The set of all tangent vectors at p is called the *tangent space* T_pM to the manifold M at p, and the set

$$TM=\bigcup_{p\in M}T_pM$$

with the projection $\pi:TM\to M$ defined by $\pi([x,v]_p)=p$ is called the *tangent bundle* of M. For each local chart (x,U) on M, we have a bijection $t_x:\pi^{-1}(U)\to U\times\mathbf{R}^n$ defined by $t_x([x,v]_q)=(q,v)$.

2.67 Proposition The tangent bundle $\pi:TM\to M$ of a smooth manifold M of dimension n is an n-dimentional vector bundle with $(t_x,\pi^{-1}(U))$ as a local trivialization for each local chart (x,U) on M.

PROOF: The maps t_x are clearly bijections satisfying the commutative diagram of Proposition 2.48. Moreover, if (x,U) and (y,V) are local charts on M, we have that

$$\begin{aligned}t_y\circ t_x^{-1}(q,v)&=t_y([x,v]_q)=t_y([y,D(y\circ x^{-1})(x(q))\,v]_q)\\&=(q,D(y\circ x^{-1})(x(q))\,v)\end{aligned}$$

for $(q,v)\in(U\cap V)\times\mathbf{R}^n$ which shows that

$$(y\times id)\circ t_y\circ t_x^{-1}\circ(x^{-1}\times id)\,(a,v)=(y\circ x^{-1}(a),D(y\circ x^{-1})(a)\,v)$$

for $(a,v)\in x(U\cap V)\times\mathbf{R}^n$. Hence the map $t_y\circ t_x^{-1}$ is a diffeomorphism from $(U\cap V)\times\mathbf{R}^n$ onto itself which is a linear isomorphism on each fibre $\{q\}\times\mathbf{R}^n$ for $q\in U\cap V$. By Proposition 2.48 it follows that $\pi:TM\to M$ is a vector bundle with $(t_x,\pi^{-1}(U))$ as a local trivialization for each local chart (x,U) on M. □

2.68 Remark We say that $(t_x,\pi^{-1}(U))$, where $t_x([x,v]_q)=(q,v)$ for $q\in U$ and $v\in\mathbf{R}^n$, is the local trivialization in the tangent bundle $\pi:TM\to M$ associated with the local chart (x,U) on M.

2.69 Definition Let $c:I\to M$ be a smooth curve on M defined on an open interval I. The *tangent vector* to the curve c at the point $c(t)$ for $t\in I$ is the vector $[x,(x\circ c)'(t)]_{c(t)}$ in $T_{c(t)}M$ for any local chart (x,U) around $c(t)$. This vector is well defined by the discussion in 2.66, i.e., if (x,U) and (y,V) are two local charts around $c(t)$ we have that $[x,(x\circ c)'(t)]_{c(t)}=[y,(y\circ c)'(t)]_{c(t)}$.

The tangent vector to c at the point $c(t)$ is also denoted simply by $c'(t)$, and we obtain a smooth curve $c':I\to TM$ on the tangent bundle of M which is a lifting of

c, i.e., with $\pi \circ c' = c$. Indeed, using the local charts $(id, \mathbf{R})$ and (x, U) on $\mathbf{R}$ and M, we have that

$$(x \times id) \circ t_x \circ c' \circ id\,(t) = (x \circ c\,(t), (x \circ c)'(t))$$

for $t \in c^{-1}(U)$ which shows that c' is smooth.

Given a smooth map $h : J \to I$ from an open interval J, the curve $\gamma = c \circ h$ is called a *reparametrization* of c. We have that

$$\gamma'(t) = [x, (x \circ \gamma)'(t)]_{\gamma(t)} = [x, h'(t)\,(x \circ c)'(h(t))]_{c(h(t))} = h'(t)\; c'(h(t))$$

for $t \in J$.

2.70 If $f : M \to N$ is a smooth map between the manifolds M^n and N^m, we want to define a map $f_* : TM \to TN$ so that (f_*, f) is a bundle map between the tangent bundles. Let p be a point on M, and let (x, U) and (y, V) be local charts around p and $f(p)$. If $[x, v]_p$ is the tangent vector to a curve c at p with $c(t_0) = p$, we define $f_*([x, v]_p)$ to be the tangent vector of the curve $f \circ c$ at the point $f(p)$, i.e.,

$$f_*\,(c'(t_0)) = (f \circ c)'(t_0) \tag{1}$$

so that

$$f_*([x, (x \circ c)'(t_0)]_p) = [y, (y \circ f \circ c)'(t_0)]_{f(p)} = [y, D(y \circ f \circ x^{-1})(x(p))\,(x \circ c)'(t_0)]_{f(p)}$$

which implies

$$f_*\,([x, v]_p) = [y, D(y \circ f \circ x^{-1})(x(p))\,v]_{f(p)}\,. \tag{2}$$

This shows in particular that $f_*([x, v]_p)$ is well defined and does not depend on the curve c having tangent vector $[x, v]_p$. The induced map $f_{*p} : T_pM \to T_{f(p)}N$ between the tangent spaces is given by the commutative diagram

$$\begin{array}{ccc} T_pM & \xrightarrow{\;f_{*p}\;} & T_{f(p)}N \\ {\scriptstyle t_{x,p}}\downarrow & & \downarrow{\scriptstyle t_{y,f(p)}} \\ \mathbf{R}^n & \xrightarrow[D(y \circ f \circ x^{-1})(x(p))]{} & \mathbf{R}^m \end{array}$$

2.71 Proposition For each smooth map $f : M \to N$ between the manifolds M^n and N^m, (f_*, f) is a bundle map between their tangent bundles.

PROOF: The pair (f_*, f) clearly satisfies the commutative diagram of Definition 2.49. We must show that f_* is smooth on TM and is linear on each fibre. For local charts (x, U) and (y, V) around p and $f(p)$, respectively, we have

$$t_y \circ f_* \circ t_x^{-1}(q,v) = t_y \circ f_*([x,v]_q) = t_y([y, D(y \circ f \circ x^{-1})(x(q))v]_{f(q)})$$
$$= (f(q), D(y \circ f \circ x^{-1})(x(q))v)$$

for $(q,v) \in (U \cap f^{-1}(V)) \times \mathbf{R}^n$ which shows that

$$(y \times id) \circ t_y \circ f_* \circ t_x^{-1} \circ (x^{-1} \times id)(a,v) = (y \circ f \circ x^{-1}(a), D(y \circ f \circ x^{-1})(a)v)$$

for $(a,v) \in x(U \cap f^{-1}(V)) \times \mathbf{R}^n$, and this completes the proof of the proposition. □

2.72 Proposition Let M^n, N^m and P^r be smooth manifolds.

(1) If the maps $f : M \to N$ and $g : N \to P$ are smooth, we have that $(g \circ f)_* = g_* \circ f_*$.

(2) If $id : M \to M$ is the identity on M, then $id_* : TM \to TM$ is the identity on TM.

(3) If $f : M \to N$ is a diffeomorphism, then so is $f_* : TM \to TN$, and we have that $(f_*)^{-1} = (f^{-1})_*$.

PROOF: (1) If $[x,v]_p$ is the tangent vector to the curve c at p with $c(t_0) = p$, then both $(g \circ f)_*([x,v]_p)$ and $(g_* \circ f_*)([x,v]_p)$ are the tangent vector to the curve $g \circ f \circ c$ at $g(f(p))$.

(2) If $[x,v]_p$ is the tangent vector to the curve c at p with $c(t_0) = p$, then $id_*([x,v]_p)$ is also the tangent vector to the same curve c at p.

(3) By (1) we have that $f_* \circ (f^{-1})_* = (f \circ f^{-1})_* = id_*$ and $(f^{-1})_* \circ f_* = (f^{-1} \circ f)_* = id_*$. Hence (2) shows that f_* is a diffeomorphism with $(f_*)^{-1} = (f^{-1})_*$. □

2.73 Remark Assertion (1) in the proposition is still true for a map g defined only on an open submanifold V of N if we identify $T(f^{-1}(V))$ with $TM|_{f^{-1}(V)}$ and TV with $TN|_V$, as the proof still holds for $p \in M \cap f^{-1}(V)$.

2.74 Proposition Let M and N be smooth manifolds of dimensions n and m, respectively. Then the tangent space of the product manifold $M \times N$ at a point (p,q) can be written as a direct sum

$$T_{(p,q)}(M \times N) = i_{q*}(T_pM) \oplus i_{p*}(T_qN)\,, \tag{1}$$

where $i_q : M \to M \times N$ and $i_p : N \to M \times N$ are the embeddings defined by $i_q(p') = (p',q)$ and $i_p(q') = (p,q')$ for $p' \in M$ and $q' \in N$. If (z_1, U) and (z_2, V) are local charts around p and q, and if $z = z_1 \times z_2$, then

$$[z,v]_{(p,q)} = i_{q*}([z_1,v_1]_p) + i_{p*}([z_2,v_2]_q) \tag{2}$$

for $v = (v_1, v_2) \in \mathbf{R}^n \times \mathbf{R}^m$.

Given a smooth map $f : M \times N \to P$, we have that

$$f_*([z,v]_{(p,q)}) = f_{q*}([z_1,v_1]_p) + f_{p*}([z_2,v_2]_q), \tag{3}$$

where $f_q : M \to P$ and $f_p : N \to P$ are the maps defined by $f_q(p') = f(p',q)$ and $f_p(q') = f(p,q')$ for $p' \in M$ and $q' \in N$.

If $g : Q \to M \times N$ is a smooth map with component maps $g_1 : Q \to M$ and $g_2 : Q \to N$, and if (x,W) is a local chart around a point q on Q, then

$$g_*([x,v]_q) = i_{g_2(q)*} \circ g_{1*}([x,v]_q) + i_{g_1(q)*} \circ g_{2*}([x,v]_q) \tag{4}$$

for $v \in \mathbf{R}^r$ where $r = \dim(Q)$.

If $h : Q \to P$ is the composed map given by $h = f \circ g$, we have that

$$h_*([x,v]_q) = f_{g_2(q)*} \circ g_{1*}([x,v]_q) + f_{g_1(q)*} \circ g_{2*}([x,v]_q) \tag{5}$$

for $v \in \mathbf{R}^r$.

PROOF : The first two formulae follow from

$$\begin{aligned}
[z,v]_{(p,q)} &= [z,(v_1,0)]_{(p,q)} + [z,(0,v_2)]_{(p,q)} \\
&= [z, D(z \circ i_q \circ z_1^{-1})(z_1(p))\, v_1]_{i_q(p)} + [z, D(z \circ i_p \circ z_2^{-1})(z_2(p))\, v_2]_{i_p(q)} \\
&= i_{q*}([z_1,v_1]_p) + i_{p*}([z_2,v_2]_q).
\end{aligned}$$

By applying f_* on both sides of formula (2) and using that $f_q = f \circ i_q$ and $f_p = f \circ i_p$, we obtain formula (3). Using formula (2) we also have that

$$\begin{aligned}
g_*([x,v]_q) &= [z, D(z \circ g \circ x^{-1})(x(q))\, v]_{g(q)} \\
&= i_{g_2(q)*}([z_1, D(z_1 \circ g_1 \circ x^{-1})(x(q))\, v]_{g_1(q)}) \\
&\quad + i_{g_1(q)*}([z_2, D(z_2 \circ g_2 \circ x^{-1})(x(q))\, v]_{g_2(q)}) \\
&= i_{g_2(q)*} \circ g_{1*}([x,v]_q) + i_{g_1(q)*} \circ g_{2*}([x,v]_q)
\end{aligned}$$

which completes the proof of formula (4). Finally, formula (5) is obtained by combining formula (3) and (4) and using Proposition 2.72 (1). □

2.75 Remark To avoid confusion when the manifolds M and N have elements in common, we denote the maps i_q, i_p, f_q and f_p by i_q^1, i_p^2, f_q^1 and f_p^2 in this case.

In particular, if $f : N \times N \to P$ is a smooth map, and $h : N \to P$ is the map given by $h(q) = f(q,q)$ for $q \in N$, then formula (5) reduces to

$$h_*([x,v]_q) = f_{q*}^1([x,v]_q) + f_{q*}^2([x,v]_q), \tag{6}$$

where $f_q^1 : N \to P$ and $f_q^2 : N \to P$ are the maps defined by $f_q^1(p) = f(p,q)$ and $f_q^2(p) = f(q,p)$ for $p \in N$.

2.76 Proposition Let $f : M \to N$ be a smooth map. Then its graph $G(f) = \{(p, f(p)) \mid p \in M\}$ is a submanifold of $M \times N$, and we have that

$$T_{(p,f(p))}G(f) = \{\, i_{f(p)*}(v) + i_{p*} \circ f_*(v) \mid v \in T_pM \,\} \tag{1}$$

for each $p \in M$. If $\pi_1 : M \times N \to M$ is the projection on the first factor, then $\pi_1|_{G(f)}$ is a diffeomorphism.

PROOF: Let $F : M \to M \times N$ be the map defined by $F(p) = (p, f(p))$ for $p \in M$. If (x, U) and (y, V) are local charts around p and $f(p)$, respectively, we have that

$$(x \times y) \circ F \circ x^{-1}(a) = (a, y \circ f \circ x^{-1}(a))$$

for $a \in x(U \cap f^{-1}(V))$, showing that F is an immersion. Since $\pi_1 \circ F = id_M$ where both F and π_1 are continuous, it follows that F is a homeomorphism onto its image $G(f)$ endowed with the subspace topology. Hence $G(f)$ is a submanifold of $M \times N$, and $\pi_1|_{G(f)}$ is a diffeomorphism. Formula (1) follows from Proposition 2.74 which implies that

$$F_*(v) = i_{f(p)*}(v) + i_{p*} \circ f_*(v)$$

for $v \in T_pM$. □

2.77 We will now see that a tangent vector to a smooth manifold M^n at a point p has an alternative description as a local derivation on the algebra $\mathscr{F}_p$ of all smooth functions, each defined on some open neighbourhood of p in M.

Let $[x, v]_p$ be the tangent vector to a curve c at p with $c(t_0) = p$, and let $f : V \to \mathbf{R}$ be a smooth function defined on an open neighbourhood V of p in M. We say that $(f \circ c)'(t_0)$ is the *derivative of f along c at p*. This derivative actually depends only on f and the tangent vector $[x, v]_p$ to c at p since we have that

$$[id, (f \circ c)'(t_0)]_{f(p)} = f_*([x, v]_p)$$

so that

$$(f \circ c)'(t_0) = (t_{id, f(p)} \circ f_*)([x, v]_p)\,,$$

where $t_{id, f(p)} : T_{f(p)}\mathbf{R} \to \mathbf{R}$ is the isomorphism defined in Remark 2.46. Hence $(f \circ c)'(t_0)$ is also called the *derivative of f at p with respect to* $[x, v]_p$. Using the curve c defined by $c(t) = x^{-1}(x(p) + (t - t_0)\, v)$, we see that

$$(f \circ c)'(t_0) = \lim_{h \to 0} \frac{f \circ x^{-1}(x(p) + hv) - f \circ x^{-1}(x(p))}{h}\,.$$

By the chain rule we now have that

$$(f \circ c)'(t_0) = \sum_{i=1}^{n} D_i(f \circ x^{-1})(x(p))\,(x^i \circ c)'(t_0) = \sum_{i=1}^{n} v^i \frac{\partial f}{\partial x^i}(p)\,.$$

Hence we may think of the tangent vector $[x,v]_p$ as an operator

$$l = \sum_{i=1}^{n} v^i \frac{\partial}{\partial x^i}\bigg|_p$$

on the algebra $\mathscr{F}_p$. By Example 2.16, l is a local derivation at p. We will see that in this way we obtain a very useful description of the tangent space T_pM as the set of all local derivations at p. We first need two lemmas.

2.78 Lemma Let l be a local derivation at a point p in the smooth manifold M. Then we have that

(1) $l(f)$ only depends on the local behavior of the function f in $\mathscr{F}_p$. More precisely, this means that if two smooth functions $f_i : V_i \to \mathbf{R}$ which are defined on open neighbourhoods V_i of p for $i = 1,2$, coincide on an open neighbourhood W of p contained in $V_1 \cap V_2$, then $l(f_1) = l(f_2)$.

(2) $l(f) = 0$ for each constant function f in $\mathscr{F}_p$.

PROOF: (1) Considering the difference $f_2 - f_1$, it is clearly enough to prove that $l(f) = 0$ for each smooth function $f : V \to \mathbf{R}$ defined on an open neighbourhood V of p and vanishing on an open neighbourhood W of p contained in V. To show this, let $h : M \to \mathbf{R}$ be a smooth function with $h = 1$ on $M - W$ and $h(p) = 0$. Such a function h exists by Corollary 2.24. Then we have that $l(f) = l(hf) = l(h)\,f(p) + h(p)\,l(f) = 0$, which proves the assertion in (1).

(2) Suppose first that $g : V \to \mathbf{R}$ has the value 1 in the open neighbourhood V of p. Then we have that $l(g) = l(g \cdot g) = l(g)\,1 + 1\,l(g)$, which shows that $l(g) = 0$. If f has the value c in V, we have that $l(f) = c\,l(g) = 0$ by the linearity of l. □

2.79 Lemma Let $f : U \to \mathbf{R}$ be a C^∞-function defined on a convex open neighbourhood U of a point a in $\mathbf{R}^n$. Then there are C^∞-functions $g_i : U \to \mathbf{R}$ for $i = 1, \ldots, n$ such that

(1) $f(u) = f(a) + \sum_{i=1}^{n} (u^i - a^i)\, g_i(u)$ for $u \in U$,

(2) $g_i(a) = D_i f(a)$.

PROOF: Let $u \in U$, and define a C^∞-function $h_u : [0,1] \to \mathbf{R}$ by $h_u(t) = f(a + t(u - a))$. Then we have that

$$\begin{aligned} f(u) - f(a) &= h_u(1) - h_u(0) = \int_0^1 h_u'(t)\,dt \\ &= \int_0^1 \sum_{i=1}^{n} D_i f(a + t(u-a))\,(u^i - a^i)\,dt = \sum_{i=1}^{n} (u^i - a^i)\, g_i(u), \end{aligned}$$

where $g_i(u) = \int_0^1 D_i f(a + t(u - a))\,dt$ for $i = 1, \ldots, n$. □

2.80 Proposition The set T'_pM of all local derivations at p is an n-dimensional vector space. If (x,U) is a local chart around p, then the set $\{\frac{\partial}{\partial x^1}\Big|_p, \ldots, \frac{\partial}{\partial x^n}\Big|_p\}$ of partial derivations at p with respect to (x,U) is a basis for T'_pM, and any local derivation l at p can be written as

$$l = \sum_{i=1}^{n} l(x^i)\, \frac{\partial}{\partial x^i}\Big|_p .$$

PROOF : Let $f : V \to \mathbf{R}$ be a smooth function defined on an open neighbourhood V of p in M, and let (x,U) be a local chart around p. Choose a convex open neighbourhood W of $x(p)$ contained in $x(U \cap V)$. By Lemma 2.79 there are C^∞-functions $g_i : W \to \mathbf{R}$ for $i = 1,\ldots,n$ such that

(1) $(f \circ x^{-1})(u) = (f \circ x^{-1})(x(p)) + \sum_{i=1}^{n} (u^i - x^i(p))\, g_i(u)$ for $u \in W$, and

(2) $g_i(x(p)) = D_i(f \circ x^{-1})(x(p)) = \frac{\partial f}{\partial x^i}(p)$.

When $u = x(q)$, assertion (1) implies that

$$f(q) = f(p) + \sum_{i=1}^{n} (x^i(q) - x^i(p))\, (g_i \circ x)(q) \ \text{ for } \ q \in x^{-1}(W),$$

and applying the derivation l on both sides gives

$$l(f) = \sum_{i=1}^{n} [\, l(x^i)\, (g_i \circ x)(p) + 0\, l(g_i \circ x)] = \sum_{i=1}^{n} l(x^i)\, \frac{\partial f}{\partial x^i}(p).$$

This shows that $\{\frac{\partial}{\partial x^1}\Big|_p, \ldots, \frac{\partial}{\partial x^n}\Big|_p\}$ spans the vector space T'_pM, and that any local derivation l at p can be written as

$$l = \sum_{i=1}^{n} l(x^i)\, \frac{\partial}{\partial x^i}\Big|_p .$$

To show that $\{\frac{\partial}{\partial x^1}\Big|_p, \ldots, \frac{\partial}{\partial x^n}\Big|_p\}$ is linearly independent, assume that

$$\sum_{i=1}^{n} v^i \frac{\partial}{\partial x^i}\Big|_p = 0.$$

By applying both sides to the functions x^j and using that $\frac{\partial}{\partial x^i}\Big|_p (x^j) = \delta_{ij}$, we see that $v^j = 0$ for $j = 1,\ldots,n$. □

2.81 We now let $T'M = \bigcup_{p \in M} T'_pM$ and define the projection $\pi' : T'M \to M$ by $\pi'\left(\sum_{i=1}^{n} v^i \frac{\partial}{\partial x^i}\bigg|_p\right) = p$. We have a bijective map $e_M : TM \to T'M$ defined by

$$e_M([x,v]_p) = \sum_{i=1}^{n} v^i \frac{\partial}{\partial x^i}\bigg|_p .$$

which is a linear isomorphism on each fibre T_pM.

If $(t_x, \pi^{-1}(U))$ are the local trivializations of the bundle T_pM defined in Proposition 2.67 for each local chart (x,U) on M, then the bijections $t'_x : \pi'^{-1}(U) \to U \times \mathbf{R}^n$ defined by $t'_x = t_x \circ e_M^{-1}$ satisfy the conditions of Proposition 2.48. Hence there is a unique topology and smooth structure on $T'M$ such that $\pi' : T'M \to M$ is an n-dimensional vector bundle and $(t'_x, \pi'^{-1}(U))$ is a local trivialization for each local chart (x,U) on M. We have that e_M is an equivalence over M between the bundles π and π' since $e_M|_{\pi'^{-1}(U)} = t_x'^{-1} \circ t_x$.

For each smooth map $f : M \to N$ we now define a map $f_\# : T'M \to T'N$ so that the diagram

$$\begin{array}{ccc} TM & \xrightarrow{f_*} & TN \\ {\scriptstyle e_M}\downarrow & & \downarrow{\scriptstyle e_N} \\ T'M & \xrightarrow{f_\#} & T'N \end{array}$$

is commutative. Since $f_\# = e_N \circ f_* \circ e_M^{-1}$ we see that $(f_\#, f)$ is a bundle map, and we want to give a more explicit description of $f_\#$. If $l = e_M([x,v]_p)$ is a local derivation in $T'M$, it follows from the diagram that $f_\#(l) = e_N(f_*([x,v]_p))$. Now let $g : V \to \mathbf{R}$ be a smooth function defined on an open neighbourhood V of $f(p)$ in N. From the discussion in 2.77 we then see that by applying the isomorphism $t_{id,g(f(p))}$ to both sides of the relation

$$g_*(f_*([x,v]_p)) = (g \circ f)_*([x,v]_p)$$

we have that

$$f_\#(l)(g) = l(g \circ f) .$$

We will from now on identify the bundle $\pi' : T'M \to M$ with $\pi : TM \to M$ and call both the tangent bundle of M. Hence we drop the apostophes and write f_* instead of $f_\#$, and we speak of both $[x,v]_p$ and the local derivation $\sum_{i=1}^{n} v^i \frac{\partial}{\partial x^i}\bigg|_p$ as a tangent vector to M at p.

2.82 Remark If $(t_x, \pi^{-1}(U))$ is the local trivialization in the tangent bundle

$\pi : TM \to M$ associated with a local chart (x,U) around a point p in M^n, and if $\mathscr{E} = \{e_1,...,e_n\}$ is the standard basis for $\mathbf{R}^n$, then we have that

$$\left.\frac{\partial}{\partial x^i}\right|_p = t_{x,p}^{-1}(e_i) \quad \text{for} \quad i = 1,\dots,n\,.$$

2.83 Remark Let $c : I \to M$ be a smooth curve on M defined on an open interval I, and let (r,I) be the standard local chart on I where $r : I \to \mathbf{R}$ is the inclusion map. Then the partial derivation $\left.\frac{\partial}{\partial r^1}\right|_t$ with respect to (r,I) at a point $t \in I$ is denoted simply by $\left.\frac{d}{dr}\right|_t$, and we have that

$$c'(t) = c_*\left(\left.\frac{d}{dr}\right|_t\right).$$

Indeed, if (x,U) is a local chart around $c(t)$, it follows from Definition 2.69 that

$$c'(t) = [x, D(x \circ c \circ r^{-1})(r(t))\, e_1\,]_{c(t)} = c_*\left([r,e_1]_t\right).$$

2.84 Lemma Let V be a vector space of dimension n. Then the tangent space T_pV at any point $p \in V$ may be identified with V by means of the linear isomorphism $\omega_p : T_pV \to V$ given by $\omega_p = x^{-1} \circ t_{x,p}$ for any linear isomorphism $x : V \to \mathbf{R}^n$, ω_p being independent of the choice of x. We will refer to this as the *canonical identification*.

If $\mathscr{E} = \{e_1,...,e_n\}$ is the standard basis for $\mathbf{R}^n$ and $\mathscr{B} = \{v_1,...,v_n\}$ is a basis for V so that $x(v_i) = e_i$ for $i = 1, ..., n$, then $\mathscr{B}^* = \{x^1,...,x^n\}$ is the dual basis of $\mathscr{B}$, and

$$\omega_p\left(\sum_{j=1}^n a^j \left.\frac{\partial}{\partial x^j}\right|_p\right) = \sum_{j=1}^n a^j v_j\,.$$

If $l \in T_pV$, then

$$x^i \circ \omega_p(l) = l(x^i)$$

for $i = 1, ..., n$, and

$$\lambda \circ \omega_p(l) = l(\lambda)$$

for any linear functional λ on V.

The map $\omega = (id_V \times x^{-1}) \circ t_x : TV \to V \times V$ is an equivalence over V so that

$$\omega\left(\sum_{j=1}^n a^j \left.\frac{\partial}{\partial x^j}\right|_p\right) = (p, x^{-1}(a))$$

and

$$\omega^{-1}(p,q) = \sum_{j=1}^n x^j(q) \left.\frac{\partial}{\partial x^j}\right|_p$$

for $p, q \in V$ and $a \in \mathbf{R}^n$.

Let $c : I \to V$ be a smooth curve on V defined on an open interval I. Then we have that

$$c(t) = \sum_{i=1}^{n} (x^i \circ c)(t)\, v_i$$

and

$$\omega_{c(t)} \circ c'(t) = \sum_{i=1}^{n} (x^i \circ c)'(t)\, v_i = \lim_{h \to 0} \frac{1}{h}\, [c(t+h) - c(t)]$$

for $t \in I$. If the curve c lies in a subspace W of V, then $\omega_{c(t)} \circ c'(t) \in W$.

If $F : V \to W$ is a linear map between the finite dimensional vector spaces V and W, and if the tangent spaces T_pV and $T_{F(p)}W$ are canonically identified with V and W by means of the linear isomorphisms $\omega_p : T_pV \to V$ and $\omega_{F(p)} : T_{F(p)}W \to W$ obtained as above from the linear isomorphisms $x : V \to \mathbf{R}^n$ and $y : W \to \mathbf{R}^m$, respectively, then

$$\omega_{F(p)} \circ F_{*p} \circ \omega_p^{-1} = F\,.$$

PROOF : We have that (x, V) is a local chart on V, and that $t_{x,p} : T_pV \to \mathbf{R}^n$ is a linear isomorphism. Hence ω_p is also a linear isomorphism which is independent of the choice of x. Indeed, if $y : V \to \mathbf{R}^n$ is another linear isomorphism, then it follows from 2.70 that

$$t_{y,p} \circ t_{x,p}^{-1} = D\,(y \circ x^{-1})(x(p)) = y \circ x^{-1}\,,$$

so that

$$y^{-1} \circ t_{y,p} = x^{-1} \circ t_{x,p}\,.$$

To prove the next part of the lemma, let $l = \sum_{j=1}^{n} a^j \left.\frac{\partial}{\partial x^j}\right|_p$. Then

$$\omega_p\,(l) = x^{-1}(a) = \sum_{j=1}^{n} a^j\, v_j\,,$$

and

$$x^i \circ \omega_p\,(l) = a^i = l\,(x^i)$$

for $i = 1, \ldots, n$. If

$$\lambda = \sum_{j=1}^{n} b_j x^j\,,$$

then

$$\lambda \circ \omega_p\,(l) = \sum_{j=1}^{n} b_j x^j \circ \omega_p\,(l) = \sum_{j=1}^{n} b_j\, l\,(x^j) = l\,(\sum_{j=1}^{n} b_j x^j) = l\,(\lambda)\,.$$

Using Definition 2.69 we have that

$$\omega_{c(t)} \circ c'(t) = x^{-1} \circ t_{x,c(t)} \circ c'(t) = x^{-1}\,(x \circ c)'(t) = \sum_{i=1}^{n} (x^i \circ c)'(t)\, v_i$$

which also implies that

$$\omega_{c(t)} \circ c'(t) = x^{-1}\big(\lim_{h\to 0} \frac{1}{h}\,[(x\circ c)(t+h) - (x\circ c)(t)]\big) = \lim_{h\to 0} \frac{1}{h}\,[c(t+h) - c(t)]$$

for $t \in I$ since x is a linear homeomorphism.

If the curve c lies in a subspace W of V, then we may assume that the basis $\mathscr{B}$ is an extension of a basis $\mathscr{C} = \{v_1, \ldots, v_m\}$ for W so that $x \circ c(I) \subset x(W) = \mathbf{R}^m \times \{0\}$, which shows that $\omega_{c(t)} \circ c'(t) \in W$.

The last part of the lemma follows from the commutative diagram in 2.70 which shows that

$$\omega_{F(p)} \circ F_{*p} \circ \omega_p^{-1} = y^{-1} \circ t_{y,F(p)} \circ F_{*p} \circ t_{x,p}^{-1} \circ x =$$
$$y^{-1} \circ D\,(y \circ F \circ x^{-1})(x(p)) \circ x = y^{-1} \circ y \circ F \circ x^{-1} \circ x = F. \qquad \square$$

Chapter 3

VECTOR FIELDS AND DIFFERENTIAL EQUATIONS

In this chapter we develop the theory of integral curves and flows of vector fields on a smooth manifold, and prove the important result that a smooth vector field has a smooth flow. We first establish the necessary facts about existence and uniqueness of solutions of differential equations.

VECTOR FIELDS

3.1 Definition By a *vector field* on an open subset V of a smooth manifold M we mean a section of its tangent bundle $\pi : TM \to M$ on V. A section of π along a smooth map $f : N \to M$ from a smooth manifold N is called a *vector field along* f.

If $X : M \to TM$ is a map with $\pi \circ X = id_M$ and p is a point in M, then $X(p)$ is often denoted by X_p. If $f : V \to \mathbf{R}$ belongs to the algebra $\mathscr{F}(V)$ of C^∞-functions on V and we think of X_p as a local derivation at p, we can define a new function $X_V(f) : V \to \mathbf{R}$ by letting X operate on f at each point $p \in V$, i.e.,

$$X_V(f)(p) = X_p(f)\,.$$

3.2 Proposition Let $\pi : TM \to M$ be the tangent bundle of a smooth manifold M^n and $X : M \to TM$ be a map with $\pi \circ X = id_M$. Then X is a vector field on M if and only if one of the following equivalent assertions are satisfied :

(1) For any local chart (x,U) on M the map $a : U \to \mathbf{R}^n$ defined by

$$X(p) = \sum_{i=1}^{n} a^i(p) \left.\frac{\partial}{\partial x^i}\right|_p$$

for $p \in U$ is smooth.

(2) Whenever V is open in M and $f \in \mathscr{F}(V)$, then $X_V(f) \in \mathscr{F}(V)$.

PROOF OF (1) : Follows from Proposition 2.54 and Remark 2.82.

PROOF OF (2): Assume first that X is a vector field on M, and let $f \in \mathscr{F}(V)$ for an open set V in M. If (x,U) is a local chart on M with $U \subset V$ and $a : U \to \mathbf{R}^n$ is the smooth map defined in (1), we have that

$$X_V(f)|_U = \sum_{i=1}^{n} a^i \frac{\partial f}{\partial x^i}$$

showing that $X_V(f)|_U \in \mathscr{F}(U)$. Since this is true for every coordinate neighbourhood U contained in V, we have that $X_V(f) \in \mathscr{F}(V)$.

Conversely, assume that X satisfies (2) and let (x,U) be a local chart on M. Then the map $a : U \to \mathbf{R}^n$ defined in (1) is given by $X_U(x^i) = a^i$ for $i = 1,...,n$ and hence is smooth, showing that X is a vector field on M. □

3.3 Example If (x,U) is a local chart on the smooth manifold M^n, then we have for each $i = 1,...,n$ a vector field $X_i = \frac{\partial}{\partial x^i}$ called the i'th *partial derivation* on U given by

$$X_i(p) = \left.\frac{\partial}{\partial x^i}\right|_p .$$

If X is an arbitrary vector field on M, we have that

$$X|_U = \sum_{i=1}^{n} a^i \frac{\partial}{\partial x^i}$$

where $a : U \to \mathbf{R}^n$ is the local representation of X on U as defined in Proposition 3.2 (1) and Definition 2.52.

3.4 Example Let V be a vector space of dimension n, and let $\omega : TV \to V \times V$ be the equivalence over V obtained from the canonical identification of the tangent space T_pV with V for each $p \in V$ as described in Lemma 2.84. Then we have a vector field $P = \omega^{-1} \circ (id_V, id_V)$ on V, called the *position vector field*, given by

$$P = \sum_{i=1}^{n} x^i \frac{\partial}{\partial x^i}$$

for any linear isomorphism $x : V \to \mathbf{R}^n$.

3.5 Definition Let $\mathscr{A}$ be an algebra. Then a linear map $F : \mathscr{A} \to \mathscr{A}$ is called a *derivation* of $\mathscr{A}$ if it satisfies the relation $F(fg) = F(f)g + fF(g)$ for every $f, g \in \mathscr{A}$.

3.6 Example If X is a vector field on an open subset V of a smooth manifold M, then the map $X_V : \mathscr{F}(V) \to \mathscr{F}(V)$ defined in Definition 3.1 is a derivation of the algebra $\mathscr{F}(V)$ of C^∞-functions on V. Indeed, we have that

$$X_V(fg)(p) = X_p(fg) = X_p(f)g(p) + f(p)X_p(g) = [X_V(f)g + fX_V(g)](p)$$

for every $p \in V$ and $f, g \in \mathscr{F}(V)$.

3.7 Lemma Let M be a smooth manifold, and let $F : \mathscr{F}(M) \to \mathscr{F}(M)$ be a derivation of the algebra $\mathscr{F}(M)$. If $f_1, f_2 \in \mathscr{F}(M)$ are smooth functions with $f_1|_V = f_2|_V$ for an open subset V of M, then we have that $F(f_1)|_V = F(f_2)|_V$.

PROOF: Considering the difference $f_2 - f_1$, it is clearly enough to prove that $F(f)|_V = 0$ for each smooth function $f \in \mathscr{F}(M)$ with $f|_V = 0$. To show this, let $p \in V$ and choose a smooth function $h \in \mathscr{F}(M)$ with $h = 1$ on $M - V$ and $h(p) = 0$. Such a function h exists by Corollary 2.24. Then we have that

$$F(f)(p) = F(hf)(p) = F(h)(p)f(p) + h(p)F(f)(p) = 0$$

which completes the proof since p was an arbitrary point in V. □

3.8 Proposition Let M^n be a smooth manifold, and let $F : \mathscr{F}(M) \to \mathscr{F}(M)$ be a derivation of $\mathscr{F}(M)$. Then there is a unique vector field X on M such that $F = X_M$.

PROOF: We first note that if $f \in \mathscr{F}_p$ is a smooth function defined on an open neighbourhood V of a point p on M, then there exists a smooth function $g \in \mathscr{F}(M)$ which coincides with f on an open neighbourhood W of p with $\overline{W} \subset V$. Indeed, there is a smooth function $h : M \to \mathbf{R}$ with $h(p) = 1$ on $\overline{W}$ and $\operatorname{supp}(h) \subset V$, and we may extend hf to a smooth function g on M by defining it to be zero outside V.

Then by Lemma 2.78, we must define

$$X_p(f) = F(g)(p),$$

and it follows from Lemma 3.7 that $X_p(f)$ does not depend on the choice of W and g. Furthermore, we see the X_p is a local derivation at p since F is a derivation on $\mathscr{F}(M)$.

It only remains to show that X is smooth on M. By Proposition 3.2 (1) we must show that the functions $a^i : U \to \mathbf{R}$ defined by

$$a^i(p) = X_p(x^i)$$

for $p \in U$ and $i = 1, \ldots, n$, are smooth for every local chart (x, U) on M. Fix p in U, and let W be an open neighbourhood of p with $\overline{W} \subset U$. In the same way as above we extend each function hx^i on U to a smooth function y^i on M so that they coincide on W. Then we have that

$$a^i(q) = F(y^i)(q)$$

for $q \in W$ showing that a^i is smooth at p. Since p was an arbitrary point in U, this completes the proof that X is smooth. □

3.9 Remark Because of Proposition 3.8 we will not distinguish between the vector field X and the map X_M, and a vector field on M can be thought of as a derivation of the algebra $\mathscr{F}(M)$ of C^∞-functions on M.

3.10 Definition Let $f : M \to N$ be a smooth map, and let X and Y be vector fields on M and N, respectively. We say that X and Y are *f-related* if the diagram

$$\begin{array}{ccc} TM & \xrightarrow{\;f_*\;} & TN \\ {\scriptstyle X}\big\uparrow & & \big\uparrow{\scriptstyle Y} \\ M & \xrightarrow{\;f\;} & N \end{array}$$

is commutative, i.e., if $f_*(X_p) = Y_{f(p)}$ for every $p \in M$.

3.11 Proposition If $f : M \to N$ is a smooth map, then the vector fields X and Y on M and N, respectively, are f-related if and only if

$$Y_N(g) \circ f = X_M(g \circ f)$$

for every smooth function $g : N \to \mathbf{R}$.

PROOF : Follows from Definition 3.10 since

$$Y_N(g)(f(p)) = Y_{f(p)}(g)$$

and

$$X_M(g \circ f)(p) = X_p(g \circ f) = f_*(X_p)(g)$$

for every $p \in M$. □

3.12 Remark If $f : M \to N$ is a smooth map and X is a vector field on M, there may be no vector field Y on N which is f-related to X. Neither is there always a vector field X on M which is f-related to a given vector field Y on N.

However, in one important special case such f-related vector fields always exist.

3.13 Definition Let $f : M \to N$ be a diffeomorphism.

(1) If X is a vector field on M, there is a unique vector field Y on N which is f-related to X given by

$$Y = f_* \circ X \circ f^{-1}.$$

Y is denoted by $f_*(X)$ and is called the *push-forward* of X by f.

(2) If Y is a vector field on N, there is a unique vector field X on M which is f-related to Y given by

$$X = (f_*)^{-1} \circ Y \circ f.$$

X is denoted by $f^*(Y)$ and is called the *pull-back* of Y by f. Note that $f^*(Y) = (f^{-1})_*(Y)$.

3.14 Proposition Let $f: M^n \to N^m$ be an immersion, and let Y be a vector field on N with $Y_{f(p)} \in f_*(T_pM)$ for each $p \in M$. Then there is a unique vector field X on M which is f-related to Y.

PROOF: For each $p \in M$ there is a unique tangent vector $X_p \in T_pM$ with $f_*(X_p) = Y_{f(p)}$. To show that the map $X: M \to TM$ is smooth, let p be an arbitrary point on M. By the rank theorem there are local charts (x, U) and (y, V) around p and $f(p)$, respectively, such that

$$y \circ f \circ x^{-1}(a^1, ..., a^n) = (a^1, ..., a^n, 0, ..., 0)$$

for $a \in x(U)$, which implies that

$$f_*\left(\left.\frac{\partial}{\partial x^i}\right|_q\right) = \left.\frac{\partial}{\partial y^i}\right|_{f(q)}$$

for $q \in U$ and $i = 1, \ldots, n$.

By Proposition 3.2 (1) the vector field Y has a local representation $b: V \to \mathbf{R}^m$ which is smooth and is given by

$$Y|_V = \sum_{i=1}^{m} b^i \frac{\partial}{\partial y^i} .$$

Hence if

$$X|_U = \sum_{i=1}^{n} a^i \frac{\partial}{\partial x^i} ,$$

we have that

$$Y_{f(q)} = f_*(X_q) = \sum_{i=1}^{n} a^i(q) \left.\frac{\partial}{\partial y^i}\right|_{f(q)}$$

so that $a^i(q) = b^i(f(q))$ for $q \in U$ and $i = 1, \ldots, n$. This shows that the local representation $a: U \to \mathbf{R}^n$ for X on U is smooth and hence that X is a vector field on M which is f-related to Y. □

INTEGRAL CURVES AND LOCAL FLOWS

3.15 Definition Let X be a vector field on a smooth manifold M^n. A smooth curve $\gamma: I \to M$ defined on an open interval I is called an *integral curve* for X if

$$\gamma'(t) = X(\gamma(t)) \tag{1}$$

for $t \in I$. If I contains 0 and $\gamma(0) = p_0$, the point p_0 is called the *starting point* or *initial condition* of γ.

The integral curve γ is called *maximal* if it has no extension to an integral curve for X on any larger open interval.

By a *local flow* for X at a point p_0 on M we mean a map

$$\alpha : I \times U_0 \to M ,$$

where I is an open interval containing 0, and U_0 is an open neighbourhood of p_0, such that the curve $\alpha_p : I \to M$ defined by $\alpha_p(t) = \alpha(t,p)$ is an integral curve for X with initial condition p for each point p in U_0.

3.16 Remark If (x,U) is a local chart around p_0, the problem of finding an integral curve $\gamma : I \to U$ for $X|_U$ with initial condition p_0 can be transformed to a similar problem on $x(U)$. The coordinate map x may be written as $x = i \circ \tilde{x}$, where $\tilde{x} : U \to x(U)$ is a diffeomorphism and $i : x(U) \to \mathbf{R}^n$ is the inclusion map. Suppose that

$$X|_U = \sum_{i=1}^{n} a^i \frac{\partial}{\partial x^i} ,$$

and let $f = a \circ x^{-1}$, $c = \tilde{x} \circ \gamma$ and $x_0 = x(p_0)$. By applying t_x to both sides of (1) in Definition 3.15 and using Definition 2.69 and the proof of Proposition 3.2 (1), we obtain

$$(\gamma(t), (x \circ \gamma)'(t)) = (\gamma(t), a(\gamma(t)))$$

which is equivalent to

$$c'(t) = f(c(t))$$

for $t \in I$, and we have that $c(0) = x_0$. We say that f is a *vector field* on $x(U)$, and c is called an *integral curve* for f with initial condition x_0.

If $\alpha : I \times U_0 \to U$ is a local flow for $X|_U$ at p_0, where I is an open interval containing 0 and U_0 is an open subset of U containing p_0, then the map

$$\beta : I \times x(U_0) \to x(U)$$

defined by $\beta = \tilde{x} \circ \alpha \circ (id \times x^{-1})$ is called a *local flow* for f at x_0. Indeed, for each point $u = x(p)$ in $x(U_0)$, we have that $\beta_u = \tilde{x} \circ \alpha_p$ is an integral curve for f with initial condition u if and only if α_p is an integral curve for X with initial condition p, so the problem of finding a local flow for $X|_U$ on U is transformed to a similar problem on $x(U)$.

3.17 Definition Let U be an open subset of a finite dimensional normed vector space E. By a *time-dependent vector field* on U we mean a continuous map $f : J \times U \to E$ where J is an open interval containing 0. A curve $c : I \to U$ of class C^1 defined on an open subinterval I of J is called an *integral curve* for f if

$$c'(t) = f(t, c(t)) \tag{1}$$

for $t \in I$. We see by induction that if the map f is smooth, so is c. If I contains 0 and $c(0) = x_0$, the point x_0 is called the *starting point* or *initial condition* of c.

The integral curve c is called *maximal* if it has no extension to an integral curve for f on any larger open subinterval of J.

3.18 Remark A continuous curve $c: I \to U$ is an integral curve for f with $c(t_0) = x_0$ if and only if it satisfies the integral equation

$$c(t) = x_0 + \int_{t_0}^{t} f(u, c(u))\, du$$

3.19 Definition Let $f: J \times U \to E$ be a time-dependent vector field, and let x_0 be a point in U. By a *local flow* for f at x_0 we mean a map

$$\alpha: I \times U_0 \to U\,,$$

where I is an open subinterval of J containing 0, and U_0 is an open subset of U containing x_0, such that the curve $\alpha_x: I \to U$ defined by $\alpha_x(t) = \alpha(t,x)$ is an integral curve for f with initial condition x, i.e., such that

$$D_1\alpha(t,x) = f(t, \alpha(t,x)) \quad \text{and} \quad \alpha(0,x) = x$$

for all $t \in I$ and $x \in U_0$.

3.20 The contraction lemma Let M be a nonempty complete metric space with distance function d, and let $f: M \to M$ be a contraction, i.e., there is a constant C with $0 \leq C < 1$ such that

$$d(f(x), f(y)) \leq C\, d(x,y)$$

for all $x, y \in M$. Then f has a unique fixed point, i.e., a point x such that $f(x) = x$. If x_0 is an arbitrary point in M, we have that $x = \lim_{n\to\infty} f^n(x_0)$.

PROOF : If x and y are two fixed points for f, we have that

$$d(x,y) = d(f(x), f(y)) \leq C\, d(x,y)$$

which implies that $d(x,y) = 0$ since $C < 1$. This shows that $x = y$ and proves the uniqueness of the fixed point for f.

To show existence, let x_0 be an arbitrary point in M, and let $x_{n+1} = f(x_n)$ for $n = 0, 1, \ldots$. By induction we have that

$$d(x_i, x_{i+1}) \leq C^i\, d(x_0, x_1)\,,$$

so if $n < m$, we have

$$d(x_n, x_m) \leq \sum_{i=n}^{m-1} d(x_i, x_{i+1}) \leq \Big(\sum_{i=n}^{m-1} C^i\Big)\, d(x_0, x_1) = \frac{C^n - C^m}{1 - C}\, d(x_0, x_1)\,.$$

As $\{C^n\}$ is a Cauchy sequence, so is $\{x_n\}$. It therefore converges to a point x in M since M is complete. Continuity of f then shows that

$$f(x) = \lim_{n\to\infty} f(x_n) = \lim_{n\to\infty} x_{n+1} = x$$

which completes the proof that f has a unique fixed point. □

3.21 Definition A mapping $f : U \to E$ defined on a subset U of a finite dimensional normed vector space E is said to be *Lipschitz* if there is a constant K such that

$$\| f(x) - f(y) \| \le K \| x - y \|$$

for all $x, y \in U$. We call K a *Lipschitz constant* for f on U. We say that f is *locally Lipschitz* if it is Lipschitz in a neighbourhood of each point in U.

If J is an open interval, then a mapping $f : J \times U \to E$ is said to be *Lipschitz* on U *uniformly with respect to* J if there is a constant K such that

$$\| f(t,x) - f(t,y) \| \le K \| x - y \|$$

for all $x, y \in U$ and $t \in J$.

We say that f is *locally Lipschitz* on U *uniformly with respect to* J if for each point (t_0, x_0) in $J \times U$ there is an open subinterval J_0 of J containing t_0 and a neighbourhood U_0 of x_0 in U such that f is Lipschitz on U_0 uniformly with respect to J_0.

3.22 Proposition If $f_1, f_2 : X \to Y$ are two continuous maps between the topological spaces X and Y, where Y is Hausdorff, then the set $D = \{x \in X \,|\, f_1(x) = f_2(x)\}$ is closed in X.

PROOF: We have that $D = f^{-1}(\Delta)$, where $f : X \to Y \times Y$ is the continuous map defined by $f(x) = (f_1(x), f_2(x))$ for $x \in X$, and $\Delta \subset Y \times Y$ is the diagonal which is closed since Y is Hausdorff. □

3.23 Proposition Let $f : J \times U \to E$ be a time-dependent vector field which is locally Lipschitz on U uniformly with respect to J, and let (t_0, x_0) be a point in $J \times U$. Choose an open subinterval J_0 of J containing t_0 and a positive real number a such that the closed ball $\overline{B}_a(x_0)$ is contained in U and there are positive constants K and L such that

(1) $\| f(t,x) - f(t,y) \| \le K \| x - y \|$ for $x, y \in \overline{B}_a(x_0)$ and $t \in J_0$,

(2) $\| f(t,x) \| \le L$ for $x \in \overline{B}_a(x_0)$ and $t \in J_0$.

If $I_b \subset J_0$ is the open interval $(t_0 - b, t_0 + b)$, where b is a positive real number with $b < 1/K$ and $b \le a/L$, then there is a unique integral curve $\alpha : I_b \to U$ for f defined on I_b with $\alpha(t_0) = x_0$.

If $\alpha_i : I_i \to U$ where $i = 1, 2$ are two integral curves for f with $\alpha_1(s) = \alpha_2(s)$ for some real number s in $I_1 \cap I_2$, then α_1 and α_2 are equal on $I_1 \cap I_2$.

PROOF: Let M be the set of all continuous curves

$$\alpha : I_b \to \overline{B}_a(x_0) .$$

Then M is a nonempty compete metric space with the usual supremum metric given by

$$d(\alpha, \beta) = \sup_{t \in I_b} \| \alpha(t) - \beta(t) \| .$$

We define the map $S : M \to M$ by

$$S(\alpha)(t) = x_0 + \int_{t_0}^{t} f(u, \alpha(u))\, du\,.$$

Indeed we have that

$$\| S(\alpha)(t) - x_0 \| = \left\| \int_{t_0}^{t} f(u, \alpha(u))\, du \right\| < bL \le a$$

for every $t \in I_b$ which shows that $S(\alpha)$ is a continuous curve in $B_a(x_0)$ and hence belongs to M. Since

$$d\,(S(\alpha), S(\beta)) \le \sup_{t \in I_b} \left\| \int_{t_0}^{t} [f(u, \alpha(u)) - f(u, \beta(u))]\, du \right\| \le bK\, d\,(\alpha, \beta)$$

and $bK < 1$, we have that S is a contraction and hence has a unique fixed point c by the contraction lemma. If $i : \overline{B}_a(x_0) \to U$ is the inclusion map, it follows by Remark 3.18 that the curve $\alpha = i \circ c$ is the unique integral curve for f defined on I_b with $\alpha(t_0) = x_0$ and $\alpha(I_b) \subset B_a(x_0)$. This shows the existence of an integral curve $\alpha : I_b \to U$ for f with $\alpha(t_0) = x_0$, and uniqueness will follow from the last part of the proposition.

Let $I = I_1 \cap I_2$ and $I_0 = \{t \in I \,|\, \alpha_1(t) = \alpha_2(t)\}$. Then I_0 is closed in I by Proposition 3.22. Since I is connected, it will follow that $I_0 = I$ if we can show that I_0 is also open and nonempty. If $t_1 \in I_0$ and $\alpha_1(t_1) = \alpha_2(t_1) = x_1$, choose I_b as in the first part of the proposition with (t_1, x_1) instead of (t_0, x_0), and let in addition b be so small that $I_b \subset I$ and $\alpha_i(I_b) \subset B_a(x_1)$ for $i = 1, 2$. Then it follows from the first part of the proof that α_1 and α_2 are equal on I_b so that I_0 is open. Since I_0 contains s, it is also nonempty, and this completes the proof that the curves α_1 and α_2 are equal on the intersection of their domains. □

3.24 Corollary Let $f : J \times U \to E$ be a time-dependent vector field which is locally Lipschitz on U uniformly with respect to J, and let (t_0, x_0) be a point in $J \times U$. Then there is a unique maximal integral curve $\alpha : I \to U$ for f defined on an open subinterval I of J containing t_0 with $\alpha(t_0) = x_0$.

PROOF : Let $\mathscr{I}$ be the set of all open subintervals I' of J containing t_0 such that there exists an integral curve $\alpha' : I' \to U$ for f with $\alpha'(t_0) = x_0$, and put $I = \bigcup_{I' \in \mathscr{I}} I'$. Define the curve $\alpha : I \to U$ as follows. If $t \in I$, choose an interval I' in $\mathscr{I}$ containing t and set $\alpha(t) = \alpha'(t)$. It follows by Proposition 3.23 that $\mathscr{I}$ is nonempty and that α is well defined and uniquely determined by the above conditions. □

3.25 Example Let $f : (-2, 2) \times \mathbf{R} \to \mathbf{R}$ be the time-dependent vector field defined by $f(t, x) = tx^2$. By separation of variables we see that the maximal integral curve for f with initial condition 2 is the curve $\alpha : (-1, 1) \to \mathbf{R}$ given by $\alpha(t) = 2/(1 - t^2)$.

3.26 Remark In Example 3.25 we see that the maximal integral curve is defined on a smaller time interval that the time-dependent vector field. The reason is that $\alpha(t) \to \infty$ when $t \to -1^+$ and $t \to 1^-$, and we see that $f(t,\alpha(t))$ is not bounded.

Another case when this can happen is when the integral curve tends to leave the open set U where the time-dependent vector field is defined. For instance will the domain of the maximal integral curve be smaller if we remove from U a point lying on the curve.

The next proposition shows that apart from these two cases, the maximal integral curve is always defined on the same time interval as the time-dependent vector field.

3.27 Proposition Let $f : (a,b) \times U \to E$ be a time-dependent vector field which is locally Lipschitz on U uniformly with respect to (a,b), and let $\alpha : (a_0, b_0) \to U$ be a maximal integral curve for f. Assume that

(1) there exists an $\varepsilon > 0$ such that $\overline{\alpha((b_0 - \varepsilon, b_0))} \subset U$,

(2) there exists a B such that $\| f(t,\alpha(t)) \| \leq B$ for $t \in (b_0 - \varepsilon, b_0)$.

Then we have that $b_0 = b$, and a similar result holds for the left endpoints.

PROOF : By Remark 3.18 it follows from (2) that

$$\| \alpha(t_1) - \alpha(t_2) \| = \left\| \int_{t_2}^{t_1} f(u,\alpha(u))\, du \right\| \leq B\, |t_1 - t_2|$$

for $t_1, t_2 \in (b_0 - \varepsilon, b_0)$. Hence $\lim_{t \to b_0^-} \alpha(t)$ exists and equals a point x_0 in U by hypothesis (1). If $b_0 < b$, it follows from Proposition 3.23 that there is an integral curve β for f defined on some open interval $(b_0 - \varepsilon_1, b_0 + \varepsilon_1)$ contained in (a,b) with $\beta(b_0) = x_0$. Since $\alpha' = \beta'$ on $(b_0 - \varepsilon_1, b_0)$, we have that α and β only differ by a constant on this interval. As their limit when $t \to {b_0}^-$ are equal, this constant must be zero. Hence α may be extended to an integral curve for f on a larger open subinterval of (a,b), and this contradiction shows that $b_0 = b$. □

3.28 Theorem Let $f : J \times U \to E$ be a time-dependent vector field which is locally Lipschitz on U uniformly with respect to J, and let x_0 be a point in U. Choose an open subinterval J_0 of J containing 0 and a positive real number a such that the closed ball $\overline{B}_{2a}(x_0)$ is contained in U and there are positive constants K and L such that

(1) $\| f(t,x) - f(t,y) \| \leq K \, \|x - y\|$ for $x, y \in \overline{B}_{2a}(x_0)$ and $t \in J_0$,

(2) $\| f(t,x) \| \leq L$ for $x \in \overline{B}_{2a}(x_0)$ and $t \in J_0$.

If $I_b \subset J_0$ is the open interval $(-b,b)$, where b is a positive real number with $b < 1/K$ and $b \leq a/L$, then there is a unique map $\alpha : I_b \times B_a(x_0) \to U$ which is a local flow for f at x_0, and α is Lipschitz on $I_b \times B_a(x_0)$.

PROOF : If x is a point in $B_a(x_0)$, then $\overline{B}_a(x) \subset \overline{B}_{2a}(x_0)$, so by Proposition 3.23 there is a unique integral curve $\alpha_x : I_b \to U$ for f defined on I_b with $\alpha_x(t_0) = x$. Hence the map $\alpha : I_b \times B_a(x_0) \to U$ defined by $\alpha(t,x) = \alpha_x(t)$ is the unique local flow for f defined on $I_b \times B_a(x_0)$.

It remains to show that α is Lipschitz on $I_b \times B_a(x_0)$. If $x, y \in B_a(x_0)$, we have that

$$\begin{aligned}
&\sup_{t\in I_b} \| \alpha(t,x) - \alpha(t,y) \| \\
&\leq \|x-y\| + \sup_{t\in I_b} \left\| \int_0^t [f(u,\alpha(u,x)) - f(u,\alpha(u,y))]\, du \right\| \\
&\leq \|x-y\| + bK \sup_{t\in I_b} \| \alpha(t,x) - \alpha(t,y) \|
\end{aligned}$$

so that

$$\sup_{t\in I_b} \| \alpha(t,x) - \alpha(t,y) \| \leq \frac{1}{1-bK} \|x-y\| ,$$

which shows that $\alpha(t,x)$ is Lipschitz in x uniformly in t. If $s,t \in I_b$ and $x \in B_a(x_0)$, we have that

$$\| \alpha(s,x) - \alpha(t,x) \| = \left\| \int_s^t f(u,\alpha(u,x))\, du \right\| \leq L \, \|s-t\| ,$$

showing that $\alpha(t,x)$ is also Lipschitz in t uniformly in x. Hence $\alpha(t,x)$ is Lipschitz in t and x jointly. □

3.29 Remark We want to prove that if a time-dependent vector field $f : J \times U \to E$ is of class C^k, then it has a local flow α at each point x_0 in U which is also of class C^k. The hard part is to prove that α is C^1 in the second variable x. Assuming this for the moment, we see from Definition 3.19 by differentiating with respect to x, that $D_2\alpha = \lambda$ must satisfy the equation

$$D_1\lambda(t,x) = D_2 f(t,\alpha(t,x)) \circ \lambda(t,x) \quad \text{with} \quad \lambda(0,x) = id_E ,$$

which is a time-dependent linear differenial equation depending on x as a parameter.

In the next proposition we show that such an equation has a unique solution λ which is continuous. Using this we will show in Theorem 3.33 that $D_2\alpha$ exists and equals λ and, by an induction argument, that the local flow α is in fact C^k. We first need a lemma.

3.30 Lemma (Gronwall's inequality) Let $f,g : [a,b) \to \mathbf{R}$ be continuous and non-negative. Suppose that $A \geq 0$ and that for $t \in [a,b)$ we have

$$f(t) \leq A + \int_a^t f(u)g(u)\, du . \tag{1}$$

Then it follows that

$$f(t) \leq A \, exp\left(\int_a^t g(u)\, du\right) \tag{2}$$

for $t \in [a,b)$. In particular, if $A = 0$, then $f(t) = 0$ for $t \in [a,b)$.

PROOF: We first suppose that $A > 0$, and let $h : [a,b) \to \mathbf{R}$ be the function defined by

$$h(t) = A + \int_a^t f(u)g(u)\,du$$

appearing on the right side of (1). Then $h(t) > 0$ and $h'(t) = f(t)g(t) \leq h(t)g(t)$ so that $h'(t)\,/\,h(t) \leq g(t)$ for $t \in [a,b)$. Integration gives

$$\log h(t) - \log A \leq \int_a^t g(u)\,du$$

and hence

$$h(t) \leq A\,exp\,(\int_a^t g(u)\,du)$$

for $t \in [a,b)$. Since $f(t) \leq h(t)$, this shows (2) in the case when $A > 0$.

If (1) is satisfied for $A = 0$, then (1) and hence (2) is satisfied for every $A > 0$ which shows that $f(t) = 0$ for $t \in [a,b)$. □

3.31 Corollary If $b < a$, then Lemma 3.30 is still valid if we replace the interval $[a,b)$ by $(b,a]$ and take absolute values of the integrals.

PROOF: Let $\overline{f},\overline{g} : [-a,-b) \to \mathbf{R}$ be the functions defined by $\overline{f}(t) = f(-t)$ and $\overline{g}(t) = g(-t)$, and suppose that

$$f(t) \leq A + |\int_a^t f(u)g(u)\,du|$$

for $t \in (b,a]$. Making the substitution $s = -u$ in the integral, this implies that

$$\overline{f}(-t) \leq A + \int_{-a}^{-t} \overline{f}(s)\overline{g}(s)\,ds\,.$$

By Lemma 1.61 we now have that

$$\overline{f}(-t) \leq A\,exp\,(\int_{-a}^{-t} \overline{g}(s)\,ds)$$

so that

$$f(t) \leq A\,exp\,(|\int_a^t g(u)\,du|)$$

for $t \in (b,a]$. □

3.32 Proposition Let J be an open interval containing 0, E and F be finite dimensional normed vector spaces, and V be an open subset of F. Let

$$g : J \times V \to L(E,E)$$

be a continuous map. Then there is a unique map

$$\lambda : J \times V \to L(E,E)$$

such that

$$D_1\lambda(t,x) = g(t,x)\circ\lambda(t,x) \quad \text{and} \quad \lambda(0,x) = id\,|_E$$

for all $t \in J$ and $x \in V$. Furthermore, this λ is continuous.

PROOF : Choose a point x in V, and let

$$f : J \times L(E,E) \to L(E,E)$$

be the time-dependent vector field defined by $f(t,v) = g(t,x)\circ v$ for $t \in J$ and $v \in L(E,E)$. From the inequality

$$\|\, f(t,v_1) - f(t,v_2)\,\| \le \|\, g(t,x)\,\|\,\|\, v_1 - v_2\,\|$$

we see that f is locally Lipschitz on $L(E,E)$ uniformly with respect to J, since $g(t,x)$ is bounded on $J_0 \times \{x\}$ for any open interval J_0 with compact closure $\overline{J_0} \subset J$.

Let $\lambda_x : I \to L(E,E)$ be the maximal integral curve for f with $\lambda_x(0) = id\,|_E$. We will show that $I = J$ using Proposition 3.27. Let $J = (a,b)$ and $I = (a_0,b_0)$, and assume that $b_0 < b$. Hypothesis (1) in Proposition 3.27 is clearly satisfied for $\varepsilon = b_0$ as the open set U now is the whole space $L(E,E)$. Since g is continuous, there is a constant C such that $\|\, g(t,x)\,\| \le C$ for $t \in [0,b_0]$. From the integral equation

$$\lambda_x(t) = id\,|_E + \int_0^t g(u,x)\circ\lambda_x(u)\,du \tag{1}$$

we therefore have that

$$\|\,\lambda_x(t)\,\| \le 1 + C\int_0^t \|\,\lambda_x(u)\,\|\;du$$

for every $t \in [0,b_0)$. By Gronwall's inequality it follows that

$$\|\,\lambda_x(t)\,\| \le e^{Ct} \le e^{Cb_0}$$

so that

$$\|\, f(t,\lambda_x(t))\,\| \le \|\, g(t,x)\,\|\,\|\,\lambda_x(t)\,\| \le Ce^{Cb_0}$$

for $t \in (0,b_0)$. This shows that hypothesis (2) in Proposition 3.27 is also satisfied with $B = Ce^{Cb_0}$. The assumption $b_0 < b$ hence leeds to a contradiction. A similar argument using Corollary 3.31 applies to a and a_0, and this completes the proof that $I = J$.

We therefore have a unique map $\lambda : J \times V \to L(E,E)$, defined by $\lambda(t,x) = \lambda_x(t)$ for $t \in J$ and $x \in V$, which satisfies the linear differetial equation and the initial value condition of the proposition. It only remains to show that λ is continuous. We know from Definition 3.17 that λ is continuous in the first variable t for each fixed value of x.

Let (t_0,x_0) be a point in $J \times V$, and let $\varepsilon > 0$. Choose an open interval J_0 containing 0 and t_0 with compact closure $\overline{J_0} \subset J$ and a compact neighbourhood V_0 of x_0 in V. Then there are positive constants C and K such that

$\| \lambda(t,x_0) \| \leq C$ for $t \in J_0$ and $\| g(t,x) \| \leq K$ for $(t,x) \in J_0 \times V_0$. Let l be the length of the interval J_0, and let $r = \varepsilon K e^{-Kl}$. Then there is an open subinterval J_1 of J_0 containing t_0 and an open neighbourhood V_1 of x_0 contained in V_0 such that

$$\| \lambda(t,x_0) - \lambda(t_0,x_0) \| < \tfrac{r}{K} \tag{2}$$

for $t \in J_1$ and

$$\| g(t,x) - g(t,x_0) \| < \tfrac{r}{C}$$

for $(t,x) \in J_0 \times V_1$.

If x is an arbitrary point in V_1 and $\psi : J_0 \to \mathbf{R}$ is the function defined by $\psi(t) = \| \lambda(t,x) - \lambda(t,x_0) \|$, it follows from the integral Equation (1) that

$$\begin{aligned} \psi(t) &\leq \int_0^t \| g(u,x) \circ \lambda(u,x) - g(u,x_0) \circ \lambda(u,x_0) \| \, du \\ &\leq \int_0^t \| g(u,x) \| \, \| \lambda(u,x) - \lambda(u,x_0) \| \, du \\ &\quad + \int_0^t \| g(u,x) - g(u,x_0) \| \, \| \lambda(u,x_0) \| \, du \\ &\leq K \int_0^t \psi(u) \, du + rt = K \int_0^t (\psi(u) + \tfrac{r}{K}) \, du \end{aligned}$$

so that

$$\psi(t) + \tfrac{r}{K} \leq \tfrac{r}{K} + K \int_0^t (\psi(u) + \tfrac{r}{K}) \, du$$

for $t \geq 0$ in J_0. By Gronwall's inequality we therefore have that

$$\psi(t) + \tfrac{r}{K} \leq \tfrac{r}{K} e^{Kt} \leq \tfrac{r}{K} e^{Kl} \,,$$

and hence

$$\| \lambda(t,x) - \lambda(t,x_0) \| \leq \tfrac{r}{K} (e^{Kl} - 1) \tag{3}$$

for $t \geq 0$ in J_0. A similar argument using Corollary 3.31 gives the same result for $t \leq 0$ in J_0. From (2) and (3) it now follows that

$$\| \lambda(t,x) - \lambda(t_0,x_0) \| \leq \| \lambda(t,x) - \lambda(t,x_0) \| + \| \lambda(t,x_0) - \lambda(t_0,x_0) \| < \varepsilon$$

for $(t,x) \in J_1 \times V_1$ which shows that λ is continuous at (t_0,x_0). □

3.33 Theorem Let $f : J \times U \to E$ be a time-dependent vector field of class C^k for an integer $k \geq 1$, and let x_0 be a point in U. Then there are positive real numbers a and b such that the local flow $\alpha : I_b \times B_a(x_0) \to U$ in Theorem 3.28 is of class C^k.

PROOF: We first show that if f is of class C^1, then so is the local flow $\alpha : I_b \times$

$B_a(x_0) \to U$ in Theorem 3.28. We allready know that α is Lipschitz so that there are positive constants A and B with

$$\| \alpha(t_1,x_1) - \alpha(t_2,x_2) \| \leq A \| t_1 - t_2 \| + B \| x_1 - x_2 \| \tag{1}$$

for $t_1, t_2 \in I_b$ and $x_1, x_2 \in B_a(x_0)$.

Let $g : I_b \times B_a(x_0) \to L(E,E)$ be the continuous map given by $g(t,x) = D_2 f(t, \alpha(t,x))$, and let $\lambda : I_b \times B_a(x_0) \to L(E,E)$ be the unique solution of the time-dependent linear differenial equation

$$D_1 \lambda(t,x) = g(t,x) \circ \lambda(t,x) \quad \text{with} \quad \lambda(0,x) = id_E. \tag{2}$$

We claim that $D_2\alpha$ exists and equals λ which is continuous by Proposition 3.32.

Let x be a point in $B_a(x_0)$, and choose positive real numbers c and d such that $B_d(x) \subset B_a(x_0)$ and $c < b$. If $\theta : I_b \times B_d(x) \to E$ is the map defined by

$$\theta(t,h) = \alpha(t, x+h) - \alpha(t,x),$$

we have that

$$\begin{aligned}
&\theta(t,h) - \lambda(t,x)h \\
&= \int_0^t [f(u,\alpha(u,x+h)) - f(u,\alpha(u,x))]\,du - [\lambda(t,x) - id_E]h \\
&= \int_0^t [f(u,\alpha(u,x+h)) - f(u,\alpha(u,x))]\,du \\
&\quad - \int_0^t D_2 f(u,\alpha(u,x))\,\lambda(u,x)\,h\,du \\
&= \int_0^t [f(u,\alpha(u,x+h)) - f(u,\alpha(u,x)) - D_2 f(u,\alpha(u,x))\,\theta(u,h)]\,du \\
&\quad + \int_0^t D_2 f(u,\alpha(u,x))\,[\theta(u,h) - \lambda(u,x)h]\,du
\end{aligned} \tag{3}$$

for $t \in (-c,c)$ and $h \in B_d(x)$.

By the continuity of g there is a constant M such that

$$\| D_2 f(u,\alpha(u,x)) \| \leq M \tag{4}$$

for every $u \in [-c,c]$.

In order to estimate the first integral in the last expression in (3), we use the mean value theorem which implies that

$$\begin{aligned}
&\| f(u,\alpha(u,x+h)) - f(u,\alpha(u,x)) - D_2 f(u,\alpha(u,x))\,\theta(u,h) \| \\
&\quad \leq \| \theta(u,h) \| \sup \| D_2 f(u,y) - D_2 f(u,\alpha(u,x)) \|,
\end{aligned} \tag{5}$$

where the sup is taken over all y on the line segment between $\alpha(u,x)$ and $\alpha(u,x+h)$. By (1) we have that $\| \theta(u,h) \| \leq B \| h \|$.

Given $\varepsilon > 0$, let $k = \varepsilon e^{-Mc}$. For each $u \in [-c,c]$ there is an open ball $B_u \subset U$ of radius δ_u centered at $\alpha(u,x)$ and an open interval $J_u \subset I_b$ of radius r_u centered at u such that $r_u < \delta_u / 2A$ and

$$\| D_2 f(s,z) - D_2 f(u,\alpha(u,x)) \| < \frac{k}{2B}$$

for all $(s,z) \in J_u \times B_u$, by the continuity of $D_2 f$ at the point $(u,\alpha(u,x))$. Since the interval $[-c,c]$ is compact, it is covered by a finite number of intervals J_{u_i} for $i = 1,...,n$. Let

$$\delta = \min\{\delta_{u_i} / 2B | 1 \le i \le n\} \cup \{d\},$$

and suppose that $\| h \| < \delta$. If $u \in [-c,c]$, then $u \in J_{u_i}$ for some i. Hence it follows from (1) that both $\alpha(u,x)$ and $\alpha(u,x+h)$ and hence every y on the line segment between $\alpha(u,x)$ and $\alpha(u,x+h)$ belong to B_{u_i}. From this it follows that

$$\| D_2 f(u,y) - D_2 f(u_i,\alpha(u_i,x)) \| < \frac{k}{2B}$$

and

$$\| D_2 f(u,\alpha(u,x)) - D_2 f(u_i,\alpha(u_i,x)) \| < \frac{k}{2B}$$

so that

$$\| D_2 f(u,y) - D_2 f(u,\alpha(u,x)) \| < \frac{k}{B} \tag{6}$$

Combining (6) with (3), (4) and (5), we obtain

$$\| \theta(t,h) - \lambda(t,x)h \| \le k \| h \| + M \int_0^t \| \theta(u,h) - \lambda(u,x)h \| \, du$$

for $t \in [0,c)$. Gronwall's inequality hence implies that

$$\| \theta(t,h) - \lambda(t,x)h \| \le \varepsilon \| h \|$$

for $t \in [0,c)$, and a similar argument using Corollary 3.31 gives the same result for $t \in (-c,0]$. Since for each point (t,x) in $I_b \times B_a(x_0)$ we may always choose a positive real number c with $|t| < c < b$, this completes the proof that $D_2\alpha$ exists and equals λ so that the local flow α is of class C^1 in the second variable x. From the equation $D_1\alpha(t,x) = f(t,\alpha(t,x))$ we see that α is also of class C^1 in the first variable t.

This shows the theorem in the case $k = 1$. Now assume inductively that it is true for $k-1$, and let f be of class C^k. We have to show that $D_1\alpha$ and $D_2\alpha$ are of class C^{k-1}. Let

$$G : I_b \times B_a(x_0) \times L(E,E) \to E \times L(E,E)$$

be the time-dependent vector field on $B_a(x_0) \times L(E,E)$ given by $G(t,x,w) = (0, g(t,x) \circ w)$, and let

$$\Lambda : I_b \times B_a(x_0) \times L(E,E) \to B_a(x_0) \times L(E,E)$$

be the map defined by $\Lambda(t,x,w) = (x, \lambda(t,x) \circ w)$. By (2) we have that

$$D_1\Lambda(t,x,w) = (0, g(t,x) \circ \lambda(t,x) \circ w) = G(t,x,\lambda(t,x) \circ w) = G(t,\Lambda(t,x,w))$$

with

$$\Lambda(0,x,w) = (x, id|_E \circ w) = (x,w)$$

showing that Λ is a local flow for G. Since G is of class C^{k-1}, it follows by the induction hypothesis that there are positive real numbers a' and b' and some open neighbourhood V of $id|_E$ in $L(E,E)$ such that Λ is of class C^{k-1} on $I_{b'} \times B_{a'}(x_0) \times V$. Since $\lambda(t,x) = pr_2 \circ \Lambda(t,x,id|_E)$, where $pr_2 : E \times L(E,E) \to L(E,E)$ is the projection on the second factor, we have that $D_2\alpha = \lambda$ is of class C^{k-1} on $I_{b'} \times B_{a'}(x_0)$. From the equation $D_1\alpha(t,x) = f(t,\alpha(t,x))$ we see that this is also the case for $D_1\alpha$, and this completes the induction step. □

3.34 Remark It follows from the equation $D_1\alpha(t,x) = f(t,\alpha(t,x))$ that the local flow α is in fact C^{k+1} in the first variable t.

Since a and b depend on k as well as on x_0, we cannot yet conclude that the theorem is true when $k = \infty$. However, this will follow from the following Remark 3.35 and Theorem 3.39.

3.35 Remark Let U be an open subset of a finite dimensional normed vector space E. By a (time-independent) *vector field* on U we mean a continuous map $f : U \to E$. With f we may associate a time-dependent vector field $g : \mathbf{R} \times U \to E$ on U defined by $g(t,x) = f(x)$. By a local flow for f at a point x_0 in U we mean a local flow for g at x_0, i.e., a map

$$\alpha : I \times U_0 \to U\ ,$$

where I is an open interval containing 0, and U_0 is an open subset of U containing x_0, such that

$$D_1\alpha(t,x) = f(\alpha(t,x)) \quad \text{and} \quad \alpha(0,x) = x$$

for all $t \in I$ and $x \in U_0$. The time-dependent vector fields are thus seen to include the time-independent ones as a special case. We see that f is of class C^k on U if and only if g is of class C^k on $\mathbf{R} \times U$, and f is locally Lipschitz on U if and only if g is locally Lipschitz on U uniformly with respect to $\mathbf{R}$.

On the other hand, with a general time-dependent vector field $g : J \times U \to E$ on U, where J is an open interval containing 0, we may associate a time-independent vector field $h : J \times U \to \mathbf{R} \times E$ on $J \times U$ defined by $h(t,x) = (1, g(t,x))$. Let

$$\beta : I \times I_0 \times U_0 \to J \times U$$

be a local flow for h, where I and I_0 are open subintervals of J containing 0, and U_0 is an open subset of U containing x_0, so that

$$D_1\beta(t,s,x) = h(\beta(t,s,x)) \quad \text{and} \quad \beta(0,s,x) = (s,x)$$

for all $t \in I$ and $(s,x) \in I_0 \times U_0$. If β_1 and β_2 are the components of β, it follows that $\beta_1(t,s,x) = s + t$, and hence we have that

$$D_1\beta_2(t,s,x) = g(s+t, \beta_2(t,s,x)) \quad \text{and} \quad \beta_2(0,s,x) = x\ .$$

Then the map

$$\alpha : I \times U_0 \to U$$

defined by $\alpha(t,x) = \beta_2(t,0,x)$ satisfies the equation

$$D_1\alpha(t,x) = g(t,\alpha(t,x)) \quad \text{with} \quad \alpha(0,x) = x$$

showing that α is a local flow for g. We see that g is of class C^k on $J \times U$ if and only if h is of class C^k, and if β is of class C^k, so is α. Hence the study of time-dependent vector fields of class C^k is reduced to the study of time-independent ones which we will consider from now on. On the other hand, there is no natural condition on h corresponding to the condition that g is locally Lipschitz on U uniformly with respect to J, which is precisely what was needed in the beginning of the proof of Proposition 3.32, and this is the reason why we have studied time-dependent vector fields explicitly in the preceding theory.

3.36 Proposition If X is a vector field on a smooth manifold M and p is a point on M, then there is an integral curve $\gamma : I \to M$ for X with initial condition p.

If $\gamma_i : I_i \to M$ where $i = 1,2$ are two integral curves for X with with $\gamma_1(s) = \gamma_2(s)$ for some real number s in $I_1 \cap I_2$, then γ_1 and γ_2 are equal on $I_1 \cap I_2$.

PROOF: The first part follows from Remark 3.16, Proposition 3.23 and Definition 3.17.

To prove the last part, let $I = I_1 \cap I_2$ and $I_0 = \{t \in I | \gamma_1(t) = \gamma_2(t)\}$. Then I_0 is closed in I by Proposition 3.22. Since I is connected, it will follow that $I_0 = I$ if we can show that I_0 is also open and nonempty. If $t_1 \in I_0$ and $\gamma_1(t_1) = \gamma_2(t_1) = x_1$, let (x,U) be a local chart around x_1. By Proposition 3.23 we have that $x \circ \gamma_1$ and $x \circ \gamma_2$ and hence γ_1 and γ_2 are equal on $\gamma_1{}^{-1}(U) \cap \gamma_2{}^{-1}(U)$ so that I_0 is open. Since I_0 contains s, it is also nonempty, and this completes the proof that the curves γ_1 and γ_2 are equal on the intersection of their domains. □

3.37 Corollary If X is a vector field on a smooth manifold M and p is a point on M, then there is a unique maximal integral curve $\gamma : I \to M$ for X with initial condition p.

PROOF: Proved in the same way as Corollary 3.24. □

GLOBAL FLOWS

3.38 Definition Let X be a vector field on a smooth manifold M. For each point p on M we denote by $\gamma_p : I(p) \to M$ the maximal integral curve for X with initial condition p. The set

$$\mathscr{D}(X) = \{(t,p) \in \mathbf{R} \times M | t \in I(p)\}$$

is called the *domain of the flow* for X, and the *(global) flow* for X is the map $\gamma: \mathscr{D}(X) \to M$ defined by $\gamma(t,p) = \gamma_p(t)$ for $p \in M$ and $t \in I(p)$.

3.39 Theorem Let X be a vector field on a smooth manifold M. Then $\mathscr{D}(X)$ is an open subset of $\mathbf{R} \times M$ containing $\{0\} \times M$, and the flow $\gamma: \mathscr{D}(X) \to M$ for X is smooth.

PROOF: It follows immediately from Definition 3.38 that $\{0\} \times M \subset \mathscr{D}(X)$. Given an integer $k \geq 1$, we will show that $\mathscr{D}(X)$ is open and that the flow γ is of class C^k on $\mathscr{D}(X)$. For each point p on M we let $I^*(p)$ be the set of points t in $I(p)$ for which there is a real number $b > 0$ and an open neighbourhood U of p such that $(t-b, t+b) \times U \subset \mathscr{D}(X)$ and γ is of class C^k on this product. Then $I^*(p)$ is clearly open in $I(p)$ and contains 0 by Theorem 3.33 and Remark 3.16. Since $I(p)$ is connected, it will follow that $I^*(p) = I(p)$ if we can show that $I^*(p)$ is also closed in $I(p)$.

Let $s \in I(p)$ be a point in the closure of $I^*(p)$. By Theorem 3.33 and Remark 3.16 there is an open neighbourhood V of $\gamma(s,p)$, a real number $a > 0$ and a unique local flow $\beta: (-a,a) \times V \to M$ for X which is of class C^k. Since the integral curve γ_p is continuous, we have that $\gamma_p^{-1}(V)$ is an open neighbourhood of s. Let $t_1 \in I^*(p) \cap \gamma_p^{-1}(V) \cap (s-a, s+a)$, and choose a real number $b > 0$ and an open neighbourhood U of p such that $(t_1 - b, t_1 + b) \times U \subset \mathscr{D}(X)$ and γ is of class C^k on this product. By choosing smaller b and U if necessary, we may assume that $\gamma((t_1 - b, t_1 + b) \times U) \subset V$.

Now let $\alpha: (t_1 - a, t_1 + a) \times U \to M$ be the map defined by

$$\alpha(t,q) = \beta(t - t_1, \gamma(t_1, q)).$$

Then α is of class C^k, and we have that $\alpha(t_1, q) = \beta(0, \gamma(t_1, q)) = \gamma(t_1, q)$ for every $q \in U$. For each such q we therefore have that α_q is an integral curve for X defined on the interval $(t_1 - a, t_1 + a)$ with $\alpha_q(t_1) = \gamma_q(t_1)$. Since γ_q is maximal, it follows that $(t_1 - a, t_1 + a) \times U \subset \mathscr{D}(X)$ and that γ coincides with α and hence is of class C^k on this product. Since $s \in (t_1 - a, t_1 + a)$, this implies that $s \in I^*(p)$ which completes the proof of the theorem. □

3.40 Proposition Let X be a vector field on a smooth manifold M, and let γ be its flow. If p is a point on M and t belongs to $I(p)$, then

$$I(\gamma(t,p)) = I(p) - t\,,$$

and we have that

$$\gamma(s, \gamma(t,p)) = \gamma(s+t, p)$$

for all $s \in I(p) - t$.

PROOF: The curves $\gamma_1: I(\gamma(t,p)) \to M$ and $\gamma_2: I(p) - t \to M$ defined by

$$\gamma_1(s) = \gamma(s, \gamma(t,p)) \quad \text{and} \quad \gamma_2(s) = \gamma(s+t, p)$$

are both maximal integral curves for X with initial condition $\gamma(t,p)$ and hence are equal. □

3.41 Definition If X is a vector field on a smooth manifold M with global flow $\gamma : \mathscr{D}(X) \to M$, and if $t \in \mathbf{R}$, we define

$$\mathscr{D}_t(X) = \{p \in M \mid (t,p) \in \mathscr{D}(X)\}$$

and let $\gamma_t : \mathscr{D}_t(X) \to M$ be the map defined by $\gamma_t(p) = \gamma(t,p)$.

3.42 Proposition For each $s,t \in \mathbf{R}$ we have that

(1) $\mathscr{D}_t(X)$ is open in M.

(2) If $p \in \mathscr{D}_t(X)$ and $\gamma_t(p) \in \mathscr{D}_s(X)$, then $p \in \mathscr{D}_{s+t}(X)$ and $\gamma_s(\gamma_t(p)) = \gamma_{s+t}(p)$. In particular the domain of $\gamma_s \circ \gamma_t$ is contained in $\mathscr{D}_{s+t}(X)$, and it equals $\mathscr{D}_{s+t}(X)$ if s and t both have the same sign.

(3) $\gamma_t(\mathscr{D}_t(X)) = \mathscr{D}_{-t}(X)$, and γ_t is a diffeomorphism onto its image with inverse γ_{-t}.

PROOF: (1) Follows from Theorem 3.39 since $\mathscr{D}_t(X) = i_t^{-1}(\mathscr{D}(X))$, where $i_t : M \to \mathbf{R} \times M$ is the continuous map defined by $i_t(p) = (t,p)$.

(2) By Definition 3.41 and 3.38 the assertions $p \in \mathscr{D}_t(X)$ and $\gamma_t(p) \in \mathscr{D}_s(X)$ are equivalent to $t \in I(p)$ and $s \in I(\gamma(t,p))$, respectively. Hence by Proposition 3.40 we have that $s+t \in I(p)$ and $\gamma(s,\gamma(t,p)) = \gamma(s+t,p)$ which is equivalent to $p \in \mathscr{D}_{s+t}(X)$ and $\gamma_s(\gamma_t(p)) = \gamma_{s+t}(p)$. This proves the first part of (2) and implies that the domain of $\gamma_s \circ \gamma_t$ is contained in $\mathscr{D}_{s+t}(X)$.

Now assume that s and t both have the same sign, and let $p \in \mathscr{D}_{s+t}(X)$ which is equivalent to $s+t \in I(p)$. Since $I(p)$ is an open interval around 0, it follows that $t \in I(p)$ and $s \in I(p) - t = I(\gamma(t,p))$ which again is equivalent to $p \in \mathscr{D}_t(X)$ and $\gamma_t(p) \in \mathscr{D}_s(X)$, showing that p is in the domain of $\gamma_s \circ \gamma_t$.

(3) Let $p \in \mathscr{D}_t(X)$ which is equivalent to $t \in I(p)$ by Definition 3.41 and 3.38. Since $I(p)$ also contains 0, we have that $-t \in I(p) - t = I(\gamma(t,p))$ which again is equivalent to $\gamma_t(p) \in \mathscr{D}_{-t}(X)$. By taking $s = -t$ in part (2), it follows that $\gamma_{-t}(\gamma_t(p)) = p$ for every $p \in \mathscr{D}_t(X)$. Replacing t with $-t$ we also have that $\gamma_t(\gamma_{-t}(p)) = p$ for every $p \in \mathscr{D}_{-t}(X)$, and this completes the proof of (3). □

3.43 Remark The domain of $\gamma_s \circ \gamma_t$ is generally not equal to $\mathscr{D}_{s+t}(X)$ in part (2) of the preceding proposition if s and t have opposite sign. For instance, when $s = -t$, the domain of $\gamma_s \circ \gamma_t$ is $\mathscr{D}_t(X)$, whereas $\mathscr{D}_{s+t}(X) = M$.

3.44 Definition A vector field X on a smooth manifold M is said to be *complete* if $\mathscr{D}(X) = \mathbf{R} \times M$, where $\mathscr{D}(X)$ is the domain of the flow for X as defined in Definition 3.38. This means that the maximal integral curve γ_p for X with initial condition p is defined on $(-\infty,\infty)$ for every $p \in M$.

In this case the maps γ_t in Definition 3.41 are defined on M for every $t \in \mathbf{R}$, and the family $\{\gamma_t \mid t \in \mathbf{R}\}$ is a group of diffeomorphisms on M parametrized by the real numbers called the 1-*parameter group* of X. When X is not complete, this family is not a group, but is referred to as the *local* 1-*parameter group* of X.

3.45 Proposition Let X be a vector field on a smooth manifold M, and let $\alpha : I \to M$ be a maximal integral curve for X. If $\alpha((a,b))$ is contained in a compact subset of M for each finite open subinterval (a,b) of I, we have that $I = \mathbf{R}$.

PROOF: It suffices to show that $a \in I$ and $b \in I$ for each such finite open subinterval (a,b) of I. Let $\{t_n\}$ be a sequence in (a,b) which converges to b. Since $\alpha((a,b))$ is contained in a compact set, there is a subsequence $\{\alpha(t_{n_i})\}$ of $\{\alpha(t_n)\}$ which converges to some point p on M. As the domain of the flow $\gamma : \mathscr{D}(X) \to M$ for X is an open subset of $\mathbf{R} \times M$ containing $\{0\} \times M$, there is a real number $\delta > 0$ and an open neighbourhood U of p such that $(-\delta, \delta) \times U \subset \mathscr{D}(X)$. Choose i so large that $t_{n_i} \in (b - \delta, b)$ and $\alpha(t_{n_i}) \in U$, and let $\beta : (t_{n_i} - \delta, t_{n_i} + \delta) \to M$ be the curve defined by $\beta(t) = \gamma(t - t_{n_i}, \alpha(t_{n_i}))$. Then β is an integral curve for X with $\beta(t_{n_i}) = \alpha(t_{n_i})$, and since α is maximal, it follows that $b \in (a, t_{n_i} + \delta) \subset I$. A similar argument holds for a. □

3.46 Corollary Let X be a vector field with compact support on a smooth manifold M. Then X is complete.

PROOF: We must show that $I(p) = \mathbf{R}$ for each point p on M. Let K be the support of X, and let γ be its flow. Suppose first that $\gamma_p(s) = q \notin K$ for some real number $s \in I(p)$. Then it follows from Proposition 3.36 that γ_p must coincide on $I(p)$ with the constant curve $c : \mathbf{R} \to M$ defined by $c(t) = q$ for all t, and since γ_p is maximal, we have that $I(p) = \mathbf{R}$.

If, on the other hand, the above assumption is not satisfied, we have that $\gamma_p(I(p)) \subset K$, in which case $I(p) = \mathbf{R}$ follows from Proposition 3.45. □

3.47 Corollary A vector field on a compact smooth manifold M is complete.

PROOF: Follows immediately from Corollary 3.46 since each such vector field has compact support. □

3.48 Proposition Let $f : M \to N$ be a smooth map, and let X and Y be vector fields on M and N, respectively, which are f-related. If $c : I \to M$ is an integral curve for X with initial condition q, then $f \circ c$ is an integral curve for Y with initial condition $f(q)$.

PROOF: If c is an integral curve for X, then

$$c'(t) = X(c(t))$$

for $t \in I$. From this it follows that

$$(f \circ c)'(t) = f_* \circ c'(t) = f_* \circ X \circ c(t) = Y \circ f \circ c(t) = Y(f \circ c(t))$$

which shows that $f \circ c$ is an integral curve for Y. If c has initial condition $c(0) = q$, then $f \circ c$ has initial condition $(f \circ c)(0) = f(q)$. □

3.49 Proposition Let $f : M \to N$ be a diffeomorphism, and let X and Y be vector fields on M and N, respectively, which are f-related. Then $c : I \to M$ is an integral curve for X with initial condition q if and only if $f \circ c$ is an integral curve for Y with initial condition $f(q)$.

If $\gamma : \mathscr{D}(X) \to M$ is the flow for X, then the flow for Y is given by $\beta : \mathscr{D}(Y) \to N$ where

$$\mathscr{D}(Y) = (id \times f)(\mathscr{D}(X)) \quad \text{and} \quad \beta = f \circ \gamma \circ (id \times f^{-1}),$$

and we have that

$$\mathscr{D}_t(Y) = f(\mathscr{D}_t(X)) \quad \text{and} \quad \beta_t = f \circ \gamma_t \circ f^{-1}$$

for $t \in \mathbf{R}$.

PROOF : The first part of the proposition follows from Proposition 3.48 applied to f and f^{-1}.

Now, let $\gamma_q : I(q) \to M$ and $\beta_p : J(p) \to N$ be the maximal integral curves for X and Y, respectively, with initial conditions q and p. By the first part of the proposition, we have that

$$J(f(q)) = I(q) \quad \text{and} \quad \beta_{f(q)} = f \circ \gamma_q \tag{1}$$

for $q \in M$.

From the first relation in (1) it follows that $(t, f(q)) \in \mathscr{D}(Y)$ if and only if $(t,q) \in \mathscr{D}(X)$, which shows that $\mathscr{D}(Y) = (id \times f)(\mathscr{D}(X))$. For a fixed $t \in \mathbf{R}$, we have that $f(q) \in \mathscr{D}_t(Y)$ if and only if $q \in \mathscr{D}_t(X)$ so that $\mathscr{D}_t(Y) = f(\mathscr{D}_t(X))$.

From the second relation in (1) we have that $\beta(t,p) = f \circ \gamma(t, f^{-1}(p))$ for $(t,p) \in \mathscr{D}(Y)$, which implies that $\beta = f \circ \gamma \circ (id \times f^{-1})$. For a fixed $t \in \mathbf{R}$, we have that $\beta_t(p) = f \circ \gamma_t(f^{-1}(p))$ for each $p \in \mathscr{D}_t(Y)$ so that $\beta_t = f \circ \gamma_t \circ f^{-1}$. □

3.50 Corollary Let $f : M \to M$ be a diffeomorphism, and let X be a vector field on M with flow γ. Then $f_*(X) = X$ if and only if $\gamma_t \circ f = f \circ \gamma_t$ for every $t \in \mathbf{R}$.

PROOF : If β is the flow for $f_*(X)$, it follows from Proposition 3.49 that $\beta_t = f \circ \gamma_t \circ f^{-1}$ for every $t \in \mathbf{R}$. Hence it is enough to show that $f_*(X) = X$ if and only if $\beta = \gamma$. Assuming that $\beta = \gamma$, we have that

$$f_*(X)(q) = f_*(X)(\beta_q(0)) = \beta_q'(0) = \gamma_q'(0) = X(\gamma_q(0)) = X(q)$$

for every $q \in M$ which shows that $f_*(X) = X$, and the converse statement is obviously true. □

Chapter 4

TENSORS

With the tangent bundle of a smooth manifold described in Chapter 2 at our disposal, we can now form new vector bundles by replacing the tangent space at each point with new vector spaces of various types, gluing them together in a smooth way. The sections in these new vector bundles will be covector fields and various types of tensor fields, including differential forms which will be investigated more fully in the next chapter. We also describe the Lie derivative of tensor fields, and the fundamental integrability theorems of Frobenius for integral manifolds of distributions.

DUAL VECTOR BUNDLES

4.1 If V is a vector space, we let V^* denote the dual space $L(V, \mathbf{R})$ consisting of all linear functionals on V. If $F : V \to W$ is a linear map, we have a linear map $F^* : W^* \to V^*$ defined by $F^*(\lambda) = \lambda \circ F$ so that the diagram

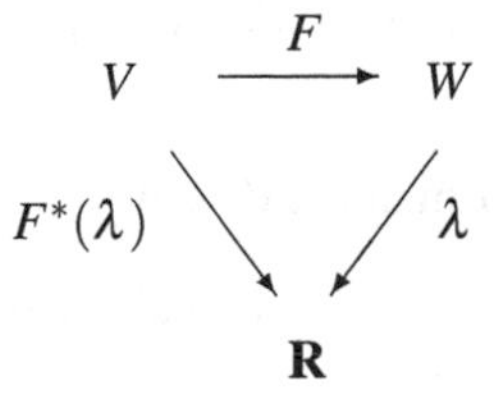

is commutative.

4.2 Proposition Let U, V and W be vector spaces.

(1) If $F : U \to V$ and $G : V \to W$ are linear maps, we have that $(G \circ F)^* = F^* \circ G^*$.

(2) If $id : V \to V$ is the identity on V, then $id^* : V^* \to V^*$ is the identity on V^*.

(3) If $F : V \to W$ is an isomorphism, so is $F^* : W^* \to V^*$ with $(F^*)^{-1} = (F^{-1})^*$.

PROOF : Follows immediately from the definition. □

4.3 Proposition If $\{v_1, ..., v_n\}$ is a basis for a vector space V, we have a basis $\{v_1^*, ..., v_n^*\}$ for the dual space V^* consisting of the linear functionals v_i^* defined by $v_i^*(v_j) = \delta_{i,j}$ for $j = 1, ..., n$. If λ is a linear functional in V^*, we have that

$$\lambda = \sum_{i=1}^{n} \lambda(v_i)\, v_i^* \, .$$

$\{v_1^*, ..., v_n^*\}$ is called the *dual basis* of $\{v_1, ..., v_n\}$.

PROOF : If $v = \sum_{i=1}^n a_i v_i$, we have that

$$\lambda(v) = \sum_{i=1}^{n} a_i \lambda(v_i) = \sum_{i=1}^{n} \lambda(v_i)\, v_i^*(v)$$

which shows the last assertion of the proposition and also that the set $\{v_1^*, ..., v_n^*\}$ spans V^*. To show that it is linearly independent, assume that $\sum_{i=1}^n a_i v_i^* = 0$. By applying both sides to v_j, we see that $a_j = 0$ for $j = 1, ..., n$. □

4.4 Proposition For each vector space V of dimension n we have a natural isomorphism $i_V : V \to V^{**}$ given by $i_V(v)(\lambda) = \lambda(v)$ for $v \in V$ and $\lambda \in V^*$, i.e., i_V is an isomorphism such that the diagram

$$\begin{array}{ccc} V & \xrightarrow{F} & W \\ {\scriptstyle i_V}\downarrow & & \downarrow{\scriptstyle i_W} \\ V^{**} & \xrightarrow{F^{**}} & W^{**} \end{array}$$

is commutative for every linear map $F : V \to W$.

PROOF : Let $\{v_1, ..., v_n\}$ be a basis for V with dual basis $\{v_1^*, ..., v_n^*\}$, and let $v = \sum_{i=1}^n a_i v_i$ be a vector in $\ker(i_V)$. Then

$$0 = i_V(v)(v_j^*) = v_j^*(v) = a_j$$

for $j = 1, ..., n$ so that $v = 0$, which shows that i_V is injective. Since $\dim V^{**} = n$ by Proposition 4.3, it follows that i_V is an isomorphism.

To show that it is natural, let $F : V \to W$ be a linear map, and let $v \in V$ and $\lambda \in V^*$. Then we have that

$$F^{**}(i_V(v))(\lambda) = i_V(v)(F^*(\lambda)) = F^*(\lambda)(v) = \lambda(F(v)) = i_W(F(v))(\lambda)$$

showing that the diagram is commutative. □

4.5 Remark We will often identify V^{**} with V by means of the natural isomorphism i_V, and we may consider a vector $v \in V$ to be a functional on V^* mapping $\lambda \in V^*$ onto $\lambda(v)$.

4.6 Proposition Let $F : V \to W$ be a linear map, and let $\mathscr{B}$ and $\mathscr{C}$ be bases for V and W, respectively, with dual bases $\mathscr{B}^*$ and $\mathscr{C}^*$. Then we have that

$$m_{\mathscr{B}^*}^{\mathscr{C}^*}(F^*) = m_{\mathscr{C}}^{\mathscr{B}}(F)^t .$$

PROOF: Let $\mathscr{B} = \{v_1, ..., v_n\}$ and $\mathscr{C} = \{w_1, ..., w_m\}$. If $m_{\mathscr{C}}^{\mathscr{B}}(F) = A$, we have that

$$F(v_j) = \sum_{i=1}^{m} A_{ij} w_i$$

for $j = 1, ..., n$. From this it follows that

$$F^*(w_j^*)(v_i) = (w_j^* \circ F)(v_i) = A_{ji}$$

which shows that

$$F^*(w_j^*) = \sum_{i=1}^{n} A_{ji} v_i^*$$

for $j = 1, ..., m$, and this completes the proof of the proposition. □

4.7 Now let $\pi : E \to M$ be an n-dimensional vector bundle over the smooth manifold M, and let $E_p = \pi^{-1}(p)$ be the fibre over p for each $p \in M$. We want to define a new bundle

$$E^* = \bigcup_{p \in M} E_p^* ,$$

called the *dual bundle* of E, by replacing each fibre E_p by its dual space. We have a projection $\pi^* : E^* \to M$ defined by $\pi^*(\lambda) = p$ for $\lambda \in E_p^*$.

4.8 Proposition $\pi^* : E^* \to M$ is an n-dimensional vector bundle over M.

PROOF: Let $\mathscr{E} = \{e_1, ..., e_n\}$ be the standard basis for $\mathbf{R}^n$, and let $\mathscr{E}^* = \{e^1, ..., e^n\}$ be its dual basis. We then have an isomorphism $r : (\mathbf{R}^n)^* \to \mathbf{R}^n$ defined by $r(e^i) = e_i$ for $i = 1, ..., n$.

Let $\{(t_\alpha, \pi^{-1}(U_\alpha)) \mid \alpha \in A\}$ be a trivializing cover of E. For each $\alpha \in A$ we have a bijection $t'_\alpha : \pi^{*-1}(U_\alpha) \to U_\alpha \times \mathbf{R}^n$ defined by $t'_\alpha(\lambda) = (p, t'_{\alpha,p}(\lambda))$ for $\lambda \in \pi^{*-1}(p)$, where $t'_{\alpha,p} = r \circ (t_{\alpha,p}^{-1})^*$. We will show that these bijections satisfy the conditions of Proposition 2.48.

For each pair of indices $\alpha, \beta \in A$, we have that $t_\alpha \circ t_\beta^{-1}$ is a diffeomorphism from $(U_\alpha \cap U_\beta) \times \mathbf{R}^n$ onto itself given by

$$(t_\alpha \circ t_\beta^{-1})(p, v) = (p, (t_{\alpha,p} \circ t_{\beta,p}^{-1})(v)) ,$$

where $t_{\alpha,p} \circ t_{\beta,p}^{-1} : \mathbf{R}^n \to \mathbf{R}^n$ is a linear isomorphism for each $p \in U_\alpha \cap U_\beta$. Since

$$t'_{\beta,p} \circ t'^{-1}_{\alpha,p} = r \circ (t_{\alpha,p} \circ t_{\beta,p}^{-1})^* \circ r^{-1} ,$$

it follows from Proposition 4.6, using the fact that $m_{\mathscr{E}}^{\mathscr{E}^*}(r)$ and $m_{\mathscr{E}^*}^{\mathscr{E}}(r^{-1})$ are both the identity matrix, that

$$m_{\mathscr{E}}^{\mathscr{E}}(t'_{\beta,p} \circ t'^{-1}_{\alpha,p}) = m_{\mathscr{E}}^{\mathscr{E}}(t_{\alpha,p} \circ t_{\beta,p}^{-1})^t .$$

Hence $t'_\beta \circ t'^{-1}_\alpha$ is also a diffeomorphism from $(U_\alpha \cap U_\beta) \times \mathbf{R}^n$ onto itself which is a linear isomorphism on each fibre. By Proposition 2.48 there is therefore a unique topology and smooth structure on E^* such that $\pi^* : E^* \to M$ is an n-dimensional vector bundle over M with $(t'_\alpha, \pi^{*-1}(U_\alpha))$ as local trivializations. □

4.9 Remark We say that $(t', \pi^{*-1}(U))$, where $t'_p = r \circ (t_p^{-1})^*$ for $p \in U$, is the local trivialization in the dual bundle $\pi^* : E^* \to M$ corresponding to the local trivialization $(t, \pi^{-1}(U))$ in the n-dimensional vector bundle $\pi : E \to M$. Here $r : (\mathbf{R}^n)^* \to \mathbf{R}^n$ is the linear isomorphism defined by $r(e^i) = e_i$ for $i = 1, ..., n$, where $\mathscr{E} = \{e_1, ..., e_n\}$ is the standard basis for $\mathbf{R}^n$ and $\mathscr{E}^* = \{e^1, ..., e^n\}$ is its dual basis.

4.10 Proposition Let $\pi : E \to M$ be an n-dimensional vector bundle over the smooth manifold M. Then the bundle map $i_E : E \to E^{**}$ over M given by $i_E(v)(\lambda) = \lambda(v)$ for $v \in E_p$ and $\lambda \in E_p^*$ where $p \in M$, is an equivalence.

PROOF: We know from Proposition 4.4 that the induced map $i_{E_p} : E_p \to E_p^{**}$ between the fibres is a natural isomorphism for each $p \in M$. It only remains to prove that i_E is a diffeomorphism.

It is enough to show that $t'' \circ i_E = t$ for each local trivialization $(t, \pi^{-1}(U))$ in a trivializing cover of E, where $(t'', \pi^{**-1}(U))$ is the corresponding local trivialization for E^{**} defined in Remark 4.9. We have that $t'_p = r \circ (t_p{}^{-1})^*$ so that $t''_p = r \circ (t'_p{}^{-1})^* = r \circ (r^{-1})^* \circ t_p{}^{**}$ for $p \in M$.

We first prove that

$$(r^{-1})^* \circ i_{\mathbf{R}^n} = r^{-1} .$$

Let $\mathscr{E} = \{e_1, ..., e_n\}$ be the standard basis for $\mathbf{R}^n$ and $\mathscr{E}^* = \{e^1, ..., e^n\}$ be its dual basis. Then we have that

$$(r^{-1})^*(i_{\mathbf{R}^n}(e_i))(e_j) = i_{\mathbf{R}^n}(e_i)(r^{-1}(e_j)) = i_{\mathbf{R}^n}(e_i)(e^j) = e^j(e_i) = \delta_{ij}$$

for $i, j = 1, ..., n$ which implies that

$$(r^{-1})^*(i_{\mathbf{R}^n}(e_i)) = e^i$$

for $i = 1, ..., n$, thereby proving the assertion.

From this it now follows that

$$t''_p \circ i_{E_p} = r \circ (r^{-1})^* \circ t_p{}^{**} \circ i_{E_p} = r \circ (r^{-1})^* \circ i_{\mathbf{R}^n} \circ t_p = t_p$$

which completes the proof of the proposition. □

4.11 Proposition Let $(t, \pi^{-1}(U))$ be a local trivialization for an n-dimensional vector bundle $\pi : E \to M$, and let $(t', \pi^{*-1}(U))$ be the corresponding local trivialization for its dual bundle $\pi^* : E^* \to M$. Furthermore, let $\mathscr{E} = \{e_1, ..., e_n\}$ be the standard basis for $\mathbf{R}^n$ and $\mathscr{E}^* = \{e^1, ..., e^n\}$ be its dual basis. If p is a point in U, then

$$\mathscr{B}_p^* = \{t_p'^{-1}(e_1), ..., t_p'^{-1}(e_n)\} = \{e^1 \circ t_p, ..., e^n \circ t_p\}$$

is a basis for the fibre $\pi^{*-1}(p)$ which is dual to the basis $\mathscr{B}_p$ for $\pi^{-1}(p)$ given in Remark 2.46. Moreover

$$\mathscr{B}_p^{**} = \{i_{E_p}(t_p^{-1}(e_1)), ..., i_{E_p}(t_p^{-1}(e_n))\} = \{e^1 \circ t_p', ..., e^n \circ t_p'\}$$

is a basis for the fibre $\pi^{**-1}(p)$ in the double dual bundle $\pi^{**} : E^{**} \to M$ which may be identified with $\mathscr{B}_p$ according to Remark 4.5.

PROOF : Since $t_p' = r \circ (t_p^{-1})^*$, we have that

$$t_p'^{-1}(e_i) = t_p^* \circ r^{-1}(e_i) = t_p^*(e^i) = e^i \circ t_p ,$$

and

$$(e^i \circ t_p)(t_p^{-1}(e_j)) = e^i(e_j) = \delta_{ij}$$

which shows that $\mathscr{B}_p^*$ is the dual basis of $\mathscr{B}_p$.

Using the notation in the proof of Proposition 4.10, we have that

$$i_{E_p}(t_p^{-1}(e_i)) = t_p''^{-1}(e_i) = e^i \circ t_p'$$

which shows the last part of the proposition. □

4.12 Proposition Let $\pi : E \to M$ be an n-dimensional vector bundle and $s : M \to E^*$ be a map with $\pi^* \circ s = id_M$, and let $\mathscr{E}^* = \{e^1, ..., e^n\}$ be the dual of the standard basis for $\mathbf{R}^n$. Then s is a section of π^* on M if and only if the map $a : U \to \mathbf{R}^n$ defined by

$$s(p) = \sum_{i=1}^{n} a_i(p)\, e^i \circ t_p$$

for $p \in U$ is smooth for every local trivialization $(t, \pi^{-1}(U))$ in a trivializing cover of E.

PROOF : If $(t', \pi^{*-1}(U))$ is the corresponding local trivialization for the dual bundle E^* as defined in Remark 4.9, it follows that

$$t_p'(e^i \circ t_p) = r \circ (t_p^{-1})^* \circ t_p^*(e^i) = r(e^i) = e_i$$

so that

$$(t' \circ s)(p) = (p, a(p))$$

for $p \in U$. Hence a is the local representation of s on U as defined in Definition 2.52. □

4.13 If M is a smooth manifold of dimension n, the dual bundle $\pi^* : T^*M \to M$ of the tangent bundle $\pi : TM \to M$ is called the *cotangent bundle* over M. The fibre T_p^*M over a point p in M is called the *cotangent space* of M at p.

If $f : V \to \mathbf{R}$ is a smooth function defined on an open neighbourhood of a point p in M, we define the *differential* of f at p to be the element $df(p)$ in T_p^*M defined by

$$df(p)(l) = l(f)$$

for each local derivation l in T_pM. By the discussion in 2.77 we know that this is the derivative of f at p with respect to the tangent vector l.

It follows from Lemma 2.78 (1) that $df(p)$ only depends on the local behavior of f around p, i.e., if $f_i : V_i \to \mathbf{R}$ for $i = 1,2$ are two smooth functions defined on open neighbourhoods V_i of p which coincide on an open neighbourhood of p contained in $V_1 \cap V_2$, then $df_1(p) = df_2(p)$.

If (x,U) is a local chart around p, we have that

$$dx^i(p)\left(\left.\frac{\partial}{\partial x^j}\right|_p\right) = \left.\frac{\partial}{\partial x^j}\right|_p (x^i) = \delta_{ij}$$

which shows that the set $\{dx^1(p),...,dx^n(p)\}$ of differentials of the coordinate functions of the local chart (x,U) at p is a basis for T_p^*M which is dual to the basis $\{\left.\frac{\partial}{\partial x^1}\right|_p,...,\left.\frac{\partial}{\partial x^n}\right|_p\}$ for T_pM of partial derivations at p with respect to (x,U).

4.14 Proposition Let $c : I \to M$ be a smooth curve on a smooth manifold M defined on an open interval I containing t, and let $f : V \to \mathbf{R}$ be a smooth function defined on an open neighbourhood V of $c(t)$. Then we have that

$$(f \circ c)'(t) = df(c(t))\,(c'(t))\,.$$

PROOF : Follows from the discussion in 2.77 and 4.13. □

4.15 Proposition If $(t_x, \pi^{-1}(U))$ is the local trivialization for the tangent bundle $\pi : TM \to M$ corresponding to a local chart (x,U) around a point p in M^n as defined in Remark 2.68, and if $\mathscr{E}^* = \{e^1,...,e^n\}$ is the dual of the standard basis for $\mathbf{R}^n$, then we have that

$$dx^i(p) = e^i \circ t_{x,p} \quad \text{for} \quad i = 1,...,n\,.$$

PROOF : Follows from Remark 2.82 and Proposition 4.11. □

4.16 Definition By a *covector field* or a *1-form* on an open subset V of a smooth manifold M we mean a section of its cotangent bundle $\pi^* : T^*M \to M$ on V.

4.17 Proposition Let $\pi^* : T^*M \to M$ be the cotangent bundle of a smooth manifold M^n and $\omega : M \to T^*M$ be a map with $\pi^* \circ \omega = id_M$. Then ω is a 1-form on M if and only if the map $a : U \to \mathbf{R}^n$ defined by

$$\omega(p) = \sum_{i=1}^{n} a_i(p)\, dx^i(p)$$

for $p \in U$ is smooth for every local chart (x,U) on M.

PROOF : Follows from Propositions 4.12 and 4.15. □

4.18 Proposition If $f : V \to \mathbf{R}$ is a smooth function defined on an open subset V of a smooth manifold M^n, then its differential $df : V \to T^*M$ is a 1-form on V. For any local chart (x,U) on M with $U \subset V$ we have that

$$df\,|_U = \sum_{i=1}^{n} \frac{\partial f}{\partial x^i}\, dx^i\,.$$

PROOF : We have that

$$df(p) = \sum_{i=1}^{n} df(p)\left(\left.\frac{\partial}{\partial x^j}\right|_p\right) dx^i(p) = \sum_{i=1}^{n} \frac{\partial f}{\partial x^i}(p)\, dx^i(p)$$

for $p \in U$. □

4.19 Definition Let S be a subset of a finite dimensional vector space V. Then the *annihilator* of S is the subspace $A(S)$ of V^* defined by

$$A(S) = \{\lambda \in V^* | \lambda(S) = \{0\}\}\,.$$

4.20 Proposition Let V_1 and V_2 be two subspaces of a finite dimensional vector space V with $V = V_1 \oplus V_2$. Then we have that

$$V^* = A(V_1) \oplus A(V_2)\,.$$

If $\pi_1 : V \to V$ and $\pi_2 : V \to V$ are the projections on V_1 and V_2, then $\pi_2^* : V^* \to V^*$ and $\pi_1^* : V^* \to V^*$ are the projections on $A(V_1)$ and $A(V_2)$, respectively.

PROOF : We have that

$$\lambda = \lambda \circ \pi_2 + \lambda \circ \pi_1$$

where $\lambda \circ \pi_2 \in A(V_1)$ and $\lambda \circ \pi_1 \in A(V_2)$ for every $\lambda \in V^*$. Since we also have that

$$A(V_1) \cap A(V_2) = \{0\}\,,$$

this completes the proof of the proposition. □

4.21 Proposition Let W be a subspace of a finite dimensional vector space V. Then we have that

$$\dim W + \dim A(W) = \dim V .$$

PROOF: Choose a basis $\mathscr{C} = \{w_1, \ldots, w_r\}$ for W, and extend it to a basis $\mathscr{B} = \{w_1, \ldots, w_n\}$ for V with dual basis $\mathscr{B}^* = \{w_1^*, \ldots, w_n^*\}$. We contend that $\{w_{r+1}^*, \ldots, w_n^*\}$ is a basis for $A(W)$.

It is clearly a linearly independent subset of $A(W)$, so we only need to show that it spans $A(W)$. This follows from the fact that every element $\lambda \in A(W)$ can be written as $\lambda = \sum_{i=1}^n a_i w_i^*$, where $a_j = \lambda(w_j) = 0$ for $j = 1, \ldots, r$, thus showing that $\dim A(W) = n - r$. □

4.22 Proposition Let $\pi' : E' \to M$ be a k-dimensional subbundle of an n-dimensional vector bundle $\pi : E \to M$. Then

$$A(E') = \bigcup_{p \in M} A(E'_p)$$

is an $(n-k)$-dimensional subbundle of the dual bundle $\pi^* : E^* \to M$, with projection $\pi'' : A(E') \to M$ sending each set $A(E'_p)$ to p.

PROOF: Given a point p in M, we choose a local trivialization $(t, \pi^{-1}(U))$ around p in E having the subbundle property

$$t(\pi^{-1}(U) \cap E') = U \times \mathbf{R}^k \times \{0\} \subset U \times \mathbf{R}^k \times \mathbf{R}^{n-k} .$$

Let $(t', \pi^{*-1}(U))$ be the corresponding local trivialization in the dual bundle $\pi^* : E^* \to M$ described in Remark 4.9, where $t'_p = r \circ (t_p^{-1})^*$ for $p \in U$. In the same way as in the proof of Proposition 4.21 we see that

$$t'_p(A(E'_p)) = r(A(\mathbf{R}^k \times \{0\})) = \{0\} \times \mathbf{R}^{n-k}$$

for $p \in U$. If $\alpha : \mathbf{R}^k \times \mathbf{R}^{n-k} \to \mathbf{R}^{n-k} \times \mathbf{R}^k$ is the linear isomorphism given by $\alpha(a,b) = (b,a)$ for $a \in \mathbf{R}^k$ and $b \in \mathbf{R}^{n-k}$, we therefore have that $(\alpha \circ t', \pi^{*-1}(U))$ is a local trivialization around p in E^* with the subbundle property

$$\alpha \circ t'(\pi^{*-1}(U) \cap A(E')) = U \times \mathbf{R}^{n-k} \times \{0\} \subset U \times \mathbf{R}^{n-k} \times \mathbf{R}^k .$$

□

TENSOR BUNDLES

4.23 Let $V_1, \ldots, V_k$ be vector spaces. Their *tensor product* $V_1 \otimes \cdots \otimes V_k$ is defined to be a vector space V with a multilinear map $\phi : V_1 \times \cdots \times V_k \to V$ satisfying the following universal property. If $F : V_1 \times \cdots \times V_k \to W$ is any multilinear map into some vector space W, then there is a unique linear map $F_* : V \to W$ which makes the diagram

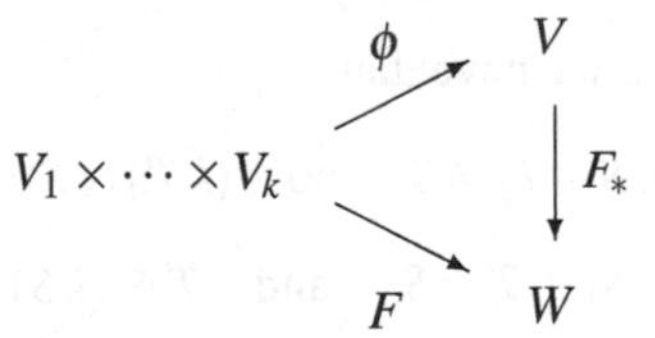

commutative. It follows from the definition that the tensor product is uniquely defined in the sense that if V' is another vector space satisfying the above condition, then there is a unique linear isomorphism between V and V'.

If $v_i \in V_i$ for $i = 1, ..., k$, we denote the element $\phi(v_1, ..., v_k)$ by $v_1 \otimes \cdots \otimes v_k$. Given a linear map $F_i : V_i \to V_i'$ for each $i = 1, ..., k$, we have a unique linear map $F_1 \otimes \cdots \otimes F_k : V_1 \otimes \cdots \otimes V_k \to V_1' \otimes \cdots \otimes V_k'$ such that the diagram

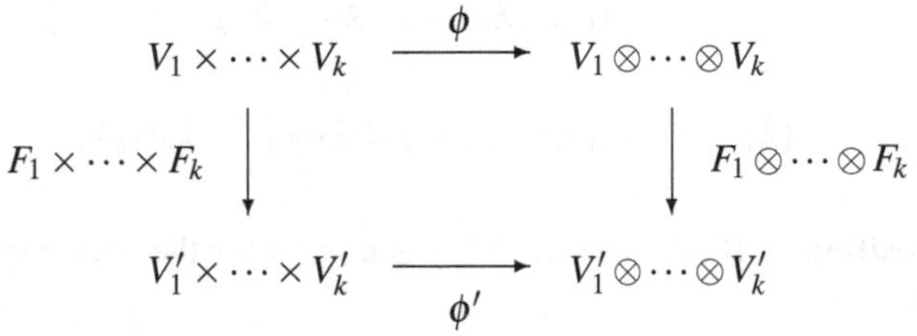

is commutative since $\phi' \circ (F_1 \times \cdots \times F_k)$ is multilinear, and we have that

$$F_1 \otimes \cdots \otimes F_k (v_1 \otimes \cdots \otimes v_k) = F_1(v_1) \otimes \cdots \otimes F_k(v_k) \, .$$

It remains to prove the existence of V. We first consider the tensor product of the dual spaces $V_1^*, ..., V_k^*$.

4.24 If $V_1, ..., V_k, W$ are vector spaces, we let $L^k(V_1, ..., V_k; W)$ be the spacc of all multilinear maps $F : V_1 \times .. \times V_k \to W$. The space $L^k(V_1, ..., V_k; \mathbf{R})$ consisting of all multilinear functionals $T : V_1 \times .. \times V_k \to \mathbf{R}$ is also denoted by $\mathscr{T}(V_1, ..., V_k)$. If $k = 1$, then $\mathscr{T}(V) = V^*$ is just the dual space of V. We want to prove in general that $\mathscr{T}(V_1, ..., V_k) = V_1^* \otimes \cdots \otimes V_k^*$ so that $V_1 \otimes \cdots \otimes V_k = \mathscr{T}(V_1^*, ..., V_k^*)$ when we identify each V_i with its double dual V_i^{**} as usual.

The components $F^j \in \mathscr{T}(V_1, ..., V_k)$ of a multilinear map $F \in L^k(V_1, ..., V_k; W)$ with respect to a basis $\mathscr{C} = \{w_1, ..., w_m\}$ for W are defined by

$$F(v_1, ..., v_k) = \sum_{j=1}^{m} F^j(v_1, ..., v_k) \, w_j$$

for $(v_1, ..., v_k) \in V_1 \times .. \times V_k$.

4.25 If $T \in \mathscr{T}(V_1, ..., V_k)$ and $S \in \mathscr{T}(V_{k+1}, ..., V_{k+l})$, we define their *tensor product* $T \otimes S \in \mathscr{T}(V_1, ..., V_{k+l})$ by

$$(T \otimes S)(v_1, ..., v_k, v_{k+1}, ..., v_{k+l}) = T(v_1, ..., v_k) S(v_{k+1}, ..., v_{k+l})$$

We see that $\otimes$ is bilinear, i.e., we have that

(1) $(T_1 + T_2) \otimes S = T_1 \otimes S + T_2 \otimes S$ and $(kT) \otimes S = k(T \otimes S)$

(2) $T \otimes (S_1 + S_2) = T \otimes S_1 + T \otimes S_2$ and $T \otimes (kS) = k(T \otimes S)$

for all multilinear functionals $T, T_1, T_2 \in \mathscr{T}(V_1, ..., V_k)$ and $S, S_1, S_2 \in \mathscr{T}(V_{k+1}, ..., V_{k+l})$ and every real number k.

Since $(U \otimes T) \otimes S = U \otimes (T \otimes S)$, we also have a well-defined tensor product with an arbitrary number of factors. In particular we have a multilinear map

$$\phi : V_1^* \times \cdots \times V_k^* \to \mathscr{T}(V_1, ..., V_k)$$

defined by

$$\phi(\lambda_1, ..., \lambda_k) = \lambda_1 \otimes \cdots \otimes \lambda_k$$

where

$$(\lambda_1 \otimes \cdots \otimes \lambda_k)(v_1, ..., v_k) = \lambda_1(v_1) \cdots \lambda_k(v_k) .$$

4.26 Proposition If $\mathscr{B}_r = \{v_1^r, ..., v_{n_r}^r\}$ is a basis for the vector space V_r for $r = 1, ..., k$, then

$$\mathscr{D} = \{v_{i_1}^{1*} \otimes \cdots \otimes v_{i_k}^{k*} \mid (i_1, ..., i_k) \in I_{n_1} \times \cdots \times I_{n_k}\} ,$$

where $I_n = \{1, ..., n\}$, is a basis for $\mathscr{T}(V_1, ..., V_k)$. If $T \in \mathscr{T}(V_1, ..., V_k)$ we have that

$$T = \sum_{i_1, ..., i_k} T(v_{i_1}^1, ..., v_{i_k}^k)\, v_{i_1}^{1*} \otimes \cdots \otimes v_{i_k}^{k*} .$$

PROOF: If $T \in \mathscr{T}(V_1, ..., V_k)$ we have that

$$\sum_{i_1, ..., i_k} T(v_{i_1}^1, ..., v_{i_k}^k)\, v_{i_1}^{1*} \otimes \cdots \otimes v_{i_k}^{k*}\, (v_{j_1}^1, ..., v_{j_k}^k) = T(v_{j_1}^1, ..., v_{j_k}^k)$$

for $1 \leq j_r \leq n_r$ and $1 \leq r \leq k$, which shows the last assertion of the proposition and also that $\mathscr{D}$ spans $\mathscr{T}(V_1, ..., V_k)$. To show that it is linearly independent, assume that

$$\sum_{i_1, ..., i_k} a_{i_1, ..., i_k}\, v_{i_1}^{1*} \otimes \cdots \otimes v_{i_k}^{k*} = 0 .$$

By applying both sides to $(v_{j_1}^1, ..., v_{j_k}^k)$, we see that $a_{j_1, ..., j_k} = 0$ for $1 \leq j_r \leq n_r$ and $1 \leq r \leq k$. □

4.27 Remark We can order the index set $I_{n_1} \times \cdots \times I_{n_k}$ for the basis $\mathscr{D}$ lexicographically so that $(i_1, ..., i_k) < (j_1, ..., j_k)$ if there is an integer $r \in I_k$ such that $i_s = j_s$ for $1 \leq s < r$ and $i_r < j_r$. We then have a unique bijection $b : I_{n_1} \times \cdots \times I_{n_k} \to I_n$ which is strictly increasing and where $n = \prod_{r=1}^{k} n_k$. It is explicitly given by

$$b(i_1, ..., i_k) = \sum_{r=1}^{k} (i_r - 1) \prod_{j=r+1}^{k} n_j + 1 \, .$$

We let $\mathscr{T}(\mathscr{B}_1, ..., \mathscr{B}_k)$ be the basis $\{v_1, ..., v_n\}$ for $\mathscr{T}(V_1, ..., V_k)$ obtained by reindexing $\mathscr{D}$ so that

$$v_{b(i_1, ..., i_k)} = v_{i_1}^{1\,*} \otimes \cdots \otimes v_{i_k}^{k\,*}$$

for $(i_1, ..., i_k) \in I_{n_1} \times \cdots \times I_{n_k}$.

4.28 Proposition If $V_1, ..., V_k$ are vector spaces, the tensor product $V_1^* \otimes \cdots \otimes V_k^*$ equals $\mathscr{T}(V_1, ..., V_k)$ with the multilinear map $\phi : V_1^* \times \cdots \times V_k^* \to \mathscr{T}(V_1, ..., V_k)$ defined in 4.25.

PROOF: Let $\mathscr{B}_r = \{v_1^r, ..., v_{n_r}^r\}$ be a basis for the vector space V_r for $r = 1, ..., k$, and let $F : V_1^* \times \cdots \times V_k^* \to W$ be a multilinear map into some vector space W. Then we have a linear map $F_* : \mathscr{T}(V_1, ..., V_k) \to W$ defined by

$$F_*(v_{i_1}^{1\,*} \otimes \cdots \otimes v_{i_k}^{k\,*}) = F(v_{i_1}^{1\,*}, ..., v_{i_k}^{k\,*}) \tag{1}$$

for $(i_1, ..., i_k) \in I_{n_1} \times \cdots \times I_{n_k}$. If $\lambda_r = \sum_{i=1}^{n_r} a_i^r v_i^{r\,*}$ is an arbitrary element in V_r^* for $r = 1, ..., k$, we have that

$$\begin{aligned} F_*(\lambda_1 \otimes \cdots \otimes \lambda_k) &= \sum_{i_1, ..., i_k} a_{i_1}^1 \cdots a_{i_k}^k \, F_*(v_{i_1}^{1\,*} \otimes \cdots \otimes v_{i_k}^{k\,*}) \\ &= \sum_{i_1, ..., i_k} a_{i_1}^1 \cdots a_{i_k}^k \, F(v_{i_1}^{1\,*}, ..., v_{i_k}^{k\,*}) = F(\lambda_1, ..., \lambda_k) \end{aligned}$$

which shows that the diagram in 4.23 is commutative.

To prove uniqueness, we see that a linear map F_* making the diagram commutative clearly must satisfy (1), and this determines F_* uniquely on $\mathscr{T}(V_1, ..., V_k)$. □

4.29 Corollary Let $V_1, ..., V_k$ be vector spaces, and let $\phi' : V_1^{**} \times \cdots \times V_k^{**} \to \mathscr{T}(V_1^*, ..., V_k^*)$ be the multilinear map defined in 4.25. Then the tensor product $V_1 \otimes \cdots \otimes V_k$ equals $\mathscr{T}(V_1^*, ..., V_k^*)$ with the multilinear map $\phi : V_1 \times \cdots \times V_k \to \mathscr{T}(V_1^*, ..., V_k^*)$ given by $\phi = \phi' \circ (i_{V_1} \times \cdots \times i_{V_k})$.

We have that $i_{V_1} \otimes \cdots \otimes i_{V_k}$ is the identity on $\mathscr{T}(V_1^*, ..., V_k^*)$, and if $v_i \in V_i$ for $i = 1, ..., k$, then $v_1 \otimes \cdots \otimes v_k$ is the element in $\mathscr{T}(V_1^*, ..., V_k^*)$ given by

$$v_1 \otimes \cdots \otimes v_k \, (\lambda_1, ..., \lambda_k) = \lambda_1(v_1) \cdots \lambda_k(v_k)$$

for $\lambda_i \in V_i^*$ where $i = 1, ..., k$.

PROOF : Let $F : V_1 \times \cdots \times V_k \to W$ be a multilinear map into some vector space W. Then we have a multilinear map $F \circ (i_{V_1} \times \cdots \times i_{V_k})^{-1} : V_1^{**} \times \cdots \times V_k^{**} \to W$, so by Proposition 4.28 there is a unique linear map $F_* : \mathscr{T}(V_1^*, ..., V_k^*) \to W$ with $F \circ (i_{V_1} \times \cdots \times i_{V_k})^{-1} = F_* \circ \phi'$. This implies that $F = F_* \circ \phi$ and completes the proof of the first part of the corollary. The last part follows from 4.23 and Remark 4.5. □

4.30 Corollary If $\mathscr{B}_r = \{v_1^r, ..., v_{n_r}^r\}$ is a basis for the vector space V_r for $r = 1, ..., k$, then

$$\mathscr{D} = \{v_{i_1}^1 \otimes \cdots \otimes v_{i_k}^k \mid (i_1, ..., i_k) \in I_{n_1} \times \cdots \times I_{n_k}\},$$

where $I_n = \{1, ..., n\}$, is a basis for $V_1 \otimes \cdots \otimes V_k$. If $T \in \mathscr{T}(V_1^*, ..., V_k^*)$ we have that

$$T = \sum_{i_1, ..., i_k} T(v_{i_1}^{1*}, ..., v_{i_k}^{k*})\, v_{i_1}^1 \otimes \cdots \otimes v_{i_k}^k .$$

The tensor product $V_1 \otimes \cdots \otimes V_k$ is generated by the set $\phi\,(V_1 \times \cdots \times V_k)$ consisting of elements of the form $v_1 \otimes \cdots \otimes v_k$, where $v_i \in V_i$ for $i = 1, ..., k$, which are called *simple* . If $F_i : V_i \to V_i'$ is a linear map for $i = 1, ..., k$, we have that

$$F_1 \otimes \cdots \otimes F_k\,(V_1 \otimes \cdots \otimes V_k) = F_1(V_1) \otimes \cdots \otimes F_k(V_k) .$$

PROOF : Follows from 4.23, Proposition 4.26 and Corollary 4.29. □

4.31 Remark By 4.23 and 4.25 we have a bilinear map

$$\mu : \mathscr{T}(V_1, ..., V_k) \times \mathscr{T}(V_{k+1}, ..., V_{k+l}) \to \mathscr{T}(V_1, ..., V_{k+l})$$

defined by

$$\mu(T,S) = T \otimes S$$

which induces a linear map

$$\mu_* : \mathscr{T}(V_1, ..., V_k) \otimes \mathscr{T}(V_{k+1}, ..., V_{k+l}) \to \mathscr{T}(V_1, ..., V_{k+l}) .$$

If the vector spaces $V_1, ..., V_{k+l}$ are finite dimensional, it follow from Proposition 4.26 and 4.28 that μ_* is a linear isomorphism since it is a surjective linear map between spaces of the same dimension.

4.32 If $F_i : V_i \to W_i$ is a linear map for $i = 1, ..., k$, we have a linear map

$$(F_1, ..., F_k)^* : \mathscr{T}(W_1, ..., W_k) \to \mathscr{T}(V_1, ..., V_k)$$

defined by

$$(F_1, ..., F_k)^*(T) = T \circ (F_1 \times \cdots \times F_k) ,$$

i.e.,

$$(F_1, ..., F_k)^*(T)\,(v_1, ..., v_k) = T\,(F_1(v_1), ..., F_k(v_k)) .$$

In particular we have that

$$(F_1, ..., F_k)^* (\lambda_1 \otimes \cdots \otimes \lambda_k) = F_1^*(\lambda_1) \otimes \cdots \otimes F_k^*(\lambda_k) .$$

for $\lambda_i \in V_i^*$ where $i = 1, ..., k$.

4.33 Proposition

(1) If $F_i : U_i \to V_i$ and $G_i : V_i \to W_i$ are linear maps for $i = 1, ..., k$, we have that $(G_1 \circ F_1, ..., G_k \circ F_k)^* = (F_1, ..., F_k)^* \circ (G_1, ..., G_k)^*$.

(2) If $id_i : V_i \to V_i$ is the identity on V_i for $i = 1, ..., k$, then $(id_1, ..., id_k)^* : \mathscr{T}(V_1, ..., V_k) \to \mathscr{T}(V_1, ..., V_k)$ is the identity on $\mathscr{T}(V_1, ..., V_k)$.

(3) If $F_i : V_i \to W_i$ are isomorphisms for $i = 1, ..., k$, then so is $(F_1, ..., F_k)^* : \mathscr{T}(W_1, ..., W_k) \to \mathscr{T}(V_1, ..., V_k)$ with $(F_1, ..., F_k)^{*-1} = (F_1^{-1}, ..., F_k^{-1})^*$.

(4) If $F_i : V_i \to W_i$ are linear maps for $i = 1, ..., k+l$, we have that $(F_1, ..., F_{k+l})^* (T \otimes S) = (F_1, ..., F_k)^*(T) \otimes (F_{k+1}, ..., F_{k+l})^*(S)$ for every $T \in \mathscr{T}(W_1, ..., W_k)$ and $S \in \mathscr{T}(W_{k+1}, ..., W_{k+l})$.

PROOF : Follows immediately from the definition. □

4.34 Proposition If $F_i : V_i \to W_i$ is a linear map for $i = 1, ..., k$, then the linear map

$$F_1 \otimes \cdots \otimes F_k : V_1 \otimes \cdots \otimes V_k \to W_1 \otimes \cdots \otimes W_k$$

defined in 4.23 is given by

$$F_1 \otimes \cdots \otimes F_k = (F_1^*, ..., F_k^*)^* .$$

PROOF : We have that

$$(F_1^*, ..., F_k^*)^* (v_1 \otimes \cdots \otimes v_k)(\lambda_1, ..., \lambda_k) = (v_1 \otimes \cdots \otimes v_k)(F_1^*(\lambda_1), ..., F_k^*(\lambda_k))$$
$$= \lambda_1(F_1(v_1)) \cdots \lambda_k(F_k(v_k)) = (F_1(v_1) \otimes \cdots \otimes F_k(v_k))(\lambda_1, ..., \lambda_k)$$

for $v_i \in V_i$ and $\lambda_i \in W_i^*$ where $i = 1, ..., k$. □

4.35 Corollary

(1) If $F_i : U_i \to V_i$ and $G_i : V_i \to W_i$ are linear maps for $i = 1, ..., k$, we have that $(G_1 \circ F_1) \otimes \cdots \otimes (G_k \circ F_k) = (G_1 \otimes \cdots \otimes G_k) \circ (F_1 \otimes \cdots \otimes F_k)$.

(2) If $id_i : V_i \to V_i$ is the identity on V_i for $i = 1, ..., k$, then $id_1 \otimes \cdots \otimes id_k : V_1 \otimes \cdots \otimes V_k \to V_1 \otimes \cdots \otimes V_k$ is the identity on $V_1 \otimes \cdots \otimes V_k$.

(3) If $F_i : V_i \to W_i$ are isomorphisms for $i = 1, ..., k$, then so is $F_1 \otimes \cdots \otimes F_k : V_1 \otimes \cdots \otimes V_k \to W_1 \otimes \cdots \otimes W_k$ with $(F_1 \otimes \cdots \otimes F_k)^{-1} = F_1^{-1} \otimes \cdots \otimes F_k^{-1}$.

PROOF : Follows from Propositions 4.2 and 4.33 or immediately from the definition in 4.23. □

4.36 Definition Let A^r be an $m_r \times n_r$ - matrix for $r = 1, ..., k$. The *tensor product* $A^1 \otimes \cdots \otimes A^k$ is the $m \times n$ - matrix A defined by

$$A_{ij} = A^1_{i_1 j_1} \cdots A^k_{i_k j_k}$$

where $m = \prod_{r=1}^k m_r$ and $n = \prod_{r=1}^k n_r$, and where $i = b(i_1, ..., i_k)$ and $j = b'(j_1, ..., j_k)$ for the bijections $b : I_{n_1} \times \cdots \times I_{n_k} \to I_n$ and $b' : I_{m_1} \times \cdots \times I_{m_k} \to I_m$ defined in Remark 4.27.

4.37 Proposition Let $F_r : V_r \to W_r$ be a linear map for $r = 1, ..., k$, and let $\mathscr{B}_r$ and $\mathscr{C}_r$ be bases for V_r and W_r, respectively. Let $\mathscr{B} = \mathscr{T}(\mathscr{B}_1, ..., \mathscr{B}_k)$ and $\mathscr{C} = \mathscr{T}(\mathscr{C}_1, ..., \mathscr{C}_k)$ be the bases for $\mathscr{T}(V_1, ..., V_k)$ and $\mathscr{T}(W_1, ..., W_k)$ defined in Remark 4.27. Then we have that

$$m^{\mathscr{C}}_{\mathscr{B}}((F_1, ..., F_k)^*) = m^{\mathscr{B}_1}_{\mathscr{C}_1}(F_1)^t \otimes \cdots \otimes m^{\mathscr{B}_k}_{\mathscr{C}_k}(F_k)^t .$$

PROOF : Let $\mathscr{B}_r = \{v^r_1, ..., v^r_{n_r}\}$ and $\mathscr{C}_r = \{w^r_1, ..., w^r_{m_r}\}$. If $m^{\mathscr{B}_r}_{\mathscr{C}_r}(F) = A^r$, we have that

$$F_r(v^r_j) = \sum_{i=1}^{m_r} A^r_{ij} w^r_i$$

for $1 \le j \le n_r$ and $1 \le r \le k$. From this it follows that

$$(F_1, ..., F_k)^* (w^{1*}_{j_1} \otimes \cdots \otimes w^{k*}_{j_k}) (v^1_{i_1}, ..., v^k_{i_k}) =$$

$$((w^{1*}_{j_1} \circ F_1) \otimes \cdots \otimes (w^{k*}_{j_k} \circ F_k)) (v^1_{i_1}, ..., v^k_{i_k}) = A^1_{j_1 i_1} \cdots A^k_{j_k i_k}$$

which shows that

$$(F_1, ..., F_k)^* (w^{1*}_{j_1} \otimes \cdots \otimes w^{k*}_{j_k}) = \sum_{i_1, ..., i_k} A^1_{j_1 i_1} \cdots A^k_{j_k i_k} v^{1*}_{i_1} \otimes \cdots \otimes v^{k*}_{i_k}$$

for $(j_1, ..., j_k) \in I_{m_1} \times \cdots \times I_{m_k}$. If $\mathscr{B} = \{v_1, ..., v_n\}$ and $\mathscr{C} = \{w_1, ..., w_m\}$, where $n = \prod_{r=1}^k n_r$ and $m = \prod_{r=1}^k m_r$, and if $(A^1)^t \otimes \cdots \otimes (A^k)^t = A$, it follows from Remark 4.27 and Definition 4.36 that

$$(F_1, ..., F_k)^* (w_j) = \sum_{i=1}^{n} A_{ij} v_i$$

which completes the proof of the proposition. □

4.38 Now let $\pi^r : E^r \to M$ be an n_r-dimensional vector bundle over the smooth manifold M for $r = 1, \ldots, k$, and let $E^r_p = \pi^{r\,-1}(p)$ be the fibre over p for each $p \in M$. We want to define a new bundle

$$\mathscr{T}(E^1, \ldots, E^k) = \bigcup_{p \in M} \mathscr{T}(E^1_p, \ldots, E^k_p)$$

with projection $\pi' : \mathscr{T}(E^1, \ldots, E^k) \to M$ defined by $\pi'(\omega) = p$ for $\omega \in \mathscr{T}(E^1_p, \ldots, E^k_p)$.

4.39 Proposition $\pi' : \mathscr{T}(E^1, \ldots, E^k) \to M$ is an n-dimensional vector bundle over M, where $n = \prod_{r=1}^k n_r$.

PROOF: Let $\mathscr{E}_r = \{e^r_1, \ldots, e^r_{n_r}\}$ be the standard basis for $\mathbf{R}^{n_r}$ for $r = 1, \ldots, k$, and let $\mathscr{B} = \mathscr{T}(\mathscr{E}_1, \ldots, \mathscr{E}_k)$ be the basis for $\mathscr{T}(\mathbf{R}^{n_1}, \ldots, \mathbf{R}^{n_k})$ defined in Remark 4.27. If $\mathscr{B} = \{v_1, \ldots, v_n\}$ and if $\mathscr{E} = \{e_1, \ldots, e_n\}$ is the standard basis for $\mathbf{R}^n$, we have a linear isomorphism $r' : \mathscr{T}(\mathbf{R}^{n_1}, \ldots, \mathbf{R}^{n_k}) \to \mathbf{R}^n$ defined by $r'(v_i) = e_i$ for $i = 1, \ldots, n$.

Let $\{U_\alpha | \alpha \in A\}$ be an open cover of M such that for each $\alpha \in A$, there are local trivializations $(t^r_\alpha, \pi^{r\,-1}(U_\alpha))$ in E^r for $r = 1, \ldots, k$. Then we have for each $\alpha \in A$ a bijection $t'_\alpha : \pi'^{\,-1}(U_\alpha) \to U_\alpha \times \mathbf{R}^n$ defined by $t'_\alpha(\omega) = (p, t'_{\alpha,p}(\omega))$ for $\omega \in \pi'^{\,-1}(p)$, where $t'_{\alpha,p} = r' \circ (t^{1\;-1}_{\alpha,p}, \ldots, t^{k\;-1}_{\alpha,p})^*$. We will show that these bijections satisfy the conditions of Proposition 2.48.

For each pair of indices $\alpha, \beta \in A$, we have that $t^r_\alpha \circ t^{r\;-1}_\beta$ is a diffeomorphism from $(U_\alpha \cap U_\beta) \times \mathbf{R}^{n_r}$ onto itself given by

$$(t^r_\alpha \circ t^{r\;-1}_\beta)(p, v) = (p, (t^r_{\alpha,p} \circ t^{r\;-1}_{\beta,p})(v))$$

for $r = 1, \ldots, k$, where $t^r_{\alpha,p} \circ t^{r\;-1}_{\beta,p} : \mathbf{R}^{n_r} \to \mathbf{R}^{n_r}$ is a linear isomorphism for each $p \in U_\alpha \cap U_\beta$. Since

$$t'_{\beta,p} \circ t'^{\,-1}_{\alpha,p} = r' \circ (t^1_{\alpha,p} \circ t^{1\;-1}_{\beta,p}, \ldots, t^k_{\alpha,p} \circ t^{k\;-1}_{\beta,p})^* \circ r'^{-1},$$

it follows from Proposition 4.37, using the fact that $m^{\mathscr{B}}_{\mathscr{E}}(r')$ and $m^{\mathscr{E}}_{\mathscr{B}}(r'^{-1})$ are both the identity matrix, that

$$m^{\mathscr{E}}_{\mathscr{E}}(t'_{\beta,p} \circ t'^{-1}_{\alpha,p}) = m^{\mathscr{E}_1}_{\mathscr{E}_1}(t^1_{\alpha,p} \circ t^{1\;-1}_{\beta,p})^t \otimes \cdots \otimes m^{\mathscr{E}_k}_{\mathscr{E}_k}(t^k_{\alpha,p} \circ t^{k\;-1}_{\beta,p})^t .$$

Hence $t'_\beta \circ t'^{-1}_\alpha$ is also a diffeomorphism from $(U_\alpha \cap U_\beta) \times \mathbf{R}^n$ onto itself which is a linear isomorphism on each fibre. By Proposition 2.48 there is therefore a unique topology and smooth structure on $\mathscr{T}(E^1, \ldots, E^k)$ such that $\pi' : \mathscr{T}(E^1, \ldots, E^k) \to M$ is an n-dimensional vector bundle over M with $(t'_\alpha, \pi'^{\,-1}(U_\alpha))$ as local trivializations.
□

4.40 Remark We say that $(t', \pi'^{\,-1}(U))$, where $t'_p = r' \circ (t^{1\;-1}_p, \ldots, t^{k\;-1}_p)^*$ for $p \in U$, is the local trivialization in the tensor bundle $\pi' : \mathscr{T}(E^1, \ldots, E^k) \to M$ corresponding to the local trivializations $(t^r, \pi^{r\,-1}(U))$ in the n_r-dimensional vector bundles $\pi^r : E^r \to M$ for $r = 1,...,k$. Here $r' : \mathscr{T}(\mathbf{R}^{n_1},...,\mathbf{R}^{n_k}) \to \mathbf{R}^n$ is the linear isomorphism defined by

$$r'(e^{i_1}_{(1)} \otimes \cdots \otimes e^{i_k}_{(k)}) = e_i \,,$$

for $(i_1,...,i_k) \in I_{n_1} \times \cdots \times I_{n_k}$, where $\mathscr{E}_r^* = \{e^1_{(r)},...,e^{n_r}_{(r)}\}$ is the dual of the standard basis in $\mathbf{R}^{n_r}$ for $r = 1,...,k$ and $\mathscr{E} = \{e_1,...,e_n\}$ is the standard basis for $\mathbf{R}^n$, and where $i = b(i_1,...,i_k)$ for the bijection $b : I_{n_1} \times \cdots \times I_{n_k} \to I_n$ defined in Remark 4.27 with $n = \prod_{r=1}^k n_r$.

4.41 Definition Let $\pi^r : E^r \to M$ be an n_r-dimensional vector bundle over the smooth manifold M for $r = 1,...,k$. If $\{U_\alpha | \alpha \in A\}$ is an open cover of M such that for each $\alpha \in A$, there are local trivializations $(t^r_\alpha, \pi^{r-1}(U_\alpha))$ in E^r for $r = 1,...,k$, then the family $\{(t^r_\alpha, \pi^{r-1}(U_\alpha)) | (r,\alpha) \in I_k \times A\}$ is called a *trivializing cover* of the bundles $E^1,...,E^k$ over M.

4.42 Definition If $\pi^r : E^r \to M$ is an n_r-dimensional vector bundle over the smooth manifold M for $r = 1,...,k$, we define their *tensor product* by

$$E^1 \otimes \cdots \otimes E^k = \mathscr{T}(E^{1*},...,E^{k*}) \,,$$

with projection $\pi' : E^1 \otimes \cdots \otimes E^k \to M$ sending each set $E^1_p \otimes \cdots \otimes E^k_p$ to p. By Proposition 4.8 and 4.39 this is an n-dimensional vector bundle over M, where $n = \prod_{r=1}^k n_r$.

4.43 Proposition Let $\pi^r_1 : E^r_1 \to M$ and $\pi^r_2 : E^r_2 \to M$ be two n_r-dimensional vector bundles for each $r = 1,\dots,k$, and let f_r be bundle maps over M from π^r_1 to π^r_2. Then we have a bundle map $(f_1,\dots,f_k)^*$ over M from $\pi'_2 : \mathscr{T}(E^1_2,\dots,E^k_2) \to M$ to $\pi'_1 : \mathscr{T}(E^1_1,\dots,E^k_1) \to M$ given by

$$(f_1,\dots,f_k)^*_p = (f_{1,p},\dots,f_{k,p})^*$$

i.e.,

$$(f_1,\dots,f_k)^*(T)(v_1,\dots,v_k) = T(f_1(v_1),\dots,f_k(v_k))$$

for all $T \in \pi'^{-1}_2(p)$ and $v_r \in \pi^{r\,-1}_1(p)$ where $p \in M$ and $r = 1,\dots,k$.

PROOF : The proof is similar to the proof of Proposition 4.39, and we use the same notation. Choose a trivializing cover $\{(t^r_{i,\alpha}, \pi^{r\,-1}_i(U_\alpha)) | (r,\alpha) \in I_k \times A\}$ of the bundles $E^1_i,\dots,E^k_i$ over M for $i = 1,2$, and let $(t'_{i,\alpha}, \pi'^{-1}_i(U_\alpha))$ be the corresponding local trivializations in the tensor bundles $\pi'_i : \mathscr{T}(E^1_i,\dots,E^k_i) \to M$.

For each index $\alpha \in A$, we have that $t^r_{2,\alpha} \circ f_r \circ t^{r\,-1}_{1,\alpha}$ is a smooth map from $U_\alpha \times \mathbf{R}^{n_r}$ to $U_\alpha \times \mathbf{R}^{n_r}$ given by

$$(t^r_{2,\alpha} \circ f_r \circ t^{r\,-1}_{1,\alpha})(p,v) = (p, (t^r_{2,\alpha,p} \circ f_r \circ t^{r\,-1}_{1,\alpha,p})(v))$$

for $r = 1, \dots, k$, where $t^r_{2,\alpha,p} \circ f_r \circ t^{r\ -1}_{1,\alpha,p} : \mathbf{R}^{n_r} \to \mathbf{R}^{n_r}$ is a linear map for each $p \in U_\alpha$. Since

$$
\begin{aligned}
& t'_{1,\alpha,p} \circ (f_1, \dots, f_k)^* \circ t'^{-1}_{2,\alpha,p} \\
&= r \circ (t^1_{2,\alpha,p} \circ f_1 \circ t^{1\ -1}_{1,\alpha,p}, \dots, t^k_{2,\alpha,p} \circ f_k \circ t^{k\ -1}_{1,\alpha,p})^* \circ r^{-1},
\end{aligned}
$$

it follows from Proposition 4.37, using the fact that $m^{\mathscr{B}}_{\mathscr{E}}(r)$ and $m^{\mathscr{E}}_{\mathscr{B}}(r^{-1})$ are both the identity matrix, that

$$
\begin{aligned}
& m^{\mathscr{E}}_{\mathscr{E}}(t'_{1,\alpha,p} \circ (f_1, \dots, f_k)^* \circ t'^{-1}_{2,\alpha,p}) \\
&= m^{\mathscr{E}_1}_{\mathscr{E}_1}(t^1_{2,\alpha,p} \circ f_1 \circ t^{1\ -1}_{1,\alpha,p})^t \otimes \cdots \otimes m^{\mathscr{E}_k}_{\mathscr{E}_k}(t^k_{2,\alpha,p} \circ f_k \circ t^{k\ -1}_{1,\alpha,p})^t .
\end{aligned}
$$

Hence $t'_{1,\alpha} \circ (f_1, \dots, f_k)^* \circ t'^{-1}_{2,\alpha}$ is a smooth map from $U_\alpha \times \mathbf{R}^n$ to $U_\alpha \times \mathbf{R}^n$ which is linear on each fibre, thus showing that $(f_1, \dots, f_k)^*$ is a bundle map over M. □

4.44 Corollary Let $\pi^r_1 : E^r_1 \to M$ and $\pi^r_2 : E^r_2 \to M$ be two n_r-dimensional vector bundles for each $r = 1, \dots, k$, and let f_r be bundle maps over M from π^r_1 to π^r_2. Then we have a bundle map $f_1 \otimes \cdots \otimes f_k$ over M from $\pi'_1 : E^1_1 \otimes \cdots \otimes E^k_1 \to M$ to $\pi'_2 : E^1_2 \otimes \cdots \otimes E^k_2 \to M$ given by

$$(f_1 \otimes \cdots \otimes f_k)_p = f_{1,p} \otimes \cdots \otimes f_{k,p}$$

i.e.,

$$f_1 \otimes \cdots \otimes f_k\, (v_1 \otimes \cdots \otimes v_k) = f_1(v_1) \otimes \cdots \otimes f_k(v_k)$$

for all $v_r \in \pi^{r\ -1}_1(p)$ where $p \in M$ and $r = 1, \dots, k$.

4.45 Proposition Let $s : M \to \mathscr{T}(E^1, \dots, E^k)$ be a map with $\pi' \circ s = id_M$, and let $\mathscr{E}^*_r = \{e^1_{(r)}, \dots, e^{n_r}_{(r)}\}$ be the dual of the standard basis in $\mathbf{R}^{n_r}$ for $r = 1, \dots, k$. Then s is a section of π' on M if and only if the functions $A_{i_1, \dots, i_k} : U \to \mathbf{R}$ defined by

$$s(p) = \sum_{i_1, \dots, i_k} A_{i_1, \dots, i_k}(p)\, (e^{i_1}_{(1)} \circ t^1_p) \otimes \cdots \otimes (e^{i_k}_{(k)} \circ t^k_p)$$

for $p \in U$ are smooth for every local trivialization $(t^r, \pi^{r\ -1}(U))$ in E^r belonging to a trivializing cover of the bundles $E^1, \dots, E^k$ over M.

PROOF: If $(t', \pi'^{-1}(U))$ is the corresponding local trivialization for the tensor bundle $\mathscr{T}(E^1, \dots, E^k)$, it follows that

$$t'_p((e^{i_1}_{(1)} \circ t^1_p) \otimes \cdots \otimes (e^{i_k}_{(k)} \circ t^k_p)) = e_i ,$$

where $i = b(i_1, \dots, i_k)$ for the bijection $b : I_{n_1} \times \cdots \times I_{n_k} \to I_n$ defined in Remark 4.27 with $n = \prod_{r=1}^k n_r$. If $a : U \to \mathbf{R}^n$ is the map defined by $a_i = A_{i_1, \dots, i_k}$, we have that

$$(t' \circ s)(p) = (p, a(p))$$

for $p \in U$, which shows that a is the local representation of s on U as defined in Definition 2.52. □

4.46 For each section T of π' on M we have a map

$$T_M : \Gamma(M;E^1) \times \cdots \times \Gamma(M;E^k) \to \mathscr{F}(M)$$

defined by

$$T_M(s_1,\dots,s_k)(p) = T(p)(s_1(p),\dots,s_k(p))$$

for each point p on M and for sections $s_r \in \Gamma(M;E^r)$ where $r = 1,\dots,k$. Indeed, if (x,U) is a local chart on M, and if $A_{i_1,\dots,i_k}$ and a_i^r are the components of the local representations of T and s_r on U for $r = 1,\dots,k$ as defined in Propositions 4.45 and 2.54, then

$$T_M(s_1,\dots,s_k)|_U = \sum_{i_1,\dots,i_k} A_{i_1,\dots,i_k} a_{i_1}^1 \cdots a_{i_k}^k$$

which shows that $T_M(s_1,\dots,s_k)$ is a smooth function on M.

We see that T_M is multilinear over $\mathscr{F}(M)$. The next proposition shows that each such multilinear map arises from a unique section of π' on M in this way.

4.47 Proposition Let M be a smooth manifold, and let

$$F : \Gamma(M;E^1) \times \cdots \times \Gamma(M;E^k) \to \mathscr{F}(M)$$

be a map which is multilinear over $\mathscr{F}(M)$. Then there is a unique section T of π' on M such that $F = T_M$.

PROOF: If $v_i \in E_p^i$, we choose a section s_i on M with $s_i(p) = v_i$ for $i = 1,\dots,k$ using Proposition 2.56 (2). Then we must define

$$T(p)(v_1,\dots,v_k) = F(s_1,\dots,s_k)(p). \tag{1}$$

For $T(p)$ to be well defined, we need to show that $T(p)(v_1,\dots,v_k)$ does not depend on the choice of $s_1,\dots,s_k$. It clearly suffices to show that

$$F(s_1,\dots,s_k)(p) = 0 \tag{2}$$

if $s_r(p) = 0$ for some r. Then it will follow that s_r in (1) may be replaced by any other section of π^r on M with the same value at p without changing the value of $F(s_1,\dots,s_k)(p)$.

Suppose that $s_r(p) = 0$, and choose a local chart (x,U) around p and a smooth function $f : M \to \mathbf{R}$ with $f(p) = 1$ and $\operatorname{supp}(f) \subset U$. If

$$s_r|_U = \sum_{i=1}^{n_r} b_i \eta_i^r ,$$

where $\eta_i^r : U \to E^r$ is the section given by $\eta_i^r(p) = t_p^{r-1}(e_i^r)$ for $p \in U$ and $i = 1,\dots,n_r$, we extend the functions fb_i and sections $f\eta_i^r$ on U to smooth functions g_i and sections θ_i on M, respectively, by defining them to be zero outside U. Then we have that

$$f^2 s_r = \sum_{i=1}^{n_r} g_i \theta_i$$

so that

$$f^2F(s_1,\ldots,s_r,\ldots,s_k) = F(s_1,\ldots,f^2s_r,\ldots,s_k) = \sum_{i=1}^{n_r} g_iF(s_1,\ldots,\theta_i,\ldots,s_k)\,.$$

Evaluating at p and using that $f(p)=1$ and $g_i(p)=0$ for $i=1,\ldots,n_r$, we obtain (2), and this compeletes the proof that $T(p)$ is well defined by formula (1).

It only remains to show that T is smooth on M. By Proposition 4.45 we must show that the functions $A_{i_1,\ldots,i_k}: U \to \mathbf{R}$ defined by

$$A_{i_1,\ldots,i_k}(p) = T(p)(\eta^1_{i_1}(p),\ldots,\eta^k_{i_k}(p))$$

for $p \in U$, are smooth for every local chart (x,U) on M. Fix p in U, and let V be an open neighbourhood of p with $\overline{V} \subset U$. By Proposition 2.56 (1) there are sections θ^r_i of π^r on M which coincide with η^r_i on $\overline{V}$ for $i=1,\ldots,n_r$ and $r=1,\ldots,k$. Then we have that

$$A_{i_1,\ldots,i_k}(q) = F(\theta^1_{i_1},\ldots,\theta^k_{i_k})(q)$$

for $q \in V$, showing that $A_{i_1,\ldots,i_k}$ is smooth at p. Since p was an arbitrary point in U, this completes the proof that T is smooth. □

4.48 Remark Because of Proposition 4.47 we will not distinguish between the section T and the map T_M, and a section in $\Gamma(M;\mathscr{T}(E^1,\ldots,E^k))$ can be thought of as an operation on k sections in $\Gamma(M;E^1)$, ... , $\Gamma(M;E^k)$, respectively, yielding a smooth funnction on M.

4.49 Proposition Let $\pi^r_1: E^r_1 \to M_1$ and $\pi^r_2: E^r_2 \to M_2$ be two vector bundles of dimensions n^r_1 and n^r_2 for each $r=1,\ldots,k$, and let $(\tilde{f}_r, f)$ be bundle maps from π^r_1 to π^r_2. If s_2 is a section of $\pi'_2: \mathscr{T}(E^1_2,\ldots,E^k_2) \to M_2$ on an open subset V_2 of M_2 and $V_1 = f^{-1}(V_2)$, then the map $s_1: V_1 \to \mathscr{T}(E^1_1,\ldots,E^k_1)$ defined by

$$s_1(p)(v_1,\ldots,v_k) = s_2(f(p))(\tilde{f}_1(v_1),\ldots,\tilde{f}_k(v_k))$$

for $p \in V_1$ and $v_r \in {\pi^r_1}^{-1}(p)$ where $r=1,\ldots,k$, is a section of $\pi'_1: \mathscr{T}(E^1_1,\ldots,E^k_1) \to M_1$ on V_1.

s_1 is called the *pull-back* of s_2 by $(\tilde{f}_1,\ldots,\tilde{f}_k)$ and is denoted by $(\tilde{f}_1,\ldots,\tilde{f}_k)^*(s_2)$. We will give two alternative proofs.

FIRST PROOF: Let p be a point in V_1, and let $(t^r_1, {\pi^r_1}^{-1}(U_1))$ and $(t^r_2, {\pi^r_2}^{-1}(U_2))$ be local trivializations of π^r_1 and π^r_2 around p and $f(p)$, respectively, with $U_2 \subset V_2$ and $U_1 \subset f^{-1}(U_2)$. Then the maps $t^r_2 \circ \tilde{f}_r \circ {t^r_1}^{-1}: U_1 \times \mathbf{R}^{n^r_1} \to U_2 \times \mathbf{R}^{n^r_2}$ are smooth and linear on each fibre.

Now let $(t'_i, {\pi'_i}^{-1}(U_i))$ be the local trivialization of π'_i defined in Remark 4.40 for $i=1,2$. We have that $t'_i: {\pi'_i}^{-1}(U_i) \to U_i \times \mathbf{R}^{n_i}$ is given by $t'_i(\omega) = (q, t'_{i,q}(\omega))$ for $q \in U_i$ and $\omega \in {\pi'_i}^{-1}(q)$, where $t'_{i,q} = r_i \circ ({t^1_{i,q}}^{-1},\ldots,{t^k_{i,q}}^{-1})^*$ and $n_i = \prod_{r=1}^k n^r_i$.

Hence $t'_i \circ s_i(q) = (q, h_i(q))$ where the local representations $h_i : U_i \to \mathbf{R}^{n_i}$ of s_i are given by

$$h_1(q) = r_1 \circ (\tilde{f}_1 \circ t^{1\;-1}_{1,q}, \ldots, \tilde{f}_k \circ t^{k\;-1}_{1,q})^* \circ s_2(f(q))$$

and

$$h_2(q) = r_2 \circ (t^{1\;-1}_{2,q}, \ldots, t^{1\;-1}_{2,q})^* \circ s_2(q)\,.$$

Combining this we have that

$$h_1(q) = r_1 \circ (t^1_{2,f(q)} \circ \tilde{f}_1 \circ t^{1\;-1}_{1,q}, \ldots, t^k_{2,f(q)} \circ \tilde{f}_k \circ t^{k\;-1}_{1,q})^* \circ r_2^{-1} \circ h_2(f(q))$$

for $q \in U_1$. If $\mathscr{E}^r_i$ and $\mathscr{B}_i$ are the standard basis for $\mathbf{R}^{n^r_i}$ and $\mathbf{R}^{n_i}$, respectively, for $r = 1, \ldots, k$ and $i = 1, 2$, we have that

$$\begin{aligned} &m^{\mathscr{B}_2}_{\mathscr{B}_1}(r_1 \circ (t^1_{2,f(q)} \circ \tilde{f}_1 \circ t^{1\;-1}_{1,q}, \ldots, t^k_{2,f(q)} \circ \tilde{f}_k \circ t^{k\;-1}_{1,q})^* \circ r_2^{-1}) \\ &= m^{\mathscr{E}^1_1}_{\mathscr{E}^1_2}(t^1_{2,f(q)} \circ \tilde{f}_1 \circ t^{1\;-1}_{1,q})^t \otimes \cdots \otimes m^{\mathscr{E}^k_1}_{\mathscr{E}^k_2}(t^k_{2,f(q)} \circ \tilde{f}_k \circ t^{k\;-1}_{1,q})^t \end{aligned}$$

which shows that s_1 is smooth in a neighbourhood of every point p in V_1 and hence is a section of π'_1 on V_1. □

SECOND PROOF: The Proposition can also be proved using Proposition 4.45 and the standard basis $\mathscr{E}^r_i = \{e^{(i,r)}_1, \ldots, e^{(i,r)}_{n^r_i}\}$ in $\mathbf{R}^{n^r_i}$ with dual basis $(\mathscr{E}^r_i)^* = \{e^1_{(i,r)}, \ldots, e^{n^r_i}_{(i,r)}\}$ for $r = 1, \ldots, k$ and $i = 1, 2$.

Again, let p be a point in V_1, and let $(t^r_1, \pi^{r\;-1}_1(U_1))$ and $(t^r_2, \pi^{r\;-1}_2(U_2))$ be local trivializations of π^r_1 and π^r_2 around p and $f(p)$, respectively, with $U_2 \subset V_2$ and $U_1 \subset f^{-1}(U_2)$. Since the maps $t^r_2 \circ \tilde{f}_r \circ t^{r\;-1}_1 : U_1 \times \mathbf{R}^{n^r_1} \to U_2 \times \mathbf{R}^{n^r_2}$ are smooth and linear on each fibre, there are smooth functions $a^r_{ji} : U_1 \to \mathbf{R}$ such that

$$t^r_{2,f(q)} \circ \tilde{f}_r \circ t^{r\;-1}_{1,q}(e^{(1,r)}_i) = \sum_{j=1}^{n^r_2} a^r_{ji}(q)\, e^{(2,r)}_j$$

for $q \in U_1$ and $i = 1, \ldots, n^r_1$.

By Proposition 4.45, there are smooth functions $A_{j_1,\ldots,j_k} : U_2 \to \mathbf{R}$ defined by

$$s_2(q') = \sum_{j_1,\ldots,j_k} A_{j_1,\ldots,j_k}(q')\,(e^{j_1}_{(2,1)} \circ t^1_{2,q'}) \otimes \cdots \otimes (e^{j_k}_{(2,k)} \circ t^k_{2,q'})$$

for $q' \in U_2$. From this it follows that

$$s_1(q) = \sum_{i_1,\ldots,i_k} B_{i_1,\ldots,i_k}(q)\,(e^{i_1}_{(1,1)} \circ t^1_{1,q}) \otimes \cdots \otimes (e^{i_k}_{(1,k)} \circ t^k_{1,q})$$

for $q \in U_1$, where

$$\begin{aligned} B_{i_1,\ldots,i_k}(q) &= s_1(q)(t^{1\;-1}_{1,q}(e^{(1,1)}_{i_1}), \ldots, t^{k\;-1}_{1,q}(e^{(1,k)}_{i_k})) \\ &= s_2(f(q))(\tilde{f}_1 \circ t^{1\;-1}_{1,q}(e^{(1,1)}_{i_1}), \ldots, \tilde{f}_k \circ t^{k\;-1}_{1,q}(e^{(1,k)}_{i_k})) \\ &= \sum_{j_1,\ldots,j_k} A_{j_1,\ldots,j_k}(f(q))\, a^1_{j_1 i_1}(q) \cdots a^k_{j_k i_k}(q). \end{aligned}$$

Since the functions $B_{i_1,\dots,i_k} : U_1 \to \mathbf{R}$ are smooth, it follows that s_1 is smooth in a neighbourhood of every point p in V_1 and hence is a section of π_1' on V_1. □

4.50 Proposition Let $\pi_1 : E_1 \to M_1$ and $\pi_2 : E_2 \to M_2$ be two vector bundles of dimension n, and let $(\tilde{f}, f)$ be a bundle map from π_1 to π_2 where $\tilde{f}$ induces a linear isomorphism between the fibres $\pi_1^{-1}(p)$ and $\pi_2^{-1}(f(p))$ for every $p \in M_1$. Then there is a bundle map $(\tilde{g}, f)$ between the dual bundles $\pi_1^* : E_1^* \to M_1$ and $\pi_2^* : E_2^* \to M_2$ which is given by

$$\tilde{g}_p = (\tilde{f}_p^{-1})^*$$

i.e.,

$$\tilde{g}(\lambda) = \lambda \circ \tilde{f}_p^{-1}$$

for $p \in M_1$ and $\lambda \in \pi_1^{-1}(p)$.

PROOF: Let $q_0 \in M_1$, and choose local trivializations $(t_1, \pi_1^{-1}(U_1))$ and $(t_2, \pi_2^{-1}(U_2))$ of π_1 and π_2 around q_0 and $f(q_0)$, respectively, with $U_1 \subset f^{-1}(U_2)$. Let $(t_1', {\pi_1^*}^{-1}(U_1))$ and $(t_2', {\pi_2^*}^{-1}(U_2))$ be the corresponding local trivializations in the dual bundles, and let $\mathscr{E}$ be the standard basis for $\mathbf{R}^n$ with dual basis $\mathscr{E}^*$.

Then $t_2 \circ \tilde{f} \circ t_1^{-1} : U_1 \times \mathbf{R}^n \to U_2 \times \mathbf{R}^n$ is a smooth map given by

$$(t_2 \circ \tilde{f} \circ t_1^{-1})(q, v) = (p, (t_{2,p} \circ \tilde{f}_q \circ t_{1,q}^{-1})(v)),$$

where $t_{2,p} \circ \tilde{f}_q \circ t_{1,q}^{-1} : \mathbf{R}^n \to \mathbf{R}^n$ is a linear isomorphism for each $q \in U_1$ and where $p = f(q)$. Since

$$t_{2,p}' \circ \tilde{g}_q \circ {t'}_{1,q}^{-1} = r \circ \{(t_{2,p} \circ \tilde{f}_q \circ t_{1,q}^{-1})^{-1}\}^* \circ r^{-1},$$

it follows from Proposition 4.6, using the fact that $m_{\mathscr{E}}^{\mathscr{E}^*}(r)$ and $m_{\mathscr{E}^*}^{\mathscr{E}}(r^{-1})$ are both the identity matrix, that

$$m_{\mathscr{E}}^{\mathscr{E}}(t_{2,p}' \circ \tilde{g}_q \circ {t'}_{1,q}^{-1}) = \{m_{\mathscr{E}}^{\mathscr{E}}(t_{2,p} \circ \tilde{f}_q \circ t_{1,q}^{-1})^{-1}\}^t.$$

Hence $t_2' \circ \tilde{g} \circ {t'}_1^{-1}$ is a smooth map from $U_1 \times \mathbf{R}^n$ to $U_2 \times \mathbf{R}^n$ which is linear on each fibre, thus showing that $(\tilde{g}, f)$ is bundle map. □

4.51 Definition Let $\pi^r : E^r \to M$ be an n_r-dimensional vector bundle over the smooth manifold M for $r = 1, \dots, k+l$. If ω and η are sections of the bundles $\mathscr{T}(E^1, \dots, E^k)$ and $\mathscr{T}(E^{k+1}, \dots, E^{k+l})$, respectively, on an open subset V of M, we define their *tensor product* $\omega \otimes \eta$ to be the section of $\mathscr{T}(E^1, \dots, E^{k+l})$ on V defined by

$$(\omega \otimes \eta)(p) = \omega(p) \otimes \eta(p)$$

for $p \in V$. If $l = 0$, then η is simply a smooth function f on V, and $f \otimes \omega$ and $\omega \otimes f$ simply mean $f\omega$.

4.52 Definition If V is a vector space, the space

$$\mathscr{T}_l^k(V) = \mathscr{T}(\underbrace{V, \dots, V}_{k}, \underbrace{V^*, \dots, V^*}_{l}) = \underbrace{V^* \otimes \cdots \otimes V^*}_{k} \otimes \underbrace{V \otimes \cdots \otimes V}_{l}$$

is called the *tensor space* of type $\binom{k}{l}$ on V, or the space of tensors on V which are *covariant* of degree k and *contravariant* of degree l. We let $\mathscr{T}_0^0(V) = \mathbf{R}$. The spaces $\mathscr{T}_0^k(V)$ and $\mathscr{T}_k^0(V)$ of purely covariant and contravariant tensors of degree k are usually denoted simply by $\mathscr{T}^k(V)$ and $\mathscr{T}_k(V)$, respectively. If $\mathscr{B}$ is a basis for V with dual basis $\mathscr{B}^*$, then the basis

$$\mathscr{T}(\underbrace{\mathscr{B}, \dots, \mathscr{B}}_{k}, \underbrace{\mathscr{B}^*, \dots, \mathscr{B}^*}_{l})$$

for $\mathscr{T}_l^k(V)$ defined in Remark 4.27 is denoted by $\mathscr{T}_l^k(\mathscr{B})$. We let $\mathscr{T}_0^0(\mathscr{B}) = \{1\}$.

If $F : V \to W$ is a linear map, we write $F^* : \mathscr{T}^k(W) \to \mathscr{T}^k(V)$ instead of $(F, \dots, F)^* : \mathscr{T}^k(W) \to \mathscr{T}^k(V)$. By Proposition 4.33 (4) we have that $F^*(T \otimes S) = F^*(T) \otimes F^*(S)$ for every covariant tensor T and S on W. If $F : V \to W$ is a linear isomorphism, we also denote the map

$$(\underbrace{F, \dots, F}_{k}, \underbrace{(F^{-1})^*, \dots, (F^{-1})^*}_{l})^* : \mathscr{T}_l^k(W) \to \mathscr{T}_l^k(V)$$

simply by $F^* : \mathscr{T}_l^k(W) \to \mathscr{T}_l^k(V)$, so that

$$F^*(T)(v_1, \dots, v_k, \lambda_1, \dots, \lambda_l) = T(F(v_1), \dots, F(v_k), \lambda_1 \circ F^{-1}, \dots, \lambda_l \circ F^{-1})$$

for $T \in \mathscr{T}_l^k(W)$, $v_1, \dots, v_k \in V$ and $\lambda_1, \dots, \lambda_l \in V^*$. When $k = 0$ and $l = 1$, the vector spaces $\mathscr{T}_1(V)$ and $\mathscr{T}_1(W)$ may be identified with V and W as in Remark 4.5, and the map $F^* : \mathscr{T}_1(W) \to \mathscr{T}_1(V)$ is then identified with $F^{-1} : W \to V$ by Proposition 4.4.

If $\pi : E \to M$ is a vector bundle over the smooth manifold M, we let

$$\mathscr{T}_l^k(E) = \mathscr{T}(\underbrace{E, \dots, E}_{k}, \underbrace{E^*, \dots, E^*}_{l})$$

and denote its projection by π_l^k. The bundles $\mathscr{T}_0^k(E)$ and $\mathscr{T}_k^0(E)$ are usually denoted simply by $\mathscr{T}^k(E)$ and $\mathscr{T}_k(E)$ with projections π^k and π_k, respectively.

4.53 Proposition For each local trivialization $(t, \pi^{-1}(U))$ in the n-dimensional vector bundle $\pi : E \to M$, we have a corresponding local trivialization $(t', (\pi_l^k)^{-1}(U))$ in the tensor bundle $\pi_l^k : \mathscr{T}_l^k(E) \to M$, where $t'_p = r'' \circ (t_p^{-1})^*$ for $p \in U$. Here $r'' : \mathscr{T}_l^k(\mathbf{R}^n) \to \mathbf{R}^m$ is the linear isomorphism defined by

$$r''(e^{i_1} \otimes \cdots \otimes e^{i_k} \otimes e_{j_1} \otimes \cdots \otimes e_{j_l}) = \widetilde{e}_s\,,$$

for $(i_1, \dots, i_k, j_1, \dots, j_l) \in I_n^{k+l}$, where $\mathscr{E} = \{e_1, \dots, e_n\}$ is the standard basis for $\mathbf{R}^n$ with dual basis $\mathscr{E}^* = \{e^1, ..., e^n\}$ and $\mathscr{B} = \{\tilde{e}_1, \dots, \tilde{e}_m\}$ is the standard basis for $\mathbf{R}^m$, and where $s = b(i_1, ..., i_k, j_1, ..., j_l)$ for the bijection $b : I_n^{k+l} \to I_m$ defined in Remark 4.27 with $m = n^{k+l}$.

We have that $t' = (id_U \times r'') \circ \tau$, where $(id_U \times r'', U \times \mathscr{T}_l^k(\mathbf{R}^n))$ is a local trivialization for the vector bundle $pr_1 : U \times \mathscr{T}_l^k(\mathbf{R}^n) \to U$, and $\tau : (\pi_l^k)^{-1}(U) \to U \times \mathscr{T}_l^k(\mathbf{R}^n)$ is the equivalence over U given by

$$\tau(T) = (p, (t_p^{-1})^*(T))$$

for $T \in \mathscr{T}_l^k(E_p)$ where $p \in U$.

PROOF : By Remarks 4.9 and 4.40 we have that

$$t'_p = r' \circ (\underbrace{t_p^{-1}, \dots, t_p^{-1}}_{k}, \underbrace{t_p^* \circ r^{-1}, \dots, t_p^* \circ r^{-1}}_{l})^* = r'' \circ (t_p^{-1})^*$$

for $p \in U$, where

$$r'' = r' \circ (\underbrace{id_{\mathbf{R}^n}, \dots, id_{\mathbf{R}^n}}_{k}, \underbrace{r^{-1}, \dots, r^{-1}}_{l})^* : \mathscr{T}_l^k(\mathbf{R}^n) \to \mathbf{R}^m$$

is given by

$$r''(e^{i_1} \otimes \cdots \otimes e^{i_k} \otimes e_{j_1} \otimes \cdots \otimes e_{j_l}) = r'(e^{i_1} \otimes \cdots \otimes e^{i_k} \otimes e^{j_1} \otimes \cdots \otimes e^{j_l}) = \tilde{e}_s$$

for $(i_1, ..., i_k, j_1, ..., j_l) \in I_n^{k+l}$. Indeed, we have that

$$\begin{aligned}(id_{\mathbf{R}^n}, \dots, id_{\mathbf{R}^n}, r^{-1}, \dots, r^{-1})^*(e^{i_1} \otimes \cdots \otimes e^{i_k} \otimes e_{j_1} \otimes \cdots \otimes e_{j_l}) \\ = e^{i_1} \otimes \cdots \otimes e^{i_k} \otimes e^{j_1} \otimes \cdots \otimes e^{j_l}\end{aligned}$$

as both sides have the same effect on $e_{r_1} \otimes \cdots \otimes e_{r_k} \otimes e_{s_1} \otimes \cdots \otimes e_{s_l}$ for $(r_1, ..., r_k, s_1, ..., s_l) \in I_n^{k+l}$. □

4.54 Remark If $V_1, ..., V_k$ are vector spaces and $\sigma \in S_k$ is a permutation, we have a linear isomorphism

$$\phi : \mathscr{T}(V_{\sigma(1)}, ..., V_{\sigma(k)}) \to \mathscr{T}(V_1, ..., V_k)$$

given by

$$\phi(T)(v_1, ..., v_k) = T(v_{\sigma(1)}, ..., v_{\sigma(k)})$$

for every $T \in \mathscr{T}(V_{\sigma(1)}, ..., V_{\sigma(k)})$ and $v_i \in V_i$ for $i = 1, \dots, k$. We will not distinguish between the isomorphic spaces obtained from $\mathscr{T}(V_1, ..., V_k)$ by interchanging the spaces $V_1, ..., V_k$.

In particular, if V is a vector space, every tensor product with k factors V^* and l factors V in any order is identified with $\mathscr{T}_l^k(V)$ with the canonical order given in

Definition 4.52 placing all the factors V^* to the left of V. Hence if $T \in \mathscr{T}^{k_1}_{l_1}(V)$ and $S \in \mathscr{T}^{k_2}_{l_2}(V)$, their tensor product $T \otimes S \in \mathscr{T}^{k_1+k_2}_{l_1+l_2}(V)$ is given by

$$(T \otimes S)(v_1,\dots,v_{k_1+k_2},\lambda_1,\dots,\lambda_{l_1+l_2})$$
$$= T(v_1,\dots,v_{k_1},\lambda_1,\dots,\lambda_{l_1})\,S(v_{k_1+1},\dots,v_{k_1+k_2},\lambda_{l_1+1},\dots,\lambda_{l_1+l_2})$$

where $v_i \in V$ for $i = 1,\dots,k_1+k_2$ and $\lambda_j \in V^*$ for $j = 1,\dots,l_1+l_2$. If $F : V \to W$ is a linear isomorphism, we have that $F^*(T \otimes S) = F^*(T) \otimes F^*(S)$.

4.55 Proposition Let $(\tilde{f}, f)$ be a bundle map from a vector bundle $\pi_1 : E_1 \to M_1$ to a vector bundle $\pi_2 : E_2 \to M_2$, and let s_2 be a section of $(\pi_2)^k_l : \mathscr{T}^k_l(E_2) \to M_2$ on M_2. If $l > 0$, we suppose that $\tilde{f}$ induces a linear isomorphism between the fibres $\pi_1^{-1}(p)$ and $\pi_2^{-1}(f(p))$ for every $p \in M_1$. Then the map $s_1 : M_1 \to \mathscr{T}^k_l(E_1)$ defined by

$$s_1(p) = (\tilde{f}_p)^* s_2(f(p))$$

for $p \in M_1$, i.e.,

$$s_1(p)(v_1,\dots,v_k,\lambda_1,\dots,\lambda_l) = s_2(f(p))(\tilde{f}_p(v_1),\dots,\tilde{f}_p(v_k),\lambda_1 \circ \tilde{f}_p^{-1},\dots,\lambda_l \circ \tilde{f}_p^{-1})$$

for all $v_1,\dots,v_k \in \pi_1{}^{-1}(p)$ and $\lambda_1,\dots,\lambda_l \in \pi_1^{*\,-1}(p)$ where $p \in M_1$, is a section of $(\pi_1)^k_l : \mathscr{T}^k_l(E_1) \to M_1$ on M_1.

s_1 is called the *pull-back* of s_2 by $\tilde{f}$ and is denoted by $\tilde{f}^*(s_2)$.

PROOF : Follows from Propositions 4.49 and 4.50. □

4.56 Remark When $k = 0$ and $l = 1$, we identify $\mathscr{T}_1(E_r) = E_r^{**}$ with E_r by means of the equivalence $i_{E_r} : E_r \to E_r{}^{**}$ defined in Proposition 4.10 for $r = 1,2$. If s_2 is a section of $\pi_2 : E_2 \to M_2$ on M_2, we have that

$$[i_{E_1} \circ \tilde{f}^*(s_2)](p)(\lambda) = i_{E_1} \circ \tilde{f}_p^{-1} \circ s_2 \circ f(p)(\lambda) = \lambda \circ \tilde{f}_p^{-1} \circ s_2(f(p))$$
$$= (i_{E_2} \circ s_2)(f(p))(\lambda \circ \tilde{f}_p^{-1}) = \tilde{f}^*(i_{E_2} \circ s_2)(p)(\lambda)$$

for $\lambda \in \pi_1^{*\,-1}(p)$ where $p \in M_1$, which shows that the above definition of the pull-back of s_2 is consistent with the one in Proposition 2.57.

4.57 Definition If M is a smooth manifold of dimension n, the bundle $T^k_l(M) = \mathscr{T}^k_l(TM)$ with projection π^k_l is called the *tensor bundle* of type $\binom{k}{l}$ over M. A section of this bundle on an open subset V of M is called a *tensor field* of type $\binom{k}{l}$ on V. The set of all tensor fields of type $\binom{k}{l}$ on V is denoted by $\mathscr{T}^k_l(V)$. In particular $\mathscr{T}^0_0(V) = \mathscr{F}(V)$ is the set of C^∞-functions on V.

The bundles $T^k_0(M)$ and $T^0_k(M)$ of purely covariant and contravariant tensors of degree k are usually denoted simply by $T^k(M)$ and $T_k(M)$ with projections π^k and π_k.

Their sections on an open subset V of M are denoted by $\mathscr{T}^k(V)$ and $\mathscr{T}_k(V)$ and are called covariant and contravariant tensor fields of degree k on V, respectively.

4.58 Proposition Let $\pi_l^k : T_l^k(M) \to M$ be the tensor bundle of type $\binom{k}{l}$ over a smooth manifold M^n and $T : M \to T_l^k(M)$ be a map with $\pi_l^k \circ T = id_M$. Then T is a tensor field of type $\binom{k}{l}$ on M if and only if the functions $A_{i_1,\dots,i_k}^{j_1,\dots,j_l} : U \to \mathbf{R}$ defined by

$$T(p) = \sum_{\substack{i_1,\dots,i_k \\ j_1,\dots,j_l}} A_{i_1,\dots,i_k}^{j_1,\dots,j_l}(p)\, dx^{i_1}(p) \otimes \cdots \otimes dx^{i_k}(p) \otimes \left.\frac{\partial}{\partial x^{j_1}}\right|_p \otimes \cdots \otimes \left.\frac{\partial}{\partial x^{j_l}}\right|_p$$

for $p \in U$ are smooth for every local chart (x,U) on M. Here each $\left.\frac{\partial}{\partial x^{j_r}}\right|_p$ for $r = 1,\dots,l$ is considered as a functional on T_p^*M as described in Remark 4.5.

PROOF : Follows from Remark 2.82 and Propositions 4.11, 4.15 and 4.45. □

4.59 For each tensor field T of type $\binom{k}{l}$ on M we have a multilinear map

$$T_M : \underbrace{\mathscr{T}_1(M) \times \cdots \times \mathscr{T}_1(M)}_{k} \times \underbrace{\mathscr{T}^1(M) \times \cdots \times \mathscr{T}^1(M)}_{l} \to \mathscr{F}(M)$$

defined by

$$T_M(X_1,\dots,X_k,\lambda_1,\dots,\lambda_l)(p) = T(p)(X_1(p),\dots,X_k(p),\lambda_1(p),\dots,\lambda_l(p))$$

for each point p on M and for vector fields $X_1,\dots,X_k \in \mathscr{T}_1(M)$ and 1-forms $\lambda_1,\dots,\lambda_l \in \mathscr{T}^1(M)$ as described in 4.46. Conversely, each such multilinear map arises from a unique tensor field in this way by Proposition 4.47. Hence we will not distinguish between the tensor field T and the map T_M, and a tensor field of type $\binom{k}{l}$ on M can be thought of as an operation on k vector fields and l 1-forms yielding a smooth funnction on M.

4.60 If $f : M \to N$ is a diffeomorphism and T is a tensor field of type $\binom{k}{l}$ on N, then we denote its pullback $(f_*)^*(T)$ simply by $f^*(T)$. If N is an open submanifold of a manifold P and T is a tensor field on P, we also denote $f^*(T|_N)$ simply by $f^*(T)$.

We use the same notation when $f : M \to N$ is a smooth map and T is a covariant tensor field of degree k on an open subset V of N. In particular, when $k = 0$ then T is simply a smooth function on V, and $f^*(T) = T \circ f$.

It follows from Proposition 4.49 that if T is covariant of degree k on V, then $f^*(T)$ is a covariant tensor field of degree k on the open subset $f^{-1}(V)$ of M, but we also want to write out the second proof of Proposition 4.49 explicitely in this case using Proposition 4.58.

4.61 Proposition If $f : M \to N$ is a smooth map and T is a covariant tensor field of degree k on an open subset V of N, then the map $f^*(T) : f^{-1}(V) \to T^k(M)$ defined by

$$f^*(T)(p) = (f_{*p})^* \, T(f(p))$$

i.e.,

$$f^*(T)(p)(v_1, \ldots, v_k) = T(f(p))(f_*(v_1), \ldots, f_*(v_k))$$

for $p \in f^{-1}(V)$ and $v_1, \ldots, v_k \in T_pM$, is a covariant tensor field of degree k on $f^{-1}(V)$.

PROOF : Let p be a point in $f^{-1}(V)$, and let (x,U) and (y,W) be local charts around p and $f(p)$, respectively, with $W \subset V$ and $U \subset f^{-1}(W)$. Then there are smooth functions $A_{j_1,\ldots,j_k} : W \to \mathbf{R}$ defined by

$$T(q') = \sum_{j_1,\ldots,j_k} A_{j_1,\ldots,j_k}(q') \, dy^{j_1}(q') \otimes \cdots \otimes dy^{j_k}(q')$$

for $q' \in W$. From this it follows that

$$f^*(T)(q) = \sum_{i_1,\ldots,i_k} B_{i_1,\ldots,i_k}(q) \, dx^{i_1}(q) \otimes \cdots \otimes dx^{i_k}(q)$$

for $q \in U$, where

$$B_{i_1,\ldots,i_k}(q) = f^*(T)(q)\left(\left.\frac{\partial}{\partial x^{i_1}}\right|_q, \ldots, \left.\frac{\partial}{\partial x^{i_k}}\right|_q \right)$$

$$= T(f(q))\left(f_*\left(\left.\frac{\partial}{\partial x^{i_1}}\right|_q\right), \ldots, f_*\left(\left.\frac{\partial}{\partial x^{i_k}}\right|_q\right) \right)$$

$$= \sum_{j_1,\ldots,j_k} A_{j_1,\ldots,j_k}(f(q)) \, dy^{j_1}(f(q))\left(f_*\left(\left.\frac{\partial}{\partial x^{i_1}}\right|_q\right)\right) \cdots dy^{j_k}(f(q))\left(f_*\left(\left.\frac{\partial}{\partial x^{i_k}}\right|_q\right)\right)$$

$$= \sum_{j_1,\ldots,j_k} A_{j_1,\ldots,j_k}(f(q)) \left(f_*\left(\left.\frac{\partial}{\partial x^{i_1}}\right|_q\right)\right)(y^{j_1}) \cdots \left(f_*\left(\left.\frac{\partial}{\partial x^{i_k}}\right|_q\right)\right)(y^{j_k})$$

$$= \sum_{j_1,\ldots,j_k} A_{j_1,\ldots,j_k}(f(q)) \, \frac{\partial(y^{j_1} \circ f)}{\partial x^{i_1}}(q) \cdots \frac{\partial(y^{j_k} \circ f)}{\partial x^{i_k}}(q) \, .$$

Since the functions $B_{i_1,\ldots,i_k} : U \to \mathbf{R}$ are smooth, it follows that $f^*(T)$ is smooth in a neighbourhood of every point p in $f^{-1}(V)$ and hence is a covariant tensor field of degree k on $f^{-1}(V)$. □

4.62 Proposition Let $f : M \to N$ be a diffeomorphism, and let T be a tensor field of type $\binom{k}{l}$ on N. If (x,U) and (y,V) are local charts on M and N, respectively, and if

$$T(p) = \sum_{\substack{r_1,\ldots,r_k \\ s_1,\ldots,s_l}} A^{s_1,\ldots,s_l}_{r_1,\ldots,r_k}(p) \, dy^{r_1}(p) \otimes \cdots \otimes dy^{r_k}(p) \otimes \left.\frac{\partial}{\partial y^{s_1}}\right|_p \otimes \cdots \otimes \left.\frac{\partial}{\partial y^{s_l}}\right|_p$$

for $p \in V$, then

$$f^*(T)(q) = \sum_{\substack{i_1,\dots,i_k \\ j_1,\dots,j_l}} B^{j_1,\dots,j_l}_{i_1,\dots,i_k}(q)\, dx^{i_1}(q) \otimes \cdots \otimes dx^{i_k}(q) \otimes \left.\frac{\partial}{\partial x^{j_1}}\right|_q \otimes \cdots \otimes \left.\frac{\partial}{\partial x^{j_l}}\right|_q$$

where

$$B^{j_1,\dots,j_l}_{i_1,\dots,i_k}(q) = \sum_{\substack{r_1,\dots,r_k \\ s_1,\dots,s_l}} A^{s_1,\dots,s_l}_{r_1,\dots,r_k}(f(q))\, \frac{\partial(y^{r_1}\circ f)}{\partial x^{i_1}}(q) \cdots \frac{\partial(y^{r_k}\circ f)}{\partial x^{i_k}}(q)$$

$$\frac{\partial(x^{j_1}\circ f^{-1})}{\partial y^{s_1}}(f(q)) \cdots \frac{\partial(x^{j_k}\circ f^{-1})}{\partial y^{s_k}}(f(q))$$

for $q \in U \cap f^{-1}(V)$.

PROOF: Since

$$dy^r(f(q))\left(f_*\left(\left.\frac{\partial}{\partial x^i}\right|_q\right)\right) = f_*\left(\left.\frac{\partial}{\partial x^i}\right|_q\right)(y^r) = \frac{\partial(y^r\circ f)}{\partial x^i}(q)$$

and

$$dx^j(q)\circ(f^{-1})_*\left(\left.\frac{\partial}{\partial y^s}\right|_{f(q)}\right) = (f^{-1})_*\left(\left.\frac{\partial}{\partial y^s}\right|_{f(q)}\right)(x^j) = \frac{\partial(x^j\circ f^{-1})}{\partial y^s}(f(q)),$$

it follows from Proposition 4.58 that

$$B^{j_1,\dots,j_l}_{i_1,\dots,i_k}(q) = f^*(T)(q)\left(\left.\frac{\partial}{\partial x^{i_1}}\right|_q, \dots, \left.\frac{\partial}{\partial x^{i_k}}\right|_q, dx^{j_1}(q), \dots, dx^{j_l}(q)\right)$$

$$= T(f(q))\left(f_*\left(\left.\frac{\partial}{\partial x^{i_1}}\right|_q\right), \dots, f_*\left(\left.\frac{\partial}{\partial x^{i_k}}\right|_q\right), dx^{j_1}(q)\circ(f^{-1})_*, \dots, dx^{j_l}(q)\circ(f^{-1})_*\right)$$

$$= \sum_{\substack{r_1,\dots,r_k \\ s_1,\dots,s_l}} A^{s_1,\dots,s_l}_{r_1,\dots,r_k}(f(q))\, \frac{\partial(y^{r_1}\circ f)}{\partial x^{i_1}}(q) \cdots \frac{\partial(y^{r_k}\circ f)}{\partial x^{i_k}}(q)$$

$$\frac{\partial(x^{j_1}\circ f^{-1})}{\partial y^{s_1}}(f(q)) \cdots \frac{\partial(x^{j_k}\circ f^{-1})}{\partial y^{s_k}}(f(q)).$$

□

CONTRACTION

4.63 Definition If $A = (a_{ij})$ is an $n \times n$-matrix, we define the *trace* of A to be

$$\operatorname{tr}(A) = \sum_{i=1}^{n} a_{ii},$$

i.e., the trace is the sum of the diagonal elements.

4.64 Proposition If A and B are $n \times n$-matrices, then

$$\operatorname{tr}(AB) = \operatorname{tr}(BA) .$$

PROOF: If $A = (a_{ij})$ and $B = (b_{ij})$, then

$$\operatorname{tr}(AB) = \sum_{i=1}^{n} \sum_{k=1}^{n} a_{ik} b_{ki} = \sum_{k=1}^{n} \sum_{i=1}^{n} b_{ki} a_{ik} = \operatorname{tr}(BA) . \qquad \square$$

4.65 Corollary If A and B are $n \times n$-matrices and B is invertible, then

$$\operatorname{tr}(BAB^{-1}) = \operatorname{tr}(A) .$$

PROOF: It follows from Proposition 4.64 that

$$\operatorname{tr}(BAB^{-1}) = \operatorname{tr}(B^{-1}BA) = \operatorname{tr}(A) . \qquad \square$$

4.66 Definition If V is an n-dimensional vector space and $F : V \to V$ is a linear map, we define the *trace* of F to be

$$\operatorname{tr}(F) = \operatorname{tr}(m^{\mathscr{B}}_{\mathscr{B}}(F))$$

where $\mathscr{B}$ is a basis for V. It follows from Corollary 4.65 that $\operatorname{tr}(F)$ is well defined and does not depend on the choice of $\mathscr{B}$. Indeed, if $\mathscr{C}$ is another basis for V, we have that

$$\operatorname{tr}(m^{\mathscr{C}}_{\mathscr{C}}(F)) = \operatorname{tr}(m^{\mathscr{B}}_{\mathscr{C}}(id)\, m^{\mathscr{B}}_{\mathscr{B}}(F)\, m^{\mathscr{C}}_{\mathscr{B}}(id)) = \operatorname{tr}(m^{\mathscr{B}}_{\mathscr{B}}(F)) .$$

4.67 If V is an n-dimensional vector space, we have a linear isomorphism

$$\phi : L(V,V) \to \mathscr{T}^1_1(V)$$

given by $\phi(F) = T$ where

$$T(v, \lambda) = \lambda(F(v))$$

for $v \in V$, $\lambda \in V^*$ and $F \in L(V,V)$.

If $\mathscr{B} = \{v_1, ..., v_n\}$ is a basis for V with dual basis $\mathscr{B}^* = \{v_1^*, ..., v_n^*\}$, then $\phi^{-1}(T) = F$ is given by

$$F(v) = \sum_{i=1}^{n} v_i^*(F(v))\, v_i = \sum_{i=1}^{n} T(v, v_i^*)\, v_i$$

for $v \in V$ and $T \in \mathscr{T}^1_1(V)$. Now if

$$T = \sum_{i,j} T_i^j v_i^* \otimes v_j ,$$

it follows in particular that

$$F(v_j) = \sum_{i=1}^{n} T(v_j, v_i^*)\, v_i = \sum_{i=1}^{n} T_j^i\, v_i$$

so that $m_{\mathscr{B}}^{\mathscr{B}}(F) = A$ where $A_{ij} = T_j^i$.

4.68 Example The tensor $\delta \in \mathscr{T}_1^1(V)$ corresponding to the identity map $id \in L(V,V)$ is called the *Kronecker delta* on the vector space V. It is given by

$$\delta(v,\lambda) = \lambda\,(v)$$

for $v \in V$ and $\lambda \in V^*$. If $\mathscr{B} = \{v_1,...,v_n\}$ is any basis for V with dual basis $\mathscr{B}^* = \{v_1^*,...,v_n^*\}$, then

$$\delta = \sum_i v_i^* \otimes v_i\,.$$

If M is a smooth manifold, we have a tensor field T of type $\binom{1}{1}$ on M given by $T(p) = \delta_p$ for every $p \in M$, where δ_p is the Kronecker delta on T_pM. We have that

$$T|_U = \sum_i dx^i \otimes \frac{\partial}{\partial x^i}$$

for every local chart (x,U) on M, showing that T is in fact a tensor field. It is called the *Kronecker delta tensor* on M and is also denoted by δ.

4.69 Definition We have a linear map

$$C : \mathscr{T}_1^1(V) \to \mathbf{R}\,,$$

called a *contraction*, defined by $C(T) = \operatorname{tr}(F)$ where $F = \phi^{-1}(T)$.

4.70 Proposition If $\mathscr{B} = \{v_1,...,v_n\}$ is a basis for V with dual basis $\mathscr{B}^* = \{v_1^*,...,v_n^*\}$, and if

$$T = \sum_{i,j} T_i^j v_i^* \otimes v_j\,,$$

then

$$C(T) = \sum_i T_i^i\,.$$

PROOF: Follows from 4.67. □

4.71 Definition We have a linear map

$$C_s^r : \mathscr{T}_l^k(V) \to \mathscr{T}_{l-1}^{k-1}(V)\,,$$

called a *contraction* in the r-th covariant and the s-th contravariant index, which is defined in the following way. Let $T \in \mathscr{T}^k_l(V)$, and fix $v_1,\dots,v_{k-1} \in V$ and $\lambda_1,\dots,\lambda_{l-1} \in V^*$. Then

$$C^r_s(T)(v_1,\dots,v_{k-1},\lambda_1,\dots,\lambda_{l-1}) = C(S)\,,$$

where $S \in \mathscr{T}^1_1(V)$ is the tensor defined by

$$S(v,\lambda) = T(v_1,\dots,v_{r-1},v,v_r,\dots,v_{k-1},\lambda_1,\dots,\lambda_{s-1},\lambda,\lambda_s,\dots,\lambda_{l-1})$$

for $v \in V$ and $\lambda \in V^*$.

4.72 Proposition If $\mathscr{B} = \{v_1,\dots,v_n\}$ is a basis for V with dual basis $\mathscr{B}^* = \{v_1^*,\dots,v_n^*\}$, and if

$$T = \sum_{\substack{i_1,\dots,i_k \\ j_1,\dots,j_l}} A^{j_1,\dots,j_l}_{i_1,\dots,i_k}\, v^*_{i_1} \otimes \cdots \otimes v^*_{i_k} \otimes v_{j_1} \otimes \cdots \otimes v_{j_l}\,,$$

then

$$C^r_s(T) = \sum_{\substack{i_1,\dots,i_{k-1} \\ j_1,\dots,j_{l-1}}} B^{j_1,\dots,j_{l-1}}_{i_1,\dots,i_{k-1}}\, v^*_{i_1} \otimes \cdots \otimes v^*_{i_{k-1}} \otimes v_{j_1} \otimes \cdots \otimes v_{j_{l-1}}$$

where

$$B^{j_1,\dots,j_{l-1}}_{i_1,\dots,i_{k-1}} = \sum_m A^{j_1,\dots,j_{s-1},m,j_s,\dots,j_{l-1}}_{i_1,\dots,i_{r-1},m,i_r,\dots,i_{k-1}}\,.$$

PROOF : Follows from Definition 4.71 and Proposition 4.70. □

4.73 Definition Let T be a tensor field of type $\binom{k}{l}$ on a smooth manifold M. Then we have a tensor field S of type $\binom{k-1}{l-1}$ on M given by

$$S(p) = C^r_s(T(p))$$

for every $p \in M$. If (x,U) is a local chart on M, and if $A^{j_1,\dots,j_l}_{i_1,\dots,i_k}$ and $B^{j_1,\dots,j_{l-1}}_{i_1,\dots,i_{k-1}}$ are the components of the local representations of T and S on U, respectively, then it follows from Proposition 4.72 that

$$B^{j_1,\dots,j_{l-1}}_{i_1,\dots,i_{k-1}} = \sum_m A^{j_1,\dots,j_{s-1},m,j_s,\dots,j_{l-1}}_{i_1,\dots,i_{r-1},m,i_r,\dots,i_{k-1}}\,,$$

showing that S is in fact a tensor field. It is called the *contraction* of T in the r-th covariant and the s-th contravariant index, and it is also denoted by $C^r_s(T)$.

4.74 Proposition Let $C : \mathscr{T}^k_l(V) \to \mathscr{T}^{k-l}_0(V)$ where $k \ge l$ be the l-fold contraction in the i-th contravariant and the i-th covariant index for $0 \le i \le l$. Then we have that

$$C(T \otimes \lambda_1 \otimes \cdots \otimes \lambda_k) = T(\lambda_1,\dots,\lambda_l)\,\lambda_{l+1} \otimes \cdots \otimes \lambda_k$$

for every $T \in \mathscr{T}_l(V)$ and $\lambda_1,\dots,\lambda_k \in \mathscr{T}^1(V)$.

PROOF: Let $\mathscr{B} = \{v_1, ..., v_n\}$ be a basis for V with dual basis $\mathscr{B}^* = \{v_1^*, ..., v_n^*\}$, and suppose that

$$T = \sum_{j_1,...,j_l} A^{j_1,...,j_l} v_{j_1} \otimes \cdots \otimes v_{j_l}$$

and

$$\lambda_r = \sum_{i_r} a_{i_r}^r v_{i_r}^*$$

for $r = 1, ..., k$. Then it follows that

$$T \otimes \lambda_1 \otimes \cdots \otimes \lambda_k = \sum_{\substack{i_1,...,i_k \\ j_1,...,j_l}} A^{j_1,...,j_l} a_{i_1}^1 \cdots a_{i_k}^k v_{i_1}^* \otimes \cdots \otimes v_{i_k}^* \otimes v_{j_1} \otimes \cdots \otimes v_{j_l}$$

so that

$$\begin{aligned} &C(T \otimes \lambda_1 \otimes \cdots \otimes \lambda_k) \\ &= \sum_{i_1,...,i_l} A^{i_1,...,i_l} a_{i_1}^1 \cdots a_{i_l}^l \sum_{i_{l+1},...,i_k} a_{i_{l+1}}^{l+1} \cdots a_{i_k}^k v_{i_{l+1}}^* \otimes \cdots \otimes v_{i_k}^* \\ &= T(\lambda_1, ..., \lambda_l) \lambda_{l+1} \otimes \cdots \otimes \lambda_k . \end{aligned}$$

□

THE LIE DERIVATIVE

4.75 Definition Let X be a vector field on a smooth manifold M with global flow $\gamma: \mathscr{D}(X) \to M$, and let T be a tensor field of type $\binom{k}{l}$ on M.

If p is a point on M, then $\gamma_t : \mathscr{D}_t(X) \to M$ is a diffeomorphism from the open neighbourhood $\mathscr{D}_t(X)$ of p onto the open neighbourhood $\mathscr{D}_{-t}(X)$ of $\gamma_t(p)$ with inverse γ_{-t} for each $t \in I(p)$, and we define the *Lie derivative* of T at p with respect to X to be

$$L_X T(p) = \lim_{t \to 0} \frac{1}{t} [\gamma_t^* T(p) - T(p)] = \frac{d}{dt}\bigg|_0 \gamma_t^* T(p) ,$$

using the vector space topology on $\mathscr{T}_l^k(T_pM)$ defined in Proposition 13.117 in the appendix.

4.76 Proposition If $c : I(p) \to \mathscr{T}_l^k(T_pM)$ is the curve defined by $c(t) = \gamma_t^* T(p)$ and we canonically identify $T_{c(0)} \mathscr{T}_l^k(T_pM)$ with $\mathscr{T}_l^k(T_pM)$ as in Lemma 2.84, then $L_X T(p) = c'(0)$, and the Lie derivative $L_X T$ is a well-defined tensor field of type $\binom{k}{l}$ on M.

Let (x, U) be a local chart on M, and let

$$X|_U = \sum_{i=1}^n a^i \frac{\partial}{\partial x^i}$$

and

$$T\,|_U = \sum_{\substack{i_1,\ldots,i_k \\ j_1,\ldots,j_l}} A^{j_1,\ldots,j_l}_{i_1,\ldots,i_k}\, dx^{i_1}\otimes\cdots\otimes dx^{i_k}\otimes\frac{\partial}{\partial x^{j_1}}\otimes\cdots\otimes\frac{\partial}{\partial x^{j_l}}\,.$$

Then we have that

$$L_X T\,|_U = \sum_{\substack{i_1,\ldots,i_k \\ j_1,\ldots,j_l}} B^{j_1,\ldots,j_l}_{i_1,\ldots,i_k}\, dx^{i_1}\otimes\cdots\otimes dx^{i_k}\otimes\frac{\partial}{\partial x^{j_1}}\otimes\cdots\otimes\frac{\partial}{\partial x^{j_l}}$$

where

$$B^{j_1,\ldots,j_l}_{i_1,\ldots,i_k} = \sum_{i=1}^{n} a^i\,\frac{\partial A^{j_1,\ldots,j_l}_{i_1,\ldots,i_k}}{\partial x^i} - \sum_{m=1}^{l}\sum_{j=1}^{n} A^{j_1,\ldots,j_{m-1},j,j_{m+1},\ldots,j_l}_{i_1,\ldots,i_k}\,\frac{\partial a^{j_m}}{\partial x^j}$$
$$+\sum_{m=1}^{k}\sum_{i=1}^{n} A^{j_1,\ldots,j_l}_{i_1,\ldots,i_{m-1},i,i_{m+1},\ldots,i_k}\,\frac{\partial a^i}{\partial x^{i_m}}\,.$$

PROOF: If p_0 is a point in U, we may choose an open neighbourhood W of p_0 contained in U and a an open interval I containing 0 such that $I\times W\subset\mathscr{D}(X)$ and $\gamma(I\times W)\subset U$. By Proposition 3.42 (3) we know that $\gamma_t : W\to\gamma_t(W)$ is a diffeomorphism for each $t\in I$ with inverse γ_{-t}. We must show that the curve $c : I\to\mathscr{T}^k_l(T_pM)$ defined by $c(t)=\gamma_t^* T(p)$ is smooth so that $L_X T(p)$ is a well-defined tensor in $\mathscr{T}^k_l(T_pM)$. By Proposition 4.62 we have that

$$\gamma_t^*(T)(q) = \sum_{\substack{i_1,\ldots,i_k \\ j_1,\ldots,j_l}} C_t{}^{j_1,\ldots,j_l}_{i_1,\ldots,i_k}(q)\, dx^{i_1}(q)\otimes\cdots\otimes dx^{i_k}(q)\otimes\left.\frac{\partial}{\partial x^{j_1}}\right|_q\otimes\cdots\otimes\left.\frac{\partial}{\partial x^{j_l}}\right|_q$$

where

$$C_t{}^{j_1,\ldots,j_l}_{i_1,\ldots,i_k}(q) = \sum_{\substack{r_1,\ldots,r_k \\ s_1,\ldots,s_l}} A^{s_1,\ldots,s_l}_{r_1,\ldots,r_k}(\gamma_t(q))\,\frac{\partial(x^{r_1}\circ\gamma_t)}{\partial x^{i_1}}(q)\cdots\frac{\partial(x^{r_k}\circ\gamma_t)}{\partial x^{i_k}}(q)$$
$$\cdot\,\frac{\partial(x^{j_1}\circ\gamma_{-t})}{\partial x^{s_1}}(\gamma_t(q))\cdots\frac{\partial(x^{j_l}\circ\gamma_{-t})}{\partial x^{s_l}}(\gamma_t(q)) \tag{1}$$

for $q\in W$.

We now transform the problem to open sets in $\mathbf{R}^n$ using the local chart (x,U). The coordinate map x can be written as $x=i\circ\tilde{x}$, where $\tilde{x} : U\to x(U)$ is a diffeomorphism and $i : x(U)\to\mathbf{R}^n$ is the inclusion map. Consider the functions $f^i = a^i\circ x^{-1}$ and $F^{j_1,\ldots,j_l}_{i_1,\ldots,i_k} = A^{j_1,\ldots,j_l}_{i_1,\ldots,i_k}\circ x^{-1}$ on $x(U)$ and $H_t{}^{j_1,\ldots,j_l}_{i_1,\ldots,i_k} = C_t{}^{j_1,\ldots,j_l}_{i_1,\ldots,i_k}\circ x^{-1}$ on $x(W)$ representing X, T and $\gamma_t^* T$, and let

$$\alpha : I\times x(W)\to x(U)$$

be the local flow for f given by $\alpha=\tilde{x}\circ\gamma\circ(id\times x^{-1})$, where γ is the local flow for X on $I\times W$. Then

$$D_1\alpha(t,u) = f(\alpha(t,u)) \quad\text{and}\quad \alpha(0,u)=u \tag{2}$$

for all $t \in I$ and $u \in x(W)$, and we have that $\alpha_t = \tilde{x} \circ \gamma_t \circ x^{-1}$. If $u = x(q)$, it follows from (1) that

$$H_t{}^{j_1,\dots,j_l}_{i_1,\dots,i_k}(u) = \sum_{\substack{r_1,\dots,r_k \\ s_1,\dots,s_l}} F^{s_1,\dots,s_l}_{r_1,\dots,r_k}(\alpha(t,u))\, D_{i_1+1}\alpha^{r_1}(t,u) \cdots D_{i_k+1}\alpha^{r_k}(t,u)$$
$$\cdot D_{s_1+1}\alpha^{j_1}(-t,\alpha(t,u)) \cdots D_{s_l+1}\alpha^{j_l}(-t,\alpha(t,u)) \tag{3}$$

for $u \in x(W)$, which shows that the curve c is smooth and that the Lie derivative $L_X T$ is a well-defined tensor field of type $\binom{k}{l}$ on M.

Using (2) we can now find the derivative of each term in (3) with respect to t at $t = 0$. First note that

$$D_{i+1}\alpha^r(0,u) = D_i u^r = \delta_{ir} \quad \text{and} \quad D_{s+1,i+1}\alpha^r(0,u) = 0\,.$$

Hence we have that

$$\frac{d}{dt}\bigg|_0 (F^{s_1,\dots,s_l}_{r_1,\dots,r_k}(\alpha(t,u))) = \sum_{i=1}^n D_i F^{s_1,\dots,s_l}_{r_1,\dots,r_k}(u)\, f^i(u) = \sum_{i=1}^n a^i(q)\, \frac{\partial A^{s_1,\dots,s_l}_{r_1,\dots,r_k}}{\partial x^i}(q)\,,$$

$$\frac{d}{dt}\bigg|_0 (D_{i+1}\alpha^r(t,u)) = D_{i+1}D_1\alpha^r(0,u) = D_i f^r(u) = \frac{\partial a^r}{\partial x^i}(q)\,,$$

and

$$\frac{d}{dt}\bigg|_0 (D_{s+1}\alpha^j(-t,\alpha(t,u))) = -D_{s+1}D_1\alpha^j(0,u) = -D_s f^j(u) = -\frac{\partial a^j}{\partial x^s}(q)\,.$$

Inserting this in (3), we obtain the formula for $B^{j_1,\dots,j_l}_{i_1,\dots,i_k}$ in the proposition. □

4.77 Remark If f is a smooth function on M, the formula for $L_X f\,|_U$ reduces to

$$L_X f\,|_U = \sum_{i=1}^n a^i \frac{\partial f}{\partial x^i}\,,$$

which shows that $L_X f = X(f)$.

If Y is a vector field on M with

$$Y\,|_U = \sum_{j=1}^n b^j \frac{\partial}{\partial x^j}\,,$$

we have that

$$L_X Y\,|_U = \sum_{i,j} \left\{ a^i \frac{\partial b^j}{\partial x^i} - b^i \frac{\partial a^j}{\partial x^i} \right\} \frac{\partial}{\partial x^j}\,.$$

4.78 Remark Let $\mathscr{B} = \{v_1, \dots, v_n\}$ be a basis for T_pM with dual basis $\mathscr{B}^* =$

$\{v^1,\dots,v^n\}$, and let $c : I \to M$ be an integral curve for X with initial condition p given by $c(t) = \gamma_t(p)$. If $v_i(t) = \gamma_{t*}(v_i)$ and $v^i(t) = v^i \circ \gamma_{-t*}$ for $i = 1,\dots,n$ and $t \in I$, then $\mathscr{B}_t = \{v_1(t),\dots,v_n(t)\}$ is a basis for $T_{c(t)}M$ with dual basis $\mathscr{B}_t^* = \{v^1(t),\dots,v^n(t)\}$, since $(\gamma_t)_{*p} : T_pM \to T_{c(t)}M$ is a linear isomorphism. We say that $v_i(t)$ and $v^i(t)$ are obtained from v_i and v^i, respectively, by *Lie transport* along the integral curve c. If

$$T(c(t)) = \sum_{\substack{i_1,\dots,i_k \\ j_1,\dots,j_l}} C_p{}^{j_1,\dots,j_l}_{i_1,\dots,i_k}(t)\, v^{i_1}(t) \otimes \cdots \otimes v^{i_k}(t) \otimes v_{j_1}(t) \otimes \cdots \otimes v_{j_l}(t)\,,$$

then

$$\begin{aligned} &\gamma_t^* T(p)(v_{i_1},\dots,v_{i_k},v^{j_1},\dots,v^{j_l}) \\ &\quad = T(\gamma_t(p))(\gamma_{t*}(v_{i_1}),\dots,\gamma_{t*}(v_{i_k}),v^{j_1}\circ\gamma_{-t*},\dots,v^{j_l}\circ\gamma_{-t*}) \\ &\quad = C_p{}^{j_1,\dots,j_l}_{i_1,\dots,i_k}(t) \end{aligned}$$

so that

$$\gamma_t^* T(p) = \sum_{\substack{i_1,\dots,i_k \\ j_1,\dots,j_l}} C_p{}^{j_1,\dots,j_l}_{i_1,\dots,i_k}(t)\, v^{i_1} \otimes \cdots \otimes v^{i_k} \otimes v_{j_1} \otimes \cdots \otimes v_{j_l}\,.$$

Hence we have that

$$L_X T(p) = \sum_{\substack{i_1,\dots,i_k \\ j_1,\dots,j_l}} \frac{d}{dt}\bigg|_0 C_p{}^{j_1,\dots,j_l}_{i_1,\dots,i_k}\, v^{i_1} \otimes \cdots \otimes v^{i_k} \otimes v_{j_1} \otimes \cdots \otimes v_{j_l}\,.$$

Let (x,U) be a local chart around p, and choose an open neighbourhood W of p contained in U and an open interval I containing 0 such that $I \times W \subset \mathscr{D}(X)$ and $\gamma(I \times W) \subset U$. If $v_i = \dfrac{\partial}{\partial x^i}\bigg|_q$ for $i = 1,\dots,n$ and $q \in W$, then the functions $C^{j_1,\dots,j_l}_{i_1,\dots,i_k} : I \times W \to \mathbf{R}$ defined by $C^{j_1,\dots,j_l}_{i_1,\dots,i_k}(t,q) = C_q{}^{j_1,\dots,j_l}_{i_1,\dots,i_k}(t)$ are smooth since $C_q{}^{j_1,\dots,j_l}_{i_1,\dots,i_k}(t)$ coincides with $C_t{}^{j_1,\dots,j_l}_{i_1,\dots,i_k}(q)$ in the proof of Proposition 4.76.

4.79 Proposition Let X be a vector field on the smooth manifold M.

(1) L_X is linear over $\mathbf{R}$.

(2) If $T \in \mathscr{T}^{k_1}_{l_1}(M)$ and $S \in \mathscr{T}^{k_2}_{l_2}(M)$, then $L_X(T \otimes S) = (L_X T) \otimes S + T \otimes (L_X S)$.

(3) $L_X \delta = 0$, where $\delta \in \mathscr{T}^1_1(M)$ is the Kronecker delta tensor.

(4) L_X commutes with contractions, i.e., if $T \in \mathscr{T}^k_l(M)$, and if $1 \le r \le k$ and $1 \le s \le l$, then $L_X(C^r_s T) = C^r_s(L_X T)$.

(5) If $T \in \mathscr{T}^k_l(M)$, and if $X_1,\dots,X_k \in \mathscr{T}_1(M)$ and $\lambda_1,\dots,\lambda_l \in \mathscr{T}^1(M)$, then

$$\begin{aligned} L_X\left[T\left(X_1,\dots,X_k,\lambda_1,\dots,\lambda_l\right)\right] &= \left(L_X T\right)\left(X_1,\dots,X_k,\lambda_1,\dots,\lambda_l\right) \\ &\quad + \sum_{i=1}^{k} T\left(X_1,\dots,L_X X_i,\dots,X_k,\lambda_1,\dots,\lambda_l\right) \\ &\quad + \sum_{i=1}^{l} T\left(X_1,\dots,X_k,\lambda_1,\dots,L_X\lambda_i,\dots,\lambda_l\right). \end{aligned}$$

PROOF: Point (1) to (4) follow directly from Remark 4.78. Point (5) follows from (2) and (4) since $T(X_1,\dots,X_k,\lambda_1,\dots,\lambda_l)$ can be obtained from $T \otimes X_1 \otimes \cdots \otimes X_k \otimes \lambda_1 \otimes \cdots \otimes \lambda_l$ by applying contractions repeatedly. □

4.80 Definition A *Lie algebra* is a vector space $\mathscr{A}$ with a bilinear map $[\ ,\ ] : \mathscr{A} \times \mathscr{A} \to \mathscr{A}$, called the *bracket product* or just *bracket*, satisfying

(1) $[X,Y] = -[Y,X]$,

(2) $[X,[Y,Z]] + [Y,[Z,X]] + [Z,[X,Y]] = 0$

for every $X,Y,Z \in \mathscr{A}$.

4.81 Remark A bilinear map satisfying (1) is called *skew symmetric*, and the identity (2) is called the *Jacobi identity*.

A bilinear map satisfies (1) if and only if it is *alternating*, i.e., if

(3) $[X,X] = 0$ for every $X \in \mathscr{A}$.

Indeed, it follows from (3) that

$$0 = [X+Y,X+Y] = [X,X]+[X,Y]+[Y,X]+[Y,Y] = 0+[X,Y]+[Y,X]+0$$

which implies (1). Conversely, if (1) is satisfied, we have that $[X,X]$ is equal to its own inverse and hence must be 0 for every $X \in \mathscr{A}$, thus implying (3).

4.82 Proposition Every associative algebra $\mathscr{A}$ is a Lie algebra, where the bracket $[X,Y] = XY - YX$ is the commutator product.

PROOF: We immediately see that the commutator product is bilinear and skew symmetric, and it only remains to prove the Jacobi identity. If $X,Y,Z \in \mathscr{A}$, we have that

$$\begin{aligned} [X,[Y,Z]] + [Y,[Z,X]] + [Z,[X,Y]] &= \{X(YZ-ZY) - (YZ-ZY)X\} \\ &\quad + \{Y(ZX-XZ) - (ZX-XZ)Y\} \\ &\quad + \{Z(XY-YX) - (XY-YX)Z\} = 0 \end{aligned}$$

since the product of every permutation of X, Y and Z occurs once with a plus and once with a minus in the formula. □

4.83 Remark As special cases we have the Lie algebra $End\,(V)$ of endomorphisms of a real or complex vector space V, and the Lie algebras $\mathfrak{gl}\,(n,\mathbf{R})$ and $\mathfrak{gl}\,(n,\mathbf{C})$ of real or complex $n \times n$-matrices, all considered as vector spaces over $\mathbf{R}$.

4.84 Definition A subspace $\mathscr{A}$ of a Lie algebra $\mathscr{B}$ which is closed under the bracket product $[\ ,\ ]$, is called a *Lie subalgebra* of $\mathscr{B}$. It is called an *ideal* in $\mathscr{B}$ if $[X,Y] \in \mathscr{A}$ for every $X \in \mathscr{B}$ and $Y \in \mathscr{A}$.

A linear map $\phi : \mathscr{A} \to \mathscr{B}$ between the Lie algebras $\mathscr{A}$ and $\mathscr{B}$ is called a *Lie algebra homomorphism* if

$$\phi\,([X,Y]) = [\phi\,(X), \phi\,(Y)]$$

for every $X,Y \in \mathscr{A}$. If ϕ is also bijective, it is called a *Lie algebra isomorphism.* A Lie algebra isomorphism of a Lie algebra with itself is called a *Lie algebra automorphism.* The linear map $\phi : \mathscr{A} \to \mathscr{B}$ is called a *Lie algebra anti-homomorphism* if

$$\phi\,([X,Y]) = -[\phi\,(X), \phi\,(Y)]$$

for every $X,Y \in \mathscr{A}$.

If $\mathscr{B} = End\,(V)$ for a vector space V, or if $\mathscr{B} = \mathfrak{gl}\,(n,\mathbf{R})$, then a Lie algebra homomorphism $\phi : \mathscr{A} \to \mathscr{B}$ is called a *representation* of the Lie algebra $\mathscr{A}$.

4.85 Proposition Let $\mathscr{A}$ be an algebra. Then the set $\mathscr{D}\,(\mathscr{A})$ of derivations of $\mathscr{A}$ is a Lie subalgebra of $End\,(\mathscr{A})$.

PROOF : If $X,Y \in \mathscr{D}\,(\mathscr{A})$, we have that

$$\begin{aligned}[X,Y]\,(fg) &= X\{Y(f)\,g + f\,Y(g)\} - Y\{X(f)\,g + f\,X(g)\} \\ &= \{XY(f)\,g + Y(f)\,X(g) + X(f)\,Y(g) + f\,XY(g)\} \\ &\quad - \{YX(f)\,g + X(f)\,Y(g) + Y(f)\,X(g) + f\,YX(g)\} \\ &= [X,Y]\,(f)\,g + f\,[X,Y]\,(g)\end{aligned}$$

for every $f,g \in \mathscr{A}$, which shows that $[X,Y]$ is also a derivation of $\mathscr{A}$. □

4.86 Proposition If X and Y are vector fields on a smooth manifold M^n, then their commutator product $[X,Y]$, defined by

$$[X,Y]\,(f) = X(Y(f)) - Y(X(f))$$

for $f \in \mathscr{F}(M)$, is a vector field on M, and we have that $[X,Y] = L_X Y$.

PROOF : The assertion that $[X,Y]$ is a vector field follows by Remark 3.9 and Proposition 4.85. To prove the last assertion, let (x,U) be a local chart on M, and let

$$X\,|_U = \sum_{i=1}^{n} a^i \frac{\partial}{\partial x^i} \quad \text{and} \quad Y\,|_U = \sum_{j=1}^{n} b^j \frac{\partial}{\partial x^j}\,.$$

Then we have that

$$X(Y(f))|_U = \sum_{i=1}^{n} a^i \frac{\partial}{\partial x^i}\left(\sum_{j=1}^{n} b^j \frac{\partial f}{\partial x^j}\right) = \sum_{i,j}\left\{a^i \frac{\partial b^j}{\partial x^i}\frac{\partial f}{\partial x^j} + a^i b^j \frac{\partial^2 f}{\partial x^i \partial x^j}\right\},$$

and we similarly obtain

$$Y(X(f))|_U = \sum_{i,j}\left\{b^i \frac{\partial a^j}{\partial x^i}\frac{\partial f}{\partial x^j} + b^i a^j \frac{\partial^2 f}{\partial x^i \partial x^j}\right\}$$

so that

$$[X,Y](f)|_U = \sum_{i,j}\left\{a^i \frac{\partial b^j}{\partial x^i} - b^i \frac{\partial a^j}{\partial x^i}\right\}\frac{\partial f}{\partial x^j}.$$

Hence it follows from Remark 4.77 that $[X,Y] = L_X Y$. □

4.87 Proposition Let M be a smooth manifold. If $X,Y \in \mathscr{T}_1(M)$ and $f,g \in \mathscr{F}(M)$, then

$$[fX, gY] = fg[X,Y] + fX(g)Y - gY(f)X.$$

PROOF: By Propositions 4.79 (2) and 4.86 we have that

$$[X, gY] = L_X(gY) = (L_X g)Y + g(L_X Y) = g[X,Y] + X(g)Y$$

which also implies

$$[fX, Y] = -[Y, fX] = -(f[Y,X] + Y(f)X) = f[X,Y] - Y(f)X.$$

Combining this, we have that

$$[fX, gY] = g[fX, Y] + fX(g)Y = fg[X,Y] + fX(g)Y - gY(f)X.$$ □

4.88 Proposition Let $f : M \to N$ be a smooth map, and let X_i and Y_i be vector fields on M and N, respectively, which are f-related for $i = 1,2$. Then the commutator products $[X_1, X_2]$ and $[Y_1, Y_2]$ are also f-related.

PROOF: By Proposition 3.11 we have that

$$Y_i(g) \circ f = X_i(g \circ f)$$

for $g \in \mathscr{F}(N)$ and $i = 1,2$, which implies that

$$\begin{aligned}[Y_1, Y_2](g) \circ f &= Y_1(Y_2(g)) \circ f - Y_2(Y_1(g)) \circ f \\ &= X_1(Y_2(g) \circ f) - X_2(Y_1(g) \circ f) \\ &= X_1(X_2(g \circ f)) - X_2(X_1(g \circ f)) = [X_1, X_2](g \circ f).\end{aligned}$$ □

4.89 Proposition Let $f : M \to N$ be a diffomorphism. If X is a vector field and T a tensor field of type $\binom{k}{l}$ on N, then

$$f^*(L_X T) = L_{f^*(X)}\, f^*(T)\,.$$

PROOF : Let γ and β be the global flows for $f^*(X)$ and X, respectively, and let $q \in M$ and $p = f(q)$. If $c : I(q) \to \mathscr{T}^k_l(T_qM)$ and $b : J(p) \to \mathscr{T}^k_l(T_pN)$ are the curves defined by

$$c(t) = \gamma_t^*\, f^*(T)(q) \quad \text{and} \quad b(t) = \beta_t^*\, T(p)\,,$$

and we canonically identify $T_{c(0)}\mathscr{T}^k_l(T_qM)$ with $\mathscr{T}^k_l(T_qM)$ and $T_{b(0)}\mathscr{T}^k_l(T_pN)$ with $\mathscr{T}^k_l(T_pN)$ as in Lemma 2.84, then

$$L_{f^*(X)}\, f^*(T)(q) = c'(0) \quad \text{and} \quad L_X T(p) = b'(0)\,.$$

Now $\beta_t = f \circ \gamma_t \circ f^{-1}$ by Proposition 3.49, so that

$$c(t) = f^*(f \circ \gamma_t \circ f^{-1})^*(T)(q) = (f_{*q})^*(\beta_t^*\, T)(p) = (f_{*q})^* \circ b(t)$$

where $(f_{*q})^* : \mathscr{T}^k_l(T_pN) \to \mathscr{T}^k_l(T_qM)$ is the linear map defined in Definition 4.52. Hence it follows from Lemma 2.84 that

$$L_{f^*(X)}\, f^*(T)(q) = c'(0) = (f_{*q})^* \circ b'(0) = f^*(L_X T)(q)\,. \qquad \square$$

4.90 Proposition Let X be a vector field on a smooth manifold M with global flow $\gamma : \mathscr{D}(X) \to M$, and let T be a tensor field of type $\binom{k}{l}$ on M. If p is a point on M and $c : I(p) \to \mathscr{T}^k_l(T_pM)$ is the curve defined by $c(t) = \gamma_t^*\, T(p)$, then we have that

$$c'(t_0) = \gamma_{t_0}^*(L_X T)(p)$$

for every $t_0 \in I(p)$ if we canonically identify $T_{c(t_0)}\mathscr{T}^k_l(T_pM)$ with $\mathscr{T}^k_l(T_pM)$ as in Lemma 2.84.

PROOF : Let $c_0 : I(p) - t_0 \to \mathscr{T}^k_l(T_pM)$ be the curve defined by $c_0(s) = c(s + t_0) = \gamma_{s+t_0}^*\, T(p)$ so that $c'(t_0) = c_0'(0)$. Consider the diffeomorphism $\gamma_{t_0} : \mathscr{D}_{t_0}(X) \to \mathscr{D}_{-t_0}(X)$, and denote the restriction of X to $\mathscr{D}_{-t_0}(X)$ by X_0. Let $J(q)$ be the domain of the maximal integral curve for X_0 with initial condition q. If α and β are the flows for X_0 and $\gamma_{t_0}^*(X_0)$, respectively, it follows from Proposition 3.49 that $\alpha_s = \gamma_{t_0} \circ \beta_s \circ \gamma_{-t_0}$ for every $s \in \mathbf{R}$ so that $\alpha_s \circ \gamma_{t_0} = \gamma_{t_0} \circ \beta_s$. Since α_s is the restriction of γ_s to $\mathscr{D}_s(X_0)$, it follows by Proposition 3.42 (2) that $\alpha_s \circ \gamma_{t_0}$ is the restriction of γ_{s+t_0} to $\gamma_{-t_0}(\mathscr{D}_s(X_0))$. Hence we have that

$$c_0(s) = \beta_s^*\,(\gamma_{t_0}^* T)\,(p)$$

for every $s \in J(\gamma_{t_0}(p))$ so that

$$c_0'(0) = L_{\gamma_{t_0}^*(X)}\, \gamma_{t_0}^*(T)\,(p) = \gamma_{t_0}^*(L_X T)\,(p)\,. \qquad \square$$

4.91 Proposition Let X and Y be vector fields on a smooth manifold M with global flows α and β, respectively, such that $[X,Y]=0$ on M. Let p be a point on M, and suppose there are open intervals I and J containing 0 such that both expressions $\alpha(t,\beta(s,p))$ and $\beta(s,\alpha(t,p))$ are defined for $(s,t)\in I\times J$. Then we have that

$$\alpha(t,\beta(s,p))=\beta(s,\alpha(t,p))$$

for every $(s,t)\in I\times J$.

PROOF: Fix $t\in J$, and consider the curves $c_i:I\to M$ for $i=1,2$ defined by $c_1(s)=\alpha(t,\beta(s,p))$ and $c_2(s)=\beta(s,\alpha(t,p))$ for $s\in I$. Then c_2 is an integral curve for Y with initial condition $\alpha(t,p)$, and we will show that this is also the case for c_1.

For each $s\in I$ we have that $c_1'(s)=\alpha_{t*}(Y(\beta(s,p)))=\alpha_{t*}(Y(q))$ where $q=\beta(s,p)\in\mathscr{D}_t(X)$. Since $L_XY=[X,Y]=0$, it follows from Proposition 4.90 that $Y(q)=\alpha_t^*Y(q)=\alpha_{-t*}\circ Y\circ\alpha_t(q)$ so that $c_1'(s)=Y(\alpha_t(q))=Y(c_1(s))$. Hence c_1 is also an integral curve for Y with initial condition $\alpha(t,p)$, which completes the proof of the proposition. □

4.92 Proposition Let X and Y be vector fields on a smooth manifold M with global flows α and β, respectively. Let p be a point on M, and suppose there are open intervals I and J containing 0 such that

$$\alpha(t,\beta(s,p))=\beta(s,\alpha(t,p))$$

for $(s,t)\in I\times J$. Then we have that $[X,Y](p)=0$.

PROOF: Keeping $t\in J$ fixed and differentiating both sides of the equation with respect to s at $s=0$, we obtain

$$\alpha_{t*}(Y(p))=Y(\alpha_t(p))\,.$$

Hence $Y(p)=\alpha_{-t*}\circ Y\circ\alpha_t(p)=\alpha_t^*Y(p)$ for $t\in J$ so that

$$[X,Y](p)=L_XY(p)=\frac{d}{dt}\bigg|_0\alpha_t^*Y(p)=0\,.$$ □

DISTRIBUTIONS AND INTEGRAL MANIFOLDS

4.93 Proposition Let X be a vector field on a smooth manifold M^n with $X(p)\neq 0$ for a point p on M. Then there exists a local chart (x,U) around p such that $X=\dfrac{\partial}{\partial x^1}$ on U.

PROOF: Let $\gamma\colon \mathscr{D}(X) \to M$ be the global flow for X, and choose an open neighbourhood W of p and an open interval I containing 0 such that $I \times W \subset \mathscr{D}(X)$. Let (z,V) be a local chart around p with $V \subset W$ and $z(p) = 0$.

Then we have that $X(p) = [z, v_1]_p$ for a vector $v_1 \neq 0$ in $\mathbf{R}^n$, and we extend $\{v_1\}$ to a basis $\mathscr{B} = \{v_1, \dots, v_n\}$ for $\mathbf{R}^n$. If $\mathscr{E} = \{e_1, \dots, e_n\}$ is the standard basis for $\mathbf{R}^n$, we have a linear isomorphism $\phi : \mathbf{R}^n \to \mathbf{R}^n$ given by $\phi(v_i) = e_i$ for $i = 1, \dots, n$. Hence if (y,V) is the local chart with $y = \phi \circ z$, we have that

$$X(p) = [z, v_1]_p = [y, D(y \circ z^{-1})(z(p))\, v_1]_p = [y, \phi(v_1)]_p = [y, e_1]_p\,.$$

Now let O be an open neighbourhood of 0 in $\mathbf{R}^{n-1}$ with $\{0\} \times O \subset y(V)$, and let $\theta : I \times O \to M$ be the map defined by

$$\theta(a_1, a_2, \dots, a_n) = \gamma(a_1, y^{-1}(0, a_2, \dots, a_n))\,.$$

Using the local chart $(t, I \times O)$ on $I \times O$, where $t : I \times O \to \mathbf{R} \times \mathbf{R}^{n-1}$ is the inclusion map, we have that

$$\theta_*([t, e_1]_0) = X(p) = [y, e_1]_p$$

and

$$\theta_*([t, e_i]_0) = [y, D(y \circ \theta \circ t^{-1})(0)\, e_i]_p = [y, e_i]_p$$

for $2 \leq i \leq n$, showing that θ_{*0} is non-singular. By the inverse function theorem, θ is a local diffeomorphism at 0 with a smooth inverse $x : U \to \mathbf{R}^n$ defined on an open neighbourhood U of p. Since

$$X(\theta(a)) = \theta_*([t, e_1]_a) = [x, D(x \circ \theta \circ t^{-1})(a)\, e_1]_{\theta(a)} = [x, e_1]_{\theta(a)}$$

for $a \in x(U)$, it follows that $X = \dfrac{\partial}{\partial x^1}$ on U. □

4.94 Definition By a *distribution* Δ on a smooth manifold M^n we mean a subbundle of its tangent bundle $\pi : TM \to M$. We say that Δ is *involutive* or *integrable* if for any pair of vector fields $X, Y \in \Delta$ defined on some open subset U of M, we have that $[X, Y] \in \Delta$.

An immersed submanifold N of M is called an *integral manifold* of Δ if $i_*(T_pN) = \Delta_p$ for every $p \in N$, where $i : N \to M$ is the inclusion map. A connected integral manifold of Δ is called *maximal* if it is not contained in any larger connected integral manifold of Δ.

4.95 Proposition Let Δ be a distribution on a smooth manifold M, and suppose that there is an integral manifold of Δ through every point on M. Then Δ is integrable.

PROOF: Let X and Y be two vector fields on some open subset U of M lying in Δ, and let $p \in U$. If N is an integral manifold of Δ passing through p, it follows from Proposition 3.14 that there are unique vector fields X' and Y' on $U \cap N$ which are i-related to X and Y, respectively, where $i : U \cap N \to U$ is the inclusion map. By Proposition 4.88 it follows that $[X',Y']$ and $[X,Y]$ are i-related so that

$$i_* [X',Y']_p = [X,Y]_p .$$

Since $[X',Y']_p \in T_pN$, it follows that $[X,Y]_p \in \Delta_p$ for every $p \in U$, thus showing that $[X,Y]$ also lies in Δ. □

4.96 Proposition A distribution Δ on a smooth manifold M is integrable if and only if given a local basis $\{X_1,\dots,X_k\}$ for Δ on an open subset U of M, there are C^∞-functions $c^r_{ij} : U \to \mathbf{R}$ such that

$$[X_i,X_j] = \sum_{r=1}^{k} c^r_{ij} X_r$$

for $i,j = 1,\dots,k$.

PROOF: If Δ is integrable, we know that $[X_i,X_j]$ belongs to Δ and hence may be written as a linear combination of $X_1,\dots,X_k$ for $i,j = 1,\dots,k$, with coefficients c^r_{ij} which are C^∞-functions on U by Corollaries 2.38 and 2.59.

Conversely, suppose that $\{X_1,\dots,X_k\}$ is a local basis for Δ on U satisfying the conditions of the proposition, and let X and Y be two vector fields on U belonging to Δ. Then there are C^∞-functions $a^i : U \to \mathbf{R}$ and $b^j : U \to \mathbf{R}$ such that

$$X = \sum_{i=1}^{k} a^i X_i \quad \text{and} \quad Y = \sum_{j=1}^{k} b^j X_j$$

so that

$$[X,Y] = \sum_{ij} [a^i X_i, b^j X_j] = \sum_{ij} \{a^i b^j \sum_r c^r_{ij} X_r + a^i X_i(b^j) X_j - b^j X_j(a^i) X_i\}$$

by Proposition 4.87. This shows that $[X,Y]$ also belongs to Δ so that Δ is integrable. □

4.97 Proposition Let Δ be a k-dimensional distribution on a smooth manifold M^n. Then the following three assertions are equivalent for a local chart (x,U) on M:

(1) The local trivialization $(t_x, \pi^{-1}(U))$ associated with (x,U) has the subbundle property $t_x(\pi^{-1}(U) \cap \Delta) = U \times \mathbf{R}^k \times \{0\} \subset U \times \mathbf{R}^k \times \mathbf{R}^{n-k}$.

(2) $\{\frac{\partial}{\partial x^1},\dots,\frac{\partial}{\partial x^k}\}$ is a local basis for Δ on U.

(3) For every $(a_0,b_0) \in x(U) \subset \mathbf{R}^k \times \mathbf{R}^{n-k}$, the set $x^{-1}(\mathbf{R}^k \times \{b_0\})$ is an integral manifold for Δ called a *slice* of U, and each slice is a submanifold of M.

PROOF : The equivalence of (1) and (2) follows from Remark 2.82.

To show the equivalence of (2) and (3), let $S = x^{-1}(\mathbf{R}^k \times \{b_0\})$ and let $t : \mathbf{R}^k \times \mathbf{R}^{n-k} \to \mathbf{R}^k \times \mathbf{R}^{n-k}$ be the translation defined by $t(a,b) = (a, b - b_0)$. Then the local chart (u,U), where $u = t \circ x$, has the submanifold property $u(U \cap S) = u(U) \cap (\mathbf{R}^k \times \{0\})$.

We have a local chart (y,S) on S with coordinate map $y = p \circ x|_S$, where $p : \mathbf{R}^k \times \mathbf{R}^{n-k} \to \mathbf{R}^k$ is the projection on the first factor. If $i : S \to M$ is the inclusion mapping, we have that

$$i_*\left(\left.\frac{\partial}{\partial y^j}\right|_q\right) = \sum_{l=1}^{n} \frac{\partial(x^l \circ i)}{\partial y^j}(q) \left.\frac{\partial}{\partial x^l}\right|_q = \left.\frac{\partial}{\partial x^j}\right|_q$$

for $q \in S$ and $j = 1, \ldots, k$. This shows that $\{ \left.\frac{\partial}{\partial x^1}\right|_q, \ldots, \left.\frac{\partial}{\partial x^k}\right|_q \}$ is a basis for $i_*(T_qS)$ and completes the proof that (2) and (3) are equivalent. □

4.98 Frobenius' Integrability Theorem (local version) Let Δ be a k-dimensional integrable distribution on a smooth manifold M^n. Then, for each point $p \in M$, there is a local chart (x,U) around p satisfying the assertions of Proposition 4.97.

PROOF : We will prove the theorem by induction on k. When $k = 1$, assertion (2) in Proposition 4.97 follows immediately from Proposition 4.93.

Now assume that the theorem is true for distributions of dimension $k-1$, and let $\{X_1, \ldots, X_k\}$ be a local basis for Δ on an open subset V of M containing p. By Proposition 4.93 there is a local chart (y,W) around p with $W \subset V$ such that

$$X_1 = \frac{\partial}{\partial y^1}$$

on W, and we may assume that $y(W) = I \times O$ for an open interval I and an open subset O of $\mathbf{R}^{n-1}$. We define a new local basis $\{Y_1, \ldots, Y_k\}$ for Δ on W by

$$Y_1 = X_1 ,$$
$$Y_j = X_j - X_j(y^1) X_1 \quad \text{for} \quad j = 2, \ldots, k .$$

Since Δ is integrable, there are C^∞-functions $c_{ij}^r : W \to \mathbf{R}$ such that

$$[Y_i, Y_j] = \sum_{r=1}^{k} c_{ij}^r Y_r \tag{1}$$

for $i, j = 1, \ldots, k$, and we have that

$$Y_j(y^1) = 0 \quad \text{for} \quad j = 2, \ldots, k .$$

By applying both sides of (1) to y^1, we see that this implies

$$c_{ij}^1 = 0 \quad \text{for} \quad i,j = 1,\ldots,k\,.$$

Now define $N = y^{-1}(\{t_0\} \times O)$, where $y(p) = (t_0, a_0) \in I \times O$, and let $i : N \to W$ be the inclusion map. By Proposition 3.14 there is for each $j = 1,\ldots,k-1$ a unique vector field Z_j on N which is i-related to Y_{j+1}, and it follows from (1) with $c_{ij}^1 = 0$ and Proposition 4.96 that the distribution Δ' on N with local basis $\{Z_1,\ldots,Z_{k-1}\}$ is integrable. By the induction hypothesis there is therefore a local chart (z, U') around p on N such that $\{\frac{\partial}{\partial z^1},\ldots,\frac{\partial}{\partial z^{k-1}}\}$ is a local basis for Δ' on U'.

Let $pr_2 : I \times O \to O$ be the projection on the second factor, and let $\rho : O \to \{t_0\} \times O$ be the diffeomorphism given by $\rho(a) = (t_0, a)$. Furthermore, let $\pi : W \to N$ be the submersion defined by $i \circ \pi = y^{-1} \circ \rho \circ pr_2 \circ y$.

Consider the local chart (x, U) around p on M with coordinate neighbourhood $U = \pi^{-1}(U')$ and coordinate map $x : U \to \mathbf{R} \times \mathbf{R}^{n-1}$ defined by $x = \{id \times (z \circ y^{-1} \circ \rho)\} \circ y$, i.e.,

$$x^j = \begin{cases} y^1 & \text{for} \quad j = 1 \\ z^{j-1} \circ \pi & \text{for} \quad j = 2,\ldots,n \end{cases}$$

so that $x \circ y^{-1}(t,a) = (t, z \circ y^{-1}(t_0, a))$. Then we have that $x(U) = I \times z(U')$, and

$$Y_1 = \frac{\partial}{\partial y^1} = \sum_{j=1}^{n} \frac{\partial x^j}{\partial y^1} \frac{\partial}{\partial x^j} = \frac{\partial}{\partial x^1} \quad \text{on } U. \tag{2}$$

We want to show that

$$Y_j(x^r) = 0 \quad \text{on } U \tag{3}$$

for $j = 1,\ldots,k$ and $r = k+1,\ldots,n$ which implies that $\{\frac{\partial}{\partial x^1},\ldots,\frac{\partial}{\partial x^k}\}$ is a local basis for Δ on U, thus completing the proof of the theorem. The case $j = 1$ follows immediately from (2), and by (1) with $c_{ij}^1 = 0$ and (2) we have that

$$\frac{\partial}{\partial x^1}(Y_j(x^r)) = Y_1(Y_j(x^r)) = [Y_1, Y_j](x^r) = \sum_{l=2}^{k} c_{1j}^l Y_l(x^r)$$

and

$$Y_j(x^r) \circ i = Z_{j-1}(x^r \circ i) = Z_{j-1}(z^{r-1}) = 0$$

for $j = 2,\ldots,k$ and $r = k+1,\ldots,n$.

Hence for each $a \in z(U')$ and $r = k+1,\ldots,n$, the curve $\gamma : I \to \mathbf{R}^{k-1}$ defined by

$$\gamma_j(t) = Y_{j+1}(x^r) \circ x^{-1}(t,a)$$

for $j = 1,\ldots,k-1$, is an integral curve for the time-independent vector field $f : \mathbf{R}^{k-1} \to \mathbf{R}^{k-1}$ given by

$$f_j(u) = \sum_{l=1}^{k-1} c_{1\,j+1}^{l+1} u_l\,,$$

and we have that $\gamma(t_0) = 0$. The uniqueness part of Proposition 3.23 and Remark 3.35 therefore implies that γ is identically zero, and this completes the proof of (3). □

4.99 Proposition Let Δ be a k-dimensional distribution on a smooth manifold M^n, and let (x,U) be a local chart on M satisfying the assertions of Proposition 4.97. If N is a connected integral manifold of Δ contained in U, then N is an open submanifold of a slice of U.

PROOF : Let $i : N \to M$ be the inclusion map, and let $p : \mathbf{R}^k \times \mathbf{R}^{n-k} \to \mathbf{R}^{n-k}$ be the projection on the second factor. For every $q \in N$ and every smooth function g defined on some open neighbourhood of $(p \circ x)(q)$ in $\mathbf{R}^{n-k}$, we have that

$$(p \circ x)_* \left(\left. \frac{\partial}{\partial x^i} \right|_q \right)(g) = \left. \frac{\partial}{\partial x^i} \right|_q (g \circ p \circ x) = D_i(g \circ p)(x(q)) = 0$$

for $i = 1, \ldots, k$, which implies that $(p \circ x \circ i)_* = 0$. Hence it follows by Proposition 2.29 that $p \circ x \circ i$ is constant, thus showing that N is contained in a slice S of U.

Since S is a submanifold of M, it follows from Corollary 2.38 that the inclusion map $i' : N \to S$ is smooth. As N and S are both integral manifolds of Δ, we have that i'_{*q} is bijective for each $q \in N$ which shows that N is an open submanifold of S. □

4.100 Proposition Let Δ be a k-dimensional integrable distribution on a smooth manifold M^n, and let N_1 and N_2 be two integral manifolds of Δ passing through a point q on M. Then there is an integral manifold N of Δ passing through q which is an open submanifold of both N_1 and N_2.

PROOF : Let (x,U) be a local chart around q on M satisfying the assertions of Proposition 4.97. Since the inclusion map $i_r : N_r \to M$ is continuous, $N_r \cap U$ is an open subset of N_r for $r = 1, 2$.

Let C_r be the connected component of $N_r \cap U$ containing q (in the topology indused by N_r). Since N_r is locally connected, C_r is an open submanifold of N_r and hence is a connected integral manifold of Δ contained in U. By Proposition 4.99, both C_1 and C_2 are therefore open submanifolds of the slice S of U passing through q. Hence $N = C_1 \cap C_2$ is an integral manifold of Δ satisfying the assertions in the proposition. □

4.101 Corollary Let Δ be a k-dimensional integrable distribution on a smooth manifold M^n, and let N_1 and N_2 be two integral manifolds of Δ. Then $N_1 \cap N_2$ is open in both N_1 and N_2, and N_1 and N_2 induce the same topology on $N_1 \cap N_2$.

PROOF : The first assertion follows immediately from the proposition, since for each $q \in N_1 \cap N_2$ there is a set N with $q \in N \subset N_1 \cap N_2$ which is open in both N_1 and N_2.

To prove the last assertion, let $\mathscr{T}_1$ and $\mathscr{T}_2$ be the topologies on $N_1 \cap N_2$ induced by N_1 and N_2, respectively. If $O \in \mathscr{T}_1$, then O is an open submanifold of N_1 and hence is an integral manifold of Δ. By the first part of the corollary, we therefore have that $O = O \cap N_2$ is open in N_2. Hence $\mathscr{T}_1 \subset \mathscr{T}_2$, and the opposite inclusion follows in the same way. □

4.102 Proposition Let N be an integral manifold of a k-dimensional integrable distribution Δ on M^n, and let $f : P \to M$ be a smooth map with $f(P) \subset N$. Then f is also smooth as a map into N.

PROOF : Let $q \in P$, and choose a local chart (x,U) around $f(q)$ on M satisfying the assertions of Proposition 4.97. Let C_1 and C_2 be the connected components of $N \cap U$ containing $f(q)$ in the topologies induced by N and M, respectively. Since the inclusion map $i : N \to M$ is continuous, we have that $C_1 \subset C_2$. We will show that we in fact have $C_1 = C_2$.

Since N is second countable and locally connected, it follows from Corollary 2.5 that $N \cap U$ has only a countable number of connected components in the topology induced by N, each contained in a slice of U by Proposition 4.99. If $p : \mathbf{R}^k \times \mathbf{R}^{n-k} \to \mathbf{R}^{n-k}$ is the projection on the second factor, we therefore have that $p \circ x(C_2)$ is a countable connected subset of $\mathbf{R}^{n-k}$. Hence $p \circ x(C_2)$ must consist of a single point, showing that C_2 is contained in the slice S of U passing through $f(q)$. By Corollary 4.101 we know that N and M induce the same topology on $N \cap S$, which implies that $C_1 = C_2$.

Since P is locally connected, the connected component D of $f^{-1}(U)$ containing q is open in P, and we have that $f(D) \subset C_1$ which is an open submanifold of S by Proposition 4.99. Since S is a submanifold of M, it follows from Corollary 2.38 that $f|_D$ is smooth considered as a map into C_1, thus showing that f is smooth at q considered as a map into N. □

4.103 Proposition Let X be a topological space which is connected and locally connected, and suppose there is a countable open cover $\{U_i \,|\, i \in \mathbf{N}\}$ of X such that each connected component of each U_i is second countable. Then X is second countable.

PROOF : We assume that the space X is nonempty, otherwise the proposition is trivially true. Let $\mathscr{C}_i$ be the family of connected components in U_i, and let $\mathscr{C} = \bigcup_{i=1}^{\infty} \mathscr{C}_i$. Since X is locally connected, the sets in $\mathscr{C}_i$ are open and disjoint. Hence it follows from Corollary 2.5 that given a set $V \in \mathscr{C}$, there is only a contable number of sets $V' \in \mathscr{C}$ with $V \cap V' \neq \emptyset$.

We now inductively define a sequence of families $\mathscr{A}_i \subset \mathscr{C}$ as follows. Fix a set $U \in \mathscr{C}$, and let $\mathscr{A}_1 = \{U\}$. For each integer $i \geq 2$, let $\mathscr{A}_i = \{V \in \mathscr{C} | V \cap V' \neq \emptyset$ for some $V' \in \mathscr{A}_{i-1}\}$. It follows by induction that each family $\mathscr{A}_i$ is countable. Hence the set $Y = \bigcup_{i=1}^{\infty} \bigcup_{V \in \mathscr{A}_i} V$ is nonempty, open and second countable. We want to show that Y is also closed in X. Since X is connected, this implies that $X = Y$ which completes the proof that X is second countable.

Let $p \in \overline{Y}$, and choose a set $V \in \mathscr{C}$ containing p. Since $V \cap Y \neq \emptyset$, there is a positive integer i and a set $V' \in \mathscr{A}_i$ such that $V \cap V' \neq \emptyset$. Then $V \in \mathscr{A}_{i+1}$ which shows that $p \in Y$ and hence that Y is a closed subset of X. □

4.104 Frobenius' Integrability Theorem (global version) Let Δ be a k-dimensional integrable distribution on a smooth manifold M^n. Then, for each point $p \in M$, there is a unique maximal connected integral manifold N_p of Δ passing through p. Every connected integral manifold of Δ passing through p is an open submanifold of N_p.

PROOF : Let $\mathscr{T}$ be the family of all subsets of M which are unions of integral manifolds of Δ. Then $\mathscr{T}$ is a new topology on M which is finer than its original topology $\mathscr{T}_0$. Indeed, $\mathscr{T}$ is clearly closed under unions, and $\emptyset$ is obtained by taking a union with an empty indexing set. By induction using Proposition 4.100, we see that $\mathscr{T}$ is also closed under finite intersections, and it contains M by Theorem 4.98.

To see that $\mathscr{T}$ is finer than $\mathscr{T}_0$, let U be an arbitrary open set in $\mathscr{T}_0$. For each point $q \in U$, there is an integral manifold N of Δ passing through q by Theorem 4.98. Since the inclusion map $i : N \to M$ is continuous, $N \cap U$ is an open submanifold of N. Hence $N \cap U$ is an integral manifold of Δ passing through q and contained in U, showing that U is open in $\mathscr{T}$ which is therefore finer than $\mathscr{T}_0$. In particular, it follows that $\mathscr{T}$ is a Hausdorff topology.

We next show that each integral manifold N of Δ is a subspace of $(M, \mathscr{T})$. Each open set U in N is an integral manifold of Δ and hence belongs to $\mathscr{T}$. Conversely, suppose that U is a subset of N belonging to $\mathscr{T}$, and let q be an arbitrary point in U. Then, by the definition of $\mathscr{T}$, there is an integral manifold N_1 of Δ passing through q and contained in U. By Proposition 4.100 there is an integral manifold N_2 of Δ passing through q which is an open submanifold of both N and N_1, and this shows U is open in N.

Since the integral manifolds of Δ form an open cover of the Hausdorff space $(M, \mathscr{T})$ consisting of subspaces which are locally Euclidean, it follows that $(M, \mathscr{T})$ is also locally Euclidean. The family $\mathscr{A}$ consisting of all local charts on every integral manifold of Δ, is an atlas on $(M, \mathscr{T})$. To see that $y \circ x^{-1}$ is C^∞ for any pair of local charts (x, U) and (y, V) in $\mathscr{A}$, let q be a point in $U \cap V$ and suppose that (x, U) and (y, V) are local charts on the integral manifolds N_1 and N_2, respectively. By Proposition 4.100 there is an integral manifold N of Δ passing through q which is an open submanifold of both N_1 and N_2, and we choose a local chart (z, W) around q on N. Then $y \circ x^{-1}$ is C^∞ at $x(q)$, since $y \circ x^{-1}|_{x(U \cap V \cap W)}$ is the composition of the C^∞ coordinate transformations $z \circ x^{-1}$ and $y \circ z^{-1}$. Hence $y \circ x^{-1}$ is C^∞ at every point in $x(U \cap V)$, thus showing that $\mathscr{A}$ is an atlas on M. We let $\mathscr{D}$ be the unique differetiable structure on M containing $\mathscr{A}$.

Now let $\{N_\alpha \mid \alpha \in A\}$ be the family of connected components in $(M, \mathscr{T})$. Since $(M, \mathscr{T})$ is locally connected, each N_α is open in $\mathscr{T}$, and $\mathscr{D}|_{N_\alpha} = \{(x, U) \in \mathscr{D} | U \subset N_\alpha\}$ is a smooth structure on N_α. We want to show that each N_α is a maximal connected integral manifold of Δ, which will complete the proof of the theorem. It only remains to show that N_α is second countable.

By Propositions 4.100 and 2.6 there is a countable atlas $\{(x_i, U_i) \mid i \in \mathbf{N}\}$ on $(M, \mathscr{T}_0)$ consisting of local charts satisfying the assertions of Proposition 4.97. For every positive integer i, it follows by Proposition 4.99 that each connected component of $N_\alpha \cap U_i$ (in the topology induced by N_α) is an open submanifold of a slice of U_i and is therefore second countable. Hence it follows from Proposition 4.103 that N_α is second countable. □

4.105 Remark The partition $\{N_\alpha \mid \alpha \in A\}$ of M into maximal connected integral manifolds of Δ is called a *foliation* of M, and each maximal connected integral manifold of Δ is called a *folium* or *leaf* of the foliation.

4.106 Proposition Let N be a maximal connected integral manifold of a k-dimensional integrable distribution Δ on M^n, and let (x,U) be a local chart on M satisfying the assertions of Proposition 4.97 with $x(U) = O_1 \times O_2$ for connected open sets $O_1 \subset \mathbf{R}^k$ and $O_2 \subset \mathbf{R}^{n-k}$. Then the connected components of $N \cap U$ (in the topology induced by N) are slices of U.

PROOF: By Proposition 4.99 it follows that a connected component C of $N \cap U$ is an open submanifold of a slice S of U. Since S is a connected integral manifold of Δ, it follows from Theorem 4.104 that $S \subset N \cap U$ so that $S = C$. □

4.107 Proposition Let $f : M_1 \to M_2$ be a diffeomorphism, and let Δ_1 and Δ_2 be distributions on M_1 and M_2, respectively, such that

$$f_*(\Delta_{1p}) = \Delta_{2f(p)}$$

for every $p \in M_1$. Suppose that N_1 is an integral manifold of Δ_1 and that $N_2 = f(N_1)$ with inclusion maps $i_1 : N_1 \to M_1$ and $i_2 : N_2 \to M_2$, and let $g : N_1 \to N_2$ be the bijective map induced by f making the diagram

$$\begin{array}{ccc} M_1 & \xrightarrow{\;f\;} & M_2 \\ {\scriptstyle i_1}\uparrow & & \uparrow{\scriptstyle i_2} \\ N_1 & \xrightarrow{\;g\;} & N_2 \end{array}$$

commutative. If N_2 is endowed with the manifold structure making g a diffeomorphism (see Remark 2.31), then N_2 is an integral manifold of Δ_2.

PROOF: We have that N_2 is an immersed submanifold of M_2 such that

$$i_{2*}(T_{f(p)}N_2) = i_{2*} \circ g_*(T_pN_1) = f_* \circ i_{1*}(T_pN_1) = f_*(\Delta_{1p}) = \Delta_{2f(p)}$$

for each $p \in N_1$. □

4.108 Proposition Let Δ_1 and Δ_2 be integrable distributions of dimensions k and $n-k$ on a smooth manifold M^n so that

$$T_pM = \Delta_{1p} \oplus \Delta_{2p}$$

for every $p \in M$. Then, for each point $p \in M$, there is a local chart (x,U) around p such that $\{\frac{\partial}{\partial x^1}, \dots, \frac{\partial}{\partial x^k}\}$ and $\{\frac{\partial}{\partial x^{k+1}}, \dots, \frac{\partial}{\partial x^n}\}$ are local bases for Δ_1 and Δ_2, respectively, on U.

PROOF : By Frobenius' integrability theorem there are local charts (y,V) and (z,W) around p such that $\{\frac{\partial}{\partial y^1},\ldots,\frac{\partial}{\partial y^k}\}$ and $\{\frac{\partial}{\partial z^{k+1}},\ldots,\frac{\partial}{\partial z^n}\}$ are local bases for Δ_1 and Δ_2 on V and W, respectively.

Since $T_pM = \Delta_{1\,p} \oplus \Delta_{2\,p}$, we have that $\{\frac{\partial}{\partial y^1}\Big|_p,\ldots,\frac{\partial}{\partial y^k}\Big|_p,\frac{\partial}{\partial z^{k+1}}\Big|_p,\ldots,\frac{\partial}{\partial z^n}\Big|_p\}$ is a basis for T_pM. Hence there are real numbers a_{rj} such that

$$\frac{\partial}{\partial y^j}\bigg|_p = \sum_{r=1}^{k} a_{rj}\,\frac{\partial}{\partial y^r}\bigg|_p + \sum_{r=k+1}^{n} a_{rj}\,\frac{\partial}{\partial z^r}\bigg|_p$$

for $j=k+1,\ldots,n$. By applying both sides of this formula to the coordinate functions y^i where $i=k+1,\ldots,n$, we see that

$$\delta_{ij} = \sum_{r=k+1}^{n} a_{rj}\,\frac{\partial y^i}{\partial z^r}(p)$$

for $i,j=k+1,\ldots,n$, which shows that the matrix $\left(\frac{\partial y^i}{\partial z^j}(p)\right)$, where $k+1 \le i,j \le n$, is non-singular. By the inverse function theorem there is therefore a local chart (x,U) around p on M with $U \subset V \cap W$ and coordinate functions

$$x^i = \begin{cases} z^i|_U & \text{for} \quad i=1,\ldots,k \\ y^i|_U & \text{for} \quad i=k+1,\ldots,n \end{cases} .$$

Since

$$\frac{\partial}{\partial y^j}\bigg|_U = \sum_{i=1}^{k} \frac{\partial z^i}{\partial y^j}\,\frac{\partial}{\partial x^i}$$

for $j=1,\ldots,k$ and

$$\frac{\partial}{\partial z^j}\bigg|_U = \sum_{i=k+1}^{n} \frac{\partial y^i}{\partial z^j}\,\frac{\partial}{\partial x^i}$$

for $j=k+1,\ldots,n$, we see the local chart (x,U) has the desired properties. □

Chapter 5

DIFFERENTIAL FORMS

EXTERIOR FORMS ON A VECTOR SPACE

5.1 Definition Let V be a vector space, and let V^k denote the cartesian product $V \times \cdots \times V$ with k factors. A multilinear map $T : V^k \to W$ into a vector space W is said to be *skew symmetric* if it changes sign whenever any pair of variables is interchanged, i.e., if

$$T(v_1, ..., v_i, ..., v_j, ..., v_k) = -T(v_1, ..., v_j, ..., v_i, ..., v_k)$$

for all $v_1, ..., v_k \in V$.

5.2 Remark A multilinear map $T : V^k \to W$ is skew symmetric if and only if it is *alternating*, which means that $T(v_1, ..., v_k) = 0$ whenever $v_i = v_j$ for some $i \neq j$. Indeed, if T is alternating, we have that

$$\begin{aligned} 0 &= T(v_1, ..., v_i + v_j, ..., v_i + v_j, ..., v_k) \\ &= T(v_1, ..., v_i, ..., v_i, ..., v_k) + T(v_1, ..., v_i, ..., v_j, ..., v_k) \\ &\quad + T(v_1, ..., v_j, ..., v_i, ..., v_k) + T(v_1, ..., v_j, ..., v_j, ..., v_k) \\ &= 0 + T(v_1, ..., v_i, ..., v_j, ..., v_k) + T(v_1, ..., v_j, ..., v_i, ..., v_k) + 0 \end{aligned}$$

showing that T is skew symmetric. Conversely, if T is skew symmetric, we have that $T(v_1, ..., v, ..., v, ..., v_k)$ is equal to its own additive inverse and hence must be 0.

5.3 Definition The skew symmetric tensors in $\mathscr{T}^k(V)$ are called *exterior forms of degree k* or simply *exterior k-forms* on V. They form a subspace of $\mathscr{T}^k(V)$ which we denote by $\Lambda^k(V)$. We let $\Lambda^0(V) = \mathbf{R}$.

More generally, we let $\mathscr{T}^k(V;W)$ be the space $L^k(V, ..., V; W)$ of all multilinear maps $F : V^k \to W$ into a vector space W, and $\Lambda^k(V;W)$ be the subspace consisting of all skew symmetric multilinear maps into W. The elements in $\Lambda^k(V;W)$ are called *vector-valued exterior k-forms* on V with values in W. We let $\Lambda^0(V;W) = W$ and $\mathscr{T}^k_l(V, W)$ denote the space

$$\mathscr{T}(\underbrace{V, ..., V}_{k}, \underbrace{W^*, ..., W^*}_{l}) .$$

If $\mathscr{B}$ is a basis for V, and $\mathscr{C}$ is a basis for W with dual basis $\mathscr{C}^*$, then the basis

$$\mathscr{T}(\underbrace{\mathscr{B}, \dots, \mathscr{B}}_{k}, \underbrace{\mathscr{C}^*, \dots, \mathscr{C}^*}_{l})$$

for $\mathscr{T}^k_l(V,W)$ defined in Remark 4.27 is denoted by $\mathscr{T}^k_l(\mathscr{B},\mathscr{C})$.

If $G: U \to V$ is a linear map, then $G^* : \mathscr{T}^k(V) \to \mathscr{T}^k(U)$ maps $\Lambda^k(V)$ into $\Lambda^k(U)$, and we also use the notation G^* for the induced map $G^* : \Lambda^k(V) \to \Lambda^k(U)$. More generally, we have a linear map $G^* : \mathscr{T}^k(V;W) \to \mathscr{T}^k(U;W)$ defined by

$$G^*(F) = F \circ G^k ,$$

i.e.,

$$G^*(F)\,(v_1, \dots, v_k) = F\,(G(v_1), \dots, G(v_k))$$

for $v_1, \dots, v_k \in U$, which maps $\Lambda^k(V;W)$ into $\Lambda^k(U;W)$. We also use the notation G^* for the induced map $G^* : \Lambda^k(V;W) \to \Lambda^k(U;W)$, and we have that

$$\lambda \circ G^*(F) = G^*(\lambda \circ F)$$

for every functional $\lambda \in W^*$.

5.4 Proposition A multilinear map $T : V^k \to W$ is skew symmetric if and only if

$$T^{\sigma} = \varepsilon(\sigma)\, T \tag{1}$$

for every permutation $\sigma \in S_k$, where $\varepsilon(\sigma)$ denotes the sign of σ.

PROOF : By Definition 5.1 we know that T is skew symmetric if and only if

$$T^{\tau} = \varepsilon(\tau)\, T$$

for every transposition τ, and this is certainly the case if (1) is satisfied for every permutation $\sigma \in S_k$.

Suppose conversely that T is skew symmetric, and let σ be an arbitrary permutation in S_k. We know by Proposition 13.7 in the appendix that σ may be expressed as a product of transpositions

$$\sigma = \tau_1 \cdots \tau_m .$$

Then (1) follows by induction on the number m of factors in σ. Assuming it is true for $\sigma' = \tau_1 \cdots \tau_{m-1}$, we have that

$$T^{\sigma} = T^{\sigma' \tau_m} = (T^{\tau_m})^{\sigma'} = \varepsilon(\tau_m)\, T^{\sigma'} = \varepsilon(\tau_m)\, \varepsilon(\sigma')\, T = \varepsilon(\sigma)\, T$$

which completes the proof of the proposition. □

5.5 Corollary A tensor $T \in \mathscr{T}^k(V)$ belongs to $\Lambda^k(V)$ if and only if

$$T^{\sigma} = \varepsilon(\sigma)T \tag{1}$$

for every permutation $\sigma \in S_k$.

5.6 Now consider the linear map $\mathscr{A} : \mathscr{T}^k(V;W) \to \mathscr{T}^k(V;W)$ defined by

$$\mathscr{A}(T) = \frac{1}{k!}\sum_{\sigma \in S_k} \varepsilon(\sigma)T^{\sigma}$$

i.e.,

$$\mathscr{A}(T)(v_1,...,v_k) = \frac{1}{k!}\sum_{\sigma \in S_k} \varepsilon(\sigma)T(v_{\sigma(1)},...,v_{\sigma(k)})$$

which is called the *alternation* operator.

5.7 Proposition If $T \in \mathscr{T}^k(V;W)$ and $\sigma \in S_k$, then $\mathscr{A}(T^{\sigma}) = \mathscr{A}(T)^{\sigma} = \varepsilon(\sigma)\mathscr{A}(T)$.

PROOF : We have that

$$\mathscr{A}(T^{\sigma}) = \frac{1}{k!}\sum_{\tau \in S_k} \varepsilon(\tau)(T^{\sigma})^{\tau} = \varepsilon(\sigma)\frac{1}{k!}\sum_{\tau \in S_k} \varepsilon(\tau\sigma)T^{\tau\sigma} = \varepsilon(\sigma)\mathscr{A}(T)$$

and

$$\mathscr{A}(T)^{\sigma} = \frac{1}{k!}\sum_{\tau \in S_k} \varepsilon(\tau)(T^{\tau})^{\sigma} = \varepsilon(\sigma)\frac{1}{k!}\sum_{\tau \in S_k} \varepsilon(\sigma\tau)T^{\sigma\tau} = \varepsilon(\sigma)\mathscr{A}(T)$$

since the maps $R_{\sigma} : S_k \to S_k$ and $L_{\sigma} : S_k \to S_k$ given by $R_{\sigma}(\tau) = \tau\sigma$ and $L_{\sigma}(\tau) = \sigma\tau$ are bijections and ε is a homomorphism. □

5.8 Proposition

(1) If $T \in \mathscr{T}^k(V;W)$, then $\mathscr{A}(T) \in \Lambda^k(V;W)$.

(2) If $\omega \in \Lambda^k(V;W)$, then $\mathscr{A}(\omega) = \omega$.

(3) If $T \in \mathscr{T}^k(V;W)$, then $\mathscr{A}(\mathscr{A}(T)) = \mathscr{A}(T)$.

PROOF: (1) Follows from Proposition 5.4 and 5.7.

(2) If $\omega \in \Lambda^k(V;W)$, then

$$\mathscr{A}(\omega) = \frac{1}{k!}\sum_{\sigma \in S_k} \varepsilon(\sigma)\omega^{\sigma} = \frac{1}{k!}\sum_{\sigma \in S_k} \varepsilon(\sigma)\varepsilon(\sigma)\omega = \omega$$

by Proposition 5.4.

(3) Follows from (1) and (2). □

5.9 Proposition If $F: U \to V$ is a linear map, and if $\mathscr{A}_U$ and $\mathscr{A}_V$ are the alternations on $\mathscr{T}^k(U;W)$ and $\mathscr{T}^k(V;W)$, respectively, then

$$F^* \circ \mathscr{A}_V = \mathscr{A}_U \circ F^* .$$

PROOF: Since $F^*(T^\sigma) = F^*(T)^\sigma$ for every $T \in \mathscr{T}^k(V;W)$ and $\sigma \in S_k$ by 13.9 in the appendix, the proposition follows from the linearity of $F^* : \mathscr{T}^k(V;W) \to \mathscr{T}^k(U;W)$. □

5.10 Proposition If $T \in \mathscr{T}^k(V)$ and $S \in \mathscr{T}^l(V)$, then

(1) $\mathscr{A}(T \otimes S) = (-1)^{kl} \mathscr{A}(S \otimes T)$,

(2) $\mathscr{A}(\mathscr{A}(T) \otimes S) = \mathscr{A}(T \otimes S) = \mathscr{A}(T \otimes \mathscr{A}(S))$.

PROOF: (1) Let $\tau \in S_{k+l}$ be the permutation defined by

$$\tau(r) = \begin{cases} r+k & \text{for} \quad 1 \le r \le l \\ r-l & \text{for} \quad l+1 \le r \le l+k \end{cases} .$$

Since τ may be obtained by moving each of the last k numbers $\{l+1, ..., l+k\}$ in succession l positions forward so that they become the first k numbers, we see that τ may be written as a product of kl transpositions and therefore has sign $\varepsilon(\tau) = (-1)^{kl}$. We now have that

$$\begin{aligned} T \otimes S(v_1, ..., v_k, v_{k+1}, ..., v_{k+l}) &= S \otimes T(v_{k+1}, ..., v_{k+l}, v_1, ..., v_k) \\ &= S \otimes T(v_{\tau(1)}, ..., v_{\tau(l)}, v_{\tau(l+1)}, ..., v_{\tau(l+k)}) . \end{aligned}$$

Hence $T \otimes S = (S \otimes T)^\tau$ which by Proposition 5.7 implies that

$$\mathscr{A}(T \otimes S) = (-1)^{kl} \mathscr{A}(S \otimes T) .$$

(2) We have an injective homomorphism $\phi : S_k \to S_{k+l}$ given by

$$\phi(\sigma)(r) = \begin{cases} \sigma(r) & \text{for} \quad 1 \le r \le k \\ r & \text{for} \quad k+1 \le r \le k+l \end{cases} ,$$

i.e., $\phi(\sigma)$ acts as σ on $\{1, ..., k\}$ and leaves $\{k+1, ..., k+l\}$ fixed. Then we have that $T^\sigma \otimes S = (T \otimes S)^{\phi(\sigma)}$ and $\varepsilon(\phi(\sigma)) = \varepsilon(\sigma)$ since ϕ maps transpositions in S_k to transpositions in S_{k+l} and hence is sign preserving. By Proposition 5.7 we have that $\mathscr{A}(T^\sigma \otimes S) = \varepsilon(\sigma) \mathscr{A}(T \otimes S)$, and from this it follows that

$$\mathscr{A}(\mathscr{A}(T) \otimes S) = \frac{1}{k!} \sum_{\sigma \in S_k} \varepsilon(\sigma) \mathscr{A}(T^\sigma \otimes S) = \frac{1}{k!} \sum_{\sigma \in S_k} \mathscr{A}(T \otimes S) = \mathscr{A}(T \otimes S)$$

since $\mathscr{A}$ is linear and $\otimes$ is bilinear.

The last equality in (2) is proved in the same way using the injective homomorphism $\psi : S_l \to S_{k+l}$ given by

$$\psi(\sigma)(r) = \begin{cases} r & \text{for } \ 1 \le r \le k \\ \sigma(r-k)+k & \text{for } \ k+1 \le r \le k+l \end{cases} .$$

Alternatively the last equality follows from the first by (1) as follows

$$\mathscr{A}(T \otimes \mathscr{A}(S)) = (-1)^{kl} \mathscr{A}(\mathscr{A}(S) \otimes T) = (-1)^{kl} \mathscr{A}(S \otimes T) = \mathscr{A}(T \otimes S) .$$

□

5.11 Definition If $\omega \in \Lambda^k(V)$ and $\eta \in \Lambda^l(V)$, we define their *wedge product* or *exterior product* $\omega \wedge \eta \in \Lambda^{k+l}(V)$ by

$$\omega \wedge \eta = \frac{(k+l)!}{k!\, l!} \mathscr{A}(\omega \otimes \eta) .$$

5.12 Proposition

(1) $\wedge$ is bilinear.

(2) If $F : V \to W$ is a linear map, then $F^*(\omega \wedge \eta) = F^*(\omega) \wedge F^*(\eta)$ for every exterior form ω and η on W.

(3) If $\omega \in \Lambda^k(V)$ and $\eta \in \Lambda^l(V)$, then $\omega \wedge \eta = (-1)^{kl} \eta \wedge \omega$.

(4) If $\omega \in \Lambda^k(V)$, $\eta \in \Lambda^l(V)$ and $\theta \in \Lambda^m(V)$, then

$$(\omega \wedge \eta) \wedge \theta = \omega \wedge (\eta \wedge \theta) = \frac{(k+l+m)!}{k!\, l!\, m!} \mathscr{A}(\omega \otimes \eta \otimes \theta) .$$

PROOF: (1) Follows from the fact that $\otimes$ is bilinear and the alternation $\mathscr{A}$ is linear.

(2) Follows from the corresponding relation for $\otimes$ and Proposition 5.9.

(3) Follows from Proposition 5.10 (1).

(4) By Proposition 5.10 (2) we have that

$$(\omega \wedge \eta) \wedge \theta = \frac{(k+l+m)!}{(k+l)!\, m!} \mathscr{A}\left(\frac{(k+l)!}{k!\, l!} \mathscr{A}(\omega \otimes \eta) \otimes \theta\right) = \frac{(k+l+m)!}{k!\, l!\, m!} \mathscr{A}(\omega \otimes \eta \otimes \theta) ,$$

and a similar calculation gives

$$\omega \wedge (\eta \wedge \theta) = \frac{(k+l+m)!}{k!\, l!\, m!} \mathscr{A}(\omega \otimes \eta \otimes \theta) .$$

□

5.13 Proposition

(1) If $\omega_1, \dots, \omega_k \in V^*$, then

$$\omega_1 \wedge \cdots \wedge \omega_k = k!\, \mathscr{A}\,(\omega_1 \otimes \cdots \otimes \omega_k) = \sum_{\sigma \in S_k} \varepsilon(\sigma)\, \omega_{\sigma(1)} \otimes \cdots \otimes \omega_{\sigma(k)}\,.$$

(2) The map $\phi : V^{*k} \to \Lambda^k(V)$ defined by $\phi(\omega_1, \dots, \omega_k) = \omega_1 \wedge \cdots \wedge \omega_k$ is multilinear and alternating.

PROOF: (1) The first equality is proved by induction on k. It is clearly true when $k = 1$. Assuming it is true for $k-1$, we have that

$$\omega_1 \wedge \cdots \wedge \omega_k = \frac{k!}{(k-1)!\,1!}\, \mathscr{A}\,((\omega_1 \wedge \cdots \wedge \omega_{k-1}) \otimes \omega_k) = k!\, \mathscr{A}\,(\omega_1 \otimes \cdots \otimes \omega_k)$$

by Proposition 5.10 (2).

The last equality follows from the definition of $\mathscr{A}$ using that

$$(*) \qquad \omega_{\sigma(1)} \otimes \cdots \otimes \omega_{\sigma(k)} = (\,\omega_1 \otimes \cdots \otimes \omega_k\,)^{\sigma^{-1}}$$

and $\varepsilon(\sigma) = \varepsilon(\sigma^{-1})$ for $\sigma \in S_k$, and that the map $I : S_k \to S_k$ given by $I(\sigma) = \sigma^{-1}$ is a bijection.

(2) ϕ is multilinear by Proposition 5.12 (1). By applying $\mathscr{A}$ to both sides of $(*)$ and using Proposition 5.7 and the first equality in (1) we have that

$$\omega_{\sigma(1)} \wedge \cdots \wedge \omega_{\sigma(k)} = \varepsilon(\sigma)\, \omega_1 \wedge \cdots \wedge \omega_k\,.$$

Hence $\phi^{\sigma} = \varepsilon(\sigma)\, \phi$ which shows that ϕ is alternating. □

5.14 Proposition If $\mathscr{B} = \{v_1, \dots, v_n\}$ is a basis for the vector space V, then

$$\mathscr{D} = \{v_{i_1}^* \wedge \cdots \wedge v_{i_k}^* \mid 1 \le i_1 < \dots < i_k \le n\}$$

is a basis for $\Lambda^k(V)$ when $k > 0$. If $\omega \in \Lambda^k(V)$ we have that

$$\omega = \sum_{i_1 < \dots < i_k} \omega(v_{i_1}, \dots, v_{i_k})\, v_{i_1}^* \wedge \cdots \wedge v_{i_k}^*\,.$$

In particular $\dim \Lambda^k(V) = \binom{n}{k}$ when $0 \le k \le n$, and $\Lambda^k(V) = \{0\}$ for $k > n$.

PROOF : By Proposition 4.26 we have that

$$\omega = \sum_{i_1, \dots, i_k} \omega(v_{i_1}, \dots, v_{i_k})\, v_{i_1}^* \otimes \cdots \otimes v_{i_k}^*\,,$$

and by applying $\mathscr{A}$ to both sides it follows by Proposition 5.8 (2) and 5.13 (1) that

$$\omega = \frac{1}{k!} \sum_{i_1, \dots, i_k} \omega(v_{i_1}, \dots, v_{i_k})\, v_{i_1}^* \wedge \cdots \wedge v_{i_k}^*\,.$$

As ω is alternating, we have that the coefficient $\omega(v_{i_1},...,v_{i_k})$ is 0 if the indices $i_1,...,i_k$ are not all different. If they are different and $\sigma \in S_k$, we have that

$$\omega(v_{i_1},...,v_{i_k})\,v^*_{i_1} \wedge \cdots \wedge v^*_{i_k} = \omega(v_{i_{\sigma(1)}},...,v_{i_{\sigma(k)}})\,v^*_{i_{\sigma(1)}} \wedge \cdots \wedge v^*_{i_{\sigma(k)}}$$

since both ω and the map ϕ from Proposition 5.13 (2) are alternating. Since there are $k!$ such rearrangements of the indices $i_1,...,i_k$, we have that

$$\omega = \sum_{i_1<...<i_k} \omega(v_{i_1},...,v_{i_k})\,v^*_{i_1} \wedge \cdots \wedge v^*_{i_k}$$

which shows the last assertion of the proposition and also that $\mathscr{D}$ spans $\Lambda^k(V)$. To show that it is linearly independent, assume that

$$\sum_{i_1<...<i_k} a_{i_1,...,i_k}\,v^*_{i_1} \wedge \cdots \wedge v^*_{i_k} = 0\,.$$

By applying both sides to $(v_{j_1},...,v_{j_k})$ where $1 \le j_1 < ... < j_k \le n$, we see that $a_{j_1,...,j_k} = 0$. □

5.15 Let V and W be vector spaces, and let $\mathscr{C} = \{w_1,...,w_m\}$ be a basis for W with dual basis $\mathscr{C}^* = \{w^*_1,...,w^*_m\}$. Then we have a linear isomorphism

$$\phi : \mathscr{T}^k(V;W) \to \mathscr{T}^k_1(V,W)$$

defined by

$$\phi(F)(v_1,...,v_k,\lambda) = \lambda(F(v_1,...,v_k))$$

for $F \in \mathscr{T}^k(V;W)$, $v_1,...,v_k \in V$ and $\lambda \in W^*$, with inverse

$$\psi : \mathscr{T}^k_1(V,W) \to \mathscr{T}^k(V;W)$$

given by

$$\psi(T)(v_1,...,v_k) = \sum_{i=1}^{m} T(v_1,...,v_k,w^*_i)\,w_i$$

for $T \in \mathscr{T}^k_1(V,W)$ and $v_1,...,v_k \in V$. Indeed, we have that

$$\phi(\psi(T))(v_1,...,v_k,w^*_j) = w^*_j(\psi(T)(v_1,...,v_k)) = T(v_1,...,v_k,w^*_j)$$

for $j = 1,...,m$ and

$$\psi(\phi(F))(v_1,...,v_k) = \sum_{i=1}^{m} \phi(F)(v_1,...,v_k,w^*_i)\,w_i$$

$$= \sum_{i=1}^{m} w^*_i(F(v_1,...,v_k))\,w_i = F(v_1,...,v_k)$$

which shows that $\psi = \phi^{-1}$. In particular, it follows that ψ does not depend on the choice of the basis $\mathscr{C}$.

If the space V is also finite dimensional, then it follows from Remarks 4.5 and 4.31 that we have a linear isomorphism

$$\mu_* : \mathscr{T}^k(V) \otimes W \to \mathscr{T}^k_1(V,W)$$

induced by the bilinear map

$$\mu : \mathscr{T}^k(V) \times W \to \mathscr{T}^k_1(V,W)$$

defined by

$$\mu\,(T,w) = T \otimes w$$

for $T \in \mathscr{T}^k(V)$ and $w \in W$, i.e.,

$$\mu\,(T,w)(v_1,...,v_k,\lambda) = T(v_1,...,v_k)\,\lambda\,(w)$$

for $v_1,...,v_k \in V$ and $\lambda \in W^*$. The composed map

$$\psi \circ \mu : \mathscr{T}^k(V) \times W \to \mathscr{T}^k(V;W)$$

is given by

$$\psi \circ \mu\,(T,w)(v_1,...,v_k) = T(v_1,...,v_k)\,w$$

for $T \in \mathscr{T}^k(V)$, $w \in W$ and $v_1,...,v_k \in V$.

Hence if $\mathscr{A}$ is the alternation operator, we have that

$$\psi \circ \mu_* \circ (\mathscr{A}, id)_* = \mathscr{A} \circ \psi \circ \mu_*$$

since

$$\psi \circ \mu_* \circ (\mathscr{A}, id)_*\,(T \otimes w)(v_1,...,v_k) = \psi \circ \mu\,(\mathscr{A}\,(T), w)(v_1,...,v_k)$$

$$= \mathscr{A}\,(T)(v_1,...,v_k)\,w = \frac{1}{k!}\sum_{\sigma \in S_k} \varepsilon(\sigma)\,T(v_{\sigma(1)},...,v_{\sigma(k)})\,w$$

$$= \frac{1}{k!}\sum_{\sigma \in S_k} \varepsilon(\sigma)\,\psi \circ \mu\,(T,w)(v_{\sigma(1)},...,v_{\sigma(k)})$$

$$= \mathscr{A}\,(\psi \circ \mu\,(T,w))(v_1,...,v_k) = \mathscr{A} \circ \psi \circ \mu_*\,(T \otimes w)(v_1,...,v_k)\,,$$

and so we have a commutative diagram

$$\begin{array}{ccc} \mathscr{T}^k(V)\otimes W & \xrightarrow{\ \psi\circ\mu_*\ } & \mathscr{T}^k(V;W) \\ \Big\downarrow (\mathscr{A},id)_* & & \Big\downarrow \mathscr{A} \\ \mathscr{T}^k(V)\otimes W & \xrightarrow[\ \psi\circ\mu_*\ ]{} & \mathscr{T}^k(V;W) \end{array}$$

By Proposition 5.8 it follows that the linear isomorphism $\psi \circ \mu_*$ maps the subspace $\Lambda^k(V) \otimes W$ of $\mathscr{T}^k(V) \otimes W$ onto $\Lambda^k(V;W)$. We let $\Lambda_1^k(V,W)$ denote the image $\mu_*(\Lambda^k(V) \otimes W)$ consisting of all tensors in $\mathscr{T}_1^k(V,W)$ which are skew symmetric in the first k variables.

5.16 Proposition If $\mathscr{B} = \{v_1, ..., v_n\}$ and $\mathscr{C} = \{w_1, ..., w_m\}$ are bases for the vector spaces V and W, respectively, then

$$\mathscr{D} = \{(v_{i_1}^* \wedge \cdots \wedge v_{i_k}^*) \otimes w_j \mid 1 \le i_1 < ... < i_k \le n, 1 \le j \le m\}$$

is a basis for $\Lambda_1^k(V,W)$ when $k > 0$. If $\omega \in \Lambda_1^k(V,W)$ we have that

$$\omega = \sum_{i_1 < ... < i_k, j} \omega(v_{i_1}, ..., v_{i_k}, w_j^*)\,(v_{i_1}^* \wedge \cdots \wedge v_{i_k}^*) \otimes w_j\,.$$

In particular $\dim \Lambda_1^k(V,W) = m\binom{n}{k}$ when $0 \le k \le n$, and $\Lambda_1^k(V,W) = \{0\}$ for $k > n$.

PROOF : By 5.15 and Proposition 5.14 it follows that $\mathscr{D}$ is a basis for $\Lambda_1^k(V,W)$ when $k > 0$, as well as the last part of the proposition. If

$$\omega = \sum_{i_1 < ... < i_k, j} a_{i_1, ..., i_k}^j (v_{i_1}^* \wedge \cdots \wedge v_{i_k}^*) \otimes w_j\,,$$

then

$$\omega(v_{r_1}, ..., v_{r_k}, w_s^*) = a_{r_1, ..., r_k}^s$$

for $1 \le r_1 < ... < r_k \le n$ and $1 \le j \le m$ which yields the formula for ω. □

5.17 Proposition Let $\{v_1, \ldots, v_n\}$ be a basis for the vector space V, and let $\omega \in \Lambda^n(V)$. If $w_j = \sum_{i=1}^n a_{ij} v_i$ for $j = 1, \ldots, n$, we have that

$$\omega(w_1, \ldots, w_n) = \det(a_{ij})\, \omega(v_1, \ldots, v_n)$$

PROOF : Let $\eta \in \mathscr{T}^n(\mathbf{R}^n)$ be the covariant tensor on $\mathbf{R}^n$ defined by

$$\eta\,((a_{11}, \ldots, a_{n1}), \ldots, (a_{1n}, \ldots, a_{nn})) = \omega\,(\sum_{i=1}^n a_{i1} v_i, \ldots, \sum_{i=1}^n a_{in} v_i)$$

Then $\eta \in \Lambda^n(\mathbf{R}^n)$ so that $\eta = c \det$ for some $c \in \mathbf{R}$, and we have that

$$c = \eta(e_1, \ldots, e_n) = \omega(v_1, \ldots, v_n) .$$

□

5.18 Corollary If $\{v_1, \ldots, v_n\}$ and $\{w_1, \ldots, w_n\}$ are two bases for a vector space V with $w_j = \sum_{i=1}^n a_{ij} v_i$ for $j = 1, \ldots, n$, then we have that

$$v_1^* \wedge \cdots \wedge v_n^* = \det(a_{ij})\, w_1^* \wedge \cdots \wedge w_n^*$$

PROOF: Using Propositions 5.14 and 5.17 we have that

$$v_1^* \wedge \cdots \wedge v_n^* = v_1^* \wedge \cdots \wedge v_n^* (w_1, \ldots w_n)\, w_1^* \wedge \cdots \wedge w_n^* = \det(a_{ij})\, w_1^* \wedge \cdots \wedge w_n^* .$$

□

DIFFERENTIAL FORMS ON A MANIFOLD

5.19 Now let $\pi : E \to M$ be an n-dimensional vector bundle over the smooth manifold M, and let $E_p = \pi^{-1}(p)$ be the fibre over p for each $p \in M$. We want to define a new bundle

$$\Lambda^k(E) = \bigcup_{p \in M} \Lambda^k(E_p)$$

of exterior k-forms on each fibre E_p. We have a projection $\pi' : \Lambda^k(E) \to M$ defined by $\pi'(\omega) = p$ for $\omega \in \Lambda^k(E_p)$.

5.20 Proposition $\pi' : \Lambda^k(E) \to M$ is a subbundle of dimension $\binom{n}{k}$ of the vector bundle $\pi^k : \mathscr{T}^k(E) \to M$.

PROOF: If $(t, \pi^{-1}(U))$ is a local trivialization for E, it follows from Propositions 4.45 and 5.13 (1) that there are sections $s_{i_1, \ldots, i_k} : U \to \mathscr{T}^k(E)$ defined by

$$s_{i_1, \ldots, i_k}(p) = (e^{i_1} \circ t_p) \wedge \cdots \wedge (e^{i_k} \circ t_p)$$

for $p \in U$ and $1 \leq i_1 < \ldots < i_k \leq n$, and $\{s_{i_1, \ldots, i_k}(p) \mid 1 \leq i_1 < \ldots < i_k \leq n\}$ is a basis for $\Lambda^k(E_p)$ by Propositions 4.11 and 5.14. The result hence follows from Proposition 2.62. □

5.21 Definition Let $\pi : E \to M$ be an n-dimensional vector bundle over the smooth manifold M. If ω and η are sections of the bundles $\Lambda^k(E)$ and $\Lambda^l(E)$, respectively, on an open subset V of M, we define their *wedge product* $\omega \wedge \eta$ to be the section of $\Lambda^{k+l}(E)$ on V defined by

$$(\omega \wedge \eta)(p) = \omega(p) \wedge \eta(p)$$

for $p \in V$. If $l = 0$, then η is simply a smooth function f on V, and $f \wedge \omega$ and $\omega \wedge f$ simply mean $f\omega$.

A section of the bundle $\pi : \Lambda^k(TM) \to M$ defined on an open subset V of M is called a *differential form* of degree k or simply a *k-form* on V. The set of all k-forms on V is denoted by $\Omega^k(V)$. In particular $\Omega^0(V) = \mathscr{F}(V)$ is the set of C^∞-functions on V. We let $\Omega(V)$ be the direct sum of $\Omega^k(V)$ for $k = 0, ..., n$.

5.22 Corollary If M is a smooth manifold and $\omega : M \to \Lambda^k(TM)$ is a map with $\pi \circ \omega = id_M$, then ω is a k-form on M if and only if the functions $\omega_{i_1,...,i_k} : U \to \mathbf{R}$ defined by

$$\omega(p) = \sum_{i_1<...<i_k} \omega_{i_1,...,i_k}(p)\, dx^{i_1}(p) \wedge \cdots \wedge dx^{i_k}(p)$$

for $p \in U$ are smooth for every local chart (x, U) on M.

PROOF : Follows from Propositions 4.58 and 5.13 (1). □

5.23 If $\pi : E \to M$ and $\widetilde{\pi} : \widetilde{E} \to M$ are vector bundles over the smooth manifold M of dimensions n and m, respectively, we let

$$\mathscr{T}_1^k(E, \widetilde{E}) = \mathscr{T}(\underbrace{E, ..., E}_{k}, \widetilde{E}^*)$$

and denote its projection by π_1^k. We want to define a subbundle

$$\Lambda_1^k(E, \widetilde{E}) = \bigcup_{p \in M} \Lambda_1^k(E_p, \widetilde{E}_p)$$

of tensors which are skew symmetric in the first k variables on each fibre E_p. We have a projection $\pi'' : \Lambda_1^k(E, \widetilde{E}) \to M$ defined by $\pi''(\omega) = p$ for $\omega \in \Lambda_1^k(E_p, \widetilde{E}_p)$.

5.24 Proposition $\pi'' : \Lambda_1^k(E, \widetilde{E}) \to M$ is a subbundle of dimension $m\binom{n}{k}$ of the vector bundle $\pi_1^k : \mathscr{T}_1^k(E, \widetilde{E}) \to M$.

PROOF : If $(t, \pi^{-1}(U))$ and $(\widetilde{t}, \widetilde{\pi}^{-1}(\widetilde{U}))$ are local trivializations for E and $\widetilde{E}$, it follows from Propositions 4.11, 4.45 and 5.13 (1) that there are sections $s_{i_1,...,i_k,j} : U \cap \widetilde{U} \to \mathscr{T}_1^k(E, \widetilde{E})$ defined by

$$s_{i_1,...,i_k,j}(p) = ((e^{i_1} \circ t_p) \wedge \cdots \wedge (e^{i_k} \circ t_p)) \otimes \widetilde{t}_p^{-1}(e_j)$$

for $p \in U \cap \widetilde{U}$, $1 \le i_1 < ... < i_k \le n$ and $1 \le j \le m$, and $\{s_{i_1,...,i_k,j}(p) \mid 1 \le i_1 < ... < i_k \le n, 1 \le j \le m\}$ is a basis for $\Lambda_1^k(E_p, \widetilde{E}_p)$ by Propositions 4.11 and 5.16. The result hence follows from Propositions 2.62. □

5.25 We now define the bundle

$$\mathscr{T}^k(E;\widetilde{E}) = \bigcup_{p\in M} \mathscr{T}^k(E_p;\widetilde{E}_p)$$

with $\widetilde{E}_p$-valued covariant tensors of degree k in each fibre E_p. We have a projection $\pi^k : \mathscr{T}^k(E;\widetilde{E}) \to M$ given by $\pi^k(\omega) = p$ for $\omega \in \mathscr{T}^k(E_p;\widetilde{E}_p)$, and a map

$$\phi : \mathscr{T}^k(E;\widetilde{E}) \to \mathscr{T}^k_1(E,\widetilde{E})$$

defined by

$$\phi(F)(v_1,...,v_k,\lambda) = \lambda(F(v_1,...,v_k))$$

for $F \in \mathscr{T}^k(E_p;\widetilde{E}_p)$, $v_1,...,v_k \in E_p$ and $\lambda \in \widetilde{E}_p^*$. From 5.15 we know that ϕ is a linear isomorphism on each fibre with $\pi^k_1 \circ \phi = \pi^k$, and we give $\mathscr{T}^k(E;\widetilde{E})$ the manifold structure such that ϕ is an equivalence over M. We also define the bundle

$$\Lambda^k(E;\widetilde{E}) = \bigcup_{p\in M} \Lambda^k(E_p;\widetilde{E}_p)$$

with $\widetilde{E}_p$-valued exterior k-forms in each fibre E_p, which is a subbundle of $\mathscr{T}^k(E;\widetilde{E})$ since

$$\Lambda^k(E;\widetilde{E}) = \phi^{-1}(\Lambda^k_1(E,\widetilde{E}))\ .$$

The projection $\pi' : \Lambda^k(E;\widetilde{E}) \to M$ induced by π^k is given by $\pi'(\omega) = p$ for $\omega \in \Lambda^k(E_p;\widetilde{E}_p)$.

5.26 Let $\pi : E \to M$ be an n-dimensional vector bundle over the smooth manifold M. If W is a vector space of dimension m, we let $\widetilde{\pi} : E_W \to M$ be the vector bundle $E_W = M \times W$ where $\widetilde{\pi}$ is the projection on the first factor. If $\mathscr{C} = \{w_1,...,w_m\}$ is a basis for W, we have a local trivialization $(\widetilde{t},\widetilde{\pi}^{-1}(M))$ for E_W given by

$$\widetilde{t}(p,w) = (p,a) \quad \text{for} \quad p \in M \ \text{ and } \ w = \sum_{j=1}^{m} a_j w_j$$

so that the projection $\rho_p : \{p\} \times W \to W$ on the second factor is a linear isomorphism for each fibre $\{p\} \times W$. From this we also obtain the linear isomorphisms

$$\rho''_p : \Lambda^k_1(E_p,\{p\} \times W) \to \Lambda^k_1(E_p,W)$$

induced by

$$(id, ... ,id,\rho_p^*)^* : \mathscr{T}^k_1(E_p,\{p\} \times W) \to \mathscr{T}^k_1(E_p,W)$$

and

$$\rho'_p : \Lambda^k(E_p;\{p\} \times W) \to \Lambda^k(E_p;W)$$

given by

$$\rho'_p(F) = \rho_p \circ F$$

for $F \in \Lambda^k(E_p;\{p\}\times W)$. If

$$\widetilde{\phi}_p : \Lambda^k(E_p;\{p\}\times W) \to \Lambda^k_1(E_p,\{p\}\times W)$$

is the linear isomorphism defined as in 5.25 and

$$\phi_p : \Lambda^k(E_p;W) \to \Lambda^k_1(E_p,W)$$

is the corresponding linear isomorphism defined by

$$\phi_p(G)(v_1,\ldots,v_k,\lambda) = \lambda\,(G(v_1,\ldots,v_k))$$

for $G \in \Lambda^k(E_p;W)$, $v_1,\ldots,v_k \in E_p$ and $\lambda \in W^*$, we have a commutative diagram

$$\begin{array}{ccc}
\Lambda^k(E_p;\{p\}\times W) & \xrightarrow{\ \widetilde{\phi}_p\ } & \Lambda^k_1(E_p,\{p\}\times W) \\
\rho'_p \downarrow & & \downarrow \rho''_p \\
\Lambda^k(E_p;W) & \xrightarrow[\ \phi_p\ ]{} & \Lambda^k_1(E_p,W)
\end{array}$$

since

$$\begin{aligned}
\phi_p \circ {\rho'}_p(F)(v_1,\ldots,v_k,\lambda) &= \lambda\,({\rho'}_p(F)(v_1,\ldots,v_k)) = \lambda \circ \rho_p \circ F(v_1,\ldots,v_k) \\
&= \widetilde{\phi}_p(F)(v_1,\ldots,v_k,\rho_p^*(\lambda)) = \rho''_p \circ \widetilde{\phi}_p(F)(v_1,\ldots,v_k,\lambda)
\end{aligned}$$

for $F \in \Lambda^k(E_p;\{p\}\times W)$, $v_1,\ldots,v_k \in E_p$ and $\lambda \in W^*$.

Using the vector spaces $\Lambda^k(E_p;W)$ and $\Lambda^k_1(E_p,W)$ where $p \in M$ as fibres, we can now construct the vector bundles

$$\pi' : \Lambda^k(E;W) \to M \quad \text{and} \quad \pi'' : \Lambda^k_1(E,W) \to M$$

such that all the maps in the corresponding commutative diagram

$$\begin{array}{ccc}
\Lambda^k(E;E_W) & \xrightarrow{\ \widetilde{\phi}\ } & \Lambda^k_1(E,E_W) \\
\rho' \downarrow & & \downarrow \rho'' \\
\Lambda^k(E;W) & \xrightarrow[\ \phi\ ]{} & \Lambda^k_1(E,W)
\end{array}$$

are equivalences over M.

5.27 Let $\pi : E \to M$ be an n-dimensional vector bundle over the smooth manifold M. If W is a vector space of dimension m, we let $\widetilde{\pi} : E_W \to M$ be the vector bundle $E_W = M \times W$ where $\widetilde{\pi}$ is the projection on the first factor. If $\mathscr{C} = \{w_1, ..., w_m\}$ is a basis for W, we have a local trivialization $(\widetilde{t}, \widetilde{\pi}^{-1}(M))$ for E_W given by

$$\widetilde{t}(p,w) = (p,a) \quad \text{for} \quad p \in M \quad \text{and} \quad w = \sum_{j=1}^{m} a_j w_j$$

so that the projection $\rho_p : \{p\} \times W \to W$ on the second factor is a linear isomorphism for each fibre $\{p\} \times W$. From this we also obtain the linear isomorphisms

$$\rho''_p : \Lambda_1^1(W, E_p) \to \Lambda_1^1(\{p\} \times W, E_p)$$

induced by

$$(\rho_p, id)^* : \mathscr{T}_1^1(W, E_p) \to \mathscr{T}_1^1(\{p\} \times W, E_p)$$

and

$$\rho'_p : \Lambda^1(W; E_p) \to \Lambda^1(\{p\} \times W; E_p)$$

given by

$$\rho'_p(F) = F \circ \rho_p$$

for $F \in \Lambda^1(W; E_p)$. If

$$\widetilde{\phi}_p : \Lambda^1(\{p\} \times W; E_p) \to \Lambda_1^1(\{p\} \times W, E_p)$$

is the linear isomorphism defined as in 5.25 and

$$\phi_p : \Lambda^1(W; E_p) \to \Lambda_1^1(W, E_p)$$

is the corresponding linear isomorphism defined by

$$\phi_p(G)(v, \lambda) = \lambda(G(v))$$

for $G \in \Lambda^1(W; E_p)$, $v \in W$ and $\lambda \in E_p^*$, we have a commutative diagram

$$\begin{array}{ccc} \Lambda^1(W; E_p) & \xrightarrow{\ \phi_p\ } & \Lambda_1^1(W, E_p) \\ \rho'_p \big\downarrow & & \big\downarrow \rho''_p \\ \Lambda^1(\{p\} \times W; E_p) & \xrightarrow[\ \widetilde{\phi}_p\ ]{} & \Lambda_1^1(\{p\} \times W, E_p) \end{array}$$

since

$$\begin{aligned} \widetilde{\phi}_p \circ \rho'_p(F)(v, \lambda) &= \lambda(\rho'_p(F)(v)) = \lambda \circ F \circ \rho_p(v) \\ &= \phi_p(F)(\rho_p(v), \lambda) = \rho''_p \circ \phi_p(F)(v, \lambda) \end{aligned}$$

for $F \in \Lambda^1(W;E_p)$, $v \in \{p\} \times W$ and $\lambda \in E_p^*$.

Using the vector spaces $\Lambda^1(W;E_p)$ and $\Lambda_1^1(W,E_p)$ where $p \in M$ as fibres, we can now construct the vector bundles

$$\pi' : \Lambda^1(W;E) \to M \quad \text{and} \quad \pi'' : \Lambda_1^1(W,E) \to M$$

such that all the maps in the corresponding commutative diagram

$$\begin{array}{ccc} \Lambda^1(W;E) & \xrightarrow{\ \phi\ } & \Lambda_1^1(W,E) \\ {\scriptstyle \rho'}\big\downarrow & & \big\downarrow{\scriptstyle \rho''} \\ \Lambda^1(E_W;E) & \xrightarrow[\ \widetilde{\phi}\ ]{} & \Lambda_1^1(E_W,E) \end{array}$$

are equivalences over M.

5.28 Proposition Let $\pi : E \to M$ and $\widetilde{\pi} : \widetilde{E} \to M$ be vector bundles over the smooth manifold M of dimensions n and m, respectively, and let $\mathscr{E} = \{e_1, ..., e_n\}$ and $\mathscr{F} = \{f_1, ..., f_m\}$ be the standard bases for $\mathbf{R}^n$ and $\mathbf{R}^m$ with dual bases $\mathscr{E}^* = \{e^1, ..., e^n\}$ and $\mathscr{F}^* = \{f^1, ..., f^m\}$. If $(t, \pi^{-1}(U))$ and $(\widetilde{t}, \widetilde{\pi}^{-1}(U))$ are local trivializations in E and $\widetilde{E}$, then we have a local trivialization $((id_U \times \beta) \circ \tau, (\pi^k)^{-1}(U))$ in the vector bundle $\pi^k : \mathscr{T}^k(E;\widetilde{E}) \to M$, where $\tau : (\pi^k)^{-1}(U) \to U \times \mathscr{T}^k(\mathbf{R}^n;\mathbf{R}^m)$ is the equivalence over U defined by

$$\tau(F) = (p, \widetilde{t}_p \circ F \circ (t_p^{-1})^k)$$

for $p \in U$ and $F \in \mathscr{T}^k(E_p;\widetilde{E}_p)$, and $\beta : \mathscr{T}^k(\mathbf{R}^n;\mathbf{R}^m) \to \mathbf{R}^s$ is the linear isomorphism given by

$$\beta^{\,b(i_1, \ldots, i_k, j)}(G) = f^j \circ G(e_{i_1}, \ldots, e_{i_k})$$

for $G \in \mathscr{T}^k(\mathbf{R}^n;\mathbf{R}^m)$ and $(i_1, ..., i_k, j) \in I_n^k \times I_m$, where $b : I_n^k \times I_m \to I_s$ is the bijection defined in Remark 4.27 with $s = n^k m$. If

$$\phi : \mathscr{T}^k(E;\widetilde{E}) \to \mathscr{T}_1^k(E,\widetilde{E})$$

is the equivalence over M defined in 5.25 and $(t', \pi''^{-1}(U))$ is the local trivialization in $\mathscr{T}_1^k(E,\widetilde{E})$ defined in Remark 4.40, then we have that

$$(id_U \times \beta) \circ \tau = t' \circ \phi\ .$$

PROOF : Let $(t'', \widetilde{\pi}^{*-1}(U))$ be the local trivialization in the dual bundle $\widetilde{\pi}^* : \widetilde{E}^* \to$

M corresponding to $(\widetilde{t}, \widetilde{\pi}^{-1}(U))$, and let $\mathscr{G} = \{g_1, ..., g_s\}$ be the standard basis for $\mathbf{R}^s$. Then we have that

$$\begin{aligned}
&(t_p^{-1}, \dots, t_p^{-1}, t_p''^{-1})^* \circ \phi(F)(e_{i_1}, \dots, e_{i_k}, f_j) \\
&= \phi(F)(t_p^{-1}(e_{i_1}), \dots, t_p^{-1}(e_{i_k}), f^j \circ \widetilde{t}_p) \\
&= f^j \circ \widetilde{t}_p \circ F \circ (t_p^{-1})^k(e_{i_1}, \dots, e_{i_k})
\end{aligned}$$

for $(i_1, ..., i_k, j) \in I_n^k \times I_m$, so that

$$\begin{aligned}
&(t_p^{-1}, \dots, t_p^{-1}, t_p''^{-1})^* \circ \phi(F) \\
&= \sum_{i_1, \dots, i_k, j} f^j \circ \widetilde{t}_p \circ F \circ (t_p^{-1})^k(e_{i_1}, \dots, e_{i_k})\, e^{i_1} \otimes \cdots \otimes e^{i_k} \otimes f^j
\end{aligned}$$

for $p \in U$. Hence it follows that

$$t_p' \circ \phi(F) = \sum_{i_1, \dots, i_k, j} f^j \circ \widetilde{t}_p \circ F \circ (t_p^{-1})^k(e_{i_1}, \dots, e_{i_k})\, g_{b(i_1, \dots, i_k, j)} = \beta \circ \tau(F)$$

for $p \in U$ and $F \in \mathscr{T}^k(E_p; \widetilde{E}_p)$, which completes the proof of the proposition. □

5.29 Corollary Let $\pi : E \to M$ be a vector bundle over the smooth manifold M of dimension m, and let W be an n-dimensional vector space. Let $\mathscr{E} = \{e_1, ..., e_n\}$ and $\mathscr{F} = \{f_1, ..., f_m\}$ be the standard bases for $\mathbf{R}^n$ and $\mathbf{R}^m$ with dual bases $\mathscr{E}^* = \{e^1, ..., e^n\}$ and $\mathscr{F}^* = \{f^1, ..., f^m\}$. If $(t, \pi^{-1}(U))$ is a local trivialization in E and $x : W \to \mathbf{R}^n$ is a linear isomorphism, then we have a local trivialization $((id_U \times \beta) \circ \tau, \pi'^{-1}(U))$ in the vector bundle $\pi' : \Lambda^1(W; E) \to M$, where $\tau : \pi'^{-1}(U) \to U \times L(\mathbf{R}^n, \mathbf{R}^m)$ is the equivalence over U defined by

$$\tau(F) = (p, t_p \circ F \circ x^{-1})$$

for $p \in U$ and $F \in \Lambda^1(W; E_p)$, and $\beta : L(\mathbf{R}^n, \mathbf{R}^m) \to \mathbf{R}^{nm}$ is the linear isomorphism given by

$$\beta^{\,b(i,j)}(G) = f^j \circ G(e_i)$$

for $G \in L(\mathbf{R}^n, \mathbf{R}^m)$, $i = 1, \dots, n$ and $j = 1, \dots, m$, where $b : I_n \times I_m \to I_{nm}$ is the bijection defined in Remark 4.27. If

$$\rho' : \Lambda^1(W; E) \to \Lambda^1(E_W; E) \quad \text{and} \quad \widetilde{\phi} : \Lambda^1(E_W; E) \to \Lambda^1_1(E_W, E)$$

are the equivalences over M defined in 5.27, using the basis $\mathscr{C} = \{x^{-1}(e_1), ..., x^{-1}(e_n)\}$ for W, and if $(t', \pi'''^{-1}(U))$ is the local trivialization in $\Lambda^1_1(E_W, E) = \mathscr{T}(E_W, E^*)$ defined in Remark 4.40, then we have that

$$(id_U \times \beta) \circ \tau = t' \circ \widetilde{\phi} \circ \rho'.$$

PROOF: Let $\widetilde{\tau} : \pi''^{-1}(U) \to U \times L(\mathbf{R}^n, \mathbf{R}^m)$ be the equivalence over U defined as in Proposition 5.28 for the vector bundle $\pi'' : \Lambda^1(E_W; E) \to M$, using the local trivialization $(\widetilde{t}, \widetilde{\pi}^{-1}(M))$ for E_W given in 5.27. Then we have that $\widetilde{t}_p \circ \rho_p^{-1} = x$ for $p \in M$, which shows that

$$\widetilde{\tau} \circ \rho'(F) = (p, t_p \circ F \circ \rho_p \circ \widetilde{t}_p^{-1}) = (p, t_p \circ F \circ x^{-1}) = \tau(F)$$

for $p \in U$ and $F \in \Lambda^1(W; E_p)$. The corollary now follows from Proposition 5.28. □

5.30 Proposition Let $\pi : E \to M$ and $\widetilde{\pi} : \widetilde{E} \to M$ be vector bundles over the smooth manifold M of dimensions n and m, and let $g : E \to \widetilde{E}$ be a bundle map over M. Then we have a section s on M in the vector bundle $\pi' : \Lambda^1(E; \widetilde{E}) \to M$, called the *section determined by the bundle map g*, defined by

$$s(p) = g_p$$

for $p \in M$. Conversely, each section s of π' on M is determined by a unique bundle map g given by

$$g(v) = s(p)(v)$$

for $p \in M$ and $v \in E_p$.

PROOF: Let $((id_U \times \beta) \circ \tau, (\pi^k)^{-1}(U))$ be the local trivialization in the vector bundle $\pi' : \Lambda^1(E; \widetilde{E}) \to M$ given in Proposition 5.28, obtained from local trivializations $(t, \pi^{-1}(U))$ and $(\widetilde{t}, \widetilde{\pi}^{-1}(U))$ in E and $\widetilde{E}$, and let $pr_2 : U \times \mathbf{R}^m \to \mathbf{R}^m$ and $pr_2' : U \times L(\mathbf{R}^n, \mathbf{R}^m) \to L(\mathbf{R}^n, \mathbf{R}^m)$ be the projections on the second factor. Then we see that s is smooth on U since

$$\beta^{b(i,j)} \circ pr_2' \circ \tau \circ s(p) = f^j \circ pr_2 \circ \widetilde{t} \circ g \circ t^{-1}(p, e_i)$$

for $p \in U$, $i = 1, \dots, n$ and $j = 1, \dots, m$, showing the first part of the proposition.

To prove the last part, let s be a section of π' on M, and let $B : L(\mathbf{R}^n, \mathbf{R}^m) \times \mathbf{R}^n \to \mathbf{R}^m$ be the bilinear map given by $B(G, a) = G(a)$ for $G \in L(\mathbf{R}^n, \mathbf{R}^m)$ and $a \in \mathbf{R}^n$. Then we see that g is smooth on $\pi^{-1}(U)$ since

$$pr_2 \circ \widetilde{t} \circ g \circ t^{-1} = B \circ \{(pr_2' \circ \tau \circ s) \times id_{\mathbf{R}^n}\}.$$

□

5.31 Proposition Let $\pi : E \to M$ and $\widetilde{\pi} : \widetilde{E} \to M$ be vector bundles over the smooth manifold M of dimensions n and m. Then we have an equivalence

$$\psi : \mathscr{T}^k(E; \Lambda^1(E; \widetilde{E})) \to \mathscr{T}^{k+1}(E; \widetilde{E})$$

over M between the vector bundles $\widetilde{\pi}^k : \mathscr{T}^k(E; \Lambda^1(E; \widetilde{E})) \to M$ and $\pi^{k+1} : \mathscr{T}^{k+1}(E; \widetilde{E}) \to M$ which is given by

$$\psi(F)(v_1, \dots, v_{k+1}) = F(v_1, \dots, v_k)(v_{k+1})$$

for $F \in \mathscr{T}^k(E_p; \Lambda^1(E_p; \widetilde{E}_p))$ and $v_1, \dots, v_{k+1} \in E_p$ where $p \in M$.

PROOF : Consider first the linear isomorphism $\phi : \mathscr{T}^k(\mathbf{R}^n;L(\mathbf{R}^n,\mathbf{R}^m)) \to \mathscr{T}^{k+1}(\mathbf{R}^n;\mathbf{R}^m)$ given by

$$\phi(G)(a_1,\dots,a_{k+1}) = G(a_1,\dots,a_k)(a_{k+1})$$

for $G \in \mathscr{T}^k(\mathbf{R}^n;L(\mathbf{R}^n,\mathbf{R}^m))$ and $a_1,\dots,a_{k+1} \in \mathbf{R}^n$. Let $r = nm$ and $s = n^{k+1}m$, and let $\mathscr{E} = \{e_1,\dots,e_n\}$, $\mathscr{F} = \{f_1,\dots,f_m\}$ and $\mathscr{G} = \{g_1,\dots,g_r\}$ be the standard bases for $\mathbf{R}^n$, $\mathbf{R}^m$ and $\mathbf{R}^r$, respectively, with dual bases $\mathscr{E}^* = \{e^1,\dots,e^n\}$, $\mathscr{F}^* = \{f^1,\dots,f^m\}$ and $\mathscr{G}^* = \{g^1,\dots,g^r\}$. Let $\beta_1 : L(\mathbf{R}^n,\mathbf{R}^m) \to \mathbf{R}^r$, $\widetilde{\beta}_k : \mathscr{T}^k(\mathbf{R}^n;\mathbf{R}^r) \to \mathbf{R}^s$ and $\beta_{k+1} : \mathscr{T}^{k+1}(\mathbf{R}^n;\mathbf{R}^m) \to \mathbf{R}^s$ be the linear isomorphisms given by

$$\beta_1^{\,b_1(i,j)}(G_1) = f^j \circ G_1(e_i) \;\;,\;\; \widetilde{\beta}_k^{\,b_2(i_1,\dots,i_k,l)}(G_2) = g^l \circ G_2(e_{i_1},\dots,e_{i_k})$$

and

$$\beta_{k+1}^{\,b_3(i_1,\dots,i_{k+1},j)}(G_3) = f^j \circ G_3(e_{i_1},\dots,e_{i_{k+1}})$$

for $i, i_1,\dots,i_{k+1} \in I_n$, $j \in I_m$ and $l \in I_r$, where $b_1 : I_n \times I_m \to I_r$, $b_2 : I_n^k \times I_r \to I_s$ and $b_3 : I_n^{k+1} \times I_m \to I_s$ are the bijections defined in Remark 4.27. Then we have that

$$\beta_{k+1}^{\,b_3(i_1,\dots,i_{k+1},j)}(\phi(G)) = f^j \circ G(e_{i_1},\dots,e_{i_k})(e_{i_{k+1}})$$
$$= \beta_1^{\,b_1(i_{k+1},j)} \circ G(e_{i_1},\dots,e_{i_k}) = g^{\,b_1(i_{k+1},j)} \circ (\beta_1 \circ G)(e_{i_1},\dots,e_{i_k})$$
$$= \widetilde{\beta}_k^{\,b_2(i_1,\dots,i_k,b_1(i_{k+1},j))}(\beta_1 \circ G) = \widetilde{\beta}_k^{\,b_3(i_1,\dots,i_{k+1},j)}(\beta_1 \circ G)$$

which shows that

$$\beta_{k+1} \circ \phi\,(G) = \widetilde{\beta}_k\,(\beta_1 \circ G) \tag{1}$$

for $G \in \mathscr{T}^k(\mathbf{R}^n;L(\mathbf{R}^n,\mathbf{R}^m))$.

Now let $(t,\pi^{-1}(U))$ and $(\widetilde{t},\widetilde{\pi}^{-1}(U))$ be local trivializations in E and $\widetilde{E}$. By Proposition 5.28 we have a local trivialization $(t_h,(\pi^h)^{-1}(U))$ in the vector bundle $\pi^h : \mathscr{T}^h(E;\widetilde{E}) \to M$, where $t_{h,p} = \beta_h \circ \tau_{h,p}$ and

$$\tau_{h,p}(F) = \widetilde{t}_p \circ F \circ \{t_p^{-1}\}^h$$

for $p \in U$ and $F \in \mathscr{T}^h(E_p;\widetilde{E}_p)$ with $h = 1, k+1$.

We also have a local trivialization $(\widetilde{t}_k,(\widetilde{\pi}^k)^{-1}(U))$ in the vector bundle $\widetilde{\pi}^k : \mathscr{T}^k(E;\Lambda^1(E;\widetilde{E})) \to M$, where $\widetilde{t}_{k,p} = \widetilde{\beta}_k \circ \widetilde{\tau}_{k,p}$ and

$$\widetilde{\tau}_{k,p}(F) = t_{1,p} \circ F \circ \{t_p^{-1}\}^k$$

for $p \in U$ and $F \in \mathscr{T}^k(E_p;\Lambda^1(E_p;\widetilde{E}_p))$.

Combining this, it follows that

$$\tau_{k+1,p} \circ \psi(F)(a_1,\dots,a_{k+1}) = \widetilde{t}_p \circ F\,(t_p^{-1}(a_1),\dots,t_p^{-1}(a_k)) \circ t_p^{-1}(a_{k+1})$$
$$= \tau_{1,p} \circ F \circ \{t_p^{-1}\}^k(a_1,\dots,a_k)(a_{k+1}) = \phi\,(\tau_{1,p} \circ F \circ \{t_p^{-1}\}^k)(a_1,\dots,a_{k+1})$$

for $a_1, \dots, a_{k+1} \in \mathbf{R}^n$, showing that

$$\tau_{k+1,p} \circ \psi(F) = \phi\,(\tau_{1,p} \circ F \circ \{t_p^{-1}\}^k) \tag{2}$$

for $p \in U$ and $F \in \mathscr{T}^k(E_p; \Lambda^1(E_p; \widetilde{E}_p))$.

From (1) and (2) it now follows that

$$t_{k+1,p} \circ \psi(F) = \beta_{k+1} \circ \phi\,(\tau_{1,p} \circ F \circ \{t_p^{-1}\}^k) = \widetilde{\beta}_k\,(t_{1,p} \circ F \circ \{t_p^{-1}\}^k) = \widetilde{t}_{k,p}(F)$$

so that

$$t_{k+1} \circ \psi \circ \widetilde{t}_k^{\,-1} = id_{U \times \mathbf{R}^s},$$

thus showing that ψ is an equivalence. □

5.32 Proposition Let $\pi_1 : E_1 \to M$, $\pi_2 : E_2 \to M$ and $\widetilde{\pi} : \widetilde{E} \to M$ be vector bundles over the smooth manifold M of dimensions n_1, n_2 and m. Then we have an equivalence

$$\psi : \Lambda^1(E_1; \Lambda^1(E_2; \widetilde{E})) \to \Lambda^1(E_2; \Lambda^1(E_1; \widetilde{E}))$$

over M between the vector bundles $\widehat{\pi}_1 : \Lambda^1(E_1; \Lambda^1(E_2; \widetilde{E})) \to M$ and $\widehat{\pi}_2 : \Lambda^1(E_2; \Lambda^1(E_1; \widetilde{E})) \to M$ which is given by

$$\psi(F)(v_2)(v_1) = F(v_1)(v_2)$$

for $F \in \Lambda^1(E_{1,p}; \Lambda^1(E_{2,p}; \widetilde{E}_p))$, $v_1 \in E_{1,p}$ and $v_2 \in E_{2,p}$ where $p \in M$.

PROOF : Consider first the isomorphism $\phi : L(\mathbf{R}^{n_1}, L(\mathbf{R}^{n_2}, \mathbf{R}^m)) \to L(\mathbf{R}^{n_2}, L(\mathbf{R}^{n_1}, \mathbf{R}^m))$ given by

$$\phi(G)(a_2)(a_1) = G(a_1)(a_2)$$

for $G \in L(\mathbf{R}^{n_1}, L(\mathbf{R}^{n_2}, \mathbf{R}^m))$, $a_1 \in \mathbf{R}^{n_1}$ and $a_2 \in \mathbf{R}^{n_2}$. Let $r_i = n_i m$ and $s = n_1 n_2 m$, and let $\mathscr{E}_i = \{e_1^i, \dots, e_{n_i}^i\}$, $\mathscr{F} = \{f_1, \dots, f_m\}$ and $\mathscr{G}_i = \{g_1^i, \dots, g_{r_i}^i\}$ be the standard bases for $\mathbf{R}^{n_i}$, $\mathbf{R}^m$ and $\mathbf{R}^{r_i}$, respectively, with dual bases $\mathscr{E}_i^* = \{e_{(i)}^1, \dots, e_{(i)}^{n_i}\}$, $\mathscr{F}^* = \{f^1, \dots, f^m\}$ and $\mathscr{G}_i^* = \{g_{(i)}^1, \dots, g_{(i)}^{r_i}\}$ for $i = 1, 2$. Let $\beta_i : L(\mathbf{R}^{n_i}, \mathbf{R}^m) \to \mathbf{R}^{r_i}$ and $\widetilde{\beta}_i : L(\mathbf{R}^{n_i}, \mathbf{R}^{r_{3-i}}) \to \mathbf{R}^s$ be the linear isomorphisms given by

$$\beta_i^{\,b_i(j,k)}(G) = f^k \circ G(e_j^i) \quad \text{and} \quad \widetilde{\beta}_i^{\,\widetilde{b}_i(j,l)}(G) = g_{(i)}^l \circ G(e_j^i)$$

for $j \in I_{n_i}$, $k \in I_m$ and $l \in I_{r_{3-i}}$, where $b_i : I_{n_i} \times I_m \to I_{r_i}$ and $\widetilde{b}_i : I_{n_i} \times I_{r_{3-i}} \to I_s$ are the bijections defined in Remark 4.27. We also have a permutation σ of I_s defined by

$$\sigma = \widehat{b}_1 \circ \rho \circ \widehat{b}_2^{-1},$$

where $\rho : I_{n_2} \times I_{n_1} \times I_m \to I_{n_1} \times I_{n_2} \times I_m$ is the bijection given by

$$\rho(j_2, j_1, k) = (j_1, j_2, k)$$

for $(j_2, j_1, k) \in I_{n_2} \times I_{n_1} \times I_m$, and $\widehat{b}_i : I_{n_i} \times I_{n_{3-i}} \times I_m \to I_s$ is the bijection defined in

Remark 4.27 for $i = 1,2$. Now we have that

$$\widetilde{\beta}_2^{\,\widehat{b}_2(j_2,j_1,k)}(\beta_1 \circ \phi(G)) = \widetilde{\beta}_2^{\,\widetilde{b}_2(j_2,b_1(j_1,k))}(\beta_1 \circ \phi(G)) = \beta_1^{\,b_1(j_1,k)} \circ \phi(G)(e_2^{j_2})$$
$$= f^k \circ \phi(G)(e_2^{j_2})(e_1^{j_1}) = f^k \circ G(e_1^{j_1})(e_2^{j_2}) = \beta_2^{\,b_2(j_2,k)} \circ G(e_1^{j_1})$$
$$= \widetilde{\beta}_1^{\,\widetilde{b}_1(j_1,b_2(j_2,k))}(\beta_2 \circ G) = \widetilde{\beta}_1^{\,\widehat{b}_1(j_1,j_2,k)}(\beta_2 \circ G) = \widetilde{\beta}_1^{\,\sigma \circ \widehat{b}_2(j_2,j_1,k)}(\beta_2 \circ G)$$

which shows that

$$\widetilde{\beta}_2\,(\beta_1 \circ \phi\,(G)) = \pi_\sigma \circ \widetilde{\beta}_1\,(\beta_2 \circ G) \tag{1}$$

for $G \in L(\mathbf{R}^{n_1}, L(\mathbf{R}^{n_2}, \mathbf{R}^m))$.

Now let $(t_i, \pi_i^{-1}(U))$ and $(\widetilde{t}, \widetilde{\pi}^{-1}(U))$ be local trivializations in E_i and $\widetilde{E}$. By Proposition 5.28 we have a local trivialization $(\widetilde{t}_i, \widetilde{\pi}_i^{-1}(U))$ in the vector bundle $\widetilde{\pi}_i : \Lambda^1(E_i;\widetilde{E}) \to M$, where $\widetilde{t}_{i,p} = \beta_i \circ \tau_{i,p}$ and

$$\tau_{i,p}(F) = \widetilde{t}_p \circ F \circ t_{i,p}^{-1}$$

for $p \in U$ and $F \in \Lambda^1(E_{i,p};\widetilde{E}_p)$ with $i = 1,2$.

We also have a local trivialization $(\widehat{t}_i, \widehat{\pi}_i^{-1}(U))$ in the vector bundle $\widehat{\pi}_i : \Lambda^1(E_i;\Lambda^1(E_{3-i};\widetilde{E})) \to M$, where $\widehat{t}_{i,p} = \widetilde{\beta}_i \circ \widetilde{\tau}_{i,p}$ and

$$\widetilde{\tau}_{i,p}(F) = \widetilde{t}_{3-i,p} \circ F \circ t_{i,p}^{-1}$$

for $p \in U$ and $F \in \Lambda^1(E_{i,p};\Lambda^1(E_{3-i,p};\widetilde{E}_p))$.

Combining this, it follows that

$$\tau_{1,p} \circ \psi(F) \circ t_{2,p}^{-1}(a_2)(a_1) = \widetilde{t}_p \circ \psi(F)\,(t_{2,p}^{-1}(a_2))\,(t_{1,p}^{-1}(a_1))$$
$$= \widetilde{t}_p \circ F\,(t_{1,p}^{-1}(a_1))\,(t_{2,p}^{-1}(a_2)) = \tau_{2,p} \circ F \circ t_{1,p}^{-1}(a_1)(a_2)$$
$$= \phi\,(\tau_{2,p} \circ F \circ t_{1,p}^{-1})(a_2)(a_1)$$

for $a_1 \in \mathbf{R}^{n_1}$ and $a_2 \in \mathbf{R}^{n_2}$, showing that

$$\tau_{1,p} \circ \psi(F) \circ t_{2,p}^{-1} = \phi\,(\tau_{2,p} \circ F \circ t_{1,p}^{-1}) \tag{2}$$

for $p \in U$ and $F \in \Lambda^1(E_{1,p};\Lambda^1(E_{2,p};\widetilde{E}_p))$.

From (1) and (2) it now follows that

$$\widehat{t}_{2,p} \circ \psi(F) = \widetilde{\beta}_2\,(\widetilde{t}_{1,p} \circ \psi(F) \circ t_{2,p}^{-1}) = \widetilde{\beta}_2\,(\beta_1 \circ \phi\,(\tau_{2,p} \circ F \circ t_{1,p}^{-1}))$$
$$= \pi_\sigma \circ \widetilde{\beta}_1\,(\beta_2 \circ \tau_{2,p} \circ F \circ t_{1,p}^{-1}) = \pi_\sigma \circ \widetilde{\beta}_1\,(\widetilde{t}_{2,p} \circ F \circ t_{1,p}^{-1}) = \pi_\sigma \circ \widehat{t}_{1,p}(F)$$

so that

$$\widehat{t}_2 \circ \psi \circ \widehat{t}_1^{\,-1} = id_U \times \pi_\sigma\,,$$

thus showing that ψ is an equivalence. □

5.33 Proposition Let $(\tilde{f}_i, f)$ be a bundle map from a vector bundle $\pi_1^i : E_1^i \to M_1$ to a vector bundle $\pi_2^i : E_2^i \to M_2$ for $i = 1,2$, and suppose that $\tilde{f}_2$ induces a linear isomorphism between the fibres $(\pi_1^2)^{-1}(p)$ and $(\pi_2^2)^{-1}(f(p))$ for every $p \in M_1$. If s_2 is a section of $\pi_2' : \Lambda^k(E_2^1; E_2^2) \to M_2$ on an open subset V_2 of M_2 and $V_1 = f^{-1}(V_2)$, then the map $s_1 : V_1 \to \Lambda^k(E_1^1; E_1^2)$ defined by

$$s_1(p)(v_1, \ldots, v_k) = \tilde{f}_{2,p}^{-1} \circ s_2(f(p))(\tilde{f}_1(v_1), \ldots, \tilde{f}_1(v_k))$$

for all $v_1, \ldots, v_k \in (\pi_1^1)^{-1}(p)$ where $p \in V_1$, is a section of $\pi_1' : \Lambda^k(E_1^1; E_1^2) \to M_1$ on V_1.

s_1 is called the *pull-back* of s_2 by $(\tilde{f}_1, \tilde{f}_2)$ and is denoted by $(\tilde{f}_1, \tilde{f}_2)^*(s_2)$.

PROOF : Follows from Proposition 4.49 since

$$\phi_1 \circ s_1 = (\tilde{f}_1, \ldots, \tilde{f}_1, \tilde{g})^*(\phi_2 \circ s_2),$$

where $\phi_i : \Lambda^k(E_i^1; E_i^2) \to \Lambda_1^k(E_i^1, E_i^2)$ is the equivalence over M_i defined in 5.25 for $i = 1,2$, and $(\tilde{g}, f)$ is the bundle map between the dual bundles $(\pi_1^2)^* : (E_1^2)^* \to M_1$ and $(\pi_2^2)^* : (E_2^2)^* \to M_2$ defined in Proposition 4.50. Indeed, we have that

$$\begin{aligned}
&\phi_1 \circ s_1(p)(v_1, \ldots, v_k, \lambda) = \lambda \circ s_1(p)(v_1, \ldots, v_k) \\
&= \lambda \circ \tilde{f}_{2,p}^{-1} \circ s_2(f(p))(\tilde{f}_1(v_1), \ldots, \tilde{f}_1(v_k)) \\
&= \phi_2 \circ s_2(f(p))(\tilde{f}_1(v_1), \ldots, \tilde{f}_1(v_k), \tilde{g}(\lambda))
\end{aligned}$$

for $v_1, \ldots, v_k \in (\pi_1^1)^{-1}(p)$ and $\lambda \in (\pi_1^2)^{*-1}(p)$. □

5.34 Corollary Let $(\tilde{f}, f)$ be a bundle map from a vector bundle $\pi_1 : E_1 \to M_1$ to a vector bundle $\pi_2 : E_2 \to M_2$, and let s_2 be a section of $\pi_2' : \Lambda^k(E_2; W) \to M_2$ on M_2. Then the map $s_1 : M_1 \to \Lambda^k(E_1; W)$ defined by

$$s_1(p) = (\tilde{f}_p)^* s_2(f(p))$$

for $p \in M_1$, i.e.,

$$s_1(p)(v_1, \ldots, v_k) = s_2(f(p))(\tilde{f}_p(v_1), \ldots, \tilde{f}_p(v_k))$$

for all $v_1, \ldots, v_k \in \pi_1^{-1}(p)$ where $p \in M_1$, is a section of $\pi_1' : \Lambda^k(E_1; W) \to M_1$ on M_1.

s_1 is called the *pull-back* of s_2 by $\tilde{f}$ and is denoted by $\tilde{f}^*(s_2)$.

PROOF : Follows from Proposition 5.33 since

$$\rho_1'^{-1} \circ s_1 = (\tilde{f}, \tilde{g})^*(\rho_2'^{-1} \circ s_2),$$

where $\rho'_i : \Lambda^k(E_i; M_i \times W) \to \Lambda^k(E_i; W)$ is the equivalence over M_i defined in 5.26 for $i = 1, 2$ and $\tilde{g} = f \times id : M_1 \times W \to M_2 \times W$. Indeed, we have that

$$\begin{aligned}\rho_1'^{-1} \circ s_1(p)(v_1, \ldots, v_k) &= (p, s_1(p)(v_1, \ldots, v_k)) \\ &= \tilde{g}_p^{-1}(f(p), s_2(f(p))(\tilde{f}_p(v_1), \ldots, \tilde{f}_p(v_k))) \\ &= \tilde{g}_p^{-1} \circ (\rho_2'^{-1} \circ s_2)(f(p))(\tilde{f}_p(v_1), \ldots, \tilde{f}_p(v_k))\end{aligned}$$

for $p \in M_1$ and $v_1, \ldots, v_k \in E_{1,p}$. □

5.35 Let $\pi : E \to M$ and $\widetilde{\pi} : \widetilde{E} \to M$ be vector bundles over the smooth manifold M, and let T be a section on M in the vector bundle $\pi^k : \mathscr{T}^k(E; \widetilde{E}) \to M$. Then we have a map

$$T_M : \underbrace{\Gamma(M;E) \times \cdots \times \Gamma(M;E)}_{k} \to \Gamma(M;\widetilde{E})$$

defined by

$$T_M(s_1, \ldots, s_k)(p) = T(p)(s_1(p), \ldots, s_k(p))$$

for each point $p \in M$ and sections $s_1, \ldots, s_k \in \Gamma(M;E)$. Indeed, let $\phi : \mathscr{T}^k(E;\widetilde{E}) \to \mathscr{T}^k_1(E, \widetilde{E})$ and $i_{\widetilde{E}} : \widetilde{E} \to \widetilde{E}^{**}$ be the equivalences over M defined in 5.25 and Proposition 4.10. For each choice of $s_1, \ldots, s_k$ we have a linear map $B : \Gamma(M;\widetilde{E}^*) \to \mathscr{F}(M)$ given by

$$B(s)(p) = \phi \circ T(p)(s_1(p), \ldots, s_k(p), s(p))$$

for $p \in M$ and $s \in \Gamma(M;\widetilde{E}^*)$, which satisfies

$$B(s)(p) = s(p)(T_M(s_1, \ldots, s_k)(p)) = i_{\widetilde{E}} \circ T_M(s_1, \ldots, s_k)(p)(s(p)) .$$

Hence $i_{\widetilde{E}} \circ T_M(s_1, \ldots, s_k)$ is the unique section S in $\Gamma(M;\widetilde{E}^{**})$ with $B = S_M$ which is given by Proposition 4.47.

We see that T_M is multilinear over $\mathscr{F}(M)$. The next proposition shows that each such multilinear map arises from a unique section of π^k in this way.

5.36 Proposition Let $\pi : E \to M$ and $\widetilde{\pi} : \widetilde{E} \to M$ be vector bundles over a smooth manifold M, and let

$$F : \underbrace{\Gamma(M;E) \times \cdots \times \Gamma(M;E)}_{k} \to \Gamma(M;\widetilde{E})$$

be a map which is multilinear over $\mathscr{F}(M)$. Then there is a unique section T on M in the vector bundle $\pi^k : \mathscr{T}^k(E;\widetilde{E}) \to M$ such that $F = T_M$.

PROOF : Let

$$G : \underbrace{\Gamma(M;E) \times \cdots \times \Gamma(M;E)}_{k} \times \Gamma(M;\widetilde{E}^*) \to \mathscr{F}(M)$$

be the multilinear map given by

$$G(s_1,\dots,s_k,s)(p) = s(p)(F(s_1,\dots,s_k)(p))$$

for $p \in M$, $s_1,\dots,s_k \in \Gamma(M;E)$ and $s \in \Gamma(M;\widetilde{E}^*)$. Then we have that $F = T_M$ for a section T of π^k if and only if $G = (\phi \circ T)_M$, where $\phi : \mathscr{T}^k(E;\widetilde{E}) \to \mathscr{T}^k_1(E,\widetilde{E})$ is the equivalence over M defined in 5.25. The result therefore follows from Proposition 4.47 applied to multilinear map G. □

5.37 Remark Because of Proposition 5.36 we will not distinguish between the section T and the map T_M, and a section in $\Gamma(M;\mathscr{T}^k(E;\widetilde{E}))$ can be thought of as an operation on k sections in $\Gamma(M;E)$ yielding a section in $\Gamma(M;\widetilde{E})$. The map T_M is skew symmetric if and only if T is a section in $\Gamma(M;\Lambda^k(E;\widetilde{E}))$.

5.38 Let $\pi : E \to M$ be a vector bundle over a smooth manifold M, and let W be a finite dimensional vector space. For each section ω on M in the vector bundle $\pi' : \Lambda^k(E;W) \to M$ we have a map

$$\omega_M : \underbrace{\Gamma(M;E) \times \cdots \times \Gamma(M;E)}_{k} \to \mathscr{F}(M;W)$$

defined by

$$\omega_M(s_1,\dots,s_k)(p) = \omega(p)(s_1(p),\dots,s_k(p))$$

for each point $p \in M$ and sections $s_1,\dots,s_k \in \Gamma(M;E)$. Indeed, if $\rho'_k : \Lambda^k(E;E_W) \to \Lambda^k(E;W)$ is the equivalence ρ' over M defined in 5.26, we have that

$$\omega_M = \rho'_0 \circ (\rho'^{-1}_k \circ \omega)_M .$$

We see that ω_M is multilinear and skew symmetric over $\mathscr{F}(M)$. The next proposition shows that each such multilinear skew symmetric map arises from a unique section of π' in this way.

5.39 Proposition Let $\pi : E \to M$ be a vector bundle over a smooth manifold M and W be a finite dimensional vector space, and let

$$F : \underbrace{\Gamma(M;E) \times \cdots \times \Gamma(M;E)}_{k} \to \mathscr{F}(M;W)$$

be a map which is multilinear and skew symmetric over $\mathscr{F}(M)$. Then there is a unique section ω on M in the vector bundle $\pi' : \Lambda^k(E;W) \to M$ such that $F = \omega_M$.

PROOF : Apply Proposition 5.36 to the map $\rho'^{-1}_0 \circ F$, using the notation in 5.38. □

5.40 Remark Because of Proposition 5.39 we will not distinguish between the section ω and the map ω_M, and a section in $\Gamma(M;\Lambda^k(E;W))$ can be thought of as an operation on k sections in $\Gamma(M;E)$ yielding a vector-valued smooth function in $\mathscr{F}(M;W)$.

5.41 A section of the bundle $\pi' : \Lambda^k(TM;W) \to M$ defined on an open subset V of the smooth manifold M is called a *vector-valued differential form* of degree k or simply a *vector-valued k-form* on V with values in W. The set of all W-valued k-forms on V is denoted by $\Omega^k(V;W)$. In particular $\Omega^0(V;W) = \mathscr{F}(V;W) = C^\infty(V,W)$ is the set of C^∞-maps from V to W.

More generally, if $\pi : E \to M$ is a vector bundle over M, then a section of the bundle $\pi^k : \mathscr{T}^k(TM;E) \to M$ defined on an open subset V of M is called a *bundle-valued covariant tensor field of degree k* on V with values in E, and the set of all E-valued covariant tensor fields on V of degree k is denoted by $\mathscr{T}^k(V;E)$.

A section of the subbundle $\pi' : \Lambda^k(TM;E) \to M$ defined on an open subset V of M is called a *bundle-valued k-form* on V with values in E, and the set of all E-valued k-forms on V is denoted by $\Omega^k(V;E)$. In particular $\Omega^0(V;E) = \Gamma(V;E)$ is the set of all sections of π on V.

5.42 Proposition Let M be a smooth manifold and $\omega : M \to \Lambda^k(TM;W)$ be a map with $\pi' \circ \omega = id_M$. Then ω is a vector-valued k-form on M with values in W if and only if one of the following equivalent assertions are satisfied:

(1) Given a basis $\mathscr{C} = \{w_1, ..., w_m\}$ for W, the functions $\omega^j_{i_1,...,i_k} : U \to \mathbf{R}$ defined by
$$\phi \circ \omega(p) = \sum_{i_1<...<i_k, j} \omega^j_{i_1,...,i_k}(p)\,(dx^{i_1}(p) \wedge \cdots \wedge dx^{i_k}(p)) \otimes w_j$$
for $p \in U$ are smooth for every local chart (x,U) on M.

(2) The components ω^j of ω with respect to the basis $\mathscr{C} = \{w_1, ..., w_m\}$ for W defined by
$$\omega(p)(v_1, ..., v_k) = \sum_{j=1}^{m} \omega^j(p)(v_1, ..., v_k)\, w_j$$
for $p \in M$ and $v_1, ..., v_k \in T_pM$, are real-valued differential forms on M.

(3) For every functional $\lambda \in W^*$, the map $\lambda \cdot \omega : M \to \Lambda^k(TM)$ defined by
$$(\lambda \cdot \omega)(p) = \lambda \circ (\omega(p))$$
for $p \in M$, is a real-valued differential form on M.

PROOF: (1) Follows from Propositions 4.15, 4.45 and 5.13 (1).

(2) Using the maps ψ and μ defined in 5.15, the components ω^j of ω are given by the formula

$$\omega(p) = \sum_{j=1}^{m} \psi \circ \mu(\omega^j(p), w_j)$$

which is equivalent to

$$\phi \circ \omega(p) = \sum_{j=1}^{m} \omega^j(p) \otimes w_j .$$

Comparing this with the formula for $\phi \circ \omega(p)$ in (1), we hence obtain that

$$\omega^j(p) = \sum_{i_1<\ldots<i_k} \omega^j_{i_1,\ldots,i_k}(p)\, dx^{i_1}(p) \wedge \cdots \wedge dx^{i_k}(p)$$

for $p \in U$ and $j = 1, \ldots, m$, and so (2) is equivalent to (1) by Corollary 5.22.

(3) (3) is equivalent to (2) since $w_j^* \cdot \omega = \omega^j$ for $j = 1, \ldots, m$, and if

$$\lambda = \sum_{j=1}^{m} c_j w_j^* ,$$

then

$$\lambda \cdot \omega = \sum_{j=1}^{m} c_j \omega^j .$$

□

5.43 Remark When $k = 0$, then $\omega(p) \in W$ may be thought of as a map from $\{0\}$ to W mapping 0 to $\omega(p)$, so that $\lambda \circ (\omega(p))$ simply means $\lambda(\omega(p))$.

5.44 Proposition Let $\pi : E \to M$ be a vector bundle over a smooth manifold M and $\omega : M \to \Lambda^k(TM;E)$ be a map with $\pi' \circ \omega = id_M$. Then ω is a bundle-valued k-form on M with values in E if and only if one of the following equivalent assertions are satisfied :

(1) The functions $\omega^j_{i_1,\ldots,i_k} : U \cap V \to \mathbf{R}$ defined by

$$\phi \circ \omega(p) = \sum_{i_1<\ldots<i_k,\, j} \omega^j_{i_1,\ldots,i_k}(p)\, (dx^{i_1}(p) \wedge \cdots \wedge dx^{i_k}(p)) \otimes s_j(p)$$

for $p \in U \cap V$ are smooth for any local basis $\mathscr{B} = \{s_1, \ldots, s_m\}$ for E on some open subset V of M and every local chart (x, U) on M.

(2) The components ω^j of ω with respect to any local basis $\mathscr{B} = \{s_1, \ldots, s_m\}$ for E on some open subset V of M defined by

$$\omega(p)(v_1, \ldots, v_k) = \sum_{j=1}^{m} \omega^j(p)(v_1, \ldots, v_k)\, s_j(p)$$

for $p \in V$ and $v_1, \ldots, v_k \in T_pM$, are real-valued differential forms on V.

(3) For any section s of the dual bundle $\pi^* : E^* \to M$ on some open subset V of M, the map $s \cdot \omega : V \to \Lambda^k(TM)$ defined by

$$(s \cdot \omega)(p) = s(p) \circ \omega(p)$$

for $p \in V$, is a real-valued differential form on V.

PROOF: (1) Follows from Corollary 2.59 and Propositions 4.11, 4.15, 4.45 and 5.13 (1).

(2) Using the maps ψ and μ defined in 5.15, the components ω^j of ω are given by the formula

$$\omega(p) = \sum_{j=1}^{m} \psi \circ \mu(\omega^j(p), s_j(p))$$

which is equivalent to

$$\phi \circ \omega(p) = \sum_{j=1}^{m} \omega^j(p) \otimes s_j(p) .$$

Comparing this with the formula for $\phi \circ \omega(p)$ in (1), we hence obtain that

$$\omega^j(p) = \sum_{i_1 < \ldots < i_k} \omega^j_{i_1,\ldots,i_k}(p)\, dx^{i_1}(p) \wedge \cdots \wedge dx^{i_k}(p)$$

for $p \in U \cap V$ and $j = 1, \ldots, m$, and so (2) is equivalent to (1) by Corollary 5.22.

(3) (3) is equivalent to (2) since $s_j^* \cdot \omega = \omega^j$ for $j = 1, \ldots, m$, where $\mathscr{B}^* = \{s_1^*, \ldots, s_m^*\}$ is the dual local basis for E^* on V, and if

$$s = \sum_{j=1}^{m} c_j s_j^* ,$$

then

$$s \cdot \omega = \sum_{j=1}^{m} c_j \omega^j .$$

□

5.45 Example Given a smooth manifold M, we have a bundle-valued 1-form $\varepsilon \in \Omega^1(M;TM)$ on M with values in the tangent bundle defined by

$$\varepsilon(p) = id_{T_pM}$$

for $p \in M$, i.e.,

$$\varepsilon(p)(v) = v$$

for $p \in M$ and $v \in T_pM$, so that $\phi \circ \varepsilon = \delta$ is the Kronecker delta tensor on M defined in Example 4.68.

5.46 Proposition Let $G : W_1 \to W_2$ be a linear map between the finite dimensional vector spaces W_1 and W_2, and let $\omega \in \Omega^k(M;W_1)$ be a vector-valued k-form on a smooth manifold M with values in W_1. Then the map $G \cdot \omega : M \to \Lambda^k(TM;W_2)$ defined by

$$(G \cdot \omega)(p) = G \circ (\omega(p))$$

for $p \in M$, is a vector-valued k-form on M with values in W_2, and we have that

$$\lambda \cdot (G \cdot \omega) = (\lambda \circ G) \cdot \omega$$

for every functional $\lambda \in W_2^*$.

PROOF: Follows from Proposition 5.42 (3). □

5.47 Proposition Let $f : M \to N$ be a smooth map and $\omega \in \Omega^k(N;W)$ be a vector-valued k-form on N with values in a finite dimensional vector space W. Then the map $f^*(\omega) : M \to \Lambda^k(TM;W)$ defined by

$$f^*(\omega)(p)(v_1, \dots, v_k) = \omega(f(p))(f_*(v_1), \dots, f_*(v_k))$$

for $p \in M$ and $v_1, \dots, v_k \in T_pM$, is a vector-valued k-form on M with values in W, and we have that

$$\lambda \cdot f^*(\omega) = f^*(\lambda \cdot \omega)$$

for every functional $\lambda \in W^*$.

PROOF: Follows from Propositions 4.61 and 5.42 (3). □

5.48 Proposition Let $f : M \to N$ be a smooth map, and let $G : W_1 \to W_2$ be a linear map between the finite dimensional vector spaces W_1 and W_2. If $\omega \in \Omega^k(N;W_1)$ is a vector-valued k-form on the smooth manifold N with values in W_1, then we have that

$$f^*(G \cdot \omega) = G \cdot f^*(\omega).$$

PROOF: For every functional $\lambda \in W_2^*$ we have that

$$\lambda \cdot f^*(G \cdot \omega) = f^*(\lambda \cdot (G \cdot \omega)) = f^*((\lambda \circ G) \cdot \omega) = (\lambda \circ G) \cdot f^*(\omega) = \lambda \cdot (G \cdot f^*(\omega))$$

by Propositions 5.47 and 5.46. □

5.49 Let M be a smooth manifold and W be a finite dimensional vector space. If $f : V \to W$ is a vector-valued smooth function defined on an open neighbourhood V of a point $p \in M$, and if $l \in T_pM$, we define $l(f)$ to be the derivative $f_*(l)$ of f at p with respect to the tangent vector l, canonically identifying $T_{f(p)}W$ with W by means of the linear isomorphism $\omega_{f(p)} : T_{f(p)}W \to W$ given in Lemma 2.84. This is analogous to the definition of $l(f)$ for real-valued functions f given in 2.77. For any linear functional λ on W, we have that

$$\lambda \circ l(f) = \lambda \circ \omega_{f(p)} \circ f_*(l) = f_*(l)(\lambda) = l(\lambda \circ f) .$$

If $\mathscr{C} = \{w_1, ..., w_m\}$ is a basis for W and $x : W \to \mathbf{R}^m$ is the linear isomorphism given by

$$x(\sum_{j=1}^{m} a^j w_j) = a$$

for $a \in \mathbf{R}^m$, then in particular we have that

$$x^j \circ l(f) = l(x^j \circ f)$$

for $j = 1, ..., m$. Hence $l(f)$ for a W-valued function f is defined by means of its components with respect to any basis $\mathscr{C}$ for W, where l acts in the usual way as a local derivation on each component function.

If X is a vector field on M and $f \in \mathscr{F}(M;W)$, we let $X(f) \in \mathscr{F}(M;W)$ be the vector-valued function defined by

$$X(f)(p) = X_p(f)$$

for $p \in M$. We have that

$$\lambda \circ X(f) = X(\lambda \circ f)$$

for every linear functional λ on W.

In particular, if $f : V \to \mathbf{C}$ is a complex-valued function defined on V with smooth real and complex part, we define

$$l(f) = l(Re\, f) + i\, l(Im\, f)$$

and

$$X(f) = X(Re\, f) + i\, X(Im\, f) .$$

5.50 Proposition Let M be a smooth manifold and W be a finite dimensional vector space. If $f : U \to \mathbf{R}$ and $g : V \to W$ are, respectively, a real-valued and a vector-valued smooth function defined on open neighbourhoods U and V of a point $p \in M$, and if $l \in T_pM$, then we have that

$$l(fg) = l(f)g(p) + f(p)\,l(g)$$

PROOF : For every functional $\lambda \in W^*$ we have that

$$\lambda \circ l(fg) = l(f\lambda \circ g) = l(f)\ \lambda \circ g(p) + f(p)\ l(\lambda \circ g) = \lambda\ (l(f)g(p) + f(p)\,l(g)) .$$

□

5.51 Let $f : V \to W$ be a vector-valued smooth function defined on an open neighbourhood V of a point p in a smooth manifold M with values in a finite dimensional vector space W. In the same way as in 4.13 we define the *differential* of f at p to be the element $df(p)$ in $\Lambda^1(T_pM; W)$ defined by

$$df(p)(l) = l(f)$$

for each local derivation l in T_pM. By the discussion in 5.49 we know that this is the derivative of f at p with respect to the tangent vector l, so that

$$df(p) = f_{*p}$$

when we canonically identify $T_{f(p)}W$ with W as in Lemma 2.84. For every functional $\lambda \in W^*$ we have that

$$\lambda \circ df(p)(l) = \lambda \circ l(f) = l(\lambda \circ f) = d(\lambda \circ f)(p)(l)$$

so that

$$\lambda \cdot df = d(\lambda \circ f) .$$

5.52 Proposition Let ω be an n-form on a smooth manifold M^n, and let (x, U) and (y, V) be two local charts around a point p on M. If

$$\omega|_U = f\, dx^1 \wedge \cdots \wedge dx^n \quad \text{and} \quad \omega|_V = g\, dy^1 \wedge \cdots \wedge dy^n ,$$

then

$$f = g \det\left(\frac{\partial y^i}{\partial x^j}\right) \quad \text{on} \quad U \cap V$$

PROOF : Since

$$\left.\frac{\partial}{\partial x^j}\right|_q = \sum_{i=1}^{n} \frac{\partial y^i}{\partial x^j}(q) \left.\frac{\partial}{\partial y^i}\right|_q$$

for $j = 1, \ldots n$ and $q \in U \cap V$, it follows from Corollary 5.18 that

$$dy^1 \wedge \cdots \wedge dy^n = \det\left(\frac{\partial y^i}{\partial x^j}\right) dx^1 \wedge \cdots \wedge dx^n$$

on $U \cap V$ which completes the proof of the proposition. □

5.53 Proposition Let $f : M \to N$ be a smooth map, and let ω and η be differential forms on N. Then we have that

$$f^*(\omega \wedge \eta) = f^*\omega \wedge f^*\eta .$$

PROOF : For each point p on M, it follows from Definition 5.21 and Proposition 5.12 (2) that

$$\begin{aligned}[f^*(\omega \wedge \eta)](p) &= (f_{*p})^*\,[(\omega \wedge \eta)(f(p))] = (f_{*p})^*\,[\omega(f(p)) \wedge \eta(f(p))] \\ &= (f_{*p})^*\,[\omega(f(p))] \wedge (f_{*p})^*\,[\eta(f(p))] = [f^*\omega](p) \wedge [f^*\eta](p) \\ &= [f^*\omega \wedge f^*\eta](p) .\end{aligned}$$

□

EXTERIOR DIFFERENTIATION OF FORMS

5.54 We want to extend the differential $d : \Omega^0(M) \to \Omega^1(M)$ of functions on a smooth manifold M, as defined in 4.13 and Proposition 4.18, to a map $d : \Omega^k(M) \to \Omega^{k+1}(M)$ called the *exterior derivative* of arbitrary k-forms on M for $k \geq 0$. We first prove a local result.

5.55 Proposition Let U be a coordinate neighbourhood in a smooth manifold M^n. Then there is a unique family of maps $d_U : \Omega^k(U) \to \Omega^{k+1}(U)$ for $k \geq 0$ satisfying the following four conditions

(1) if $f \in \Omega^0(U)$, then $d_U(f) = df$ which is the ordinary differential of f ,

(2) $d_U(\omega + \eta) = d_U\,\omega + d_U\,\eta$ for every $\omega, \eta \in \Omega^k(U)$,

(3) $d_U(\omega \wedge \eta) = (d_U\,\omega) \wedge \eta + (-1)^k \omega \wedge (d_U\,\eta)$ whenever $\omega \in \Omega^k(U)$ and $\eta \in \Omega^l(U)$,

(4) $d_U \circ d_U = 0$.

If $x : U \to \mathbf{R}^n$ is a coordinate map and the form $\omega \in \Omega^k(U)$, where $k > 0$, is given by

$$\omega = \sum_{i_1 < \ldots < i_k} \omega_{i_1,\ldots,i_k}\, dx^{i_1} \wedge \cdots \wedge dx^{i_k} ,$$

then

$$\begin{aligned} d_U(\omega) &= \sum_{i_1 < \ldots < i_k} d\,\omega_{i_1,\ldots,i_k} \wedge dx^{i_1} \wedge \cdots \wedge dx^{i_k} \\ &= \sum_{i_1 < \ldots < i_k} \sum_{i=1}^{n} \frac{\partial\, \omega_{i_1,\ldots,i_k}}{\partial x^i}\, dx^i \wedge dx^{i_1} \wedge \cdots \wedge dx^{i_k} . \end{aligned}$$

PROOF : We first prove uniqueness of d_U. If ω is the k-form given in the last part of the proposition, it follows from (1) and (3) that

$$\begin{aligned} & d_U(\omega_{i_1,\ldots,i_k}\, dx^{i_1} \wedge \cdots \wedge dx^{i_k}) = d\,\omega_{i_1,\ldots,i_k} \wedge dx^{i_1} \wedge \cdots \wedge dx^{i_k} \\ & + (-1)^0\, \omega_{i_1,\ldots,i_k} \wedge d_U(dx^{i_1} \wedge \cdots \wedge dx^{i_k}) , \end{aligned}$$

and we will prove by induction that $d_U(dx^{i_1} \wedge \cdots \wedge dx^{i_k}) = 0$. In fact, for $k = 1$ we have that $d_U(dx^{i_1}) = d_U(d_U(x^{i_1})) = 0$ by (1) and (4). If the formula holds for $k-1$, it follows from (3) that

$$\begin{aligned} & d_U(dx^{i_1} \wedge \cdots \wedge dx^{i_k}) = d_U(dx^{i_1}) \wedge dx^{i_2} \wedge \cdots \wedge dx^{i_k} \\ & + (-1)^1\, dx^{i_1} \wedge d_U(dx^{i_2} \wedge \cdots \wedge dx^{i_k}) = 0 . \end{aligned}$$

The uniqueness of d_U as well as the formula for $d_U(\omega)$ now follows from (2).

To prove existence, we must show that the map d_U given by (1) and the last part of the proposition really satisfies all four conditions. We immediately see that (2) is satisfied. Note that we have

$$d_U\,(f\,dx^{i_1}\wedge\cdots\wedge dx^{i_k}) = df\wedge dx^{i_1}\wedge\cdots\wedge dx^{i_k}$$

for arbitrary indices $i_1,...,i_k$ in I_n. If some of the indices are equal, then both sides are equal to zero. Otherwise there is a permutation $\sigma\in S_k$ such that $i_{\sigma(1)}<...<i_{\sigma(k)}$, and by Propositions 5.4 and 5.13 (2) we have

$$\begin{aligned} d_U\,(f\,dx^{i_1}\wedge\cdots\wedge dx^{i_k}) &= d_U\,(\varepsilon(\sigma)\,f\,dx^{i_{\sigma(1)}}\wedge\cdots\wedge dx^{i_{\sigma(k)}}) \\ &= \varepsilon(\sigma)\,df\wedge dx^{i_{\sigma(1)}}\wedge\cdots\wedge dx^{i_{\sigma(k)}} = df\wedge dx^{i_1}\wedge\cdots\wedge dx^{i_k}. \end{aligned}$$

In wiew of (2), it is enough to prove (3) when $\omega = f\,dx^{i_1}\wedge\cdots\wedge dx^{i_k}$ and $\eta = g\,dx^{j_1}\wedge\cdots\wedge dx^{j_l}$, in which case we have

$$\begin{aligned} d_U\,(\omega\wedge\eta) &= (g\,df + f\,dg)\wedge dx^{i_1}\wedge\cdots\wedge dx^{i_k}\wedge dx^{j_1}\wedge\cdots\wedge dx^{j_l} \\ &= (df\wedge dx^{i_1}\wedge\cdots\wedge dx^{i_k})\wedge(g\,dx^{j_1}\wedge\cdots\wedge dx^{j_l}) \\ &\quad + (-1)^k(f\,dx^{i_1}\wedge\cdots\wedge dx^{i_k})\wedge(dg\wedge dx^{j_1}\wedge\cdots\wedge dx^{j_l}) \\ &= (d_U\,\omega)\wedge\eta + (-1)^k\,\omega\wedge(d_U\,\eta). \end{aligned}$$

Similarly, it is enough to prove (4) for the case where $\omega = f\,dx^{i_1}\wedge\cdots\wedge dx^{i_k}$. We then have that

$$\begin{aligned} d_U\,(d_U\,\omega) &= d_U\,(\sum_{i=1}^{n}\frac{\partial f}{\partial x^i}\,dx^i\wedge dx^{i_1}\wedge\cdots\wedge dx^{i_k}) \\ &= \sum_{i=1}^{n}\sum_{j=1}^{n}\frac{\partial^2 f}{\partial x^j\,\partial x^i}\,dx^j\wedge dx^i\wedge dx^{i_1}\wedge\cdots\wedge dx^{i_k} \end{aligned}$$

where the terms with $i=j$ vanish since $dx^i\wedge dx^i = 0$, and the other terms cancel in pairs since $\dfrac{\partial^2 f}{\partial x^i\,\partial x^j} = \dfrac{\partial^2 f}{\partial x^j\,\partial x^i}$ and $dx^i\wedge dx^j = -\,dx^j\wedge dx^i$.

The arguments above also hold for 0-forms if we make the usual convention that a wedge product with no factors is replaced by the constant function 1 on U. □

5.56 Remark Let ω be the k-form given in the last part of the proposition, where $x: U\to\mathbf{R}^m$ is any smooth map defined on an open subset U of M with m components. Then $d_U(\omega)$ is still uniquely determined by the first part of the given formula, provided that d_U satisfies the four conditions in the proposition. In fact the first part of the proof is still valid for the given ω even if (x,U) is not a coordinate system.

5.57 Proposition If V and U are coordinate neighbourhoods in a smooth manifold M with $V \subset U$, then

$$d_V(\omega|_V) = (d_U\omega)|_V$$

for every $\omega \in \Omega^k(U)$ where $k \geq 0$.

PROOF: The case $k = 0$ follows from 4.13, for if $f \in C^\infty(U)$ and $p \in V$ we have that $d(f|_V)(p) = df(p)$ since $f|_V$ and f coincide in a neighbourhood of p. This shows that $d(f|_V) = (df)|_V$.

If $k > 0$ it is enough to prove the assertion when $\omega = f\,dx^{i_1} \wedge \cdots \wedge dx^{i_k}$ for a coordinate map $x : U \to \mathbf{R}^n$. Then $x|_V$ is a coordinate map on V, and using the first part of the proof we have that

$$\begin{aligned} d_V(\omega|_V) &= d_V(f|_V\, d(x|_V)^{i_1} \wedge \cdots \wedge d(x|_V)^{i_k}) \\ &= d(f|_V) \wedge d(x|_V)^{i_1} \wedge \cdots \wedge d(x|_V)^{i_k}) = (d_U\,\omega)|_V\,. \end{aligned}$$

□

5.58 Proposition Let M^n be a smooth manifold. Then there is a unique family of maps $d : \Omega^k(M) \to \Omega^{k+1}(M)$ for $k \geq 0$ satisfying the following four conditions

(1) if $f \in \Omega^0(M)$, then df is the ordinary differential of f,

(2) $d(\omega + \eta) = d\omega + d\eta$ for every $\omega, \eta \in \Omega^k(M)$,

(3) $d(\omega \wedge \eta) = (d\omega) \wedge \eta + (-1)^k \omega \wedge (d\eta)$ whenever $\omega \in \Omega^k(M)$ and $\eta \in \Omega^l(M)$,

(4) $d \circ d = 0$.

If (x, U) is a local chart on M and the form $\omega \in \Omega^k(M)$, where $k > 0$, is given locally by

$$\omega|_U = \sum_{i_1 < \ldots < i_k} \omega_{i_1,\ldots,i_k}\, dx^{i_1} \wedge \cdots \wedge dx^{i_k}\,,$$

then

$$\begin{aligned} (d\omega)|_U &= \sum_{i_1 < \ldots < i_k} d\omega_{i_1,\ldots,i_k} \wedge dx^{i_1} \wedge \cdots \wedge dx^{i_k} \\ &= \sum_{i_1 < \ldots < i_k} \sum_{i=1}^{n} \frac{\partial \omega_{i_1,\ldots,i_k}}{\partial x^i}\, dx^i \wedge dx^{i_1} \wedge \cdots \wedge dx^{i_k}\,. \end{aligned}$$

PROOF: We first prove existence of d. If $\omega \in \Omega^k(M)$ and p is a point in M, we choose a coordinate neighbourhood U of p in M and define

$$d\omega(p) = d_U(\omega|_U)(p)\,.$$

Then d is well defined, for if V is another coordinate neighbourhood of p it follows from Proposition 5.57 that

$$d_U(\omega|_U)(p) = d_{U\cap V}(\omega|_{U\cap V})(p) = d_V(\omega|_V)(p)\,.$$

We have to show that d satisfies the four conditions of the proposition.

To prove (1), note that the first part of the proof of Proposition 5.57 also holds for every $f \in C^\infty(M)$. Hence

$$d_U(f|_U) = d(f|_U) = (df)|_U$$

where d now denotes the ordinary differential of the functions $f|_U$ and f, and this completes the proof of (1).

To prove (2) we use that the same condition is satisfied for $d|_U$ so that

$$\begin{aligned} d(\omega+\eta)(p) &= d_U(\omega|_U + \eta|_U)(p) = d_U(\omega|_U)(p) + d_U(\eta|_U)(p) \\ &= d\omega(p) + d\eta(p)\,, \end{aligned}$$

and the last two conditions follow in the same way.

Finally, the formula for $(d\omega)|_U$ follows from Proposition 5.55 since

$$(d\omega)|_U = d_U(\omega|_U)$$

by the definition of d.

We now prove that if $\omega \in \Omega^k(M)$ and p is a point in M, then $d\omega(p)$ is uniquely determined by the four conditions in the proposition. We first show that $d\omega(p)$ only depends on the local behavior of the k-form ω , i.e., if η is another k-form on M which coincides with ω on an open neighbourhood V of p, then $d\omega(p) = d\eta(p)$. To see this let W be an open neighbourhood of p with $\overline{W} \subset V$, and choose a smooth function $h : M \to \mathbf{R}$ with $h = 1$ on $\overline{W}$ and $\operatorname{supp}(h) \subset V$. Such a function h exists by Corollary 2.24. Then $h\omega = h\eta$, and by (1) and (3) we have that

$$d(h\omega)(p) = dh(p)\wedge\omega(p) + h(p)\wedge d\omega(p) = d\omega(p)$$

so that

$$d\omega(p) = d(h\omega)(p) = d(h\eta)(p) = d\eta(p)\,.$$

The uniqueness of $d\omega(p)$ follows from (1) in the case when $k = 0$. If $k > 0$ we choose a local chart (x,U) around p and suppose that $\omega|_U$ has the local expression given in the last part of the proposition. If V is an open neighbourhood of p with $\overline{V} \subset U$, we may again choose a smooth function $g : M \to \mathbf{R}$ with $g = 1$ on $\overline{V}$ and $\operatorname{supp}(g) \subset U$. We now extend the functions $g\,\omega_{i_1,\dots,i_k}$ and gx^i on U to smooth functions $\eta_{i_1,\dots,i_k}$ and y^i on M, respectively, by defining them to be zero outside U, and we let

$$\eta = \sum_{i_1<\dots<i_k} \eta_{i_1,\dots,i_k}\, dy^{i_1}\wedge\cdots\wedge dy^{i_k}\,.$$

Then η is a k-form on M which coincides with ω on V, and by Remark 5.56 we see that $d\,\eta$ is uniquely determined by the four conditions in the proposition and is given by

$$d\,\eta = \sum_{i_1<\ldots<i_k} d\,\eta_{i_1,\ldots,i_k} \wedge dy^{i_1} \wedge \cdots \wedge dy^{i_k} .$$

Combining this we now see that $d\,\omega\,(p)$ is uniquely determined and is given by

$$d\,\omega\,(p) = \sum_{i_1<\ldots<i_k} d\,\omega_{i_1,\ldots,i_k}(p) \wedge dx^{i_1}(p) \wedge \cdots \wedge dx^{i_k}(p)$$

thus proving the uniqueness of d. □

5.59 Proposition Let $f : M \to N$ be a smooth map, and let ω be a k-form on N. Then we have that

$$f^*(d\,\omega) = d\,(f^*\,\omega)\,.$$

In particular, if $g : N \to \mathbf{R}$ is a smooth function on N, then

$$f^*(dg) = d\,(g \circ f)\,.$$

PROOF: If l is a local derivation at a point p on M, it follows from 2.81, 4.13 and Proposition 4.61 that

$$(f^*(dg))\,(p)\,(l) = dg\,(f(p))\,(f_*\,l) = (f_*\,l)\,(g) = l\,(g \circ f) = d\,(g \circ f)\,(p)\,(l)$$

which proves the last relation in our proposition.

Using this, we can now show the first relation. Let (y,V) be a local chart on N, and let U be a coordinate neighbourhood contained in $f^{-1}(V)$. If

$$\omega\,|_V = \sum_{j_1<\ldots<j_k} \omega_{j_1,\ldots,j_k}\, dy^{j_1} \wedge \cdots \wedge dy^{j_k}\,,$$

it follows by induction from Proposition 5.53 and Remark 5.56 that

$$\begin{aligned} f^*(d\,\omega)\,|_U &= \sum_{j_1<\ldots<j_k} d\,(\omega_{j_1,\ldots,j_k} \circ f\,|_U) \wedge d(y^{j_1} \circ f\,|_U) \wedge \cdots \wedge d(y^{j_k} \circ f\,|_U) \\ &= d\,|_U\,(f^*\,\omega\,|_U) = d\,(f^*\,\omega)\,|_U . \end{aligned}$$

□

5.60 Definition The *exterior derivative* $d\,\omega$ of a vector-valued k-form $\omega \in \Omega^k(M;W)$ is defined by means of the components with respect to a basis $\mathscr{C} = \{w_1,\ldots,w_m\}$ for W, i.e., $d\,\omega \in \Omega^{k+1}(M;W)$ is the vector-valued $(k+1)$-form defined by

$$(d\,\omega)^j = d\,(\omega^j)$$

for $j = 1,\ldots,m$.

5.61 Proposition If $\omega \in \Omega^k(M;W)$ is a vector-valued k-form on M, then

$$\lambda \cdot d\,\omega = d\,(\lambda \cdot \omega)$$

for every functional $\lambda \in W^*$. In particular, the exterior derivative $d\,\omega$ does not depend on the choice of basis $\mathscr{C}$ for W in Definition 5.60.

PROOF : If $\mathscr{C}^* = \{w_1^*, ..., w_m^*\}$ is the dual of the basis $\mathscr{C}$ and

$$\lambda = \sum_{j=1}^{m} c_j w_j^* ,$$

then

$$\lambda \cdot d\,\omega = \sum_{j=1}^{m} c_j (d\,\omega)^j = \sum_{j=1}^{m} c_j d\,(\omega^j) = d\,(\sum_{j=1}^{m} c_j \omega^j) = d\,(\lambda \cdot \omega)\,. \qquad \square$$

5.62 Proposition Let $G : W_1 \to W_2$ be a linear map between the finite dimensional vector spaces W_1 and W_2, and let $\omega \in \Omega^k(M; W_1)$ be a vector-valued k-form on a smooth manifold M with values in W_1. Then we have that

$$d\,(G \cdot \omega) = G \cdot d\,\omega\,.$$

PROOF : For every functional $\lambda \in W_2^*$ we have that

$$\lambda \cdot d\,(G \cdot \omega) = d\,(\lambda \cdot (G \cdot \omega)) = d\,((\lambda \circ G) \cdot \omega) = (\lambda \circ G) \cdot d\,\omega = \lambda \cdot (G \cdot d\,\omega)$$

by Propositions 5.61 and 5.46. $\square$

5.63 Proposition Let $f : M \to N$ be a smooth map, and let $\omega \in \Omega^k(N; W)$ be a vector-valued k-form on N with values in a finite dimensional vector space W. Then we have that

$$f^*\,(d\,\omega) = d\,(f^*\,\omega)\,.$$

In particular, if $g : N \to W$ is a smooth vector-valued function on N, then

$$f^*\,(dg) = d\,(g \circ f)\,.$$

PROOF : For every functional $\lambda \in W^*$ we have that

$$\lambda \cdot f^*(d\omega) = f^*(\lambda \cdot d\omega) = f^*(d(\lambda \cdot \omega)) = d(f^*(\lambda \cdot \omega)) = d(\lambda \cdot f^*\omega) = \lambda \cdot d(f^*\omega)$$

by Propositions 5.47, 5.61 and 5.59. $\square$

5.64 Proposition Let M^n be a smooth manifold and W be a finite dimensional vector space. Then the family of maps $d : \Omega^k(M; W) \to \Omega^{k+1}(M; W)$ for $k \geq 0$ satisfies the following conditions:

(1) if $f \in \Omega^0(M; W)$, then df is the ordinary differential of f defined in 5.51,

(2) $d\,(\omega + \eta) = d\,\omega + d\,\eta$ for every $\omega, \eta \in \Omega^k(M; W)$,

(3) $d \circ d = 0$.

PROOF : Condition (1) follows from 5.51. Using Propositions 5.58 (2) and (4) we have that

$$\lambda \cdot d(\omega+\eta) = d(\lambda \cdot \omega + \lambda \cdot \eta) = d(\lambda \cdot \omega) + d(\lambda \cdot \eta) = \lambda \cdot (d\omega + d\eta)$$

and

$$\lambda \cdot d(d\omega) = d(\lambda \cdot d\omega) = d(d(\lambda \cdot \omega)) = 0$$

for every $\omega, \eta \in \Omega^k(M;W)$ and every functional $\lambda \in W^*$, thereby showing (2) and (3). □

WEDGE PRODUCT OF VECTOR-VALUED FORMS

5.65 Let W_1, W_2 and W be finite dimensional vector spaces, and let

$$\nu : W_1 \times W_2 \to W$$

be a bilinear map. If $T \in \mathscr{T}^k(V;W_1)$ and $S \in \mathscr{T}^l(V;W_2)$ for a vector space V, we define their *tensor product* $T \otimes_\nu S \in \mathscr{T}^{k+l}(V;W)$ with respect to ν by

$$(T \otimes_\nu S)(v_1, \ldots, v_k, v_{k+1}, \ldots, v_{k+l}) = \nu\,(T(v_1, \ldots, v_k), S(v_{k+1}, \ldots, v_{k+l}))\,.$$

If $\omega \in \Lambda^k(V;W_1)$ and $\eta \in \Lambda^l(V;W_2)$, their *wedge product* $\omega \wedge_\nu \eta \in \Lambda^{k+l}(V;W)$ with respect to ν is defined by

$$\omega \wedge_\nu \eta = \frac{(k+l)!}{k!\,l!}\,\mathscr{A}(\omega \otimes_\nu \eta)\,.$$

We will often omit the reference to ν in the notation if it is clear from the context.

Since ν is bilinear and $\mathscr{A}$ is linear, it follows that $\otimes_\nu$ and $\wedge_\nu$ are bilinear. If $F : U \to V$ is a linear map from a vector space U, we see that

$$F^*(T \otimes_\nu S) = F^*(T) \otimes_\nu F^*(S)\,,$$

and hence

$$F^*(\omega \wedge_\nu \eta) = F^*(\omega) \wedge_\nu F^*(\eta)$$

by Proposition 5.9.

5.66 Proposition If ω^i and η^j are the components of $\omega \in \Lambda^k(V;W_1)$ and $\eta \in \Lambda^l(V;W_2)$ with respect to the bases $\mathscr{B} = \{e_1, \ldots, e_r\}$ and $\mathscr{C} = \{f_1, \ldots, f_s\}$ for W_1 and W_2, respectively, and if $g_{ij} = \nu(e_i, f_j)$ for $i = 1, \ldots, r$ and $j = 1, \ldots, s$, then

$$(\omega \wedge_\nu \eta)(v_1, \ldots, v_{k+l}) = \sum_{ij} (\omega^i \wedge \eta^j)(v_1, \ldots, v_{k+l})\; g_{ij} \qquad (1)$$

for $v_1, \ldots, v_{k+l} \in V$, and

$$\lambda \circ (\omega \wedge_\nu \eta) = \sum_{ij} \lambda(g_{ij})\ \omega^i \wedge \eta^j \qquad (2)$$

for every linear functional $\lambda \in W^*$.

PROOF : Since $\nu : W_1 \times W_2 \to W$ is bilinear, we have that

$$\begin{aligned}
&(\omega \otimes_\nu \eta)(v_1, \ldots, v_k, v_{k+1}, \ldots, v_{k+l}) \\
&= \nu(\omega(v_1, \ldots, v_k), \eta(v_{k+1}, \ldots, v_{k+l})) \\
&= \sum_{ij} \omega^i(v_1, \ldots, v_k)\ \eta^j(v_{k+1}, \ldots, v_{k+l})\ \nu(e_i, f_j) \\
&= \sum_{ij} (\omega^i \otimes \eta^j)(v_1, \ldots, v_k, v_{k+1}, \ldots, v_{k+l})\ g_{ij}
\end{aligned}$$

for every $v_1, \ldots, v_{k+l} \in V$. By appying the alternation operator $\mathscr{A}$ on both sides, we obtain formula (1), and formula (2) is obtained from (1) by applying λ on both sides. □

5.67 Proposition Let V, W_1 and W be finite dimensional vector spaces, and let $\nu : W_1 \times W_1 \to W$ be a bilinear map so that

$$\nu(u, v) = (-1)^n\ \nu(v, u)$$

for $u, v \in W_1$, where n is a fixed integer. If $\omega \in \Lambda^k(V; W_1)$ and $\eta \in \Lambda^l(V; W_1)$, then

$$\omega \wedge \eta = (-1)^{kl+n}\ \eta \wedge \omega\,.$$

PROOF : Follows from Propositions 5.66 and 5.12 (3). □

5.68 Proposition If $\omega \in \Omega^k(V; W_1)$ and $\eta \in \Omega^l(V; W_2)$ are vector-valued differential forms on an open subset V of a smooth manifold M, we have a vector-valued differential form $\omega \wedge_\nu \eta \in \Omega^{k+l}(V; W)$, called the *wedge product* of ω and η with respect to ν, defined by

$$(\omega \wedge_\nu \eta)(p) = \omega(p) \wedge_\nu \eta(p)$$

for $p \in V$. If ω^i and η^j are the components of ω and η with respect to the bases $\mathscr{B} = \{e_1, \ldots, e_r\}$ and $\mathscr{C} = \{f_1, \ldots, f_s\}$ for W_1 and W_2, respectively, and if $g_{ij} = \nu(e_i, f_j)$ for $i = 1, \ldots, r$ and $j = 1, \ldots, s$, then

$$(\omega \wedge_\nu \eta)(p)(v_1, \ldots, v_{k+l}) = \sum_{ij} (\omega^i \wedge \eta^j)(p)(v_1, \ldots, v_{k+l})\ g_{ij} \qquad (1)$$

for $p \in V$ and $v_1, \ldots, v_{k+l} \in T_pM$, and

$$\lambda \cdot (\omega \wedge_\nu \eta) = \sum_{ij} \lambda(g_{ij})\ \omega^i \wedge \eta^j \qquad (2)$$

for every linear functional $\lambda \in W^*$.

PROOF : Follows from Propositions 5.66 and 5.42 (3). □

5.69 Proposition Let V be an open subset of a smooth manifold M. If W_1, W_2 and W are finite dimensional vector spaces and $\nu : W_1 \times W_2 \to W$ is a bilinear map, then

$$d(\omega \wedge_\nu \eta) = (d\omega) \wedge_\nu \eta + (-1)^k \omega \wedge_\nu (d\eta)$$

whenever $\omega \in \Omega^k(V;W_1)$ and $\eta \in \Omega^l(V;W_2)$.

PROOF : Follows from the same relation for real-valued forms given in proposition 5.58 (3), together with Definition 5.60 and Propositions 5.61 and 5.68. □

5.70 Proposition Let $f : M \to N$ be a smooth map. If W_1, W_2 and W are finite dimensional vector spaces and $\nu : W_1 \times W_2 \to W$ is a bilinear map, then

$$f^*(\omega \wedge_\nu \eta) = f^*\omega \wedge_\nu f^*\eta$$

whenever $\omega \in \Omega^k(N;W_1)$ and $\eta \in \Omega^l(N;W_2)$.

PROOF : For each point p on M, it follows from 5.65 and Proposition 5.68 that

$$[f^*(\omega \wedge_\nu \eta)](p) = (f_{*p})^*[(\omega \wedge_\nu \eta)(f(p))] = (f_{*p})^*[\omega(f(p)) \wedge_\nu \eta(f(p))]$$
$$= (f_{*p})^*[\omega(f(p))] \wedge_\nu (f_{*p})^*[\eta(f(p))] = [f^*\omega](p) \wedge_\nu [f^*\eta](p)$$
$$= [f^*\omega \wedge_\nu f^*\eta](p).$$

□

INTERIOR PRODUCT

5.71 Definition If V is a vector space, we define the *interior product* of a vector $v \in V$ and a covariant tensor $T \in \mathscr{T}^k(V)$ to be the covariant tensor $i_v T \in \mathscr{T}^{k-1}(V)$ given by

$$i_v T(v_1, \dots, v_{k-1}) = T(v, v_1, \dots, v_{k-1})$$

for $v_1, \dots, v_{k-1} \in V$. We let $i_v T = 0$ when $k = 0$. If T is alternating, so is $i_v T$.

We define the *kernel* of a k-form $\omega \in \Lambda^k(V)$ by

$$\ker(\omega) = \{v \in V \mid i_v \omega = 0\},$$

which equals the kernel of the linear map $\phi : V \to \Lambda^{k-1}(V)$ given by $\phi(v) = i_v \omega$ for $v \in V$.

5.72 Proposition Let $\pi : E \to M$ be an n-dimensional vector bundle over a smooth manifold M, and let $s : M \to \mathscr{T}^k(E)$ be a section in the vector bundle $\pi^k : \mathscr{T}^k(E) \to M$. Then we have a bundle map $\phi : E \to \mathscr{T}^{k-1}(E)$ over M given by

$$\phi(v) = i_v s(p)$$

for $p \in M$ and $v \in E_p$.

PROOF : Let $(t, \pi^{-1}(U))$ be a local trivialization in the vector bundle $\pi : E \to M$, and let $\tau : (\pi^k)^{-1}(U) \to U \times \mathscr{T}^k(\mathbf{R}^n)$ and $\tau' : (\pi^{k-1})^{-1}(U) \to U \times \mathscr{T}^{k-1}(\mathbf{R}^n)$ be the corresponding equivalences over U defined in Proposition 4.53. Then

$$\tau \circ s(p) = (p, h(p))$$

for $p \in U$, where $h : U \to \mathscr{T}^k(\mathbf{R}^n)$ is the smooth map given by $h(p) = (t_p^{-1})^* s(p)$. From this it follows that

$$\tau' \circ \phi \circ t^{-1}(p,a) = (p, B(a, h(p)))$$

for $(p,a) \in U \times \mathbf{R}^n$, where $B : \mathbf{R}^n \times \mathscr{T}^k(\mathbf{R}^n) \to \mathscr{T}^{k-1}(\mathbf{R}^n)$ is the bilinear map defined by $B(a,T) = i_a T$. Indeed, we have that

$$(t_p^{-1})^* \{i_{t_p^{-1}(a)} s(p)\}(a_1, \dots, a_{k-1}) = h(p)(a, a_1, \dots, a_{k-1}) = i_a h(p)(a_1, \dots, a_{k-1})$$

for $p \in U$ and $a, a_1, \dots, a_{k-1} \in \mathbf{R}^n$. □

5.73 Remark If $\omega : M \to \Lambda^k(E)$ is a section in the vector bundle $\pi' : \Lambda^k(E) \to M$, then the bundle map obtained from ω as described in Proposition 5.72 also induces a bundle map $\phi : E \to \Lambda^{k-1}(E)$ over M.

Suppose that there is an integer r such that $\dim\ker \omega(p) = \dim\ker(\phi_p) = r$ for every $p \in M$. Then it follows from Proposition 2.65 that $\ker(\phi)$ is an r-dimensional subbundle of E which we also call the *kernel* of ω and denote by $\ker(\omega)$.

5.74 Proposition If $v \in V$ and $\lambda_1, \dots, \lambda_k \in V^*$, we have that

$$i_v(\lambda_1 \wedge \cdots \wedge \lambda_k) = \sum_{i=1}^{k} (-1)^{i+1} i_v(\lambda_i)\, \lambda_1 \wedge \cdots \wedge \lambda_{i-1} \wedge \lambda_{i+1} \wedge \cdots \wedge \lambda_k .$$

PROOF : If $v_1, \dots, v_k \in V$, it follows from Proposition 5.13 (1) that

$$i_{v_1}(\lambda_1\wedge\cdots\wedge\lambda_k)(v_2,\dots,v_k)=(\lambda_1\wedge\cdots\wedge\lambda_k)(v_1,\dots,v_k)$$

$$=\sum_{\sigma\in S_k}\varepsilon(\sigma)\,\lambda_{\sigma(1)}(v_1)\cdots\lambda_{\sigma(k)}(v_k)=\begin{vmatrix}\lambda_1(v_1) & \lambda_1(v_2) & \cdots & \lambda_1(v_k)\\ \lambda_2(v_1) & \lambda_2(v_2) & \cdots & \lambda_2(v_k)\\ \vdots & \vdots & & \vdots\\ \lambda_k(v_1) & \lambda_k(v_2) & \cdots & \lambda_k(v_k)\end{vmatrix}$$

$$=\sum_{i=1}^{k}(-1)^{i+1}\lambda_i(v_1)\,(\lambda_1\wedge\cdots\wedge\lambda_{i-1}\wedge\lambda_{i+1}\wedge\cdots\wedge\lambda_k)(v_2,\dots,v_k)$$

$$=\sum_{i=1}^{k}(-1)^{i+1}i_{v_1}(\lambda_i)(\lambda_1\wedge\cdots\wedge\lambda_{i-1}\wedge\lambda_{i+1}\wedge\cdots\wedge\lambda_k)\,(v_2,\dots,v_k)$$

where we have expanded the determinant along the first column. □

5.75 Proposition If $\omega\in\Lambda^k(V)$ and $\eta\in\Lambda^l(V)$, we have that

$$i_v(\omega\wedge\eta)=(i_v\,\omega)\wedge\eta+(-1)^k\,\omega\wedge(i_v\,\eta)\,.$$

PROOF: If $\omega=\omega_1\wedge\cdots\wedge\omega_k$ and $\eta=\eta_1\wedge\cdots\wedge\eta_l$, where $\omega_1,\dots,\omega_k,\eta_1,\dots,\eta_l\in V^*$, it follows from Proposition 5.74 that

$$\begin{aligned}i_v(\omega\wedge\eta)&=\sum_{i=1}^{k}(-1)^{i+1}i_v(\omega_i)\,\omega_1\wedge\cdots\wedge\omega_{i-1}\wedge\omega_{i+1}\wedge\cdots\wedge\omega_k\wedge\eta\\&\quad+\sum_{j=1}^{l}(-1)^{k+j+1}i_v(\eta_j)\,\omega\wedge\eta_1\wedge\cdots\wedge\eta_{j-1}\wedge\eta_{j+1}\wedge\cdots\wedge\eta_l\\&=(i_v\,\omega)\wedge\eta+(-1)^k\,\omega\wedge(i_v\,\eta)\,.\end{aligned}$$

The general result now follows since every form $\omega\in\Lambda^k(V)$ and $\eta\in\Lambda^l(V)$ can be expressed as a linear combination of forms of the above type. □

5.76 Proposition If $F:V\to W$ is a linear map, and if $v\in V$ and $T\in\mathscr{T}^k(W)$, then we have that

$$i_v\,F^*(T)=F^*(i_{F(v)}T)\,.$$

PROOF: We have that

$$i_v\,F^*(T)\,(v_1,\dots,v_{k-1})=F^*(T)\,(v,v_1,\dots,v_{k-1})=T\,(F(v),F(v_1),\dots,F(v_{k-1}))$$
$$=i_{F(v)}T\,(F(v_1),\dots,F(v_{k-1}))=F^*(i_{F(v)}T\,)\,(v_1,\dots,v_{k-1})$$

for $v_1,\dots,v_{k-1}\in V$. □

5.77 Proposition If X is a vector field and ω a k-form on a smooth manifold M, we have a $(k-1)$-form $i_X\omega$ on M, called the *interior product* of X and ω, defined by

$$(i_X\,\omega)(p) = i_{X(p)}\;\omega\,(p)$$

for each $p \in M$.

PROOF : We have that $i_X\omega = \phi \circ X$, where $\phi : TM \to \Lambda^{k-1}(TM)$ is the bundle map over M obtained from ω as described in Remark 5.73. □

THE LIE DERIVATIVE OF FORMS

5.78 Definition If $T \in \mathscr{T}^k(M)$ is a covariant tensor field of degree k on a smooth manifold M, and if $\sigma \in S_k$ is a permutation, we define T^σ by $T^\sigma(p) = T(p)^\sigma$ for every $p \in M$.

We define a map $\mathscr{A} : \mathscr{T}^k(M) \to \mathscr{T}^k(M)$, called the *alternation* operator on $\mathscr{T}^k(M)$, by

$$\mathscr{A}\,(T)(p) = \mathscr{A}\,(T(p))$$

for each $p \in M$. We see that $\mathscr{A}$ is linear over $\mathscr{F}(M)$.

5.79 Proposition If $f : M \to N$ is a smooth map, and if $\mathscr{A}_M$ and $\mathscr{A}_N$ are the alternations on $\mathscr{T}^k(M)$ and $\mathscr{T}^k(N)$, respectively, then

$$f^*(\mathscr{A}_N(T)) = \mathscr{A}_M(f^*(T))$$

for every $T \in \mathscr{T}^k(N)$.

PROOF : For each point p on M, it follows from Definition 5.78 and Proposition 5.9 that

$$\begin{aligned}[f^*(\mathscr{A}_N(T))]\,(p) &= (f_{*p})^*\,[\,\mathscr{A}_N(T)\,(f(p))] = (f_{*p})^*\,[\,\mathscr{A}_{T_{f(p)}N}\,(T(f(p)))]\\ &= \mathscr{A}_{T_pM}\,[(f_{*p})^*\,(T(f(p)))] = \mathscr{A}_{T_pM}\,[f^*(T)(p)] = [\mathscr{A}_M(f^*(T))]\,(p)\end{aligned}$$

where $\mathscr{A}_{T_pM}$ and $\mathscr{A}_{T_{f(p)}N}$ are the alternations on $\mathscr{T}^k(T_pM)$ and $\mathscr{T}^k(T_{f(p)}N)$, respectively. □

5.80 Proposition If X is a vector field on the smooth manifold M, then the Lie derivative L_X commutes with the alternation operator $\mathscr{A}$ on $\mathscr{T}^k(M)$.

PROOF : Let γ be the global flow for X, and let $T \in \mathscr{T}^k(M)$ and $p \in M$. If $c : I(p) \to \mathscr{T}^k_l(T_pM)$ and $b : I(p) \to \mathscr{T}^k_l(T_pM)$ are the curves defined by

$$c(t) = \gamma_t^*(\mathscr{A}(T))(p) \quad \text{and} \quad b(t) = \gamma_t^* T(p) ,$$

and we canonically identify $T_{c(0)}\mathscr{T}^k_l(T_pM)$ and $T_{b(0)}\mathscr{T}^k_l(T_pM)$ with $\mathscr{T}^k_l(T_pM)$ as in Lemma 2.84, then

$$L_X(\mathscr{A}(T))(p) = c'(0) \quad \text{and} \quad L_X T(p) = b'(0) .$$

Since $c(t) = \mathscr{A} \circ b(t)$ for $t \in I(p)$ by Proposition 5.79, it now follows from Lemma 2.84 that

$$L_X(\mathscr{A}(T))(p) = c'(0) = \mathscr{A} \circ b'(0) = \mathscr{A}(L_X T)(p) .$$

□

5.81 Proposition Let X be a vector field on the smooth manifold M.

(1) If ω is a k-form on M, so is $L_X\,\omega$.

(2) If $\omega \in \Lambda^k(V)$ and $\eta \in \Lambda^l(V)$, then $L_X(\omega \wedge \eta) = (L_X\,\omega) \wedge \eta + \omega \wedge (L_X\,\eta)$.

PROOF: (1) It follows from Propositions 5.8 and 5.80 that

$$L_X\,\omega = L_X\,(\mathscr{A}\,(\omega)) = \mathscr{A}\,(L_X\,\omega)$$

which shows that $L_X\,\omega$ is a k-form on M.

(2) Follows from Propositions 4.79 (2) and 5.80. □

5.82 Proposition If X is a vector field and ω a k-form on the smooth manifold M, then we have that

$$L_X(d\,\omega) = d\,(L_X\omega) .$$

PROOF : Let p_0 be a point on M^n and (x,U) be a local chart around p_0, and choose an open neighbourhood W of p_0 contained in U and a real number $\varepsilon > 0$ such that $<-\varepsilon,\varepsilon> \times W \subset \mathscr{D}(X)$ and $\gamma(<-\varepsilon,\varepsilon> \times W) \subset U$. Let $c_q : I \to M$ be an integral curve for X with initial condition $q \in W$, and let $v^i(t)$ be obtained from $v^i = dx^i(q)$ for $i = 1,\dots,n$ by Lie transport along c_q as defined in Remark 4.78. Then we have that

$$v^i(t) = dx^i(q) \circ \gamma_{-t\,*} = \gamma_{-t}^*\,dx^i(q) = d\,(x^i \circ \gamma_{-t})\,(\gamma_t(q)) = d\,y^i(p) ,$$

where $p = c_q(t)$ and $(y, \gamma_t(W))$ is a local chart on M with $y = x \circ \gamma_{-t}$. Then if

$$\begin{aligned}\omega\,(c_q(t)) &= \sum_{i_1,\dots,i_k} C_{q\,i_1,\dots,i_k}(t)\, v^{i_1}(t) \wedge \cdots \wedge v^{i_k}(t) \\ &= \sum_{i_1,\dots,i_k} D_{t\,i_1,\dots,i_k}(p)\, d\,y^{i_1}(p) \wedge \ldots \wedge d\,y^{i_k}(p)\end{aligned}$$

where $D_{t\,i_1,\ldots,i_k} = C_{t\,i_1,\ldots,i_k} \circ \gamma_{-t}$, we have that

$$d\,\omega(c_q(t)) = \sum_{i_1,\ldots,i_k} \sum_{i=1}^{n} \frac{\partial}{\partial y^i}\bigg|_p D_{t\,i_1,\ldots,i_k}\, dy^i(p) \wedge dy^{i_1}(p) \wedge \cdots \wedge dy^{i_k}(p)$$
$$= \sum_{i_1,\ldots,i_k} \sum_{i=1}^{n} \frac{\partial}{\partial x^i}\bigg|_q C_{t\,i_1,\ldots,i_k}\, v^i(t) \wedge v^{i_1}(t) \wedge \cdots \wedge v^{i_k}(t)$$

so that

$$L_X(d\,\omega)(q) = \sum_{i_1,\ldots,i_k} \sum_{i=1}^{n} \frac{d}{dt}\bigg|_0 \frac{\partial}{\partial x^i}\bigg|_q C_{i_1,\ldots,i_k}\, dx^i(q) \wedge dx^{i_1}(q) \wedge \cdots \wedge dx^{i_k}(q)$$
$$= d\,(L_X\omega)(q)$$

for every $q \in W$. □

5.83 Proposition The Lie derivative of a k-form ω with respect to a vector field X on a smooth manifold M is given by the *Cartan formula*

$$L_X\,\omega = d\,(i_X\,\omega) + i_X\,(d\,\omega)\,.$$

PROOF: Follows since both L_X and $d \circ i_X + i_X \circ d$ are derivations on $\Omega^k(U)$ for each open subset U of M, which commute with d and restriction to open subsets, and have the same effect on functions. Indeed, whenever $\omega \in \Omega^k(U)$ and $\eta \in \Omega^l(U)$, it follows from Propositions 5.55 and 5.75 that

$$(d \circ i_X)(\omega \wedge \eta) = d\,[(i_X\,\omega) \wedge \eta + (-1)^k\,\omega \wedge (i_X\,\eta)]$$
$$= (d \circ i_X)(\omega) \wedge \eta + (-1)^{k-1}(i_X\,\omega) \wedge (d\,\eta)$$
$$+ (-1)^k (d\,\omega) \wedge (i_X\,\eta) + \omega \wedge (d \circ i_X)(\eta)$$

and

$$(i_X \circ d)(\omega \wedge \eta) = i_X\,[(d\,\omega) \wedge \eta + (-1)^k\,\omega \wedge (d\,\eta)]$$
$$= (i_X \circ d)(\omega) \wedge \eta + (-1)^{k+1}(d\,\omega) \wedge (i_X\,\eta)$$
$$+ (-1)^k (i_X\,\omega) \wedge (d\,\eta) + \omega \wedge (i_X \circ d)(\eta)$$

so that

$$(d \circ i_X + i_X \circ d)(\omega \wedge \eta) = (d \circ i_X + i_X \circ d)(\omega) \wedge \eta + \omega \wedge (d \circ i_X + i_X \circ d)(\eta).$$

Moreover, we have that

$$(d \circ i_X + i_X \circ d) \circ d = d \circ i_X \circ d = d \circ (d \circ i_X + i_X \circ d)$$

and

$$L_X(f) = X(f) = df(X) = i_X(df) = (d \circ i_X + i_X \circ d)(f)\,.$$

The result hence follows from the expressions for ω in local coordinates. □

5.84 Proposition Let ω be a k-form on the smooth manifold M, and let $X_1, \dots, X_{k+1}$ be vector fields on M. Then we have that

$$d\,\omega\,(X_1, \dots, X_{k+1}) = \sum_{i=1}^{k+1} (-1)^{i+1} X_i\,(\omega\,(X_1, \dots, \hat{X}_i, \dots, X_{k+1})) + \sum_{1 \le i < j \le k+1} (-1)^{i+j}\,\omega\,([X_i, X_j], X_1, \dots, \hat{X}_i, \dots, \hat{X}_j, \dots, X_{k+1})\,,$$

where a circumflex over a term means that it is omitted.

PROOF: We prove the formula by induction on k. It is clearly true for $k = 0$ since $d\,\omega\,(X_1) = X_1\,(\omega)$ when ω is a smooth function on M. Assuming that the formula is true for $k-1$, it follows from Propositions 4.79 (5) and 5.83 that

$$\begin{aligned}
d\,\omega\,(X_1, \dots, X_{k+1}) &= i_{X_1}(d\,\omega)\,(X_2, \dots, X_{k+1}) \\
&= L_{X_1}\,\omega\,(X_2, \dots, X_{k+1}) - d\,(i_{X_1}\,\omega)\,(X_2, \dots, X_{k+1}) \\
&= X_1\,(\omega\,(X_2, \dots, X_{k+1})) - \sum_{j=2}^{k+1} \omega\,(X_2, \dots, [X_1, X_j], \dots, X_{k+1}) \\
&\quad - \sum_{i=2}^{k+1} (-1)^i X_i\,(\omega\,(X_1, X_2, \dots, \hat{X}_i, \dots, X_{k+1})) \\
&\quad - \sum_{2 \le i < j \le k+1} (-1)^{i+j}\,\omega\,(X_1, [X_i, X_j], X_2, \dots, \hat{X}_i, \dots, \hat{X}_j, \dots, X_{k+1}) \\
&= \sum_{i=1}^{k+1} (-1)^{i+1} X_i\,(\omega\,(X_1, \dots, \hat{X}_i, \dots, X_{k+1})) \\
&\quad + \sum_{1 \le i < j \le k+1} (-1)^{i+j}\,\omega\,([X_i, X_j], X_1, \dots, \hat{X}_i, \dots, \hat{X}_j, \dots, X_{k+1})\,.
\end{aligned}$$

□

5.85 Remark Proposition 5.84 is also valid for vector-valued k-forms ω on M with values in a finite dimensional vector space W. To see this, we simply apply Proposition 5.84 to $\lambda \cdot \omega$ for an arbitrary linear functional λ on W, and use 5.49 and Proposition 5.61.

5.86 Remark The case $k = 1$ of Proposition 5.84 and Remark 5.85 is particularly useful. If ω is a real-valued or a vector-valued 1-form on the smooth manifold M, and if X and Y are vector fields on M, then we have that

$$d\,\omega(X,Y) = X\,(\,\omega\,(Y)) - Y\,(\,\omega\,(X)) - \omega\,([X,Y])\,.$$

5.87 Proposition Let ω be a k-form on a smooth manifold M, and suppose that there is an integer r such that $\dim\ker\omega(p) = r$ for every $p \in M$. Then $\ker(\omega)$ is an r-dimensional distribution on M. A vector field X on M belongs to $\ker(\omega)$ if and only if $i_X\,\omega = 0$. If X and Y are vector fields belonging to $\ker(\omega)$, then we have that

$$i_X\, i_Y\, d\omega = i_{[X,Y]}\;\omega, \tag{1}$$

and the distribution $\ker(\omega)$ is integrable when ω is a closed form.

PROOF : The first part of the proposition follows from Remark 5.73, and formula (1) follows from Proposition 5.84. □

5.88 Corollary Let ω be a nowhere vanishing 1-form on a smooth manifold M^n. Then $\ker(\omega)$ is an $(n-1)$-dimensional distribution on M. It is integrable if and only if $i^*d\omega = 0$, where $i : \ker(\omega) \to TM$ is the inclusion map.

PROOF : Follows from Propositions 5.87 and 2.56. □

Chapter 6

INTEGRATION ON MANIFOLDS

MANIFOLDS WITH BOUNDARY

6.1 Let A be any subset of $\mathbf{R}^n$ such that $O \subset A \subset \overline{O}$ for an open set O in $\mathbf{R}^n$. A map $f : A \to \mathbf{R}^m$ is said to be *smooth* at a point $p \in A$ if there is an open neighbourhood V_1 of p in $\mathbf{R}^n$ and a smooth map $f_1 : V_1 \to \mathbf{R}^m$ such that $f_1|_{A\cap V_1} = f|_{A\cap V_1}$.

The *derivative* $Df(p)$ is defined to be $Df_1(p)$. We see that it is well defined, for if $f_2 : V_2 \to \mathbf{R}^m$ is another smooth map defined on some open neighbourhood V_2 of p with $f_2|_{A\cap V_2} = f|_{A\cap V_2}$, then $Df_1(p) = Df_2(p)$. Indeed, this is clearly satisfied for $p \in O$, and if $p \in A - O$ we can choose a sequence $\{p_k\}$ in $O \cap V_1 \cap V_2$ converging to p so that

$$Df_1(p) = \lim_{k\to\infty} Df_1(p_k) = \lim_{k\to\infty} Df_2(p_k) = Df_2(p)\,.$$

f is said to be smooth on A if it is smooth at every point in A.

6.2 Proposition Let A be a subset of $\mathbf{R}^n$ such that $O \subset A \subset \overline{O}$ for an open set O in $\mathbf{R}^n$. Then a map $f : A \to \mathbf{R}^m$ is smooth on A if and only if it can be extended to a smooth map $\tilde{f} : V \to \mathbf{R}^m$ defined on some open set V in $\mathbf{R}^n$ containing A.

PROOF : If f has such an extension $\tilde{f}$, it follows directly from the definition that f is smooth on A.

Conversely, suppose that f is smooth on A, and choose for each point p in A an open neighbourhood V_p of p in $\mathbf{R}^n$ and a smooth map $g_p : V_p \to \mathbf{R}^m$ such that $g_p|_{A\cap V_p} = f|_{A\cap V_p}$. Let $V = \bigcup_{p\in A} V_p$, and let $\{\phi_p | p \in A\}$ be a partition of unity on V subordinate to the open cover $\{V_p | p \in A\}$. For each point p in A, let f_p be the extension of the map $\phi_p g_p$ to V defining it to be zero outside V_p. Then $\tilde{f} = \sum_{p\in A} f_p$ is a smooth map which is an extension of f to the open set V containing A. □

6.3 Let $\mathbf{H}^n = \{x \in \mathbf{R}^n | x^n \geq 0\}$ denote the *upper half space* in $\mathbf{R}^n$, and let $\operatorname{Int}\mathbf{H}^n = \{x \in \mathbf{R}^n | x^n > 0\}$ and $\partial\mathbf{H}^n = \{x \in \mathbf{R}^n | x^n = 0\}$ be its *interior* and its *boundary*, respectively.

A second countable Hausdorff space M is called a *topological manifold with boundary* if each point p in M has an open neighbourhood U homeomorphic to an open set in $\mathbf{H}^n$ for some integer $n \geq 0$. A map $x : U \to \mathbf{H}^n$ sending U homeomorphically onto an open subset $x(U)$ of $\mathbf{H}^n$, is called a *coordinate map* at p, and the pair

(x,U) is called a *local coordinate system* or a *local chart*. We can now define an atlas and a smooth structure on M in the same way as in Definition 2.1. If M has a smooth structure, it is called a *smooth manifold with boundary*.

A point p in M^n is called a *boundary point* if there is a local chart (x,U) around p with $x(p) \in \partial\mathbf{H}^n$. By the inverse function theorem it then follows that $y(p) \in \partial\mathbf{H}^n$ for every local chart (y,V) around p. The set of boundary points in M is called the *boundary* of M and is denoted by ∂M. Its complement $M - \partial M$ is called the *interior* of M and is denoted by $\operatorname{Int} M$.

If (x,U) is a local chart on M, we have that $x(U \cap \partial M) = x(U) \cap \partial\mathbf{H}^n$ and $x(U \cap \operatorname{Int} M) = x(U) \cap \operatorname{Int}\mathbf{H}^n$. Hence $\operatorname{Int} M$ is open in M, and it follows from Example 2.9 (c) and Proposition 2.36 that $\operatorname{Int} M$ and ∂M are smooth submanifolds of M without boundary of dimensions n and $n-1$, respectively. For each local chart (x,U) on M with $x(U) \cap \partial\mathbf{H}^n \neq \emptyset$, we have an induced local chart $(\tilde{x},\tilde{U})$ on ∂M defined by $\tilde{U} = U \cap \partial M$ and $\tilde{x} = p \circ x|_{\tilde{U}}$ where $p : \mathbf{R}^{n-1} \times \mathbf{R} \to \mathbf{R}^{n-1}$ is the projection on the first factor. These charts form an induced atlas on ∂M.

6.4 Let M^n be a smooth manifold with boundary, and let U be an open neighbourhood of a point p in M. Then a map $x : U \to \mathbf{R}^n$, sending U homeomorphically onto some subset $x(U)$ of $\mathbf{R}^n$, is called a *generalized coordinate map* at p if there is open set O in $\mathbf{R}^n$ with $O \subset x(U) \subset \overline{O}$, and if $y \circ x^{-1}$ and $x \circ y^{-1}$ are C^∞ for every local chart (y,V) in the smooth structure on M.

Note that the domain $x(U \cap V)$ of $y \circ x^{-1}$ is an open subset of $x(U)$ so that $x(U \cap V) = x(U) \cap W$ for some open set W in $\mathbf{R}^n$, and we have that $O \cap W \subset x(U) \cap W \subset \overline{O \cap W}$. To show the last inclusion, let Q be any open neighbourhood of a point $q \in x(U) \cap W$. Then $Q \cap W$ is also an open neighbourhood of q so that $Q \cap W \cap O \neq \emptyset$ which shows that $q \in \overline{O \cap W}$. Hence it is meaningful to require that $y \circ x^{-1}$ is C^∞ in the sense defined in 6.1. The same is true for $x \circ y^{-1}$ which is defined on the open subset $y(U \cap V)$ of the half space $\mathbf{H}^n$.

The pair (x,U) is called a *generalized local chart* around p. If $f : V \to \mathbf{R}$ is a smooth function defined on some open neighbourhood V of p, we define the i-th *partial derivative* of f at p with respect to (x,U) to be

$$\frac{\partial f}{\partial x^i}(p) = D_i(f \circ x^{-1})(x(p)) \, ,$$

i.e., the i-th partial derivative of $f \circ x^{-1}$ at $x(p)$.

EXACT FORMS

6.5 A k-form ω on a smooth manifold M is said to be *closed* if $d\omega = 0$. It is called *exact* if there is a $(k-1)$-form η on M such that $\omega = d\eta$. Since $d \circ d = 0$, every exact form is closed, but the converse statement is not true in general. We are going to show that it is in fact true for smoothly contractible manifolds, and therefore also true locally on an arbitrary smooth manifold.

6.6 Lemma Let $V = V_1 \oplus V_2$ where $\dim V_1 = n$ and $\dim V_2 = 1$, and let $\pi_i : V \to V$ be the projection with range V_i for $i = 1,2$. If $\lambda \in \Lambda^1(V)$ is a non-zero 1-form on V with $\lambda = \pi_2^*(\lambda)$, then every k-form $\omega \in \Lambda^k(V)$ can be written uniquely as

$$\omega = \alpha + (\lambda \wedge \beta)$$

where $\alpha \in \Lambda^k(V)$ and $\beta \in \Lambda^{k-1}(V)$ are forms with $\alpha = \pi_1^*(\alpha)$ and $\beta = \pi_1^*(\beta)$.

If v is a basis vector for V_2 with $\lambda(v) = 1$, then α and β are given by $\alpha = \pi_1^*(\omega)$ and $\beta(u_1,...,u_{k-1}) = \omega(v, \pi_1(u_1),...,\pi_1(u_{k-1}))$ for $u_1,...,u_{k-1} \in V$.

PROOF : The last part of the lemma follows immediately from the first, and it only remains to show the existence of the forms α and β. Let $\mathscr{B}_1 = \{v_1,...,v_n\}$ be a basis for V_1, and let $v_0 = v$. Then $\mathscr{B} = \{v_0,...,v_n\}$is a basis for V with dual basis $\mathscr{B}^* = \{v_0^*,...,v_n^*\}$ where $\lambda = v_0^*$. By Proposition 5.14 we have that

$$\omega = \sum_{0 \le i_1 < ... < i_k \le n} \omega(v_{i_1},...,v_{i_k})\, v_{i_1}^* \wedge \cdots \wedge v_{i_k}^*$$

which shows that

$$\omega = \alpha + (\lambda \wedge \beta)$$

where

$$\alpha = \sum_{1 \le i_1 < ... < i_k \le n} \omega(v_{i_1},...,v_{i_k})\, v_{i_1}^* \wedge \cdots \wedge v_{i_k}^*$$

and

$$\beta = \sum_{1 \le i_1 < ... < i_{k-1} \le n} \omega(v_0, v_{i_1},...,v_{i_{k-1}})\, v_{i_1}^* \wedge \cdots \wedge v_{i_{k-1}}^* ,$$

and we see that $\alpha = \pi_1^*(\alpha)$ and $\beta = \pi_1^*(\beta)$. □

6.7 Remark The condition $\omega = \pi_j^*(\omega)$ for a form $\omega \in \Lambda^r(V)$ in the above lemma, where $j = 1,2$, is equivalent to the condition that

$$\omega(v_1,...,v_r) = 0 \quad \text{if some} \quad v_i \in V_{3-j} .$$

6.8 Proposition Let M^n be a smooth manifold and $I = [0,1]$, and let $i_t : M \to M \times I$ be the map defined by $i_t(p) = (p,t)$ for $p \in M$ and $t \in I$. Then there is a family of linear maps

$$h : \Omega^k(M \times I) \to \Omega^{k-1}(M)$$

for $k > 0$ satisfying

$$d \circ h + h \circ d = i_1^* - i_0^* .$$

PROOF: Let $\rho_1 : M \times I \to M$ and $\rho_2 : M \times I \to I$ be the projections on the first and second factor, and let $i_p : I \to M \times I$ be the map defined by $i_p(t) = (p,t)$ for $t \in I$ and $p \in M$. Then the tangent space of $M \times I$ at a point (p,t) can be written as a direct sum

$$T_{(p,t)}(M \times I) = i_{t*}(T_pM) \oplus i_{p*}(T_tI) .$$

The summands are the ranges of the projections $\pi_{1*(p,t)}$ and $\pi_{2*(p,t)}$, where $\pi_1 = i_t \circ \rho_1$ and $\pi_2 = i_p \circ \rho_2$. Let $s : M \times I \to \mathbf{R}$ be the function defined by $s = i \circ \rho_2$ where $i : I \to \mathbf{R}$ is the inclusion map. Then we have a non-zero 1-form $ds(p,t)$ on $T_{(p,t)}(M \times I)$ with $ds(p,t) = \pi_2^*(ds(p,t))$ since $s \circ \pi_2 = i \circ \rho_2 \circ i_p \circ \rho_2 = i \circ \rho_2 = s$.

By Lemma 6.6 it now follows that each k-form $\omega \in \Omega^k(M \times I)$ can be written uniquely as

$$\omega = \alpha + (ds \wedge \beta)$$

where $\alpha \in \Omega^k(M \times I)$ and $\beta \in \Omega^{k-1}(M \times I)$ are forms with $\alpha(p,t) = \pi_1^*(\alpha(p,t))$ and $\beta(p,t) = \pi_1^*(\beta(p,t))$ for all $(p,t) \in M \times I$, and we define

$$(h\omega)(p)(v_1,\ldots,v_{k-1}) = \int_0^1 (i_t^*\beta)(p)(v_1,\ldots,v_{k-1})\,dt$$

for $p \in M$ and $v_1,\ldots,v_{k-1} \in T_pM$.

Having given an invariant definition of h, we can now prove the formula in the proposition locally using a coordinate system. Let (x,U) be a local chart in M, and let $\overline{x} = x \circ \rho_1$. Then $(\overline{x}^1,\ldots,\overline{x}^n,s)$ is a generalized coordinate map, in the sense given in 6.4, defined on the open coordinate neighbourhood $U \times I$ in $M \times I$.

If

$$\begin{aligned}\omega|_{U\times I} = & \sum_{i_1<\ldots<i_k} \alpha_{i_1,\ldots,i_k}\, d\overline{x}^{i_1} \wedge \ldots \wedge d\overline{x}^{i_k} \\ & + \sum_{j_1<\ldots<j_{k-1}} \beta_{j_1,\ldots,j_{k-1}}\, ds \wedge d\overline{x}^{j_1} \wedge \ldots \wedge d\overline{x}^{j_{k-1}} ,\end{aligned}$$

then

$$(h\omega)(p) = \sum_{j_1<\ldots<j_{k-1}} \left(\int_0^1 \beta_{j_1,\ldots,j_{k-1}}(p,t)\,dt \right) dx^{j_1}(p) \wedge \ldots \wedge dx^{j_{k-1}}(p)$$

for $p \in U$, so that

$$\begin{aligned} & d(h\omega)(p) \\ = & \sum_{j_1<\ldots<j_{k-1}} \sum_{j=1}^{n} \left(\int_0^1 \frac{\partial \beta_{j_1,\ldots,j_{k-1}}}{\partial x^j}(p,t)\,dt \right) dx^j(p) \wedge dx^{j_1}(p) \wedge \ldots \wedge dx^{j_{k-1}}(p) .\end{aligned}$$

Furthermore, we have that

$$(d\,\omega)\,|_{U\times I} = \sum_{i_1<\ldots<i_k}\sum_{i=1}^{n}\frac{\partial\,\alpha_{i_1,\ldots,i_k}}{\partial\overline{x}^i}\,d\,\overline{x}^i\wedge d\,\overline{x}^{i_1}\wedge\ldots\wedge d\,\overline{x}^{i_k}$$
$$+\sum_{i_1<\ldots<i_k}\frac{\partial\,\alpha_{i_1,\ldots,i_k}}{\partial s}\,d\,s\wedge d\,\overline{x}^{i_1}\wedge\ldots\wedge d\,\overline{x}^{i_k}$$
$$+\sum_{j_1<\ldots<j_{k-1}}\sum_{j=1}^{n}\frac{\partial\,\beta_{j_1,\ldots,j_{k-1}}}{\partial\overline{x}^j}\,d\,\overline{x}^j\wedge d\,s\wedge d\,\overline{x}^{j_1}\wedge\ldots\wedge d\,\overline{x}^{j_{k-1}}\,,$$

which implies

$$h(d\,\omega)(p)$$
$$=\sum_{i_1<\ldots<i_k}\left(\int_0^1\frac{\partial\,\alpha_{i_1,\ldots,i_k}}{\partial s}(p,t)\,dt\right)d\,x^{i_1}(p)\wedge\ldots\wedge d\,x^{i_k}(p)$$
$$-\sum_{j_1<\ldots<j_{k-1}}\sum_{j=1}^{n}\left(\int_0^1\frac{\partial\,\beta_{j_1,\ldots,j_{k-1}}}{\partial x^j}(p,t)\,dt\right)d\,x^j(p)\wedge d\,x^{j_1}(p)\wedge\ldots\wedge d\,x^{j_{k-1}}(p)\,.$$

Combining this, it follows that

$$d\,(h\omega)(p)+h(d\,\omega)(p)$$
$$=\sum_{i_1<\ldots<i_k}\left(\int_0^1\frac{\partial\,\alpha_{i_1,\ldots,i_k}}{\partial s}(p,t)\,dt\right)d\,x^{i_1}(p)\wedge\ldots\wedge d\,x^{i_k}(p)$$
$$=\sum_{i_1<\ldots<i_k}\left[\alpha_{i_1,\ldots,i_k}(p,1)-\alpha_{i_1,\ldots,i_k}(p,0)\right]d\,x^{i_1}(p)\wedge\ldots\wedge d\,x^{i_k}(p)$$
$$=\; i_1^*\,\omega\,(p)-i_0^*\,\omega\,(p)$$

which completes the proof of the proposition. □

6.9 Poincaré Lemma If M is a smoothly contractible manifold, then every closed form ω on M is exact.

PROOF: Let $F: M\times I\to M$ be a smooth contraction of M to a point $p\in M$, and let $i_t: M\to M\times I$ be the map defined by $i_t(q)=(q,t)$ for $q\in M$ and $t\in I$. Then $F\circ i_0 = id\,|_M$ is the identity map, and $F\circ i_1$ the constant map on M with value p. Hence it follows from Proposition 6.8 that

$$\omega=(F\circ i_0)^*\,\omega-(F\circ i_1)^*\,\omega=-(d\circ h+h\circ d)\,F^*\omega=d(-h\,F^*\omega)$$

since $d(F^*\omega) = F^*(d\omega) = 0$. □

6.10 Proposition Let (x,U) be a local chart around a point p on a smooth manifold M where $x(U)$ is star-shaped with respect to $x(p)$. Then U is smoothly contractible to p.

PROOF : We have a smooth contraction $F : U \times I \to U$ of U to p defined by

$$F(q,t) = x^{-1}(t\,x(p) + (1-t)\,x(q))$$

for $q \in U$ and $t \in I$. □

ORIENTATION

6.11 Definition We say that two ordered bases $(v_1,\dots,v_n)$ and $(w_1,\dots,w_n)$ for a vector space V are *equally oriented*, and write $(v_1,\dots,v_n) \sim (w_1,\dots,w_n)$, if

$$w_j = \sum_{i=1}^{n} a_{ij} v_i \quad \text{for} \quad j = 1,\dots,n \tag{1}$$

with $\det(a_{ij}) > 0$. If $\det(a_{ij}) < 0$, the bases are said to be *oppositely oriented*.

We see that $\sim$ is an equivalence relation in the family of ordered bases for V, and the equivalence class of $(v_1,\dots,v_n)$ is denoted by $[\,v_1,\dots,v_n\,]$ and is called an *orientation* of V. Since each ordered basis in V is equivalent to either $(v_1,\dots,v_n)$ or $(-v_1,\dots,v_n)$, we see that V has exactly two orientations. If μ denotes one orientation of V, the other is denoted by $-\mu$. A vector space V with an orientation μ is called an *oriented vector space* and is denoted by (V,μ) if we want to indicate the orientation explicitly. An ordered basis $(v_1,\dots,v_n)$ in (V,μ) is said to be *positively oriented* if $[\,v_1,\dots,v_n\,] = \mu$ and *negatively oriented* if $[\,v_1,\dots,v_n\,] = -\mu$.

Now let (V,μ) and (W,ν) be two oriented vector spaces of dimension n, and let $F : V \to W$ be a linear isomorphism. By applying F to both sides of (1), we see that $(v_1,\dots,v_n) \sim (w_1,\dots,w_n)$ implies that $(F(v_1),\dots,F(v_n)) \sim (F(w_1),\dots,F(w_n))$. From this and the same argument for F^{-1} we see that $\{(F(v_1),\dots,F(v_n)) \mid (v_1,\dots,v_n) \in \mu\}$ is an orientation of W which we denote by $\overline{F}(\mu)$. We say that F is *orientation preserving* if $\overline{F}(\mu) = \nu$ and *orientation reversing* if $\overline{F}(\mu) = -\nu$.

6.12 Example The *standard orientation* of $\mathbf{R}^n$ is $[\,e_1,\dots,e_n\,]$ where $(e_1,\dots,e_n)$ is the standard basis.

If M is a smooth manifold, we can introduce the *standard orientation* on the product bundle $M \times \mathbf{R}^n$ by choosing the orientation $[\,(p,e_1),\dots,(p,e_n)\,]$ on the fibre $\{p\} \times \mathbf{R}^n$ for each point p in M.

6.13 Definition Let $\pi : E \to M$ be an n-dimensional vector bundle, and choose an orientation μ_p on each fibre $\pi^{-1}(p)$. The family $\mu = \{\mu_p | p \in M\}$ is said to be an *orientation* of E if for each point p in M, there is an open neighbourhood U around p and n sections $s_1, ..., s_n$ of π on U such that $[\, s_1(q), ..., s_n(q)\,] = \mu_q$ for all $q \in U$.

The vector bundle E is said to be *orientiable* if it has an orientation, and *non-orientiable* otherwise. A vector bundle E with an orientation μ is called an *oriented vector bundle* and is denoted by (E, μ) if we want to indicate the orientation explicitly in the notation.

6.14 Proposition Let V be an n-dimensional vector space, and let $\omega \in \Lambda^n(V)$ be a non-zero n-form on V. Then there is a unique orientation μ of V such that

$$[\, v_1, ..., v_n\,] = \mu \quad \text{if and only if} \quad \omega(v_1, ..., v_n) > 0\,.$$

PROOF : Follows directly from Proposition 5.17. □

6.15 Proposition Let $\pi : E \to M$ be an n-dimensional vector bundle, and let ω be a nowhere vanishing section of the bundle $\pi' : \Lambda^n(E) \to M$ on M. Then there is a unique orientation μ of E such that

$$[\, v_1, ..., v_n\,] = \mu_p \quad \text{if and only if} \quad \omega(p)(v_1, ..., v_n) > 0$$

for every $p \in M$.

In the last two propositions μ is called the orientation determined by ω, and ω is called a *volume element* compatible with the orientation μ.

PROOF : Given a point p on M, let $(t, \pi^{-1}(U))$ be a local trivialization around p with U connected. Then we have a nowhere vanishing smooth function $f : U \to \mathbf{R}$ defined by

$$f(q) = \omega(q)(t_q^{-1}(e_1), ..., t_q^{-1}(e_n))$$

for $q \in U$, which must be either positive or negative on U, and we can define n sections $s_1, ..., s_n$ of π on U by $s_1(q) = \operatorname{sgn}(f(q))\, t_q^{-1}(e_1)$ and $s_i(q) = t_q^{-1}(e_i)$ for $i = 2, ..., n$ so that $[\, s_1(q), ..., s_n(q)\,] = \mu_q$ for all $q \in U$. This completes the proof that μ is an orientation of E. □

6.16 Proposition If μ is an orientation of the n-dimensional vector bundle $\pi : E \to M$, then there is a nowhere vanishing section ω of the bundle $\pi' : \Lambda^n(E) \to M$ on M which is compatible with μ.

PROOF : Let $\{U_\alpha | \alpha \in A\}$ be an open cover of M such that for each $\alpha \in A$, there are n sections $s_{\alpha,1}, ..., s_{\alpha,n}$ of π on U_α with $[\, s_{\alpha,1}(p), ..., s_{\alpha,n}(p)\,] = \mu_p$ for every $p \in U_\alpha$. Let $s^*_{\alpha,1}, ..., s^*_{\alpha,n}$ be the sections of the dual bundle $\pi^* : E^* \to M$ on U_α so that $\{s^*_{\alpha,1}(p), ..., s^*_{\alpha,n}(p)\}$ is the dual basis of $\{s_{\alpha,1}(p), ..., s_{\alpha,n}(p)\}$ for $p \in U_\alpha$.

To see that they are sections of π^*, let $\mathscr{E} = \{e_1, ..., e_n\}$ be the standard basis for $\mathbf{R}^n$ and $\mathscr{E}^* = \{e^1, ..., e^n\}$ be its dual basis, and let $(t_\alpha, \pi^{-1}(U_\alpha))$ be the local trivialization in E associated with the sections $s_{\alpha,1},...,s_{\alpha,n}$. For each $p \in U_\alpha$ and $i = 1,...,n$, we have that $s_{\alpha,i}(p) = t_{\alpha,p}^{-1}(e_i)$ so that $s_{\alpha,i}^*(p) = e^i \circ t_{\alpha,p}$ by Proposition 4.11. From Proposition 4.12 it therefore follows that $s_{\alpha,i}^*$ is a section of π^* on U_α for $i = 1,...,n$.

Now let $\{\phi_\alpha | \alpha \in A\}$ be a partition of unity on M subordinate to the open cover $\{U_\alpha | \alpha \in A\}$, and for each $\alpha \in A$, let ω_α be the extension of the section $\phi_\alpha s_{\alpha,1}^* \wedge \ldots \wedge s_{\alpha,n}^*$ to M defining it to be zero outside U_α. Then $\omega = \sum_{\alpha \in A} \omega_\alpha$ is a nowhere vanishing section of $\pi' : \Lambda^n(E) \to M$ which is compatible with μ. □

6.17 Corollary The n-dimensional vector bundle $\pi : E \to M$ is orientiable if and only if one of the following equivalent assertions are satisfied :

(1) The bundle $\pi' : \Lambda^n(E) \to M$ has a nowhere vanishing section.
(2) The bundle $\pi' : \Lambda^n(E) \to M$ is trivial.

PROOF : (1) follows from Proposition 6.15 and 6.16, and (2) from Proposition 2.58. □

6.18 Definition A smooth manifold M is said to be *orientiable* if its tangent bundle is orientiable, and *non-orientiable* otherwise. An orientation of TM is also called an *orientation* of M.

A smooth manifold M with an orientation μ is called an *oriented smooth manifold* and is denoted by (M, μ) if we want to indicate the orientation explicitly. A local chart (x, U) on M is said to be *positively oriented* with respect to μ if

$$[\left.\frac{\partial}{\partial x^1}\right|_p, ..., \left.\frac{\partial}{\partial x^n}\right|_p] = \mu_p$$

for every $p \in U$. It is said to be *negatively oriented* with respect to μ if

$$[\left.\frac{\partial}{\partial x^1}\right|_p, ..., \left.\frac{\partial}{\partial x^n}\right|_p] = -\mu_p$$

for every $p \in U$.

6.19 Remark If $(t_x, \pi^{-1}(U))$ is the local trivialization in the tangent bundle assosiated with the local chart (x, U) on M, then it follows by Remark 2.82 that (x, U) is positively (negatively) oriented if and only if the equivalence $t_x : \pi^{-1}(U) \to U \times \mathbf{R}^n$ is orientation preserving (orientation reversing) on each fibre when $U \times \mathbf{R}^n$ is given the standard orientation.

6.20 Proposition If (M^n, μ) is an oriented smooth manifold, then a local chart (x, U) on M with U connected is either positively or negatively oriented.

PROOF : If ω is a nowhere vanishing n-form on M which is compatible with the orientation μ, we have a nowhere vanishing smooth function $f : U \to \mathbf{R}$ defined by

$$f(q) = \omega(q)\left(\frac{\partial}{\partial x^1}\bigg|_q, \ldots, \frac{\partial}{\partial x^n}\bigg|_q \right)$$

for $q \in U$, which must be either positive or negative on U since U is connected. Hence the result follows from Definition 6.18. □

6.21 Remark If the local chart (x, U) on M^n is negatively oriented and $\phi : \mathbf{R}^n \to \mathbf{R}^n$ is the reflection given by

$$\phi(a_1, a_2, \ldots, a_n) = (-a_1, a_2, \ldots, a_n) ,$$

then $(\phi \circ x, U)$ is a local chart on M which is positively oriented.

6.22 Proposition A smooth manifold M^n is orientable if and only if it has an atlas $\mathscr{A}$ such that

$$\det\left(\frac{\partial y^i}{\partial x^j}\right) > 0 \quad \text{on} \quad U \cap V$$

for every pair of local charts (x, U) and (y, V) in $\mathscr{A}$.

PROOF : Assume first that M is orientable, and let μ be an orientation of M. Then it follows from Proposition 6.20 and Remark 6.21 that M has an atlas $\mathscr{A}$ consisting of positively oriented charts. If (x, U) and (y, V) are two local charts in $\mathscr{A}$, the inequality in the proposition follows from the relations

$$\frac{\partial}{\partial x^j}\bigg|_p = \sum_{i=1}^{n} \frac{\partial y^i}{\partial x^j}(p) \frac{\partial}{\partial y^i}\bigg|_p$$

for $j = 1, \ldots, n$ and $p \in U \cap V$.

Conversely, assume that M has an atlas $\mathscr{A}$ satisfying the property in the proposition, and let p be a point on M. To define the orientation μ_p on the tangent space T_pM, we choose a local chart (x, U) around p belonging to $\mathscr{A}$ and let

$$\mu_p = \left[\frac{\partial}{\partial x^1}\bigg|_p, \ldots, \frac{\partial}{\partial x^n}\bigg|_p \right] .$$

It follows from the inequality in the proposition the μ_p is well defined and does not depend on the choice of local chart (x, U).

The family $\mu = \{\mu_p | p \in M\}$ is an orientation of M since for each point p on M, there is a local chart (x, U) around p belonging to $\mathscr{A}$, and the partial derivations $X_j = \frac{\partial}{\partial x^j}$ for $j = 1, \ldots, n$ are n vector fields on U such that $[X_1(q), \ldots, X_n(q)] = \mu_q$ for all $q \in U$. □

6.23 Remark An atlas satisfying the property of Proposition 6.22 is called an *oriented atlas* on M. Such an atlas $\mathscr{A}$ uniquely determines an orientation μ of M such that every local chart in $\mathscr{A}$ is positively oriented with respect to μ.

6.24 Proposition Let M^n be an orientable smooth manifold, and let (x,U) and (y,V) be local charts on M with U and V connected. Then $\det\left(\frac{\partial y^i}{\partial x^j}\right)$ cannot change sign on $U \cap V$.

PROOF: Let μ be an orientation of M. Then it follows from Proposition 6.20 that each of the local charts (x,U) and (y,V) is either positively or negatively oriented. From the relations

$$\left.\frac{\partial}{\partial x^j}\right|_p = \sum_{i=1}^{n} \frac{\partial y^i}{\partial x^j}(p) \left.\frac{\partial}{\partial y^i}\right|_p$$

for $j = 1,...,n$ and $p \in U \cap V$, we see that $\det\left(\frac{\partial y^i}{\partial x^j}\right)$ is positive on $U \cap V$ if the local charts (x,U) and (y,V) are equally oriented, and it is negative on $U \cap V$ if they are oppositely oriented. □

6.25 Lemma Let M^n be a smooth manifold with boundary. If $p \in \partial M$ and (x,U) and (y,V) are two local charts around p, we have that

$$\frac{\partial y^n}{\partial x^j}(p) = 0 \quad \text{for} \quad j = 1,...,n-1 \quad \text{and} \quad \frac{\partial y^n}{\partial x^n}(p) > 0\,.$$

PROOF: Since

$$x(U \cap V \cap \partial M) = x(U \cap V) \cap \partial \mathbf{H}^n$$

and

$$y(U \cap V \cap \partial M) = y(U \cap V) \cap \partial \mathbf{H}^n\,,$$

we have that

$$(y \circ x^{-1})\,(x(U \cap V) \cap \partial \mathbf{H}^n) = y(U \cap V) \cap \partial \mathbf{H}^n$$

which implies the first assertion in the lemma.

The second assertion follows from the inclusion

$$(y \circ x^{-1})\,(x(U \cap V)) = y(U \cap V) \subset \mathbf{H}^n$$

and the fact that

$$\det\left(\frac{\partial y^i}{\partial x^j}(p)\right) \neq 0\,.$$

□

6.26 Proposition If M^n is an orientable smooth manifold with boundary, then ∂M is an orientable smooth submanifold of M^n without boundary of dimension $n-1$.

PROOF: By 6.3 we only need to show that ∂M is orientable. Let $\mathscr{A}$ be an oriented atlas on M, and let (x,U) and (y,V) be two local charts in $\mathscr{A}$ intersecting ∂M. If $(\tilde{x},\widetilde{U})$ and $(\tilde{y},\widetilde{V})$ are the induced local charts on ∂M, it follows by Lemma 6.25 that

$$\det D(y\circ x^{-1}) = \det D(\tilde{y}\circ\tilde{x}^{-1})\cdot\frac{\partial y^n}{\partial x^n}$$

on $\widetilde{U}\cap\widetilde{V}$ where $\frac{\partial y^n}{\partial x^n}>0$. Hence $\det D(y\circ x^{-1})>0$ on $U\cap V$ implies that $\det D(\tilde{y}\circ\tilde{x}^{-1})>0$ on $\widetilde{U}\cap\widetilde{V}$, showing that the induced atlas on ∂M is also oriented.

□

6.27 Let M^n be a smooth manifold with boundary, and let $p\in\partial M$. If (x,U) is a local chart around p, we say that the tangent vector $\sum_{i=1}^{n} v^i \frac{\partial}{\partial x^i}\Big|_p$ is *inward pointing* if $v^n>0$ and *outward pointing* if $v^n<0$.

This does not depend on the choice of local chart, for if $\sum_{i=1}^{n} w^i \frac{\partial}{\partial y^i}\Big|_p$ is the representation of the tangent vector with respect to another local chart (y,V) around p, we have that

$$w = D(y\circ x^{-1})(x(p))\,v$$

so that

$$w^n = \frac{\partial y^n}{\partial x^n}(p)\,v^n$$

by Lemma 6.25 where $\frac{\partial y^n}{\partial x^n}(p)>0$. Hence $w^n>0$ (resp., $w^n<0$) if and only if $v^n>0$ (resp., $v^n<0$).

6.28 Proposition Let (M^n,μ) be an oriented smooth manifold with boundary, and let $i:\partial M\to M$ be the inclusion map. Then we have an induced orientation $\partial\mu$ on ∂M such that given an ordered basis $(v_1,\dots,v_{n-1})$ in $T_p(\partial M)$, we have that

$$[\,v_1,\dots,v_{n-1}\,] = (\partial\mu)_p \quad \text{if and only if} \quad [\,u, i_* v_1,\dots,i_* v_{n-1}\,] = \mu_p$$

for any outward pointing vector $u\in T_pM$.

If $\mathscr{A}$ is an atlas on M consisting of positively oriented local charts with respect to μ, then the orientation determined by the induced atlas on ∂M is $(-1)^n\,\partial\mu$.

PROOF: We first note that $(u,i_* v_1,\dots,i_* v_{n-1})$ is an ordered basis for T_pM since $u\notin i_*T_p(\partial M)$. If u' is another outward pointing vector in T_pM, then

$$u' = a u + \sum_{i=1}^{n-1} a_i\, i_* v_i\,,$$

where $a>0$. Indeed, if $(t_x,\pi^{-1}(U))$ is the local trivialization assosiated with a local chart (x,U) around p, then $t_{x,p}(u')^n = a t_{x,p}(u)^n$ where $t_{x,p}(u)^n > 0$ and $t_{x,p}(u')^n > 0$ since u and u' are outward pointing. Hence

$$[u', i_* v_1, \dots, i_* v_{n-1}] = [u, i_* v_1, \dots, i_* v_{n-1}]$$

which shows that $(\partial\mu)_p$ does not depend on the choice of u.

If $(v_1, \dots, v_{n-1})$ and $(w_1, \dots, w_{n-1})$ are two ordered bases for $T_p(\partial M)$, we clearly have that $(v_1, \dots, v_{n-1}) \sim (w_1, \dots, w_{n-1})$ if and only if $(u, i_* v_1, \dots, i_* v_{n-1}) \sim (u, i_* w_1, \dots, i_* w_{n-1})$. For suppose that

$$w_j = \sum_{i=1}^{n-1} A_{ij}\, v_i \quad \text{for} \quad j = 1, \dots, n-1,$$

and that

$$u = B_{00}\, u + \sum_{i=1}^{n-1} B_{i0}\, i_* v_i \quad \text{and}$$

$$i_* w_j = B_{0j}\, u + \sum_{i=1}^{n-1} B_{ij}\, i_* v_i \quad \text{for} \quad j = 1, \dots, n-1.$$

Then

$$B = \begin{pmatrix} 1 & 0 \\ 0 & A \end{pmatrix}$$

which implies that $\det B = \det A$, showing that $(\partial\mu)_p$ is well defined by the conditions in the proposition.

Now let (x,U) be a local chart around $p \in \partial M$ which is positively oriented with respect to μ, and let $(\tilde{x}, \widetilde{U})$ be the induced local chart on ∂M. Then $-\left.\frac{\partial}{\partial x^n}\right|_p$ is an outward pointing vector in T_pM, and we have that

$$[\,-\left.\frac{\partial}{\partial x^n}\right|_p, i_*\left(\left.\frac{\partial}{\partial \tilde{x}^1}\right|_p\right), \dots, i_*\left(\left.\frac{\partial}{\partial \tilde{x}^{n-1}}\right|_p\right)] = [\,-\left.\frac{\partial}{\partial x^n}\right|_p, \left.\frac{\partial}{\partial x^1}\right|_p, \dots, \left.\frac{\partial}{\partial x^{n-1}}\right|_p]$$

$$= (-1)^n\, [\left.\frac{\partial}{\partial x^1}\right|_p, \dots, \left.\frac{\partial}{\partial x^n}\right|_p] = (-1)^n\, \mu_p$$

so that

$$[\left.\frac{\partial}{\partial \tilde{x}^1}\right|_p, \dots, \left.\frac{\partial}{\partial \tilde{x}^{n-1}}\right|_p] = (-1)^n\, (\partial\mu)_p\,.$$

This shows the last part of the proposition and also that $\partial\mu$ is an orientation of ∂M. □

6.29 Let (N, ν) and (M, μ) be two oriented smooth manifolds where N^{n-1} is an immersed submanifold of M^n, and let $i : N \to M$ be the inclusion map and $p \in N$.

Then a tangent vector $u \in T_pM - i_*(T_pN)$ is said to be *positive* if given an ordered basis $(v_1, \dots, v_{n-1})$ in T_pN, we have that

$$[v_1, \dots, v_{n-1}] = v_p \quad \text{if and only if} \quad [u, i_* v_1, \dots, i_* v_{n-1}] = \mu_p .$$

A tangent vector $u \in T_pM - i_*(T_pN)$ is said to be *negative* if $-u$ is positive.

The sign of u is well defined and does not depend on the choice of ordered basis $(v_1, \dots, v_{n-1})$, for let $(w_1, \dots, w_{n-1})$ be another ordered basis in T_pN. Then we clearly have that $(v_1, \dots, v_{n-1}) \sim (w_1, \dots, w_{n-1})$ if and only if $(u, i_* v_1, \dots, i_* v_{n-1}) \sim (u, i_* w_1, \dots, i_* w_{n-1})$.

Two tangent vectors $u, u' \in T_pM - i_*(T_pN)$ have the same sign if

$$[u, i_* v_1, \dots, i_* v_{n-1}] = [u', i_* v_1, \dots, i_* v_{n-1}]$$

for an ordered basis $(v_1, \dots, v_{n-1})$ in T_pN, which is satisfied if $u' = au + w$ where $a > 0$ and $w \in i_*(T_pN)$. They have opposite sign if $a < 0$.

A vector field $X : N \to TM$ along the inclusion map $i : N \to M$ is said to be *positive* if $X(p)$ is a positive vector for every $p \in N$.

6.30 Remark In the case where (M, μ) is an oriented smooth manifold with boundary and (N, v) is the oriented manifold $(\partial M, \partial \mu)$, the positive (negative) tangent vectors are the outward (inward) pointing vectors.

6.31 Proposition Let N^{n-1} be an immersed submanifold of an oriented smooth manifold (M^n, μ), and let $X : N \to TM$ be a vector field along the inclusion map $i : N \to M$ with $X(p) \in T_pM - i_*(T_pN)$ for every $p \in N$. Then we have a unique induced orientation v of N such that X is positive, i.e., given an ordered basis $(v_1, \dots, v_{n-1})$ in the tangent space T_pN at an arbitrary point $p \in N$, we have that

$$[v_1, \dots, v_{n-1}] = v_p \quad \text{if and only if} \quad [X(p), i_* v_1, \dots, i_* v_{n-1}] = \mu_p .$$

PROOF: If $(v_1, \dots, v_{n-1})$ and $(w_1, \dots, w_{n-1})$ are two ordered bases in T_pN, we clearly have that $(v_1, \dots, v_{n-1}) \sim (w_1, \dots, w_{n-1})$ if and only if $(X(p), i_* v_1, \dots, i_* v_{n-1}) \sim (X(p), i_* w_1, \dots, i_* w_{n-1})$, which shows that v_p is well defined by the conditions in the proposition.

Now let ω be a volume element on M compatible with μ, and let $s_1, \dots, s_{n-1}$ be a local basis for TN on an open neighbourhood U of a point $p \in N$ such that

$$[s_1(p), \dots, s_{n-1}(p)] = v_p$$

which implies that

$$\omega(p)(X(p), i_* \circ s_1(p), \dots, i_* \circ s_{n-1}(p)) > 0 .$$

By continuity there is therefore a neighbourhood V of p contained in U so that

$$\omega(q)(X(q), i_* \circ s_1(q), \dots, i_* \circ s_{n-1}(q)) > 0$$

for $q \in V$, which implies that

$$[s_1(q), \dots, s_{n-1}(q)] = v_q$$

for $q \in V$. This shows that v is an orientation of N. □

INTEGRATION OF DIFFERENTIAL FORMS

6.32 We want to define the integral of an n-form ω with compact support on an oriented smooth manifold (M^n, μ). Suppose first that there is a positively oriented local chart (x, U) on M with $\text{supp}(\omega) \subset U$. Then we have a smooth function $f : U \to \mathbf{R}$ such that

$$\omega|_U = f\, dx^1 \wedge \ldots \wedge dx^n$$

with $\text{supp}(f) = \text{supp}(\omega)$, and we define

$$\int_M \omega = \int_{x(U)} f \circ x^{-1}$$

where the last integral is the Riemann integral over any rectangle containing the support of $f \circ x^{-1}$. We extend the integrand to $\mathbf{R}^n$ by defining it to be zero outside $x(U)$. The next lemma shows that $\int_M \omega$ is well defined and does not depend on the choice of local chart (x, U) containing $\text{supp}(\omega)$.

6.33 Lemma Let (x, U) and (y, V) be positively oriented local charts on M with $\text{supp}(\omega) \subset U \cap V$. If

$$\omega|_U = f\, dx^1 \wedge \ldots \wedge dx^n \quad \text{and} \quad \omega|_V = g\, dy^1 \wedge \ldots \wedge dy^n\,,$$

then

$$\int_{x(U)} f \circ x^{-1} = \int_{y(V)} g \circ y^{-1}\,.$$

PROOF: Since $\text{supp}(f)$ and $\text{supp}(g)$ are both contained in $U \cap V$, we may replace $x(U)$ and $y(V)$ in the above integrals by $x(U \cap V)$ and $y(U \cap V)$, respectively. By Proposition 5.52 we have that

$$f \circ x^{-1} = (g \circ y^{-1}) \circ (y \circ x^{-1}) \det D(y \circ x^{-1})$$

on $x(U \cap V)$, and the coordinate transformation $y \circ x^{-1}$ is a diffeomorphism mapping $x(U \cap V)$ onto $y(U \cap V)$. Since the local charts (x, U) and (y, V) are positively oriented, we have that $\det D(y \circ x^{-1}) > 0$ on $x(U \cap V)$. Hence

$$\int_{x(U \cap V)} f \circ x^{-1} = \int_{x(U \cap V)} (g \circ y^{-1}) \circ (y \circ x^{-1})\, |\det D(y \circ x^{-1})| = \int_{y(U \cap V)} g \circ y^{-1}$$

by the change of variable formula for multiple integrals. □

6.34 Lemma Let ω and (x, U) be as above, and let $\mathscr{A} = \{(x_\alpha, U_\alpha) \,|\, \alpha \in A\}$ be an atlas on M consisting of positively oriented local charts. Let $\{\phi_\alpha \,|\, \alpha \in A\}$ be a partition of unity on M subordinate to the open cover $\{U_\alpha \,|\, \alpha \in A\}$. Then we have that

$$\int_M \omega = \sum_{\alpha\in A} \int_M \phi_\alpha\,\omega$$

where only a finite number of terms in the sum are non-zero.

PROOF : Each $\phi_\alpha\,\omega$ is an n-form on M with compact support contained in U, and since $\operatorname{supp}(\omega)$ is compact, $\phi_\alpha\,\omega$ is non-zero only for a finite number of indices $\alpha \in A$. If

$$\omega\,|_U = f\,dx^1 \wedge \ldots \wedge dx^n\,,$$

we have that

$$\int_M \omega = \int_{x(U)} f \circ x^{-1} = \sum_{\alpha\in A} \int_{x(U)} (\phi_\alpha f) \circ x^{-1} = \sum_{\alpha\in A} \int_M \phi_\alpha\,\omega\,.$$

□

6.35 We now want to define the integral of an n-form ω with compact support on an oriented smooth manifold (M^n, μ) in the case where $\operatorname{supp}(\omega)$ is not necessarily contained in a single coordinate neighbourhood. Let $\mathscr{A} = \{(x_\alpha, U_\alpha) \,|\, \alpha \in A\}$ be an atlas on M consisting of positively oriented local charts with respect to μ, and let $\{\phi_\alpha \,|\, \alpha \in A\}$ be a partition of unity on M subordinate to the open cover $\{U_\alpha \,|\, \alpha \in A\}$. Then we define

$$\int_M \omega = \sum_{\alpha\in A} \int_M \phi_\alpha\,\omega\,.$$

Since $\operatorname{supp}(\omega)$ is compact, only a finite number of terms in the sum are non-zero. The next lemma shows that $\int_M \omega$ is well defined and does not depend on the choice of atlas and partition of unity on M.

6.36 Lemma Suppose we have two atlases $\mathscr{A} = \{(x_\alpha, U_\alpha) \,|\, \alpha \in A\}$ and $\mathscr{B} = \{(y_\beta, V_\beta) \,|\, \beta \in B\}$ on M consisting of positively oriented local charts, and let $\{\phi_\alpha \,|\, \alpha \in A\}$ and $\{\psi_\beta \,|\, \beta \in B\}$ be partitions of unity on M subordinate to the open covers $\{U_\alpha \,|\, \alpha \in A\}$ and $\{V_\beta \,|\, \beta \in B\}$, respectively. Then we have that

$$\sum_{\alpha\in A} \int_M \phi_\alpha\,\omega = \sum_{\beta\in B} \int_M \psi_\beta\,\omega\,.$$

PROOF : By Lemma 6.34 we have that

$$\int_M \phi_\alpha\,\omega = \sum_{\beta\in B} \int_M \psi_\beta\,\phi_\alpha\,\omega$$

for each $\alpha \in A$, and

$$\int_M \psi_\beta\,\omega = \sum_{\alpha\in A} \int_M \phi_\alpha\,\psi_\beta\,\omega$$

for each $\beta \in B$. Hence it follows that

$$\sum_{\alpha\in A}\int_M \phi_\alpha\,\omega = \sum_{\alpha\in A}\sum_{\beta\in B}\int_M \psi_\beta\,\phi_\alpha\,\omega = \sum_{\beta\in B}\sum_{\alpha\in A}\int_M \phi_\alpha\,\psi_\beta\,\omega = \sum_{\beta\in B}\int_M \psi_\beta\,\omega\,.$$

□

6.37 Remark If there is a positively oriented local chart (x,U) on M with $\operatorname{supp}(\omega)\subset U$, we now have two definitions of $\int_M \omega$ given in 6.32 and 6.35. However, by Lemma 6.34, these two definitions coincide.

6.38 Stokes' Theorem Let M^n be an oriented smooth manifold with boundary, and let $i:\partial M \to M$ be the inclusion map. If ω is an $(n-1)$ - form on M with compact support, then

$$\int_M d\,\omega = \int_{\partial M} i^*\omega\,.$$

PROOF : Let μ be the orientation of M, and let $\partial\mu$ be the induced orientation on ∂M given in Proposition 6.28. Suppose first that there is a positively oriented local chart (x,U) on M with $\operatorname{supp}(\omega)\subset U$. If

$$\omega|_U = \sum_{i=1}^n (-1)^{i-1} f_i\,dx^1\wedge\ldots\wedge dx^{i-1}\wedge dx^{i+1}\wedge\ldots\wedge dx^n\,,$$

we have that

$$d\,\omega|_U = \left(\sum_{i=1}^n \frac{\partial f_i}{\partial x^i}\right) dx^1\wedge\ldots\wedge dx^n\,.$$

Choose a closed cube $C_r = \{a\in\mathbf{R}^n \mid |a^i|\le r \text{ for } 1\le i\le n\}$ containing the support of $f_i\circ x^{-1}$ for $i=1,\ldots,n$. We now distinguish two types of local charts (x,U).

Consider first the case where $x(U)\cap\partial\mathbf{H}^n=\emptyset$, and let g_i be the extension of $f_i\circ x^{-1}$ to $\mathbf{R}^n$ defining it to be zero outside $x(U)$ for $i=1,\ldots,n$. Then we have that

$$\begin{aligned}\int_M d\,\omega &= \sum_{i=1}^n \int_{-r}^r\cdots\int_{-r}^r D_i g_i\,dx^1\ldots dx^n\\ &= \sum_{i=1}^n \int_{-r}^r\cdots\int_{-r}^r [g_i(x^1,\ldots,x^{i-1},r,x^{i+1},\ldots,x^n)\\ &\quad - g_i(x^1,\ldots,x^{i-1},-r,x^{i+1},\ldots,x^n)]\,dx^1\ldots dx^{i-1}\,dx^{i+1}\ldots dx^n = 0\,.\end{aligned}$$

Since $i^*\omega=0$, this completes the proof of Stokes' theorem in this case.

Consider next the case where $x(U)\cap\partial\mathbf{H}^n\neq\emptyset$, and let $(\tilde{x},\widetilde{U})$ be the induced local chart on ∂M. Let g_i be the extension of $f_i\circ x^{-1}$ to $\mathbf{H}^n$, defining it to be zero outside $x(U)$ for $i=1,\ldots,n$. Then we have that

$$\int_M d\omega = \sum_{i=1}^{n} \int_0^r \int_{-r}^{r} \cdots \int_{-r}^{r} D_i g_i \, dx^1 \ldots dx^n$$

$$= -\int_{-r}^{r} \cdots \int_{-r}^{r} g_n(x^1, \ldots, x^{n-1}, 0) \, dx^1 \ldots dx^{n-1} .$$

Since $i^* dx^n = d(x^n \circ i) = 0$, it follows that

$$(i^* \omega)|_{\widetilde{U}} = (-1)^{n-1} (f_n \circ i) \, d\tilde{x}^1 \wedge \ldots \wedge d\tilde{x}^{n-1} .$$

By Proposition 6.28, the local chart $(\tilde{x}, \widetilde{U})$ on ∂M is positively (resp., negatively) oriented with respect to $\partial\mu$ when n is even (resp., odd). Hence using this chart, we must include an extra factor $(-1)^n$ in the formula for $\int_{\partial M} i^* \omega$, and we obtain

$$\int_{\partial M} i^* \omega = -\int_{-r}^{r} \cdots \int_{-r}^{r} g_n(x^1, \ldots, x^{n-1}, 0) \, dx^1 \ldots dx^{n-1}$$

which completes the proof of Stokes' theorem in the case where $x(U) \cap \partial \mathbf{H}^n \neq \emptyset$.

Now consider the general case where $\operatorname{supp}(\omega)$ is not necessarily contained in a single coordinate neighbourhood. Let $\mathscr{A} = \{(x_\alpha, U_\alpha) \mid \alpha \in A\}$ be an atlas on M consisting of positively oriented local charts with respect to μ, and let $\{\phi_\alpha | \alpha \in A\}$ be a partition of unity on M subordinate to the open cover $\{U_\alpha | \alpha \in A\}$. By the first part of the proof we have that

$$\int_M d(\phi_\alpha \omega) = \int_{\partial M} i^* (\phi_\alpha \omega)$$

for every $\alpha \in A$, and the integrals are non-zero only for a finite number of indices α since $\operatorname{supp}(\omega)$ is compact. Hence we have that

$$\int_M d\omega = \sum_{\alpha \in A} \int_M d(\phi_\alpha \omega) = \sum_{\alpha \in A} \int_{\partial M} i^* (\phi_\alpha \omega) = \int_{\partial M} i^* \omega$$

which completes the proof of Stokes' theorem in the general case. □

Chapter 7

METRIC AND SYMPLECTIC STRUCTURES

COVARIANT TENSORS OF DEGREE 2

7.1 Definition Let V be a finite dimensional vector space. A bilinear functional $g : V \times V \to \mathbf{R}$ is said to be *non-degenerate* if

$$g(v,w) = 0 \text{ for every } w \in V \text{ implies that } v = 0\,,$$

and it is said to be *positive definite* (resp., *negative definite*) if it satisfies the stronger condition

$$g(v,v) > 0 \text{ (resp. } g(v,v) < 0) \text{ for every } v \neq 0\,.$$

The *transpose* g^t of g is the bilinear functional $g^t : V \times V \to \mathbf{R}$ defined by

$$g^t(v,w) = g(w,v)$$

for $v,w \in V$. We say that g is *symmetric* if $g^t = g$ and *skew symmetric* if $g^t = -g$.

A bilinear functional $g \in \mathscr{T}^2(V)$ which is symmetric and non-degenerate, is called a *metric* on V. If it is also positive definite, it is called a *Riemannian metric* or an *inner product* on V. A metric is said to be *definite* if it is either positive or negative definite, otherwise it is said to be *indefinite*. We also use $<v,w>$ as an alternative notation for $g(v,w)$ when $v,w \in V$. If V and W are vector spaces with metrics g and h, respectively, then a linear map $F : V \to W$ is called a *linear isometry* if

$$h(F(v),F(w)) = g(v,w)$$

for all $v,w \in V$.

A bilinear functional $g \in \mathscr{T}^2(V)$ which is skew symmetric and non-degenerate, is called a *symplectic form* on V.

7.2 If V is a vector space, we have a linear isomorphism

$$\phi : \mathscr{T}^2(V) \to L(V,V^*)$$

given by $\phi(g) = G$, where

$$G(v)(w) = g(v,w)$$

for $v, w \in V$ and $g \in \mathscr{T}^2(V)$. The inverse map

$$\phi^{-1} : L(V,V^*) \to \mathscr{T}^2(V)$$

is given by $\phi^{-1}(G) = g$, where

$$g(v,w) = G(v)(w)$$

for $v, w \in V$ and $G \in L(V,V^*)$.

If V is finite dimensional and $\mathscr{B} = \{v_1, ..., v_n\}$ is a basis for V with dual basis $\mathscr{B}^* = \{v_1^*, ..., v_n^*\}$, then

$$g = \sum_{i,j} g_{ij} v_i^* \otimes v_j^*$$

where $g_{ij} = g(v_i, v_j)$ for $i, j = 1, ..., n$. These numbers are called the *components* of g with respect to the basis $\mathscr{B}$, and the matrix (g_{ij}) is also denoted by $m_{\mathscr{B}}(g)$. We see that $m_{\mathscr{B}}(g^t) = m_{\mathscr{B}}(g)^t$.

Now if $\phi(g) = G$, we have that

$$G(v_j) = \sum_{i=1}^{n} G(v_j)(v_i)\, v_i^* = \sum_{i=1}^{n} g_{ji} v_i^*$$

so that $m_{\mathscr{B}^*}^{\mathscr{B}}(G) = (g_{ij})^t$. Hence if

$$v = \sum_{i=1}^{n} a^i v_i ,$$

then

$$G(v) = \sum_{i=1}^{n} a_i v_i^* \quad \text{where} \quad a_i = \sum_{j=1}^{n} g_{ji} a^j .$$

We say that $G(v)$ is obtained from v by *lowering indices*, and the map G is denoted by $g^\flat$. We have that $g^\flat(v) = i_v g$. The linear functional $g^\flat(v)$ is also denoted by $v^\flat$.

By the *rank* of the bilinear map g we mean the rank of the matrix (g_{ij}), which is equal to $\dim G(V)$ and therefore independent of the choice of the basis $\mathscr{B}$. We immediately see that g and its transpose g^t always have the same rank.

We see that g is non-degenerate if and only if $\phi(g) = G$ is a linear isomorphism so that the matrix (g_{ij}) is invertible, i.e., $\det(g_{ij}) \neq 0$. The inverse of (g_{ij}) is denoted by (g^{ij}). If

$$\lambda = \sum_{i=1}^{n} a_i v_i^* ,$$

then

$$G^{-1}(\lambda) = \sum_{i=1}^{n} a^i v_i \quad \text{where} \quad a^i = \sum_{j=1}^{n} g^{ji} a_j .$$

We say that $G^{-1}(\lambda)$ is obtained from λ by *raising indices*, and the map G^{-1} is denoted by $g^\sharp$. We have that $i_{g^\sharp(\lambda)}\, g = \lambda$. The vector $g^\sharp(\lambda)$ is also denoted by $\lambda^\sharp$.

More generally, we have linear isomorphisms

$$L_s^r : \mathscr{T}_l^k(V) \to \mathscr{T}_{l-1}^{k+1}(V) \quad \text{and} \quad R_s^r : \mathscr{T}_l^k(V) \to \mathscr{T}_{l+1}^{k-1}(V)\,,$$

also called *lowering* and *raising* of indices, respectively, which are defined by

$$L_s^r(T)(w_1,\ldots,w_{k+1},\lambda_1,\ldots,\lambda_{l-1})$$

$$= T(w_1,\ldots,w_{r-1},w_{r+1},\ldots,w_{k+1},\lambda_1,\ldots,\lambda_{s-1},G(w_r),\lambda_s,\ldots,\lambda_{l-1})$$

and

$$R_s^r(T)(w_1,\ldots,w_{k-1},\lambda_1,\ldots,\lambda_{l+1})$$

$$= T(w_1,\ldots,w_{r-1},G^{-1}(\lambda_s),w_r,\ldots,w_{k-1},\lambda_1,\ldots,\lambda_{s-1},\lambda_{s+1},\ldots,\lambda_{l+1})$$

for $w_1,\ldots,w_{k+1} \in V$ and $\lambda_1,\ldots,\lambda_{l+1} \in V^*$.

Finally, we see that g is symmetric if and only if the matrix (g_{ij}) is symmetric, i.e., we have that $g_{ij} = g_{ji}$ for $i,j = 1,\ldots,n$.

7.3 Proposition Let $\mathscr{B} = \{v_1,\ldots,v_n\}$ is a basis for a vector space V with dual basis $\mathscr{B}^* = \{v_1^*,\ldots,v_n^*\}$, and let g be a metric on V and $T \in \mathscr{T}_l^k(V)$ be a tensor so that

$$g = \sum_{i,j} g_{ij} v_i^* \otimes v_j^* \quad \text{and} \quad T = \sum_{\substack{i_1,\ldots,i_k \\ j_1,\ldots,j_l}} A_{i_1,\ldots,i_k}^{j_1,\ldots,j_l} v_{i_1}^* \otimes \cdots \otimes v_{i_k}^* \otimes v_{j_1} \otimes \cdots \otimes v_{j_l}\,.$$

Then the tensors $L_s^r(T)$ and $R_s^r(T)$ obtained by lowering and raising indices in T are given by

$$L_s^r(T) = \sum_{\substack{i_1,\ldots,i_{k+1} \\ j_1,\ldots,j_{l-1}}} B_{i_1,\ldots,i_{k+1}}^{j_1,\ldots,j_{l-1}} v_{i_1}^* \otimes \cdots \otimes v_{i_{k+1}}^* \otimes v_{j_1} \otimes \cdots \otimes v_{j_{l-1}}$$

and

$$R_s^r(T) = \sum_{\substack{i_1,\ldots,i_{k-1} \\ j_1,\ldots,j_{l+1}}} C_{i_1,\ldots,i_{k-1}}^{j_1,\ldots,j_{l+1}} v_{i_1}^* \otimes \cdots \otimes v_{i_{k-1}}^* \otimes v_{j_1} \otimes \cdots \otimes v_{j_{l+1}}\,,$$

where

$$B_{i_1,\ldots,i_{k+1}}^{j_1,\ldots,j_{l-1}} = \sum_m g_{i_r m} A_{i_1,\ldots,i_{r-1},i_{r+1},\ldots,i_{k+1}}^{j_1,\ldots,j_{s-1},m,j_s,\ldots,j_{l-1}}$$

and

$$C_{i_1,\ldots,i_{k-1}}^{j_1,\ldots,j_{l+1}} = \sum_m g^{j_s m} A_{i_1,\ldots,i_{r-1},m,i_r,\ldots,i_{k-1}}^{j_1,\ldots,j_{s-1},j_{s+1},\ldots,j_{l+1}}\,.$$

PROOF: Follows from 7.2. □

7.4 If $\psi : \mathscr{T}^1_1(V) \to \mathscr{T}^1(V;V)$ is the linear isomorphism defined in 5.15 and $T \in \mathscr{T}^2(V)$, then we have that

$$T(v,w) = g(v, \psi \circ R^1_1(T)(w)) = g^t(\psi \circ R^2_1(T)(v), w)$$

for every $v, w \in V$. Indeed, if $\mathscr{B} = \{v_1, ..., v_n\}$ is a basis for V with dual basis $\mathscr{B}^* = \{v_1^*, ..., v_n^*\}$, then $\mathscr{C} = \{(g^t)^\sharp(v_1^*), ..., (g^t)^\sharp(v_n^*)\}$ is also a basis for V with dual basis $\mathscr{C}^* = \{g^\flat(v_1), ..., g^\flat(v_n)\}$ since

$$g^\flat(v_i)\,\{(g^t)^\sharp(v_j^*)\} = g(v_i, (g^t)^\sharp(v_j^*)) = g^t((g^t)^\sharp(v_j^*), v_i) = v_j^*(v_i) = \delta_{i,j}\,.$$

Hence we have that

$$\psi \circ R^1_1(T)(w) = \sum_{j=1}^n R^1_1(T)(w, g^\flat(v_j))\,(g^t)^\sharp(v_j^*) = \sum_{j=1}^n T(v_j, w)\,(g^t)^\sharp(v_j^*)$$

and

$$\psi \circ R^2_1(T)(v) = \sum_{j=1}^n R^2_1(T)(v, g^\flat(v_j))\,(g^t)^\sharp(v_j^*) = \sum_{j=1}^n T(v, v_j)\,(g^t)^\sharp(v_j^*)$$

which completes the proof of the assertion.

7.5 Definition A covariant tensor field $g \in \mathscr{T}^2(M)$ of degree 2 on a smooth manifold M is said to be *non-degenerate* if $g(p)$ is non-degenerate for every $p \in M$. Let T be a tensor field of type $\binom{k}{l}$ on M. Then we have tensor fields S_1 and S_2 on M of type $\binom{k+1}{l-1}$ and $\binom{k-1}{l+1}$, respectively, given by

$$S_1(p) = L^r_s(T(p)) \quad \text{and} \quad S_2(p) = R^r_s(T(p))$$

for every $p \in M$. If (x, U) is a local chart on M, and if g_{ij}, $A^{j_1,...,j_l}_{i_1,...,i_k}$, $B^{j_1,...,j_{l-1}}_{i_1,...,i_{k+1}}$ and $C^{j_1,...,j_{l+1}}_{i_1,...,i_{k-1}}$ are the components of the local representations of g, T, S_1 and S_2 on U, then it follows from Proposition 7.3 that

$$B^{j_1,...,j_{l-1}}_{i_1,...,i_{k+1}} = \sum_m g_{i_r m} A^{j_1,...,j_{s-1},m,j_s,...,j_{l-1}}_{i_1,...,i_{r-1},i_{r+1},...,i_{k+1}}$$

and

$$C^{j_1,...,j_{l+1}}_{i_1,...,i_{k-1}} = \sum_m g^{j_s m} A^{j_1,...,j_{s-1},j_{s+1},...,j_{l+1}}_{i_1,...,i_{r-1},m,i_r,...,i_{k-1}}\,,$$

showing that S_1 and S_2 are in fact tensor fields. We say that they are obtained by *lowering* and *raising* of indices in T, and they are denoted by $L^r_s(T)$ and $R^r_s(T)$, respectively. If X is a vector field on M, then the covector field $L^1_1(X)$ is usually denoted by $g^\flat(X)$ or $X^\flat$. Similarly, if ω is a covector field on M, then the vector field $R^1_1(\omega)$ is usually denoted by $g^\sharp(\omega)$ or $\omega^\sharp$.

PSEUDO-RIEMANNIAN MANIFOLDS

7.6 Now consider the linear map $\mathscr{S} : \mathscr{T}^k(V) \to \mathscr{T}^k(V)$ defined by

$$\mathscr{S}(T) = \frac{1}{k!} \sum_{\sigma \in S_k} T^{\sigma}$$

i.e.,

$$\mathscr{S}(T)(v_1, \ldots, v_k) = \frac{1}{k!} \sum_{\sigma \in S_k} T(v_{\sigma(1)}, \ldots, v_{\sigma(k)})$$

which is called the *symmetrization* operator. If $\alpha, \beta \in V^*$, we define their *symmetric product* $\alpha\beta$ by

$$\alpha\beta = \mathscr{S}(\alpha \otimes \beta) = \frac{1}{2}(\alpha \otimes \beta + \beta \otimes \alpha).$$

In particular, we have that $\alpha\alpha = \alpha \otimes \alpha$ which is usually denoted by α^2.

If $g : V \times V \to \mathbf{R}$ is a symmetric bilinear functional, and if $\mathscr{B} = \{v_1, \ldots, v_n\}$ is a basis for V with dual basis $\mathscr{B}^* = \{v_1^*, \ldots, v_n^*\}$, then

$$\begin{aligned} g &= \sum_{i,j} g_{ij} v_i^* \otimes v_j^* = \sum_{i,j} \frac{1}{2}(g_{ij} + g_{ji})\, v_i^* \otimes v_j^* \\ &= \frac{1}{2} \sum_{i,j} g_{ij} v_i^* \otimes v_j^* + \frac{1}{2} \sum_{i,j} g_{ij} v_j^* \otimes v_i^* = \sum_{i,j} g_{ij}\, v_i^* v_j^* . \end{aligned}$$

7.7 Definition Let V be a finite dimensional vector space with a metric g. Then two vectors $v, w \in V$ are said to be *orthogonal*, and we write $v \perp w$, if $g(v,w) = 0$. If S is a subset of V, then $v \perp S$ means that $v \perp w$ for every $w \in S$. We have a subspace $S^{\perp}$ of V, called the *orthogonal space of* S, which is given by

$$S^{\perp} = \{ v \in V \,|\, v \perp S \} = g^{\sharp}(A(S)) ,$$

where $A(S)$ is the annihilator of S defined in Definition 4.19. When $S = \{w\}$, then $S^{\perp}$ is denoted simply by $w^{\perp}$ and is called the *orthogonal space of* w.

The *norm* of a vector $v \in V$ is defined by $\|v\| = |g(v,v)|^{1/2}$. A non-zero vector v is called a *unit vector* if $\|v\| = 1$ and a *null vector* if $\|v\| = 0$. A set of mutually orthogonal vectors is called *orthogonal*, and it is called *orthonormal* if all the vectors are unit vectors.

If W is a subspace of V, then the restriction $g|_{W \times W}$ is a symmetric bilinear functional on $W \times W$ which is denoted simply by $g|_W$. We say that W is a *non-degenerate* subspace of V if $g|_W$ is non-degenerate.

7.8 Proposition Let W be a subspace of a finite dimensional vector space V with a metric g. Then we have that

(1) $\dim W + \dim W^{\perp} = \dim V$

(2) $(W^{\perp})^{\perp} = W$

PROOF: (1) Follows from Proposition 4.21.

(2) Follows since $W \subset (W^{\perp})^{\perp}$, where

$$\dim W = \dim (W^{\perp})^{\perp} = \dim V - \dim W^{\perp}$$

according to (1). □

7.9 Proposition A subspace W of a finite dimensional vector space V with a metric g is non-degenerate if and only if

$$V = W \oplus W^{\perp}.$$

We have that W is non-degenerate if and only if $W^{\perp}$ is non-degenerate.

PROOF: From Definition 7.7 we see that W is non-degenerate if and only if $W \cap W^{\perp} = \{0\}$. The first part of the proposition therefore follows from Proposition 7.8 (1) using the formula

$$\dim (W + W^{\perp}) + \dim (W \cap W^{\perp}) = \dim W + \dim W^{\perp} = \dim V.$$

The last part of the proposition follows from the first part and Proposition 7.8 (2). □

7.10 Proposition An n-dimensional vector space V with a metric g always has an orthonormal basis.

PROOF: We prove the proposition by induction on n. It is obviously true for $n = 0$, and we will prove it for $n > 0$ assuming that it is true for every vector space V of dimension $n - 1$.

There is a vector $v \in V$ with $a = g(v, v) \neq 0$. Otherwise, if $g(v, v) = 0$ for every $v \in V$, we would have that

$$g(v, w) = \frac{1}{2}\left[g(v + w, v + w) - g(v, v) - g(w, w)\right] = 0$$

for every $v, w \in V$, contradicting the fact that g is non-degenerate.

We let $v_n = |a|^{-1/2} v$ so that $g(v_n, v_n) = \pm 1$, and consider the subspace

$$W = \{w \in V \,|\, g(v_n, w) = 0\}$$

which is of dimension $n - 1$ since it is the kernel of the non-zero functional $F : V \to \mathbf{R}$ defined by $F(v) = g(v_n, v)$.

Now g is non-degenerate on W, for suppose that $u \in W$ with $g(u, w) = 0$ for every $w \in W$. Since $g(u, v_n) = 0$, it follows that $g(u, v) = 0$ for every $v \in V$ which implies that $u = 0$. By the induction hypothesis there is therefore an orthonormal basis $\{v_1, \ldots, v_{n-1}\}$ for W, and $\mathscr{B} = \{v_1, \ldots, v_n\}$ is a basis for V with the desired properties. □

7.11 Remark Let $\mathscr{B} = \{v_1, ..., v_n\}$ be an orthonormal basis for a vector space V with a metric g. Then the subspaces $V_r = L(v_1, ..., v_r)$ are non-degenerate for $r = 1, \ldots, n$. Indeed, if $v \in V_r$ with $g(v, w) = 0$ for every $w \in V_r$, then $v = \sum_{i=1}^{r} a_i v_i$ where $a_j = g(v, v_j)/g(v_j, v_j) = 0$ for $j = 1, \ldots, r$.

7.12 Proposition Let g be a metric on an n-dimensional vector space V, and let k be the number of vectors v_i in an orthonormal basis $\mathscr{B} = \{v_1, ..., v_n\}$ for which $g(v_i, v_i) = -1$. Then k does not depend on the choice of orthonormal basis $\mathscr{B}$, and it is called the *index* of the metric g. It is also called the index of the vector space V and is denoted by $\operatorname{ind} V$.

PROOF: Since g is negative definite on the subspace $\{0\}$ of V, we may choose a subspace W of V of maximal dimension such that g is negative definite on W. We will prove that k is the dimension of W.

We may assume that the vectors in $\mathscr{B}$ are ordered in such a way that $g(v_i, v_i) = -1$ for $1 \leq i \leq k$ and $g(v_i, v_i) = 1$ for $k+1 \leq i \leq n$. Then $V = V_1 \oplus V_2$, where V_1 and V_2 are the subspaces of V spanned by $\{v_1, ..., v_k\}$ and $\{v_{k+1}, ..., v_n\}$, respectively, and g is negative definite on V_1 and positive definite on V_2. Indeed, if

$$0 \neq v = \sum_{i=1}^{k} a^i v_i$$

we have that

$$g(v, v) = \sum_{i,j=1}^{k} g(v_i, v_j) a^i a^j = -\sum_{i=1}^{k} (a^i)^2 < 0,$$

showing that g is negative definite on V_1, and a similar argument applies to V_2. In particular, it follows that $\dim(W) \geq \dim(V_1) = k$.

To show that $\dim(W) \leq k$, we consider the linear map $F : V \to V_1$ defined by $F(v) = v_1$ for $v = v_1 + v_2$ with $v_1 \in V_1$ and $v_2 \in V_2$, and we let $G = F|_W : W \to V_1$. Then $\ker(G) \subset W \cap V_2$ which implies that $\ker(G) = \{0\}$, since g is negative definite on W and positive definite on V_2, and this completes the proof that $\dim(W) \leq \dim(V_1) = k$. □

7.13 Remark Let $\mathscr{B} = \{v_1, ..., v_n\}$ be an orthonormal basis for an n-dimensional vector space V with a metric g of index k. Then the n-tuple $\varepsilon = (\varepsilon_1, \ldots, \varepsilon_n) \in \{-1, 1\}^n$, where $\varepsilon_i = g(v_i, v_i)$ for $i = 1, \ldots, n$, is called the *signature* of $\mathscr{B}$. An n-tuple $\varepsilon \in \{-1, 1\}^n$ having k components which are -1 and $n-k$ components which are 1, is called a signature compatible with g.

Unless otherwise specified, we will assume that the vectors in an orthonormal basis $\mathscr{B}$ are ordered so that

$$\varepsilon_i = \begin{cases} -1 & \text{for} \quad 1 \leq i \leq k \\ 1 & \text{for} \quad k+1 \leq i \leq n \end{cases}.$$

We call this ε the *canonical signature* for the metric g.

The metric g is positive definite, negative definite or indefinite when $k=0, k=n$ or $0<k<n$, respectively. If the index of g is 1 and $n \geq 2$, g is called a *Lorentz* metric.

7.14 Proposition Let W be a non-degenerate subspace of a finite dimensional vector space V with a metric g. Then we have that

$$\text{ind } V = \text{ind } W + \text{ind } W^{\perp}.$$

PROOF: By Propositions 7.9 and 7.10 there are orthonormal bases $\mathscr{B}=\{v_1,\ldots,v_r\}$ and $\mathscr{C}=\{v_{r+1},\ldots,v_n\}$ for W and $W^{\perp}$ where the number of vectors v_i for which $g(v_i,v_i)=-1$ are ind W and ind $W^{\perp}$, respectively. The proposition now follows since $\{v_1,\ldots,v_n\}$ is an orthonormal basis for V. □

7.15 Definition Given an n-tuple $\varepsilon=(\varepsilon_1,\ldots,\varepsilon_n)\in\{-1,1\}^n$, there is a unique metric g^n_ε in $\mathbf{R}^n$ such that the standard basis $\mathscr{E}=\{e_1,\ldots,e_n\}$ is orthonormal with signature ε. It is given by

$$g^n_\varepsilon(a,b)=\sum_{i=1}^{n}\varepsilon_i\, a^i\, b^i$$

for $a,b\in\mathbf{R}^n$. The vector space $\mathbf{R}^n$ equipped with this metric is denoted by $\mathbf{R}^n_\varepsilon$ and is called the *pseudo-Euclidean* space of dimension n and signature ε. If the index of the metric g^n_ε is k and ε is its canonical signature, then g^n_ε and $\mathbf{R}^n_\varepsilon$ are also denoted by g^n_k and $\mathbf{R}^n_k$, respectively. The matrix $S=m_{\mathscr{E}}(g^n_\varepsilon)$ is the $n\times n$ diagonal matrix with diagonal elements $S_{ii}=\varepsilon_i$ for $i=1,\ldots,n$, and it is called the *signature matrix* for the metric g^n_ε. We have that

$$g^n_\varepsilon(a,b)=a^t S\, b$$

for column vectors a and b in $\mathbf{R}^n$. The space $\mathbf{R}^n_1$ where $n\geq 2$ is called the n-dimensional *Minkowski space*, and the space $\mathbf{R}^n$ equipped with the metric g^n_0 is called the n-dimensional *Euclidean space*.

7.16 Definition By a *fibre metric* in a vector bundle $\pi: E\to M$ over a smooth manifold M we mean a section g in the vector bundle $\pi^2:\mathscr{T}^2(E)\to M$ so that $g(p)$ is a metric in the fibre E_p for each $p\in M$. If the metric $g(p)$ is Riemannian for each $p\in M$, then g is called a *Riemannian* fibre metric in $\pi: E\to M$.

If $F_i: N\to E$ for $i=1,2$ are liftings of a smooth map $f: N\to M$ from a smooth manifold N, we often use $<F_1,F_2>$ as an alternative notation for the function $h: N\to\mathbf{R}$ defined by $h(p)=g(f(p))\,(F_1(p),F_2(p))$ for $p\in N$.

By a *metric* on a smooth manifold M we mean a fibre metric in its tangent bundle $\pi: TM\to M$, i.e., a covariant tensor field $g\in\mathscr{T}^2(M)$ on M of degree 2 so that $g(p)$ is a metric in the tangent space T_pM for each $p\in M$. A manifold with a metric is called a *pseudo-Riemannian* manifold. A vector field X along a smooth map $f: N\to M$ from a smooth manifold N with $\|X(p)\|=1$ for every $p\in N$ is called a *unit vector field along* f. If N is an open subset of M and f is the inclusion map, it is called a *unit vector field* on N.

An immersion or an embedding $f : M_1 \to M_2$ between two pseudo-Riemannian manifolds M_1 and M_2 with metrics g_1 and g_2 is said to be *isometric* if $f^* g_2 = g_1$. If $f : M_1 \to M_2$ is also a diffeomorphism, it is called an *isometry*. An (immersed) submanifold M_1 of a pseudo-Riemannian manifold M_2 with metric g_2 is called a *pseudo-Riemannian (immersed) submanifold* if $i^* g_2$ is a metric on M_1, where $i : M_1 \to M_2$ is the inclusion map. If the codimension of M_1 is 1, M_1 is called a *pseudo-Riemannian hypersurface* of M_2.

If the metric $g(p)$ is Riemannian (resp., Lorentz) for each $p \in M$, then g is called a *Riemannian (resp. Lorentz)* metric on M, and M is called a *Riemannian (resp. Lorentz)* manifold.

7.17 Proposition Every vector bundle $\pi : E \to M$ has a Riemannian fibre metric.

PROOF: Let $\{(t_\alpha, \pi^{-1}(U_\alpha)) \mid \alpha \in A\}$ be a trivializing cover of E, and let $\{\phi_\alpha \mid \alpha \in A\}$ be a partition of unity on M subordinate to the open cover $\{U_\alpha \mid \alpha \in A\}$. For each $\alpha \in A$, we let g_α be the section in $\pi^2 : \mathscr{T}^2(E) \to M$ defined by

$$g_\alpha(p) = \begin{cases} \phi_\alpha(p)\, t^*_{\alpha,p}\, g^n_0 & \text{for} \quad p \in U_\alpha \\ 0 & \text{for} \quad p \notin U_\alpha \end{cases},$$

where n is the dimension of the vector bundle and g^n_0 is the Riemannian metric in $\mathbf{R}^n$ defined in Definition 7.15. Then $g = \sum_{\alpha\in A} g_\alpha$ is a Riemannian fibre metric in $\pi : E \to M$. □

7.18 Proposition A vector bundle $\pi : E \to M$ with a fibre metric g over a smooth manifold M^n is equivalent over M to its dual bundle $\pi^* : E^* \to M$, with the equivalence $e : E \to E^*$ given by

$$e(v)(w) = g(p)(v, w)$$

for $p \in M$ and $v, w \in E_p$.

PROOF: Let $\mathscr{E} = \{e_1, \ldots, e_n\}$ be the standard basis for $\mathbf{R}^n$, and let $\mathscr{E}^* = \{e^1, \ldots, e^n\}$ be its dual basis. For every local trivialization $(t, \pi^{-1}(U))$ in E, the functions $g_{ij} : U \to \mathbf{R}$ defined by

$$g(p) = \sum_{i,j} g_{ij}(p)\, (e^i \circ t_p) \otimes (e^j \circ t_p)$$

for $p \in U$, are smooth by Proposition 4.45. If $(t', \pi^{*-1}(U))$, where $t'_p = r \circ (t_p^{-1})^*$ for $p \in U$, is the corresponding local trivialization in the dual bundle $\pi^* : E^* \to M$ defined in Remark 4.9, we have that

$$(t_p^{-1})^* \circ e_p \circ t_p^{-1}(e_i)(e_j) = g(p)(t_p^{-1}(e_i), t_p^{-1}(e_j)) = g_{ij}(p)\,,$$

so that

$$t'_p \circ e_p \circ t_p^{-1}(e_i) = \sum_j g_{ij}(p)\, e_j$$

for $p \in U$ and $i = 1, \ldots, n$. Hence it follows that

$$t' \circ e \circ t^{-1}(p, v) = (p, t'_p \circ e_p \circ t_p^{-1}(v))$$

for $p \in U$ and $v \in \mathbf{R}^n$, where

$$m_{\mathscr{E}}^{\mathscr{E}}(t'_p \circ e_p \circ t_p^{-1}) = (g_{ij}(p)),$$

showing that e is an equivalence. □

7.19 Definition If α and β are 1-forms on an open subset V of a smooth manifold M, we define their *symmetric product* $\alpha\beta$ to be the covariant tensor field on V of degree 2 defined by

$$(\alpha\beta)(p) = \alpha(p)\,\beta(p)$$

for $p \in V$.

7.20 Example Consider the smooth manifold $M = \mathbf{R}^2$. A point $p \in M$ with cartesian coordinates (x, y) in the open set

$$V = \{(x, y) \in \mathbf{R}^2 \,|\, y \neq 0 \text{ or } x > 0\}$$

can also be described by polar coordinates (r, ϕ) in the open set

$$U = \{(r, \phi) \in \mathbf{R}^2 \,|\, r > 0 \text{ and } -\pi < \phi < \pi\}.$$

The connection between these coordinates is given by

$$x = r\cos\phi$$
$$y = r\sin\phi$$

and

$$r = \sqrt{x^2 + y^2}$$

$$\phi = \begin{cases} \cot^{-1}(x/y) - \pi & \text{for } y < 0 \\ 0 & \text{for } y = 0 \text{ and } x > 0 \\ \cot^{-1}(x/y) & \text{for } y > 0 \end{cases}$$

As well as being real numbers, we can also think of x, y and r, ϕ as functions of p, so that they are the coordinate functions of local charts (id, M) and (ψ, V) on M. The meaning will always be clear from the context. Here $\psi = i \circ \rho^{-1}$, where $\rho : U \to V$ is the diffeomorphism given by

$$\rho\,(r, \phi) = (r\cos\phi\,,\, r\sin\phi)\,,$$

and $i : U \to \mathbf{R}^2$ is the inclusion map. We have that

$$D\rho\,(r, \phi) = \begin{pmatrix} \cos\phi & -r\sin\phi \\ \sin\phi & r\cos\phi \end{pmatrix}$$

so that

$$\det D\rho\ (r,\phi) = r > 0$$

for $(r,\phi) \in U$. Proposition 2.80 now implies that

$$\frac{\partial}{\partial r} = \frac{\partial x}{\partial r}\frac{\partial}{\partial x} + \frac{\partial y}{\partial r}\frac{\partial}{\partial y} = \cos\phi\ \frac{\partial}{\partial x} + \sin\phi\ \frac{\partial}{\partial y}$$

and

$$\frac{\partial}{\partial \phi} = \frac{\partial x}{\partial \phi}\frac{\partial}{\partial x} + \frac{\partial y}{\partial \phi}\frac{\partial}{\partial y} = -r\sin\phi\ \frac{\partial}{\partial x} + r\cos\phi\ \frac{\partial}{\partial y}\,,$$

and by Proposition 4.18 we also have that

$$dx = \frac{\partial x}{\partial r}dr + \frac{\partial x}{\partial \phi}d\phi = \cos\phi\ dr - r\sin\phi\ d\phi$$

and

$$dy = \frac{\partial y}{\partial r}dr + \frac{\partial y}{\partial \phi}d\phi = \sin\phi\ dr + r\cos\phi\ d\phi$$

on V. The standard metric on $\mathbf{R}^2$ is given by

$$g = dx^2 + dy^2$$

in cartesian coordinates. Using polar coordinates, we therefore have that

$$g|_V = dr^2 + r^2\ d\phi^2\,.$$

7.21 Example Consider the smooth manifold $M = \mathbf{R}^3$. A point $p \in M$ with cartesian coordinates (x,y,z) in the open set

$$V = \{(x,y,z) \in \mathbf{R}^3 | y \neq 0 \text{ or } x > 0\}$$

can also be described by spherical coordinates (r,θ,ϕ) in the open set

$$U = \{(r,\theta,\phi) \in \mathbf{R}^3 | r > 0\,,\, 0 < \theta < \pi \text{ and } -\pi < \phi < \pi\}\,.$$

The connection between these coordinates is given by

$$\begin{aligned} x &= r\sin\theta\ \cos\phi \\ y &= r\sin\theta\ \sin\phi \\ z &= r\cos\theta \end{aligned}$$

and

$$r = \sqrt{x^2+y^2+z^2}$$
$$\theta = \cos^{-1}(z/\sqrt{x^2+y^2+z^2})$$
$$\phi = \begin{cases} \cot^{-1}(x/y)-\pi & \text{for } y<0 \\ 0 & \text{for } y=0 \text{ and } x>0 \\ \cot^{-1}(x/y) & \text{for } y>0 \end{cases}$$

As well as being real numbers, we can also think of x,y,z and r,θ,ϕ as functions of p, so that they are the coordinate functions of local charts (id,M) and (ψ,V) on M. The meaning will always be clear from the context. Here $\psi = i\circ\rho^{-1}$, where $\rho : U \to V$ is the diffeomorphism given by

$$\rho\,(r,\theta,\phi) = (r\sin\theta\,\cos\phi\,,\, r\sin\theta\,\sin\phi\,,\, r\cos\theta)\,,$$

and $i : U \to \mathbf{R}^3$ is the inclusion map. We have that

$$D\rho\,(r,\theta,\phi) = \begin{pmatrix} \sin\theta\,\cos\phi & r\cos\theta\,\cos\phi & -r\sin\theta\,\sin\phi \\ \sin\theta\,\sin\phi & r\cos\theta\,\sin\phi & r\sin\theta\,\cos\phi \\ \cos\theta & -r\sin\theta & 0 \end{pmatrix}$$

so that

$$\det D\rho\,(r,\theta,\phi) = (\cos\theta)(r^2\cos\theta\,\sin\theta)+(r\sin\theta)(r\sin^2\theta) = r^2\sin\theta > 0$$

for $(r,\theta,\phi)\in U$. Proposition 2.80 now implies that

$$\frac{\partial}{\partial r} = \frac{\partial x}{\partial r}\frac{\partial}{\partial x}+\frac{\partial y}{\partial r}\frac{\partial}{\partial y}+\frac{\partial z}{\partial r}\frac{\partial}{\partial z} = \sin\theta\,\cos\phi\,\frac{\partial}{\partial x}+\sin\theta\,\sin\phi\,\frac{\partial}{\partial y}+\cos\theta\,\frac{\partial}{\partial z}\,,$$

$$\frac{\partial}{\partial \theta} = \frac{\partial x}{\partial \theta}\frac{\partial}{\partial x}+\frac{\partial y}{\partial \theta}\frac{\partial}{\partial y}+\frac{\partial z}{\partial \theta}\frac{\partial}{\partial z} = r\cos\theta\,\cos\phi\,\frac{\partial}{\partial x}+r\cos\theta\,\sin\phi\,\frac{\partial}{\partial y}-r\sin\theta\,\frac{\partial}{\partial z}$$

and

$$\frac{\partial}{\partial \phi} = \frac{\partial x}{\partial \phi}\frac{\partial}{\partial x}+\frac{\partial y}{\partial \phi}\frac{\partial}{\partial y}+\frac{\partial z}{\partial \phi}\frac{\partial}{\partial z} = -r\sin\theta\,\sin\phi\,\frac{\partial}{\partial x}+r\sin\theta\,\cos\phi\,\frac{\partial}{\partial y}\,,$$

and by Proposition 4.18 we also have that

$$dx = \frac{\partial x}{\partial r}dr+\frac{\partial x}{\partial \theta}d\theta+\frac{\partial x}{\partial \phi}d\phi = \sin\theta\,\cos\phi\,dr+r\cos\theta\,\cos\phi\,d\theta-r\sin\theta\,\sin\phi\,d\phi,$$

$$dy = \frac{\partial y}{\partial r}dr+\frac{\partial y}{\partial \theta}d\theta+\frac{\partial y}{\partial \phi}d\phi = \sin\theta\,\sin\phi\,dr+r\cos\theta\,\sin\phi\,d\theta+r\sin\theta\,\cos\phi\,d\phi$$

and

$$dz = \frac{\partial z}{\partial r}\,dr + \frac{\partial z}{\partial \theta}\,d\theta + \frac{\partial z}{\partial \phi}\,d\phi \;=\; \cos\theta\; dr - r\,\sin\theta\; d\theta$$

on V. The standard metric on $\mathbf{R}^3$ is given by

$$g = dx^2 + dy^2 + dz^2$$

in cartesian coordinates. Using spherical coordinates, we therefore have that

$$g|_V = dr^2 + r^2 (d\theta^2 + \sin^2\theta \; d\phi^2)\,.$$

7.22 Example Consider the smooth manifold $M = \mathbf{R}^4$. A point $p \in M$ with cartesian coordinates (w,x,y,z) in the open set

$$V = \{(w,x,y,z) \in \mathbf{R}^4 | y \neq 0 \text{ or } x > 0\}$$

can also be described by hyperspherical coordinates (r,χ,θ,ϕ) in the open set

$$U = \{(r,\chi,\theta,\phi) \in \mathbf{R}^4 | r > 0\,,\, 0 < \chi < \pi\,,\, 0 < \theta < \pi \text{ and } -\pi < \phi < \pi\}\,.$$

The connection between these coordinates is given by

$$\begin{aligned} w &= r\,\cos\chi \\ x &= r\,\sin\chi\;\sin\theta\;\cos\phi \\ y &= r\,\sin\chi\;\sin\theta\;\sin\phi \\ z &= r\,\sin\chi\;\cos\theta \end{aligned}$$

and

$$\begin{aligned} r &= \sqrt{w^2+x^2+y^2+z^2} \\ \chi &= \cos^{-1}(w\,/\,\sqrt{w^2+x^2+y^2+z^2}\,) \\ \theta &= \cos^{-1}(z\,/\,\sqrt{x^2+y^2+z^2}\,) \end{aligned}$$

$$\phi = \begin{cases} \cot^{-1}(x/y) - \pi & \text{for } \; y < 0 \\ 0 & \text{for } \; y = 0 \text{ and } x > 0 \\ \cot^{-1}(x/y) & \text{for } \; y > 0 \end{cases}$$

As well as being real numbers, we can also think of w,x,y,z and r,χ,θ,ϕ as functions of p, so that they are the coordinate functions of local charts (id,M) and (ψ,V) on M. The meaning will always be clear from the context. Here $\psi = i \circ \rho^{-1}$, where $\rho : U \to V$ is the diffeomorphism given by

$$\rho\,(r,\chi,\theta,\phi) = (r\,\cos\chi\,,\, r\,\sin\chi\;\sin\theta\;\cos\phi\,,\, r\,\sin\chi\;\sin\theta\;\sin\phi\,,\, r\,\sin\chi\;\cos\theta)\,,$$

and $i : U \to \mathbf{R}^4$ is the inclusion map. We have that

$Dp\,(r,\chi,\theta,\phi) =$

$$\begin{pmatrix} \cos\chi & -r\sin\chi & 0 & 0 \\ \sin\chi\,\sin\theta\,\cos\phi & r\cos\chi\,\sin\theta\,\cos\phi & r\sin\chi\,\cos\theta\,\cos\phi & -r\sin\chi\,\sin\theta\,\sin\phi \\ \sin\chi\,\sin\theta\,\sin\phi & r\cos\chi\,\sin\theta\,\sin\phi & r\sin\chi\,\cos\theta\,\sin\phi & r\sin\chi\,\sin\theta\,\cos\phi \\ \sin\chi\,\cos\theta & r\cos\chi\,\cos\theta & -r\sin\chi\,\sin\theta & 0 \end{pmatrix}$$

so that

$$\det Dp\,(r,\chi,\theta,\phi) = (\cos\chi)(r^3\cos\chi\,\sin^2\chi\,\sin\theta) - (-r\sin\chi)(r^2\sin^3\chi\,\sin\theta)$$
$$= r^3\sin^2\chi\,\sin\theta > 0$$

for $(r,\chi,\theta,\phi)\in U$. Proposition 4.18 now implies that

$$dw = \cos\chi\;dr - r\sin\chi\;d\chi\,,$$
$$dx = \sin\chi\,\sin\theta\,\cos\phi\;dr + r\cos\chi\,\sin\theta\,\cos\phi\;d\chi + r\sin\chi\,\cos\theta\,\cos\phi\;d\theta - r\sin\chi\,\sin\theta\,\sin\phi\;d\phi\,,$$
$$dy = \sin\chi\,\sin\theta\,\sin\phi\;dr + r\cos\chi\,\sin\theta\,\sin\phi\;d\chi + r\sin\chi\,\cos\theta\,\sin\phi\;d\theta + r\sin\chi\,\sin\theta\,\cos\phi\;d\phi$$

and

$$dz = \sin\chi\,\cos\theta\;dr + r\cos\chi\,\cos\theta\;d\chi - r\sin\chi\,\sin\theta\;d\theta$$

on V. The standard metric on $\mathbf{R}^4$ is given by

$$g = dw^2 + dx^2 + dy^2 + dz^2$$

in cartesian coordinates. Using hyperspherical coordinates, we therefore have that

$$g|_V = dr^2 + r^2(d\chi^2 + \sin^2\chi\,(d\theta^2 + \sin^2\theta\;d\phi^2))\,.$$

By Proposition 2.41 we know that the 3-sphere

$$\mathbf{S}^3_a = \{(w,x,y,z)\in\mathbf{R}^4 \,|\, w^2+x^2+y^2+z^2 = a^2\}$$

with radius a is a closed submanifold of the Euclidean space $\mathbf{R}^4$. Let $i:\mathbf{S}^3_a\to\mathbf{R}^4$ be the inclusion map and $p:\mathbf{R}\times\mathbf{R}^3\to\mathbf{R}^3$ be the projection on the last factor, and let $V_a = V\cap\mathbf{S}^3_a$. Then $(p\circ\psi\circ i, V_a)$ is a local chart on $\mathbf{S}^3_a$ with coordinate functions $\chi\circ i$, $\theta\circ i$ and $\phi\circ i$ which we also denote simply by χ, θ and ϕ. The pull-back to V_a of the standard metric g on $\mathbf{R}^4$ is given by

$$i^*g|_{V_a} = a^2(d\chi^2 + \sin^2\chi\,(d\theta^2 + \sin^2\theta\;d\phi^2))\,.$$

7.23 Example Consider the smooth manifold

$$M = \{(w,x,y,z) \in \mathbf{R}^4 | w > \sqrt{x^2+y^2+z^2}\,\}\,.$$

A point $p \in M$ with cartesian coordinates (w,x,y,z) in the open set

$$V = \{(w,x,y,z) \in M | y \neq 0 \text{ or } x > 0\}$$

can also be described by pseudospherical coordinates (r,χ,θ,ϕ) in the open set

$$U = \{(r,\chi,\theta,\phi) \in \mathbf{R}^4 | r > 0\,,\, \chi > 0\,,\, 0 < \theta < \pi \text{ and } -\pi < \phi < \pi\}\,.$$

The connection between these coordinates is given by

$$\begin{aligned} w &= r\cosh\chi \\ x &= r\sinh\chi\,\sin\theta\,\cos\phi \\ y &= r\sinh\chi\,\sin\theta\,\sin\phi \\ z &= r\sinh\chi\,\cos\theta \end{aligned}$$

and

$$\begin{aligned} r &= \sqrt{w^2-x^2-y^2-z^2} \\ \chi &= \cosh^{-1}(\,w\,/\,\sqrt{w^2-x^2-y^2-z^2}\,) \\ \theta &= \cos^{-1}(\,z\,/\,\sqrt{x^2+y^2+z^2}\,) \\ \phi &= \begin{cases} \cot^{-1}(x/y) - \pi & \text{for } \ y < 0 \\ 0 & \text{for } \ y = 0 \text{ and } x > 0 \\ \cot^{-1}(x/y) & \text{for } \ y > 0 \end{cases} \end{aligned}$$

As well as being real numbers, we can also think of w,x,y,z and r,χ,θ,ϕ as functions of p, so that they are the coordinate functions of local charts (id,M) and (ψ,V) on M. The meaning will always be clear from the context. Here $\psi = i \circ \rho^{-1}$, where $\rho : U \to V$ is the diffeomorphism given by

$$\rho\,(r,\chi,\theta,\phi) = (r\cosh\chi\,,\, r\sinh\chi\,\sin\theta\,\cos\phi\,,\, r\sinh\chi\,\sin\theta\,\sin\phi\,,\, r\sinh\chi\,\cos\theta)\,,$$

and $i : U \to \mathbf{R}^4$ is the inclusion map. We have that

$$D\rho\,(r,\chi,\theta,\phi) =$$

$$\begin{pmatrix} \cosh\chi & r\sinh\chi & 0 & 0 \\ \sinh\chi\sin\theta\cos\phi & r\cosh\chi\sin\theta\cos\phi & r\sinh\chi\cos\theta\cos\phi & -r\sinh\chi\sin\theta\sin\phi \\ \sinh\chi\sin\theta\sin\phi & r\cosh\chi\sin\theta\sin\phi & r\sinh\chi\cos\theta\sin\phi & r\sinh\chi\sin\theta\cos\phi \\ \sinh\chi\cos\theta & r\cosh\chi\cos\theta & -r\sinh\chi\sin\theta & 0 \end{pmatrix}$$

so that

$$\det D\rho\,(r,\chi,\theta,\phi) = (\cosh\chi)(r^3\cosh\chi\,\sinh^2\chi\,\sin\theta) - (r\sinh\chi)(r^2\sinh^3\chi\,\sin\theta)$$
$$= r^3\sinh^2\chi\,\sin\theta > 0$$

for $(r,\chi,\theta,\phi)\in U$. Proposition 4.18 now implies that

$$dw = \cosh\chi\; dr + r\sinh\chi\; d\chi\,,$$

$$dx = \sinh\chi\,\sin\theta\,\cos\phi\; dr + r\cosh\chi\,\sin\theta\,\cos\phi\; d\chi + r\sinh\chi\,\cos\theta\,\cos\phi\; d\theta - r\sinh\chi\,\sin\theta\,\sin\phi\; d\phi\,,$$

$$dy = \sinh\chi\,\sin\theta\,\sin\phi\; dr + r\cosh\chi\,\sin\theta\,\sin\phi\; d\chi + r\sinh\chi\,\cos\theta\,\sin\phi\; d\theta + r\sinh\chi\,\sin\theta\,\cos\phi\; d\phi$$

and

$$dz = \sinh\chi\,\cos\theta\; dr + r\cosh\chi\,\cos\theta\; d\chi - r\sinh\chi\,\sin\theta\; d\theta$$

on V. The Lorentz metric on $\mathbf{R}^4_1$ is given by

$$g^4_1 = -\,dw^2 + dx^2 + dy^2 + dz^2$$

in cartesian coordinates. Using pseudospherical coordinates, we therefore have that

$$g^4_1|_V = -\,dr^2 + r^2\,(d\chi^2 + \sinh^2\chi\,(d\theta^2 + \sin^2\theta\; d\phi^2))\,.$$

By Proposition 2.41 we know that the 3-hyperboloid

$$\mathbf{H}^3_a = \{(w,x,y,z)\in\mathbf{R}^4\,|\,w^2 - x^2 - y^2 - z^2 = a^2\}$$

where $a > 0$ is a closed submanifold of the Minkowski space $\mathbf{R}^4_1$. Let $i : \mathbf{H}^3_a \to \mathbf{R}^4$ be the inclusion map and $p : \mathbf{R}\times\mathbf{R}^3 \to \mathbf{R}^3$ be the projection on the last factor, and let $V_a = V \cap \mathbf{H}^3_a$. Then $(p\circ\psi\circ i, V_a)$ is a local chart on the upper sheet of $\mathbf{H}^3_a$ with coordinate functions $\chi\circ i$, $\theta\circ i$ and $\phi\circ i$ which we also denote simply by χ, θ and ϕ. The pull-back to V_a of the Lorentz metric g^4_1 on $\mathbf{R}^4_1$ is given by

$$i^* g^4_1|_{V_a} = a^2\,(d\chi^2 + \sinh^2\chi\,(d\theta^2 + \sin^2\theta\; d\phi^2))\,.$$

THE HODGE STAR OPERATOR

7.24 Let V be a finite dimensional vector space with a metric g. Identifying V^{**} with V as in Remark 4.5 by means of the natural isomorphism $i_V : V \to V^{**}$ given by $i_V(v)(\lambda) = \lambda(v)$ for $v\in V$ and $\lambda\in V^*$, we have an isomorphism $H = i_V\circ G^{-1}$:

$V^* \to V^{**}$. Hence we obtain a non-degenerate bilinear functional $h : V^* \times V^* \to \mathbf{R}$ given by

$$h(\lambda,\mu) = H(\lambda)(\mu) = \mu(G^{-1}(\lambda))$$

for $\lambda,\mu \in V^*$. If $\mathscr{B} = \{v_1,...,v_n\}$ is a basis for V with dual basis $\mathscr{B}^* = \{v_1^*,...,v_n^*\}$, then the components of h with respect to $\mathscr{B}^*$ are given by

$$h(v_i^*,v_j^*) = v_j^*(\sum_{k=1}^{n} g^{ik} v_k) = g^{ij}$$

for $i,j = 1,...,n$, so that

$$h = \sum_{i,j} g^{ij} v_i \otimes v_j\,.$$

If g is symmetric, so is h, and therefore h is a metric on V^* which is said to be induced by the metric g on V. We have that

$$h(G(v),G(w)) = G(w)(v) = g(w,v) = g(v,w)$$

for $v,w \in V$, which shows that $G : V \to V^*$ is a linear isometry with respect to the metrics g and h. Hence if g is positive definite (resp., negative definite), the same is true for h.

The metric h may be extended to a symmetric bilinear functional $h_k : \mathscr{T}^k(V) \times \mathscr{T}^k(V) \to \mathbf{R}$ so that

$$h_k(\lambda_1 \otimes \cdots \otimes \lambda_k, \mu_1 \otimes \cdots \otimes \mu_k) = \frac{1}{k!}\, h(\lambda_1,\mu_1) \cdots h(\lambda_k,\mu_k) \tag{1}$$

for $\lambda_1, \ldots, \lambda_k, \mu_1, \ldots, \mu_k \in V^*$. Since this implies that

$$h_k(\lambda_1 \otimes \cdots \otimes \lambda_k, \mu_1 \otimes \cdots \otimes \mu_k) = \frac{1}{k!}\, \mu_1 \otimes \cdots \otimes \mu_k (G^{-1}(\lambda_1), \ldots, G^{-1}(\lambda_k)), \tag{2}$$

it follows that

$$h_k(\lambda_1 \otimes \cdots \otimes \lambda_k, \sum_{i=1}^{r} c_i \mu_1^i \otimes \cdots \otimes \mu_k^i) = 0$$

for every $\lambda_1, \ldots, \lambda_k \in V^*$ if and only if

$$\sum_{i=1}^{r} c_i\, \mu_1^i \otimes \cdots \otimes \mu_k^i = 0\,,$$

so that h_k is well defined and non-degenerate.

From (2) and Proposition 4.74 it also follows that

$$h_k(\lambda_1 \otimes \cdots \otimes \lambda_k, \mu_1 \otimes \cdots \otimes \mu_k) = \frac{1}{k!}\, C(R(\mu_1 \otimes \cdots \otimes \mu_k) \otimes \lambda_1 \otimes \cdots \otimes \lambda_k)\,,$$

where $C : \mathscr{T}_k^k(V) \to \mathbf{R}$ is the k-fold contraction in the i-th contravariant and the i-th covariant index for $0 \leq i \leq k$, and $R : \mathcal{T}_0^k(V) \to \mathcal{T}_k^0(V)$ is the linear isomorphism given by

$$R(T)(\lambda_1,\dots,\lambda_k) = T(G^{-1}(\lambda_1),\dots,G^{-1}(\lambda_k))$$

for $T \in \mathcal{T}_0^k(V)$ and $\lambda_1,\dots,\lambda_k \in V^*$, raising all the indices of T as defined above. Hence we have that

$$h_k(T,S) = \frac{1}{k!}\, C(R(T) \otimes S)$$

for $T,S \in \mathcal{T}^k(V)$. If

$$T = \sum_{i_1,\dots,i_k} A_{i_1,\dots,i_k} v_{i_1}^* \otimes \cdots \otimes v_{i_k}^*$$

then

$$R(T) = \sum_{i_1,\dots,i_k} A^{i_1,\dots,i_k} v_{i_1} \otimes \cdots \otimes v_{i_k} \quad \text{where } A^{i_1,\dots,i_k} = \sum_{j_1,\dots,j_k} g^{i_1 j_1} \cdots g^{i_k j_k} A_{j_1,\dots,j_k}\,.$$

Hence, if

$$S = \sum_{j_1,\dots,j_k} B_{j_1,\dots,j_k} v_{j_1}^* \otimes \cdots \otimes v_{j_k}^*\,,$$

we have that

$$h_k(T,S) = \frac{1}{k!} \sum_{i_1,\dots,i_k} A^{i_1,\dots,i_k} B_{i_1,\dots,i_k}\,.$$

We will show that the restriction $\widetilde{h}_k : \Lambda^k(V) \times \Lambda^k(V) \to \mathbf{R}$ of h_k is non-degenerate, and therefore is a metric on $\Lambda^k(V)$. By (1) and Proposition 5.13 (1) we have that

$$h_k(\lambda_1 \wedge \cdots \wedge \lambda_k, \mu_1 \otimes \cdots \otimes \mu_k) = \frac{1}{k!} \sum_{\sigma \in S_k} \varepsilon(\sigma)\, h(\lambda_{\sigma(1)}, \mu_1) \cdots h(\lambda_{\sigma(k)}, \mu_k)$$

$$= \frac{1}{k!} \begin{vmatrix} h(\lambda_1,\mu_1) & h(\lambda_1,\mu_2) & \cdots & h(\lambda_1,\mu_k) \\ h(\lambda_2,\mu_1) & h(\lambda_2,\mu_2) & \cdots & h(\lambda_2,\mu_k) \\ \vdots & \vdots & & \vdots \\ h(\lambda_k,\mu_1) & h(\lambda_k,\mu_2) & \cdots & h(\lambda_k,\mu_k) \end{vmatrix}$$

for $\lambda_1, \dots, \lambda_k, \mu_1, \dots, \mu_k \in V^*$. Since the last expression is skew symmetric as a function of $\mu_1,\dots,\mu_k$, it follows from Proposition 5.8 (2) that

$$\widetilde{h}_k(\lambda_1 \wedge \cdots \wedge \lambda_k, \mu_1 \wedge \cdots \wedge \mu_k) = \begin{vmatrix} h(\lambda_1,\mu_1) & h(\lambda_1,\mu_2) & \cdots & h(\lambda_1,\mu_k) \\ h(\lambda_2,\mu_1) & h(\lambda_2,\mu_2) & \cdots & h(\lambda_2,\mu_k) \\ \vdots & \vdots & & \vdots \\ h(\lambda_k,\mu_1) & h(\lambda_k,\mu_2) & \cdots & h(\lambda_k,\mu_k) \end{vmatrix}. \quad (3)$$

Now let $\mathcal{B} = \{v_1,\dots,v_n\}$ be an orthonormal basis for V with dual basis $\mathcal{B}^* =$

$\{v_1^*, ..., v_n^*\}$. Then $\mathscr{B}^*$ is an orthonormal basis for V^*, since G is a linear isometry and $G(v_i) = g(v_i, v_i)\ v_i^*$ for $i = 1, ..., n$. From (3) it follows that

$$\mathscr{D} = \{v_{i_1}^* \wedge \cdots \wedge v_{i_k}^* \mid 1 \leq i_1 < ... < i_k \leq n\}$$

is an othonormal basis for $\Lambda^k(V)$, which completes the proof that $\widetilde{h}_k$ is non-degenerate and therefore is a metric on $\Lambda^k(V)$.

7.25 Proposition Let (V, μ) be an n-dimensional oriented vector space with a metric g. Then there is a unique volume element ω in V compatible with μ such that $\omega = v_1^* \wedge \cdots \wedge v_n^*$ for any positively oriented orthonormal basis $(v_1, ..., v_n)$ in V. ω is called the *metric volume element* in V compatible with μ.

If $(w_1, ..., w_n)$ is any positively oriented basis in V, we have that

$$\omega = |\,\overline{g}\,|^{1/2}\ w_1^* \wedge \cdots \wedge w_n^*$$

where $\overline{g} = \det(g(w_i, w_j))$.

PROOF: Let $\{v_1, \ldots, v_n\}$ and $\{w_1, \ldots, w_n\}$ be two positively oriented bases for V with $w_j = \sum_{i=1}^n a_{ij}\, v_i$ for $j = 1, \ldots, n$. Then we have that

$$g(w_i, w_j) = \sum_{rs} a_{ri}\, a_{sj}\, g(v_r, v_s)$$

for $i, j = 1, \ldots, n$, which means that

$$C = A^t BA$$

for the matrices $A = (a_{ij})$, $B = (g(v_i, v_j))$ and $C = (g(w_i, w_j))$. Since the bases are equally oriented, we have that $\det(A) > 0$, and

$$v_1^* \wedge \cdots \wedge v_n^* = \det(A)\ w_1^* \wedge \cdots \wedge w_n^*$$

by Corollary 5.18.

Suppose now that both bases are orthonormal. Then $\det(B) = \det(C) = (-1)^k$, where k is the index of the metric g, and it follows that $\det(A) = 1$ which proves the first part of the proposition.

If on the other hand only the basis $\{v_1, \ldots, v_n\}$ is assumed to be orthonormal, then we obtain $\det(A) = |\det(C)|^{1/2}$ which proves the last part. □

7.26 The antisymmetric *Levi–Civita symbol* is defined by

$$\varepsilon_{i_1 \ldots i_n} = \begin{cases} 1 & \text{if } (i_1, \ldots, i_n) \text{ is an even permutation} \\ -1 & \text{if } (i_1, \ldots, i_n) \text{ is an odd permutation} \\ 0 & \text{otherwise} \end{cases}$$

Let V be an n-dimensional oriented vector space with a metric g of index k. If

$(v_1,...,v_n)$ is a positively oriented basis in V, it follows from Proposition 5.13 (1) and 7.25 that the metric volume element ω in V is given by

$$\omega = \sum_{i_1,...,i_n} a_{i_1,...,i_n}\, v^*_{i_1} \otimes \cdots \otimes v^*_{i_n}$$

with components

$$a_{i_1 \dots i_n} = |\,\overline{g}\,|^{1/2}\ \varepsilon_{i_1 \dots i_n}\,,$$

where $\overline{g} = \det(g(v_i, v_j))$. Raising the indices we have that

$$a^{i_1 \dots i_n} = |\,\overline{g}\,|^{-1/2}\ \varepsilon^{i_1 \dots i_n}$$

where $\varepsilon^{i_1 \dots i_n} = (-1)^k\, \varepsilon_{i_1 \dots i_n}$.

7.27 Let (V,μ) be an n-dimensional oriented vector space with a metric g, and let ε be the metric volume element in V compatible with μ. If $0 \le k \le n$, we have a linear isomorphism $R : \mathscr{T}^k_0(V) \to \mathscr{T}^0_k(V)$ given by

$$R(T)(\lambda_1,\dots,\lambda_k) = T(g^\sharp(\lambda_1),\dots,g^\sharp(\lambda_k))$$

for $T \in \mathscr{T}^k_0(V)$ and $\lambda_1,\dots,\lambda_k \in V^*$, raising all the indices of T as defined in 7.24. We see that R maps $\Lambda^k(V)$ onto $\Lambda^k(V^*)$. Furthermore, let $C : \mathscr{T}^n_k(V) \to \mathscr{T}^{n-k}_0(V)$ be the k-fold contraction in the i-th contravariant and the i-th covariant index for $0 \le i \le k$. We can now define a linear map

$$* : \Lambda^k(V) \to \Lambda^{n-k}(V)\,,$$

called the *Hodge star operator*, by

$$*\,\omega = \frac{1}{k!}\ C(R(\omega) \otimes \varepsilon)$$

for $\omega \in \Lambda^k(V)$. Let $\mathscr{B} = \{v_1,...,v_n\}$ be a basis for V with dual basis $\mathscr{B}^* = \{v^*_1,...,v^*_n\}$, and suppose that

$$\omega = \sum_{i_1<\dots<i_k} a_{i_1,...,i_k}\, v^*_{i_1} \wedge \cdots \wedge v^*_{i_k}$$

and

$$\varepsilon = c_{1,...,n}\, v^*_1 \wedge \cdots \wedge v^*_n\,.$$

It will be useful to express the forms ω and ε in terms of tensor products of the basis vectors from $\mathscr{B}^*$, instead of wedge products. By Proposition 4.26 we have that

$$\omega = \sum_{i_1,...,i_k} a_{i_1,...,i_k}\, v^*_{i_1} \otimes \cdots \otimes v^*_{i_k}$$

and

$$\varepsilon = \sum_{i_1,...,i_n} c_{i_1,...,i_n}\, v^*_{i_1} \otimes \cdots \otimes v^*_{i_n}\,,$$

where the new coefficients $a_{i_1,\dots,i_k} = \omega(v_{i_1},\dots,v_{i_k})$ and $c_{i_1,\dots,i_n} = \varepsilon(v_{i_1},\dots,v_{i_n})$ are skew symmetric in their indices, and agree with the old ones when the indices are strictly increasing. Now we have that

$$R(\omega) = \sum_{i_1,\dots,i_k} a^{i_1,\dots,i_k} v_{i_1} \otimes \cdots \otimes v_{i_k} \quad \text{where} \quad a^{i_1,\dots,i_k} = \sum_{j_1,\dots,j_k} g^{i_1 j_1} \cdots g^{i_k j_k} a_{j_1,\dots,j_k} ,$$

so that

$$*\,\omega = \sum_{j_1,\dots,j_{n-k}} b_{j_1,\dots,j_{n-k}} v^*_{j_1} \otimes \cdots \otimes v^*_{j_{n-k}}$$

with

$$b_{j_1,\dots,j_{n-k}} = \frac{1}{k!} \sum_{i_1,\dots,i_k} a^{i_1,\dots,i_k} c_{i_1,\dots,i_k,j_1,\dots,j_{n-k}} .$$

As the coefficients $a^{i_1,\dots,i_k}$ and $b_{j_1,\dots,j_{n-k}}$ are also skew symmetric in their indices, it follows from Proposition 5.13 (1) that

$$*\,\omega = \sum_{j_1<\dots<j_{n-k}} b_{j_1,\dots,j_{n-k}} v^*_{j_1} \wedge \cdots \wedge v^*_{j_{n-k}}$$

where

$$b_{j_1,\dots,j_{n-k}} = \sum_{i_1<\dots<i_k} a^{i_1,\dots,i_k} c_{i_1,\dots,i_k,j_1,\dots,j_{n-k}} .$$

In particular, if $\omega = g^\flat(v)$ for a vector $v = \sum_{i=1}^{n} a^i v_i$, then

$$\begin{aligned} *\,\omega &= \sum_{j_1<\dots<j_{n-1}} \left(\sum_{i=1}^{n} a^i c_{i,j_1,\dots,j_{n-1}} \right) v^*_{j_1} \wedge \cdots \wedge v^*_{j_{n-1}} \\ &= \sum_{i=1}^{n} (-1)^{i+1} c_{1,\dots,n}\, a^i\, v^*_1 \wedge \cdots \wedge v^*_{i-1} \wedge v^*_{i+1} \wedge \cdots \wedge v^*_n \\ &= c_{1,\dots,n} \sum_{i=1}^{n} (-1)^{i+1} i_v(v^*_i)\, v^*_1 \wedge \cdots \wedge v^*_{i-1} \wedge v^*_{i+1} \wedge \cdots \wedge v^*_n = i_v(\varepsilon) \end{aligned}$$

by Proposition 5.74.

If $\mathscr{C} = \{w_1,\dots,w_n\}$ is an orthonormal basis for V with dual basis $\mathscr{C}^* = \{w^*_1,\dots,w^*_n\}$, we have that

$$*(w^*_{i_1} \wedge \cdots \wedge w^*_{i_k}) = g(w_{i_1},w_{i_1}) \cdots g(w_{i_k},w_{i_k})\, w^*_{j_1} \wedge \cdots \wedge w^*_{j_{n-k}} \tag{1}$$

if

$$w^*_{i_1} \wedge \cdots \wedge w^*_{i_k} \wedge w^*_{j_1} \wedge \cdots \wedge w^*_{j_{n-k}} = \varepsilon. \tag{2}$$

Indeed, if

$$v_r = \begin{cases} w_{i_r} & \text{for } 1 \le r \le k \\ w_{j_{r-k}} & \text{for } k+1 \le r \le n \end{cases} ,$$

then $\mathscr{B} = \{v_1, ..., v_n\}$ is a positively oriented orthonormal basis for V, and we have that

$$*(v_1^* \wedge \cdots \wedge v_k^*) = b_{k+1,...,n} \, v_{k+1}^* \wedge \cdots \wedge v_n^*$$

where

$$b_{k+1,...,n} = a^{1,...,k} = g(v_1, v_1) \cdots g(v_k, v_k) .$$

In particular, we have that

$$*1 = \varepsilon \quad \text{and} \quad *\varepsilon = (-1)^s$$

where s is the index of the metric g.

By Proposition 5.12 (3) it follows from (2) that

$$w_{j_1}^* \wedge \cdots \wedge w_{j_{n-k}}^* \wedge (-1)^{k(n-k)} \, w_{i_1}^* \wedge \cdots \wedge w_{i_k}^* = \varepsilon ,$$

so that (1) implies that

$$**(w_{i_1}^* \wedge \cdots \wedge w_{i_k}^*) = (-1)^{k(n-k)+s} \, w_{i_1}^* \wedge \cdots \wedge w_{i_k}^* .$$

Hence we have that

$$** \, \omega = (-1)^{k(n-k)+s} \, \omega$$

for $\omega \in \Lambda^k(V)$.

From (1) it follows that

$$(w_{i_1}^* \wedge \cdots \wedge w_{i_k}^*) \wedge *(w_{i_1}^* \wedge \cdots \wedge w_{i_k}^*) = < w_{i_1}^* \wedge \cdots \wedge w_{i_k}^*, w_{i_1}^* \wedge \cdots \wedge w_{i_k}^* > \varepsilon ,$$

and we have that

$$(w_{r_1}^* \wedge \cdots \wedge w_{r_k}^*) \wedge *(w_{i_1}^* \wedge \cdots \wedge w_{i_k}^*) = 0$$

and

$$< w_{r_1}^* \wedge \cdots \wedge w_{r_k}^*, w_{i_1}^* \wedge \cdots \wedge w_{i_k}^* > = 0$$

when $\{r_1, ..., r_k\} \neq \{i_1, ..., i_k\}$. Hence it follows from Proposition 5.14 that

$$\omega \wedge * \eta = < \omega, \eta > \varepsilon$$

for $\omega, \eta \in \Lambda^k(V)$.

7.28 Definition Let M^n be an oriented pseudo-Riemannian manifold with a metric g of index s. Then we have a linear map

$$* : \Omega^k(M) \to \Omega^{n-k}(M) ,$$

called the *Hodge star operator*, defined by

$$(*\omega)(p) = *\omega(p)$$

for $p \in M$, $\omega \in \Omega^k(M)$ and $k = 0,...,n$. We let $\Omega^k(M) = \{0\}$ and $* : \Omega^k(M) \to \Omega^{n-k}(M)$ be the zero map when $k < 0$ or $k > n$.

We now define the *codifferential operator* $\delta : \Omega^k(M) \to \Omega^{k-1}(M)$ by

$$\delta = (-1)^{n(k+1)+s+1} * d * ,$$

and the *Laplace–Beltrami operator* $\Delta : \Omega^k(M) \to \Omega^k(M)$ by

$$\Delta = \delta d + d \delta$$

for $k = 0,...,n$.

7.29 Let M^n be a compact oriented pseudo-Riemannian manifold. Then we have a metric on the vector space $\Omega^k(M)$ of k-forms on M defined by

$$< \omega, \eta > \ = \int_M \omega \wedge * \eta$$

for $\omega, \eta \in \Omega^k(M)$ and $k = 0,...,n$. The metric may also be extended to $\Omega(M) = \oplus_{k=0}^n \Omega^k(M)$ by requiring the various $\Omega^k(M)$ to be orthogonal.

7.30 Proposition Let M^n be a compact oriented pseudo-Riemannian manifold with a metric g of index s. Then δ is the adjoint of d, and Δ is self-adjoint on $\Omega(M)$, i.e.,

$$< d\omega, \eta > \ = \ < \omega, \delta \eta >$$

and

$$< \Delta\omega, \eta > \ = \ < \omega, \Delta\eta >$$

for $\omega, \eta \in \Omega(M)$.

PROOF: To prove the first formula, we only need to consider the case where $\omega \in \Omega^{k-1}(M)$ and $\eta \in \Omega^k(M)$ for $k = 1,...,n$. By Proposition 5.58 we have that

$$d(\omega \wedge * \eta) = d\omega \wedge * \eta + (-1)^{k-1} \omega \wedge d * \eta .$$

Since $d * \eta \in \Omega^{n-k+1}(M)$, it follows from 7.27 that

$$\begin{aligned}
&(-1)^{k-1} d * \eta = (-1)^{k-1} (-1)^{(n-k+1)(k-1)+s} ** d * \eta \\
&= (-1)^{n(k-1)-k(k-1)+s} ** d * \eta = (-1)^{n(k+1)+s} ** d * \eta \\
&= - * \delta\eta .
\end{aligned}$$

Combining this and using Stokes' theorem, we conclude that

$$0 = \int_M d(\omega \wedge * \eta) = \int_M (d\omega \wedge * \eta - \omega \wedge * \delta\eta) = < d\omega, \eta > - < \omega, \delta\eta >$$

which completes the proof of the first formula in the proposition. The second formula follows immediately from the first. □

7.31 Let $f : M_1 \to M_2$ be a smooth map between two n-dimensional smooth manifolds M_1 and M_2 with volume elements ω_1 and ω_2. Then the smooth function $J : M_1 \to \mathbf{R}$ defined by

$$f^*\omega_2 = J\,\omega_1$$

is called the *Jacobian function* of f. For each point $p \in M_1$, the number $|J(p)|$ measures the rate of change of volume near p under the mapping f. The map f is said to be *volume preserving* if $J = 1$ so that $f^*\omega_2 = \omega_1$.

7.32 Let M^n be a smooth manifold with a volume element ω. Then a vector field X on M with global flow $\gamma : \mathscr{D}(X) \to M$ is called *incompressible* if the diffeomorphism $\gamma_t : \mathscr{D}_t(X) \to \mathscr{D}_{-t}(X)$ is volume preserving for each $t \in \mathbf{R}$. By Proposition 4.90 this is equivalent to the assertion that

$$L_X\,\omega = 0\,.$$

In order to measure the deviation of a vector field X on M from being incompressible, we define the *ω-divergence* $\operatorname{div}_\omega X$ of X with respect to the volume element ω by

$$L_X\,\omega = (\operatorname{div}_\omega X)\,\omega\,.$$

If (x,U) is a local chart on M and

$$\omega\,|_U = h\;dx^1\wedge\ldots\wedge dx^n \quad\text{and}\quad X\,|_U = \sum_{i=1}^n a^i\,\frac{\partial}{\partial x^i}\,,$$

then

$$(\operatorname{div}_\omega X)\,|_U = \frac{1}{h}\sum_{i=1}^n \frac{\partial}{\partial x^i}(ha^i)$$

since

$$\begin{aligned}(L_X\omega)|_U &= X(h)dx^1\wedge\ldots\wedge dx^n + \sum_{i=1}^n h dx^1\wedge\ldots\wedge L_X(dx^i)\wedge\ldots\wedge dx^n\\ &= \sum_{i=1}^n a^i\,\frac{\partial h}{\partial x^i}dx^1\wedge\ldots\wedge dx^n + \sum_{i=1}^n h\frac{\partial a^i}{\partial x^i}dx^1\wedge\ldots\wedge dx^n\\ &= \sum_{i=1}^n \frac{\partial}{\partial x^i}(ha^i)dx^1\wedge\ldots\wedge dx^n\end{aligned}$$

by Proposition 4.76 and 4.79 (2).

7.33 Proposition Let (M^n,μ) be an oriented pseudo-Riemannian manifold with a metric g, and let ε be the metric volume element on M compatible with μ. If X is a vector field on M, we have that

$$\operatorname{div}_\varepsilon X = *\,d*(X^\flat) = -\,\delta\,(X^\flat)\,.$$

In this case $\operatorname{div}_\varepsilon X$ is usually denoted simply by $\operatorname{div} X$.

PROOF : We have that

$$L_X \varepsilon = d(i_X \varepsilon) = d{*}(X^\flat)$$

by 7.27 and Proposition 5.83. □

7.34 Definition Let M^n be a pseudo-Riemannian manifold with a metric g. Then we define the *gradient* of a smooth function $f : M \to \mathbf{R}$ by

$$\operatorname{grad} f = (df)^\sharp .$$

If (x, U) is a local chart on M, we have that

$$df|_U = \sum_{i=1}^{n} \frac{\partial f}{\partial x^i}\, dx^i$$

so that

$$\operatorname{grad} f|_U = \sum_{i,j} g^{ij} \frac{\partial f}{\partial x^j} \frac{\partial}{\partial x^i} .$$

7.35 Remark Let M^n be an oriented pseudo-Riemannian manifold with a metric g, and let $f : M \to \mathbf{R}$ be a smooth function. Then we have that

$$\Delta f = \delta(df) = -\operatorname{div}(\operatorname{grad} f) .$$

Hence the Laplace-Beltrami operator has the opposite sign of the ordinary Laplacian on functions

$$\nabla^2 f = \operatorname{div}(\operatorname{grad} f) .$$

7.36 Let (M, μ) be a 3-dimensional oriented Riemannian manifold with a metric g, and let ε be the metric volume element on M compatible with μ. Then we define the *curl* of a vector field X on M by

$$\operatorname{curl} X = \{ * d(X^\flat) \}^\sharp$$

so that we have a commutative diagram

$$\begin{array}{ccccccc}
\mathscr{F}(M) & \xrightarrow{\text{grad}} & \mathscr{T}_1(M) & \xrightarrow{\text{curl}} & \mathscr{T}_1(M) & \xrightarrow{\text{div}} & \mathscr{F}(M) \\
\phi_1 \downarrow & & \phi_2 \downarrow & & \phi_3 \downarrow & & \phi_4 \downarrow \\
\Omega^0(M) & \xrightarrow{d} & \Omega^1(M) & \xrightarrow{d} & \Omega^2(M) & \xrightarrow{d} & \Omega^3(M)
\end{array}$$

where $\phi_1 : \mathscr{F}(M) \to \Omega^0(M)$, $\phi_2 : \mathscr{T}_1(M) \to \Omega^1(M)$, $\phi_3 : \mathscr{T}_1(M) \to \Omega^2(M)$ and $\phi_4 : \mathscr{F}(M) \to \Omega^3(M)$ are the linear isomorphisms defined by

$$\phi_1(f) = f \ , \ \phi_2(X) = X^\flat \ , \ \phi_3(X) = *(X^\flat) \ \text{ and } \ \phi_4(f) = *f$$

for $f \in \mathscr{F}(M)$ and $X \in \mathscr{T}_1(M)$. Since $d \circ d = 0$, it follows in particular that

$$\text{curl}(\text{grad}\, f) = 0 \ \text{ and } \ \text{div}(\text{curl}\, X) = 0$$

for every $f \in \mathscr{F}(M)$ and $X \in \mathscr{T}_1(M)$. By 7.27 we also have that

$$i_{\,\text{curl}X}\ \varepsilon = d(X^\flat)\ .$$

7.37 Example Consider the smooth manifold $M = \mathbf{R}^3$ with its standard orientation μ so that the local chart (id, M) is positively oriented, and with its standard metric

$$g = dx^2 + dy^2 + dz^2$$

in cartesian coordinates x, y and z which are the coordinate functions of id. Then the metrical volume element is given by

$$\varepsilon = dx \wedge dy \wedge dz\ .$$

For a smooth function $f : M \to \mathbf{R}$ we have that

$$df = \frac{\partial f}{\partial x}\, dx + \frac{\partial f}{\partial y}\, dy + \frac{\partial f}{\partial z}\, dz\ .$$

Since (g_{ij}) is the identity matrix, it follows that

$$\text{grad}\, f = (df)^\sharp = \frac{\partial f}{\partial x}\frac{\partial}{\partial x} + \frac{\partial f}{\partial y}\frac{\partial}{\partial y} + \frac{\partial f}{\partial z}\frac{\partial}{\partial z}\ .$$

Let X be a vector field on M with

$$X = a_x \frac{\partial}{\partial x} + a_y \frac{\partial}{\partial y} + a_z \frac{\partial}{\partial z}$$

so that

$$X^\flat = a_x\, dx + a_y\, dy + a_z\, dz\ .$$

Then we have that

$$d(X^\flat) = \left(\frac{\partial a_z}{\partial y} - \frac{\partial a_y}{\partial z}\right) dy \wedge dz + \left(\frac{\partial a_x}{\partial z} - \frac{\partial a_z}{\partial x}\right) dz \wedge dx + \left(\frac{\partial a_y}{\partial x} - \frac{\partial a_x}{\partial y}\right) dx \wedge dy$$

which implies that

$$*d(X^\flat) = \left(\frac{\partial a_z}{\partial y} - \frac{\partial a_y}{\partial z}\right) dx + \left(\frac{\partial a_x}{\partial z} - \frac{\partial a_z}{\partial x}\right) dy + \left(\frac{\partial a_y}{\partial x} - \frac{\partial a_x}{\partial y}\right) dz\ ,$$

thus showing that

$$\operatorname{curl} X = \{*d(X^\flat)\}^\sharp = \left(\frac{\partial a_z}{\partial y} - \frac{\partial a_y}{\partial z}\right)\frac{\partial}{\partial x} + \left(\frac{\partial a_x}{\partial z} - \frac{\partial a_z}{\partial x}\right)\frac{\partial}{\partial y} + \left(\frac{\partial a_y}{\partial x} - \frac{\partial a_x}{\partial y}\right)\frac{\partial}{\partial z}\,.$$

Now identifying the tangent space of $\mathbf{R}^3$ at an arbtrary point with $\mathbf{R}^3$ as described in Lemma 2.84, we obtain the usual expressions

$$\operatorname{grad} f = \left(\frac{\partial f}{\partial x}, \frac{\partial f}{\partial y}, \frac{\partial f}{\partial z}\right)$$

and

$$\operatorname{curl} X = \left(\frac{\partial a_z}{\partial y} - \frac{\partial a_y}{\partial z}, \frac{\partial a_x}{\partial z} - \frac{\partial a_z}{\partial x}, \frac{\partial a_y}{\partial x} - \frac{\partial a_x}{\partial y}\right) = \begin{vmatrix} \mathbf{i} & \mathbf{j} & \mathbf{k} \\ \dfrac{\partial}{\partial x} & \dfrac{\partial}{\partial y} & \dfrac{\partial}{\partial z} \\ a_x & a_y & a_z \end{vmatrix}\,.$$

By 7.32 we have that

$$\operatorname{div} X = \frac{\partial a_x}{\partial x} + \frac{\partial a_y}{\partial y} + \frac{\partial a_z}{\partial z}$$

and

$$\nabla^2 f = \frac{\partial^2 a_x}{\partial x^2} + \frac{\partial^2 a_y}{\partial y^2} + \frac{\partial^2 a_z}{\partial z^2}\,.$$

7.38 Example Let (M^3, μ) be an oriented Riemannian manifold with a metric g, and suppose that (ψ, O) is a positively oriented local chart on M with coordinate functions u, v and w so that

$$g|_O = \frac{1}{U^2}\,du^2 + \frac{1}{V^2}\,dv^2 + \frac{1}{W^2}\,dw^2\,,$$

where U, V and W are positive smooth functions on O. Then the vector fields

$$\mathbf{e}_u = U\frac{\partial}{\partial u}\quad,\quad \mathbf{e}_v = V\frac{\partial}{\partial v}\quad\text{and}\quad \mathbf{e}_w = W\frac{\partial}{\partial w}$$

form an orthonormal local basis $\mathscr{B}$ for TM on O, with dual basis $\mathscr{B}^*$ consisting of

$$\mathbf{e}^u = \frac{1}{U}\,du\quad,\quad \mathbf{e}^v = \frac{1}{V}\,dv\quad\text{and}\quad \mathbf{e}^w = \frac{1}{W}\,dw\,.$$

The metrical volume element is given by

$$\varepsilon = \mathbf{e}^u \wedge \mathbf{e}^v \wedge \mathbf{e}^w = \frac{1}{UVW}\,du \wedge dv \wedge dw\,.$$

For a smooth function $f : O \to \mathbf{R}$ we have that

$$df = \frac{\partial f}{\partial u}\,du + \frac{\partial f}{\partial v}\,dv + \frac{\partial f}{\partial w}\,dw = U\frac{\partial f}{\partial u}\,\mathbf{e}^u + V\frac{\partial f}{\partial v}\,\mathbf{e}^v + W\frac{\partial f}{\partial w}\,\mathbf{e}^w\,.$$

Since $m_{\mathscr{B}}(g)$ is the identity matrix at each point, it follows that

$$\operatorname{grad} f = (df)^{\sharp} = U\,\frac{\partial f}{\partial u}\,\mathbf{e}_u + V\,\frac{\partial f}{\partial v}\,\mathbf{e}_v + W\,\frac{\partial f}{\partial w}\,\mathbf{e}_w\,.$$

Let X be a vector field on O with

$$X = a_u\,\mathbf{e}_u + a_v\,\mathbf{e}_v + a_w\,\mathbf{e}_w = a_u\,U\,\frac{\partial}{\partial u} + a_v\,V\,\frac{\partial}{\partial v} + a_w\,W\,\frac{\partial}{\partial w}$$

so that

$$X^{\flat} = a_u\,\mathbf{e}^u + a_v\,\mathbf{e}^v + a_w\,\mathbf{e}^w = \frac{a_u}{U}\,du + \frac{a_v}{V}\,dv + \frac{a_w}{W}\,dw\,.$$

Then we have that

$$d(X^{\flat}) = \left[\frac{\partial}{\partial v}\left(\frac{a_w}{W}\right) - \frac{\partial}{\partial w}\left(\frac{a_v}{V}\right)\right] dv\wedge dw + \left[\frac{\partial}{\partial w}\left(\frac{a_u}{U}\right) - \frac{\partial}{\partial u}\left(\frac{a_w}{W}\right)\right] dw\wedge du$$

$$+\left[\frac{\partial}{\partial u}\left(\frac{a_v}{V}\right) - \frac{\partial}{\partial v}\left(\frac{a_u}{U}\right)\right] du\wedge dv = VW\left[\frac{\partial}{\partial v}\left(\frac{a_w}{W}\right) - \frac{\partial}{\partial w}\left(\frac{a_v}{V}\right)\right] \mathbf{e}^v\wedge\mathbf{e}^w$$

$$+\,WU\left[\frac{\partial}{\partial w}\left(\frac{a_u}{U}\right) - \frac{\partial}{\partial u}\left(\frac{a_w}{W}\right)\right] \mathbf{e}^w\wedge\mathbf{e}^u + UV\left[\frac{\partial}{\partial u}\left(\frac{a_v}{V}\right) - \frac{\partial}{\partial v}\left(\frac{a_u}{U}\right)\right] \mathbf{e}^u\wedge\mathbf{e}^v$$

which implies that

$$*\,d(X^{\flat}) = VW\left[\frac{\partial}{\partial v}\left(\frac{a_w}{W}\right) - \frac{\partial}{\partial w}\left(\frac{a_v}{V}\right)\right] \mathbf{e}^u + WU\left[\frac{\partial}{\partial w}\left(\frac{a_u}{U}\right) - \frac{\partial}{\partial u}\left(\frac{a_w}{W}\right)\right] \mathbf{e}^v$$

$$+\,UV\left[\frac{\partial}{\partial u}\left(\frac{a_v}{V}\right) - \frac{\partial}{\partial v}\left(\frac{a_u}{U}\right)\right] \mathbf{e}^w\,,$$

thus showing that

$$\operatorname{curl} X = \{*\,d(X^{\flat})\}^{\sharp} = VW\left[\frac{\partial}{\partial v}\left(\frac{a_w}{W}\right) - \frac{\partial}{\partial w}\left(\frac{a_v}{V}\right)\right] \mathbf{e}_u$$

$$+\,WU\left[\frac{\partial}{\partial w}\left(\frac{a_u}{U}\right) - \frac{\partial}{\partial u}\left(\frac{a_w}{W}\right)\right] \mathbf{e}_v + UV\left[\frac{\partial}{\partial u}\left(\frac{a_v}{V}\right) - \frac{\partial}{\partial v}\left(\frac{a_u}{U}\right)\right] \mathbf{e}_w\,.$$

By 7.32 we have that

$$\operatorname{div} X = UVW\left[\frac{\partial}{\partial u}\left(\frac{a_u}{VW}\right) + \frac{\partial}{\partial v}\left(\frac{a_v}{WU}\right) + \frac{\partial}{\partial w}\left(\frac{a_w}{UV}\right)\right]$$

which also implies that

$$\nabla^2 f = \operatorname{div}(\operatorname{grad} f) = UVW\left[\frac{\partial}{\partial u}\left(\frac{U}{VW}\,\frac{\partial f}{\partial u}\right) + \frac{\partial}{\partial v}\left(\frac{V}{WU}\,\frac{\partial f}{\partial v}\right) + \frac{\partial}{\partial w}\left(\frac{W}{UV}\,\frac{\partial f}{\partial w}\right)\right]\,.$$

7.39 Example We are going to apply the formulae in Example 7.38 to the oriented Riemannian manifold $M = \mathbf{R}^3$ and the spherical coordinate system (ψ, O) with coordinate functions r, θ and ϕ. By Example 7.21 this coordinate system is positively oriented with respect to the standard orientation in $\mathbf{R}^3$, and we have that

$$g|_O = dr^2 + r^2\, d\theta^2 + r^2 \sin^2\theta\; d\phi^2$$

so that

$$U = 1 \;\;,\;\; V = \frac{1}{r} \;\;\text{and}\;\; W = \frac{1}{r\sin\theta}\,.$$

Given a smooth function $f : O \to \mathbf{R}$ and a vector field X on O with

$$X = a_r\,\mathbf{e}_r + a_\theta\,\mathbf{e}_\theta + a_\phi\,\mathbf{e}_\phi\;,$$

we have that

$$\operatorname{grad} f = \frac{\partial f}{\partial r}\,\mathbf{e}_r + \frac{1}{r}\,\frac{\partial f}{\partial \theta}\,\mathbf{e}_\theta + \frac{1}{r\sin\theta}\,\frac{\partial f}{\partial \phi}\,\mathbf{e}_\phi,$$

$$\operatorname{curl} X = \frac{1}{r\sin\theta}\left[\frac{\partial}{\partial\theta}(\sin\theta\; a_\phi) - \frac{\partial a_\theta}{\partial\phi}\right]\mathbf{e}_r$$
$$+ \frac{1}{r\sin\theta}\left[\frac{\partial a_r}{\partial\phi} - \sin\theta\,\frac{\partial}{\partial r}(r a_\phi)\right]\mathbf{e}_\theta + \frac{1}{r}\left[\frac{\partial}{\partial r}(r a_\theta) - \frac{\partial a_r}{\partial\theta}\right]\mathbf{e}_\phi,$$

$$\operatorname{div} X = \frac{1}{r^2}\,\frac{\partial}{\partial r}(r^2 a_r) + \frac{1}{r\sin\theta}\,\frac{\partial}{\partial\theta}(\sin\theta\; a_\theta) + \frac{1}{r\sin\theta}\,\frac{\partial a_\phi}{\partial\phi} \qquad \text{and}$$

$$\nabla^2 f = \frac{1}{r^2}\,\frac{\partial}{\partial r}\left(r^2\,\frac{\partial f}{\partial r}\right) + \frac{1}{r^2\sin\theta}\,\frac{\partial}{\partial\theta}\left(\sin\theta\;\frac{\partial f}{\partial\theta}\right) + \frac{1}{r^2\sin^2\theta}\,\frac{\partial^2 f}{\partial\phi^2}.$$

7.40 Proposition Let (M,μ) be an oriented pseudo-Riemannian manifold with a metric g, and let P^{n-1} be an immersed pseudo-Riemannian submanifold of M^n with orientation ν and metric $i^*(g)$, where $i : P \to M$ is the inclusion map. Then there is a unique vector field $N : P \to TM$ along i, called the *positive unit normal field* on P, so that $N(p)$ is a positive unit normal vector to $i_*(T_pP)$ in T_pM for each $p \in P$.

PROOF : The map $N : P \to TM$ is clearly uniquely defined, so we only need to show that it is smooth. Let ω and η be volume elements on M and P compatible with the orientations μ and ν, respectively. Now let $p_0 \in P$, and choose an orthonormal basis $\mathscr{C} = \{v_1,\dots,v_n\}$ for $T_{p_0}M$ so that $v_j = i_*(w_j)$ for $j = 1, \dots, n-1$, where $\mathscr{B} = \{w_1,\dots,w_{n-1}\}$ is an orthonormal basis for $T_{p_0}P$. We can also assume that $\omega(p_0)(v_n,v_1,\dots,v_{n-1}) > 0$ and $\eta(p_0)(w_1,\dots,w_{n-1}) > 0$ so that $N(p_0) = v_n$.

By Corollary 2.35 there are local charts (x,U) and (y,V) around p_0 on P and M, respectively, such that $U = \{q \in V \,|\, y^n(q) = 0\}$ and $x^1(q) = y^1(q)\;, \dots, x^{n-1}(q) = y^{n-1}(q)$ for $q \in U$. Hence $y \circ i = j_1 \circ x$, where $j_1 : \mathbf{R}^{n-1} \to \mathbf{R}^{n-1} \times \mathbf{R}$ is the linear map given by $j_1(a) = (a,0)$ for $a \in \mathbf{R}^{n-1}$. This implies that

$$t_{y,p} \circ i_{*p} = j_1 \circ t_{x,p} \quad\text{ and }\quad i_{*p} \circ t_{x,p}^{-1} = t_{y,p}^{-1} \circ j_1$$

for $p \in U$ by the commutative diagram in 2.70, where $(t_x, \pi'^{-1}(U))$ and $(t_y, \pi^{-1}(V))$

are the local trivializations in the tangent bundles $\pi' : TP \to P$ and $\pi : TM \to M$ associated with the local charts (x,U) and (y,V), respectively.

We have vector fields $Y_1', \dots, Y_n'$ along $i|_U$ where $\{Y_1'(p), \dots, Y_n'(p)\}$ is a basis for T_pM for each $p \in U$ and $Y_i'(p_0) = v_i$ for $i = 1, \dots, n$, given by

$$Y_i'(p) = t_{y,p}^{-1} \circ t_{y,p_0}(v_i)$$

for $p \in U$ and $i = 1, \dots, n$. We also have that $Y_j' = i_* \circ X_j'$ for $j = 1, \dots, n-1$, where

$$X_i'(p) = t_{x,p}^{-1} \circ t_{x,p_0}(w_i)$$

for $p \in U$ and $i = 1, \dots, n-1$, so that $\{X_1'(p), \dots, X_{n-1}'(p)\}$ is a basis for T_pP for each $p \in U$ and $X_i'(p_0) = w_i$ for $i = 1, \dots, n-1$.

Let W be an open neighbourhood of p_0 on P contained in U so that the $i \times i$-matrix $\left(g(p)(Y_\alpha'(p), Y_\beta'(p)) \right)$, where $1 \leq \alpha, \beta \leq i$, is non-singular for each $p \in W$ and $1 \leq i \leq n$. Then the subspace $V_i(p)$ of T_pM spanned by $\{ Y_1'(p), \dots, Y_i'(p) \}$ is non-degenerate for $p \in W$ and $1 \leq i \leq n$. Using this, we are going to construct vector fields $Y_1, \dots, Y_n$ along $i|_W$ so that $\{Y_1(p), \dots, Y_n(p)\}$ is an orthonormal basis for T_pM for each $p \in W$ and $Y_i(p_0) = v_i$ for $i = 1, \dots, n$.

By applying the Gram-Schmidt process to $\{ Y_1'(p), \dots, Y_n'(p) \}$, we obtain an orthogonal basis $\{ Y_1''(p), \dots, Y_n''(p) \}$ for T_pM with $Y_i''(p_0) = v_i$ for $i = 1, \dots, n$, given by

$$Y_1''(p) = Y_1'(p)$$

and

$$Y_i''(p) = Y_i'(p) - \sum_{j=1}^{i-1} \frac{g(p)\,(Y_i'(p), Y_j''(p))}{g(p)\,(Y_j''(p), Y_j''(p))}\, Y_j''(p)$$

for $p \in W$ and $i = 2, \dots, n$. Indeed, assuming inductively that $\{ Y_1''(p), \dots, Y_{i-1}''(p) \}$ is an orthogonal basis for $V_{i-1}(p)$, we have that $Y_j''(p) \notin V_{i-1}(p)^\perp$ so that $g(p)(Y_j''(p), Y_j''(p)) \neq 0$ for $j = 1, \dots, i-1$ since $V_{i-1}(p)$ is a non-degenerate subspace of T_pM, thus showing that $Y_i''(p)$ is well defined. Furthermore, $g(p)(Y_i''(p), Y_r''(p)) = 0$ for $r = 1, \dots, i-1$ which shows that $\{ Y_1''(p), \dots, Y_i''(p) \}$ is an orthogonal basis for $V_i(p)$. Hence we have an orthonormal basis $\{ Y_1(p), \dots, Y_n(p) \}$ for T_pM with $Y_i(p_0) = v_i$ for $i = 1, \dots, n$, given by

$$Y_i(p) = Y_i''(p) \,/\, \|Y_i''(p)\|$$

for $p \in W$ and $i = 1, \dots, n$. We also have that $Y_j = i_* \circ X_j$ for $j = 1, \dots, n-1$, where the orthonormal basis $\{ X_1(p), \dots, X_{n-1}(p) \}$ for T_pP is obtained in the same way by using the Gram-Schmidt process on $\{ X_1'(p), \dots, X_{n-1}'(p) \}$.

Now there is an open neighbourhood O of p_0 contained in W so that $\omega(p)(Y_n(p), Y_1(p), \dots, Y_{n-1}(p)) > 0$ and $\eta(p)(X_1(p), \dots, X_{n-1}(p)) > 0$ for $p \in O$. This implies that $N(p) = Y_n(p)$ for $p \in O$, proving that N is smooth at p_0. As p_0 was an arbitrary point on P, this shows that N is a vector field along i, thus completing the proof of the proposition. □

7.41 Proposition Let (M,μ) be a 1-dimensional oriented Riemannian manifold. Then there is a unique vector field τ on M, called the *positive unit tangent field*, so that $(\tau(p))$ is a positively oriented orthonormal basis in T_pM for each $p \in M$.

PROOF: The map $\tau : M \to TM$ is clearly uniquely defined, and it is also smooth. For if (x,U) is a positively oriented local chart on M, then we have that

$$\tau|_U = \tfrac{\partial}{\partial x} \,/\, \|\tfrac{\partial}{\partial x}\| \,. \qquad \square$$

7.42 Proposition Let P^{n-1} be a Riemannian immersed submanifold of an oriented Riemannian manifold (M^n,μ), and let $i : P \to M$ be the inclusion map. If $N : P \to TM$ is the positive unit normal field on P with respect to an orientation ν of P and $\omega \in \Omega^1(M)$ is a 1-form on M, we have that

$$i^*(*\omega) = (\omega \circ i)(N)\ \varepsilon$$

where ε is the metric volume element on P compatible with ν.

PROOF: Let $\mathscr{B} = (v_2,\dots,v_n)$ be a positively oriented orthonormal basis for the tangent space T_pP at a point $p \in P$. Then $\mathscr{C} = (N_p, i_* v_2,\dots,i_* v_n)$ is a positively oriented orthonormal basis for T_pM, and we let $\mathscr{C}^* = (\omega_1,\dots,\omega_n)$ be the dual basis in T_p^*M. Using the dual basis $\mathscr{B}^* = ((i_{*p})^*\,\omega_2,\dots,(i_{*p})^*\,\omega_n)$ of $\mathscr{B}$ in T_p^*P, we have that

$$\varepsilon(p) = (i_{*p})^*(\omega_2 \wedge \cdots \wedge \omega_n)\,.$$

Since $(i_{*p})^*(\omega_1) = 0$, this implies that

$$(i_{*p})^*(*\omega_k) = \delta_{1k}\,\varepsilon(p) = \omega_k(N_p)\,\varepsilon(p)$$

for $k = 1,\dots,n$, which shows the formula in the proposition. $\square$

7.43 Gauss' Divergence Theorem Let (M^n,μ) be an oriented Riemannian manifold with boundary, and let ε and $\widetilde{\varepsilon}$ be the metric volume elements on M and ∂M compatible with μ and $\partial\mu$, respectively. Let $N : \partial M \to TM$ be the outward unit normal field on ∂M, and $i : \partial M \to M$ be the inclusion map. If X is a vector field on M with compact support, then we have that

$$\int_M (\operatorname{div} X)\ \varepsilon = \int_{\partial M} < X \circ i, N > \widetilde{\varepsilon}\,.$$

PROOF: By Theorem 6.38 and Propositions 7.33 and 7.42 we have that

$$\begin{aligned}\int_M (\operatorname{div} X)\ \varepsilon &= \int_M *d*(X^\flat)\ \varepsilon = \int_M d*(X^\flat) = \int_{\partial M} i^*(*X^\flat)\\ &= \int_{\partial M} (X \circ i)^\flat(N)\ \widetilde{\varepsilon} = \int_{\partial M} < X \circ i, N > \widetilde{\varepsilon}.\end{aligned} \qquad \square$$

7.44 The Classical Stokes' Theorem Let M^2 be a compact manifold with boundary in $\mathbf{R}^3$ with its standard metric $<,>$ and orientation, and let $N : M \to T\mathbf{R}^3$ be the positive unit normal field on M with respect to an orientation μ of M. Let dA and dS be the metric volume elements on M and ∂M compatible with μ and $\partial\mu$, respectively, and let $i : M \to \mathbf{R}^3$ and $j : \partial M \to \mathbf{R}^3$ be the inclusion maps. If $T : \partial M \to T\mathbf{R}^3$ is the vector field along j defined by $T = j_* \circ \tau$, where τ is the positive unit tangent field on ∂M, and X is a vector field on an open subset U of $\mathbf{R}^3$ containing M, then we have that

$$\int_M < \mathrm{curl} X \circ i, N > dA = \int_{\partial M} < X \circ j, T > dS .$$

PROOF : By Theorem 6.38 and Proposition 7.42 we have that

$$\begin{aligned}\int_M < \mathrm{curl} X \circ i, N > dA &= \int_M (* d(X^\flat) \circ i)(N)\, dA = \int_M i^*(** d(X^\flat)) \\ &= \int_M i^*(d(X^\flat)) = \int_M d(i^*(X^\flat)) = \int_{\partial M} j^*(X^\flat) = \int_{\partial M} j^*(X^\flat)(\tau)\, dS \\ &= \int_{\partial M} (X \circ j)^\flat (T)\, dS = \int_{\partial M} < X \circ j, T > dS.\end{aligned}$$

□

TIME DEPENDENT VECTOR FIELDS

7.45 Definition Let $\pi : TM \to M$ be the tangent bundle of a smooth manifold M, and let V be an open subset of M. By a *time-dependent vector field* on V we mean a smooth map $X : J \times V \to TM$ such that $\pi \circ X = \pi_2$, where J is an open interval and $\pi_2 : J \times V \to M$ is the map given by $\pi_2(t,p) = p$ for $(t,p) \in J \times V$. This means that the map $X_t : V \to TM$ defined by $X_t(p) = X(t,p)$ for $p \in V$, is a vector field on V for each $t \in J$.

With X we may associate a (time independent) vector field $\widetilde{X} : J \times V \to T(J \times M)$, called the *suspension* of X, given by

$$\widetilde{X}(t,p) = i_{p\,*}\left(\left.\frac{d}{dr}\right|_t \right) + i_{t\,*} \circ X(t,p)$$

for $(t,p) \in J \times V$, where $i_p : J \to J \times M$ and $i_t : M \to J \times M$ are the embeddings defined by $i_p(t') = (t',p)$ and $i_t(p') = (t,p')$ for $t' \in J$ and $p' \in V$, and (r,J) is the standard local chart on J where $r : J \to \mathbf{R}$ is the inclusion map. We see that

$$\pi_{2\,*} \circ \widetilde{X} = X .$$

7.46 Definition Let X be a time-dependent vector field on a smooth manifold

M over the open time interval J. A smooth curve $\gamma : I \to M$ defined on an open subinterval I of J is called an *integral curve* for X if

$$\gamma'(t) = X(t, \gamma(t)) \tag{1}$$

for $t \in I$. If I contains 0 and $\gamma(0) = p_0$, the point p_0 is called the *starting point* or *initial condition* of γ. The integral curve γ is called *maximal* if it has no extension to an integral curve for X on any larger open interval.

7.47 Proposition Let X be a time-dependent vector field on a smooth manifold M over the open time interval J, and let $\widetilde{X} : J \times M \to T(J \times M)$ be its suspension. Then a smooth curve $\gamma : I \to M$ defined on an open subinterval I of J containing s is an integral curve for X with $\gamma(s) = p$ if and only if the curve $\beta : I - s \to J \times M$ given by

$$\beta(t-s) = (t, \gamma(t))$$

for $t \in I$, is an integral curve for $\widetilde{X}$ with initial condition (s, p).

PROOF : Follows from Remark 2.83 and Proposition 2.74 which implies that

$$\beta'(t-s) = i_{\gamma(t)\,*}\left(\left.\frac{d}{dr}\right|_t\right) + i_{t\,*} \circ \gamma'(t)$$

and

$$\widetilde{X}(\beta(t-s)) = i_{\gamma(t)\,*}\left(\left.\frac{d}{dr}\right|_t\right) + i_{t\,*} \circ X(t, \gamma(t))$$

for $t \in I$. □

7.48 Definition Let X be a time-dependent vector field on a smooth manifold M over the open time interval J. For each point (s, p) on $J \times M$ we denote by $\gamma_{(s,p)} : I(s,p) \to M$ the maximal integral curve for X with $\gamma_{(s,p)}(s) = p$. The set

$$\mathscr{D}(X) = \{(t, s, p) \in J \times J \times M \,|\, t \in I(s,p)\}$$

is called the *domain of the time-dependent flow* for X, and the *(global) time-dependent flow* for X is the map $\gamma : \mathscr{D}(X) \to M$ defined by $\gamma(t, s, p) = \gamma_{(s,p)}(t)$ for $(s, p) \in J \times M$ and $t \in I(s,p)$.

7.49 Corollary Let X be a time-dependent vector field on a smooth manifold M over the open time interval J, and let $\widetilde{X} : J \times M \to T(J \times M)$ be its suspension. If $\gamma : \mathscr{D}(X) \to M$ is the global time-dependent flow for X and $\beta : \mathscr{D}(\widetilde{X}) \to J \times M$ the global flow for $\widetilde{X}$, and if $\Lambda : \mathbf{R} \times \mathbf{R} \times M \to \mathbf{R} \times \mathbf{R} \times M$ is the diffeomorphism given by $\Lambda(t, s, p) = (t - s, s, p)$ for $t, s \in \mathbf{R}$ and $p \in M$, then

$$\Lambda(\mathscr{D}(X)) = \mathscr{D}(\widetilde{X})$$

and

$$\beta \circ \Lambda(t,s,p) = (t, \gamma(t,s,p))$$

for $(t,s,p) \in \mathscr{D}(X)$. In particular, $\mathscr{D}(X)$ is an open subset of $J \times J \times M$ containing $\Delta_J \times M$, where $\Delta_J \subset J \times J$ is the diagonal, and the time-dependent flow γ for X is smooth. If $\pi_2 : J \times M \to M$ is the projection on the second factor, we have that

$$\pi_2 \circ \beta \circ \Lambda = \gamma .$$

7.50 Definition If X is a time-dependent vector field on a smooth manifold M over the open time interval J with global time-dependent flow $\gamma : \mathscr{D}(X) \to M$, and if $t,s \in J$, we define

$$\mathscr{D}_{t,s}(X) = \{p \in M | (t,s,p) \in \mathscr{D}(X)\}$$

and let $\gamma_{t,s} : \mathscr{D}_{t,s}(X) \to M$ be the map defined by $\gamma_{t,s}(p) = \gamma(t,s,p)$.

7.51 Corollary Let X be a time-dependent vector field on a smooth manifold M over the open time interval J, and let $\widetilde{X} : J \times M \to T(J \times M)$ be its suspension. If $\gamma : \mathscr{D}(X) \to M$ is the global time-dependent flow for X and $\beta : \mathscr{D}(\widetilde{X}) \to J \times M$ the global flow for $\widetilde{X}$, and if $i_s : M \to J \times M$ is the embedding given by $i_s(p) = (s,p)$ for $p \in M$, then

$$i_s(\mathscr{D}_{t,s}(X)) = \mathscr{D}_{t-s}(\widetilde{X}) \cap (\{s\} \times M)$$

and

$$\beta_{t-s} \circ i_s = i_t \circ \gamma_{t,s}$$

for $t,s \in J$. If $\pi_2 : J \times M \to M$ is the projection on the second factor, we have that

$$\pi_2 \circ \beta_{t-s} \circ i_s = \gamma_{t,s} .$$

7.52 Proposition Let X be a time-dependent vector field on a smooth manifold M over the open time interval J. For each $s,t,u \in J$ we have that

(1) $\mathscr{D}_{t,s}(X)$ is open in M.

(2) If $p \in \mathscr{D}_{t,s}(X)$ and $\gamma_{t,s}(p) \in \mathscr{D}_{u,t}(X)$, then $p \in \mathscr{D}_{u,s}(X)$ and $\gamma_{u,t}(\gamma_{t,s}(p)) = \gamma_{u,s}(p)$. In particular the domain of $\gamma_{u,t} \circ \gamma_{t,s}$ is contained in $\mathscr{D}_{u,s}(X)$, and it equals $\mathscr{D}_{u,s}(X)$ if t is between s and u.

(3) $\gamma_{t,s}(\mathscr{D}_{t,s}(X)) = \mathscr{D}_{s,t}(X)$, and $\gamma_{t,s}$ is a diffeomorphism onto its image with inverse $\gamma_{s,t}$.

PROOF: Follows from Corollary 7.51 and Proposition 3.42. □

7.53 Definition Let X be a time-dependent vector field on a smooth manifold M over an open time interval J containing 0. For each point p on M we denote by $\alpha_p : I(p) \to M$ the maximal integral curve for X with initial condition p. The set

$$\mathscr{D}'(X) = \{(t,p) \in J \times M | t \in I(p)\}$$

is called the *domain of the flow* for X, and the *(global) flow* for X is the map $\alpha : \mathscr{D}'(X) \to M$ defined by $\alpha(t,p) = \alpha_p(t)$ for $p \in M$ and $t \in I(p)$.

7.54 Remark If $\gamma : \mathscr{D}(X) \to M$ is the global time-dependent flow for X and $i : J \times M \to J \times J \times M$ is the embedding given by $i(t,p) = (t,0,p)$ for $t \in J$ and $p \in M$, then

$$i(\mathscr{D}'(X)) = \mathscr{D}(X) \cap (J \times \{0\} \times M) \quad \text{and} \quad \alpha = \gamma \circ i\,.$$

In particular, $\mathscr{D}'(X)$ is an open subset of $J \times M$ containing $\{0\} \times M$, and the flow α for X is smooth.

If β is the global flow for the suspension $\widetilde{X} : J \times M \to T(J \times M)$ of X and $\Lambda : \mathbf{R} \times \mathbf{R} \times M \to \mathbf{R} \times \mathbf{R} \times M$ is the diffeomorphism defined in Corollary 7.49, then $i = \Lambda \circ i$ so that

$$i(\mathscr{D}'(X)) = \mathscr{D}(\widetilde{X}) \cap (J \times \{0\} \times M)$$

and

$$\beta \circ i(t,p) = (t, \alpha(t,p))$$

for $(t,p) \in \mathscr{D}'(X)$. If $\pi_2 : J \times M \to M$ is the projection on the second factor, we have that

$$\pi_2 \circ \beta \circ i = \alpha\,.$$

7.55 Definition Let X be a time-dependent vector field on a smooth manifold M over an open time interval J containing 0. If $\alpha : \mathscr{D}'(X) \to M$ is the global flow for X, and if $t \in J$, we define

$$\mathscr{D}_t(X) = \{p \in M | (t,p) \in \mathscr{D}'(X)\}$$

and let $\alpha_t : \mathscr{D}_t(X) \to M$ be the map defined by $\alpha_t(p) = \alpha(t,p)$.

7.56 Remark If $\gamma : \mathscr{D}(X) \to M$ is the global time-dependent flow for X, then

$$\mathscr{D}_t(X) = \mathscr{D}_{t,0}(X) \quad \text{and} \quad \alpha_t = \gamma_{t,0}$$

for $t \in J$. Hence, if β is the global flow for the suspension $\widetilde{X} : J \times M \to T(J \times M)$ of X and $i_s : M \to J \times M$ is the embedding given by $i_s(p) = (s,p)$ for $s \in J$ and $p \in M$, then

$$i_0(\mathscr{D}_t(X)) = \mathscr{D}_t(\widetilde{X}) \cap (\{0\} \times M)$$

and

$$\beta_t \circ i_0 = i_t \circ \alpha_t$$

for $t \in J$. If $\pi_2 : J \times M \to M$ is the projection on the second factor, we have that

$$\pi_2 \circ \beta_t \circ i_0 = \alpha_t \; .$$

7.57 Definition Let $\pi_l^k : T_l^k(M) \to M$ be the tensor bundle of type $\binom{k}{l}$ over a smooth manifold M. By a *time-dependent tensor field* of type $\binom{k}{l}$ on M we mean a smooth map $T : J \times M \to T_l^k(M)$ such that $\pi_l^k \circ T = \pi_2$, where J is an open interval containing 0, and $\pi_2 : J \times M \to M$ is the projection on the second factor. This means that the map $T_t : M \to T_l^k(M)$ defined by $T_t(p) = T(t,p)$ for $p \in M$, is a tensor field of type $\binom{k}{l}$ on M for each $t \in J$.

7.58 Definition Let X be a time-dependent vector field on a smooth manifold M over the open time interval J with global time-dependent flow $\gamma : \mathscr{D}(X) \to M$, and let T be a covariant tensor field of degree k on M.

If p is a point on M and $s \in J$, then we define the *Lie derivative* of T at (s,p) with respect to X to be

$$L_X T(s,p) = \lim_{t \to s} \frac{1}{t-s} \left[\gamma_{t,s}^* T(p) - T(p) \right] = \frac{d}{dt}\bigg|_s \gamma_{t,s}^* T(p) \, ,$$

using the vector space topology on $\mathscr{T}^k(T_pM)$ defined in Proposition 13.117 in the appendix.

7.59 Proposition Let X be a time-dependent vector field on a smooth manifold M over the open time interval J with suspension $\widetilde{X} : J \times M \to T(J \times M)$, and let T be a covariant tensor field of degree k on M. Then the Lie derivative $L_X T$ is a well-defined covariant time-dependent tensor field of degree k on M given by

$$(L_X T)_s = i_s^* \left(L_{\widetilde{X}} \left(\pi_2^* T \right) \right)$$

for $s \in J$, where $i_s : M \to J \times M$ is the embedding defined by $i_s(p) = (s,p)$ for $p \in M$, and $\pi_2 : J \times M \to M$ is the projection on the second factor.

If $\gamma : \mathscr{D}(X) \to M$ is the global time-dependent flow for X and $c : I(s,p) \to \mathscr{T}^k(T_pM)$ is the curve defined by $c(t) = \gamma_{t,s}^* T(p)$, then we have that

$$c'(t) = \gamma_{t,s}^* (L_X T)_t (p)$$

for every $t \in I(s,p)$ if we canonically identify $T_{c(t)} \mathscr{T}^k(T_pM)$ with $\mathscr{T}^k(T_pM)$ as in Lemma 2.84.

PROOF : Let $\beta : \mathscr{D}(\widetilde{X}) \to J \times M$ be the global flow for $\widetilde{X}$, and let $b : I(s,p) - s \to \mathscr{T}^k(T_{(s,p)}(J \times M))$ be the curve defined by

$$b(t-s) = \beta_{t-s}^{*}(\pi_2^{*} T)(s,p)$$

for $t \in I(s,p)$. Since $\gamma_{t,s} = \pi_2 \circ \beta_{t-s} \circ i_s$ by Corollary 7.51, it follows that

$$c(t) = i_s^{*}(\beta_{t-s}^{*}(\pi_2^{*} T))(p) = (i_{s*p})^{*}(\beta_{t-s}^{*}(\pi_2^{*} T))(s,p) = (i_{s*p})^{*} \circ b(t-s)$$

for $t \in I(s,p)$, where $(i_{s*p})^{*} : \mathscr{T}^k(T_{(s,p)}(J \times M)) \to \mathscr{T}^k(T_pM)$ is the linear map Defined in Definition 4.52. This shows that the curve c is smooth, and if we canonically identify $T_{c(s)}\mathscr{T}^k(T_pM)$ with $\mathscr{T}^k(T_pM)$ and $T_{b(0)}\mathscr{T}^k(T_{(s,p)}(J \times M))$ with $\mathscr{T}^k(T_{(s,p)}(J \times M))$ as in Lemma 2.84, we have that

$$(L_X T)_s(p) = c'(s) = (i_{s*p})^{*} \circ b'(0) = i_s^{*}(L_{\widetilde{X}}(\pi_2^{*} T))(p) .$$

Since

$$b'(t-s) = \beta_{t-s}^{*}(L_{\widetilde{X}}(\pi_2^{*} T))(s,p)$$

and $\beta_{t-s} \circ i_s = i_t \circ \gamma_{t,s}$ by Proposition 4.90 and Corollary 7.51, it also follows that

$$\begin{aligned} c'(t) &= (i_{s*p})^{*} \circ b'(t-s) = i_s^{*}(\beta_{t-s}^{*}(L_{\widetilde{X}}(\pi_2^{*} T)))(p) \\ &= \gamma_{t,s}^{*}(i_t^{*}(L_{\widetilde{X}}(\pi_2^{*} T)))(p) = \gamma_{t,s}^{*}(L_X T)_t(p). \end{aligned}$$

□

7.60 Definition Let X be a time-dependent vector field on a smooth manifold M over the open time interval J, and let ω be a k-form on M. Then we have a time-dependent $(k-1)$-form $i_X \omega$ on M, called the *interior product* of X and ω, defined by

$$(i_X \omega)(t,p) = i_{X(t,p)}\, \omega(p)$$

for each $p \in M$ and $t \in J$. We see that

$$(i_X \omega)_t = i_{X_t} \omega$$

for each $t \in J$.

7.61 Proposition Let X be a time-dependent vector field on a smooth manifold M over the open time interval J with suspension $\widetilde{X} : J \times M \to T(J \times M)$, and let ω be a k-form on M. Then we have that

$$i_{X_t} \omega = i_t^{*}(i_{\widetilde{X}}(\pi_2^{*} \omega))$$

for $t \in J$, where $i_t : M \to J \times M$ is the embedding defined by $i_t(p) = (t,p)$ for $p \in M$, and $\pi_2 : J \times M \to M$ is the projection on the second factor.

PROOF : Using Proposition 5.76 we have that

$$i_t^*\,(i_{\widetilde{X}}\,(\pi_2^*\,\omega))(p) = (i_{t\,*\,p})^*\,(i_{\widetilde{X}}\,(\pi_2^*\,\omega))(t,p)$$
$$= (i_{t\,*\,p})^*\,i_{\widetilde{X}(t,p)}\,(\pi_{2\,*\,(t,p)})^*\,\omega(p) = (i_{t\,*\,p})^*(\pi_{2\,*\,(t,p)})^*\,i_{\pi_{2*}\circ\,\widetilde{X}(t,p)}\;\omega(p)$$
$$= ((\pi_2\circ i_t)_{\,*\,p})^*\,i_{X(t,p)}\;\omega(p) = (i_{X_t}\,\omega)(p)$$

for every $p \in M$. □

7.62 Proposition Let X be a time-dependent vector field on a smooth manifold M over the open time interval J, and let ω be a k-form on M. Then we have that

$$(L_X\,\omega)_t = d\,(i_{X_t}\,\omega) + i_{X_t}\,(d\,\omega)$$

for $t \in J$.

PROOF : If $\widetilde{X} : J \times M \to T\,(J \times M)$ is the suspension of X, it follows from Propositions 7.59, 7.61, 5.83 and 5.59 that

$$(L_X\,\omega)_t = i_t^*\,(L_{\widetilde{X}}\,(\pi_2^*\,\omega)) = i_t^*\,\{d\,(i_{\widetilde{X}}\,(\pi_2^*\,\omega)) + i_{\widetilde{X}}\,(d\,(\pi_2^*\,\omega))\}$$
$$= d\,(i_t^*\,(i_{\widetilde{X}}\,(\pi_2^*\,\omega))) + i_t^*\,(i_{\widetilde{X}}\,(\pi_2^*\,(d\,\omega))) = d\,(i_{X_t}\,\omega) + i_{X_t}\,(d\,\omega)$$

for $t \in J$. □

SYMPLECTIC MANIFOLDS

7.63 Definition A non-degenerate, closed 2-form $\omega \in \Omega^2(M)$ is called a *symplectic form* on M. A smooth manifold M with a symplectic form ω is called a *symplectic manifold*, and it is also denoted by (M,ω). If (M_1,ω_1) and (M_2,ω_2) are two sympectic manifolds, then a smooth map $f : M_1 \to M_2$ is said to be *symplectic* or to be a *canonical transformation* if $f^*\omega_2 = \omega_1$.

7.64 Proposition Let (M_1,ω_1) and (M_2,ω_2) be sympectic manifolds, and let $\pi_1 : M_1 \times M_2 \to M_1$ and $\pi_2 : M_1 \times M_2 \to M_2$ be the projections on the first and second factor. Then

$$\Omega = \pi_1^*\,\omega_1 - \pi_2^*\,\omega_2$$

is a symplectic form on $M_1 \times M_2$. A smooth map $f : M_1 \to M_2$ with graph $G(f)$ is symplectic if and only if

$$i^*\,\Omega = 0\,,$$

where $i : G(f) \to M_1 \times M_2$ is the inclusion map.

PROOF : As

$$\Omega(p,q)\,(i_{q*}(v_1)+i_{p*}(v_2)\,,i_{q*}(w_1)+i_{p*}(w_2))$$
$$=\omega_1(p)(v_1,w_1)-\omega_2(q)(v_2,w_2)$$

for $v_1,w_1\in T_pM_1$ and $v_2,w_2\in T_qM_2$, the 2-form Ω is non-degenerate. By Proposition 5.59 it is also closed, and Ω is therefore a symplectic form on $M_1\times M_2$. The last part of the proposition follows since

$$i^*\,\Omega(p,f(p))\,(i_{f(p)*}(v)+i_{p*}\circ f_*(v)\,,i_{f(p)*}(w)+i_{p*}\circ f_*(w))$$
$$=(\omega_1-f^*\omega_2)(p)(v,w)$$

for $v,w\in T_pM_1$. □

7.65 Proposition Let V be an n-dimensional vector space, and let $\omega\in\Lambda^2(V)$ be a skew symmetric bilinear functional of rank r. Then $r=2\,m$ for an integer m, and there is a basis $\mathscr{B}=\{v_1,\ldots,v_n\}$ for V such that

$$m_{\mathscr{B}}(\omega)=\begin{pmatrix}0 & I_m & 0\\ -I_m & 0 & 0\\ 0 & 0 & 0\end{pmatrix},$$

where I_m is the $m\times m$ identity matrix. If $\mathscr{B}^*=\{v_1^*,\ldots,v_n^*\}$ is the dual basis of $\mathscr{B}$, then we have that

$$\omega=\sum_{i=1}^{m}v_i^*\wedge v_{i+m}^*\,.$$

PROOF : We prove the first part of the propsition by induction on n. It is true for $n=0$ and $n=1$ since $\omega=0$ in this case, so that $m=0$ and we can let $\mathscr{B}$ be any basis for V. We will now prove it for $n>1$ assuming that it is true for every vector space V of dimension less than n.

The assertion is obviously true if $\omega=0$. Otherwise, there are vectors $v_m,v_{2m}\in V$ with $a=\omega\,(v_m,v_{2m})\neq 0$. Dividing v_m by a, we may assume that $\omega\,(v_m,v_{2m})=1$. Let $V_1=L(v_m,v_{2m})$ be the subspace of V spanned by v_m and v_{2m}, and let

$$W_1=\{w\in V\mid \omega\,(w,v)=0\text{ for every }v\in V_1\}\,.$$

Then we have that $V=V_1\oplus W_1$. Indeed, if $v\in V$, then

$$v-\omega\,(v,v_{2m})\,v_m+\omega\,(v,v_m)\,v_{2m}\in W_1$$

which shows that $V=V_1+W_1$, and we clearly have that $V_1\cap W_1=\{0\}$.

The restriction $\omega_1:W_1\times W_1\to\mathbf{R}$ of ω is a skew symmetric bilinear functional of rank $r-2$. By the induction hypothesis there is therefore a basis $\mathscr{B}_1=\{v_1,\ldots,v_{m-1},v_{m+1},\ldots,v_{2m-1},v_{2m+1},\ldots,v_n\}$ for W_1 with

$$m_{\mathscr{B}_1}(\omega_1)=\begin{pmatrix}0 & I_{m-1} & 0\\ -I_{m-1} & 0 & 0\\ 0 & 0 & 0\end{pmatrix},$$

and $\mathscr{B} = \{v_1, ..., v_n\}$ is then a basis for V satisfying the first part of the proposition.

To prove the last part of the proposition, we use that

$$v_i^* \wedge v_{i+m}^* \, (v_j, v_k) = v_i^*(v_j) \, v_{i+m}^*(v_k) - v_{i+m}^*(v_j) \, v_i^*(v_k) = \delta_{i,j} \, \delta_{i+m,k} - \delta_{i+m,j} \, \delta_{i,k}$$

for $i = 1, ..., m$ and $j, k = 1, ..., n$, which follows from Proposition 5.13 (1). This implies that

$$\sum_{i=1}^{m} v_i^* \wedge v_{i+m}^* \, (v_j, v_k) = \begin{cases} \delta_{j+m,k} - \delta_{k+m,j} & \text{for } 1 \leq j, k \leq 2m \\ 0 & \text{otherwise} \end{cases}$$

which completes the proof of the last formula in the proposition. □

7.66 Corollary Let V be an n-dimensional vector space, and let $\omega \in \Lambda^2(V)$ be a skew symmetric bilinear functional. Then ω is a symplectic form on V if and only if $n = 2k$ for an integer k, and the k-th exterior power $\omega^k = \omega \wedge \cdots \wedge \omega \neq 0$.

PROOF : If $r = 2\,m$ is the rank of ω, and $\mathscr{B} = \{v_1, ..., v_n\}$ is the basis for V given in Proposition 7.65, then we have that

$$\begin{aligned} \omega^m &= \sum_{i_1,...,i_m=1}^{m} v_{i_1}^* \wedge v_{i_1+m}^* \wedge \cdots \wedge v_{i_m}^* \wedge v_{i_m+m}^* \\ &= \sum_{\sigma \in S_m} v_{\sigma(1)}^* \wedge v_{\sigma(1)+m}^* \wedge \cdots \wedge v_{\sigma(m)}^* \wedge v_{\sigma(m)+m}^* \\ &= (-1)^{m(m-1)/2} \sum_{\sigma \in S_m} v_{\sigma(1)}^* \wedge \cdots \wedge v_{\sigma(m)}^* \wedge v_{\sigma(1)+m}^* \wedge \cdots \wedge v_{\sigma(m)+m}^* \\ &= m!\,(-1)^{m(m-1)/2} \, v_1^* \wedge \cdots \wedge v_{2m}^* \neq 0. \end{aligned}$$

Each term in the third sum is obtained by moving successively the basis vectors $v_{\sigma(2)}^*, ..., v_{\sigma(m)}^*$ in the dual basis leftwords and collecting them on the left hand side of the exterior product, resulting in $1 + ... + (m-1) = m(m-1)/2$ transpositions as each basis vector $v_{\sigma(j+1)}^*$ is moved successively to the left of the j basis vectors $v_{\sigma(j)+m}^*, ..., v_{\sigma(1)+m}^*$ for $j = 1, ..., m-1$.

Now, if ω is symplectic, then $r = n$ and the last assertion in the proposition is satisfied with $k = m$. On the other hand, if ω is not symplectic, then $r < n$. In this case, either n is odd, or $n = 2k$ for an integer $k > m$ so that $\omega^k = 0$ which follows from the above formula for ω^m. □

7.67 Proposition Let $\omega \in \Omega^2(M)$ be a 2-form on a smooth manifold M^n. Then ω is non-degenerate if and only if $n = 2k$ for an integer k, and the k-th exterior power $\omega^k = \omega \wedge \cdots \wedge \omega$ is a nowhere vanishing n-form on M. In particular, a manifold M with a non-degenerate 2-form ω must be orientable.

PROOF : Follows from Corollary 7.66 and Proposition 6.15. □

7.68 Darboux' Theorem Let $\omega \in \Omega^2(M)$ be a non-degenerate 2-form on a smooth manifold M^{2k}. Then $d\omega = 0$ if and only if there is a local chart (x,U) around each point $p \in M$ with $x(p) = 0$ and

$$\omega|_U = \sum_{i=1}^{k} dx^i \wedge dx^{i+k} .$$

PROOF : Let (y,V) be a local chart around p with $y(p) = 0$, and suppose that the form ω is closed and is given locally by

$$\omega|_V = \sum_{i<j} \omega_{ij}\, dy^i \wedge dy^j .$$

We let $\eta_0 = \omega|_V$ and η_1 be the 2-form with constant local representation on V given by

$$\eta_1(q) = \sum_{i<j} \omega_{ij}(p)\, dy^i(q) \wedge dy^j(q)$$

for $q \in V$, and we let η be the time-dependent 2-form on V defined on an open time interval J containing $[0,1]$ by

$$\eta_t = \eta_0 + t\,\widetilde{\eta}$$

for $t \in J$, where $\widetilde{\eta} = \eta_1 - \eta_0$.

Since $\eta_t(p) = \omega(p)$ which is non-degenerate for $t \in J$, there is for each such t an open subinterval I_t of J containing t and an open neighbourhood U_t of p contained in V such that η is non-degenerate on $I_t \times U_t$. As the interval $[0,1]$ is compact, it is covered by a finite number of intervals I_{t_i} for $i = 1,...,n$. Let $I' = \bigcup_{i=1}^{n} I_{t_i}$, and choose an open ball $B_r(0)$ contained in $y(\bigcap_{i=1}^{n} U_{t_i})$. By the Poincaré lemma, there is a 1-form θ on $U' = y^{-1}(B_r(0))$ such that $\widetilde{\eta} = d\,\theta$, and we may assume that $\theta(p) = 0$.

Now, since η is non-degenerate on $I' \times U'$, we have a time-dependent vector field X on U' over the time interval I', where $X_t = -(\eta_t)^\sharp(\theta)$ is obtained from $-\theta$ by raising indices for each $t \in I'$. Let $\alpha : \mathscr{D}'(X) \to M$ be the flow for X. As $X_t(p) = 0$ for $t \in I'$, we have that $\alpha(t,p) = p$ for every $t \in I'$ which shows that $I' \times \{p\} \subset \mathscr{D}'(X)$. Using that $\mathscr{D}'(X)$ is open and the interval $[0,1]$ is compact, it follows in the same way as above that there is an open subinterval I of I' containing $[0,1]$ and an open neighbourhood U of p contained in U' with $I \times U \subset \mathscr{D}'(X)$.

Fix a $q \in U$, and consider the curve $c : I \to \Lambda^2(T_qM)$ given by $c(t) = \alpha_t^*\, \eta_t(q)$ for $t \in I$. If we canonically identify $T_{c(t)}\Lambda^2(T_qM)$ with $\Lambda^2(T_qM)$ as in Lemma 2.84 and use Remark 2.75 and Propositions 7.59 and 7.62, we have that

$$\begin{aligned} c'(t) &= \alpha_t^*\,(L_X\, \eta_t)_t\,(q) + \alpha_t^*\, \widetilde{\eta}\,(q) = \alpha_t^*\, d\,(i_{X_t}\, \eta_t)(q) + \alpha_t^*\, \widetilde{\eta}\,(q) \\ &= \alpha_t^*\,(-d\,\theta + \widetilde{\eta}\,)(q) = 0 \end{aligned}$$

for $t \in I$. If (z,U) is the local chart around p with $z = y \circ \alpha_1|_U$, we therefore have that

$$\omega(q) = \alpha_1^* \eta_1(q) = \sum_{i<j} \omega_{ij}(p)\, dz^i(q) \wedge dz^j(q)$$

for $q \in U$ by Propositions 5.53 and 5.59. Hence it follows from Proposition 4.15 that

$$\omega(q) = t_{z,q}^* \beta$$

for $q \in U$, where $(t_z, \pi^{-1}(U))$ is the local trivialization for the tangent bundle $\pi : TM \to M$ corresponding to the local chart (z,U), and $\beta \in \Lambda^2(\mathbf{R}^{2k})$ is the symplectic 2-form on $\mathbf{R}^{2k}$ given by

$$\beta = \sum_{i<j} \omega_{ij}(p)\, e^i \wedge e^j ,$$

where $\mathscr{E} = \{e_1, ..., e_{2k}\}$ is the standard basis for $\mathbf{R}^{2k}$ and $\mathscr{E}^* = \{e^1, ..., e^{2k}\}$ is the dual basis.

By Proposition 7.65 there is a basis $\mathscr{B} = \{v_1, ..., v_{2k}\}$ for $\mathbf{R}^{2k}$ with dual basis $\mathscr{B}^* = \{v^1, ..., v^{2k}\}$ so that

$$\beta = \sum_{i=1}^{k} v^i \wedge v^{i+k} .$$

Now let $x = \Lambda \circ z$, where $\Lambda : \mathbf{R}^{2k} \to \mathbf{R}^{2k}$ is the linear isomorphism with $\Lambda(v_i) = e_i$ for $i = 1, ..., 2k$. By Proposition 4.6 it follows that $\Lambda^*(e^i) = v^i$ for $i = 1, ..., 2k$, so that

$$\beta = \Lambda^* \left(\sum_{i=1}^{k} e^i \wedge e^{i+k} \right) .$$

Since $t_{x,q} = \Lambda \circ t_{z,q}$ by the commutative diagram in 2.70, this finally shows that

$$\omega(q) = t_{x,q}^* \left(\sum_{i=1}^{k} e^i \wedge e^{i+k} \right) = \sum_{i=1}^{k} dx^i(q) \wedge dx^{i+k}(q)$$

for $q \in U$, which completes the proof of the "only if" part of the theorem. The "if" part is immediately seen to be true. □

7.69 Definition If (M, ω) is a symplectic manifold, then a local chart (x,U) having the properties described in Darboux' theorem, is called a *symplectic chart* on M. Its coordinate map x and coordinate functions x^i are said to be *canonical*.

7.70 Proposition Let $\pi : T^*M \to M$ be the cotangent bundle of a smooth manifold M^n. Then we have a 1-form θ on T^*M defined by

$$\theta(v) = v \circ \pi_{*v}$$

for $v \in T^*M$, called the *canonical* 1-*form* or the *Cartan form* on T^*M.

If (x,U) is a local chart on M and $(t'_x, \pi^{-1}(U))$ is the corresponding local trivialization in the cotangent bundle defined in Remark 4.9, we obtain a local chart (z,W) on T^*M where $W = \pi^{-1}(U)$ and $z = (x \times id) \circ t'_x$ so that

$$z\left(\sum_{i=1}^{n} a_i \, dx^i(u)\right) = (x(u), a)$$

for $u \in U$ and $a \in \mathbf{R}^n$. Denoting the two component maps of $z : W \to \mathbf{R}^n \times \mathbf{R}^n$ by q and p, we have that

$$\theta|_W = \sum_{i=1}^{n} p_i \, dq^i \, .$$

The 2-form $\omega = -\, d\theta$ is a symplectic form on T^*M which is given locally by

$$\omega|_W = \sum_{i=1}^{n} dq^i \wedge dp_i \, .$$

It is called the *canonical 2-form* on T^*M.

PROOF : Let $(t_x, \pi'^{-1}(U))$ and $(t_z, \pi''^{-1}(W))$ be the local trivializations for the tangent bundles $\pi' : TM \to M$ and $\pi'' : T(T^*M) \to T^*M$ corresponding to the local charts (x,U) and (z,W), and let $\mathscr{E} = \{e_1, ..., e_n\}$ and $\mathscr{F} = \{f_1, ..., f_{2n}\}$ be the standard bases for $\mathbf{R}^n$ and $\mathbf{R}^{2n}$, respectively, with dual bases $\mathscr{E}^* = \{e^1, ..., e^n\}$ and $\mathscr{F}^* = \{f^1, ..., f^{2n}\}$.

Now let $v = \sum_{i=1}^{n} a_i \, dx^i(u)$ be a point in W. By Proposition 4.15 we have that

$$v = t^*_{x,u}(\lambda) \quad \text{where} \quad \lambda = \sum_{i=1}^{n} a_i e^i \, .$$

Since

$$x \circ \pi \circ z^{-1} = pr_1|_{x(U) \times \mathbf{R}^n} \, ,$$

it follows that $D(x \circ \pi \circ z^{-1})(z(v)) = pr_1$, where $pr_1 : \mathbf{R}^n \times \mathbf{R}^n \to \mathbf{R}^n$ is the projection on the first factor. By 2.70 we therefore have a commutative diagram

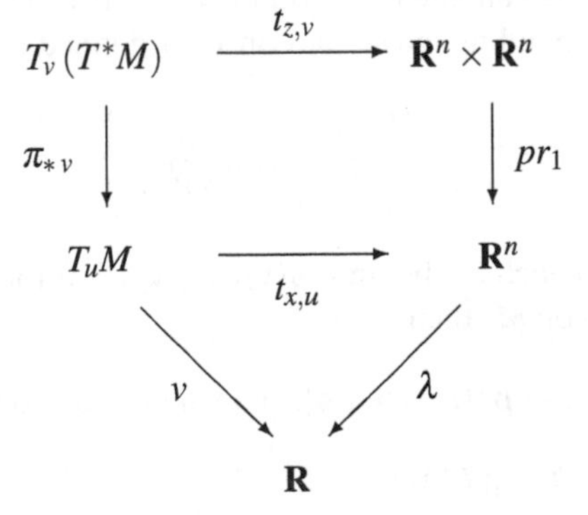

Using that

$$\lambda \circ pr_1 = \sum_{i=1}^{2n} \lambda \circ pr_1(f_i)\, f^i = \sum_{i=1}^{n} \lambda(e_i)\, f^i = \sum_{i=1}^{n} a_i f^i \,,$$

we finally have that

$$\theta(v) = t_{z,v}^*(\lambda \circ pr_1) = \sum_{i=1}^{n} a_i\,(f^i \circ t_{z,v}) = \sum_{i=1}^{n} z^{n+i}(v)\; dz^i(v)\,.$$

As in the proof of Corollary 7.66 we see that

$$\omega^n|_V = n!\,(-1)^{n(n-1)/2}\; dz^1 \wedge \cdots \wedge dz^{2n}$$

which shows that ω^n is a nowhere vanishing $2n$-form on T^*M. Hence it follows from Proposition 7.67 that the 2-form ω is symplectic. □

7.71 Proposition Let $\pi : T^*M \to M$ be the cotangent bundle of a smooth manifold M, and let θ and ω be the canonical 1- and 2-forms on T^*M. Then we have that

$$\beta^*\theta = \beta \quad \text{and} \quad \beta^*\omega = -\,d\beta$$

for every 1-form β on M.

PROOF : We have that

$$\begin{aligned}\beta^*\theta(u)(v) &= \theta(\beta(u))(\beta_{*u}(v)) = \beta(u) \circ \pi_{*\,\beta(u)}(\beta_{*u}(v)) \\ &= \beta(u) \circ (\pi \circ \beta)_{*u}(v) = \beta(u)(v)\end{aligned}$$

for every $u \in M$ and $v \in T_uM$. The last formula in the proposition follows from the first one since $\omega = -\,d\theta$. □

7.72 Proposition Let $\pi : T^*M \to M$ be the cotangent bundle of a smooth manifold M, and let ω be the canonical 2-form on T^*M. If (x,U) is a local chart on M, and (z,W) is the corresponding local chart on T^*M defined in Proposition 7.70, then

$$\mathscr{B} = \left\{ \left.\frac{\partial}{\partial p^1}\right|_v, \ldots, \left.\frac{\partial}{\partial p^n}\right|_v \right\}$$

is a basis for $\ker \pi_{*\,v}$ for each $v \in W$, and $\omega(v)(w_1,w_2) = 0$ for every $w_1, w_2 \in \ker \pi_{*\,v}$. If β is a closed 1-form on M, then

$$(\omega \circ \beta)(u)(w_1,w_2) = (\omega \circ \beta)(u)((\beta \circ \pi)_*(w_1), w_2) + (\omega \circ \beta)(u)(w_1, (\beta \circ \pi)_*(w_2))$$

for $u \in M$ and $w_1, w_2 \in T_{\beta(u)}(T^*M)$.

PROOF : Since

$$\pi_* \left(\frac{\partial}{\partial z^k} \bigg|_v \right)(x^i) = \frac{\partial (x^i \circ \pi)}{\partial z^k}(v) = \delta_{ik}$$

for $i = 1,...,n$ and $k = 1,...,2n$, we have that

$$\pi_* \left(\sum_{i=1}^{2n} a_i \frac{\partial}{\partial z^i} \bigg|_v \right) = \sum_{i=1}^{n} a_i \frac{\partial}{\partial x^i} \bigg|_u$$

for $u \in U$, $v \in T_u^*M$ and $a \in \mathbf{R}^{2n}$, which proves the first part of the proposition.

Using that $w - (\beta \circ \pi)_*(w) \in \ker \pi_{*\,\beta(u)}$ when $u \in M$ and $w \in T_{\beta(u)}(T^*M)$, and that $\beta^*\omega = 0$ when β is closed by Proposition 7.71, we now obtain that

$$\begin{aligned} 0 &= (\omega \circ \beta)(u)(w_1 - (\beta \circ \pi)_*(w_1), w_2 - (\beta \circ \pi)_*(w_2)) \\ &= (\omega \circ \beta)(u)(w_1, w_2) - (\omega \circ \beta)(u)((\beta \circ \pi)_*(w_1), w_2) \\ &\quad - (\omega \circ \beta)(u)(w_1, (\beta \circ \pi)_*(w_2)) \end{aligned}$$

which shows the last part of the proposition. □

HAMILTONIAN SYSTEMS

7.73 Definition Let (M, ω) be a symplectic manifold, and let $H \in \mathscr{F}(M)$ be a smooth function on M. Then the vector field $X_H = \omega^\sharp(dH)$ obtained by raising indices in the 1-form dH, is called the *Hamiltonian vector field* with energy function H. The function H is also called the *Hamiltonian*.

7.74 Proposition Let X_H be a Hamiltonian vector field with energy function H on a symplectic manifold (M, ω) of dimension $2n$, and let (x, U) be a symplectic chart on M. Denoting the two component maps of $x : U \to \mathbf{R}^n \times \mathbf{R}^n$ by q and p so that

$$\omega|_U = \sum_{i=1}^{n} dq^i \wedge dp_i\,,$$

we have that

$$X_H|_U = \sum_{i=1}^{n} \left(\frac{\partial H}{\partial p_i} \frac{\partial}{\partial q^i} - \frac{\partial H}{\partial q^i} \frac{\partial}{\partial p_i} \right).$$

A smooth curve $\gamma : I \to U$ is an integral curve for X_H if and only if it satisfies the *Hamiltonian equations*

$$(q^i \circ \gamma)'(t) = \frac{\partial H}{\partial p_i}(\gamma(t)) \quad \text{and} \quad (p_i \circ \gamma)'(t) = -\frac{\partial H}{\partial q^i}(\gamma(t))$$

for $t \in I$ and $i = 1,...,n$.

PROOF : Using the basis

$$\mathscr{B} = \left\{ \frac{\partial}{\partial q^1}\bigg|_u, \ldots, \frac{\partial}{\partial q^n}\bigg|_u, \frac{\partial}{\partial p^1}\bigg|_u, \ldots, \frac{\partial}{\partial p^n}\bigg|_u \right\}$$

for T_uM with dual basis

$$\mathscr{B}^* = \{dq^1(u), \ldots, dq^n(u), dp_1(u), \ldots, dp_n(u)\}$$

where $u \in U$, we have that

$$m_{\mathscr{B}}(\omega(u)) = \begin{pmatrix} 0 & I_n \\ -I_n & 0 \end{pmatrix}$$

so that

$$m_{\mathscr{B}}^{\mathscr{B}^*}(\omega(u)^\sharp) = \{m_{\mathscr{B}}(\omega(u))^t\}^{-1} = \begin{pmatrix} 0 & I_n \\ -I_n & 0 \end{pmatrix},$$

where I_n is the $n \times n$ identity matrix. The proposition now follows from the formula

$$dH|_U = \sum_{i=1}^{n} \left(\frac{\partial H}{\partial q^i}\, dq^i + \frac{\partial H}{\partial p_i}\, dp_i \right). \qquad \square$$

7.75 Definition Let (M, ω) be a symplectic manifold. To each smooth function $f \in \mathscr{F}(M)$ we associate a vector field $X_f = \omega^\sharp(df)$ obtained by raising indices in the 1-form df. If $f, g \in \mathscr{F}(M)$, their *Poisson bracket* $\{f, g\}$ is the differeniable function on M defined by

$$\{f, g\} = \omega(X_f, X_g).$$

7.76 Proposition Let (M, ω) be a symplectic manifold, and consider two smooth functions $f, g \in \mathscr{F}(M)$. If $\gamma : I \to M$ is an integral curve for X_g, then we have that

$$(f \circ \gamma)'(t) = \{f, g\} \circ \gamma(t)$$

for $t \in I$. In particular, the function $f \circ \gamma$ is constant on I if and only if $\{f, g\}$ vanishes on $\gamma(I)$.

PROOF : By Proposition 4.14 we have that

$$\begin{aligned}(f \circ \gamma)'(t) &= df(\gamma(t))(\gamma'(t)) = df(\gamma(t))(X_g(\gamma(t))) \\ &= \omega(\gamma(t))(X_f(\gamma(t)), X_g(\gamma(t))) = \{f, g\} \circ \gamma(t)\end{aligned}$$

for $t \in I$. $\square$

7.77 Corollary Let X_H be a Hamiltonian vector field with energy function H on a symplectic manifold (M, ω), and let $\gamma : I \to M$ be an integral curve for X_H. Then the function $H \circ \gamma$ is constant on I.

PROOF : Follows from Proposition 7.76 since $\{H, H\}$ vanishes on M. $\square$

LAGRANGIAN SYSTEMS

7.78 Definition Let $\pi : E \to M$ and $\widetilde{\pi} : \widetilde{E} \to M$ be two vector bundles over a smooth manifold M. A smooth map $f : E \to \widetilde{E}$ is said to be *fibre preserving* if the diagram

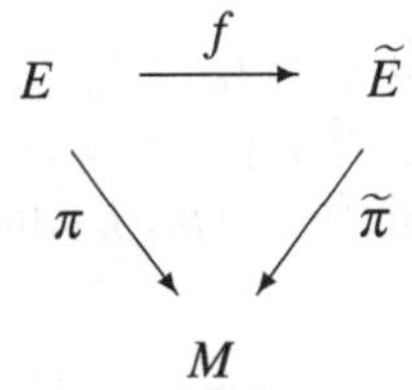

is commutative. We let f_p denote the induced smooth map $f_p : \pi^{-1}(p) \to \widetilde{\pi}^{-1}(p)$ between the fibres for each $p \in M$.

7.79 Proposition Let $\pi : E \to M$ and $\widetilde{\pi} : \widetilde{E} \to M$ be two vector bundles over a smooth manifold M of dimensions n and m, respectively, and let $f : E \to \widetilde{E}$ be a smooth map which is fibre preserving. Then we have a smooth and fibre preserving map

$$F(f) : E \to \Lambda^1(E\,;\widetilde{E}) ,$$

called the *fibre derivative* of f, which is defined by

$$F(f)(v) = (f_p)_{*v}$$

for $p \in M$ and $v \in E_p$, where we canonically identify $T_v E_p$ with E_p and $T_{f(v)} \widetilde{E}_p$ with $\widetilde{E}_p$ as in Lemma 2.84.

PROOF: Let $(t, \pi^{-1}(U))$ and $(\widetilde{t}, \widetilde{\pi}^{-1}(U))$ be local trivializations in E and $\widetilde{E}$. Then we have a smooth map $h : U \times \mathbf{R}^n \to \mathbf{R}^m$ so that

$$\widetilde{t} \circ f \circ t^{-1}(p,a) = (p, h(p,a))$$

for $(p,a) \in U \times \mathbf{R}^n$. For each $p \in U$ and $v \in E_p$ we have a commutative diagram

$$\begin{array}{ccc} T_vE_p & \xrightarrow{(f_p)_{*v}} & T_{f(v)}\widetilde{E}_p \\ {\scriptstyle \omega_v}\downarrow & & \downarrow{\scriptstyle \widetilde{\omega}_{f(v)}} \\ E_p & \xrightarrow[F(f)(v)]{} & \widetilde{E}_p \end{array}$$

where $\omega_v = t_p^{-1} \circ t_{t_p,v}$ and $\widetilde{\omega}_{f(v)} = \widetilde{t}_p^{-1} \circ t_{\widetilde{t}_p,f(v)}$ are the canonical identifications given in Lemma 2.84. If $\tau : \pi'^{-1}(U) \to U \times L(\mathbf{R}^n, \mathbf{R}^m)$ is the equivalence over U defined in Proposition 5.28 and $t(v) = (p,a)$, it therefore follows from the commutative diagram in 2.70 that

$$\tau \circ F(f) \circ t^{-1}(p,a) = \tau \circ F(f)(v) = (p, \widetilde{t}_p \circ F(f)(v) \circ t_p^{-1})$$

$$= (p, \widetilde{t}_p \circ \widetilde{\omega}_{f(v)} \circ (f_p)_{*v} \circ \omega_v^{-1} \circ t_p^{-1}) = (p, t_{\widetilde{t}_p, f(v)} \circ (f_p)_{*v} \circ t_{t_p,v}^{-1})$$

$$= (p, D(\widetilde{t}_p \circ f_p \circ t_p^{-1})(t_p(v))) = (p, D_2 h(p,a))$$

which shows that the fibre derivative $F(f)$ is smooth and fibre preserving. □

7.80 Proposition Let $\pi : E \to M$ be an n-dimensional vector bundle over a smooth manifold M, and let $f : E \to W$ be a smooth map into an m-dimensional vector space W. We let $f_p = f|_{E_p}$ denote the restriction of f to the fibre E_p for each $p \in M$. Then we have a smooth and fibre preserving map

$$F(f) : E \to \Lambda^1(E\,;W)\,,$$

called the *fibre derivative* of f, which is defined by

$$F(f)(v) = (f_p)_{*v}$$

for $p \in M$ and $v \in E_p$, where we canonically identify $T_v E_p$ with E_p and $T_{f(v)} W$ with W as in Lemma 2.84.

PROOF: Let $\widetilde{\pi} : E_W \to M$ be the vector bundle $E_W = M \times W$ defined in 5.26, and consider the smooth and fibre preserving map $\widetilde{f} : E \to E_W$ given by

$$\widetilde{f}(v) = (p, f(v))$$

for $p \in M$ and $v \in E_p$. For each $p \in M$ we have that

$$\rho_p \circ \widetilde{f}_p = f_p\,,$$

where $\rho_p : \{p\} \times W \to W$ is the projection on the second factor. Since ρ_p is a linear isomorphism, it follows from Lemma 2.84 that

$$\rho_p \circ F(\widetilde{f})(v) = F(f)(v)$$

for $p \in M$ and $v \in E_p$. If $\rho' : \Lambda^1(E; E_W) \to \Lambda^1(E; W)$ is the equivalence over M defined in 5.26, we therefore have that

$$\rho' \circ F(\widetilde{f}) = F(f)\,,$$

which shows that the fibre derivative $F(f)$ is smooth and fibre preserving. □

7.81 Proposition Let $\pi : TM \to M$ and $\pi^* : T^*M \to M$ be the tangent and cotangent bundle of a smooth manifold M^n, and let $L : TM \to \mathbf{R}$ be a smooth function. If (x, U) is a local chart on M, and if $(t_x, \pi^{-1}(U))$ and $(t'_x, \pi'^{-1}(U))$ are the corresponding local trivializations in the tangent and cotangent bundle defined in Remark 2.68 and 4.9, we obtain local charts (y, V) and (z, W) on TM and T^*M, respectively, where $V = \pi^{-1}(U)$, $W = \pi^{*-1}(U)$, $y = (x \times id) \circ t_x$ and $z = (x \times id) \circ t'_x$ so that

$$y\left(\sum_{i=1}^{n} a_i \left.\frac{\partial}{\partial x^i}\right|_u \right) = (x(u), a)$$

and

$$z\left(\sum_{i=1}^{n} a_i \, dx^i(u) \right) = (x(u), a)$$

for $u \in U$ and $a \in \mathbf{R}^n$. Denoting the two component maps of $y : V \to \mathbf{R}^n \times \mathbf{R}^n$ by q and $\dot{q}$, and the two component maps of $z : W \to \mathbf{R}^n \times \mathbf{R}^n$ by q and p, the fibre derivative $F(L) : TM \to T^*M$, which is called the *Legendre transformation* corresponding to the *Lagrangian function* L, is given by

$$q^i \circ F(L) = q^i \quad \text{and} \quad p_i \circ F(L) = \frac{\partial L}{\partial \dot{q}^i}$$

for $i = 1, \ldots, n$. The function $p_i \circ F(L)$ is called the *conjugate momentum* of the coordinate function q^i on V. We have that

$$D(z \circ F(L) \circ y^{-1})(y(v)) = \begin{pmatrix} I_n & 0 \\ B(v) & A(v) \end{pmatrix},$$

where I_n is the $n \times n$ identity matrix and

$$B(v) = \left(\frac{\partial^2 L}{\partial \dot{q}^i \, \partial q^j}(v) \right) \quad \text{and} \quad A(v) = \left(\frac{\partial^2 L}{\partial \dot{q}^i \, \partial \dot{q}^j}(v) \right)$$

for $v \in V$. In particular, we see that $F(L)$ is a local diffeomorphism in V if and only if the matrix $A(v)$ is non-singular for every $v \in V$.

PROOF : For each $u \in U$ and $v \in T_uM$ we have a commutative diagram

$$\begin{array}{ccc} T_v(T_uM) & \xrightarrow{(L_u)_{*v}} & T_{L(v)}\,\mathbf{R} \\ \omega_v \big\downarrow & & \big\downarrow \widetilde{\omega}_{L(v)} \\ T_uM & \xrightarrow[F(L)(v)]{} & \mathbf{R} \end{array}$$

where $\omega_v = t_{x,u}^{-1} \circ t_{t_{x,u},v}$ and $\widetilde{\omega}_{L(v)} = id^{-1} \circ t_{id,L(v)}$ are the canonical identifications given in Lemma 2.84. Let $r : (\mathbf{R}^n)^* \to \mathbf{R}^n$ be the linear isomorphism defined by $r(e^i) = e_i$ for $i = 1,...,n$, where $\mathscr{E} = \{e_1,...,e_n\}$ is the standard basis for $\mathbf{R}^n$ and $\mathscr{E}^* = \{e^1,...,e^n\}$ is its dual basis. By Remark 4.9 we have that

$$z \circ F(L)(v) = (x(u), r \circ (t_{x,u}^{-1})^* \circ F(L)(v))$$

where

$$(t_{x,u}^{-1})^* \circ F(L)(v) = F(L)(v) \circ t_{x,u}^{-1} = \widetilde{\omega}_{L(v)} \circ (L_u)_{*v} \circ \omega_v^{-1} \circ t_{x,u}^{-1}$$

$$= t_{id,L(v)} \circ (L_u)_{*v} \circ t_{t_{x,u},v}^{-1} = D(L_u \circ t_{x,u}^{-1})(t_{x,u}(v))\;.$$

Hence it follows that

$$z^i \circ F(L)(v) = y^i(v)$$

and

$$z^{n+i} \circ F(L)(v) = D_i(L_u \circ t_{x,u}^{-1})(t_{x,u}(v)) = \frac{\partial L}{\partial y^{n+i}}(v)$$

for $i = 1,...,n$. Indeed, if $\mathscr{F} = \{f_1,...,f_{2n}\}$ is the standard basis for $\mathbf{R}^{2n}$, the final term in the last formula equals $c'(0)$, where $c : \mathbf{R} \to \mathbf{R}$ is the curve defined by

$$c(t) = L \circ y^{-1}(y(v) + t f_{n+i})$$

for $t \in \mathbf{R}$. If $t_x(v) = (u,a)$, the last equality in the formula follows since

$$c(t) = L \circ y^{-1}(x(u), a + te_i) = L \circ t_x^{-1}(u, a + te_i) = L_u \circ t_{x,u}^{-1}(a + te_i)\;.$$

□

7.82 Proposition Let $\pi : TM \to M$ and $\pi^* : T^*M \to M$ be the tangent and cotangent bundle of a smooth manifold M^n, and let $L : TM \to \mathbf{R}$ be a smooth function. If θ and ω are the canonical 1- and 2-forms on T^*M, we let

$$\theta_L = F(L)^*\,\theta \quad \text{and} \quad \omega_L = F(L)^*\,\omega$$

be the pull-backs to TM by the fibre derivative of L. We have that $\omega_L = -\, d\theta_L$, and ω_L is called the *Lagrange 2-form* corresponding to the Lagrangian function L.

If (x,U) is a local chart on M, and (y,V) is the corresponding local chart on TM defined in Proposition 7.81, we have that

$$\theta_L|_V = \sum_i \frac{\partial L}{\partial \dot{q}^i}\, dq^i$$

and

$$\omega_L|_V = \sum_{ij} \left(\frac{\partial^2 L}{\partial \dot{q}^i \partial q^j}\, dq^i \wedge dq^j + \frac{\partial^2 L}{\partial \dot{q}^i \partial \dot{q}^j}\, dq^i \wedge d\dot{q}^j \right).$$

The matrix of $\omega_L(v)$ with respect to the basis

$$\mathscr{B} = \left\{ \left.\frac{\partial}{\partial q^1}\right|_v, \ldots, \left.\frac{\partial}{\partial q^n}\right|_v, \left.\frac{\partial}{\partial \dot{q}^1}\right|_v, \ldots, \left.\frac{\partial}{\partial \dot{q}^n}\right|_v \right\}$$

for $T_v(TM)$ is therefore

$$m_{\mathscr{B}}(\omega_L(v)) = \begin{pmatrix} B(v) - B(v)^t & A(v) \\ -A(v) & 0 \end{pmatrix}$$

for $v \in V$, where $A(v)$ and $B(v)$ are the matrices defined in Proposition 7.81.

PROOF: By Propositions 7.70 and 7.81 we have that

$$\theta_L|_V = \sum_i \{p_i \circ F(L)\}\, d\{q^i \circ F(L)\} = \sum_i \frac{\partial L}{\partial \dot{q}^i}\, dq^i,$$

and using that

$$d\left(\frac{\partial L}{\partial \dot{q}^i}\right) = \sum_j \left(\frac{\partial^2 L}{\partial \dot{q}^i \partial q^j}\, dq^j + \frac{\partial^2 L}{\partial \dot{q}^i \partial \dot{q}^j}\, d\dot{q}^j \right),$$

we obtain the formula for $\omega_L|_V$. □

7.83 Definition Let M be a smooth manifold, and let $L : TM \to \mathbf{R}$ be a smooth function. Then L is said to be a *regular Lagrangian* if the fibre derivative $F(L)$ is a local diffeomorphism. It is called a *hyperregular Lagrangian* if $F(L)$ is a diffeomorphism.

7.84 Remark We see that the Lagrange 2-form ω_L is symplectic if and only if L is a regular Lagrangian.

7.85 Definition Let M be a smooth manifold, and let $L : TM \to \mathbf{R}$ be a regular Lagrangian. Then we define the *action* $A : TM \to \mathbf{R}$ by

$$A(v) = F(L)(v)(v)$$

for $v \in TM$, and the *energy* $E : TM \to \mathbf{R}$ by $E = A - L$. The vector field $X_E = (\omega_L)^\sharp(dE)$ obtained by raising indices in the 1-form dE, is called the *Lagrangian vector field* for L.

7.86 Remark If (x,U) is a local chart on M, and (y,V) and (z,W) are the corresponding local charts on TM and T^*M defined in Proposition 7.81, we have that

$$A|_V = \sum_i \dot{q}^i \frac{\partial L}{\partial \dot{q}^i},$$

since

$$\lambda(v) = \sum_i \dot{q}^i(v)\, p_i(\lambda)$$

for $v \in T_uM$ and $\lambda \in T_u^*M$ when $u \in U$.

7.87 Proposition Let M be a smooth manifold, and let X_E be the Lagrangian vector field for a regular Lagrangian $L : TM \to \mathbf{R}$. If $\gamma : I \to M$ is an integral curve for X_E, then the function $E \circ \gamma$ is constant on I.

PROOF : By Proposition 4.14 we have that

$$\begin{aligned}(E \circ \gamma)'(t) &= dE(\gamma(t))(\gamma'(t)) = dE(\gamma(t))(X_E(\gamma(t))) \\ &= \omega_L(\gamma(t))(X_E(\gamma(t)), X_E(\gamma(t))) = 0\end{aligned}$$

for $t \in I$. □

7.88 Definition Let $\pi : TM \to M$ be the tangent bundle of a smooth manifold M. Then a vector field X on TM satisfying

$$\pi_* \circ X = id_{TM}$$

is called a *second order equation* on M.

7.89 Proposition Let $\pi : TM \to M$ be the tangent bundle of a smooth manifold M. Then a vector field X on TM is a second order equation on M if and only if

$$(\pi \circ \gamma)' = \gamma$$

for every integral curve $\gamma : I \to TM$ of X.

PROOF : If $\gamma : I \to TM$ is an integral curve of X, then

$$(\pi \circ \gamma)'(t) = \pi_* \circ \gamma'(t) = \pi_* \circ X(\gamma(t))$$

for $t \in I$ by 2.70. The proposition therefore follows from Proposition 3.36. □

7.90 Proposition Let (x,U) be a local chart on a smooth manifold M^n, and let (y,V) be the corresponding local chart on TM defined in Proposition 7.81. Furthermore, let X_E be a Lagrangian vector field for a regular Lagrangian $L: TM \to \mathbf{R}$. Then X_E is a second order equation on M, and a smooth curve $\gamma: I \to V$ is an integral curve for X_E if and only if it satisfies the *Lagrange's equations*

$$(q^i \circ \gamma)'(t) = \dot{q}^i(\gamma(t)) \quad \text{and} \quad \left(\frac{\partial L}{\partial \dot{q}^i} \circ \gamma\right)'(t) = \frac{\partial L}{\partial q^i}(\gamma(t))$$

for $t \in I$ and $i = 1, \ldots, n$.

PROOF : Using the basis

$$\mathscr{B} = \left\{ \left.\frac{\partial}{\partial q^1}\right|_v, \ldots, \left.\frac{\partial}{\partial q^n}\right|_v, \left.\frac{\partial}{\partial \dot{q}^1}\right|_v, \ldots, \left.\frac{\partial}{\partial \dot{q}^n}\right|_v \right\}$$

for $T_v(TM)$ with dual basis

$$\mathscr{B}^* = \{dq^1(v), \ldots, dq^n(v), d\dot{q}^1(v), \ldots, d\dot{q}^n(v)\}$$

where $v \in V$, it follows from Proposition 7.82 that

$$m^{\mathscr{B}}_{\mathscr{B}^*}(\omega_L(v)^\flat) = \begin{pmatrix} B(v)^t - B(v) & -A(v) \\ A(v) & 0 \end{pmatrix}$$

so that

$$m^{\mathscr{B}^*}_{\mathscr{B}}(\omega_L(v)^\sharp) = \begin{pmatrix} 0 & A(v)^{-1} \\ -A(v)^{-1} & A(v)^{-1}\{B(v)^t - B(v)\}A(v)^{-1} \end{pmatrix}.$$

By Remark 7.86 we have that

$$E|_V = \sum_j \dot{q}^j \frac{\partial L}{\partial \dot{q}^j} - L|_V,$$

so that

$$dE|_V = \sum_i (a_i\, dq^i + b_i\, d\dot{q}^i)$$

where

$$a_i = \frac{\partial E}{\partial q^i} = \sum_j \dot{q}^j \frac{\partial^2 L}{\partial q^i \partial \dot{q}^j} - \frac{\partial L}{\partial q^i}$$

and

$$b_i = \frac{\partial E}{\partial \dot{q}^i} = \sum_j \dot{q}^j \frac{\partial^2 L}{\partial \dot{q}^i \partial \dot{q}^j}.$$

Writing a, b, e and $\dot{q}$ as column vectors, where $e_i = \frac{\partial L}{\partial q^i}$, this means that

$$a = B^t \dot{q} - e \quad \text{and} \quad b = A\dot{q}.$$

Now we have that

$$X_E|v = (\omega_L)^\sharp(dE)|v = \sum_{i=1}^{n}\left(c_i \frac{\partial}{\partial q^i} + d_i \frac{\partial}{\partial \dot{q}^i}\right)$$

where

$$c = \dot{q} \quad \text{and} \quad Ad = -(B^t\dot{q} - e) + (B^t - B)\,\dot{q} = e - B\dot{q}\,.$$

Hence γ is an integral curve for X_E if and only if

$$(q^i \circ \gamma)'(t) = \dot{q}^i(\gamma(t))$$

and

$$\begin{aligned}
\left(\frac{\partial L}{\partial \dot{q}^i} \circ \gamma\right)'(t) &= d\left(\frac{\partial L}{\partial \dot{q}^i}\right)(\gamma(t))\,(\gamma'(t)) \\
&= \sum_j \left(\frac{\partial^2 L}{\partial \dot{q}^i\, \partial q^j}\,(\gamma(t))\; dq^j(\gamma(t))\,(\gamma'(t)) + \frac{\partial^2 L}{\partial \dot{q}^i\, \partial \dot{q}^j}\,(\gamma(t))\; d\dot{q}^j(\gamma(t))\,(\gamma'(t)) \right) \\
&= \sum_j \left(\frac{\partial^2 L}{\partial \dot{q}^i\, \partial q^j}\,(\gamma(t))\; (q^j \circ \gamma)'(t) + \frac{\partial^2 L}{\partial \dot{q}^i\, \partial \dot{q}^j}\,(\gamma(t))\; (\dot{q}^j \circ \gamma)'(t) \right) \\
&= \sum_j \left(\frac{\partial^2 L}{\partial \dot{q}^i\, \partial q^j}\,(\gamma(t))\; \dot{q}^j(\gamma(t)) + \frac{\partial^2 L}{\partial \dot{q}^i\, \partial \dot{q}^j}\,(\gamma(t))\; (\dot{q}^j \circ \gamma)'(t) \right) = \frac{\partial L}{\partial q^i}\,(\gamma(t))
\end{aligned}$$

for $t \in I$ and $i = 1,\dots,n$.

To show that X_E is a second order equation on M, let $v = \sum_{i=1}^{n} w_i \left.\frac{\partial}{\partial x^i}\right|_u$. Then we have that

$$\pi_*\left(\left.\frac{\partial}{\partial y^k}\right|_v\right)(x^i) = \frac{\partial (x^i \circ \pi)}{\partial y^k}\,(v) = \delta_{ik}$$

for $i = 1,\dots,n$ and $k = 1,\dots,2n$, which implies that

$$\pi_* \circ X_E(v) = \sum_{i=1}^{n} \dot{q}^i(v) \left.\frac{\partial}{\partial x^i}\right|_u = v\,.$$

□

7.91 Remark If the regular Lagrangian L is independent of the coordinate q^i, then q^i is called a *cyclic coordinate*. From the Lagrange's equations it follows that its conjugate momentum $p_i \circ F(L) = \frac{\partial L}{\partial \dot{q}^i}$ is a *constant of the motion*, i.e., it is constant on $\gamma(I)$ for every curve $\gamma : I \to V$ which is an integral curve for X_E.

7.92 Proposition Let M be a smooth manifold, and let X_E be the Lagrangian vector field for a hyperregular Lagrangian $L: TM \to \mathbf{R}$. Let X_H be the Hamiltonian vector field with energy function $H = E \circ F(L)^{-1}$, where E is the energy of L. Then X_E and X_H are $F(L)$-related, so that the diagram

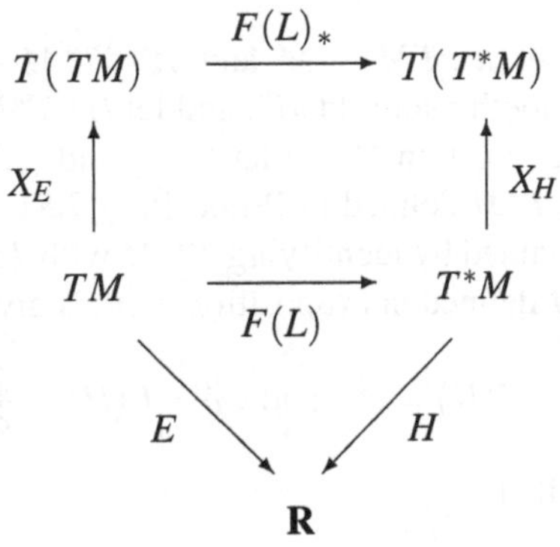

is commutative. If θ is the canonical 1-form on T^*M and $\theta_L = F(L)^*\,\theta$, then

$$\theta(X_H) = A \circ F(L)^{-1} \quad \text{and} \quad \theta_L(X_E) = A$$

where A is the action of L.

PROOF: Let ω be the canonical 2-form on T^*M, and let $\omega_L = F(L)^*\,\omega$ be the Lagrange 2-form on TM. Then we have that

$$\begin{aligned}
&\omega(F(L)(v))\,(F(L)_* \circ X_E(v), F(L)_*(w)) = \omega_L(v)\,(X_E(v), w) \\
&= dE(v)(w) = d(H \circ F(L))(v)(w) = F(L)^*\,dH(v)(w) \\
&= dH(F(L)(v))\,(F(L)_*(w)) = \omega(F(L)(v))\,(X_H \circ F(L)(v)), F(L)_*(w))
\end{aligned}$$

for every $v \in TM$ and $w \in T_v(TM)$. Since $F(L)_{*v} : T_v(TM) \to T_{F(L)(v)}(T^*M)$ is a linear isomorphism for each $v \in TM$, this implies that

$$F(L)_* \circ X_E = X_H \circ F(L)\,,$$

showing that X_E and X_H are $F(L)$-related.

Now let $v \in TM$ and $w = F(L)(v)$, and let π and π' denote the projections in the tangent and cotangent bundle of M, respectively. Then we have that

$$\begin{aligned}
&\theta(X_H)(w) = \theta(w)\,(X_H(w)) = w \circ \pi'_* \circ X_H \circ F(L)(v) \\
&= w \circ \pi'_* \circ F(L)_* \circ X_E\,(v) = w \circ \pi_* \circ X_E\,(v) = w(v) = A(v)
\end{aligned}$$

and

$$\begin{aligned}\theta_L(X_E)(v) = \theta_L(v)(X_E(v)) &= \theta(w)(F(L)_* \circ X_E(v)) \\ &= \theta(w)(X_H \circ F(L)(v)) = \theta(X_H)(w)\end{aligned}$$

which completes the proof of the proposition. □

7.93 Proposition Let $\pi : TM \to M$ and $\pi^* : T^*M \to M$ be the tangent and cotangent bundle of a smooth manifold M^n, and let $H : T^*M \to \mathbf{R}$ be a smooth function. Given a local chart (x,U) on M, we let (y,V) and (z,W) be the corresponding local charts on TM and T^*M defined in Proposition 7.81. Then the fibre derivative $F(H) : T^*M \to TM$, obtained by identifying $T^{**}M$ with TM as usual by the equivalence $i_{TM} : TM \to T^{**}M$ defined in Proposition 4.10, is given by

$$q^i \circ F(H) = q^i \quad \text{and} \quad \dot{q}^i \circ F(H) = \frac{\partial H}{\partial p_i}$$

for $i = 1,...,n$. We have that

$$D(y \circ F(H) \circ z^{-1})(z(v)) = \begin{pmatrix} I_n & 0 \\ B(v) & A(v) \end{pmatrix},$$

where I_n is the $n \times n$ identity matrix and

$$B(v) = \left(\frac{\partial^2 H}{\partial p_i \partial q^j}(v)\right) \quad \text{and} \quad A(v) = \left(\frac{\partial^2 H}{\partial p_i \partial p_j}(v)\right)$$

for $v \in W$. In particular, we see that $F(H)$ is a local diffeomorphism in W if and only if the matrix $A(v)$ is non-singular for every $v \in W$.

PROOF : For each $u \in U$ and $v \in T_u^*M$ we have a commutative diagram

$$\begin{array}{ccc} T_v(T_u^*M) & \xrightarrow{(H_u)_{*v}} & T_{H(v)}\mathbf{R} \\ \omega_v \big\downarrow & & \big\downarrow \widetilde{\omega}_{H(v)} \\ T_u^*M & \xrightarrow[i_{TM} \circ F(H)(v)]{} & \mathbf{R} \end{array}$$

where $\omega_v = t'^{-1}_{x,u} \circ t_{t'_{x,u},v}$ and $\widetilde{\omega}_{H(v)} = id^{-1} \circ t_{id,H(v)}$ are the canonical identifications given in Lemma 2.84. Let $r : (\mathbf{R}^n)^* \to \mathbf{R}^n$ be the linear isomorphism defined by $r(e^i) = e_i$ for $i = 1,...,n$, where $\mathscr{E} = \{e_1,...,e_n\}$ is the standard basis for $\mathbf{R}^n$ and $\mathscr{E}^* = \{e^1,...,e^n\}$ is its dual basis. By Remark 4.9 we have that

$$y \circ F(H)(v) = (x(u), t_{x,u} \circ F(H)(v)) = (x(u), r \circ (t'^{-1}_{x,u})^* \circ i_{TM} \circ F(H)(v))$$

where

$$(t_{x,u}^{\prime-1})^* \circ i_{TM} \circ F(H)(v) = i_{TM} \circ F(H)(v) \circ t_{x,u}^{\prime-1}$$

$$= \widetilde{\omega}_{H(v)} \circ (H_u)_{*v} \circ \omega_v^{-1} \circ t_{x,u}^{\prime-1} = t_{id,H(v)} \circ (H_u)_{*v} \circ t_{t'_{x,u},v}^{-1}$$

$$= D(H_u \circ t_{x,u}^{\prime-1})(t'_{x,u}(v)) \,.$$

Hence it follows that

$$y^i \circ F(H)(v) = z^i(v)$$

and

$$y^{n+i} \circ F(H)(v) = D_i(H_u \circ t_{x,u}^{\prime-1})(t'_{x,u}(v)) = \frac{\partial H}{\partial z^{n+i}}(v)$$

for $i = 1,...,n$. Indeed, if $\mathscr{F} = \{f_1,...,f_{2n}\}$ is the standard basis for $\mathbf{R}^{2n}$, the final term in the last formula equals $c'(0)$, where $c : \mathbf{R} \to \mathbf{R}$ is the curve defined by

$$c(t) = H \circ z^{-1}(z(v) + t f_{n+i})$$

for $t \in \mathbf{R}$. If $t'_x(v) = (u,a)$, the last equality in the formula follows since

$$c(t) = H \circ z^{-1}(x(u), a + te_i) = H \circ t_x^{\prime-1}(u, a + te_i) = H_u \circ t_{x,u}^{\prime-1}(a + te_i) \,.$$

□

7.94 Definition Let M be a smooth manifold, and let $H : T^*M \to \mathbf{R}$ be a smooth function. Then we define the *action* $G : T^*M \to \mathbf{R}$ of H by

$$G = \theta(X_H) \,,$$

where θ is the canonical 1-form on T^*M, and X_H is the Hamiltonian vector field on T^*M with energy function H.

7.95 Remark If (x,U) is a local chart on M, and (z,W) is the corresponding local chart on T^*M defined in Proposition 7.81, we have that

$$G|_W = \sum_i p_i \frac{\partial H}{\partial p_i} \,,$$

by Proposition 7.70 and 7.74.

7.96 Definition Let M be a smooth manifold, and let $H : T^*M \to \mathbf{R}$ be a smooth function. Then H is said to be a *regular Hamiltonian* if the fibre derivative $F(H)$ is a local diffeomorphism. It is called a *hyperregular Hamiltonian* if $F(H)$ is a diffeomorphism.

7.97 Proposition Let M^n be a smooth manifold, and let $H : T^*M \to \mathbf{R}$ be a hyperregular Hamiltonian with action G. Then the function $L : TM \to \mathbf{R}$ given by $L = A - E$, where

$$A = G \circ F(H)^{-1} \quad \text{and} \quad E = H \circ F(H)^{-1} \,,$$

is a hyperregular Lagrangian, and we have that

$$F(L) = F(H)^{-1} \,.$$

PROOF: Let (x,U) be a local chart on M, and let (y,V) and (z,W) be the corresponding local charts on TM and T^*M defined in Proposition 7.81. By Proposition 7.93 we have that

$$q^i \circ F(H)^{-1} = q^i \quad \text{and} \quad \frac{\partial H}{\partial p_i} \circ F(H)^{-1} = \dot{q}^i$$

for $i = 1, ..., n$. Hence it follows from Remark 7.95 that

$$L|_V = \sum_j \{p_j \circ F(H)^{-1}\}\, \dot{q}^j - H|_W \circ F(H)^{-1} \,,$$

and using the chain rule on the last term, we obtain

$$p_i \circ F(L) = \frac{\partial L}{\partial \dot{q}^i} = \sum_j \frac{\partial \{p_j \circ F(H)^{-1}\}}{\partial \dot{q}^i}\, \dot{q}^j + \{p_i \circ F(H)^{-1}\}$$

$$- \sum_j \{\frac{\partial H}{\partial p_j} \circ F(H)^{-1}\}\, \frac{\partial \{p_j \circ F(H)^{-1}\}}{\partial \dot{q}^i} = p_i \circ F(H)^{-1}$$

for $i = 1, ..., n$. The proposition now follows since we also have that

$$q^i \circ F(L) = q^i = q^i \circ F(H)^{-1}$$

for $i = 1, ..., n$. □

7.98 Remark Since

$$A = G \circ F(H)^{-1} = \theta(X_H) \circ F(L) = \theta(X_{E \circ F(L)^{-1}}) \circ F(L) \,,$$

it follows from Proposition 7.92 that A is the action of L, and hence that $E = A - L$ is the energy of L. Therefore $H = E \circ F(L)^{-1}$ can be recovered from L in the way described in Proposition 7.92.

7.99 Proposition Let M^n be a smooth manifold, and let $L : TM \to \mathbf{R}$ be a hyperregular Lagrangian with energy E. Then the function $H : T^*M \to \mathbf{R}$ given by $H = E \circ F(L)^{-1}$ is a hyperregular Hamiltonian, and we have that

$$F(H) = F(L)^{-1} \,.$$

PROOF : Let (x,U) be a local chart on M, and let (y,V) and (z,W) be the corresponding local charts on TM and T^*M defined in Proposition 7.81. By the same proposition we also have that

$$q^i \circ F(L)^{-1} = q^i \quad \text{and} \quad \frac{\partial L}{\partial \dot{q}^i} \circ F(L)^{-1} = p_i$$

for $i = 1,...,n$. Hence it follows from Remark 7.86 that

$$H|_W = \sum_j \{\dot{q}^j \circ F(L)^{-1}\}\, p_j - L|_V \circ F(L)^{-1} ,$$

and using the chain rule on the last term, we obtain

$$\dot{q}^i \circ F(H) = \frac{\partial H}{\partial p_i} = \sum_j \frac{\partial \{\dot{q}^j \circ F(L)^{-1}\}}{\partial p_i}\, p_j + \{\dot{q}^i \circ F(L)^{-1}\}$$

$$-\sum_j \{\frac{\partial L}{\partial \dot{q}^j} \circ F(L)^{-1}\}\, \frac{\partial \{\dot{q}^j \circ F(L)^{-1}\}}{\partial p_i} = \dot{q}^i \circ F(L)^{-1}$$

for $i = 1,...,n$. The proposition now follows since we also have that

$$q^i \circ F(H) = q^i = q^i \circ F(L)^{-1}$$

for $i = 1,...,n$. □

7.100 Remark By Proposition 7.92 and Definition 7.94 we have that

$$H = E \circ F(L)^{-1} = (A - L) \circ F(L)^{-1} = G - L \circ F(H) ,$$

so that

$$L = G \circ F(H)^{-1} - H \circ F(H)^{-1} .$$

Hence L can be recovered from H in the way described in Proposition 7.97.

CONSERVATIVE SYSTEMS

7.101 Proposition Let M be a pseudo-Riemannian manifold with a metric g, and let $V \in \mathscr{F}(M)$ be a smooth function on M. Then the map $L : TM \to \mathbf{R}$ defined by $L = T - V \circ \pi$, where

$$T(v) = \tfrac{1}{2}\, g(u)(v,v)$$

for $u \in M$ and $v \in T_uM$, is a hyperregular Lagrangian whose fibre derivative $F(L) : TM \to T^*M$ is given by

$$F(L)_u = g(u)^\flat$$

for $u \in M$. The action and energy of L are given by $A = 2T$ and $E = T + V \circ \pi$.

PROOF : Follows from Remark 2.75 and Lemma 2.84. □

7.102 Example Consider a system of n particles with masses $m_1, \ldots, m_n$ moving in $\mathbf{R}^3$. Then the *state* of the system is specified by a pair $(q,v) \in \mathbf{R}^{3n} \times \mathbf{R}^{3n}$, where $(q^{3k-2}, q^{3k-1}, q^{3k})$ is the position and $(v^{3k-2}, v^{3k-1}, v^{3k})$ the velocity of the kth particle for $k = 1, \ldots, n$. The point q moves in the *configuration space* $\mathbf{R}^{3n}$, and $\mathbf{R}^{3n} \times \mathbf{R}^{3n}$ is called the *state space* of the system. Let $\mathscr{E} = \{e_1, \ldots, e_{3n}\}$ be the standard basis for $\mathbf{R}^{3n}$ with dual basis $\mathscr{E}^* = \{e^1, \ldots, e^{3n}\}$, where e^i is the ith component of the identity map $id : \mathbf{R}^{3n} \to \mathbf{R}^{3n}$ for $i = 1, \ldots, 3n$. By Lemma 2.84 the state space may be identified with the total space in the tangent bundle $\pi : T\mathbf{R}^{3n} \to \mathbf{R}^{3n}$ by means of the equivalence $\omega : T\mathbf{R}^{3n} \to \mathbf{R}^{3n} \times \mathbf{R}^{3n}$ over $\mathbf{R}^{3n}$ given by

$$\omega\left(\sum_{i=1}^{3n} v^i \frac{\partial}{\partial e^i}\bigg|_q \right) = (q,v)$$

for $(q,v) \in \mathbf{R}^{3n} \times \mathbf{R}^{3n}$. As well as being real numbers, we also let $q^1, \ldots, q^{3n}, v^1, \ldots, v^{3n}$ denote the coordinate functions of the local chart $(\omega, T\mathbf{R}^{3n})$ on $T\mathbf{R}^{3n}$.

We have a metric g on $\mathbf{R}^{3n}$ defined by

$$g(q)(v,w) = \sum_{i=1}^{3n} M_i v^i w^i$$

for $q \in \mathbf{R}^{3n}$ and $v, w \in T_q\mathbf{R}^{3n}$, where $M_{3k-2} = M_{3k-1} = M_{3k} = m_k$ for $k = 1, \ldots, n$. Let $(F_{3k-2}, F_{3k-1}, F_{3k})$ be the force acting on the kth particle for $k = 1, \ldots, n$. Assuming that the forces are conservative, there is a *potential energy function* $V \in \mathscr{F}(\mathbf{R}^{3n})$ such that

$$F_i = -\frac{\partial V}{\partial q^i} \circ \pi$$

for $i = 1, \ldots, 3n$.

By Proposition 7.101 we have a hyperregular Lagrangian $L : T\mathbf{R}^{3n} \to \mathbf{R}$ defined by $L = T - V \circ \pi$, where $T : T\mathbf{R}^{3n} \to \mathbf{R}$ is the *kinetic energy function* given by

$$T \circ \omega^{-1}(q,v) = \sum_{i=1}^{3n} \tfrac{1}{2} M_i (v^i)^2$$

for $(q,v) \in \mathbf{R}^{3n} \times \mathbf{R}^{3n}$. The energy of L, called the *total energy function*, is therefore $E = T + V \circ \pi$, i.e.,

$$E \circ \omega^{-1}(q,v) = \sum_{i=1}^{3n} \tfrac{1}{2} M_i (v^i)^2 + V(q)$$

for $(q,v) \in \mathbf{R}^{3n} \times \mathbf{R}^{3n}$. The Lagrange's equations for a curve $\gamma : I \to T\mathbf{R}^{3n}$ now take the form

$$(q^i \circ \gamma)'(t) = (v^i \circ \gamma)(t) \quad \text{and} \quad M_i (v^i \circ \gamma)'(t) = -\frac{\partial V}{\partial q^i} \circ \pi \circ \gamma(t)$$

for $t \in I$ and $i = 1,...,3n$. Combining these equations, we obtain *Newton's second law*

$$M_i (q^i \circ \gamma)''(t) = F_i \circ \gamma(t)$$

for $t \in I$ and $i = 1,...,3n$.

The system can also be described in the *phase space* $\mathbf{R}^{3n} \times (\mathbf{R}^{3n})^*$, which may be identified with the total space in the cotangent bundle $\pi : T^*\mathbf{R}^{3n} \to \mathbf{R}^{3n}$ by means of the equivalence $\widetilde{\omega} : T^*\mathbf{R}^{3n} \to \mathbf{R}^{3n} \times (\mathbf{R}^{3n})^*$ over $\mathbf{R}^{3n}$ given by $\widetilde{\omega}_q = (\omega_q^*)^{-1}$ for $q \in \mathbf{R}^{3n}$, i.e.,

$$\widetilde{\omega}\,\Big(\sum_{i=1}^{3n} p_i de^i(q)\Big) = \left(q, \sum_{i=1}^{3n} p_i e^i\right)$$

for $(q,p) \in \mathbf{R}^{3n} \times \mathbf{R}^{3n}$. As well as being real numbers, we also let $q^1, \ldots, q^{3n}, p_1, \ldots, p_{3n}$ denote the coordinate functions of the local chart $((id \times r) \circ \widetilde{\omega}, T^*\mathbf{R}^{3n})$ on $T^*\mathbf{R}^{3n}$, where $r : (\mathbf{R}^n)^* \to \mathbf{R}^n$ is the linear isomorphism defined by $r(e^i) = e_i$ for $i = 1,...,3n$.

The Legendre transformation $F(L) : T\mathbf{R}^{3n} \to T^*\mathbf{R}^{3n}$ is given by

$$\widetilde{\omega} \circ F(L) \circ \omega^{-1}\left(q, \sum_{i=1}^{3n} v^i e_i\right) = \left(q, \sum_{i=1}^{3n} M_i v^i e^i\right)$$

for $(q,v) \in \mathbf{R}^{3n} \times \mathbf{R}^{3n}$, with inverse

$$\omega \circ F(L)^{-1} \circ \widetilde{\omega}^{-1}\left(q, \sum_{i=1}^{3n} p_i e^i\right) = \left(q, \sum_{i=1}^{3n} \frac{p_i}{M_i} e_i\right)$$

for $(q,p) \in \mathbf{R}^{3n} \times \mathbf{R}^{3n}$. By Proposition 7.92 the Hamiltonian of the system is $H = E \circ F(L)^{-1}$, i.e.,

$$H \circ \widetilde{\omega}^{-1}\left(q, \sum_{i=1}^{3n} p_i e^i\right) = \sum_{i=1}^{3n} \frac{p_i^2}{2M_i} + V(q)$$

for $(q,p) \in \mathbf{R}^{3n} \times \mathbf{R}^{3n}$. The Hamilton equations for a curve $\gamma : I \to T^*\mathbf{R}^{3n}$ therefore take the form

$$(q^i \circ \gamma)'(t) = \frac{1}{M_i}(p_i \circ \gamma)(t) \quad \text{and} \quad (p_i \circ \gamma)'(t) = -\frac{\partial V}{\partial q^i} \circ \pi \circ \gamma(t)$$

for $t \in I$ and $i = 1,...,3n$. Combining these equations, we again obtain Newton's second law.

TIME DEPENDENT SYSTEMS

7.103 Theorem Let $\omega \in \Omega^2(M)$ be a closed 2-form of rank $2k$ on a smooth manifold M^n. Then there is a local chart (x,U) around each point $p \in M$ with $x(p) = 0$

and

$$\omega|_U = \sum_{i=1}^{k} dx^i \wedge dx^{i+k} .$$

PROOF: By Proposition 5.87 we know that $\ker(\omega)$ is an integrable distribution on M of dimension $n-2k$. Hence there is a local chart (y,V) around p so that $\{\frac{\partial}{\partial y^{2k+1}}, \dots, \frac{\partial}{\partial y^n}\}$ is a local basis for $\ker(\omega)$ on V by Frobenius' integrability theorem. We may assume that $y(p)=0$ and that $y(V)=O_1 \times O_2$ for connected open sets $O_1 \subset \mathbf{R}^{2k}$ and $O_2 \subset \mathbf{R}^{n-2k}$. Then the form ω is given locally by

$$\omega|_V = \sum_{1 \le i < j \le 2k} a_{ij}\, dy^i \wedge dy^j ,$$

where the skew symmetric matrix $(\omega_{ij}(q))$, given by $\omega_{ij}(q) = a_{ij}(q)$ for $1 \le i < j \le 2k$, is of rank $2k$ for every $q \in V$. Using that ω is closed, we have that

$$d\omega|_V = \sum_{1 \le i < j \le 2k} \sum_{r=1}^{n} \frac{\partial a_{ij}}{\partial y^r}\, dy^r \wedge dy^i \wedge dy^j = 0$$

which implies that

$$\frac{\partial a_{ij}}{\partial y^r} = 0 \qquad (1)$$

for $1 \le i < j \le 2k$ and $r = 2k+1, \dots, n$.

Now define $N = y^{-1}(O_1 \times \{0\})$, and let $\alpha : N \to V$ be the inclusion map. Then $\alpha^*(\omega)$ is a symplectic form on N given by

$$\alpha^*(\omega) = \sum_{1 \le i < j \le 2k} b_{ij}\, du^i \wedge du^j ,$$

where $b_{ij} = a_{ij} \circ \alpha$ and (u,N) is the local chart on N defined by $u^i = y^i \circ \alpha$ for $i = 1, \dots, 2k$. Let $pr_1 : \mathbf{R}^{2k} \times \mathbf{R}^{n-2k} \to \mathbf{R}^{2k}$ be the projection on the first factor, and let $\pi : V \to N$ be the submersion defined by $\pi = u^{-1} \circ pr_1 \circ y$. Then $u \circ \pi = pr_1 \circ y$ so that $u^i \circ \pi = y^i$ for $i = 1, \dots, 2k$. From (1) it also follows that $b_{ij} \circ \pi = a_{ij}$ which implies that $\omega|_V = (\alpha \circ \pi)^*(\omega)$.

By Darboux' theorem there is now a local chart (z,W) around p on N with $z(p) = 0$ and

$$\alpha^*(\omega)|_W = \sum_{i=1}^{k} dz^i \wedge dz^{i+k} .$$

Using this, we obtain the local chart (x,U) around p on M with coordinate neighbourhood $U = \pi^{-1}(W)$ and coordinate map $x : U \to \mathbf{R}^{2k} \times \mathbf{R}^{n-2k}$ defined by $x = \{(z \circ u^{-1}) \times id\} \circ y$, i.e.,

$$x^j = \begin{cases} z^j \circ \pi & \text{for} \quad j = 1, \dots, 2k \\ y^j & \text{for} \quad j = 2k+1, \dots, n \end{cases} ,$$

which satisfies the conditions $x(p) = 0$ and

$$\omega|_U = \sum_{i=1}^{k} dx^i \wedge dx^{i+k} .$$

$\square$

7.104 Theorem Let $\pi : T^*M \to M$ be the cotangent bundle of a connected smooth manifold M^n, and let X_H be a Hamiltonian vector field with energy function H on T^*M, and $S : M \to R$ be a smooth function on M. Then the following two assertions are equivalent :

(1) If $\gamma : I \to M$ is an integral curve for the vector field $\pi_* \circ X_H \circ dS$ on M, then $dS \circ \gamma$ is an integral curve for X_H.

(2) S satisfies the Hamilton-Jacobi equation $H \circ dS = E$, where E is a constant.

PROOF : Since dS and X_H are sections of the vector bundles $\pi : T^*M \to M$ and $\pi'' : T(T^*M) \to T^*M$, respectively, it follows from the commutative diagram

$$\begin{array}{ccc} T(T^*M) & \xrightarrow{\pi_*} & TM \\ \pi'' \downarrow & & \downarrow \pi' \\ T^*M & \xrightarrow[\pi]{} & M \end{array}$$

that $\pi_* \circ X_H \circ dS$ is a section of $\pi' : TM \to M$.

Let $\alpha : I \to T^*M$ be the smooth curve on T^*M given by $\alpha = dS \circ \gamma$. If $\gamma : I \to M$ is an integral curve for $\pi_* \circ X_H \circ dS$, we have that

$$\alpha'(t) = (dS)_* \circ \gamma'(t) = (dS \circ \pi)_* \circ X_H(\alpha(t))$$

for $t \in I$. From Proposition 7.72 it follows that

$$\omega(\alpha(t))((dS \circ \pi)_* \circ X_H(\alpha(t)), w) = \omega(\alpha(t))(X_H(\alpha(t)), w)$$
$$- \omega(\alpha(t))(X_H(\alpha(t)), (dS \circ \pi)_*(w))$$

for $t \in I$ and $w \in T_{\alpha(t)}(T^*M)$, where

$$\omega(\alpha(t))(X_H(\alpha(t)), (dS \circ \pi)_*(w)) = dH(\alpha(t))((dS \circ \pi)_*(w))$$
$$= (dS)^* dH(\gamma(t))(\pi_*(w)) = d(H \circ dS)(\gamma(t))(\pi_*(w)) .$$

Hence we conclude that

$$(dS \circ \pi)_* \circ X_H(\alpha(t)) = X_H(\alpha(t))$$

if and only if

$$d(H \circ dS)(\gamma(t)) = 0$$

for $t \in I$, showing that the assertions (1) and (2) are equivalent. □

7.105 Definition An odd dimensional smooth manifold M with a closed 2-form ω of maximal rank is called a *contact manifold*, and it is also denoted by (M, ω).

7.106 Definition Let (M, ω) be a symplectic manifold, and let $H \in \mathscr{F}(\mathbf{R} \times M)$ be a smooth function on $\mathbf{R} \times M$. Then the time dependent vector field X_H on M over the time interval $\mathbf{R}$ defined by $(X_H)_s = X_{H_s}$ for $s \in \mathbf{R}$, is called the *Hamiltonian vector field* with energy function H.

7.107 Proposition Let (M, ω) be a symplectic manifold, and let $t : \mathbf{R} \times M \to \mathbf{R}$ and $\pi_2 : \mathbf{R} \times M \to M$ be the projections on the first and second factor. If $H \in \mathscr{F}(\mathbf{R} \times M)$ is a smooth function on $\mathbf{R} \times M$, we have a contact manifold $(\mathbf{R} \times M, \omega_H)$, where

$$\omega_H = \pi_2^* \, \omega + dH \wedge dt \, .$$

The suspension $\widetilde{X}_H$ of the Hamiltonian vector field with energy function H is a local basis for $\ker \omega_H$.

PROOF : We see that

$$\begin{aligned}
&\omega_H(s,p)\left(i_{p\,*}\left(a \left.\frac{d}{dr}\right|_s \right) + i_{s\,*}(v) \, , i_{p\,*}\left(b \left.\frac{d}{dr}\right|_s \right) + i_{s\,*}(w) \right) \\
&= \omega(p)(v,w) - a\; dH_s(p)(w) + b\; dH_s(p)(v) \\
&= \omega(p)(v - a\,X_H(s,p), w) \,+\, b\;\; \omega(p)(X_H(s,p), v) = 0
\end{aligned}$$

for every $b \in \mathbf{R}$ and $w \in T_pM$ if and only if $v = a\,X_H(s,p)$. This shows that $\widetilde{X}_H(s,x) = i_{p\,*}\left(\left.\frac{d}{dr}\right|_s \right) + i_{s\,*} \circ X_H(s,p)$ forms a basis for $\ker \omega_H(s,p)$ for each $(s,p) \in \mathbf{R} \times M$. □

7.108 Theorem (Hamilton-Jacobi) Let $\pi : T^*M \to M$ be the cotangent bundle of a smooth manifold M^n, and let X_H be a time dependent Hamiltonian vector field on T^*M over the open time interval J with energy function H, and $W : J \times M \to \mathbf{R}$ be a smooth function on $J \times M$. Then the following two assertions are equivalent :

(1) If $\gamma : I \to M$ is an integral curve for the time dependent vector field Y on M given by $Y_s = \pi_* \circ X_{H_s} \circ dW_s$ for $s \in J$, then the curve $\alpha : I \to T^*M$ defined by $\alpha(s) = dW_s(\gamma(s))$ for $s \in I$ is an integral curve for X_H.

(2) W satisfies the Hamilton-Jacobi equation

$$H_s \circ dW_s + \left(\frac{\partial W}{\partial t}\right)_s = const$$

on M for each $s \in J$.

PROOF : Let (r,J) be the standard local chart on J where $r : J \to \mathbf{R}$ is the inclusion map, and let (x,U) be a local chart on M and (z,W) be the corresponding local chart on T^*M defined in Proposition 7.70. We denote by t and y the two component maps of the coordinate map $r \times x : J \times U \to \mathbf{R} \times \mathbf{R}^n$ on $J \times M$.

Now let $F : J \times M \to T^*M$ be the smooth map defined by $F(s,u) = dW_s(u)$ for $s \in J$ and $u \in M$, which is given locally by

$$F(s,u) = \sum_{i=1}^{n} \frac{\partial W}{\partial y^i}(s,u)\, dx^i(u)$$

so that

$$F_{u*}\left(\left.\frac{d}{dr}\right|_s\right) = \sum_{i=1}^{2n} \frac{\partial(z^i \circ F)}{\partial t}(s,u) \left.\frac{\partial}{\partial z^i}\right|_{F(s,u)} = \sum_{i=1}^{n} \frac{\partial^2 W}{\partial t\, \partial y^i}(s,u) \left.\frac{\partial}{\partial p^i}\right|_{F(s,u)}$$

for $s \in J$ and $u \in U$. Using the local expression for ω given in Proposition 7.74, we see that

$$\omega(F(s,u))\left(F_{u*}\left(\left.\frac{d}{dr}\right|_s\right), w\right) = -\sum_{i=1}^{n} \frac{\partial^2 W}{\partial t\, \partial y^i}(s,u)\, dq^i(F(s,u))(w)$$

$$= -\sum_{i=1}^{n} \left.\frac{\partial}{\partial x^i}\right|_u \left(\frac{\partial W}{\partial t}\right)_s dx^i(u)(\pi_*(w)) = -d\left(\frac{\partial W}{\partial t}\right)_s(u)(\pi_*(w))$$

for each $w \in T_{F(s,u)}(T^*M)$ where $s \in J$ and $u \in U$.

If $\gamma : I \to M$ is an integral curve for Y, we have that

$$\alpha'(s) = F_{\gamma(s)*}\left(\left.\frac{d}{dr}\right|_s\right) + (dW_s)_*(\gamma'(s)) = F_{\gamma(s)*}\left(\left.\frac{d}{dr}\right|_s\right) + (dW_s \circ \pi)_*(X_{H_s}(\alpha(s)))$$

for $s \in I$. From Proposition 7.72 it follows that

$$\omega(\alpha(s))((dW_s \circ \pi)_*(X_{H_s}(\alpha(s))), w) = \omega(\alpha(s))(X_{H_s}(\alpha(s)), w)$$
$$- \omega(\alpha(s))(X_{H_s}(\alpha(s)), (dW_s \circ \pi)_*(w))$$

for $s \in I$ and $w \in T_{\alpha(s)}(T^*M)$, where

$$\omega(\alpha(s))\,(X_{H_s}(\alpha(s)),(dW_s \circ \pi)_*(w)) = dH_s(\alpha(s))\,((dW_s \circ \pi)_*(w))$$
$$= (dW_s)^* dH_s(\gamma(s))\,(\pi_*(w)) = d(H_s \circ dW_s)(\gamma(s))\,(\pi_*(w))\,.$$

Hence we conclude that

$$F_{\gamma(s)\,*}\left(\frac{d}{dr}\bigg|_s\right) + (dW_s \circ \pi)_*(X_{H_s}(\alpha(s))) = X_{H_s}(\alpha(s))$$

if and only if

$$d\left(H_s \circ dW_s + \left(\frac{\partial W}{\partial t}\right)_s\right)(\gamma(s)) = 0$$

for each $s \in I$, showing that the assertions (1) and (2) are equivalent. □

Chapter 8

LIE GROUPS

LIE GROUPS AND THEIR LIE ALGEBRAS

8.1 Definition A *Lie group* G is a group which is at the same time a smooth manifold such that the maps $\mu : G \times G \to G$ and $\nu : G \to G$ given by the group operations $\mu(g,h) = gh$ and $\nu(g) = g^{-1}$ are smooth. The unit element of G is denoted by e.

8.2 Remark Smoothness of μ and ν is equivalent to assuming that the map $\rho : G \times G \to G$ given by $\rho(g,h) = gh^{-1}$ is smooth.

8.3 If G is a Lie group and $g \in G$, we define the *left* and *right translations* $L_g : G \to G$ and $R_g : G \to G$ by

$$L_g(h) = gh \quad \text{and} \quad R_g(h) = hg$$

for $h \in G$. Both L_g and R_g are diffeomorphisms with inverses $L_{g^{-1}}$ and $R_{g^{-1}}$, respectively. In particular, it follows that a Lie group has the same dimension at all points.

A vector field X on G is called *left invariant* if $L_{g*}(X) = X$ for every $g \in G$, which means that X is L_g-related to itself for every $g \in G$. It follows from Propositions 4.86 and 4.88 that the left invariant vector fields form a Lie subalgebra of the Lie algebra of all vector fields on G.

8.4 Proposition If G is a Lie group and $X \in T_eG$, then there is a unique left invariant vector field $\widetilde{X}$ on G with $\widetilde{X}_e = X$ which is given by

$$\widetilde{X}_g = L_{g*}(X)$$

for $g \in G$. It is called the left invariant vector field determined by X.

PROOF : Since $L_{gh} = L_g \circ L_h$, we have that

$$\widetilde{X}_{gh} = L_{g*}(\widetilde{X}_h)$$

for every $g,h \in G$. To complete the proof, we only need to show that $\widetilde{X}$ is smooth in some open neighbourhood V of e. Since L_g maps V diffeomorphically onto an open neighbourhood $L_g(V)$ of g, and we have that

$$\widetilde{X}|_{L_g(V)} = L_{g*}(\widetilde{X}|_V)\,,$$

it then follows that $\widetilde{X}$ is smooth at every point $g \in G$ and therefore is a left invariant vector field on G with $\widetilde{X}_e = X$.

Let (z,U) be a local chart around e, and let n be the dimension of G. As G is a Lie group, the map $\mu : G \times G \to G$ given by $\mu(g,h) = gh$ is smooth. Choose an open neighbourhood V of e so that $\mu(V \times V) \subset U$, and let $x = z|_V$. Then the map $f : x(V) \times x(V) \to \mathbf{R}^n$ defined by $f = z \circ \mu \circ (x^{-1} \times x^{-1})$ is smooth, and we have that

$$f(a,b) = z \circ L_{x^{-1}(a)} \circ x^{-1}(b)$$

for $a,b \in x(V)$.

Now let $(t_z, \pi^{-1}(U))$ be the local trivialization in the tangent bundle $\pi : TG \to G$ associated with the local chart (z,U) on G. If $X = [x,v]_e$, we have that

$$\begin{aligned}\widetilde{X} \circ x^{-1}(a) = L_{x^{-1}(a)*}(X) &= [z, D(z \circ L_{x^{-1}(a)} \circ x^{-1})(x(e))\,v]_{x^{-1}(a)} \\ &= [z, D_2 f(a,x(e))\,v]_{x^{-1}(a)}\end{aligned}$$

so that

$$(z \times id) \circ t_z \circ \widetilde{X} \circ x^{-1}(a) = (a, D_2 f(a,x(e))\,v)$$

for $a \in x(V)$, thus showing that $\widetilde{X}$ is smooth in V. □

8.5 We thus have a linear isomorphism ϕ from the tangent space T_eG to the Lie algebra of all left invariant vector fields on G given by $\phi(X) = \widetilde{X}$, and we introduce a bracket product in T_eG which makes ϕ a Lie algebra isomorphism by defining

$$[X,Y] = [\widetilde{X},\widetilde{Y}](e)$$

for $X,Y \in T_eG$. The vector space T_eG with this bracket product is called the *Lie algebra of G* and is denoted by $\mathscr{L}(G)$ or $\mathfrak{g}$.

8.6 Definition A smooth homomorphism $\phi : G \to H$ between the Lie groups G and H is called a *Lie group homomorphism* . If ϕ is also a diffeomorphism, it is called a *Lie group isomorphism.* A Lie group isomorphism of a Lie group with itself is called a *Lie group automorphism.*

If G is a Lie group, then a Lie group homomorphism $\phi : \mathbf{R} \to G$ is called a *one-parameter subgroup* of G.

8.7 Proposition Let $\phi : G \to H$ be a Lie group homomorphism, and suppose that $X \in T_eG$ and $X' = \phi_*(X)$. Then the left invariant vector fields $\widetilde{X}$ and $\widetilde{X}'$ on G and H determined by X and X', respectively, are ϕ-related.

PROOF : For each $g \in G$, we have that

$$\phi \circ L_g = L_{\phi(g)} \circ \phi$$

so that

$$\phi_* (\widetilde{X}_g) = \phi_* L_{g*} (X) = L_{\phi(g)*} (X') = \widetilde{X}'_{\phi(g)} \, .$$

□

8.8 Proposition If $\phi : G \to H$ is a Lie group homomorphism, then $\phi_{*e} : T_eG \to T_eH$ is a Lie algebra homomorphism.

PROOF : Let $X, Y \in T_eG$, and suppose that $X' = \phi_* (X)$ and $Y' = \phi_* (Y)$. Then it follows from Propositions 8.7 and 4.88 that $[\widetilde{X}, \widetilde{Y}]$ and $[\widetilde{X}', \widetilde{Y}']$ are ϕ-related, which implies that

$$\phi_* ([X,Y]) = [\phi_* (X), \phi_* (Y)] \, .$$

□

8.9 Remark We will usually denote ϕ_{*e} simply by $\phi_* : \mathfrak{g} \to \mathfrak{h}$.

8.10 Examples

(a) Let $\mathfrak{gl}(n,\mathbf{R})$ be the Lie algebra of real $n \times n$-matrices with bracket product $[A,B] = AB - BA$ as defined in Remark 4.83. It is a vector space of dimension n^2, and we let $y^{ij} : \mathfrak{gl}(n,\mathbf{R}) \to \mathbf{R}$ be the linear funtional assigning to each matrix A its ij-th entry A_{ij}. Then we have a linear isomorphism $z : \mathfrak{gl}(n,\mathbf{R}) \to \mathbf{R}^{n^2}$ given by $z^{b(j,i)} = y^{ij}$ for $1 \leq i,j \leq n$, where $b : I_n \times I_n \to I_{n^2}$ is the bijection defined in Remark 4.27. We give $\mathfrak{gl}(n,\mathbf{R})$ the manifold structure defined in Example 2.9 (b) so that z is a coordinate map. If $\mathscr{E} = \{e_1, \ldots, e_{n^2}\}$ is the standard basis for $\mathbf{R}^{n^2}$, we have a corresponding basis $\mathscr{C} = \{E^i_j | 1 \leq i,j \leq n\}$ for $\mathfrak{gl}(n,\mathbf{R})$, where E^i_j is the $n \times n$-matrix with all the entries equal to zero except the ij-th entry which is 1. We have that $z(E^i_j) = e_{b(j,i)}$ for $1 \leq i,j \leq n$, and

$$E^i_j E^r_s = \delta_{jr} E^i_s$$

for $1 \leq i,j,r,s \leq n$.

Now let $Gl(n,\mathbf{R})$ be the group of non-singular real $n \times n$-matrices. As the determinant function $\det : \mathfrak{gl}(n,\mathbf{R}) \to \mathbf{R}$ is given by

$$\det(A) = \sum_{\sigma \in S_n} \varepsilon(\sigma) A_{\sigma(1)\,1} \cdots A_{\sigma(n)\,n}$$

which is a polynomial map and therefore is continuous, it follows that $Gl(n,\mathbf{R}) = \det^{-1}(\mathbf{R} - \{0\})$ is an open submanifold of $\mathfrak{gl}(n,\mathbf{R})$. If α :

$Gl(n,\mathbf{R}) \to \mathfrak{gl}(n,\mathbf{R})$ is the inclusion map, then $(z \circ \alpha, Gl(n,\mathbf{R}))$ is a local chart on $Gl(n,\mathbf{R})$ with coordinate functions $x^{ij} = y^{ij} \circ \alpha$. Moreover, since

$$(AB)_{ij} = \sum_{k=1}^{n} A_{ik} B_{kj}$$

and

$$(A^{-1})_{ij} = (-1)^{i+j} \, det(A^{ji}) / det(A),$$

where A^{ij} is the matrix obtained from A by deleting the i-th row and the j-th column, it follows that $x^{ij}(AB^{-1})$ is a rational function of $x^{kl}(A)$ and $x^{kl}(B)$ with non-zero denominator. Hence the group operations in $Gl(n,\mathbf{R})$ are smooth so that $Gl(n,\mathbf{R})$ is a Lie group called the *general linear group*.

We will show that the Lie algebra $\mathfrak{g}$ of $Gl(n,\mathbf{R})$ may be identified with $\mathfrak{gl}(n,\mathbf{R})$ using Lemma 2.84. We have that $\mathfrak{g} = T_e Gl(n,\mathbf{R}) = T_e \mathfrak{gl}(n,\mathbf{R})$ which we canonically identify with $\mathfrak{gl}(n,\mathbf{R})$ by means of the linear isomorphism $\omega : \mathfrak{g} \to \mathfrak{gl}(n,\mathbf{R})$ given by $\omega = z^{-1} \circ t_{z,e}$. It only remains to show that

$$\omega([X,Y]) = [\omega(X), \omega(Y)] \tag{1}$$

for $X, Y \in \mathfrak{g}$, so that ω is a Lie algebra isomorphism.

As usual we let $\widetilde{X}$ and $\widetilde{Y}$ be the left invariant vector fields on $Gl(n,\mathbf{R})$ determined by X and Y. By Lemmas 2.84 and 2.78 (1) we have that

$$\omega(X)_{ij} = X(x^{ij})$$

for every $X \in \mathfrak{g}$ and $1 \le i,j \le n$, so that

$$\omega([X,Y])_{ij} = [X,Y](x^{ij}) = X(\widetilde{Y}(x^{ij})) - Y(\widetilde{X}(x^{ij})).$$

To compute the last expression, we use that

$$x^{ij} \circ L_A(B) = x^{ij}(AB) = \sum_{k=1}^{n} x^{ik}(A)\, x^{kj}(B)$$

for $A, B \in Gl(n,\mathbf{R})$, so that

$$\widetilde{Y}(x^{ij})(A) = \widetilde{Y}_A(x^{ij}) = L_{A*}(Y)(x^{ij}) = Y(x^{ij} \circ L_A) = \sum_{k=1}^{n} x^{ik}(A)\, Y(x^{kj})$$

and hence

$$X(\widetilde{Y}(x^{ij})) = \sum_{k=1}^{n} X(x^{ik})\, Y(x^{kj}) = \sum_{k=1}^{n} \omega(X)_{ik}\, \omega(Y)_{kj} = (\omega(X)\, \omega(Y))_{ij}$$

which implies (1).

(b) Let V be a finite dimensional real vector space and $\mathrm{End}\,(V)$ be the Lie algebra of linear endomorphisms of V with bracket product $[F,G] = F \circ G - G \circ F$. If $\mathscr{B} = \{v_1, ..., v_n\}$ is a basis for V, we have a Lie algebra isomorphism

$$\phi : \mathrm{End}\,(V) \to \mathfrak{gl}\,(n, \mathbf{R})$$

given by

$$\phi\,(F) = m_{\mathscr{B}}^{\mathscr{B}}(F)$$

for $F \in \mathrm{End}(V)$. We give $\mathrm{End}\,(V)$ the manifold structure defined in Example 2.9 (b) so that $\tilde{z} = z \circ \phi$ is a coordinate map and hence ϕ is a diffeomorphism. Corresponding to the basis $\mathscr{C}$ for $\mathfrak{gl}\,(n,\mathbf{R})$ defined in (a), we have a basis $\mathscr{D} = \{F_j^i | 1 \le i, j \le n\}$ for $\mathrm{End}\,(V)$, where $F_j^i : V \to V$ is the linear map given by $F_j^i(v_k) = \delta_{jk} v_i$ for $k = 1, \dots, n$, so that $\phi\,(F_j^i) = E_j^i$ for $1 \le i, j \le n$. Hence we have that

$$F_j^i \circ F_s^r = \delta_{jr} F_s^i$$

for $1 \le i, j, r, s \le n$.

Now let $\mathrm{Aut}\,(V)$ be the group of non-singular endomorphisms of V, where the group product is composition of endomorphisms. Then

$$\mathrm{Aut}\,(V) = \phi^{-1}(Gl\,(n, \mathbf{R}))$$

is an open submanifold of $\mathrm{End}\,(V)$ and hence a Lie group since

$$G \circ F^{-1} = \phi^{-1}\,(\phi\,(G)\,\phi\,(F)^{-1})$$

for $F, G \in \mathrm{Aut}\,(V)$, and the map

$$\psi : \mathrm{Aut}\,(V) \to Gl\,(n, \mathbf{R})$$

induced by ϕ is a Lie group isomorphism.

We will show that the Lie algebra $\mathfrak{h}$ of $\mathrm{Aut}\,(V)$ may be identified with $\mathrm{End}\,(V)$ using Lemma 2.84. We have that $\mathfrak{h} = T_e \mathrm{Aut}\,(V) = T_e \mathrm{End}\,(V)$ which we canonically identify with $\mathrm{End}\,(V)$ by means of the linear isomorphism $\widetilde{\omega} : \mathfrak{h} \to \mathrm{End}\,(V)$ given by $\widetilde{\omega} = \tilde{z}^{-1} \circ t_{\tilde{z},e}$. Since $D(z \circ \psi \circ \tilde{z}^{-1})(\tilde{z}(e)) = id$, it follows from the commutative diagram in 2.70 that $t_{\tilde{z},e} = t_{z,e} \circ \psi_*$ which shows that $\widetilde{\omega} = \phi^{-1} \circ \omega \circ \psi_*$ is a Lie algebra isomorphism. We also have a commutative diagram

$$\begin{array}{ccc} \mathfrak{h} & \xrightarrow{\psi_*} & \mathfrak{g} \\ \widetilde{\omega} \downarrow & & \downarrow \omega \\ \mathrm{End}\,(V) & \xrightarrow{\phi} & \mathfrak{gl}\,(n, \mathbf{R}) \end{array}$$

which shows that ψ_* is identified with ϕ when we identify the Lie algebras of Aut(V) and $Gl(n,\mathbf{R})$ with End(V) and $\mathfrak{gl}(n,\mathbf{R})$ in the way described above.

We often denote End(V) and Aut(V) by $\mathfrak{gl}(V)$ and $Gl(V)$, respectively. When $V=\mathbf{R}^n$, we always use the standard basis $\mathscr{E}=\{e_1,\dots,e_n\}$ to define the isomorphisms ϕ and ψ and the basis $\mathscr{D}$.

(c) Let $\mathfrak{gl}(n,\mathbf{C})$ be the Lie algebra of complex $n\times n$-matrices with bracket product $[A,B]=AB-BA$ as defined in Remark 4.83. It is a real vector space of dimension $2n^2$, and we let $y^{ij}:\mathfrak{gl}(n,\mathbf{C})\to\mathbf{C}$ be the complex linear funtional assigning to each matrix A its ij-th entry A_{ij}. Then we have a linear isomorphism $z:\mathfrak{gl}(n,\mathbf{C})\to\mathbf{R}^{2n^2}$ given by $z^{b(1,j,i)}=Re\,y^{ij}$ and $z^{b(2,j,i)}=Im\,y^{ij}$ for $1\le i,j\le n$, where $b:I_2\times I_n\times I_n\to I_{2n^2}$ is the bijection defined in Remark 4.27. We give $\mathfrak{gl}(n,\mathbf{C})$ the manifold structure defined in Example 2.9 (b) so that z is a coordinate map. If $\mathscr{E}=\{e_1,\dots,e_{2n^2}\}$ is the standard basis for $\mathbf{R}^{2n^2}$, we have a corresponding basis $\mathscr{C}=\{E^i_j\,|\,1\le i,j\le n\}\cup\{\sqrt{-1}\,E^i_j\,|\,1\le i,j\le n\}$ for $\mathfrak{gl}(n,\mathbf{C})$, where E^i_j is the $n\times n$-matrix with all the entries equal to zero except the ij-th entry which is 1. We have that $z(E^i_j)=e_{b(1,j,i)}$ and $z(\sqrt{-1}\,E^i_j)=e_{b(2,j,i)}$ for $1\le i,j\le n$.

Now let $Gl(n,\mathbf{C})$ be the group of non-singular complex $n\times n$-matrices. As the determinant function $\det:\mathfrak{gl}(n,\mathbf{C})\to\mathbf{C}$ is given by

$$\det(A)=\sum_{\sigma\in S_n}\varepsilon(\sigma)A_{\sigma(1)\,1}\cdots A_{\sigma(n)\,n}$$

which is a polynomial map and therefore is continuous, it follows that $Gl(n,\mathbf{C})=\det^{-1}(\mathbf{C}-\{0\})$ is an open submanifold of $\mathfrak{gl}(n,\mathbf{C})$. If $\alpha: Gl(n,\mathbf{C})\to\mathfrak{gl}(n,\mathbf{C})$ is the inclusion map, then $(z\circ\alpha, Gl(n,\mathbf{C}))$ is a local chart on $Gl(n,\mathbf{C})$ whose coordinate functions are the real and imaginary parts of $x^{ij}=y^{ij}\circ\alpha$. In the same way as in (a) we see that the group operations in $Gl(n,\mathbf{C})$ are smooth so that $Gl(n,\mathbf{C})$ is a Lie group called the *complex general linear group*. Using 5.49 we also see that the Lie algebra of $Gl(n,\mathbf{C})$ may be identified with $\mathfrak{gl}(n,\mathbf{C})$ in the same way as in (a).

(d) Let V be a finite dimensional complex vector space and End(V) be the Lie algebra of linear endomorphisms of V with bracket product $[F,G]=F\circ G-G\circ F$. If $\mathscr{B}=\{v_1,\dots,v_n\}$ is a basis for V, we have a Lie algebra isomorphism

$$\phi:\mathrm{End}(V)\to\mathfrak{gl}(n,\mathbf{C})$$

given by

$$\phi(F)=m^{\mathscr{B}}_{\mathscr{B}}(F)$$

for $F\in\mathrm{End}(V)$. We give End(V) the manifold structure defined in Example 2.9 (b) so that $\tilde{z}=z\circ\phi$ is a coordinate map and hence ϕ is a diffeomorphism. Corresponding to the basis $\mathscr{C}$ for $\mathfrak{gl}(n,\mathbf{C})$ defined in (c), we have a basis $\mathscr{D}=\{F^i_j\,|\,1\le i,j\le n\}\cup\{\sqrt{-1}\,F^i_j\,|\,1\le i,j\le n\}$ for End(V), where $F^i_j:V\to V$ is the linear map given by $F_j^i(v_k) = \delta_{jk} v_i$ for $k = 1, \dots, n$, so that $\phi(F_j^i) = E_j^i$ for $1 \le i, j \le n$.

Now let Aut (V) be the group of non-singular endomorphisms of V, where the group product is composition of endomorphisms. Then

$$\mathrm{Aut}(V) = \phi^{-1}(Gl(n, \mathbf{C}))$$

is an open submanifold of End (V) and hence a Lie group since

$$G \circ F^{-1} = \phi^{-1}(\phi(G)\, \phi(F)^{-1})$$

for $F, G \in \mathrm{Aut}(V)$, and the map

$$\psi : \mathrm{Aut}(V) \to Gl(n, \mathbf{C})$$

induced by ϕ is a Lie group isomorphism.

In the same way as in (b) we see that the Lie algebra of Aut (V) may be identified with End (V) using Lemma 2.84, and ψ_* is then identified with ϕ. We often denote End (V) and Aut (V) by $\mathfrak{gl}(V)$ and $Gl(V)$, respectively. When $V = \mathbf{C}^n$, we always use the standard basis $\mathscr{E} = \{e_1, \dots, e_n\}$ to define the isomorphisms ϕ and ψ and the basis $\mathscr{D}$.

GROUP REPRESENTATIONS

8.11 Definition A Lie group homomorphism $\phi : G \to H$, where $H = \mathrm{Aut}(V)$ for a vector space V or $H = Gl(n, \mathbf{R})$, is called a *representation* of the Lie group G. The representation ϕ is said to be *faithful* if it is injective.

We say that the representations $\phi_i : G \to \mathrm{Aut}(V_i)$ for $i = 1, 2$ are *equivalent* if there is a linear isomorphism $S : V_1 \to V_2$ so that

$$S \circ \phi_1(g) = \phi_2(g) \circ S$$

for every $g \in G$. Two representations $\phi_i : G \to Gl(n, \mathbf{R})$ for $i = 1, 2$ are said to be *equivalent* if there is a matrix $A \in Gl(n, \mathbf{R})$ so that

$$A\, \phi_1(g) = \phi_2(g)\, A$$

for every $g \in G$.

8.12 Proposition Let $\psi : G \to \mathrm{Aut}(V)$ be a representation of a Lie Group G on a finite dimensional vector space V. Then we also have a representation $\psi^* : G \to \mathrm{Aut}(V^*)$ of G on the dual space V^*, called the *contragradient representation*, which is defined by

$$\psi^*(g) = \psi(g^{-1})^*$$

for $g \in G$.

PROOF : We immediately see that ψ^* is a group homomorphism, so it only remains to show that it is smooth. This follows since

$$i \circ \psi^* = \alpha \circ \psi \circ \phi ,$$

where $i : \text{Aut}(V^*) \to \text{End}(V^*)$ is the inclusion map, $\phi : G \to G$ is the smooth map given by $\phi(g) = g^{-1}$ for $g \in G$, and $\alpha : \text{End}(V) \to \text{End}(V^*)$ is the linear map given by $\alpha(F) = F^*$ for $F \in \text{End}(V)$. □

8.13 Proposition If V is a finite dimensional vector space, we have a representation $\rho^k_l : Gl(V) \to Aut(\mathscr{T}^k_l(V))$ of the Lie group $Gl(V)$ on the vector space $\mathscr{T}^k_l(V)$ given by

$$\rho^k_l(F) = (F^{-1})^*$$

for $F \in Gl(V)$ as defined in Definition 4.52, i.e.,

$$\rho^k_l(F)(T)(v_1, \ldots, v_k, \lambda_1, \ldots, \lambda_l) = T(F^{-1}(v_1), \ldots, F^{-1}(v_k), \lambda_1 \circ F, \ldots, \lambda_l \circ F)$$

for $T \in \mathscr{T}^k_l(V)$, $v_1, \ldots, v_k \in V$ and $\lambda_1, \ldots, \lambda_l \in V^*$. In particular, we have that $\rho^0_0(F) = id_{\mathbf{R}}$ for every $F \in Gl(V)$.

PROOF : We immediately see that ρ is a group homomorphism, so it only remains to show that it is smooth. If $\mathscr{B}$ is a basis for V and $\mathscr{C} = \mathscr{T}^k_l(\mathscr{B})$ is the corresponding basis for $\mathscr{T}^k_l(V)$, it follows from Propositions 4.6 and 4.37 that

$$m^{\mathscr{C}}_{\mathscr{C}}((F^{-1})^*) = \underbrace{\{m^{\mathscr{B}}_{\mathscr{B}}(F)^{-1}\}^t \otimes \cdots \otimes \{m^{\mathscr{B}}_{\mathscr{B}}(F)^{-1}\}^t}_{k} \otimes \underbrace{m^{\mathscr{B}}_{\mathscr{B}}(F) \otimes \cdots \otimes m^{\mathscr{B}}_{\mathscr{B}}(F)}_{l} .$$

Hence if $\psi_{\mathscr{B}} : Gl(V) \to Gl(n, \mathbf{R})$ and $\psi_{\mathscr{C}} : \text{Aut}(\mathscr{T}^k_l(V)) \to Gl(n^{k+l}, \mathbf{R})$, where $n = dim(V)$, are the Lie group isomorphisms defined in Example 8.10 (b), we have that

$$\psi_{\mathscr{C}} \circ \rho^k_l \circ \psi^{-1}_{\mathscr{B}}(A) = \underbrace{\{A^{-1}\}^t \otimes \cdots \otimes \{A^{-1}\}^t}_{k} \otimes \underbrace{A \otimes \cdots \otimes A}_{l}$$

for $A \in Gl(n, \mathbf{R})$, thus showing that ρ is smooth and completing the proof of the proposition. □

LIE SUBGROUPS

8.14 Definition Let G be a Lie group. A Lie group H is called a *Lie subgroup* of G if it is a subgroup and an immersed submanifold of G.

8.15 Proposition If G is a Lie group, then the connected component H of G containing e is an open Lie subgroup of G.

PROOF: By Remark 2.7 we know that H is an open submanifold of G. Since the map $\rho : G \times G \to G$ given by $\rho(g,h) = gh^{-1}$ is smooth, it follows that $\rho(H \times H)$ is a connected set containing e. Hence we have that $\rho(H \times H) \subset H$, which shows that H is a subgroup of G.

To prove that H is a Lie subgroup of G, we must show that the map $\omega : H \times H \to H$ given by $\omega(g,h) = gh^{-1}$ is smooth. Since $i \circ \omega = \rho \circ (i \times i)$ is smooth, where $i : H \to G$ is the inclusion map, this follows from Corollary 2.38. □

8.16 Proposition If H is a Lie subgroup of G, then $\mathfrak{h}$ is a Lie subalgebra of $\mathfrak{g}$.

PROOF: Since H is a Lie subgroup of G, the inclusion map $i : H \to G$ is a Lie group homomorphism. Hence we may use Proposition 8.8 to conclude that $\mathfrak{h}$ is a Lie subalgebra of $\mathfrak{g}$. □

8.17 Proposition Let G be a Lie group, and let $\mathfrak{h}$ be a Lie subalgebra of $\mathfrak{g}$. Then we have an integrable distribution Δ on G given by

$$\Delta_g = \{\widetilde{X}(g) | X \in \mathfrak{h}\}$$

for $g \in G$. If H is a Lie subgroup of G with Lie algebra $\mathfrak{h}$, then H is an integral manifold for Δ.

PROOF: We first show that Δ is an integrable distribution on G. If $\{X_1, \ldots, X_k\}$ is a basis for $\mathfrak{h}$, then the left invariant vector fields $\widetilde{X_1}, \ldots, \widetilde{X_k}$ form a local basis for Δ on G, showing that Δ is a distribution on G. Since $\mathfrak{h}$ is a Lie algebra, there are real numbers c_{ij}^r such that

$$[X_i, X_j] = \sum_{r=1}^{k} c_{ij}^r X_r \tag{1}$$

for $i, j = 1, \ldots, k$. By applying L_{g*} to both sides of (1) for each $g \in G$, we obtain

$$[\widetilde{X}_i, \widetilde{X}_j] = \sum_{r=1}^{k} c_{ij}^r \widetilde{X}_r$$

which by Proposition 4.96 implies that the distribution Δ is integrable.

To prove the last part of the propositioon, let H be a Lie subgroup of G with Lie algebra $\mathfrak{h}$, and let $i : H \to G$ be the inclusion map. Since

$$i_*(T_h H) = i_* \circ L_{h*}(T_e H) = L_{h*} \circ i_*(T_e H) = L_{h*}(\mathfrak{h}) = \Delta_h$$

for every $h \in H$, we see that H is an integral manifold of Δ. □

8.18 Theorem Let G be a Lie group, and let $\mathfrak{h}$ be a Lie subalgebra of $\mathfrak{g}$. Then there is a unique connected Lie subgroup H of G with Lie algebra $\mathfrak{h}$.

PROOF : Let Δ be the integrable distribution on G determined by the Lie subalgebra $\mathfrak{h}$ as defined in Proposition 8.17, and let H be the maximal connected integral manifold for Δ passing through e. Since

$$L_{g*}(\Delta_h) = \Delta_{gh}$$

for every $g, h \in G$, it follows from Proposition 4.107 that $L_{g^{-1}}(H)$ is also a maximal connected integral manifold of Δ passing through e and therefore coincides with H for each $g \in H$. Hence we have that $g, h \in H$ implies $g^{-1}h \in H$, which shows that H is a subgroup of G.

To prove that H is a Lie subgroup of G, we must show that the map $\rho : H \times H \to H$ given by $\rho(g,h) = gh^{-1}$ is smooth. This follows from Proposition 4.102 as the map $i \circ \rho : H \times H \to G$ is smooth, where $i : H \to G$ is the inclusion map.

Finally, it follows from Proposition 8.16 that the Lie algebra of H is Δ_e with bracket product inherited from $\mathfrak{g}$. Hence the Lie algebra of H is $\mathfrak{h}$, and this completes the existence part of the theorem.

To prove the uniqueness part, let H' be any connected Lie subgroup of G with Lie algebra $\mathfrak{h}$. By Proposition 8.17 we know that H' is a connected integral manifold of Δ passing through e. Hence the Frobenius theorem implies that H' is an open submanifold of H. Since both H' and H are subgroups of G, H' is also a subgroup of H, and each left coset of H' in H is open in H. As H is connected, this implies that $H' = H$. □

8.19 Example We have that $Gl\,(n,\mathbf{R})$ is a submanifold and a Lie subgroup of $Gl\,(n,\mathbf{C})$ since

$$z(Gl\,(n,\mathbf{R})) = z(Gl\,(n,\mathbf{C})) \cap (\mathbf{R}^{n^2} \times \{0\}),$$

where $z : \mathfrak{gl}\,(n,\mathbf{C}) \to \mathbf{R}^{n^2} \times \mathbf{R}^{n^2}$ is the linear isomorphism defined in Example 8.10 (c). If $\tilde{z} : \mathfrak{gl}\,(n,\mathbf{R}) \to \mathbf{R}^{n^2}$ is the corresponding linear isomorphism defined in Example 8.10 (a), we have a commutative diagram

$$\begin{array}{ccc} \mathfrak{gl}\,(n,\mathbf{R}) & \xrightarrow{\ \phi\ } & \mathfrak{gl}\,(n,\mathbf{C}) \\ {\scriptstyle \tilde{z}}\downarrow & & \downarrow{\scriptstyle z} \\ \mathbf{R}^{n^2} & \xrightarrow{\ i\ } & \mathbf{R}^{n^2} \times \mathbf{R}^{n^2} \end{array}$$

where $\phi : \mathfrak{gl}\,(n,\mathbf{R}) \to \mathfrak{gl}\,(n,\mathbf{C})$ is the inclusion map, and $i : \mathbf{R}^{n^2} \to \mathbf{R}^{n^2} \times \mathbf{R}^{n^2}$ is the map given by $i(a) = (a,0)$.

As described in Example 8.10, the Lie algebras $\tilde{\mathfrak{g}}$ and $\mathfrak{g}$ of $Gl\,(n,\mathbf{R})$ and $Gl\,(n,\mathbf{C})$ may be identified with $\mathfrak{gl}\,(n,\mathbf{R})$ and $\mathfrak{gl}\,(n,\mathbf{C})$, respectively, by means of the linear isomorphisms $\tilde{\omega} : \tilde{\mathfrak{g}} \to \mathfrak{gl}\,(n,\mathbf{R})$ and $\omega : \mathfrak{g} \to \mathfrak{gl}\,(n,\mathbf{C})$ given by $\tilde{\omega} = \tilde{z}^{-1} \circ t_{\tilde{z},e}$ and

$\omega = z^{-1} \circ t_{z,e}$. If $\psi : Gl(n, \mathbf{R}) \to Gl(n, \mathbf{C})$ is the inclusion map induced by ϕ, we have by 2.70 a commutative diagram

$$\begin{array}{ccc} \widetilde{\mathfrak{g}} & \xrightarrow{\psi_*} & \mathfrak{g} \\ {\scriptstyle t_{\tilde{z},e}}\big\downarrow & & \big\downarrow{\scriptstyle t_{z,e}} \\ \mathbf{R}^{n^2} & \xrightarrow{i} & \mathbf{R}^{n^2} \times \mathbf{R}^{n^2} \end{array}$$

since $D(z \circ \psi \circ \tilde{z}^{-1})(\tilde{z}(e)) = i$. Combining these two diagrams, we obtain the commutative diagram

$$\begin{array}{ccc} \widetilde{\mathfrak{g}} & \xrightarrow{\psi_*} & \mathfrak{g} \\ {\scriptstyle \widetilde{\omega}}\big\downarrow & & \big\downarrow{\scriptstyle \omega} \\ \mathfrak{gl}(n, \mathbf{R}) & \xrightarrow{\phi} & \mathfrak{gl}(n, \mathbf{C}) \end{array}$$

which shows that ψ_* is identified with the inclusion map ϕ when we identify the Lie algebras of $Gl(n, \mathbf{R})$ and $Gl(n, \mathbf{C})$ with $\mathfrak{gl}(n, \mathbf{R})$ and $\mathfrak{gl}(n, \mathbf{C})$ in the way described Example 8.10.

COVERINGS

8.20 Definition A covering $p : \widetilde{M} \to M$ is said to be *smooth* if M and $\widetilde{M}$ are smooth manifolds of the same dimension and the covering projection p is smooth of maximal rank. The manifold $\widetilde{M}$ is called a *covering manifold* of M.

8.21 Remark It follows from the rank theorem that a covering $p : \widetilde{M} \to M$ is smooth if and only if, for each evenly covered neighbourhood U in M and each sheet $\widetilde{U}_\alpha$ over U, the homeomorphism $p_\alpha : \widetilde{U}_\alpha \to U$ induced by p is in fact a diffeomorphism.

8.22 Proposition Let $p : \widetilde{M} \to M$ be a covering of a smooth manifold M^n with a connected covering space $\widetilde{M}$. Then $\widetilde{M}$ is locally Euclidean and second countable, and there is a unique smooth structure on $\widetilde{M}$ such that the covering is smooth.

PROOF : We first show that $\widetilde{M}$ is second countable. For each point $u \in M$ there is an open neighbourhood U around u which is evenly covered by p. By eventually choosing a smaller U using Propositions 13.38 and 13.81 in the appendix, we may assume that U is also connected. Since M is second countable and therefore is a Lindelöf space, M is covered by a countable family $\{U_i \mid i \in \mathbf{N}\}$ of such neighbourhoods.

Now it follows from Proposition 13.82 in the appendix that the connected components of $p^{-1}(U_i)$ are the sheets over U_i which are second countable since they are homeomorphic to U_i. We also have that $\widetilde{M}$ is locally connected since this is true for the manifold M and p is a local homeomorphism. By applying Proposition 4.103 to the countable open cover $\{p^{-1}(U_i) \mid i \in \mathbf{N}\}$ of $\widetilde{M}$, it therefore follows that $\widetilde{M}$ is second countable.

We next show that $\widetilde{M}$ is locally Euclidean. Since p is a local homeomorphism, each point $\tilde{u}$ on $\widetilde{M}$ has an open neighbourhood $\widetilde{U}_1$ which is mapped homeomorphically by p onto an open neighbourhood U_1 of $p(\tilde{u})$. If $p_1 : \widetilde{U}_1 \to U_1$ is the induced homeomorphism and (x_2, U_2) is a local chart around $p(\tilde{u})$, then $\tilde{x} = x_2 \circ p_1$ maps the open neighbourhood $\widetilde{U} = p_1^{-1}(U_1 \cap U_2)$ of $\tilde{u}$ homeomorphically onto an open set in $\mathbf{R}^n$. Since $\widetilde{M}$ is also Hausdorff by Proposition 13.76 in the appendix, we therefore have that $\widetilde{M}$ is locally Euclidean.

If the covering is smooth, then it follows from the rank theorem that p_1 is a diffeomorphism. Hence $(\tilde{x}, \widetilde{U})$ must be a local chart around $\tilde{u}$ in the differetiable structure on $\widetilde{M}$. As these local charts around each point on $\widetilde{M}$ form an atlas on $\widetilde{M}$, the uniqueness of the smooth structure follows from Proposition 2.2.

To prove the existence, let $(\tilde{y}, \widetilde{V})$ be another local chart on $\widetilde{M}$ obtained in the same way as $(\tilde{x}, \widetilde{U})$ from a homeomorphism $q_1 : \widetilde{V}_1 \to V_1$ induced by p and a local chart (y_2, V_2) on M. Then the coordinate transformation $\tilde{y} \circ \tilde{x}^{-1} = y_2 \circ q_1 \circ p_1^{-1} \circ x_2^{-1}$ is the restriction of $y_2 \circ x_2^{-1}$ to $x_2(U_1 \cap U_2 \cap V_1 \cap V_2)$, showing that the local charts $(\tilde{x}, \widetilde{U})$ and $(\tilde{y}, \widetilde{V})$ are C^∞-related. Hence we obtain a smooth structure on $\widetilde{M}$ such that the covering is smooth, since $y_2 \circ p \circ \tilde{x}^{-1} = \tilde{y} \circ \tilde{x}^{-1}$. □

8.23 Corollary Let M be a connected differeniable manifold, and let $u \in M$. If H is a subgroup of $\pi_1(M, u)$, then there is a smooth covering $p : (\widetilde{M}, \tilde{u}) \to (M, u)$ with a connected covering manifold $\widetilde{M}$ such that $p_* \pi_1(\widetilde{M}, \tilde{u}) = H$.

PROOF : Since a connected smooth manifold is pathwise connected, locally pathwise connected and semi-locally simply connected, the result follows from Theorem 13.104 in the appendix and Proposition 8.22. □

8.24 Proposition If $p : \widetilde{M} \to M$ is a smooth covering, then a lifting $\tilde{f} : N \to \widetilde{M}$ of a smooth map $f : N \to M$ is smooth.

PROOF: Let $u \in N$, and choose an evenly covered neighbourhood U of $f(u)$. Let $\widetilde{U}_\alpha$ be the sheet over U containing $\tilde{f}(u)$, and let $s_\alpha : U \to \widetilde{U}_\alpha$ be the inverse of the diffeomorphism induced by p and $i_\alpha : \widetilde{U}_\alpha \to \widetilde{M}$ be the inclusion map. Then $V = \tilde{f}^{-1}(\widetilde{U}_\alpha)$ is an open neighbourhood of u, and we have that $\tilde{f}|_V = i_\alpha \circ s_\alpha \circ f|_V$ which shows that $\tilde{f}$ is smooth at u. □

8.25 Proposition Let G be a connected Lie group with unit element e, and let $p : (\widetilde{G}, \tilde{e}) \to (G, e)$ be a smooth covering of (G, e) with a connected covering manifold $\widetilde{G}$. Then there is a unique Lie group structure on $\widetilde{G}$ with $\tilde{e}$ as unit element such that the covering projection p is a Lie group homomorphism.

PROOF: Let $\rho : (G \times G, (e,e)) \to (G, e)$ be the smooth map given by $\rho(g,h) = gh^{-1}$ for $g, h \in G$. Then it follows from Proposition 13.109 in the appendix that

$$\rho_* \circ (p \times p)_* \, \pi_1(\widetilde{G} \times \widetilde{G}, (\tilde{e}, \tilde{e})) \subset p_* \pi_1(\widetilde{G}, \tilde{e})$$

since

$$\begin{aligned} \rho_* \circ (p \times p)_* ([(\tilde{\sigma}_1, \tilde{\sigma}_2)]) &= [\rho \circ (p \circ \tilde{\sigma}_1, p \circ \tilde{\sigma}_2)] = [p \circ \tilde{\sigma}_1] * [p \circ \tilde{\sigma}_2]^{-1} \\ &= p_*([\tilde{\sigma}_1] * [\tilde{\sigma}_2]^{-1}) \in p_* \pi_1(\widetilde{G}, \tilde{e}) \end{aligned}$$

for every pair of closed paths $\tilde{\sigma}_1$ and $\tilde{\sigma}_2$ in $\widetilde{G}$ at $\tilde{e}$.

By the lifting theorem and Proposition 8.24 there is therefore a unique smooth map $\tilde{\rho} : (\widetilde{G} \times \widetilde{G}, (\tilde{e}, \tilde{e})) \to (\widetilde{G}, \tilde{e})$ which is a lifting of $\rho \circ (p \times p)$, so that the diagram

$$\begin{array}{ccc} (\widetilde{G} \times \widetilde{G}, (\tilde{e}, \tilde{e})) & \xrightarrow{\ \tilde{\rho}\ } & (\widetilde{G}, \tilde{e}) \\ {\scriptstyle p \times p}\downarrow & & \downarrow{\scriptstyle p} \\ (G \times G, (e,e)) & \xrightarrow[\ \rho\ ]{} & (G, e) \end{array}$$

is commutative.

We now let $\tilde{g}^{-1} = \tilde{\rho}(\tilde{e}, \tilde{g})$ and $\tilde{g}\tilde{h} = \tilde{\rho}(\tilde{g}, \tilde{h}^{-1})$ for $\tilde{g}, \tilde{h} \in \widetilde{G}$. To prove that this defines a Lie group structure on $\widetilde{G}$ with $\tilde{e}$ as unit element, we must show that

$$\widetilde{g}\,\tilde{e} = \tilde{e}\,\widetilde{g} = \widetilde{g} \text{ and } \widetilde{g}\,\widetilde{g}^{-1} = \widetilde{g}^{-1}\,\widetilde{g} = \tilde{e} \text{ for every } \widetilde{g} \in \widetilde{G}, \tag{1}$$

and that

$$(\widetilde{g}\widetilde{h})\widetilde{k} = \widetilde{g}(\widetilde{h}\widetilde{k}) \text{ for every } \widetilde{g}, \widetilde{h}, \widetilde{k} \in \widetilde{G}. \tag{2}$$

To show the first part of (1), we use the fact that G is a group so that the two maps $\mu_1, \mu_2 : (G,e) \to (G,e)$ given by $\mu_1(g) = g\,e$ and $\mu_2(g) = e\,g$ and the identity map $id : (G,e) \to (G,e)$ are all equal. Hence the same is true for the maps $\widetilde{\mu}_1, \widetilde{\mu}_2 : (\widetilde{G}, \tilde{e}) \to (\widetilde{G}, \tilde{e})$ given by $\widetilde{\mu}_1(\widetilde{g}) = \widetilde{g}\,\widetilde{e}$ and $\widetilde{\mu}_2(\widetilde{g}) = \widetilde{e}\,\widetilde{g}$ and the identity map $\widetilde{id} : (\widetilde{G}, \tilde{e}) \to (\widetilde{G}, \tilde{e})$, since they are the unique liftings of $\mu_1 \circ p$, $\mu_2 \circ p$ and $id \circ p$, respectively. Indeed, we have that

$$p \circ \widetilde{\mu}_1(\widetilde{g}) = p \circ \widetilde{\rho}\,(\widetilde{g}, \widetilde{\rho}\,(\widetilde{e}, \widetilde{e})) = \rho\,(p\,(\widetilde{g}), \rho\,(e,e)) = \mu_1 \circ p\,(\widetilde{g})$$

and

$$p \circ \widetilde{\mu}_2(\widetilde{g}) = p \circ \widetilde{\rho}\,(\widetilde{e}, \widetilde{\rho}\,(\widetilde{e}, \widetilde{g})) = \rho\,(e, \rho\,(e, p\,(\widetilde{g}))) = \mu_2 \circ p\,(\widetilde{g})$$

for every $\widetilde{g} \in \widetilde{G}$, and we also have that $p \circ \widetilde{id} = id \circ p$. This completes the proof of the first part of (1), and the second part of (1) and (2) are proved in the same way. Finally, we have that

$$p\,(\widetilde{g}\widetilde{h}) = p \circ \widetilde{\rho}\,(\widetilde{g}, \widetilde{\rho}\,(\widetilde{e}, \widetilde{h})) = \rho\,(p\,(\widetilde{g}), \rho\,(e, p\,(\widetilde{h}))) = p\,(\widetilde{g})\,p\,(\widetilde{h})$$

for every $\widetilde{g}, \widetilde{h} \in \widetilde{G}$, which shows that the covering projection p is a Lie group homomorphism, and this completes the proof of the existence part of the proposition.

To show the uniqueness part, suppose that $\widetilde{G}$ has any Lie group structure with $\tilde{e}$ as unit element such that the covering projection p is a Lie group homomorphism, and let $\widetilde{\lambda} : \widetilde{G} \times \widetilde{G} \to \widetilde{G}$ be the smooth map defined by $\widetilde{\lambda}\,(\widetilde{g}, \widetilde{h}) = \widetilde{g}\widetilde{h}^{-1}$ for $\widetilde{g}, \widetilde{h} \in \widetilde{G}$. Then we have that

$$p \circ \widetilde{\lambda}\,(\widetilde{g}, \widetilde{h}) = p\,(\widetilde{g})\,p\,(\widetilde{h})^{-1} = \rho \circ (p \times p)(\widetilde{g}, \widetilde{h})$$

for every $\widetilde{g}, \widetilde{h} \in \widetilde{G}$, which shows that $\widetilde{\lambda}$ is a lifting of $\rho \circ (p \times p)$ with $\widetilde{\lambda}\,(\widetilde{e}, \widetilde{e}) = \widetilde{e}$. Hence $\widetilde{\lambda} = \widetilde{\rho}$ by the unique lifting theorem, thus showing that the Lie group structure on $\widetilde{G}$ must coincide with the one defined in the first part of the proof and therefore is uniquely determined by the conditions in the proposition. □

8.26 Corollary Let G be a connected Lie group with unit element e, and let H be a subgroup of $\pi_1(G,e)$. Then there is a smooth covering $p : (\widetilde{G}, \tilde{e}) \to (G,e)$ such that $p_*\,\pi_1(\widetilde{G}, \tilde{e}) = H$, where the covering manifold $\widetilde{G}$ is a Lie group with unit element $\tilde{e}$ and the covering projection p is a Lie group homomorphism.

PROOF : Follows from Corollary 8.23 and Proposition 8.25. □

8.27 Proposition Let $\phi : G \to H$ be a Lie group homomorphism between the connected Lie groups G and H such that $\phi_* : \mathfrak{g} \to \mathfrak{h}$ is a Lie algebra isomorphism. Then ϕ is a smooth covering projection. If H is simply connected, then ϕ is a diffeomorphism.

PROOF : We have that ϕ_{*g} is an isomorphism for each $g \in G$, since the left translations L_g and $L_{\phi(g)}$ are diffeomorphisms in G and H, respectively, with $L_{\phi(g)} \circ \phi \circ L_g^{-1} = \phi$. By the inverse function theorem it follows that ϕ is a local diffeomorphism which therefore has a discrete kernel and is open by Proposition 13.86 in the appendix. Since $\phi(G)$ contains an open neighbourhood of the unit element e in H, it follows from Proposition 13.112 in the appendix that ϕ is a surjection. The proposition now follows from Propositions 13.111 and 13.99 in the appendix. □

8.28 Theorem Let G and H be Lie groups, and let $\Phi : \mathfrak{g} \to \mathfrak{h}$ be a Lie algebra homomorphism. If G is simply connected, then there is a unique Lie group homomorphism $\phi : G \to H$ with $\phi_* = \Phi$.

PROOF : We consider the graph $\mathfrak{k}$ of Φ given by $\mathfrak{k} = \{(X, \Phi(X)) | X \in \mathfrak{g}\}$. As Φ is a Lie algebra homomorphism, $\mathfrak{k}$ is a Lie subalgebra of $\mathfrak{g} \times \mathfrak{h} = \mathscr{L}(G \times H)$, so by Theorem 8.18 there is a unique connected Lie subgroup K of $G \times H$ whose Lie algebra is $\mathfrak{k}$.

Now let $\pi_1 : G \times H \to G$ and $\pi_2 : G \times H \to H$ be the projections on the first and second factor, and let $\omega = \pi_1|_K$. Then $\omega : K \to G$ is a Lie group homomorphism, and $\omega_* : \mathfrak{k} \to \mathfrak{g}$ is a Lie algebra isomorphism since

$$\omega_* (X, \Phi(X)) = X$$

for $X \in \mathfrak{g}$. By Proposition 8.27 it follows that ω is a diffeomorphism, and we have a Lie group homomorphism $\phi : G \to H$ defined by $\phi = \pi_2 \circ \omega^{-1}$. Since

$$\phi_* (X) = \pi_{2*} (X, \Phi(X)) = \Phi(X)$$

for $X \in \mathfrak{g}$, we have that $\phi_* = \Phi$, showing that ϕ satisfies the conditions of the theorem.

To show uniqueness of ϕ, suppose that $\psi : G \to H$ is another Lie group homomorphism with $\psi_* = \Phi$. Then the map $\theta : G \to G \times H$ defined by

$$\theta(g) = (g, \psi(g))$$

for $g \in G$ is a Lie group homomorphism which is a one-to-one immersion. By Remark 2.31 its image G' is a connected Lie subgroup of $G \times H$ with Lie algebra $\mathfrak{k}$, since

$$\theta_*(X) = (X, \Phi(X))$$

for $X \in \mathfrak{g}$. Hence we have that $G' = K$ by Theorem 8.18, showing that $\psi(g) = \phi(g)$ for every $g \in G$. □

THE EXPONENTIAL MAP

8.29 Proposition Let G be a Lie group, and let $X \in \mathfrak{g}$. Then there is a unique one-parameter subgroup $\phi_X : \mathbf{R} \to G$ with $\phi_X'(0) = X$.

The left invariant vector field $\widetilde{X}$ on G determined by X is complete, with flow $\gamma : \mathbf{R} \times G \to G$ given by

$$\gamma(t,g) = L_g \circ \phi_X(t) .$$

In particular, ϕ_X is the maximal integral curve for $\widetilde{X}$ with initial condition $\phi_X(0) = e$.

PROOF : We use the local chart $(r, \mathbf{R})$ on the Lie group $\mathbf{R}$, where $r : \mathbf{R} \to \mathbf{R}$ is the identity map. Then the tangent vector $\left.\frac{d}{dr}\right|_0$ is a basis for the Lie algebra of $\mathbf{R}$, which we may also identify with $\mathbf{R}$, and we have a unique Lie algebra homomorphism $\Phi_X : \mathbf{R} \to \mathfrak{g}$ such that

$$\Phi_X \left(\left.\frac{d}{dr}\right|_0 \right) = X .$$

By Theorem 8.28 there is a unique Lie group homomorphism $\phi_X : \mathbf{R} \to G$ with $\phi_{X*} = \Phi_X$, and hence the first part of the proposition follows from Remark 2.83 which implies that

$$\phi_X'(0) = \phi_{X*} \left(\left.\frac{d}{dr}\right|_0 \right) .$$

To prove the last part of the proposition, let $L_s : \mathbf{R} \to \mathbf{R}$ be the left translation on $\mathbf{R}$ defined by $L_s(t) = s + t$. Then

$$L_{s*} \left(\left.\frac{d}{dr}\right|_t \right) = \left.\frac{d}{dr}\right|_{L_s(t)}$$

for every $s, t \in \mathbf{R}$, which shows that $\frac{d}{dr}$ is the left invariant vector field on $\mathbf{R}$ determined by $\left.\frac{d}{dr}\right|_0$. Hence it follows from Proposition 8.7 that the vector fields $\frac{d}{dr}$ and $\widetilde{X}$ are ϕ_X-related, so that

$$\phi_X'(t) = \phi_{X*} \left(\left.\frac{d}{dr}\right|_t \right) = \widetilde{X}(\phi_X(t))$$

for every $t \in \mathbf{R}$. Since $\widetilde{X}$ is left invariant, the last formula is still true if ϕ_X is replaced by $L_g \circ \phi_X$, and this completes the proof of the proposition. □

8.30 Definition If G is a Lie group, we define the *exponential map* $exp : \mathfrak{g} \to G$ by

$$exp\,(X) = \phi_X\,(1)$$

for $X \in \mathfrak{g}$.

8.31 Proposition The one-parameter subgroup $\phi_X : \mathbf{R} \to G$ defined in Proposition 8.29 is given by

$$\phi_X\,(t) = exp\,(tX)$$

for $t \in \mathbf{R}$.

PROOF: Using the Lie group homomorphism $\mu_t : \mathbf{R} \to \mathbf{R}$ defined by $\mu_t\,(s) = t\,s$ for $s \in \mathbf{R}$, we obtain a one-parameter subgroup $\phi_X \circ \mu_t : \mathbf{R} \to G$ satisfying

$$(\phi_X \circ \mu_t)\,'(0) = t\,\phi_X'\,(0) = tX$$

by Definition 2.69. Hence we have that

$$\phi_X \circ \mu_t = \phi_{tX}\,,$$

which implies that

$$\phi_X\,(t) = \phi_{tX}\,(1) = exp\,(tX)\,.$$

□

8.32 Corollary Let G be a Lie group, and let $X \in \mathfrak{g}$. Then

(1) $exp\,(t_1 + t_2)X = (exp\;t_1X)(exp\;t_2X)$ for $t_1, t_2 \in \mathbf{R}$,

(2) $exp\,(-tX) = (exp\;tX)^{-1}$ for $t \in \mathbf{R}$.

8.33 Corollary Let G be a Lie group, and let $\phi : G \to G$ be the smooth map given by $\phi\,(g) = g^{-1}$ for $g \in G$. Then we have that

$$\phi_{*\,g} = -\,(R_{g^{-1}} \circ L_{g^{-1}})_{*\,g}$$

for $g \in G$. In particular, the linear map $\phi_{*\,e} : \mathfrak{g} \to \mathfrak{g}$ is given by $\phi_{*\,e}(X) = -X$ for $X \in \mathfrak{g}$.

PROOF: By Proposition 8.31 and Corollary 8.32 (2) we have that

$$\phi \circ L_g \circ \phi_X\,(t) = exp\,(-tX)\,g^{-1} = R_{g^{-1}} \circ \phi_{-X}\,(t)$$

for $t \in \mathbf{R}$. Hence it follows from 2.70 that

$$\phi_* \circ (L_g)_*(X) = (\phi \circ L_g \circ \phi_X)'(0) = -\,(R_{g^{-1}})_*(X)$$

for every $X \in \mathfrak{g}$. □

8.34 Proposition Let G be a Lie group, and let $X, Y \in \mathfrak{g}$. Then we have that $[X,Y] = 0$ if and only if

$$exp\,(tX)\; exp\,(sY) = exp\,(sY)\; exp\,(tX)$$

for all $s,t \in \mathbf{R}$.

PROOF: Let α and β be the flows of the left invariant vector fields $\widetilde{X}$ and $\widetilde{Y}$ on G determined by X and Y, respectively. Then we have that

$$\alpha\,(t, \beta\,(s,e)) = exp\,(sY)\; exp\,(tX)$$

and

$$\beta\,(s, \alpha\,(t,e)) = exp\,(tX)\; exp\,(sY)$$

for every $s,t \in \mathbf{R}$ by Propositions 8.29 and 8.31. The result therefore follows from Propositions 4.91 and 4.92. □

8.35 Proposition Let G be a Lie group, and let $X, Y \in \mathfrak{g}$ with $[X,Y] = 0$. Then we have that

$$exp\; t\,(X+Y) = exp\,(tX)\; exp\,(tY)$$

for every $t \in \mathbf{R}$.

PROOF: Using Corollary 8.32 (1) and Proposition 8.34 we see that the smooth curve $\gamma : \mathbf{R} \to G$ given by

$$\gamma(t) = exp\,(tX)\; exp\,(tY)$$

for $t \in \mathbf{R}$, is a one-parameter subgroup of G, and we have that $\gamma'(0) = X + Y$ by Remark 2.75. The result therefore follows from Proposition 8.31. □

8.36 Lemma Let V be a vector space of dimension n, and let $v \in V$. If $\psi_v : \mathbf{R} \to V$ is the curve on V defined by $\psi_v(t) = tv$ for $t \in \mathbf{R}$, and we canonically identify T_0V with V as in Lemma 2.84, then we have that $\psi_v{}'(0) = v$.

PROOF: Let $x : V \to \mathbf{R}^n$ be a linear isomorphism, and let $x(v) = a$. We identify T_0V with V by means of the linear isomorphism $\omega_0 : T_0V \to V$ given by $\omega_0 = x^{-1} \circ t_{x,0}$. Now $x \circ \psi_v(t) = ta$ for $t \in \mathbf{R}$ so that

$$\psi_v{}'(0) = [x, (x \circ \psi_v)'(0)]_{\psi_v(0)} = [x, a]_0$$

which implies that

$$\omega_0 \circ \psi_v{}'(0) = x^{-1}(a) = v\,.$$

□

8.37 Proposition A finite dimensional vector space V is an abelian Lie group with addition as its group operation, and its Lie algebra may be identified with V. The exponential map $exp : V \to V$ is then identified with $id_V : V \to V$.

PROOF: Since that map $\psi_v : \mathbf{R} \to V$, defined as in Lemma 8.36 by $\psi_v(t) = tv$ for $t \in \mathbf{R}$, is a one-parameter subgroup of V with $\psi_v'(0) = v$, we have that

$$exp\,(v) = \psi_v(1) = v$$

for $v \in V$. □

8.38 Proposition The exponential map $exp : \mathfrak{g} \to G$ is smooth, and $exp_{*0} : \mathfrak{g} \to \mathfrak{g}$ is the identity map. Hence exp maps an open neighbourhood U of $0 \in \mathfrak{g}$ diffeomorphically onto an open neighbourhood V of $e \in G$.

We let $log : V \to U$ denote the inverse of the diffeomorphism induced by exp.

PROOF: Let Y be the vector field on $G \times \mathfrak{g}$ defined by

$$Y(g,X) = (\widetilde{X}(g), 0) = (L_{g*}(X), 0)\,.$$

By Proposition 8.29 the flow $\gamma : \mathbf{R} \times G \times \mathfrak{g} \to G \times \mathfrak{g}$ of Y is given by

$$\gamma(t,g,X) = (L_g \circ \phi_X(t), X)\,,$$

and if $\pi : G \times \mathfrak{g} \to G$ is the projection on the first factor, we have that

$$exp\,(X) = \pi \circ \gamma(1,e,X)$$

which shows that exp is smooth.

Using the curves ϕ_X and ψ_X on G and $\mathfrak{g}$, respectively, defined in Proposition 8.29 and Lemma 8.36, we have that

$$exp_*(X) = (exp \circ \psi_X)'(0) = \phi_X'(0) = X$$

for every $X \in \mathfrak{g}$ by 2.70 and Proposition 8.31, showing that $exp_{*0} = id$. The last part of the proposition now follows from the inverse function theorem. □

8.39 Remark If $x : \mathfrak{g} \to \mathbf{R}^n$ is a linear isomorphism and $y = x \circ log$, then (y, V) is a local chart around e on G, called a *canonical chart* or a *canonical coordinate system* on G. Let $\mathscr{E} = \{e_1, ..., e_n\}$ be the standard basis for $\mathbf{R}^n$ and $\mathscr{C} = \{X_1, ..., X_n\}$ be a basis for $\mathfrak{g}$ so that $x(X_i) = e_i$ for $i = 1, ..., n$. Then the coordinate map y is given by

$$y(exp\,(\sum_{i=1}^{n} a_i X_i)) = a$$

for $a \in x(U)$.

8.40 Proposition If $\phi : G \to H$ is a Lie group homomorphism, then

$$exp \circ \phi_* = \phi \circ exp$$

so that we have a commutative diagram

$$\begin{array}{ccc} \mathfrak{g} & \xrightarrow{\phi_*} & \mathfrak{h} \\ {\scriptstyle exp}\downarrow & & \downarrow{\scriptstyle exp} \\ G & \xrightarrow{\phi} & H \end{array}$$

PROOF : If $X \in \mathfrak{g}$, then it follows from 2.70 and Proposition 8.29 that $\phi \circ \phi_X : \mathbf{R} \to H$ is a one-parameter subgroup of H with

$$(\phi \circ \phi_X)'(0) = \phi_* X .$$

Hence we have that

$$exp\,(\phi_* X) = \phi \circ \phi_X\,(1) = \phi\,(exp\,X) .$$

□

8.41 Proposition Let H be a Lie subgroup of G, and let $X \in \mathfrak{g}$. Then $X \in \mathfrak{h}$ if and only if $exp\,(tX) \in H$ for every $t \in \mathbf{R}$.

PROOF : Since H is a Lie subgroup of G, the inclusion map $i : H \to G$ is a Lie group homomorphism, and we have that

$$exp \circ i_* = i \circ exp$$

by Proposition 8.40. Hence it follows that $exp\,(tX) \in H$ whenever $X \in \mathfrak{h}$ and $t \in \mathbf{R}$.

Assuming conversely that $exp\,(tX) \in H$ for every $t \in \mathbf{R}$, it follows by Propositions 8.17 and 4.102 that the map $\phi : \mathbf{R} \to H$, defined by

$$i \circ \phi\,(t) = exp\,(tX)$$

for $t \in \mathbf{R}$, is smooth and therefore a one-parameter subgroup of H with $i_* \circ \phi\,'(0) = X$. This shows that $X \in \mathfrak{h}$. □

8.42 Proposition Let H be a subgroup of a Lie group G, and let $\mathfrak{h}$ be a subspace of $\mathfrak{g}$. Suppose that $\psi : U \to V$ is a diffeomorphism from an open neighbourhood U of $0 \in \mathfrak{g}$ where exp is one-to-one onto an open neighbourhood V of $e \in G$, and that

$$exp\,(U \cap \mathfrak{h}) = \psi(U \cap \mathfrak{h}) = V \cap H . \qquad (1)$$

Then H is a submanifold and a Lie subgroup of G, $\mathfrak{h}$ is a Lie subalgebra of $\mathfrak{g}$, and $\mathfrak{h}$ is the Lie algebra of H.

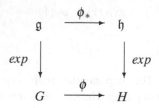

PROOF : Choose a basis $\mathscr{B} = \{X_1,...,X_k\}$ for $\mathfrak{h}$, and extend $\mathscr{B}$ to a basis $\mathscr{C} = \{X_1,...,X_n\}$ for $\mathfrak{g}$. Let $x : \mathfrak{g} \to \mathbf{R}^n$ be the linear isomorphism given by $x(X_i) = e_i$ for $i = 1, ..., n$, where $\mathscr{E} = \{e_1,...,e_n\}$ is the standard basis for $\mathbf{R}^n$, and let $y = x \circ \psi^{-1}$ so that

$$y(\psi(\sum_{i=1}^{n} a_i X_i)) = a$$

for $a \in x(U)$. Then (y,V) is a local chart around e on G having the submanifold property

$$y(V \cap H) = y(V) \cap (\mathbf{R}^k \times \{0\}) .$$

Hence the same is true for the local chart $(y \circ L_{g^{-1}}, gV)$ around g for each $g \in H$, which shows that H is a submanifold and a Lie subgroup of G.

Now it follows from Proposition 8.16 that the Lie algebra $\mathscr{L}(H)$ of H is a Lie subalgebra of $\mathfrak{g}$, and it only remains to prove that $\mathscr{L}(H) = \mathfrak{h}$. If $X \in \mathscr{L}(H)$, then $exp\,(tX) \in H$ for every $t \in \mathbf{R}$ by Proposition 8.41. We choose a $t \neq 0$ with $tX \in U$ and $exp\,(tX) \in V$. Then it follows from (1) and the fact that exp is one-to-one on U that $tX \in U \cap \mathfrak{h}$, which shows that $X \in \mathfrak{h}$. Since $\mathscr{L}(H)$ and $\mathfrak{h}$ are subspaces of $\mathfrak{g}$ having the same dimension k with $\mathscr{L}(H) \subset \mathfrak{h}$, we conclude that $\mathscr{L}(H) = \mathfrak{h}$. □

CLOSED SUBGROUPS

8.43 Theorem Let G be a Lie group, and let H be a closed subgroup of G. Then H is a submanifold and a Lie subgroup of G.

Let $\mathfrak{h}$ denote the Lie algebra of H and $\mathfrak{h}'$ be a subspace of $\mathfrak{g}$ so that $\mathfrak{g} = \mathfrak{h} \oplus \mathfrak{h}'$, and let $\phi : \mathfrak{g} \to G$ be the map given by

$$\phi\,(X + X') = exp\,(X')\; exp\,(X)$$

for $X \in \mathfrak{h}$ and $X' \in \mathfrak{h}'$. Then there are open neighbourhoods W and W' of 0 in $\mathfrak{h}$ and $\mathfrak{h}'$, respectively, and an open neighbourhood V of e in G so that ϕ induces a diffeomorphism $\psi : U \to V$ having the properties in Proposition 8.42 where $U = W + W'$.

PROOF : We will prove the theorem by showing that the set

$$\mathfrak{h} = \{X \in \mathfrak{g} \,|\, exp\,(tX) \in H \text{ for all } t \in \mathbf{R}\}$$

is a subspace of $\mathfrak{g}$ which together with the map $\psi : U \to V$ described above satisfies formula (1) in Proposition 8.42.

Given a sequence $\{X_i\}_{i=1}^{\infty}$ of elements in $\mathfrak{g}$ which converges to an element $X \in \mathfrak{g}$, and a sequence $\{t_i\}_{i=1}^{\infty}$ of positive real numbers which converges to 0 so that

$exp\,(t_iX_i) \in H$ for all i, we claim that $X \in \mathfrak{h}$. If $t \in \mathbf{R}$, we let k_i be the largest integer $\leq t/t_i$ so that

$$t/t_i - 1 < k_i \leq t/t_i$$

for all i. Then the sequence $\{k_i t_i X_i\}_{i=1}^{\infty}$ converges to tX, and

$$exp\,(k_i t_i X_i) = exp\,(t_i X_i)^{k_i} \in H$$

which shows that $exp\,(tX) \in H$, since H is closed. As this is true for all $t \in \mathbf{R}$, we have that $X \in \mathfrak{h}$.

Using this, we can now show that $\mathfrak{h}$ is a subspace of $\mathfrak{g}$. It is clearly closed under multiplication with scalars, so we only need to show that it is closed under addition. Suppose that U' is an open neighbourhood of $0 \in \mathfrak{g}$ which is mapped diffeomorphically by exp onto an open neighbourhood V' of $e \in G$. Let $Y, Z \in \mathfrak{h}$, and consider the smooth curve $\gamma : \mathbf{R} \to G$ given by

$$\gamma(t) = exp\,(tY)\; exp\,(tZ)$$

for $t \in \mathbf{R}$. We choose an open interval I with $0 \in I \subset \gamma^{-1}(V')$, and let $\alpha : I \to U'$ be the smooth curve given by $\alpha(t) = log \circ \gamma(t)$ for $t \in I$. Then it follows from Remark 2.75 and Proposition 8.38 that $\alpha'(0) = Y + Z$, and we have that $\alpha(0) = 0$ and $exp\,(\alpha(t)) = \gamma(t) \in H$ for $t \in I$.

Now let $\{t_i\}_{i=1}^{\infty}$ be a sequence of positive real numbers in I which converges to 0, and let $X_i = \alpha(t_i)/t_i$ for $i = 1, 2, 3....$ Then the sequence $\{X_i\}_{i=1}^{\infty}$ converges to $Y + Z$, and $exp\,(t_iX_i) = exp\,(\alpha(t_i)) \in H$ for all i. This shows that $Y + Z \in \mathfrak{h}$ and completes the proof that $\mathfrak{h}$ is a subspace of $\mathfrak{g}$.

Choose a basis $\mathscr{B} = \{X_1, ..., X_k\}$ for $\mathfrak{h}$, and extend $\mathscr{B}$ to a basis $\mathscr{C} = \{X_1, ..., X_n\}$ for $\mathfrak{g}$. Then

$$\mathfrak{g} = \mathfrak{h} \oplus \mathfrak{h}'\,,$$

where $\mathfrak{h}' = L(X_{k+1}, ..., X_n)$ is the subspace of $\mathfrak{g}$ spanned by $\{X_{k+1}, ..., X_n\}$. Let $\phi : \mathfrak{g} \to G$ be the smooth map given by

$$\phi\,(X + X') = exp\,(X')\; exp\,(X)$$

for $X \in \mathfrak{h}$ and $X' \in \mathfrak{h}'$. Using the curves ϕ_X and ψ_X on G and $\mathfrak{g}$, respectively, defined in Proposition 8.29 and Lemma 8.36, we have that

$$\phi_*\,(X_i) = (\phi \circ \psi_{X_i})'(0) = \phi_{X_i}'(0) = X_i$$

for $i = 1, ..., n$ by 2.70 and Proposition 8.31, showing that $\phi_{*0} = id_{\mathscr{G}}$. Hence there are open neighbourhoods W and W' of 0 in $\mathfrak{h}$ and $\mathfrak{h}'$, respectively, with $W + W' \subset U'$, such that ϕ is a diffeomorphism from $W + W'$ onto an open neighbourhood V of e.

We claim that W' can be chosen so that $exp\,(X') \notin H$ when $X' \in W' - \{0\}$. Otherwise, there would be a sequence $\{X_i'\}_{i=1}^{\infty}$ in $\mathfrak{h}' - \{0\}$ which converges to 0 with $exp\,(X_i') \in H$ for all i. We have a norm on $\mathfrak{g}$ so that the linear isomorphism $x : \mathfrak{g} \to \mathbf{R}^n$ given by $x(X_i) = e_i$ for $i = 1, ..., n$ is an isometry, where $\mathscr{E} = \{e_1, ..., e_n\}$

is the standard basis for $\mathbf{R}^n$. Let $t_i = \|X_i'\|$ and $Y_i' = X_i'/t_i$ for $i = 1,2,3....$ As the set $S' = \{Y' \in \mathfrak{h}' \mid \|Y'\| = 1\}$ is compact, there is a subsequence $\{Y_{i_r}'\}_{r=1}^{\infty}$ of $\{Y_i'\}_{i=1}^{\infty}$ which converges to an element $Y \in S'$, and $\{t_{i_r}\}_{r=1}^{\infty}$ is a sequence of positive real numbers which converges to 0 so that $exp\ (t_{i_r} Y_{i_r}) \in H$ for all r. This shows that $Y \in \mathfrak{h}$, which is impossible since $\mathfrak{h} \cap S' = \emptyset$.

We now clearly have that $\phi\ (U \cap \mathfrak{h}) \subset V \cap H$, where $U = W + W'$. To prove the reverse inclusion, let $g \in V \cap H$. Then there are elements $X \in W$ and $X' \in W'$ with $g = exp\ (X')\ exp\ (X)$. Since g, $exp\ (X) \in H$, we have that $exp\ (X') \in H$ which implies that $X' = 0$. Hence it follows that $g \in \phi\ (U \cap \mathfrak{h})$, which completes the proof of the theorem. □

8.44 Theorem Let H be a closed subgroup of a Lie group G. Then the set $G/H = \{gH | g \in G\}$ of left cosets has a unique manifold structure so that the natural projection $\pi : G \to G/H$ is smooth, and given a map $f : G/H \to M$ into a smooth manifold M, then f is smooth if and only if $f \circ \pi$ is smooth.

For each $g \in G$ there is an open neighbourhood U_g of $\pi(g)$ in G/H and a smooth map $s_g : U_g \to G$ so that $\pi \circ s_g = \tilde{\imath}_g$, where $\tilde{\imath}_g : U_g \to G/H$ is the inclusion map.

PROOF: We give G/H the quotient topology, consisting of the sets $O \subset G/H$ such that $\pi^{-1}(O)$ are open in G. Then the projection π is continuous, and it is also an open map since

$$\pi^{-1}(\pi(U)) = UH = \bigcup_{h \in H} Uh$$

is open for every open set U in G.

We next show that G/H is Hausdorff. Let $\rho : G \times G \to G$ be the continuous map given by $\rho(g_1, g_2) = g_2^{-1} g_1$. Then $g_1 H \neq g_2 H$ if and only if $(g_1, g_2) \notin \rho^{-1}(H)$. Given two distinct cosets $g_1 H$ and $g_2 H$ in G/H, there are therefore open neighbourhoods U_1 and U_2 of g_1 and g_2 in G, respectively, so that $(U_1 \times U_2) \cap \rho^{-1}(H) = \emptyset$, as $\rho^{-1}(H)$ is closed in $G \times G$. This implies that $\pi(U_1)$ and $\pi(U_2)$ are disjoint open neighbourhoods of $g_1 H$ and $g_2 H$ in G/H.

The space G/H is also second countable. If $\{O_i \mid i \in \mathbf{N}\}$ is a countable basis for the topology in G, then $\{\pi(O_i) \mid i \in \mathbf{N}\}$ is a countable basis for the topology in G/H. For each $g \in G$ we let $\tilde{L}_g : G/H \to G/H$ denote the left translation on G/H induced by the left translation $L_g : G \to G$ on G. It is the homeomorphism defined by $\tilde{L}_g(g'H) = gg'H$ for $g' \in G$ so that $\pi \circ L_g = \tilde{L}_g \circ \pi$.

Now let $\psi : W + W' \to V$ be the diffeomorphism given in Theorem 8.43, and choose open neighbourhoods W_0 and W_0' of 0 contained in W and W', respectively, such that ψ maps $W_0 + W_0'$ onto an open neighbourhood V_e of e with $V_e^{-1} V_e \subset V$. We contend that $\pi \circ \psi$ induces a homeomorphism $\phi : W_0' \to \pi(V_e)$. For each $gH \in \pi(V_e)$ there are $X \in W_0$ and $X' \in W_0'$ with $\psi(X + X') = g$. Using that $exp\ (X) \in H$ it follows that $\phi(X') = gH$, thus showing that ϕ is surjective.

To show that it is injective, assume that $\phi(X') = \phi(Y')$ where $X', Y' \in W_0'$. Then we have that

$$exp\ (-Y')\ exp\ (X') = h \in H \cap V\ ,$$

and there is an $X \in W$ with $exp\ (X) = h$. From this it follows that

$$\psi(X') = \psi(X + Y')$$

which shows that $X' = Y'$.

The bijection ϕ is clearly continuous. It is also open since

$$\phi(O) = \phi(W_0 + O)$$

is open for every open set O in W_0'. This completes the proof that ϕ is a homeomorphism.

Let $x : \mathfrak{h}' \to \mathbf{R}^{n-k}$ be a linear isomorphism, and consider for each $g \in G$ the open neighbourhood $U_g = \pi(V_g)$ of $\pi(g)$, where $V_g = L_g(V_e)$, and the map $y_g : U_g \to \mathbf{R}^{n-k}$ given by $y_g = x \circ \phi^{-1} \circ \tilde{L}_{g^{-1}}$. We contend that $\{(y_g, U_g) \mid g \in G\}$ is an atlas on G/H.

To see this, let $\pi(g_1\ exp\ (X_1')) = \pi(g_2\ exp\ (X_2'))$, where $X_1', X_2' \in W_0'$, be a point in $U_{g_1} \cap U_{g_2}$. Then there is an $h \in H$ so that

$$g_2^{-1} g_1\ exp\ (X_1')\, h = exp\ (X_2')\ ,$$

and we can choose an open neighbourhood W_1' of X_1' contained in W_0' such that

$$g_2^{-1} g_1\ exp\ (W_1')\, h \subset V_e\ .$$

We have a smooth map $\alpha : W_1' \to W_0'$ given by

$$\alpha(X') = \rho \circ \psi^{-1}(g_2^{-1} g_1\ exp\ (X')\, h)$$

for $X' \in W_1'$, where $\rho : \mathfrak{g} \to \mathfrak{h}'$ is the projection defined by $\rho(X + X') = X'$ for $X \in \mathfrak{h}$ and $X' \in \mathfrak{h}'$. Since

$$\phi \circ \alpha(X') = \pi(g_2^{-1} g_1\ exp\ (X')) = \tilde{L}_{g_2^{-1}} \circ \tilde{L}_{g_1} \circ \phi\ (X')$$

for every $X' \in W_1'$, it follows that

$$\phi^{-1} \circ \tilde{L}_{g_2^{-1}} \circ \tilde{L}_{g_1} \circ \phi\ |_{W_1'} = \alpha\ ,$$

which shows that $y_{g_2} \circ y_{g_1}^{-1}$ is smooth. Hence G/H is a smooth manifold, and the projection $\pi : G \to G/H$ is smooth since

$$\phi^{-1} \circ \tilde{L}_{g^{-1}} \circ \pi\,(g') = \rho \circ \psi^{-1} \circ L_{g^{-1}}\,(g')$$

for every $g \in G$ and $g' \in V_g$.

For each $g \in G$, we let $s_g : U_g \to G$ be the smooth map given by

$$s_g = i_g \circ L_g \circ \psi \circ \phi^{-1} \circ \tilde{L}_{g^{-1}}\ ,$$

where $i_g : V_g \to G$ is the inclusion map. Then we have that $\pi \circ s_g = \tilde{i}_g$, which proves the last part of the theorem. Suppose that $f : G/H \to M$ is a map into a smooth manifold M so that $f \circ \pi$ is smooth. Then f is also smooth since

$$f\,|_{U_g} = f \circ \pi \circ s_g$$

for every $g \in G$.

To prove the uniqueness of the manifold structure, let $(G/H)_1$ and $(G/H)_2$ denote G/H with two manifold structures satisfying the first part of the theorem. Then we immediately see that the identity map $id : (G/H)_1 \to (G/H)_2$ and its inverse are smooth, since $\pi \circ id = \pi$ and $\pi \circ id^{-1} = \pi$. This shows that the two manifold structures must coincide. □

8.45 Proposition Let $\phi : G \to H$ be a Lie group homomorphism. Then $\ker\phi$ is a closed submanifold and a Lie subgroup of G with Lie algebra $\ker\phi_*$.

PROOF : Since $\ker\phi$ is a closed subgroup of G, it follows from Theorem 8.43 that it is a closed submanifold and a Lie subgroup of G. Let $\mathfrak{k}$ be its Lie algebra, and suppose that $X \in \mathfrak{g}$. Then it follows from Propositions 8.40 and 8.41 that $X \in \mathfrak{k}$ if and only if

$$exp\,(t\,\phi_*(X)) = \phi \circ exp\,(tX) = e$$

for every $t \in \mathbf{R}$, which is equivalent to $\phi_*(X) = 0$. Hence we have that $\mathfrak{k} = \ker\phi_*$.
□

MATRIX GROUPS

8.46 Let V be a finite dimensional complex vector space, and let $\mathscr{B} = \{v_1, ..., v_n\}$ be a basis for V. Then we have a linear isomorphism $\psi : V \to \mathbf{C}^n$ given by

$$\psi(v) = a \quad \text{when} \quad v = \sum_{j=1}^{n} a_j v_j\,,$$

where a is written as a column vector. We define a norm in V by letting $\|v\| = \|a\|$ so that ψ is an isometric isomorphism, using the norm in $\mathbf{C}^n$ given by

$$\|a\| = \left(\sum_{j=1}^{n} |a_j|^2\right)^{1/2}.$$

Let $\phi : \mathfrak{gl}\,(V) \to \mathfrak{gl}\,(n, \mathbf{C})$ be the linear isomorphism defined by

$$\phi\,(F) = m_{\mathscr{B}}^{\mathscr{B}}(F)$$

for $F \in \mathfrak{gl}(V)$. If $\phi(F) = A$, then $\psi \circ F \circ \psi^{-1}$ is the linear map $T \in \mathfrak{gl}(\mathbf{C}^n)$ given by $T(a) = A\,a$ for $a \in \mathbf{C}^n$, so that

$$\|F\| = \|T\| = \max\{\|T(a)\| \mid \|a\| \leq 1\}$$

$$= \max\left\{\left(\sum_{i=1}^{n}\left|\sum_{j=1}^{n} A_{ij}\,a_j\right|^2\right)^{1/2} \;\middle|\; \sum_{j=1}^{n}|a_j|^2 \leq 1\right\} \tag{1}$$

by Propositions 13.120 and 13.123 in the appendix. We also define a norm in $\mathfrak{gl}(n,\mathbf{C})$ by letting $\|A\| = \|F\|$, so that ϕ is an isometric isomorphism. If $\mathscr{C} = \{e_1,...,e_n\}$ is the standard basis for $\mathbf{C}^n$ and A_s is the s-th column in A, we see that

$$|A_{rs}| \leq \|A_s\| \leq \|A\|$$

for all r and s by taking $a = e_s$ in formula (1).

8.47 We have an inner product in $\mathbf{C}^n$ given by

$$<a,b> = a^t\,\bar{b} = \sum_{j=1}^{n} a_j\,\bar{b}_j$$

for column vectors a and b in $\mathbf{C}^n$, so that $\|a\| = <a,a>^{1/2}$. The *adjoint* of a matrix $A \in \mathfrak{gl}(n,\mathbf{C})$ is the unique matrix $A^* \in \mathfrak{gl}(n,\mathbf{C})$ which satisfies the relation

$$<A\,a,b> = <a,A^*\,b>$$

for all $a,b \in \mathbf{C}^n$. Since

$$<A\,a,b> = (A\,a)^t\,\bar{b} = a^t A^t\,\bar{b} = a^t\,\overline{\bar{A}^t\,b} = <a,\bar{A}^t\,b>,$$

we have that $A^* = \bar{A}^t$.

Using the Cauchy Schwartz inequality we see that

$$\|A\,a\|^2 = <A\,a,A\,a> = <a,A^*A\,a> \leq \|A^*A\|\,\|a\|^2$$

so that

$$\|A\|^2 \leq \|A^*A\| \leq \|A^*\|\,\|A\|.$$

Hence we have that

$$\|A\| \leq \|A^*\| \leq \|(A^*)^*\| = \|A\|$$

which implies that $\|A^*\| = \|A\|$.

A matrix $A \in \mathfrak{gl}(n,\mathbf{C})$ is said to be *unitary* if it satisfies the relation

$$<A\,a,A\,b> = <a,b>$$

for all $a,b \in \mathbf{C}^n$, which is equivalent to assertion that $A^{-1} = A^*$.

8.48 Definition An $n \times n$-matrix A is said to be *upper triangular* if $A_{ij} = 0$ for $1 \leq j < i \leq n$, so that all its elements below the diagonal are zero.

8.49 Remark We see that the product AB of two upper triangular $n \times n$-matrices A and B is an upper triangular $n \times n$-matrix with diagonal elements

$$(AB)_{ii} = A_{ii} B_{ii}$$

for $i = 1, ..., n$.

8.50 Proposition Let V be an n-dimensional complex vector space, and let $T \in \mathfrak{gl}(V)$. Then there is a basis $\mathscr{B} = \{v_1, ..., v_n\}$ for V so that $m_{\mathscr{B}}^{\mathscr{B}}(T)$ is an upper triangular matrix.

PROOF: We prove the proposition by induction on n. It is obviously true for $n = 1$, and we will prove it for $n > 1$ assuming that it is true for every complex vector space V of dimension $n - 1$.

Let v_1 be an eigenvector for T with eigenvalue a_{11}, and extend $\{v_1\}$ to a basis $\{v_1, w_2, ... w_n\}$ for V. Let $V_1 = L(v_1)$ and $W = L(w_2, ... w_n)$ so that $V = V_1 \oplus W$, and consider the linear map $S = \pi \circ T\,|_W$, where $\pi : V \to W$ is the projection on W. As $\dim(W) = n - 1$, there is by the induction hypothesis a basis $\mathscr{C} = \{v_2, ..., v_n\}$ for W so that the matrix $m_{\mathscr{C}}^{\mathscr{C}}(S)$ is upper triangular, i.e.,

$$S(v_j) = \sum_{i=2}^{j} a_{ij}\, v_i$$

for $2 \leq j \leq n$ and complex numbers a_{ij}. Since

$$T(v_j) = a_{1j}\, v_1 + S(v_j)$$

for $2 \leq j \leq n$ and complex numbers a_{1j}, this completes the proof of the proposition.

□

8.51 Corollary If $A \in \mathfrak{gl}(n, \mathbf{C})$, there is a matrix $B \in Gl(n, \mathbf{C})$ so that BAB^{-1} is upper triangular.

PROOF: Let $T \in \mathfrak{gl}(\mathbf{C}^n)$ be the linear map so that $m_{\mathscr{E}}^{\mathscr{E}}(T) = A$, where $\mathscr{E}$ is the standard basis for $\mathbf{C}^n$, and choose a basis $\mathscr{B}$ for $\mathbf{C}^n$ so that $m_{\mathscr{B}}^{\mathscr{B}}(T)$ is an upper triangular matrix C. If we let $B = m_{\mathscr{B}}^{\mathscr{E}}(id)$, where $id : \mathbf{C}^n \to \mathbf{C}^n$ is the identity map, we have that

$$BAB^{-1} = m_{\mathscr{B}}^{\mathscr{E}}(id)\; m_{\mathscr{E}}^{\mathscr{E}}(T)\; m_{\mathscr{E}}^{\mathscr{B}}(id) \;=\; m_{\mathscr{B}}^{\mathscr{B}}(T) \;=\; C\,.$$

□

8.52 Proposition The exponential map $exp : \mathfrak{gl}(n, \mathbf{C}) \to Gl(n, \mathbf{C})$ is given by

$$exp(A) = \sum_{k=0}^{\infty} \frac{A^k}{k!}, \tag{1}$$

and we have that

$$\det(exp(A)) = e^{tr(A)} \tag{2}$$

for $A \in \mathfrak{gl}(n, \mathbf{C})$.

PROOF: By 8.46 and Proposition 13.126 in the appendix we have a continuous group homomorphism $\phi : \mathbf{R} \to Gl(n, \mathbf{C})$ given by

$$\phi(t) = \sum_{k=0}^{\infty} \frac{(tA)^k}{k!}$$

for $t \in \mathbf{R}$. Using the local chart $(z \circ \alpha, Gl(n, \mathbf{C}))$ on $Gl(n, \mathbf{C})$ defined in Example 8.10 (c), we see that

$$z^i \circ \phi(t) = \sum_{k=0}^{\infty} \frac{z^i(A^k)\, t^k}{k!}$$

is a power series with infinite radius of convergence for $i = 1, ..., 2n^2$. By differentiating the series term by term at $t = 0$, we have that $(z^i \circ \phi)'(0) = z^i(A)$ for each i. This shows that ϕ is a one-parameter subgroup of $Gl(n, \mathbf{C})$ with $\phi'(0) = A$, so that

$$exp(A) = \phi(1) = \sum_{k=0}^{\infty} \frac{A^k}{k!}$$

by Definition 8.30.

Using Corollary 8.51 we now choose a $B \in Gl(n, \mathbf{C})$ so that $BAB^{-1} = C$ is an upper triangular matrix. Then we have that

$$\det(exp(A)) = \det(B\, exp(A)\, B^{-1}) = \det(exp(BAB^{-1}))$$

$$= \prod_{i=1}^{n} e^{C_{ii}} = e^{tr(C)} = e^{tr(A)}$$

by Proposition 13.126 in the appendix, Remark 8.49 and Corollary 4.65. □

8.53 Remark By Example 8.19 and Proposition 8.40 the exponential map $exp : \mathfrak{gl}(n, \mathbf{R}) \to Gl(n, \mathbf{R})$ is induced by $exp : \mathfrak{gl}(n, \mathbf{C}) \to Gl(n, \mathbf{C})$ and satisfies the same two formulas for every $A \in \mathfrak{gl}(n, \mathbf{R})$.

8.54 Corollary Let V be a finite dimensional real or complex vector space. Then the exponential map $exp : \mathfrak{gl}(V) \to Gl(V)$ is given by

$$exp(F) = \sum_{k=0}^{\infty} \frac{F^k}{k!}, \tag{1}$$

and we have that

$$\det(exp\,(F)) = e^{tr\,(F)} \tag{2}$$

for $F \in \mathfrak{gl}\,(V)$.

PROOF : If $\mathscr{B} = \{v_1, ..., v_n\}$ is a basis for V, we have an isometric algebra isomorphism

$$\phi : \mathfrak{gl}\,(V) \to \mathfrak{gl}\,(n, K)$$

where K is either $\mathbf{R}$ or $\mathbf{C}$, which is given by

$$\phi\,(F) = m_{\mathscr{B}}^{\mathscr{B}}(F)$$

for $F \in \mathfrak{gl}\,(V)$. Moreover, ϕ induces a Lie group isomorphism

$$\psi : Gl\,(V) \to Gl\,(n, K)$$

so that $\psi_* = \phi$ when we identify the Lie algebras of $Gl\,(V)$ and $Gl\,(n, K)$ with $\mathfrak{gl}\,(V)$ and $\mathfrak{gl}\,(n, K)$ as described in Example 8.10. Using Proposition 8.40 we therefore have a commutative diagram

$$\begin{array}{ccc} \mathfrak{gl}\,(V) & \xrightarrow{\ \phi\ } & \mathfrak{gl}\,(n, K) \\ {\scriptstyle exp}\downarrow & & \downarrow{\scriptstyle exp} \\ Gl\,(V) & \xrightarrow{\ \psi\ } & Gl\,(n, K) \end{array}$$

which implies formula (1) and (2) since

$$\psi \circ exp\,(F) = exp \circ \phi\,(F) = \sum_{k=0}^{\infty} \frac{\phi\,(F)^k}{k!} = \phi\left(\sum_{k=0}^{\infty} \frac{F^k}{k!}\right)$$

and

$$\det(exp\,(F)) = \det(\psi \circ exp\,(F)) = \det(exp \circ \phi\,(F)) = e^{tr\,(\phi\,(F))} = e^{tr\,(F)}$$

for $F \in \mathfrak{gl}\,(V)$. □

8.55 Examples

(a) Consider the subgroup

$$Sl(n, \mathbf{R}) = \{A \in Gl\,(n, \mathbf{R}) \,|\, \det(A) = 1\}$$

of matrices in $Gl\,(n, \mathbf{R})$ with determinant 1, called the *special linear group*, and the subspace

$$\mathfrak{sl}(n, \mathbf{R}) = \{A \in \mathfrak{gl}\,(n, \mathbf{R}) | tr\,(A) = 0\}$$

of matrices in $\mathfrak{gl}(n,\mathbf{R})$ with trace 0. If U is an open neighbourhood of $0 \in \mathfrak{gl}(n,\mathbf{R})$ which is mapped diffeomorphically by exp onto an open neighbourhood V of $I \in Gl(n,\mathbf{R})$, we have that

$$exp\,(U \cap \mathfrak{sl}(n,\mathbf{R})) = V \cap Sl(n,\mathbf{R})$$

by Remark 8.53. Using Proposition 8.42 we therefore conclude that $Sl(n,\mathbf{R})$ is a submanifold and a closed Lie subgroup of $Gl(n,\mathbf{R})$, having the Lie subalgebra $\mathfrak{sl}(n,\mathbf{R})$ of $\mathfrak{gl}(n,\mathbf{R})$ as its Lie algebra.

(b) Consider the subgroup

$$Sl(n,\mathbf{C}) = \{A \in Gl(n,\mathbf{C}) \,|\, \det(A) = 1\}$$

of matrices in $Gl(n,\mathbf{C})$ with determinant 1, called the *complex special linear group*, and the subspace

$$\mathfrak{sl}(n,\mathbf{C}) = \{A \in \mathfrak{gl}(n,\mathbf{C}) \,|\, tr(A) = 0\}$$

of matrices in $\mathfrak{gl}(n,\mathbf{C})$ with trace 0. Let W be an open neighbourhood of $0 \in \mathfrak{gl}(n,\mathbf{C})$ which is mapped diffeomorphically by exp onto an open neighbourhood of $I \in Gl(n,\mathbf{C})$, and let

$$U = \{A \in W \,|\, |tr(A)| < 2\pi\}$$

and $V = exp\,(U)$. We contend that

$$exp\,(U \cap \mathfrak{sl}(n,\mathbf{C})) = V \cap Sl(n,\mathbf{C})$$

which by Proposition 8.42 implies that $Sl(n,\mathbf{C})$ is a submanifold and a closed Lie subgroup of $Gl(n,\mathbf{C})$, having the Lie subalgebra $\mathfrak{sl}(n,\mathbf{C})$ of $\mathfrak{gl}(n,\mathbf{C})$ as its Lie algebra. To prove the assertion, suppose that $A \in U \cap \mathfrak{sl}(n,\mathbf{C})$. Then we have that

$$\det(\,exp\,(A)) = e^{tr(A)} = 1$$

which implies that $exp\,(A) \in V \cap Sl(n,\mathbf{C})$. Conversely, assuming that $A \in U$ with $exp\,(A) \in V \cap Sl(n,\mathbf{C})$, we have that $|tr(A)| < 2\pi$ and

$$e^{tr(A)} = \det(\,exp\,(A)) = 1\,.$$

This shows that $tr(A) = 0$ so that $A \in U \cap \mathfrak{sl}(n,\mathbf{C})$, and completes the proof of the assertion.

8.56 Proposition Let $\psi : \mathfrak{gl}(n,K) \to \mathfrak{gl}(n,K)$, where K is either $\mathbf{R}$ or $\mathbf{C}$, be a bounded linear map with $\psi^{-1} = \psi$ and $\psi(AB) = \psi(A)\,\psi(B)$ whenever $AB = BA$. Then

$$G = \{A \in Gl(n,K) \,|\, A^{-1} = \psi(A)\}$$

is a submanifold and a closed Lie subgroup of $Gl\,(n,K)$, having the Lie subalgebra

$$\mathfrak{g} = \{A \in \mathfrak{gl}\,(n,K) \,|\, \psi(A) = -A\}$$

of $\mathfrak{gl}\,(n,K)$ as its Lie algebra. There is an open neighbourhood U of $0 \in \mathfrak{gl}\,(n,K)$ which is mapped diffeomorphically by exp onto an open neighbourhood V of $I \in Gl\,(n,K)$ so that

$$exp\,(U \cap \mathfrak{g}) = V \cap G\ . \tag{1}$$

Suppose in addition that $|tr\,(\psi(A))| = |tr\,(A)|$ for all $A \in \mathfrak{gl}\,(n,K)$ in the case where $K = \mathbf{C}$. Then we have that $G \cap Sl(n,K)$ is a submanifold and a closed Lie subgroup of $Gl\,(n,K)$ with Lie algebra $\mathfrak{g} \cap \mathfrak{sl}(n,K)$, and the above neighbourhoods U and V can be chosen so that

$$exp\,(U \cap \mathfrak{g} \cap \mathfrak{sl}(n,K)) = V \cap G \cap Sl(n,K)\ . \tag{2}$$

PROOF : We first observe that

$$\psi \circ exp = exp \circ \psi$$

which follows from the relation

$$\psi\left(\sum_{k=0}^{n} \frac{A^k}{k!}\right) = \sum_{k=0}^{n} \frac{\psi(A)^k}{k!}$$

when $n \to \infty$. Let W be an open neighbourhood of $0 \in \mathfrak{gl}\,(n,K)$ which is mapped diffeomorphically by exp onto an open neighbourhood of $I \in Gl\,(n,K)$, and let

$$U = W \cap \psi(W) \cap (-W) \cap (-\psi(W)) \tag{3}$$

and $V = exp\,(U)$. Suppose that $A \in U \cap \mathfrak{g}$. Then we have that

$$\psi \circ exp\,(A) = exp \circ \psi(A) = exp\,(-A) = exp\,(A)^{-1}$$

which implies that $exp\,(A) \in V \cap G$. Conversely, assuming that $A \in U$ with $exp\,(A) \in V \cap G$, we have that $\psi(A), -A \in U$ and

$$exp \circ \psi(A) = \psi \circ exp\,(A) = exp\,(A)^{-1} = exp\,(-A)\ .$$

This shows that $\psi(A) = -A$ so that $A \in U \cap \mathfrak{g}$, and completes the proof of formula (1). Formula (2) follows by the same arguments combined with Example 8.55, just replacing (3) by

$$U = W \cap \psi(W) \cap (-W) \cap (-\psi(W)) \cap \{A \in \mathfrak{gl}\,(n,\mathbf{C}) \,|\, |tr\,(A)| < 2\pi\} \tag{3'}$$

in the case where $K = \mathbf{C}$. The proposition now follows from Proposition 8.42. □

8.57 Using Proposition 8.56 we obtain the following closed Lie subgroups of $Gl(n,\mathbf{R})$ or $Gl(n,\mathbf{C})$ in the relative topology

(1) The *unitary group* $U(n) = \{A \in Gl(n,\mathbf{C}) | A^{-1} = A^*\}$
(2) The *orthogonal group* $O(n) = \{A \in Gl(n,\mathbf{R}) | A^{-1} = A^t\}$
(3) The *complex orthogonal group* $O(n,\mathbf{C}) = \{A \in Gl(n,\mathbf{C}) | A^{-1} = A^t\}$

(4) The *special unitary group* $SU(n) = U(n) \cap Sl(n,\mathbf{C})$
(5) The *special orthogonal group* $SO(n) = O(n) \cap Sl(n,\mathbf{R})$
(6) The *special complex orthogonal group* $SO(n,\mathbf{C}) = O(n,\mathbf{C}) \cap Sl(n,\mathbf{C})$

with Lie algebras consisting of

(1') *skew hermitian* matrices $\mathfrak{u}(n) = \{A \in \mathfrak{gl}(n,\mathbf{C}) | A^* = -A\}$
(2') *skew symmetric* real matrices $\mathfrak{o}(n) = \{A \in \mathfrak{gl}(n,\mathbf{R}) | A^t = -A\}$
(3') *skew symmetric* complex matrices $\mathfrak{o}(n,\mathbf{C}) = \{A \in \mathfrak{gl}(n,\mathbf{C}) | A^t = -A\}$

(4') matrices in $\mathfrak{u}(n)$ with *trace* 0 $\mathfrak{su}(n) = \mathfrak{u}(n) \cap \mathfrak{sl}(n,\mathbf{C})$
(5') matrices in $\mathfrak{o}(n)$ $\mathfrak{so}(n) = \mathfrak{o}(n)$
(6') matrices in $\mathfrak{o}(n,\mathbf{C})$ $\mathfrak{so}(n,\mathbf{C}) = \mathfrak{o}(n,\mathbf{C})$

respectively.

8.58 Examples

(a) Let S be the signature matrix for the metric g^n_ε in $\mathbf{R}^n$ defined in Definition 7.15. Then we have a closed Lie subgroup

$$O_\varepsilon(n) = \{A \in Gl(n,\mathbf{R}) | A^{-1} = SA^tS\}$$

of $Gl(n,\mathbf{R})$, called the *pseudo-orthogonal group of signature* ε, with the Lie subalgebra

$$\mathfrak{o}_\varepsilon(n) = \{A \in \mathfrak{gl}(n,\mathbf{R}) | SA^tS = -A\}$$

of $\mathfrak{gl}(n,\mathbf{R})$ as its Lie algebra by Proposition 8.56. If the index of the metric g^n_ε is k and ε is its canonical signature, then $O_\varepsilon(n)$ and $\mathfrak{o}_\varepsilon(n)$ are also denoted by $O_k(n)$ and $\mathfrak{o}_k(n)$, or by $O(k,n-k)$ and $\mathfrak{o}(k,n-k)$, respectively. $\mathfrak{o}_k(n)$ consists of matrices of the form

$$\begin{pmatrix} a & x \\ x^t & b \end{pmatrix}$$

where $a \in \mathfrak{o}(k)$, $b \in \mathfrak{o}(n-k)$ and x is an arbitrary $k \times (n-k)$ matrix. Indeed, if

$$A = \begin{pmatrix} a & x \\ y & b \end{pmatrix},$$

we have that

$$SA^tS = \begin{pmatrix} -I_k & 0 \\ 0 & I_{n-k} \end{pmatrix} \begin{pmatrix} a^t & y^t \\ x^t & b^t \end{pmatrix} \begin{pmatrix} -I_k & 0 \\ 0 & I_{n-k} \end{pmatrix} = \begin{pmatrix} a^t & -y^t \\ -x^t & b^t \end{pmatrix}$$

which equals

$$-A = \begin{pmatrix} -a & -x \\ -y & -b \end{pmatrix}$$

if and only if $a^t = -a$, $b^t = -b$ and $y = x^t$.

(b) Let V be a finite dimensional real vector space with a metric g of index k. If $\mathscr{B} = \{v_1, ..., v_n\}$ is an orthonormal basis for V with $g(v_i, v_i) = \varepsilon_i$ for $i = 1, .., n$, the linear isomorphism $\psi : V \to \mathbf{R}^n$ given by

$$\psi(v) = a \quad \text{when} \quad v = \sum_{j=1}^{n} a_j v_j \,,$$

is an isometry from V to the pseudo-Euclidean space $\mathbf{R}^n_\varepsilon$. Consider the subgroup

$$O_\varepsilon(V) = \{F \in Gl\,(V) | F^* g = g\}$$

of $Gl(V)$ consisting of the linear isometries in V. If $\phi : \mathfrak{gl}\,(V) \to \mathfrak{gl}\,(n, \mathbf{R})$ is the linear isomorphism defined by

$$\phi\,(F) = m^{\mathscr{B}}_{\mathscr{B}}(F)$$

for $F \in \mathfrak{gl}\,(V)$, we have that $O_\varepsilon(V) = \phi^{-1}(O_\varepsilon(n))$. Indeed, if $\phi\,(F) = A$, then $F^* g = g$ if and only if

$$g^n_\varepsilon(a,b) = g^n_\varepsilon(Aa, Ab) = a^t A^t SA\, b = g^n_\varepsilon(a, (SA^tS)Ab)$$

for every $a, b \in \mathbf{R}^n$, which is equivalent to $A^{-1} = SA^tS$. Hence $O_\varepsilon(V)$ is a closed Lie subgroup of $Gl(V)$ in the relative topology, with the Lie subalgebra

$$\mathfrak{o}_\varepsilon(V) = \phi^{-1}(\mathfrak{o}_\varepsilon(n))$$

of $\mathfrak{gl}\,(V)$ as its Lie algebra. We see that $F \in \mathfrak{o}_\varepsilon(V)$ if and only if

$$g(F(v), w) = -\,g(v, F(w)) \tag{1}$$

for every $v, w \in V$, which follows from the fact that $A = \phi\,(F) \in \mathfrak{o}_\varepsilon(n)$ if and only if

$$g^n_\varepsilon(Aa, b) = a^t A^t S\, b = g^n_\varepsilon(a, (SA^tS)\, b) = -\,g^n_\varepsilon(a, Ab)$$

for every $a, b \in \mathbf{R}^n$. If ε is the canonical signature of g, then $O_\varepsilon(V)$ and $\mathfrak{o}_\varepsilon(V)$ are also denoted by $O_k(V)$ and $\mathfrak{o}_k(V)$, respectively. If the index is 0, they are denoted simply by $O(V)$ and $\mathfrak{o}(V)$.

8.59 Proposition For each matrix $A \in SO(3)$ there is a unit column vector $w \in \mathbf{R}^3$ with $Aw = w$. If $F : \mathbf{R}^3 \to \mathbf{R}^3$ is the linear map having A as its standard matrix, there is a positively oriented orthonormal basis $\mathscr{B} = \{u, v, w\}$ in $\mathbf{R}^3$ so that

$$m_{\mathscr{B}}^{\mathscr{B}}(F) = \begin{pmatrix} \cos\theta & -\sin\theta & 0 \\ \sin\theta & \cos\theta & 0 \\ 0 & 0 & 1 \end{pmatrix}$$

where $0 \leq \theta < 2\pi$. The linear map F, also denoted by $R_w(\theta)$, is a rotation by an angle θ about the line through the origin with direction vector w. We have that

$$R_w(\theta) = R_{-w}(2\pi - \theta)$$

when $0 < \theta < 2\pi$, and $R_w(0) = id$ for every unit vector w.

PROOF : Since

$$\begin{aligned} \det(I - A) &= \det(I - A^t) = \det(I - A^{-1}) \\ &= \det(A^{-1})\,\det(A - I) = -\det(I - A) \end{aligned}$$

so that $\det(I - A) = 0$, it follows that 1 is an eigenvalue of A. Let w be an eigenvector of A with eigenvalue 1 and $\|w\| = 1$, and choose vectors u and v so that $\mathscr{B} = \{u, v, w\}$ is a positively oriented orthonormal basis in $\mathbf{R}^3$. Then it follows from Example 8.58 (b) that

$$m_{\mathscr{B}}^{\mathscr{B}}(F) = \begin{pmatrix} a_1 & a_2 & 0 \\ b_1 & b_2 & 0 \\ 0 & 0 & 1 \end{pmatrix}$$

where

$$a_1^2 + b_1^2 = 1 \quad , \quad a_2^2 + b_2^2 = 1 \quad \text{and} \quad a_1 b_2 - b_1 a_2 = 1$$

so that

$$(a_1 - b_2)^2 + (b_1 + a_2)^2 = (a_1^2 + b_1^2) + (a_2^2 + b_2^2) - 2(a_1 b_2 - b_1 a_2) = 0\,,$$

which implies that

$$a_1 = b_2 = \cos\theta \quad \text{and} \quad b_1 = -a_2 = \sin\theta \quad \text{where} \quad 0 \leq \theta < 2\pi\,.$$

□

8.60 Example The 3×3 matrices

$$\mathscr{L}_1 = \begin{pmatrix} 0 & 0 & 0 \\ 0 & 0 & -1 \\ 0 & 1 & 0 \end{pmatrix}, \quad \mathscr{L}_2 = \begin{pmatrix} 0 & 0 & 1 \\ 0 & 0 & 0 \\ -1 & 0 & 0 \end{pmatrix} \quad \text{and} \quad \mathscr{L}_3 = \begin{pmatrix} 0 & -1 & 0 \\ 1 & 0 & 0 \\ 0 & 0 & 0 \end{pmatrix}$$

form a basis for the Lie algebra $\mathfrak{so}(3)$ of skew symmetric matrices, satisfying the commutation relations

$$[\mathscr{L}_1, \mathscr{L}_2] = \mathscr{L}_3 \quad , \quad [\mathscr{L}_2, \mathscr{L}_3] = \mathscr{L}_1 \quad \text{and} \quad [\mathscr{L}_3, \mathscr{L}_1] = \mathscr{L}_2 \, .$$

The corresponding one-parameter subgroups $\phi_{\mathscr{L}_i} : \mathbf{R} \to SO(3)$ are given by $\phi_{\mathscr{L}_i}(\theta) = exp\,(\theta \mathscr{L}_i)$ for $\theta \in \mathbf{R}$ and $i = 1,2,3$, where

$$exp\,(\theta \mathscr{L}_1) = \begin{pmatrix} 1 & 0 & 0 \\ 0 & \cos\theta & -\sin\theta \\ 0 & \sin\theta & \cos\theta \end{pmatrix} , \quad exp\,(\theta \mathscr{L}_2) = \begin{pmatrix} \cos\theta & 0 & \sin\theta \\ 0 & 1 & 0 \\ -\sin\theta & 0 & \cos\theta \end{pmatrix}$$

and

$$exp\,(\theta \mathscr{L}_3) = \begin{pmatrix} \cos\theta & -\sin\theta & 0 \\ \sin\theta & \cos\theta & 0 \\ 0 & 0 & 1 \end{pmatrix}$$

are the standard matrices for rotations by the angle θ about the $x-$, $y-$ and $z-$axis, respectively.

8.61 Let g be the Lorentz metric in $\mathbf{R}^4$ given by

$$g(a,b) = a^t S\, b$$

for column vectors a and b in $\mathbf{R}^4$, where

$$S = \begin{pmatrix} -1 & 0 & 0 & 0 \\ 0 & 1 & 0 & 0 \\ 0 & 0 & 1 & 0 \\ 0 & 0 & 0 & 1 \end{pmatrix}$$

is the signature matrix. The vector space $\mathbf{R}^4$ equipped with this metric is called the *Minkowski space*. A vector $a \in \mathbf{R}^4$ is said to be *spacelike*, *lightlike* or *timelike* if $g(a,a) > 0$, $g(a,a) = 0$ or $g(a,a) < 0$, respectively. We denote the components of a with respect to the standard basis $\mathscr{E} = \{e_0, e_1, e_2, e_3\}$ for $\mathbf{R}^4$ by a_0, a_1, a_2 and a_3. The pseudo-orthogonal group

$$O_1(4) = \{A \in Gl\,(4,\mathbf{R}) | A^t S A = S\}$$

described in Example 8.58 (a) is called the *Lorentz group* and is also denoted by $\mathscr{L}$. If $A \in \mathscr{L}$, then $\det(A)^2 = 1$, and

$$-A_{00}^2 + \sum_{r=1}^{3} A_{r0}^2 = -1$$

so that

$$A_{00}^2 = 1 + \sum_{r=1}^{3} A_{r0}^2 \geq 1 \, .$$

A matrix $A \in \mathscr{L}$ is called *forward-timelike* if $A_{00} \geq 1$ and *backward-timelike* if $A_{00} \leq -1$. Hence $\mathscr{L}$ is the union of four disjoint subsets which are both open and closed :

$$\mathscr{L}^{\uparrow +} = \{A \in \mathscr{L} | A_{00} \geq 1, \det(A) = 1\}, \quad \mathscr{L}^{\uparrow -} = \{A \in \mathscr{L} | A_{00} \geq 1, \det(A) = -1\},$$
$$\mathscr{L}^{\downarrow +} = \{A \in \mathscr{L} | A_{00} \leq -1, \det(A) = 1\}, \mathscr{L}^{\downarrow -} = \{A \in \mathscr{L} | A_{00} \leq -1, \det(A) = -1\}.$$

For every $A \in \mathscr{L}$ we have that $A^t \in \mathscr{L}$ so that

$$\sum_{r=1}^{3} A_{r0}^2 = \sum_{r=1}^{3} A_{0r}^2 = A_{00}^2 - 1 \tag{1}$$

If $A, B \in \mathscr{L}$ are forward-timelike, then so is AB since

$$\begin{aligned}\left| \sum_{r=1}^{3} A_{0r} B_{r0} \right| &\leq \left(\sum_{r=1}^{3} A_{0r}^2 \right)^{1/2} \left(\sum_{r=1}^{3} B_{r0}^2 \right)^{1/2} = (A_{00}^2 - 1)^{1/2} (B_{00}^2 - 1)^{1/2} \\ &= \{(A_{00} - 1)(B_{00} + 1)\}^{1/2} \{(A_{00} + 1)(B_{00} - 1)\}^{1/2} \\ &= \{(A_{00} B_{00} - 1)^2 - (A_{00} - B_{00})^2\}^{1/2} \leq A_{00} B_{00} - 1\end{aligned}$$

which implies that

$$(AB)_{00} = \sum_{r=0}^{3} A_{0r} B_{r0} \geq 1 .$$

Furthermore, if $A \in \mathscr{L}$ is forward-timelike, then so is A^{-1} since

$$(A^{-1})_{00} = (SA^t S)_{00} = A_{00} .$$

This shows that $\mathscr{L}^{\uparrow +}$ is a closed subgroup and therefore a Lie subgroup of $\mathscr{L}$, called the *proper Lorentz group*. The matrices S, $P = -S$ and $-I$ are the standard matrices for *time inversion*, *space inversion* (or *parity transformation*) and *total inversion*, respectively, and we have that

$$\mathscr{L}^{\uparrow -} = P \mathscr{L}^{\uparrow +}, \quad \mathscr{L}^{\downarrow +} = (-I) \mathscr{L}^{\uparrow +} \quad \text{and} \quad \mathscr{L}^{\downarrow -} = S \mathscr{L}^{\uparrow +} .$$

8.62 We have an injective Lie algebra homomorphism $\widetilde{\alpha} : \mathfrak{gl}(3, \mathbf{R}) \to \mathfrak{gl}(4, \mathbf{R})$ given by

$$\widetilde{\alpha}(A) = \begin{pmatrix} 1 & 0 \\ 0 & A \end{pmatrix}$$

for $A \in \mathfrak{gl}(3, \mathbf{R})$, which induces a Lie group homomorphism $\alpha : SO(3) \to \mathscr{L}^{\uparrow +}$ and a corresponding Lie algebra homomorphism $\alpha_* : \mathfrak{so}(3) \to \mathfrak{o}_1(4)$. By

Example 8.58 (a) the Lie algebra $\mathfrak{o}_1(4)$ consists of matrices of the form

$$\begin{pmatrix} 0 & b_1 & b_2 & b_3 \\ b_1 & 0 & -a_3 & a_2 \\ b_2 & a_3 & 0 & -a_1 \\ b_3 & -a_2 & a_1 & 0 \end{pmatrix}$$

for arbitrary real parameters a_i and b_i. The matrices $\mathscr{R}_i = \alpha_*(\mathscr{L}_i)$ for $i = 1,2,3$ are obtained by choosing $a_i = 1$ and all other parameters equal to 0. Let $\mathscr{B}_i$ for $i = 1,2,3$ be the matrices obtained by choosing $b_i = 1$ and all other parameters equal to 0. These six matrices form a basis for the Lie algebra $\mathfrak{o}_1(4)$, satisfying the commutation relations

$$[\mathscr{R}_{\sigma(1)}, \mathscr{R}_{\sigma(2)}] = \varepsilon(\sigma)\mathscr{R}_{\sigma(3)}\ , \quad [\mathscr{B}_{\sigma(1)}, \mathscr{B}_{\sigma(2)}] = -\varepsilon(\sigma)\mathscr{R}_{\sigma(3)}$$

and

$$[\mathscr{B}_{\sigma(1)}, \mathscr{R}_{\sigma(2)}] = \varepsilon(\sigma)\mathscr{B}_{\sigma(3)}$$

for every permutation $\sigma \in S_3$.

The corresponding one-parameter subgroups $\phi_{\mathscr{R}_i} : \mathbf{R} \to \mathscr{L}^{\uparrow +}$ and $\phi_{\mathscr{B}_i} : \mathbf{R} \to \mathscr{L}^{\uparrow +}$ are given by $\phi_{\mathscr{R}_i} = \alpha \circ \phi_{\mathscr{L}_i}$ for $i = 1,2,3$, and by $\phi_{\mathscr{B}_i}(\chi) = exp(\chi\mathscr{B}_i)$ for $\chi \in \mathbf{R}$ and $i = 1,2,3$, where

$$exp(\chi\mathscr{B}_1) = \begin{pmatrix} \cosh\chi & \sinh\chi & 0 & 0 \\ \sinh\chi & \cosh\chi & 0 & 0 \\ 0 & 0 & 1 & 0 \\ 0 & 0 & 0 & 1 \end{pmatrix}, \ exp(\chi\mathscr{B}_2) = \begin{pmatrix} \cosh\chi & 0 & \sinh\chi & 0 \\ 0 & 1 & 0 & 0 \\ \sinh\chi & 0 & \cosh\chi & 0 \\ 0 & 0 & 0 & 1 \end{pmatrix}$$

and

$$exp(\chi\mathscr{B}_3) = \begin{pmatrix} \cosh\chi & 0 & 0 & \sinh\chi \\ 0 & 1 & 0 & 0 \\ 0 & 0 & 1 & 0 \\ \sinh\chi & 0 & 0 & \cosh\chi \end{pmatrix}$$

are the standard matrices for *velocity transformations* or *boosts* with velocity parameter χ along the $x-$, $y-$ and $z-$axis, respectively.

8.63 Remark By formula (1) in 8.61 we see that a matrix $A \in \mathscr{L}^{\uparrow +}$ belongs to $\alpha(SO(3))$ if and only if $A_{00} = 1$, which is also equivalent to the assertion that $A e_0 = e_0$.

Hence if $A, B \in \mathscr{L}^{\uparrow +}$, then $A e_0 = B e_0$ if and only if $A = BR$ for a matrix $R \in \alpha(SO(3))$.

8.64 Example Let J be the skew symmetric $2n \times 2n$ matrix

$$J = \begin{pmatrix} 0 & I \\ -I & 0 \end{pmatrix}$$

where I is the $n \times n$ identity matrix. Then we have a closed Lie subgroup

$$Sp(n,\mathbf{C}) = \{A \in Gl(2n,\mathbf{C}) | A^{-1} = J^{-1}A^t J\}$$

of $Gl(2n,\mathbf{C})$, called the *complex symplectic group*, with the Lie subalgebra

$$\mathfrak{sp}(n,\mathbf{C}) = \{A \in \mathfrak{gl}(2n,\mathbf{C}) | J^{-1}A^t J = -A\}$$

of $\mathfrak{gl}(2n,\mathbf{C})$ as its Lie algebra by Proposition 8.56. $\mathfrak{sp}(n,\mathbf{C})$ consists of matrices of the form

$$\begin{pmatrix} a & x \\ y & -a^t \end{pmatrix}$$

where x and y are symmetric $n \times n$ matrices, and a is an arbitrary $n \times n$ matrix. Indeed, if

$$A = \begin{pmatrix} a & x \\ y & b \end{pmatrix},$$

we have that

$$J^{-1}A^t J = \begin{pmatrix} 0 & -I \\ I & 0 \end{pmatrix} \begin{pmatrix} a^t & y^t \\ x^t & b^t \end{pmatrix} \begin{pmatrix} 0 & I \\ -I & 0 \end{pmatrix} = \begin{pmatrix} b^t & -x^t \\ -y^t & a^t \end{pmatrix}$$

which equals

$$-A = \begin{pmatrix} -a & -x \\ -y & -b \end{pmatrix}$$

if and only if $x^t = x$, $y^t = y$ and $b = -a^t$.

Consider the non-degenerate skew symmetric bilinear form on $\mathbf{C}^{2n}$ given by

$$<a,b> = a^t J b = \sum_{i=1}^{n} (a_i b_{n+i} - a_{n+i} b_i)$$

for column vectors a and b in $\mathbf{C}^{2n}$. A matrix $A \in \mathfrak{gl}(2n,\mathbf{C})$ satisfies the relation

$$<Aa, Ab> = <a,b>$$

for all $a, b \in \mathbf{C}^{2n}$ if and only if $A^t J A = J$, which is equivalent to $A \in Sp(n,\mathbf{C})$.

In the case $n = 1$ we have that

$$Sp(1,\mathbf{C}) = Sl(2,\mathbf{C})$$

since $A^t J A = \det(A)\, J$ when $A \in \mathfrak{gl}(2,\mathbf{C})$.

THE ALGEBRA OF QUATERNIONS

8.65 The algebra $\mathbf{H}$ of *quaternions* is the real vector space $\mathbf{H} = \mathbf{R} \times \mathbf{R}^3$ with a product defined by

$$(a,u)(b,v) = (ab - u \cdot v, av + bu + u \times v)$$

for $(a,u),(b,v) \in \mathbf{H}$. This product is associative since

$$\begin{aligned}
\{(a,u)(b,v)\}(c,w) &= (ab-u\cdot v, av+bu+u\times v)\,(c,w)\\
&= ((ab-u\cdot v)\,c-(av+bu+u\times v)\cdot w,\\
&\quad (ab-u\cdot v)\,w+c\,(av+bu+u\times v)+(av+bu+u\times v)\times w)\\
&= (abc-av\cdot w-bu\cdot w-cu\cdot v-u\times v\cdot w,\\
&\quad ab\,w+ac\,v+bc\,u+a\,v\times w+b\,u\times w+c\,u\times v\\
&\quad -(v\cdot w)\,u+(u\cdot w)\,v-(u\cdot v)\,w)\\
&= (a\,(bc-v\cdot w)-u\cdot(bw+cv+v\times w),\\
&\quad a\,(bw+cv+v\times w)+(bc-v\cdot w)\,u+u\times(bw+cv+v\times w))\\
&= (a,u)\,(bc-v\cdot w, bw+cv+v\times w) = (a,u)\{(b,v)(c,w)\}
\end{aligned}$$

for $(a,u),(b,v),(c,w) \in \mathbf{H}$, but it is not commutative. We call the components a and u of the quaternion (a,u) its *scalar* and *vector* part, respectively. The quaternions with zero scalar part form a subspace $\mathbf{H}_0 = \{0\}\times\mathbf{R}^3$ of $\mathbf{H}$ called the subspace of *pure* quaternions. The injection $\rho:\mathbf{R}\to\mathbf{H}$ defined by $\rho(a)=(a,0)$ is an algebra homomorphism, and we identify the subalgebra $\mathbf{R}\times\{0\}$ of $\mathbf{H}$ with $\mathbf{R}$, and the quaternion $(a,0)$ with the real number a. If $\{i,j,k\}$ is the standard basis for $\mathbf{R}^3$, we also denote the quaternions $(0,i)$, $(0,j)$ and $(0,k)$ simply by i, j and k, so that the quaternion $q=(a,(b,c,d))$ can be written as $q=a+bi+cj+dk$. We have that

$$i^2=j^2=k^2=-1$$

and

$$ij=-ji=k \ , \ jk=-kj=i \ \text{ and } \ ki=-ik=j\ .$$

The *conjugate* of a quaternion $q=(a,u)$ is the quaternion $\bar{q}=(a,-u)$, and we have that

$$q\bar{q}=\bar{q}q=a^2+\|u\|^2=\|q\|^2\ ,$$

where the non-negative real number

$$\|q\|=\sqrt{a^2+\|u\|^2}$$

is called the *norm* of $q=(a,u)$. Hence each quaternion $q\neq 0$ has a multiplicative inverse which is given by $q^{-1}=\bar{q}\,/\,\|q\|^2$, showing that $\mathbf{H}$ is a division algebra. We also have that

$$\overline{p\,q}=\bar{q}\,\bar{p} \ \text{ and } \ \|p\,q\|=\|p\|\|q\|$$

when $p,q\in\mathbf{H}$.

The injection $\sigma : \mathbf{C} \to \mathbf{H}$ defined by $\sigma(a,b) = (a,(0,0,b))$ is an algebra homomorphism, and we identify the subalgebra $\mathbf{R} \times (\{0\} \times \{0\} \times \mathbf{R})$ of $\mathbf{H}$ with $\mathbf{C}$, and the quaternion $(a,(0,0,b)) = a+bk$ with the complex number $(a,b) = a+bi$. Since

$$a+bi+cj+dk = (a+dk)+j(c+bk)\,,$$

$\mathbf{H}$ can be considered to be a 2-dimensional complex vector space with a basis $\mathscr{E} = \{1, j\}$, and with multiplication with complex scalars from the right. Using the left translations $L_q : \mathbf{H} \to \mathbf{H}$ in $\mathbf{H}$ defined by $L_q(p) = qp$ for $p,q \in \mathbf{H}$, we have an injective algebra homomorphism $\phi : \mathbf{H} \to \mathfrak{gl}(2,\mathbf{C})$ given by $\phi(q) = m^{\mathscr{E}}_{\mathscr{E}}(L_q)$ for $q \in \mathbf{H}$, so that

$$\phi(z_1 + jz_2) = \begin{pmatrix} z_1 & -\bar{z}_2 \\ z_2 & \bar{z}_1 \end{pmatrix}$$

for $z_1, z_2 \in \mathbf{C}$. Indeed, using that $jz = \bar{z}j$ when $z \in \mathbf{C}$, we have that $(z_1 + jz_2)j = -\bar{z}_2 + j\bar{z}_1$. In particular it follows that

$$\phi(1) = \begin{pmatrix} 1 & 0 \\ 0 & 1 \end{pmatrix}, \phi(i) = \begin{pmatrix} 0 & i \\ i & 0 \end{pmatrix}, \phi(j) = \begin{pmatrix} 0 & -1 \\ 1 & 0 \end{pmatrix} \text{ and } \phi(k) = \begin{pmatrix} i & 0 \\ 0 & -i \end{pmatrix}.$$

The last three matrices are skew hermitian with trace 0, and we denote them by τ_1, τ_2 and τ_3, respectively. We have that $\tau_r = i\,\sigma_r$ for $r = 1,2,3$, where

$$\sigma_1 = \begin{pmatrix} 0 & 1 \\ 1 & 0 \end{pmatrix}, \sigma_2 = \begin{pmatrix} 0 & i \\ -i & 0 \end{pmatrix} \text{ and } \sigma_3 = \begin{pmatrix} 1 & 0 \\ 0 & -1 \end{pmatrix}$$

are the *Pauli spin matrices* which are hermitian with trace 0. We also introduce the hermitian and skew hermitian matrix

$$\sigma_0 = \begin{pmatrix} 1 & 0 \\ 0 & 1 \end{pmatrix} \text{ and } \tau_0 = \begin{pmatrix} i & 0 \\ 0 & i \end{pmatrix}$$

so that $\tau_0 = i\,\sigma_0$. We have that $\mathscr{B} = \{\tau_1, \tau_2, \tau_3\}$, $\mathscr{C} = \{\tau_0, \tau_1, \tau_2, \tau_3\}$ and $\mathscr{D} = \{\tau_1, \tau_2, \tau_3, \sigma_1, \sigma_2, \sigma_3\}$ are bases for the Lie algebras $\mathfrak{su}(2)$, $\mathfrak{u}(2)$ and $\mathfrak{sl}(2,\mathbf{C})$, respectively. Finally, we note that

$$\det(\phi(q)) = \|q\|$$

when $q \in \mathbf{H}$.

8.66 Now consider the module $\mathbf{H}^n$ over $\mathbf{H}$ with multiplication with quaternionic scalars from the right. We define the *inner product* of two quaternionic vectors $a, b \in \mathbf{H}^n$ by

$$<a,b> = \sum_{i=1}^{n} \bar{a}_i b_i\,.$$

This product is additive in a and b, and we have that

$$<b,a> = \overline{<a,b>}\;, \quad <aq,b> = \bar{q}<a,b> \quad \text{and} \quad <a,bq> = <a,b>q$$

when $q \in \mathbf{H}$. The non-negative real number

$$\|a\| = \left(\sum_{i=1}^{n} \|a_i\|^2 \right)^{1/2} = <a,a>^{1/2}$$

is called the *length* of a. A quaternionic linear map $F : \mathbf{H}^n \to \mathbf{H}^n$ is said to be *symplectic* if

$$\|F(a)\| = \|a\|$$

for every $a \in \mathbf{H}^n$, which is equivalent to the assertion that

$$<F(a),F(b)> = <a,b>$$

for every $a,b \in \mathbf{H}^n$. Indeed, assuming that F preserves the length of every quaternionic vector and using that

$$\|a+bq\|^2 = \|a\|^2 + <a,b>q + \overline{<a,b>q} + \|bq\|^2 ,$$

we see that $<F(a),F(b)>q$ and $<a,b>q$ have equal scalar parts for every quaternion q. By choosing $q = 1, i, j$ and k, this shows that F preserves the inner product.

Let $\mathscr{E} = \{e_1, \dots, e_n\}$ be the standard basis for $\mathbf{H}^n$, where e_i is the vector with all components equal to 0 except the i-th component which is 1. Writing $a \subset \mathbf{H}^n$ as a column vector, we have that $F(a) = Aa$, where $A = m^{\mathscr{E}}_{\mathscr{E}}(F)$ is the quaternionic $n \times n$ matrix given by

$$F(e_j) = \sum_{i=1}^{n} e_i A_{ij}$$

for $j = 1, \dots, n$. We see that F is symplectic if and only if $\bar{A}^t A = I$, where I is the quaternionic $n \times n$ identity matrix.

Using only complex scalars, we have a complex vector space isomorphism $\psi : \mathbf{H}^n \to \mathbf{C}^{2n}$ defined by $\psi(a) = x$ where $a_i = x_i + jx_{n+i}$ for $i = 1, \dots, n$. The inner product of the quaternionic vectors a and b, where $\psi(a) = x$ and $\psi(b) = y$, is given by

$$<a,b> = \sum_{i=1}^{n} \bar{a}_i b_i = \sum_{i=1}^{n} (\bar{x}_i - \bar{x}_{n+i} j)(y_i + jy_{n+i})$$
$$= \sum_{i=1}^{2n} \bar{x}_i y_i + j \sum_{i=1}^{n} (x_i y_{n+i} - x_{n+i} y_i) .$$

By 8.47, 8.57 and Example 8.64 the quaternionic linear map $F : \mathbf{H}^n \to \mathbf{H}^n$ is symplectic if and only if the matrix $m^{\mathscr{B}}_{\mathscr{B}}(\psi \circ F \circ \psi^{-1})$, where $\mathscr{B}$ is the standard basis in $\mathbf{C}^{2n}$, belongs to $U(2n) \cap Sp(n, \mathbf{C})$. Indeed, we need to show that a complex linear map F preserving the inner product in $\mathbf{H}^n$ actually is quaternionic linear, i.e., that $F(aq) = F(a)q$ for every $a \in \mathbf{H}^n$ and $q \in \mathbf{H}$. But this follows since

$$<F(b),F(aq) - F(a)q> = <F(b),F(aq)> - <F(b),F(a)>q$$
$$= <b,aq> - <b,a>q = 0$$

for every $b \in \mathbf{H}^n$. If $m_{\mathscr{E}}^{\mathscr{E}}(F) = a + jb$ for complex $n \times n$ matrices a and b, then

$$m_{\mathscr{B}}^{\mathscr{B}}(\psi \circ F \circ \psi^{-1}) = \begin{pmatrix} a & -\bar{b} \\ b & \bar{a} \end{pmatrix}.$$

By Propositions 8.42 and 8.56 we know that

$$Sp\,(n) = U(2n) \cap Sp\,(n, \mathbf{C})$$

is a closed Lie subgroup of $Gl\,(2n, \mathbf{C})$, called the *symplectic group*, with the Lie subalgebra

$$\mathfrak{sp}\,(n) = \mathfrak{u}(2n) \cap \mathfrak{sp}\,(n, \mathbf{C})$$

of $\mathfrak{gl}\,(2n, \mathbf{C})$ as its Lie algebra. $\mathfrak{sp}\,(n)$ consists of matrices of the form

$$\begin{pmatrix} a & -\bar{b} \\ b & \bar{a} \end{pmatrix}$$

where a is a skew hermitian and b a symmetric $n \times n$ matrix. Indeed, if

$$A = \begin{pmatrix} a & c \\ b & -a^t \end{pmatrix}$$

is a matrix in $\mathfrak{sp}\,(n, \mathbf{C})$, where b and c are symmetric as described in Example 8.64, then we have that

$$A^* = \begin{pmatrix} a^* & \bar{b} \\ \bar{c} & -\bar{a} \end{pmatrix}$$

which equals

$$-A = \begin{pmatrix} -a & -c \\ -b & a^t \end{pmatrix}$$

if and only if $a^* = -a$ and $c = -\bar{b}$.

In the case $n = 1$ we have that

$$Sp\,(1) = SU(2)$$

since $Sp\,(1, \mathbf{C}) = Sl(2, \mathbf{C})$ by Example 8.64.

8.67 Example Let $\mathbf{H}$ be the Lie algebra of quaternions with bracket product $[p,q] = pq - qp$ as defined in Remark 4.83. It is a real vector space of dimension 4, and the identity map $z : \mathbf{H} \to \mathbf{R}^4$ is a real linear isomorphism. We give $\mathbf{H}$ the manifold structure defined in Example 2.9 (b) so that z is a coordinate map.

Now let $\mathbf{H}^* = \mathbf{H} - \{0\}$ be the group of non-zero quaternions which is an open submanifold of $\mathbf{H}$. If $\alpha : \mathbf{H}^* \to \mathbf{H}$ is the inclusion map, then $(z \circ \alpha, \mathbf{H}^*)$ is a local chart on $\mathbf{H}^*$. By 8.65 the group operations in $\mathbf{H}^*$ are smooth so that $\mathbf{H}^*$ is a Lie group.

We will show that the Lie algebra $\mathfrak{g}$ of $\mathbf{H}^*$ may be identified with $\mathbf{H}$ using Lemma 2.84. We have that $\mathfrak{g} = T_e \mathbf{H}^* = T_e \mathbf{H}$ which we canonically identify with

$\mathbf{H}$ by means of the linear isomorphism $\omega : \mathfrak{g} \to \mathbf{H}$ given by $\omega = z^{-1} \circ t_{z,e}$. It only remains to show that

$$\omega([X,Y]) = [\omega(X), \omega(Y)] \tag{1}$$

for $X, Y \in \mathfrak{g}$, so that ω is a Lie algebra isomorphism.

As usual we let $\widetilde{X}$ and $\widetilde{Y}$ be the left invariant vector fields on $\mathbf{H}^*$ determined by X and Y. If $h : \mathbf{H} \to \mathbf{H}$ is the identity map, it follows from 5.49 that

$$\omega(X) = \omega \circ h_*(X) = X(h)$$

for every $X \in \mathfrak{g}$, so that

$$\omega([X,Y]) = [X,Y](h) = X(\widetilde{Y}(h)) - Y(\widetilde{X}(h)) .$$

To compute the last expression, we will use the real linear maps $L_p : \mathbf{H} \to \mathbf{H}$ and $R_p : \mathbf{H} \to \mathbf{H}$ defined by $L_p(q) = pq$ and $R_p(q) = qp$ for $p, q \in \mathbf{H}$. Indentifying the tangent space $T_p\mathbf{H}$ with $\mathbf{H}$ by means of the linear isomorphism $\omega_p : T_p\mathbf{H} \to \mathbf{H}$ given by $\omega_p = z^{-1} \circ t_{z,p}$, we have that

$$\begin{aligned}\widetilde{Y}(h)(p) &= \widetilde{Y}_p(h) = L_{p*}(Y)(h) = \omega_p \circ h_* \circ L_{p*}(Y) \\ &= \omega_p \circ L_{p*}(Y) = L_p \circ \omega(Y) = R_{\omega(Y)}(p)\end{aligned}$$

for $p \in \mathbf{H}^*$. This implies that

$$X(\widetilde{Y}(h)) = \omega_{\omega(Y)} \circ R_{\omega(Y)*}(X) = R_{\omega(Y)} \circ \omega(X) = \omega(X)\ \omega(Y)$$

which completes the proof of (1).

The exponential map $exp : \mathbf{H} \to \mathbf{H}^*$ is given by

$$exp(q) = \sum_{k=0}^{\infty} \frac{q^k}{k!} \tag{2}$$

for $q \in \mathbf{H}$, which is proved in the same way as in Proposition 8.52.

Now consider the injective Lie algebra homomorphism $\phi : \mathbf{H} \to \mathfrak{gl}(2, \mathbf{C})$ defined in 8.65, and the induced map $\psi : \mathbf{H}^* \to Gl(2, \mathbf{C})$ which is a Lie group homomorphism. Identifying the Lie algebra of $Gl(2, \mathbf{C})$ with $\mathfrak{gl}(2, \mathbf{C})$ as usual by means of the linear isomorphism $\widetilde{\omega} = \widetilde{z}^{-1} \circ t_{\widetilde{z},e}$, where $\widetilde{z} : \mathfrak{gl}(2, \mathbf{C}) \to \mathbf{R}^8$ is the linear isomorphism defined in Example 8.10 (c), the Lie algebra homomorphism ψ_* is identified with ϕ in the same way as in Example 8.19. Indeed, $\widetilde{z} \circ \phi \circ z^{-1} = \lambda : \mathbf{R}^4 \to \mathbf{R}^8$ is the linear map given by $\lambda(a,b,c,d) = (a,c,-c,a,d,b,b,-d)$ for $(a,b,c,d) \in \mathbf{R}^4$, so that $t_{\widetilde{z},e} \circ \psi_* \circ t_{z.e}^{-1} = D(\widetilde{z} \circ \psi \circ z^{-1})(z(e)) = \lambda$ by the commutative diagram in 2.70. This implies that $\widetilde{\omega} \circ \psi_* \circ \omega^{-1} = \phi$.

Using Proposition 8.40 we therefore have a commutative diagram

$$\begin{array}{ccc} \mathbf{H} & \xrightarrow{\ \phi\ } & \mathfrak{gl}(2,\mathbf{C}) \\ {\scriptstyle exp}\downarrow & & \downarrow{\scriptstyle exp} \\ \mathbf{H}^* & \xrightarrow{\ \psi\ } & Gl(2,\mathbf{C}) \end{array}$$

where $\phi(\mathbf{H}_0) = \mathfrak{sp}(1)$ and $\psi(\mathbf{S}^3) = Sp(1)$ by 8.66. Hence it follows from Propositions 8.42 and 8.56 that the group $\mathbf{S}^3$ of unit quaternions is a closed Lie subgroup of $\mathbf{H}^*$, with the Lie subalgebra $\mathbf{H}_0$ of $\mathbf{H}$ consisting of the pure quaternions as its Lie algebra. Furthermore, ϕ induces a Lie group isomorphism from $\mathbf{S}^3$ to $Sp(1) = SU(2)$, which we also denote by ϕ, and a corresponding Lie algebra isomorphism ϕ_* from $\mathbf{H}_0$ to $\mathfrak{sp}(1) = \mathfrak{su}(2)$.

LEFT INVARIANT FORMS

8.68 Definition A k-form ω on G is called *left invariant* if $L_g^*(\omega) = \omega$ for every $g \in G$, which means that $\omega(h) = L_g^*(\omega(gh))$ for every $g, h \in G$. Left invariant 1-forms on G are also known as *Maurer-Cartan forms*.

8.69 Proposition If G is a Lie group and $\omega \in \Lambda^k(\mathfrak{g})$, then there is a unique left invariant k-form $\widetilde{\omega}$ on G with $\widetilde{\omega}(e) = \omega$ which is given by

$$\widetilde{\omega}(g) = L_{g^{-1}}^*(\omega)$$

for $g \in G$. It is called the left invariant k-form determined by ω.

PROOF: Since $L_{h^{-1}} = L_{(gh)^{-1}} \circ L_g$, we have that

$$\widetilde{\omega}(h) = L_g^*(\widetilde{\omega}(gh))$$

for every $g, h \in G$.

To complete the proof, we only need to show that $\widetilde{\omega}$ is smooth. If $\{X_1, \dots, X_n\}$ is a basis for $\mathfrak{g}$, then the left invariant vector fields $\widetilde{X_1}, \dots, \widetilde{X_n}$ form a local basis for TG on G. If $\{\omega^1, \dots, \omega^n\}$ is the dual local basis for $\Lambda^1(TG)$ on G, then

$$\widetilde{\omega} = \sum_{i_1 < \dots < i_k} \omega_{i_1, \dots, i_k}\, \omega^{i_1} \wedge \dots \wedge \omega^{i_k}$$

where

$$\omega_{i_1, \dots, i_k}(g) = \widetilde{\omega}(g)\,((\widetilde{X_1})_g, \dots, (\widetilde{X_k})_g) = \omega(X_1, \dots, X_k)$$

for $g \in G$, thus showing that $\widetilde{\omega}$ is smooth. □

8.70 Proposition A k-form ω on G is left invariant if and only if $\omega(X_1, \ldots, X_k)$ is a constant function on G for any choice of left invariant vector fields $X_1, \ldots, X_k$ on G.

PROOF: If the vector fields $X_1, \ldots, X_k$ are left invariant, then

$$L_g^*(\omega)(h)((X_1)_h, \ldots, (X_k)_h) = \omega(gh)(L_{g*}(X_1)_h, \ldots, L_{g*}(X_k)_h)$$
$$= \omega(gh)((X_1)_{gh}, \ldots, (X_k)_{gh})$$

for every $g, h \in G$. Assuming that ω is left invariant, the last assertion of the proposition follows by choosing $h = e$.

The converse statement follows from the fact that any tangent vector $w \in T_hG$ equals $\widetilde{X}_h$ for the left invariant vector field $\widetilde{X}$ determined by $X = (L_{h^{-1}})_*(w)$. □

8.71 Definition A vector-valued k-form ω on G with values in a finite dimensional vector space W is called *left invariant* if $L_g^*(\omega) = \omega$ for every $g \in G$, which means that $\omega(h) = L_g^*(\omega(gh))$ for every $g, h \in G$. Vector-valued left invariant 1-forms on G are also known as vector-valued *Maurer-Cartan forms*.

8.72 Proposition A vector-valued k-form ω on G with values in a finite dimensional vector space W is left invariant if and only if the real-valued k-form $\lambda \cdot \omega$ is left invariant for every functional $\lambda \in W^*$.

PROOF: Follows from Proposition 5.47. □

8.73 Proposition Let G is a Lie group and W be a finite dimensional vector space. If $\omega \in \Lambda^k(\mathfrak{g}; W)$, then there is a unique left invariant W-valued k-form $\widetilde{\omega}$ on G with $\widetilde{\omega}(e) = \omega$ which is given by

$$\widetilde{\omega}(g) = L_{g^{-1}}^*(\omega)$$

for $g \in G$. It is called the left invariant W-valued k-form determined by ω.

PROOF: Follows from Propositions 5.42 (3) and 8.72 since $\lambda \cdot \widetilde{\omega} = \widetilde{\lambda \circ \omega}$ for every functional $\lambda \in W^*$ by Definition 5.3. □

8.74 Proposition A vector-valued k-form ω on G with values in a finite dimensional vector space W is left invariant if and only if $\omega(X_1, \ldots, X_k)$ is a constant vector-valued function on G for any choice of left invariant vector fields $X_1, \ldots, X_k$ on G.

PROOF: Follows from Propositions 8.70 and 8.72 since

$$\lambda \circ \omega(X_1, \ldots, X_k) = (\lambda \cdot \omega)(X_1, \ldots, X_k)$$

for every functional $\lambda \in W^*$. □

8.75 Definition By the *canonical Maurer–Cartan form* or the *canonical 1-form* on a Lie group G we mean the unique $\mathfrak{g}$-valued left invariant 1-form θ on G determined by the 1-form $id : \mathfrak{g} \to \mathfrak{g}$ in $\Lambda^1(\mathfrak{g};\mathfrak{g})$. It is given by

$$\theta(g)(w) = (L_{g^{-1}})_*(w)$$

for $g \in G$ and $w \in T_gG$.

8.76 Remark By Proposition 8.74 we have that

$$\theta(g)(\widetilde{X}_g) = X$$

for every $g \in G$ and $X \in \mathfrak{g}$, where $\widetilde{X}$ is the left invariant vector field on G determined by X.

8.77 Theorem (Maurer-Cartan's structure equation) If θ is the canonical Maurer-Cartan form on a Lie group G, then we have that

$$d\theta(g)(w_1,w_2) = -[\,\theta(g)(w_1),\, \theta(g)(w_2)\,]$$

for $g \in G$ and $w_1, w_2 \in T_gG$.

PROOF : Suppose that $w_1, w_2 \in T_gG$, and let $X_i = \theta(g)(w_i) = (L_{g^{-1}})_*(w_i)$ so that $\widetilde{X}_{i\,g} = L_{g*}(X_i) = w_i$ for $i = 1,2$. For each $X \in \mathfrak{g}$ we know that $\theta(\widetilde{X})$ is a constant function on G with value X. Hence it follows from Remark 5.86 and Lemma 2.78 (2) that

$$\begin{aligned} d\theta(g)(w_1,w_2) &= d\theta(\widetilde{X}_1,\widetilde{X}_2)(g) \\ &= \widetilde{X}_1(\theta(\widetilde{X}_2))(g) - \widetilde{X}_2(\theta(\widetilde{X}_1))(g) - \theta([\widetilde{X}_1,\widetilde{X}_2])(g) \\ &= -[X_1,X_2] = -[\,\theta(g)(w_1),\, \theta(g)(w_2)\,]\,. \end{aligned}$$

□

8.78 Remark The formula in Theorem 8.77 can also be written in the form

$$d\theta = -\tfrac{1}{2}\; \theta \wedge \theta$$

where the wedge product is with respect to the bilinear map

$$\nu : \mathfrak{g} \times \mathfrak{g} \to \mathfrak{g}$$

given by $\nu(X,Y) = [X,Y]$.

Chapter 9

GROUP ACTIONS

INTRODUCTION

9.1 Definition Let M be a smooth manifold and G be a Lie group. By an *operation* or an *action of G on M on the left* we mean an operation $\mu : G \times M \to M$ of G on M as Defined in Definition 13.12 in the appendix, where μ is smooth. An operation of G on M *on the right* is defined similarly as in Remark 13.13 in the appendix.

9.2 Proposition Let G be a Lie group and V a finite dimensional vector space, and let $\mu : G \times V \to V$ and $\phi : G \to \mathrm{Aut}\,(V)$ be maps so that

$$\mu(g,v) = \phi(g)(v)$$

for $g \in G$ and $v \in V$. Then μ is an operation of G on V on the left which is linear in its second argument if and only if ϕ is a representation.

μ is called the operation determined by ϕ.

PROOF : Assume that ϕ is smooth. Then we see that μ is smooth since

$$\mu = B \circ (\phi \times id_V)\,,$$

where $B : \mathrm{End}\,(V) \times V \to V$ is the bilinear map defined by $B(F,v) = F(v)$ for $F \in \mathrm{End}\,(V)$ and $v \in V$.

Conversely, assume that μ is smooth, and let $\mathscr{B} = \{v_1,\ldots,v_n\}$ be a basis for V with dual basis $\mathscr{B}^* = \{v_1^*,\ldots,v_n^*\}$. Then ϕ is smooth since

$$v_i^* \circ \phi(g)(v_j) = v_i^* \circ \mu(g,v_j)$$

for $g \in G$ and $i,j = 1,\ldots,n$.

Since

$$\phi(e)(v) = \mu(e,v)$$

for $v \in V$ where $e \in G$ is the unit element, and

$$\phi(g) \circ \phi(h)(v) = \mu(g,\mu(h,v)) \quad \text{and} \quad \phi(gh)(v) = \mu(gh,v)$$

for $g,h \in G$ and $v \in V$, we see that μ is an operation if and only if ϕ is a representation of G on V. □

9.3 Definition By an operation or an action of a Lie group G on a vector bundle $\pi : E \to M$ on the left we mean a pair $(\widetilde{\mu}, \mu)$ of operations $\widetilde{\mu} : G \times E \to E$ and $\mu : G \times M \to M$ of G on E and M, respectively, such that the corresponding pair $(\tilde{L}_g, L_g)$ of left translations forms a bundle map for each $g \in G$. This implies that the projection $\pi : E \to M$ is equivariant with respect to the actions $\widetilde{\mu}$ and μ of G on E and M.

9.4 Proposition Let $\mu : G \times M \to M$ be an operation of a Lie group G on a smooth manifold M^n on the left. Then we have an operation $(\widetilde{\mu}, \mu)$ of G on the tangent bundle $\pi : TM \to M$ such that the left translations L_g and $\tilde{L}_g$ corresponding to μ and $\widetilde{\mu}$, respectively, are related by

$$\tilde{L}_g = (L_g)_*$$

for $g \in G$.

PROOF: By Proposition 2.71 we know that $(\tilde{L}_g, L_g)$ is a bundle map for each $g \in G$, and it only remains to prove that $\widetilde{\mu}$ is smooth at every point $(h, v) \in G \times T_qM$ where $q \in M$. Let (z, U) be a local chart around $\mu(h, q)$ on M, and choose local charts (x, V) and (y, W) around h and q on G and M, respectively, so that $\mu(V \times W) \subset U$. Then the map $f : x(V) \times y(W) \to \mathbf{R}^n$ defined by $f = z \circ \mu \circ (x \times y)^{-1}$ is smooth, and we have that

$$f(a,b) = z \circ L_{x^{-1}(a)} \circ y^{-1}(b)$$

for $a \in x(V)$ and $b \in y(W)$.

If $(t_y, \pi^{-1}(W))$ and $(t_z, \pi^{-1}(U))$ are the local trivializations in the tangent bundle associated with the local charts (y, W) and (z, U), we have that

$$\begin{aligned} & t_z \circ \widetilde{\mu} \circ (id_V \times t_y)^{-1}(g, p, v) = t_z \circ L_{g*}([y, v]_p) \\ & = t_z([z, D(z \circ L_g \circ y^{-1})(y(p))v]_{L_g(p)}) = (L_g(p), D(z \circ L_g \circ y^{-1})(y(p))v) \end{aligned}$$

for $g \in V$ and $(p, v) \in W \times \mathbf{R}^n$. This shows that

$$\begin{aligned} & (z \times id_{\mathbf{R}^n}) \circ t_z \circ \widetilde{\mu} \circ (id_V \times t_y)^{-1} \circ (x \times y \times id_{\mathbf{R}^n})^{-1}(a, b, v) \\ & = (z \circ L_{x^{-1}(a)} \circ y^{-1}(b), D(z \circ L_{x^{-1}(a)} \circ y^{-1})(b)v) = (f(a,b), D_2 f(a,b)v) \end{aligned}$$

for $(a, b, v) \in x(V) \times y(W) \times \mathbf{R}^n$, which completes the proof of the proposition. □

9.5 Corollary Let G be a Lie group acting on the smooth manifold M on the left, and suppose that $p \in M$ is a fixed point, i.e., that $gp = p$ for every $g \in G$. For each $g \in G$, we let $L_g : M \to M$ be the diffeomorphism defined by $L_g(q) = gq$ for $q \in M$. Then we have a representation $\phi : G \to \text{Aut}(T_pM)$ of G defined by

$$\phi(g) = (L_g)_{*p}$$

for $g \in G$.

PROOF : Using that the tangent space T_pM is a submanifold of TM, we obtain an action of G on T_pM on the left which is linear in its second argument by Proposition 9.4. The result therefore follows from Proposition 9.2. □

9.6 Proposition Let $(\widetilde{\mu},\mu)$ be an operation of a Lie group G on a vector bundle $\pi : E \to M$ of dimension n. Then there is an operation $(\hat{\mu},\mu)$ of G on the dual bundle $\pi^* : E^* \to M$ such that the left translations $\tilde{L}_g$ and $\hat{L}_g$ corresponding to $\widetilde{\mu}$ and $\hat{\mu}$, respectively, are related by

$$(\hat{L}_g)_p = (\tilde{L}_{g^{-1}})_p^*$$

for each $p \in M$ and $g \in G$.

PROOF : Using Proposition 4.50, we only need to prove that $\hat{\mu}$ is smooth at every point $(h,v) \in G \times E_q^*$ where $q \in M$. Choose local trivializations $(t_1, \pi_1^{-1}(U_1))$ and $(t_2, \pi_2^{-1}(U_2))$ of π around q and hq, respectively, and a coordinate neighbourhood V around h on G, with $\mu(V \times U_1) \subset U_2$. Let $(t_1', {\pi_1^*}^{-1}(U_1))$ and $(t_2', {\pi_2^*}^{-1}(U_2))$ be the corresponding local trivializations in the dual bundle, and let $\mathscr{E}$ be the standard basis for $\mathbf{R}^n$ with dual basis $\mathscr{E}^*$.

Then $t_2 \circ \widetilde{\mu} \circ (id_V \times t_1^{-1}) : V \times U_1 \times \mathbf{R}^n \to U_2 \times \mathbf{R}^n$ is a smooth map given by

$$(t_2 \circ \widetilde{\mu} \circ (id_V \times t_1^{-1}))(g,p,v) = (gp, (t_{2,gp} \circ \tilde{L}_{g,p} \circ t_{1,p}^{-1})(v)),$$

where $t_{2,gp} \circ \tilde{L}_{g,p} \circ t_{1,p}^{-1} : \mathbf{R}^n \to \mathbf{R}^n$ is a linear isomorphism for each $g \in V$ and $p \in U_1$. Since

$$t'_{2,gp} \circ \hat{L}_{g,p} \circ {t'}_{1,p}^{-1} = r \circ \{(t_{2,gp} \circ \tilde{L}_{g,p} \circ t_{1,p}^{-1})^{-1}\}^* \circ r^{-1},$$

it follows from Proposition 4.6, using the fact that $m_{\mathscr{E}}^{\mathscr{E}^*}(r)$ and $m_{\mathscr{E}^*}^{\mathscr{E}}(r^{-1})$ are both the identity matrix, that

$$m_{\mathscr{E}}^{\mathscr{E}}(t'_{2,gp} \circ \hat{L}_{g,p} \circ {t'}_{1,p}^{-1}) = \{m_{\mathscr{E}}^{\mathscr{E}}(t_{2,gp} \circ \tilde{L}_{g,p} \circ t_{1,p}^{-1})^{-1}\}^t .$$

Hence $t_2' \circ \hat{\mu} \circ (id_V \times {t'}_1^{-1})$ is a smooth map from $V \times U_1 \times \mathbf{R}^n$ to $U_2 \times \mathbf{R}^n$, thus showing that $(\hat{\mu},\mu)$ is an action of G on $\pi^* : E^* \to M$. □

THE ADJOINT REPRESENTATION

9.7 Let G be a Lie group. For each $g \in G$, we have an automorphism $i_g : G \to G$ defined by

$$i_g(h) = ghg^{-1}$$

for $h \in G$, called the *inner automorphism* of G by the element g. Defining $Ad_g : \mathfrak{g} \to \mathfrak{g}$ by $Ad_g = i_{g*}$, we have a commutative diagram

$$\begin{array}{ccc} \mathfrak{g} & \xrightarrow{Ad_g} & \mathfrak{g} \\ {\scriptstyle exp}\downarrow & & \downarrow{\scriptstyle exp} \\ G & \xrightarrow{i_g} & G \end{array}$$

by Proposition 8.40, so that

$$exp\ t Ad_g(X) = g\,(exp\ tX)\,g^{-1}$$

for $X \in \mathfrak{g}$ and $t \in \mathbf{R}$.

Now the map $a : G \times G \to G$ defined by $a(g,h) = ghg^{-1} = i_g(h)$ for $g,h \in G$, is an operation of G on itself on the left with fixed point e. By Corollary 9.5 we thus have a representation

$$Ad : G \to Gl(\mathfrak{g})$$

of G defined by $Ad(g) = Ad_g$ for $g \in G$, which is called the *adjoint representation*. If we define

$$ad : \mathfrak{g} \to \mathfrak{gl}(\mathfrak{g})$$

by $ad = Ad_*$ and again use Proposition 8.40, we have the commutative diagram

$$\begin{array}{ccc} \mathfrak{g} & \xrightarrow{ad} & \mathfrak{gl}(\mathfrak{g}) \\ {\scriptstyle exp}\downarrow & & \downarrow{\scriptstyle exp} \\ G & \xrightarrow{Ad} & Gl(\mathfrak{g}) \end{array}$$

The contragradient $Ad^* : G \to Gl(\mathfrak{g}^*)$ of the adjoint representation is called the *coadjoint representation*.

9.8 Examples

(a) For each non-singular matrix $A \in Gl(n,\mathbf{R})$ the inner automorphism $i_A : Gl(n,\mathbf{R}) \to Gl(n,\mathbf{R})$ is induced by the linear isomorphism $e_A : \mathfrak{gl}(n,\mathbf{R}) \to \mathfrak{gl}(n,\mathbf{R})$ given by

$$e_A(B) = ABA^{-1}$$

for $B \in \mathfrak{gl}(n,\mathbf{R})$. Hence it follows from Lemma 2.84 that $Ad(A) = e_{A*} = e_A$ when we canonically identify $T_e\,\mathfrak{gl}(n,\mathbf{R})$ with $\mathfrak{gl}(n,\mathbf{R})$ as in Example 8.10 (a). The adjoint representation

$$Ad : Gl(n,\mathbf{R}) \to Gl(\mathfrak{gl}(n,\mathbf{R}))$$

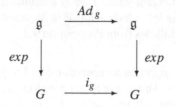

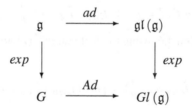

is therefore given by

$$Ad\,(A)(B) = ABA^{-1}$$

for $A \in Gl\,(n,\mathbf{R})$ and $B \in \mathfrak{gl}\,(n,\mathbf{R})$.

(b) In the same way we see that the adjoint representation

$$Ad : Gl\,(n,\mathbf{C}) \to Gl\,(\mathfrak{gl}\,(n,\mathbf{C}))$$

is given by

$$Ad\,(A)(B) = ABA^{-1}$$

for $A \in Gl\,(n,\mathbf{C})$ and $B \in \mathfrak{gl}\,(n,\mathbf{C})$.

(c) We also see that the adjoint representation

$$Ad : Gl\,(V) \to Gl\,(\mathfrak{gl}\,(V))\,,$$

where V is a finite dimensional real or complex vector space, is given by

$$Ad\,(F)(G) = F \circ G \circ F^{-1}$$

for $F \in Gl\,(V)$ and $G \in \mathfrak{gl}\,(V)$.

(d) The adjoint representation

$$Ad : \mathbf{H}^* \to Gl\,(\mathbf{H})$$

of the Lie group $\mathbf{H}^*$ of non-zero quaternions described in Example 8.67 is given by

$$Ad\,(p)(q) = p\,q\,p^{-1}$$

for $p \in \mathbf{H}^*$ and $q \in \mathbf{H}$.

9.9 Proposition Let G be a Lie group. Then we have that

$$ad\,(X)(Y) = [X,Y]$$

for $X,Y \in \mathfrak{g}$.

PROOF : Let $\widetilde{X}$ and $\widetilde{Y}$ be the left invariant vector fields on G determined by X and Y. According to Proposition 4.86 we have that

$$[X,Y] = [\widetilde{X},\widetilde{Y}]\,(e) = L_{\widetilde{X}}\,\widetilde{Y}\,(e)\,,$$

and the flow $\gamma : \mathbf{R} \times G \to G$ of the vector field $\widetilde{X}$ is given by

$$\gamma(t,g) = g\ exp\,(tX)$$

for $t \in \mathbf{R}$ and $g \in G$ by Propositions 8.29 and 8.31.

Now let $c : \mathbf{R} \to \mathfrak{g}$ be the curve defined by

$$\begin{aligned} c(t) &= [\gamma_t^* \widetilde{Y}]_e = [R^*_{exp\,(tX)} \widetilde{Y}]_e = (R_{exp\,(-tX)})_* \widetilde{Y}_{exp\,(tX)} \\ &= (R_{exp\,(-tX)})_* (L_{exp\,(tX)})_* \widetilde{Y}_e = i_{exp\,(tX)\,*} Y = Ad\,(exp\,(tX))\,Y \\ &= \Lambda \circ Ad \circ \phi_X(t) \end{aligned}$$

for $t \in \mathbf{R}$, where $\phi_X : \mathbf{R} \to G$ is the one-parameter subgroup defined in Proposition 8.29, and $\Lambda : \mathrm{End}\,(\mathfrak{g}) \to \mathfrak{g}$ is the linear map given by $\Lambda(F) = F(Y)$ for $F \in \mathrm{End}\,(\mathfrak{g})$. Using 2.70 and Proposition 4.76 it follows that

$$L_{\widetilde{X}} \widetilde{Y}(e) = c'(0) = \Lambda \circ (Ad \circ \phi_X)'(0) = ad\,(X)(Y)$$

if we canonically identify $T_{c(0)}\,\mathfrak{g}$ with $\mathfrak{g}$ as in Lemma 2.84. □

9.10 Definition The Lie subalgebra

$$Z(\mathfrak{g}) = \{X \in \mathfrak{g} \,|\, [X,Y] = 0 \text{ for all } Y \in \mathfrak{g}\}$$

of a Lie algebra $\mathfrak{g}$, which is the kernel of $ad : \mathfrak{g} \to \mathfrak{gl}\,(\mathfrak{g})$, is called the *center* of $\mathfrak{g}$. We say that $\mathfrak{g}$ is *abelian* if $Z(\mathfrak{g}) = \mathfrak{g}$, so that $[X,Y] = 0$ for every $X,Y \in \mathfrak{g}$.

9.11 Definition The subgroup

$$Z(G) = \{g \in G \,|\, gh = hg \text{ for all } h \in G\}$$

of a Lie group G is called the *center* of G.

9.12 Proposition The center $Z(G)$ of a connected Lie group G is a closed submanifold and a Lie subgroup of G with Lie algebra $Z(\mathfrak{g})$, and it is the kernel of the adjoint representation $Ad : G \to Gl\,(\mathfrak{g})$.

PROOF: The first assertion follows from the last by Proposition 8.45. If $g \in Z(G)$, we have for each $X \in \mathfrak{g}$ that

$$exp\ t\,Ad_g\,(X) = g\,(exp\ tX\,)\,g^{-1} = exp\ tX$$

when $t \in \mathbf{R}$, so that $Ad_g\,(X) = X$ which shows that $g \in \ker Ad$.

Conversely, suppose that $g \in \ker Ad$, and let U be an open neighbourhood of $0 \in \mathfrak{g}$ which is mapped diffeomorphically by exp onto an open neighbourhood V of $e \in G$. Then

$$g\,(exp\,X\,)\,g^{-1} = exp\ Ad_g\,(X) = exp\,X$$

for every $X \in U$, which shows that $gh = hg$ for every $h \in V$. Using that G is connected, it follows from Proposition 13.112 in the appendix that $g \in Z(G)$. □

9.13 Corollary A connected Lie group G is abelian if and only if its Lie algebra $\mathfrak{g}$ is abelian.

9.14 Proposition Let $\phi : G_1 \to G_2$ be a Lie group homomorphism. Then we have that

$$Ad(\phi(g)) \circ \phi_* = \phi_* \circ Ad(g)$$

for every $g \in G_1$.

PROOF : Follows from the relation

$$i_{\phi(g)} \circ \phi = \phi \circ i_g$$

for the inner automorphisms i_g and $i_{\phi(g)}$ in G_1 and G_2. □

9.15 Proposition If θ is the canonical Maurer-Cartan form on a Lie group G, then

$$R_g^*\,\theta = Ad\,(g^{-1}) \cdot \theta$$

for every $g \in G$.

PROOF : We have that

$$R_g^*\,\theta\,(h)\,(w) = \theta\,(hg)\,((R_g)_*\,w) = (L_{g^{-1}})_* \circ (L_{h^{-1}})_* \circ (R_g)_*\,(w)$$

$$= (L_{g^{-1}})_* \circ (R_g)_* \circ (L_{h^{-1}})_*\,(w) = Ad\,(g^{-1})\;\theta\,(h)(w)$$

for $g, h \in G$ and $w \in T_hG$. □

THE GROUPS $SO(3)$ AND $\mathbf{S}^3$

9.16 Consider the adjoint representation $Ad : \mathbf{S}^3 \to Gl\,(\mathbf{H}_0)$ which by Example 9.8 (d) and Proposition 9.14 is given by

$$Ad\,(q)(p) = q\,p\,q^{-1} = q\,p\,\bar{q}$$

for $q \in \mathbf{S}^3$ and $p \in \mathbf{H}_0$. Using the basis $\mathscr{C} = \{i, j, k\}$ for $\mathbf{H}_0$, we also have a representation $\widetilde{\rho} : \mathbf{S}^3 \to Gl\,(3, \mathbf{R})$ defined by

$$\widetilde{\rho}\,(q) = m_{\mathscr{C}}^{\mathscr{C}}(Ad\,(q))$$

for $q \in \mathbf{S}^3$. Since $\|Ad\,(q)(p)\| = \|p\|$ for $p \in \mathbf{H}_0$ when $q \in \mathbf{S}^3$, and $\widetilde{\rho}$ is a continuous map from the connected set $\mathbf{S}^3$, it follows that $\widetilde{\rho}\,(\mathbf{S}^3) \subset SO(3)$. We want to show that $\widetilde{\rho}\,(\mathbf{S}^3) = SO(3)$, and that the homomorphism $\rho : \mathbf{S}^3 \to SO(3)$ induced by $\widetilde{\rho}$ is a double covering of $SO(3)$.

Every unit quaternion $q \in \mathbf{S}^3$ can be written in the form

$$q = \cos\tfrac{\theta}{2} + \sin\tfrac{\theta}{2}\, t$$

where $t = (0, w) \in \mathbf{S}^3 \cap \mathbf{H}_0$ is a pure unit quaternion and $0 \leq \theta < 4\pi$. We contend that $\widetilde{\rho}\,(q)$ is the standard matrix for a rotation $R_w(\theta)$ by the angle θ about the axis through the origin with direction vector w, as defined in Proposition 8.59. To show this, we choose pure unit quaternions $r = (0, u)$ and $s = (0, v)$ so that $\mathscr{B} = \{u, v, w\}$ is a positively oriented orthonormal basis in $\mathbf{R}^3$. Then we have that

$$r^2 = s^2 = t^2 = -1$$

and

$$rs = -sr = t \quad , \quad st = -ts = r \quad \text{and} \quad tr = -rt = s\,,$$

so that

$$\begin{aligned} Ad\,(q)(r) &= \cos\theta\ r + \sin\theta\ s\,, \\ Ad\,(q)(s) &= -\sin\theta\ r + \cos\theta\ s\,, \\ Ad\,(q)(t) &= t\,. \end{aligned}$$

This implies that

$$m_{\mathscr{B}}^{\mathscr{B}}(\alpha^{-1} \circ Ad\,(q) \circ \alpha) = \begin{pmatrix} \cos\theta & -\sin\theta & 0 \\ \sin\theta & \cos\theta & 0 \\ 0 & 0 & 1 \end{pmatrix}$$

where $\alpha : \mathbf{R}^3 \to \mathbf{H}_0$ is the linear isomorphism given by $\alpha(\xi) = (0, \xi)$ for $\xi \in \mathbf{R}^3$, showing that $\alpha^{-1} \circ Ad\,(q) \circ \alpha = R_w(\theta)$. We have that $\ker\rho = \{1, -1\}$ since each $q \in \ker\rho$ commutes with every pure quaternion $p \in \mathbf{H}_0$ and therefore is real of norm 1. As this is a discrete subgroup of $\mathbf{S}^3$, the result now follows from Proposition 13.111 in the appendix.

Since

$$\cos\tfrac{\theta}{2} + \sin\tfrac{\theta}{2}\, t = exp\,(\tfrac{\theta}{2}\, t)\,,$$

we see that the standard matrix for the rotation $R_w(\theta)$ is given by

$$\rho\,(\cos\tfrac{\theta}{2} + \sin\tfrac{\theta}{2}\, t) = exp\,(\theta\, \rho_*(\tfrac{t}{2}))\,.$$

By Example 8.60 it follows that the Lie algebra isomorphism $\rho_* : \mathbf{H}_0 \to \mathfrak{so}(3)$ maps the basis $\{i/2, j/2, k/2\}$ for $\mathbf{H}_0$ onto the basis $\{\mathscr{L}_1, \mathscr{L}_2, \mathscr{L}_3\}$ for $\mathfrak{so}(3)$.

9.17 Remark The representation of rotations in $\mathbf{R}^3$ by unit quaternions described in 9.16 is very useful when combining rotations. For instance, to find the composition $R_i(\frac{\pi}{2}) \circ R_j(\frac{\pi}{2})$ we simply compute

$$(\cos\tfrac{\pi}{4}+\sin\tfrac{\pi}{4}\,i)(\cos\tfrac{\pi}{4}+\sin\tfrac{\pi}{4}\,j)=\tfrac{1}{\sqrt{2}}\,(1+i)\cdot\tfrac{1}{\sqrt{2}}\,(1+j)=\tfrac{1}{2}\,(1+i+j+k)$$

$$=\tfrac{1}{2}+\tfrac{\sqrt{3}}{2}\cdot\tfrac{1}{\sqrt{3}}\,(i+j+k)=\cos\tfrac{\pi}{3}+\sin\tfrac{\pi}{3}\cdot\tfrac{1}{\sqrt{3}}\,(i+j+k)\,,$$

giving the rotation $R_v(\frac{2\pi}{3})$ where $v=\frac{1}{\sqrt{3}}\,(i+j+k)$.

9.18 Remark The bracket product in $\mathbf{H}_0$ is given by

$$[(0,u),(0,v)]=(0,2\,u\times v)$$

for $u,v\in\mathbf{R}^3$. Hence we have a Lie algebra isomorphism $\alpha_2=\frac{1}{2}\,\alpha:\mathbf{R}^3\to\mathbf{H}_0$ from the Lie algebra $\mathbf{R}^3$ with bracket product $[u,v]=u\times v$, which is given by $\alpha_2(u)=(0,u/2)$ for $u\in\mathbf{R}^3$. By 9.16 we also have a Lie algebra isomorphism $\lambda=\rho_*\circ\alpha_2:\mathbf{R}^3\to\mathfrak{so}(3)$ which maps the standard basis $\mathscr{E}=\{i,j,k\}$ for $\mathbf{R}^3$ onto the basis $\{\mathscr{L}_1,\mathscr{L}_2,\mathscr{L}_3\}$ for $\mathfrak{so}(3)$. Since

$$\lambda\,(u)=m^{\mathscr{E}}_{\mathscr{E}}(\alpha^{-1}\circ ad\,(\alpha_2(u))\circ\alpha)\,,$$

we have that

$$\lambda\,(u)\,v=\alpha_2^{-1}([\,\alpha_2(u),\alpha_2(v)\,])=u\times v$$

for column vectors $u,v\in\mathbf{R}^3$.

If $A=\rho(q)$ where $q\in\mathbf{S}^3$, it follows from Proposition 9.14 that

$$Ad\,(q)=\rho_*^{-1}\circ Ad\,(A)\circ\rho_*$$

so that

$$A=m^{\mathscr{E}}_{\mathscr{E}}(\alpha^{-1}\circ Ad\,(q)\circ\alpha)=m^{\mathscr{E}}_{\mathscr{E}}(\lambda^{-1}\circ Ad\,(A)\circ\lambda\,)\,.$$

This shows that

$$Ad\,(A)(\lambda\,(u))=\lambda\,(Au)$$

for every matrix $A\in SO(3)$ and column vector $u\in\mathbf{R}^3$.

9.19 Consider the smooth map $q:\mathbf{R}^3\to\mathbf{S}^3$ given by

$$q(\phi,\theta,\psi)=exp\left(\tfrac{\phi}{2}\,k\right)\,exp\left(\tfrac{\theta}{2}\,i\right)\,exp\left(\tfrac{\psi}{2}\,k\right)$$

for $(\phi,\theta,\psi)\in\mathbf{R}^3$, which corresponds to performing in succession a rotation by an angle ψ about the z-axis, by an angle θ about the x-axis, and finally by an angle ϕ about the z-axis. We say that $q(\phi,\theta,\psi)$ is the quaternion determined by the *Euler angles* ϕ, θ and ψ. Since $p\,i=i\bar{p}$ for every quaternion p, we have that

$$\begin{aligned}
q(\phi,\theta,\psi)&=exp\left(\tfrac{\phi}{2}\,k\right)\,(\cos\tfrac{\theta}{2}+\sin\tfrac{\theta}{2}\,i)\,exp\left(\tfrac{\psi}{2}\,k\right)\\
&=exp\left(\tfrac{\phi+\psi}{2}\,k\right)\cos\tfrac{\theta}{2}+i\,exp\left(\tfrac{\psi-\phi}{2}\,k\right)\sin\tfrac{\theta}{2}\\
&=exp\left(\tfrac{\phi+\psi}{2}\,k\right)\cos\tfrac{\theta}{2}+j\,exp\left(\tfrac{\psi-\phi+\pi}{2}\,k\right)\sin\tfrac{\theta}{2}\\
&=a+jb
\end{aligned}$$

for complex numbers

$$a = exp\left(\tfrac{\phi+\psi}{2}\, i\right)\cos\tfrac{\theta}{2} \quad\text{and}\quad b = exp\left(\tfrac{\psi-\phi+\pi}{2}\, i\right)\sin\tfrac{\theta}{2}$$

so that

$$|a| = \cos\tfrac{\theta}{2}\ ,\quad \arg a = \tfrac{\phi+\psi}{2}\ ,\quad |b| = \sin\tfrac{\theta}{2} \quad\text{and}\quad \arg b = \tfrac{\psi-\phi+\pi}{2}\,.$$

This shows that q maps the set

$$U = \{(\phi,\theta,\psi) \mid 0 \le \phi < 2\pi,\, 0 \le \theta \le \pi,\, -2\pi \le \psi < 2\pi\}$$

onto $\mathbf{S}^3$. Furthermore, q induces a bijection $\tilde{q}$ between the subsets

$$\{(\phi,\theta,\psi) \in U \mid 0 < \theta < \pi\} \quad\text{and}\quad \{a + jb \in \mathbf{S}^3 \mid ab \neq 0\}$$

of U and $\mathbf{S}^3$, so that $\tilde{q}(\phi,\theta,\psi) = a + jb$ if and only if

$$\phi = \arg a - \arg b + \tfrac{\pi}{2}\ ,\quad \theta = 2\ \arg(\,|a| + |b|\,i\,) \quad\text{and}\quad \psi = \arg a + \arg b - \tfrac{\pi}{2}\,.$$

By restricting the Euler angles to the open set

$$U^\circ = \{(\phi,\theta,\psi) \mid 0 < \phi < 2\pi,\, 0 < \theta < \pi,\, -2\pi < \psi < 2\pi\}\,,$$

they therefore are the coordinate functions of a local chart on $\mathbf{S}^3$.

Combining q with the double covering projection $\rho : \mathbf{S}^3 \to SO(3)$ defined in 9.16, we also obtain a smooth map $R = \rho \circ q : \mathbf{R}^3 \to SO(3)$ given by

$$R(\phi,\theta,\psi) = exp(\phi\ \mathscr{L}_3)\ exp(\theta\ \mathscr{L}_1)\ exp(\psi\ \mathscr{L}_3)$$

$$= \begin{pmatrix} \cos\phi & -\sin\phi\cos\theta & \sin\phi\sin\theta \\ \sin\phi & \cos\phi\cos\theta & -\cos\phi\sin\theta \\ 0 & \sin\theta & \cos\theta \end{pmatrix} \begin{pmatrix} \cos\psi & -\sin\psi & 0 \\ \sin\psi & \cos\psi & 0 \\ 0 & 0 & 1 \end{pmatrix} =$$

$$\begin{pmatrix} \cos\phi\cos\psi - \sin\phi\sin\psi\cos\theta & -\cos\phi\sin\psi - \sin\phi\cos\psi\cos\theta & \sin\phi\sin\theta \\ \sin\phi\cos\psi + \cos\phi\sin\psi\cos\theta & -\sin\phi\sin\psi + \cos\phi\cos\psi\cos\theta & -\cos\phi\sin\theta \\ \sin\psi\sin\theta & \cos\psi\sin\theta & \cos\theta \end{pmatrix}$$

for $(\phi,\theta,\psi) \in \mathbf{R}^3$, mapping the set

$$V = \{(\phi,\theta,\psi) \mid 0 \le \phi < 2\pi,\, 0 \le \theta \le \pi,\, 0 \le \psi < 2\pi\}$$

onto $SO(3)$. By restricting the Euler angles to the open set

$$V^\circ = \{(\phi,\theta,\psi) \mid 0 < \phi < 2\pi,\, 0 < \theta < \pi,\, 0 < \psi < 2\pi\}\,,$$

they are the coordinate functions of a local chart on $SO(3)$.

9.20 Remark The quaternion $q = q(\phi + \frac{\pi}{2}, \theta, 0)$ corresponds to performing a rotation by an angle θ about the x-axis, followed by a rotation by an angle $\phi + \frac{\pi}{2}$ about the z-axis. Hence it will transform the vector k to the vector $w = (\sin\theta\cos\phi, \sin\theta\sin\phi, \cos\theta)$ in agreement with the direct calculation

$$\begin{aligned}
Ad\,(q)(k) &= q k q^{-1} \\
&= exp\left(\tfrac{\phi+\pi/2}{2}\,k\right) exp\left(\tfrac{\theta}{2}\,i\right) k \; exp\left(-\tfrac{\theta}{2}\,i\right) exp\left(-\tfrac{\phi+\pi/2}{2}\,k\right) \\
&= exp\left(\tfrac{\phi+\pi/2}{2}\,k\right) exp(\theta\, i)\, k \; exp\left(-\tfrac{\phi+\pi/2}{2}\,k\right) \\
&= exp\left(\tfrac{\phi+\pi/2}{2}\,k\right) (\cos\theta + \sin\theta\; i) \; exp\left(-\tfrac{\phi-\pi/2}{2}\,k\right) \qquad (1) \\
&= exp\left(\tfrac{\pi}{2}\,k\right)\cos\theta + \; exp(\phi\; k)\,\sin\theta\; i \\
&= \cos\theta\; k + (\cos\phi + \sin\phi\; k)\,\sin\theta\; i \\
&= \sin\theta\cos\phi\;\; i + \sin\theta\sin\phi\;\; j + \cos\theta\;\, k = t
\end{aligned}$$

where $t = (0, w)$. From this it also follows from 9.7 that

$$exp\left(\tfrac{\psi}{2}\,t\right) = exp \circ Ad\,(q)\left(\tfrac{\psi}{2}\,k\right) = i_q \circ exp\left(\tfrac{\psi}{2}\,k\right) = q \; exp\left(\tfrac{\psi}{2}\,k\right) q^{-1}\,.$$

This shows that a rotation by an angle ψ about the axis through the origin with direction vector w can be performed as a composition of rotations about the x- and z-axis in the following way :

(1) First rotate by the angle $-(\phi + \frac{\pi}{2})$ about the z-axis so that w is transformed to the vector $(0, -\sin\theta, \cos\theta)$ in the yz-plane.

(2) Rotate by the angle $-\theta$ about the x-axis so that this vector is transformed to the vector k along the z-axis.

(3) Now perform the rotation by the angle ψ about the z-axis.

(4) Rotate by the angle θ about the x-axis so that k is transformed back to the vector $(0, -\sin\theta, \cos\theta)$.

(5) Finally rotate by the angle $\phi + \frac{\pi}{2}$ about the z-axis so that this vector is transformed back to w.

THE LORENTZ GROUP AND $Sl(2, \mathbf{C})$

9.21 Let $\widetilde{R} = \alpha \circ R : \mathbf{R}^3 \to \mathscr{L}^{\uparrow +}$ be the composition of R with the injective Lie group homomorphism $\alpha : SO(3) \to \mathscr{L}^{\uparrow +}$ described in 8.62, and consider the smooth map $\Lambda : \mathbf{R}^6 \to \mathscr{L}^{\uparrow +}$ given by

$$\Lambda(\phi_1, \theta_1, \chi, \phi_2, \theta_2, \psi_2) = \widetilde{R}(\phi_1 + \tfrac{\pi}{2}, \theta_1, 0)\ exp\,(\chi \mathscr{B}_3)\ \widetilde{R}(\phi_2, \theta_2, \psi_2)$$

for $(\phi_1, \theta_1, \chi, \phi_2, \theta_2, \psi_2) \in \mathbf{R}^6$. We contend that Λ maps the set $W \times \mathbf{R} \times V$ onto $\mathscr{L}^{\uparrow +}$, where

$$W = \{(\phi, \theta) \mid -\tfrac{\pi}{2} \le \phi < \tfrac{3\pi}{2}, 0 \le \theta \le \pi\}$$

and V is the set defined in 9.19.

To show this, suppose that $A \in \mathscr{L}^{\uparrow +}$, and let

$$r = (A_{00}^2 - 1)^{1/2} \quad \text{and} \quad \chi = \log\,(A_{00} + r)$$

so that

$$\sum_{r=1}^{3} A_{r0}^2 = r^2 \ , \ A_{00}^2 - r^2 = 1 \ , \ \cosh \chi = A_{00} \ \text{ and } \ \sinh \chi = r \, .$$

Now choose $(\phi_1, \theta_1) \in W$ so that

$$A_{10} = r \sin \theta_1 \cos \phi_1 \ , \ A_{20} = r \sin \theta_1 \sin \phi_1 \ \text{ and } \ A_{30} = r \cos \theta_1 \, ,$$

using spherical coordinates for the vector (A_{10}, A_{20}, A_{30}). Then we have that

$\widetilde{R}(\phi_1 + \frac{\pi}{2}, \theta_1, 0)\ exp\,(\chi \mathscr{B}_3)\ e_0$

$$= \begin{pmatrix} 1 & 0 & 0 & 0 \\ 0 & -\sin\phi_1 & -\cos\theta_1\cos\phi_1 & \sin\theta_1\cos\phi_1 \\ 0 & \cos\phi_1 & -\cos\theta_1\sin\phi_1 & \sin\theta_1\sin\phi_1 \\ 0 & 0 & \sin\theta_1 & \cos\theta_1 \end{pmatrix} \begin{pmatrix} \cosh\chi & 0 & 0 & \sinh\chi \\ 0 & 1 & 0 & 0 \\ 0 & 0 & 1 & 0 \\ \sinh\chi & 0 & 0 & \cosh\chi \end{pmatrix} \begin{pmatrix} 1 \\ 0 \\ 0 \\ 0 \end{pmatrix}$$

$$= \begin{pmatrix} 1 & 0 & 0 & 0 \\ 0 & -\sin\phi_1 & -\cos\theta_1\cos\phi_1 & \sin\theta_1\cos\phi_1 \\ 0 & \cos\phi_1 & -\cos\theta_1\sin\phi_1 & \sin\theta_1\sin\phi_1 \\ 0 & 0 & \sin\theta_1 & \cos\theta_1 \end{pmatrix} \begin{pmatrix} A_{00} \\ 0 \\ 0 \\ r \end{pmatrix} = \begin{pmatrix} A_{00} \\ A_{10} \\ A_{20} \\ A_{30} \end{pmatrix} = A\, e_0 \, .$$

By 9.19 and Remark 8.63 there are therefore Euler angles $(\phi_2, \theta_2, \psi_2) \in V$ so that $\Lambda(\phi_1, \theta_1, \chi, \phi_2, \theta_2, \psi_2) = A$, thus completing the proof of the assertion. By restricting the parameters to the open set $W^\circ \times \mathbf{R}^* \times V^\circ$, where $\mathbf{R}^* = \mathbf{R} - \{0\}$ and

$$W^\circ = \{(\phi, \theta) \mid -\tfrac{\pi}{2} < \phi < \tfrac{3\pi}{2}, 0 < \theta < \pi\}\, ,$$

they are the coordinate functions of a local chart on $\mathscr{L}^{\uparrow +}$.

As the proper Lorentz group $\mathscr{L}^{\uparrow +}$ is the image of the connected set $W \times \mathbf{R} \times V$ by the continuous map Λ, it must be connected. Hence it follows from Proposition 13.37 in the appendix that the sets $\mathscr{L}^{\uparrow +}$, $\mathscr{L}^{\uparrow -}$, $\mathscr{L}^{\downarrow +}$ and $\mathscr{L}^{\downarrow -}$ defined in 8.61 are the connected components of the Lorentz group $\mathscr{L}$.

9.22 Let $A = \phi \circ q : \mathbf{R}^3 \to SU(2)$ be the composition of the smooth map $q : \mathbf{R}^3 \to \mathbf{S}^3$ with the Lie group isomorphism $\phi : \mathbf{S}^3 \to SU(2)$ described in 9.19 and Example 8.67, and consider the smooth map $S : \mathbf{R}^6 \to Sl(2,\mathbf{C})$ given by

$$S(\phi_1,\theta_1,\chi,\phi_2,\theta_2,\psi_2) = A(\phi_1 + \tfrac{\pi}{2},\theta_1,0)\ exp\,(\tfrac{\chi}{2}\,\sigma_3)\ A(\phi_2,\theta_2,\psi_2)$$

for $(\phi_1,\theta_1,\chi,\phi_2,\theta_2,\psi_2) \in \mathbf{R}^6$. We contend that S maps the set $W \times \mathbf{R} \times U$ onto $Sl(2,\mathbf{C})$, where W and U are the sets defined in 9.19 and 9.21.

To show this, suppose that $B \in Sl(2,\mathbf{C})$, and let $x \in \mathbf{R}^4$ be the vector given by

$$iBB^* = B\,\tau_0\,B^* = \sum_{r=0}^{3} x_r\,\tau_r = \begin{pmatrix} i(x_0+x_3) & -x_2+ix_1 \\ x_2+ix_1 & i(x_0-x_3) \end{pmatrix}$$

satisfying

$$-x_0^2+x_1^2+x_2^2+x_3^2 = \det(iBB^*) = -1\,,$$

where $\mathscr{C} = \{\tau_0,\tau_1,\tau_2,\tau_3\}$ is the basis for $\mathfrak{u}(2)$ introduced in 8.65. In the same way as in 9.21 we let

$$r = (x_0^2-1)^{1/2} \quad \text{and} \quad \chi = \log\,(x_0+r)$$

so that

$$\sum_{r=1}^{3} x_r^2 = r^2\,,\quad x_0^2 - r^2 = 1\,,\quad \cosh\chi = x_0 \quad \text{and} \quad \sinh\chi = r\,.$$

We choose $(\phi_1,\theta_1) \in W$ so that

$$x_1 = r\sin\theta_1\cos\phi_1\,,\quad x_2 = r\sin\theta_1\sin\phi_1 \quad \text{and} \quad x_3 = r\cos\theta_1\,,$$

using spherical coordinates for the vector (x_1,x_2,x_3). Now we have that

$$iA(\phi_1+\tfrac{\pi}{2},\theta_1,0)\ exp\,(\chi\,\sigma_3)\,A(\phi_1+\tfrac{\pi}{2},\theta_1,0)^*$$

$$= exp\left(\tfrac{\phi+\pi/2}{2}\,\tau_3\right)\, exp\,(\tfrac{\theta}{2}\,\tau_1)\,(\cosh\chi\ \tau_0 + \sinh\chi\ \tau_3)\, exp\,(-\tfrac{\theta}{2}\,\tau_1)\ exp\left(-\tfrac{\phi+\pi/2}{2}\,\tau_3\right)$$

$$= x_0\,\tau_0 + r\ exp\left(\tfrac{\phi+\pi/2}{2}\,\tau_3\right)\, exp\,(\tfrac{\theta}{2}\,\tau_1)\,\tau_3\ exp\,(-\tfrac{\theta}{2}\,\tau_1)\ exp\left(-\tfrac{\phi+\pi/2}{2}\,\tau_3\right)$$

$$= x_0\,\tau_0 + r\,(\sin\theta\,\cos\phi\ \ \tau_1 + \sin\theta\,\sin\phi\ \ \tau_2 + \cos\theta\ \ \tau_3) = \sum_{r=0}^{3} x_r\,\tau_r = iBB^*$$

where the third equality is obtained by applying the Lie group isomorphism $\phi : \mathbf{S}^3 \to SU(2)$ to formula (1) in Remark 9.20. This implies that

$$\{A(\phi_1+\tfrac{\pi}{2},\theta_1,0)\ exp\,(\tfrac{\chi}{2}\,\sigma_3)\}^{-1}\,B \in SU(2)\,.$$

By 9.19 there are therefore Euler angles $(\phi_2,\theta_2,\psi_2) \in U$ so that $S(\phi_1,\theta_1,\chi,\phi_2,\theta_2,\psi_2) = B$, thus completing the proof of the assertion. By restricting the parameters to the open set $W^\circ \times \mathbf{R}^* \times U^\circ$, they are the coordinate functions of a local chart on $Sl(2,\mathbf{C})$.

It also follows that $Sl(2,\mathbf{C})$ is connected, as it is the image of the connected set $W \times \mathbf{R} \times U$ by the continuous map S.

9.23 Consider the representation $\widehat{\sigma} : Sl(2,\mathbf{C}) \to Gl\,(\mathfrak{u}(2))$ defined by

$$\widehat{\sigma}\,(B)(A) = BAB^* = (iB)A\,(iB)^*$$

for $B \in Sl(2,\mathbf{C})$ and $A \in \mathfrak{u}(2)$. Using the linear isomorphism $\beta : \mathbf{R}^4 \to \mathfrak{u}(2)$ given by

$$\beta(x) = \sum_{r=0}^{3} x_r \tau_r = \begin{pmatrix} i(x_0+x_3) & -x_2+ix_1 \\ x_2+ix_1 & i(x_0-x_3) \end{pmatrix}$$

for $x \in \mathbf{R}^4$, we also have a representation $\widetilde{\sigma} : Sl(2,\mathbf{C}) \to Gl\,(4,\mathbf{R})$ defined by

$$\widetilde{\sigma}\,(B) = m^{\mathscr{E}}_{\mathscr{E}}(\beta^{-1} \circ \widehat{\sigma}\,(B) \circ \beta)$$

for $B \in Sl(2,\mathbf{C})$, where $\mathscr{E}$ is the standard basis in $\mathbf{R}^4$.

Since $\det(\widehat{\sigma}\,(B)(A)) = \det(A)$ for $A \in \mathfrak{u}(2)$ when $B \in Sl(2,\mathbf{C})$, and

$$\det(\beta(x)) = -x_0^2 + x_1^2 + x_2^2 + x_3^2$$

for $x \in \mathbf{R}^4$, it follows that $\widetilde{\sigma}\,(Sl(2,\mathbf{C})) \subset O_1(4)$. Actually, $\widetilde{\sigma}\,(Sl(2,\mathbf{C})) \subset L^{\uparrow+}$ as $\widetilde{\sigma}$ is a continuous map from the connected set $Sl(2,\mathbf{C})$. We want to show that $\widetilde{\sigma}\,(Sl(2,\mathbf{C})) = L^{\uparrow+}$, and that the homomorphism $\sigma : Sl(2,\mathbf{C}) \to L^{\uparrow+}$ induced by $\widetilde{\sigma}$ is a double covering of $L^{\uparrow+}$.

Let $\mathscr{B} = \{u,v,w\}$ be a positively oriented orthonormal basis in $\mathbf{R}^3$, and let $r = \beta(0,u)$, $s = \beta(0,v)$ and $t = \beta(0,w)$ be the corresponding matrices in $\mathfrak{su}(2) \subset \mathfrak{u}(2)$. Since β coincides with the algebra homomorphism $\phi : \mathbf{H} \to \mathfrak{gl}(2,\mathbf{C})$ described in 8.65 on $\{0\} \times \mathbf{R}^3$, we have that

$$r^2 = s^2 = t^2 = -\sigma_0$$

and

$$rs = -sr = t \quad , \quad st = -ts = r \quad \text{and} \quad tr = -rt = s\,.$$

Now let $B = exp\,(\frac{\chi}{2}\,b)$ where $b = w_1\,\sigma_1 + w_2\,\sigma_2 + w_3\,\sigma_3$ and $\chi \in \mathbf{R}$. Since $t = ib$, we have that $b^2 = \sigma_0$ so that

$$iB = \cosh\tfrac{\chi}{2}\;\tau_0 + \sinh\tfrac{\chi}{2}\;t \quad \text{and} \quad (iB)^* = -iB\,.$$

This implies that

$$\begin{aligned} \widehat{\sigma}\,(B)(\tau_0) &= \cosh\chi\;\tau_0 + \sinh\chi\;t\,, \\ \widehat{\sigma}\,(B)(r) &= r\,, \\ \widehat{\sigma}\,(B)(s) &= s\,, \\ \widehat{\sigma}\,(B)(t) &= \sinh\chi\;\tau_0 + \cosh\chi\;t\,, \end{aligned}$$

so that

$$m_{\mathscr{C}}^{\mathscr{C}}(\beta^{-1} \circ \widehat{\sigma}(B) \circ \beta) = \begin{pmatrix} \cosh\chi & 0 & 0 & \sinh\chi \\ 0 & 1 & 0 & 0 \\ 0 & 0 & 1 & 0 \\ \sinh\chi & 0 & 0 & \cosh\chi \end{pmatrix}$$

using the basis $\mathscr{C} = \{e_0, (0,u), (0,v), (0,w)\}$ in $\mathbf{R}^4$. Hence $\beta^{-1} \circ \widehat{\sigma}(B) \circ \beta$ is a velocity transformation or boost with velocity parameter χ along the axis through the origin with direction vector w. In particular, we have that

$$\sigma \circ exp\,(\tfrac{\chi}{2}\,\sigma_3) = exp\,(\chi\,\mathscr{B}_3)$$

for $\chi \in \mathbf{R}$. Since we also have that $\sigma \circ A = \widetilde{R}$, it follows from 9.21 and 9.22 that $\widetilde{\sigma}\,(Sl(2,\mathbf{C})) = L^{\uparrow +}$. If $B \in \ker\sigma$, then $B \in SU(2)$ since $\widehat{\sigma}\,(B)(\tau_0) = \tau_0$. Hence it follows from 9.16 that $\ker\sigma = \{\sigma_0, -\sigma_0\}$ which is a discrete subgroup of $Sl(2,\mathbf{C})$. By Proposition 13.111 in the appendix we therefore conclude that the homomorphism $\sigma : Sl(2,\mathbf{C}) \to L^{\uparrow +}$ is a double covering of the proper Lorentz group $L^{\uparrow +}$.

Since

$$\sigma_*|_{\mathfrak{su}\,(2)} = \alpha_* \circ \rho_* \circ \phi_*^{-1}$$

and

$$\sigma \circ exp\,(\tfrac{\chi}{2}\,b) = exp\,\big(\chi\,\sigma_*(\tfrac{b}{2})\big)$$

for $\chi \in \mathbf{R}$, it follows from 8.62, 8.65, 9.16 and Example 8.67 that the Lie algebra isomorphism $\sigma_* : \mathfrak{sl}(2,\mathbf{C}) \to \mathfrak{o}_1(4)$ maps the basis $\{\tau_1/2, \tau_2/2, \tau_3/2, \sigma_1/2, \sigma_2/2, \sigma_3/2\}$ for $\mathfrak{sl}(2,\mathbf{C})$ onto the basis $\{\mathscr{R}_1, \mathscr{R}_2, \mathscr{R}_3, \mathscr{B}_1, \mathscr{B}_2, \mathscr{B}_3\}$ for $\mathfrak{o}_1(4)$.

9.24 Remark The matrix $A = A(\phi + \frac{\pi}{2}, \theta, 0)$ in $SU(2)$ corresponds to performing a rotation by an angle θ about the x-axis, followed by a rotation by an angle $\phi + \frac{\pi}{2}$ about the z-axis. Hence it will transform the vector k to the vector $w = (\sin\theta\cos\phi, \sin\theta\sin\phi, \cos\theta)$. By applying the Lie group isomorphism $\phi : \mathbf{S}^3 \to SU(2)$ to formula (1) in Remark 9.20 and dividing by i, we obtain

$$Ad\,(A)(\sigma_3) = w_1\,\sigma_1 + w_2\,\sigma_2 + w_3\,\sigma_3 = b$$

so that

$$exp\,(\tfrac{\chi}{2}\,b) = exp \circ Ad\,(A)\,(\tfrac{\chi}{2}\,\sigma_3) = i_A \circ exp\,(\tfrac{\chi}{2}\,\sigma_3) = A\;exp\,(\tfrac{\chi}{2}\,\sigma_3)\,A^{-1}\,.$$

Hence a boost with velocity parameter χ along the axis through the origin with direction vector w can be performed as a composition of the five transformations described in Remark 9.20, where the rotation (3) is replaced by a boost with velocity parameter χ along the z-axis.

9.25 Remark The physical interpretation of the coordinate x_0 is that $x_0 = ct$, where t is time and c is the velocity of light in vacuum. The velocity v is given by

$$\tanh\chi = v/c\,,$$

where χ is the velocity parameter. From this it follows that

$$\cosh\chi = (1 - v^2/c^2)^{-1/2} = \gamma \quad\text{and}\quad \sinh\chi = \gamma v/c\,.$$

The boost $exp\,(\chi\mathscr{B}_3)$ gives the connection between the coordinates (ct,x,y,z) and (ct',x',y',z') with respect to two inertial systems S and S' with parallel coordinate axes, where S' is moving with constant velocity v with respect to S in the positive z-direction so that their origins coincide at the time $t = t' = 0$. Then we have that

$$\begin{pmatrix} ct \\ x \\ y \\ z \end{pmatrix} = \begin{pmatrix} \gamma & 0 & 0 & \gamma v/c \\ 0 & 1 & 0 & 0 \\ 0 & 0 & 1 & 0 \\ \gamma v/c & 0 & 0 & \gamma \end{pmatrix} \begin{pmatrix} ct' \\ x' \\ y' \\ z' \end{pmatrix}$$

which implies that

$$x = x'\,,\ y = y'\,,\ z = \gamma(z' + vt')\,,\ t = \gamma(t' + z'v/c^2)\,.$$

SEMIDIRECT PRODUCTS

9.26 Proposition Let $F_i : V_i \to V_i$ be a linear map in the finite dimensional vector space V_i for $i = 1,2,3$, and let $B : V_1 \times V_2 \to V_3$ be a bilinear map. Then we have that

$$F_3(B(u,v)) = B(F_1(u),v) + B(u,F_2(v)) \tag{1}$$

for $(u,v) \in V_1 \times V_2$ if and only if

$$exp\,(t\,F_3)(B(u,v)) = B(exp\,(t F_1)(u), exp\,(t F_2)(v)) \tag{2}$$

for $(u,v) \in V_1 \times V_2$ and $t \in \mathbf{R}$.

PROOF : By 4.23 the relation (1) implies that

$$t\,F_3 \circ B_* = B_* \circ (t\,F_1 \otimes id_{V_2} + id_{V_1} \otimes t\,F_2)$$

for $t \in \mathbf{R}$, where $B_* : V_1 \otimes V_2 \to V_3$ is the linear map induced by B. Hence we have that

$$(t\,F_3)^k \circ B_* = B_* \circ (t\,F_1 \otimes id_{V_2} + id_{V_1} \otimes t\,F_2)^k$$

for every positive integer k, which implies that

$$\left(\sum_{k=0}^{n} \frac{(t\,F_3)^k}{k!}\right) \circ B_* = B_* \circ \sum_{k=0}^{n} \frac{(t\,F_1 \otimes id_{V_2} + id_{V_1} \otimes t\,F_2)^k}{k!}$$

for every positive integer n. By letting $n \to \infty$, we thus obtain

$$exp\,(t F_3) \circ B_* = B_* \circ exp\,(t F_1 \otimes id_{V_2} + id_{V_1} \otimes t F_2)\,.$$

Now using the relation

$$\sum_{k=0}^{n} \frac{(t F_1 \otimes id_{V_2})^k}{k!} = \left(\sum_{k=0}^{n} \frac{(t F_1)^k}{k!} \right) \otimes id_{V_2}$$

and letting $n \to \infty$, we obtain the formula

$$exp\,(t F_1 \otimes id_{V_2}) = exp\,(t F_1) \otimes id_{V_2}\,,$$

and similarly we have that

$$exp\,(id_{V_1} \otimes t F_2) = id_{V_1} \otimes exp\,(t F_2)\,.$$

Since

$$(t F_1 \otimes id_{V_2}) \circ (id_{V_1} \otimes t F_2) = (id_{V_1} \otimes t F_2) \circ (t F_1 \otimes id_{V_2})\,,$$

it therefore follows from Proposition 13.126 in the appendix that

$$exp\,(t F_1 \otimes id_{V_2} + id_{V_1} \otimes t F_2) = exp\,(t F_1 \otimes id_{V_2}) \circ exp\,(id_{V_1} \otimes t F_2)$$
$$= \{exp\,(t F_1) \otimes id_{V_2}\} \circ \{id_{V_1} \otimes exp\,(t F_2)\} = exp\,(t F_1) \otimes exp\,(t F_2)$$

so that

$$exp\,(t F_3) \circ B_* = B_* \circ \{exp\,(t F_1) \otimes exp\,(t F_2)\}$$

which completes the proof of (2).

Conversely, fix $(u,v) \in V_1 \times V_2$ and assume that (2) is satisfied for every $t \in \mathbf{R}$. Let $c : \mathbf{R} \to V_3$ be the curve defined by

$$c(t) = B(exp\,(tF_1)(u), exp\,(tF_2)(v))$$

for $t \in \mathbf{R}$. If we canonically identify $T_{c(0)}V_3$ with V_3 as in Lemma 2.84 and use Remark 2.75 and Proposition 8.29 and 8.31, we have that

$$F_3(B(u,v)) = c'(0) = B(F_1(u),v) + B(u,F_2(v))$$

which shows that (1) is satisfied. □

9.27 Proposition The set of Lie algebra automorphisms

$$\mathrm{Aut}(\mathfrak{g}) = \{\phi \in Gl\,(\mathfrak{g}) | \phi\,([X,Y]) = [\phi\,(X), \phi\,(Y)] \text{ for } X,Y \in \mathfrak{g}\}$$

in a Lie algebra $\mathfrak{g}$ is a closed Lie subgroup of $Gl\,(\mathfrak{g})$ in the relative topology with the Lie subalgebra

$$\partial\,(\mathfrak{g}) = \{\psi \in \mathfrak{gl}\,(\mathfrak{g}) | \psi([X,Y]) = [\psi(X),Y] + [X,\psi(Y)] \text{ for } X,Y \in \mathfrak{g}\}$$

of $\mathfrak{gl}(\mathfrak{g})$, consisting of the derivations in $\mathfrak{g}$, as its Lie algebra.

PROOF: Since $\mathrm{Aut}(\mathfrak{g})$ is a closed subgroup of $Gl(\mathfrak{g})$, it follows from Theorem 8.43 that it is a closed Lie subgroup of $Gl(\mathfrak{g})$ in the relative topology. Let $\mathfrak{h}$ be its Lie algebra.

By Proposition 8.41 we know that an element $\psi \in \mathfrak{gl}(\mathfrak{g})$ belongs to $\mathfrak{h}$ if and only if $exp\,(t\,\psi) \in \mathrm{Aut}(\mathfrak{g})$ for every $t \in \mathbf{R}$. Hence we conclude that $\mathfrak{h} = \partial(\mathfrak{g})$ by Proposition 9.26, using the bilinear map $B : \mathfrak{g} \times \mathfrak{g} \to \mathfrak{g}$ given by $B(X,Y) = [X,Y]$ for $X,Y \in \mathfrak{g}$. □

9.28 Let $\mathfrak{g}$ and $\mathfrak{h}$ be Lie algebras, and let $\sigma : \mathfrak{g} \to \partial(\mathfrak{h})$ be a Lie algebra homomorphism of $\mathfrak{g}$ into the Lie algebra $\partial(\mathfrak{h})$ of derivations in $\mathfrak{h}$. Then we have a Lie algebra structure on $\mathfrak{g} \times \mathfrak{h}$ given by

$$[(X_1,Y_1),(X_2,Y_2)] = ([X_1,X_2],\, [Y_1,Y_2] + \sigma(X_1)(Y_2) - \sigma(X_2)(Y_1))$$

for $X_1,X_2 \in \mathfrak{g}$ and $Y_1,Y_2 \in \mathfrak{h}$, satisfying the Jacobi identity since

$$\begin{aligned}
&[(X_1,Y_1),[(X_2,Y_2),(X_3,Y_3)]] + [(X_2,Y_2),[(X_3,Y_3),(X_1,Y_1)]] \\
&+ [(X_3,Y_3),[(X_1,Y_1),(X_2,Y_2)]] \\
&= (0,\, [Y_1,\sigma(X_2)(Y_3) - \sigma(X_3)(Y_2)] \\
&\quad + \sigma(X_1)\{[Y_2,Y_3] + \sigma(X_2)(Y_3) - \sigma(X_3)(Y_2)\} - \sigma([X_2,X_3])(Y_1) \\
&\quad + [Y_2,\sigma(X_3)(Y_1) - \sigma(X_1)(Y_3)] \\
&\quad + \sigma(X_2)\{[Y_3,Y_1] + \sigma(X_3)(Y_1) - \sigma(X_1)(Y_3)\} - \sigma([X_3,X_1])(Y_2) \\
&\quad + [Y_3,\sigma(X_1)(Y_2) - \sigma(X_2)(Y_1)] \\
&\quad + \sigma(X_3)\{[Y_1,Y_2] + \sigma(X_1)(Y_2) - \sigma(X_2)(Y_1)\} - \sigma([X_1,X_2])(Y_3)) \\
&= (0,\, \{-[Y_2,\sigma(X_1)(Y_3)] - [\sigma(X_1)(Y_2),Y_3] + \sigma(X_1)([Y_2,Y_3])\} \\
&\quad + \{\sigma(X_2)\circ\sigma(X_3) - \sigma(X_3)\circ\sigma(X_2) - \sigma([X_2,X_3])\}(Y_1) \\
&\quad + \{-[Y_3,\sigma(X_2)(Y_1)] - [\sigma(X_2)(Y_3),Y_1] + \sigma(X_2)([Y_3,Y_1])\} \\
&\quad + \{\sigma(X_3)\circ\sigma(X_1) - \sigma(X_1)\circ\sigma(X_3) - \sigma([X_3,X_1])\}(Y_2) \\
&\quad + \{-[Y_1,\sigma(X_3)(Y_2)] - [\sigma(X_3)(Y_1),Y_2] + \sigma(X_3)([Y_1,Y_2])\} \\
&\quad + \{\sigma(X_1)\circ\sigma(X_2) - \sigma(X_2)\circ\sigma(X_1) - \sigma([X_1,X_2])\}(Y_3)) = (0,0)
\end{aligned}$$

for $X_1,X_2,X_3 \in \mathfrak{g}$ and $Y_1,Y_2,Y_3 \in \mathfrak{h}$. We see that $\mathfrak{g} \times \mathfrak{h}$ is a Lie algebra with this Lie algebra structure. It is called the *semidirect product* of $\mathfrak{g}$ and $\mathfrak{h}$ with respect to σ, and it is denoted by $\mathfrak{g} \times_\sigma \mathfrak{h}$.

The Lie algebra homomorphisms $i_1 : \mathfrak{g} \to \mathfrak{g} \times_\sigma \mathfrak{h}$ and $i_2 : \mathfrak{h} \to \mathfrak{g} \times_\sigma \mathfrak{h}$, given by

$i_1(X) = (X,0)$ and $i_2(Y) = (0,Y)$ for $X \in \mathfrak{g}$ and $Y \in \mathfrak{h}$, are injections so that $i_1(\mathfrak{g})$ is a Lie subalgebra and $i_2(\mathfrak{h})$ is an ideal in $\mathfrak{g} \times_\sigma \mathfrak{h}$ with

$$\mathfrak{g} \times_\sigma \mathfrak{h} = i_1(\mathfrak{g}) \oplus i_2(\mathfrak{h}) .$$

We also have a Lie algebra homomorphism $\psi_1 : \mathfrak{g} \times_\sigma \mathfrak{h} \to \mathfrak{g}$ given by $\psi_1(X,Y) = X$ for $X \in \mathfrak{g}$ and $Y \in \mathfrak{h}$.

9.29 Proposition Let $\mathfrak{k}$ be a Lie algebra, and let $\mathfrak{g}$ be a Lie subalgebra and $\mathfrak{h}$ be an ideal in $\mathfrak{k}$ so that

$$\mathfrak{k} = \mathfrak{g} \oplus \mathfrak{h} .$$

Then the map $\sigma : \mathfrak{g} \to \partial(\mathfrak{h})$ given by $\sigma(X)(Y) = [X,Y]$ for $X \in \mathfrak{g}$ and $Y \in \mathfrak{h}$, is a Lie algebra homomorphism of $\mathfrak{g}$ into the Lie algebra $\partial(\mathfrak{h})$ of derivations in $\mathfrak{h}$, and the map $\psi : \mathfrak{g} \times_\sigma \mathfrak{h} \to \mathfrak{k}$ given by $\psi(X,Y) = X + Y$ for $X \in \mathfrak{g}$ and $Y \in \mathfrak{h}$, is a Lie algebra isomorphism.

PROOF: Since $\mathfrak{h}$ is an ideal in $\mathfrak{k}$, we have that $\sigma(X)(\mathfrak{h}) \subset \mathfrak{h}$ for every $X \in \mathfrak{g}$, and $\sigma(X)$ is a derivation in $\mathfrak{h}$ by the Jacobi identity which implies that

$$\begin{aligned}
&\sigma(X)([Y_1,Y_2]) = [X,[Y_1,Y_2]] = [[X,Y_1],Y_2] + [Y_1,[X,Y_2]] \\
&= [\sigma(X)(Y_1),Y_2] + [Y_1,\sigma(X)(Y_2)]
\end{aligned}$$

for every $X \in \mathfrak{g}$ and $Y_1,Y_2 \in \mathfrak{h}$. It also implies that

$$\begin{aligned}
&\sigma([X_1,X_2])(Y) = [[X_1,X_2],Y] = [X_1,[X_2,Y]] - [X_2,[X_1,Y]] \\
&= \{\sigma(X_1) \circ \sigma(X_2) - \sigma(X_2) \circ \sigma(X_1)\}(Y)
\end{aligned}$$

for every $X_1,X_2 \in \mathfrak{g}$ and $Y \in \mathfrak{h}$, which shows that $\sigma : \mathfrak{g} \to \partial(\mathfrak{h})$ is a Lie algebra homomorphism. Finally, we see that $\psi : \mathfrak{g} \times_\sigma \mathfrak{h} \to \mathfrak{k}$ is a linear isomorphism which is also a Lie algebra isomorphism since

$$\begin{aligned}
&\psi([(X_1,Y_1),(X_2,Y_2)]) = \psi([X_1,X_2],\, [Y_1,Y_2] + \sigma(X_1)(Y_2) - \sigma(X_2)(Y_1)) \\
&= [X_1,X_2] + [Y_1,Y_2] + [X_1,Y_2] + [Y_1,X_2] = [X_1 + Y_1, X_2 + Y_2] \\
&= [\psi(X_1,Y_1), \psi(X_2,Y_2)]
\end{aligned}$$

for $X_1,X_2 \in \mathfrak{g}$ and $Y_1,Y_2 \in \mathfrak{h}$. □

9.30 Let G and H be Lie groups, and let $\rho : G \to \mathrm{Aut}(H)$ be a homomorphism of G into the group $\mathrm{Aut}(H)$ of Lie group automorphisms of H, so that the map $\phi : G \times H \to H$ given by $\phi(g,h) = \rho(g)(h)$ is smooth. Then we have a group structure on $G \times H$ given by

$$(g_1,h_1)(g_2,h_2) = (g_1 g_2, h_1\, \rho(g_1)(h_2))$$

for $g_1, g_2 \in G$ and $h_1, h_2 \in H$, satisfying the associative law since

$$\{(g_1,h_1)(g_2,h_2)\}(g_3,h_3) = (g_1g_2, h_1\,\rho(g_1)(h_2))(g_3,h_3)$$
$$= ((g_1g_2)g_3, h_1\,\rho(g_1)(h_2)\,\rho(g_1g_2)(h_3)) = (g_1(g_2g_3), h_1\,\rho(g_1)(h_2\,\rho(g_2)(h_3)))$$
$$= (g_1,h_1)(g_2g_3, h_2\,\rho(g_2)(h_3)) = (g_1,h_1)\{(g_2,h_2)(g_3,h_3)\}$$

for $g_1, g_2, g_3 \in G$ and $h_1, h_2, h_3 \in H$. The unit element is (e_G, e_H), where e_G and e_H are the unit elements in G and H, and we have that

$$(g,h)^{-1} = (g^{-1}, \rho(g^{-1})(h^{-1}))$$

for $g \in G$ and $h \in H$. We see that $G \times H$ is a Lie group with this group structure and the product manifold structure. It is called the *semidirect product* of G and H with respect to ρ, and it is denoted by $G \times_\rho H$.

The Lie group homomorphisms $i_1 : G \to G \times_\rho H$ and $i_2 : H \to G \times_\rho H$, given by $i_1(g) = (g, e_H)$ and $i_2(h) = (e_G, h)$ for $g \in G$ and $h \in H$, are embeddings so that $i_1(G)$ and $i_2(H)$ are closed Lie subgroups of $G \times_\rho H$, and we have that

$$(g_1,h_1)(g_2,h_2)(g_1,h_1)^{-1} = (g_1 g_2 g_1^{-1}, h_1\,\rho(g_1)(h_2)\,\rho(g_1 g_2 g_1^{-1})(h_1^{-1}))$$

for $g_1, g_2 \in G$ and $h_1, h_2 \in H$. In particular, we have that

$$i_1(g)\, i_2(h)\, i_1(g)^{-1} = i_2 \circ \rho(g)(h) \tag{1}$$

for $g \in G$ and $h \in H$. We also have a Lie group homomorphism $\psi_1 : G \times_\rho H \to G$ given by $\psi_1(g,h) = g$ for $g \in G$ and $h \in H$.

Let $\mathfrak{g}$ and $\mathfrak{h}$ be the Lie algebras of G and H, respectively, and let $\mathfrak{k}$ be the Lie algebra of $G \times_\rho H$. By Propositions 8.8 and 9.27 and Corollaries 9.5 and 2.38 we have a Lie group homomorphism $\phi : G \to \mathrm{Aut}(\mathfrak{h})$ given by $\phi(g) = \rho(g)_*$ for $g \in G$, and $\sigma = \phi_* : \mathfrak{g} \to \partial(\mathfrak{h})$ is therefore a Lie algebra homomorphism. We want to show that the map

$$\psi : \mathfrak{g} \times_\sigma \mathfrak{h} \to \mathfrak{k}$$

given by

$$\psi(X,Y) = i_{1*}(X) + i_{2*}(Y)$$

for $X \in \mathfrak{g}$ and $Y \in \mathfrak{h}$, is a Lie algebra isomorphism. By Proposition 2.74 we know that ψ is a linear isomorphism since

$$\mathfrak{k} = i_{1*}(\mathfrak{g}) \oplus i_{2*}(\mathfrak{h}),$$

where $i_{1*} : \mathfrak{g} \to \mathfrak{k}$ and $i_{2*} : \mathfrak{h} \to \mathfrak{k}$ are injective Lie algebra homomorphisms. Now it follows from (1) that

$$i_{i_1(g)} \circ i_2 = i_2 \circ \rho(g)$$

for $g \in G$, where $i_{i_1(g)} : G \times_\rho H \to G \times_\rho H$ is the inner automorphism of $G \times_\rho H$ by $i_1(g)$ as defined in 9.7, and this implies that

$$Ad_{i_1(g)} \circ i_{2*} = i_{2*} \circ \phi(g)$$

so that

$$\Lambda_1 \circ Ad \circ i_1 = \Lambda_2 \circ \phi ,$$

where $\Lambda_1 : \mathfrak{gl}(\mathfrak{k}) \to L(\mathfrak{h},\mathfrak{k})$ and $\Lambda_2 : \mathfrak{gl}(\mathfrak{h}) \to L(\mathfrak{h},\mathfrak{k})$ are the linear maps given by $\Lambda_1(F) = F \circ i_{2*}$ and $\Lambda_2(G) = i_{2*} \circ G$ for $F \in \mathfrak{gl}(\mathfrak{k})$ and $G \in \mathfrak{gl}(\mathfrak{h})$. Hence we have that

$$\Lambda_1 \circ ad \circ i_{1*} = \Lambda_2 \circ \sigma ,$$

which means that

$$[i_{1*}(X), i_{2*}(Y)] = i_{2*}(\sigma(X)(Y))$$

for $X \in \mathfrak{g}$ and $Y \in \mathfrak{h}$. This implies that ψ is a Lie algebra isomorphism since

$$\psi([(X_1,Y_1),(X_2,Y_2)]) = \psi([X_1,X_2], [Y_1,Y_2] + \sigma(X_1)(Y_2) - \sigma(X_2)(Y_1))$$
$$= i_{1*}([X_1,X_2]) + i_{2*}([Y_1,Y_2] + \sigma(X_1)(Y_2) - \sigma(X_2)(Y_1))$$
$$= [i_{1*}(X_1), i_{1*}(X_2)] + [i_{2*}(Y_1), i_{2*}(Y_2)] + [i_{1*}(X_1), i_{2*}(Y_2)] + [i_{2*}(Y_1), i_{1*}(X_2)]$$
$$= [i_{1*}(X_1) + i_{2*}(Y_1), i_{1*}(X_2) + i_{2*}(Y_2)] = [\psi(X_1,Y_1), \psi(X_2,Y_2)]$$

for $X_1,X_2 \in \mathfrak{g}$ and $Y_1,Y_2 \in \mathfrak{h}$.

AFFINE SPACES

9.31 Definition A set E on which a vector space V acts freely and transitively is called an *affine space*, and the action of a vector $v \in V$ on a point $p \in E$ is denoted by $p+v$. We say that E is *modelled* on the vector space V. Each vector $v \in V$ can be identified with a bijection $T_v : E \to E$ given by $T_v(p) = p+v$ for $p \in E$. This T_v is called the *translation* on E by v, and V is called the vector space of translations on E. For each pair of points $p,q \in E$ there is a unique vector $v \in V$ with $p+v = q$ which is denoted by $q-p$.

A map $F : E_1 \to E_2$ between the affine spaces E_1 and E_2 is called *affine* if there is a linear map $T : V_1 \to V_2$ between the corresponding vector spaces of translations so that

$$F(p+v) = F(p) + T(v) \tag{1}$$

for every $p \in E_1$ and $v \in V_1$. The map T is called the *linear part* of F. If $G : E_2 \to E_3$ is an affine map into the affine space E_3 with linear part $S : V_2 \to V_3$, then the composition $G \circ F$ is again an affine map with linear part $S \circ T$.

A subset E_1 of an affine space E_2 is called an *affine subspace* of E_2 if it is an affine space having the same vector space V of translations as E_2, so that the inclusion map $i : E_1 \to E_2$ is an affine map with the identity map $id_V : V \to V$ as its linear part. This means that each translation on E_2 by a vector $v \in V$ maps E_1 onto itself and induces the translation on E_1 by v.

9.32 Proposition Let $F : E_1 \to E_2$ be an affine map with $T : V_1 \to V_2$ as its linear part. Then F is bijective if and only if T is bijective, and the inverse map $F^{-1} : E_2 \to E_1$ is affine with $T^{-1} : V_2 \to V_1$ as its linear part.

PROOF : The first part of the propositiom follows from the relation

$$T(p_2 - p_1) = F(p_2) - F(p_1)$$

for $p_1, p_2 \in E_1$. This also implies that

$$T^{-1}(q_2 - q_1) = F^{-1}(q_2) - F^{-1}(q_1)$$

for $q_1, q_2 \in E_2$, completing the proof of the last part of the propsition. □

9.33 Example A vector space V acts freely and transitively on itself and is therefore an affine space. A map $F : V_1 \to V_2$ between the vector spaces V_1 and V_2 is affine if it is the composition of a linear map $T : V_1 \to V_2$ and a translation T_w on V_2 by a vector $w \in V_2$, i.e.,

$$F = T_w \circ T$$

so that

$$F(v) = T(v) + w$$

for every $v \in V_1$. This follows from formula (1) in Definition 9.31 by taking $p = 0$ and $w = F(0)$. Conversely, if F satisfies the above formula, then

$$F(u+v) = T(u+v) + w = T(u) + T(v) + w = F(u) + T(v)$$

for every $u, v \in V_1$, showing that F is affine with T as its linear part.

9.34 Example Let E be an affine space modelled on a vector space V, and fix a point $o \in E$. Then we have a bijection $S_o : V \to E$ given by $S_o(v) = o + v$ for $v \in V$, which is an affine map with the identity map $id_V : V \to V$ as its linear part. The inverse map $S_o^{-1} : E \to V$ is given by $S_o^{-1}(p) = p - o$ for $p \in E$, and the vector $p - o$ is called the *position vector* of p. There is a unique vector space structure on E such that S_o is a linear isomorphism, which is given by

$$a p + b q = S_o(a S_o^{-1}(p) + b S_o^{-1}(q))$$

i.e.,

$$a p + b q = o + a(p - o) + b(q - o)$$

for $a, b \in \mathbf{R}$ and $p, q \in E$. In this vector space structure o becomes the zero vector.

9.35 Example Let $\rho : G \to Gl(V)$ be a representation of a Lie group G in a finite dimensional vector space V, and consider the semidirect product $G \times_\rho V$. Let $GA(V)$ be the group of invertible affine maps $F : V \to V$ given by $F(w) = T(w) + v$

for $w \in V$, where $T \in Gl\,(V)$ and $v \in V$, and where the group product is composition. Then we have a homomorphism $\sigma : G \times_\rho V \to GA\,(V)$ given by

$$\sigma(g,v)(w) = \rho(g)(w) + v$$

for $(g,v) \in G \times_\rho V$ and $w \in V$. Indeed, we have that

$$\begin{aligned}\sigma((g_1,v_1)(g_2,v_2))(w) &= \sigma(g_1g_2, \rho(g_1)(v_2)+v_1)(w) = \rho(g_1g_2)(w) + \rho(g_1)(v_2) + v_1 \\ &= \rho(g_1)(\rho(g_2)(w)+v_2) + v_1 = \sigma(g_1,v_1) \circ \sigma(g_2,v_2)(w)\end{aligned}$$

for every $(g_1,v_1),(g_2,v_2) \in G \times_\rho V$ and $w \in V$. In the special case where $\rho = id : Gl\,(V) \to Gl\,(V)$, σ is a group isomorphism, and we give $GA\,(V)$ the manifold structure so that σ is a Lie group isomorphism.

We now have a homomorphism $\psi : GA\,(V) \to Gl\,(V \times \mathbf{R})$ which is given by

$$\psi \circ \sigma\,(T,v)(w,t) = (\,T(w) + t\,v,\ t)$$

for $(T,v) \in Gl\,(V) \times_{id} V$ and $(w,t) \in V \times \mathbf{R}$. Indeed, we have that

$$\begin{aligned}&\psi(\,\sigma\,(T_1,v_1) \circ \sigma\,(T_2,v_2))\,(w,t) = \psi \circ \sigma\,((\,T_1,v_1)(\,T_2,v_2))\,(w,t) \\ &= \psi \circ \sigma\,(\,T_1 \circ T_2, T_1(v_2)+v_1)(w,t) = (\,T_1 \circ T_2(w) + T_1(t\,v_2) + t\,v_1, t) \\ &= (\,T_1(\,T_2(w) + t\,v_2) + t\,v_1, t) = \{\psi \circ \sigma\,(\,T_1,v_1)\} \circ \{\psi \circ \sigma\,(\,T_2,v_2)\}\,(w,t)\end{aligned}$$

for every $(T_1,v_1),(T_2,v_2) \in Gl\,(V) \times_{id} V$ and $(w,t) \in V \times \mathbf{R}$. If $\alpha : V \to V \times \mathbf{R}$ is the affine map defined by $\alpha(w) = (w,1)$ for $w \in V$, then

$$\psi(F) \circ \alpha = \alpha \circ F$$

for every $F \in GA\,(V)$. Now since $\alpha(V) = V \times \{1\}$, we obtain a map $\psi_1 : GA\,(V) \to GA\,(V \times \{1\})$ which is given by

$$\psi_1 \circ \sigma\,(T,v)(w,1) = (\,T(w) + v,\ 1)$$

for $(T,v) \in Gl\,(V) \times_{id} V$ and $w \in V$. If we let $\alpha_1 : V \to V \times \{1\}$ be the invertible affine map induced by α, it follows that

$$F = \alpha_1^{-1} \circ \psi_1(F) \circ \alpha_1$$

for every $F \in GA\,(V)$. Given a basis $\mathscr{B} = \{v_1, ..., v_n\}$ for V, we have that $\mathscr{C} = \{(v_1,0), ..., (v_n,0), (0,1)\}$ is a basis for $V \times \mathbf{R}$. If $m^{\mathscr{B}}_{\mathscr{B}}(T) = A$ and $[v]_{\mathscr{B}} = a$, then

$$m^{\mathscr{C}}_{\mathscr{C}}(\psi \circ \sigma\,(T,v)) = \begin{pmatrix} A & a \\ 0 & 1 \end{pmatrix}.$$

Let $\phi : Gl\,(V \times \mathbf{R}) \to Gl\,(n+1, \mathbf{R})$ be the Lie group isomorphism given by $\phi\,(S) = m^{\mathscr{C}}_{\mathscr{C}}(S)$ for $S \in Gl\,(V \times \mathbf{R})$, and let $i_1 : Gl\,(V) \to Gl\,(V) \times_{id} V$ and $i_2 : V \to$

$Gl(V) \times_{id} V$ be the Lie group homomorphisms defined as in 9.30 by $i_1(T) = (T, 0)$ and $i_2(v) = (I_n, v)$ for $T \in Gl(V)$ and $v \in V$. Then $\phi \circ \psi \circ \sigma \circ i_1(Gl(V)) = G$, $\phi \circ \psi \circ \sigma \circ i_2(V) = H$ and $\phi \circ \psi \circ \sigma(Gl(V) \times_{id} V) = K$ are the closed Lie subgroups of $Gl(n+1, \mathbf{R})$ given by

$$G = \{\begin{pmatrix} A & 0 \\ 0 & 1 \end{pmatrix} | A \in Gl(n, \mathbf{R})\} \quad , \quad H = \{\begin{pmatrix} I_n & a \\ 0 & 1 \end{pmatrix} | a \in \mathbf{R}^n\}$$

and

$$K = \{\begin{pmatrix} A & a \\ 0 & 1 \end{pmatrix} | A \in Gl(n, \mathbf{R}) \text{ and } a \in \mathbf{R}^n\} .$$

For every $B \in \mathfrak{gl}(n, \mathbf{R})$ and $a \in \mathbf{R}^n$ we have that

$$exp\ t\begin{pmatrix} B & 0 \\ 0 & 0 \end{pmatrix} = \begin{pmatrix} exp\,(tB) & 0 \\ 0 & 1 \end{pmatrix}$$

and

$$exp\ t\begin{pmatrix} 0 & a \\ 0 & 0 \end{pmatrix} = \begin{pmatrix} I_n & ta \\ 0 & 1 \end{pmatrix}$$

for $t \in \mathbf{R}$, which follows from the relation

$$\sum_{k=0}^{n} \frac{1}{k!} \begin{pmatrix} tB & 0 \\ 0 & 0 \end{pmatrix}^k = \begin{pmatrix} \sum_{k=0}^{n} \frac{(tB)^k}{k!} & 0 \\ 0 & 1 \end{pmatrix}$$

when $n \to \infty$, and the fact that

$$\begin{pmatrix} 0 & ta \\ 0 & 0 \end{pmatrix}^k = 0$$

when $k \geq 2$. Hence the Lie algebras of G, H and K are the Lie subalgebras of $\mathfrak{gl}(n+1, \mathbf{R})$ given by

$$\mathfrak{g} = \{\begin{pmatrix} B & 0 \\ 0 & 0 \end{pmatrix} | B \in \mathfrak{gl}(n, \mathbf{R})\} \quad , \quad \mathfrak{h} = \{\begin{pmatrix} 0 & a \\ 0 & 0 \end{pmatrix} | a \in \mathbf{R}^n\}$$

and

$$\mathfrak{k} = \mathfrak{g} \oplus \mathfrak{h} = \{\begin{pmatrix} B & a \\ 0 & 0 \end{pmatrix} | B \in \mathfrak{gl}(n, \mathbf{R}) \text{ and } a \in \mathbf{R}^n\} .$$

The bracket product in $\mathfrak{gl}(V) \times_{id} V$ is defined as in 9.28 and 9.30 by

$$[(T_1, v_1), (T_2, v_2)] = (T_1 \circ T_2 - T_2 \circ T_1, T_1(v_2) - T_2(v_1))$$

for $(T_1, v_1), (T_2, v_2) \in \mathfrak{gl}(V) \times_{id} V$.

9.36 Lemma Let E be an affine space modelled on a vector space V of dimension n. For each point $p \in E$ we have a bijective affine map $S_p : V \to E$ defined by $S_p(v) =$

$p+v$ for $v \in V$. We give E the manifold structure so that S_p is a diffeomorphism, which is independent of the choice of p. Then the tangent space T_pE at any point $p \in E$ may be identified with V by means of the linear isomorphism $\omega_p : T_pE \to V$ given by $\omega_p = x^{-1} \circ t_{x',p}$ for any bijective affine map $x' : E \to \mathbf{R}^n$ with linear part $x : V \to \mathbf{R}^n$, ω_p being independent of the choice of x'. We will refer to this as the *canonical identification.*

If $\mathscr{E} = \{e_1, ..., e_n\}$ is the standard basis for $\mathbf{R}^n$ and $\mathscr{B} = \{v_1, ..., v_n\}$ is a basis for V so that $x(v_i) = e_i$ for $i = 1, ..., n$, then $\mathscr{B}^* = \{x^1, ..., x^n\}$ is the dual basis of $\mathscr{B}$, and

$$\omega_p \left(\sum_{j=1}^{n} a^j \frac{\partial}{\partial x'^j}\bigg|_p \right) = \sum_{j=1}^{n} a^j v_j \,.$$

If $l \in T_pE$, then

$$x^i \circ \omega_p (l) = l\,(x'^{\,i})$$

for $i = 1, ..., n$, and

$$\lambda \circ \omega_p (l) = l\,(\lambda \circ S_p^{-1})$$

for any linear functional λ on V.

The map $\omega = (id_E \times x^{-1}) \circ t_{x'} : TE \to E \times V$ is an equivalence over E so that

$$\omega \left(\sum_{j=1}^{n} a^j \frac{\partial}{\partial x'^j}\bigg|_p \right) = (p, x^{-1}(a))$$

and

$$\omega^{-1}(p,v) = \sum_{j=1}^{n} x^j(v) \, \frac{\partial}{\partial x'^j}\bigg|_p$$

for $p \in E$, $v \in V$ and $a \in \mathbf{R}^n$.

If $F : E_1 \to E_2$ is an affine map with $T : V_1 \to V_2$ as its linear part, and if the tangent spaces T_pE_1 and $T_{F(p)}E_2$ are canonically identified with V_1 and V_2 by means of the linear isomorphisms $\omega_p : T_pE_1 \to V_1$ and $\omega_{F(p)} : T_{F(p)}E_2 \to V_2$, then we have that

$$\omega_{F(p)} \circ F_{*p} \circ \omega_p^{-1} = T \,.$$

PROOF : Since $S_q = S_p \circ T_v$ when $q = S_p(v)$ and $T_v : V \to V$ is translation by v, the manifold structure on E is independent of the choice of p. We have that (x', E) is a local chart on E since $x' \circ S_p(v) = x'(p) + x(v)$ for $v \in V$, which implies that $x' = T_{x'(p)} \circ x \circ S_p^{-1}$ where $T_{x'(p)} : \mathbf{R}^n \to \mathbf{R}^n$ is translation by $x'(p)$. As $t_{x',p} : T_pV \to \mathbf{R}^n$ is a linear isomorphism, the same is true for the map ω_p. To show that it is independent of the choice of x', let $y' : E \to \mathbf{R}^n$ be another bijective affine map with linear part $y : V \to \mathbf{R}^n$. Then it follows from 2.70 that

$$t_{y',p} \circ t_{x',p}^{-1} = D\,(y' \circ x'^{-1})(x'(p)) = D\,(y \circ x^{-1})(0) = y \circ x^{-1} \,,$$

so that

$$y^{-1} \circ t_{y',p} = x^{-1} \circ t_{x',p} \,.$$

To prove the next part of the lemma, let $l = \sum_{j=1}^{n} a^j \frac{\partial}{\partial x'^j}\Big|_p$. Then

$$\omega_p(l) = x^{-1}(a) = \sum_{j=1}^{n} a^j v_j ,$$

and

$$x^i \circ \omega_p(l) = a^i = l(x'^i)$$

for $i = 1, ..., n$. If

$$\lambda = \sum_{j=1}^{n} b_j x^j ,$$

then

$$\lambda \circ \omega_p(l) = \sum_{j=1}^{n} b_j x^j \circ \omega_p(l) = \sum_{j=1}^{n} b_j l(x'^j) = l(\sum_{j=1}^{n} b_j x'^j) = l(\lambda \circ S_p^{-1}) .$$

Now assume that the linear isomorphisms $\omega_p : T_pE_1 \to V_1$ and $\omega_{F(p)} : T_{F(p)}E_2 \to V_2$ are obtained from the bijective affine maps $x' : E_1 \to \mathbf{R}^n$ and $y' : E_2 \to \mathbf{R}^m$ with linear parts $x : V_1 \to \mathbf{R}^n$ and $y : V_2 \to \mathbf{R}^m$, respectively. Then the last part of the lemma follows from the commutative diagram in 2.70 which shows that

$$\begin{aligned}
&\omega_{F(p)} \circ F_{*p} \circ \omega_p^{-1} = y^{-1} \circ t_{y',F(p)} \circ F_{*p} \circ t_{x',p}^{-1} \circ x \\
&= y^{-1} \circ D(y' \circ F \circ x'^{-1})(x(p)) \circ x \\
&= y^{-1} \circ D(y \circ S_{F(p)}^{-1} \circ F \circ S_p \circ x^{-1})(0) \circ x \\
&= y^{-1} \circ y \circ T \circ x^{-1} \circ x = T
\end{aligned}$$

since

$$F \circ S_p(v) = F(p+v) = F(p) + T(v) = S_{F(p)} \circ T(v)$$

for $v \in V$. □

INFINITESIMAL GROUP ACTIONS

9.37 Proposition Let G be a Lie group acting on the smooth manifold M on the right. For each $p \in M$, we let $\sigma_p : G \to M$ be the smooth map defined by $\sigma_p(g) = pg$ for $g \in G$. Then we have a Lie algebra homomorphism $\sigma : \mathfrak{g} \to \mathscr{T}_1 M$ defined by

$$\sigma(X)_p = \sigma_{p*}(X)$$

for $X \in \mathfrak{g}$ and $p \in M$, and $\sigma(X)$ is σ_p-related to the left invariant vector field $\widetilde{X}$ on G determined by X. The homomorphism σ is called the *infinitesimal action* of G on M on the right.

PROOF: We first need to show that $\sigma(X)$ is smooth on M so that it is a vector field on M. Let (z,U) be a local chart around a point p on M, and let n be the dimension of M at p. If $\nu : M \times G \to M$ is the operation of G on M, there is an open neighbourhood V of p on M and a local chart (y,W) around e on G so that $\nu(V \times W) \subset U$, and we let $x = z|_V$. Then the map $f : x(V) \times y(W) \to \mathbf{R}^n$ defined by $f = z \circ \nu \circ (x^{-1} \times y^{-1})$ is smooth, and we have that

$$f(a,b) = z \circ \sigma_{x^{-1}(a)} \circ y^{-1}(b)$$

for $a \in x(V)$ and $b \in y(W)$.

Now let $(t_z, \pi^{-1}(U))$ be the local trivialization in the tangent bundle $\pi : TM \to M$ assosiated with the local chart (z,U) on M. If $X = [y,v]_e$, it follows that

$$\begin{aligned}\sigma(X) \circ x^{-1}(a) &= \sigma_{x^{-1}(a)*}(X) = [z, D(z \circ \sigma_{x^{-1}(a)} \circ y^{-1})(y(e))\,v]_{x^{-1}(a)} \\ &= [z, D_2 f(a, y(e))\,v]_{x^{-1}(a)}\end{aligned}$$

so that

$$(z \times id) \circ t_z \circ \sigma(X) \circ x^{-1}(a) = (a, D_2 f(a, y(e))\,v)$$

for $a \in x(V)$, thus showing that $\sigma(X)$ is smooth in V.

To see that the vector field $\sigma(X)$ is σ_p-related to $\widetilde{X}$ for every $X \in \mathfrak{g}$ and $p \in M$, we first note that

$$\sigma_p \circ L_g = \sigma_{pg}$$

for every $g \in G$, since

$$\sigma_p \circ L_g(h) = \sigma_p(gh) = p(gh) = (pg)h = \sigma_{pg}(h)$$

for $h \in G$, and this implies that

$$\sigma_{p*}(\widetilde{X}_g) = \sigma_{p*} \circ L_{g*}(X) = (\sigma_p \circ L_g)_*(X) = \sigma_{pg*}(X) = \sigma(X)_{\sigma_p(g)}\,.$$

From this it now follows that σ is a Lie algebra homomorphism. Indeed, we see immediately from the definition that σ is linear, and if $X, Y \in \mathfrak{g}$, then $[\sigma(X), \sigma(Y)]$ is σ_p-related to $[\widetilde{X}, \widetilde{Y}]$ by Proposition 4.88, so that

$$\sigma([X,Y])_p = \sigma_{p*}([\widetilde{X}, \widetilde{Y}]_e) = [\sigma(X), \sigma(Y)]_p$$

for every $p \in M$. □

9.38 Proposition If G is a Lie group acting on the smooth manifold M on the right, then

$$(R_g)_*\,\sigma(X) = \sigma(Ad(g^{-1})\,X)$$

for every $X \in \mathfrak{g}$ and $g \in G$.

PROOF: Since

$$R_g \circ \sigma_p(h) = phg = pg\, g^{-1}hg = \sigma_{pg} \circ i_{g^{-1}}(h)$$

for $p \in M$ and $h \in G$, we have that

$$((R_g)_* \, \sigma(X))_{pg} = (R_g)_* \, \sigma(X)_p = (R_g)_* \circ \sigma_{p*}(X)$$

$$= \sigma_{pg*} \circ Ad\,(g^{-1})\,(X) = \sigma\,(Ad\,(g^{-1})\,X)_{pg}\,. \qquad \square$$

9.39 Proposition Let G be a Lie group acting on the smooth manifold M on the right. Then the vector field $\sigma(X)$ defined in Proposition 9.37 is complete, with flow $\gamma : \mathbf{R} \times M \to M$ given by

$$\gamma(t,p) = p\; exp\,(tX)\,.$$

In particular, we have that

$$\sigma(X)_p = \gamma_p{}'(0)\,.$$

If G acts effectively on M, then σ is injective, and if G acts freely on M, then $\sigma(X)$ is nowhere zero when $X \neq 0$.

PROOF: From Proposition 9.37 we know that $\sigma(X)$ is σ_p-related to the left invariant vector field $\widetilde{X}$ on G determined by X, and the one-parameter subgroup $\phi_X : \mathbf{R} \to G$ defined in Proposition 8.29 is an integral curve for $\widetilde{X}$ with initial condition e. Hence it follows from Proposition 3.48 that $\gamma_p = \sigma_p \circ \phi_X$ is an integral curve for $\sigma(X)$ with initial condition $\sigma_p(e) = p$, showing that γ is the flow for $\sigma(X)$.

Now suppose that G acts effectively on M and that $\sigma(X) = 0$. Then $\gamma(t,p) = p$ for every $p \in M$ and $t \in \mathbf{R}$. Since G acts effectively, this implies that $\phi_X(t) = e$ for every $t \in \mathbf{R}$ which shows that $X = 0$.

Finally, suppose that G acts freely on M and that $\sigma(X)_p = 0$ for a point $p \in M$. For each $g \in G$, we let $R_g : M \to M$ be the diffeomorphism defined by $R_g(p) = pg$ for $p \in M$. Since

$$\gamma_p(s+t) = R_{\,exp\,(tX)} \circ \gamma_p(s)$$

for $s,t \in \mathbf{R}$, we have that

$$\gamma_p{}'(t) = R_{\,exp\,(tX)\,*} \circ \gamma_p{}'(0) = 0$$

for $t \in \mathbf{R}$. This implies that $\gamma_p(t) = p$ and therefore $\phi_X(t) = e$ for every $t \in \mathbf{R}$, since G acts freely on M, thus showing that $X = 0$. $\square$

9.40 Proposition Let G be a Lie group acting on the smooth manifold M on the left. For each $p \in M$, we let $\sigma_p : G \to M$ be the smooth map defined by $\sigma_p(g) = gp$ for $g \in G$. Then we have a Lie algebra anti-homomorphism $\sigma : \mathfrak{g} \to \mathscr{T}_1 M$ defined by

$$\sigma(X)_p = \sigma_{p*}(X)$$

for $X \in \mathfrak{g}$ and $p \in M$, satisfying

$$(L_g)^* \, \sigma(X) = \sigma(Ad\,(g^{-1})\,X)$$

for every $X \in \mathfrak{g}$ and $g \in G$. The vector field $\sigma(X)$ is complete, with flow $\gamma : \mathbf{R} \times M \to M$ given by

$$\gamma(t,p) = exp\,(tX)\; p\,.$$

In particular, we have that

$$\sigma(X)_p = \gamma_p{}'(0)\,.$$

PROOF : The map $\widetilde{\mu} : M \times G \to M$ defined by

$$\widetilde{\mu}(p,g) = g^{-1}p$$

for $p \in M$ and $g \in G$, is an action of G on M on the right. We let $\widetilde{\sigma} : \mathfrak{g} \to \mathscr{T}_1 M$ be the corresponding Lie algebra homomorphism defined in Proposition 9.37. Then we have that

$$\sigma_p = \widetilde{\sigma}_p \circ \phi$$

for $p \in M$, where $\phi : G \to G$ is the smooth map given by $\phi(g) = g^{-1}$ for $g \in G$. By Corollary 8.33 it follows that

$$\sigma = -\widetilde{\sigma}\,,$$

and the proposition now follows from Propositions 9.37, 9.38 and 9.39. □

9.41 Proposition Let $\mu : G \times M \to M$ and $\mu' : G \times M' \to M'$ be two operations of a Lie group G on the smooth manifolds M and M' on the left, and let $\sigma : \mathfrak{g} \to \mathscr{T}_1 M$ and $\sigma' : \mathfrak{g} \to \mathscr{T}_1 M'$ be the corresponding Lie algebra anti-homomorphisms defined in Proposition 9.40. If $f : M \to M'$ is a smooth map which is equivariant with respect to the actions μ and μ', then the vector fields $\sigma(X)$ and $\sigma'(X)$ are f-related for every $X \in \mathfrak{g}$.

PROOF : We have that

$$f \circ \sigma_p(g) = f \circ L_g(p) = L'_g \circ f(p) = \sigma'_{f(p)}(g)$$

for $g \in G$ and $p \in M$, which implies that

$$f_* \circ \sigma(X)(p) = f_* \circ \sigma_{p\,*}(X) = (f \circ \sigma_p)_*(X) = \sigma'_{f(p)\,*}(X) = \sigma'(X) \circ f(p)$$

for $p \in M$ and $X \in \mathfrak{g}$. □

HAMILTONIAN SYSTEMS WITH SYMMETRY

9.42 Definition An action $\mu : G \times M \to M$ of a Lie group G on a symplectic manifold (M, ω) on the left is called *symplectic* if the map L_g is symplectic for every $g \in G$.

9.43 Definition Let $\mu : G \times M \to M$ be a symplectic action of a Lie group G on a symplectic manifold (M, ω) on the left. Given a smooth map $J : M \to \mathfrak{g}^*$, we have for each $\xi \in \mathfrak{g}$ a smooth function $\hat{J}(\xi) : M \to \mathbf{R}$ defined by $\hat{J}(\xi)(p) = J(p)(\xi)$ for $p \in M$. The map J is called a *momentum mapping* for μ if

$$\sigma(\xi) = X_{\hat{J}(\xi)} ,$$

which means that $\sigma(\xi)$ is the Hamiltonian vector field with energy function $\hat{J}(\xi)$, for every $\xi \in \mathfrak{g}$.

9.44 Proposition Let J be a momentum mapping for a symplectic action of a Lie group G on a symplectic manifold (M, ω) on the left, and let X_H be a Hamiltonian vector field on M with energy function H which is invariant under the action, i.e.,

$$H(p) = H(gp)$$

for every $p \in M$ and $g \in G$. If $c : I \to M$ is an integral curve for X_H, then the function $J \circ c$ is constant on I.

PROOF: If $\gamma : \mathbf{R} \times M \to M$ is the flow for $\sigma(\xi)$ defined in Proposition 9.40 for a given $\xi \in \mathfrak{g}$, we have that $H \circ \gamma_p$ is constant for each $p \in M$. Hence it follows from Proposition 7.76 that $\{H, \hat{J}(\xi)\} = 0$ on M which implies that $\hat{J}(\xi) \circ c$ is constant on I. Since this is true for every $\xi \in \mathfrak{g}$, it follows that the function $J \circ c$ is constant on I. □

9.45 Definition Let $\mu : G \times M \to M$ be a symplectic action of a Lie group G on a symplectic manifold (M, ω) on the left with a momentum mapping J. We say that J is *Ad^*-equivariant* if it is equivariant with respect to μ and the action of G on $\mathfrak{g}^*$ determined by the coadjoint representation. This means that

$$J \circ L_g = Ad\,(g^{-1})^* \circ J$$

for every $g \in G$.

9.46 Proposition Let J be an Ad^*-equivariant momentum mapping for a symplectic action of a Lie group G on a symplectic manifold (M, ω) on the left. Then we have that

$$\{\hat{J}(\xi), \hat{J}(\eta)\} = \hat{J}([\,\xi, \eta\,])$$

for every $\xi, \eta \in \mathfrak{g}$.

PROOF: Let $p \in M$ and $\xi, \eta \in \mathfrak{g}$. Using that J is Ad^*-equivariant, we have that

$$\hat{J}(\xi)(exp\,(t\eta)\,p) = J(exp\,(t\eta)\,p)(\xi)$$
$$= J(p) \circ Ad\,(exp\,(-t\eta))\,(\xi) = \alpha \circ Ad \circ exp\,(-t\eta)$$

for every $t \in \mathbf{R}$, where $\alpha : \mathrm{End}\,(\mathfrak{g}) \to \mathbf{R}$ is the linear map given by

$$\alpha(F) = J(p) \circ F(\xi)$$

for $F \in \mathrm{End}\,(\mathfrak{g})$. Taking the derivative with respect to t at 0, it follows from Propositions 7.76, 9.9 and 9.40 that

$$\{\hat{J}(\xi), \hat{J}(\eta)\}(p) = -J(p) \circ ad\,(\eta)\,(\xi) = \hat{J}([\,\xi, \eta\,])(p)\,.$$

Since this is true for every $p \in M$, we obtain the formula in the proposition for $\xi, \eta \in \mathfrak{g}$. □

9.47 Proposition Let (M, ω) be a symplectic manifold where $\omega = -d\theta$ for a 1-form θ on M, and let G be a Lie group acting on M on the left so that $L_g^*(\theta) = \theta$ for every $g \in G$. Then the map $J : M \to \mathfrak{g}^*$ given by

$$J(p)(\xi) = \theta(\sigma\,(\xi))(p)$$

for $p \in M$ and $\xi \in \mathfrak{g}$, is an Ad^*-equivariant momentum mapping for the action. We have that

$$\hat{J}(\xi) = \theta(\sigma\,(\xi))$$

for every $\xi \in \mathfrak{g}$.

PROOF: Let $\xi \in \mathfrak{g}$, and consider the flow $\gamma : \mathbf{R} \times M \to M$ for $\sigma\,(\xi)$ defined in Proposition 9.40. Since

$$\gamma_t^*\,\theta = L_{exp\,(t\,\xi)}^*\,\theta = \theta$$

for every $t \in \mathbf{R}$, it follows that the Lie derivative $L_{\sigma\,(\xi)}\,\theta = 0$. Using Proposition 5.83 we now have that

$$d\hat{J}(\xi) = d(i_{\sigma\,(\xi)}\,\theta) = -\,i_{\sigma\,(\xi)}(d\theta) = i_{\sigma\,(\xi)}(\omega)$$

which implies that

$$\sigma\,(\xi) = X_{\hat{J}(\xi)}\,,$$

thus showing that J is a momentum mapping for the action.

J is also Ad^*-equivariant, since

$$J \circ L_g(p)(\xi) = \theta(\sigma\,(\xi))\,(L_g(p)) = \theta(L_g(p)) \circ (L_g)_* \circ (L_g)_*^{-1} \circ \sigma\,(\xi) \circ L_g(p)$$
$$= L_g^*(\theta)\,\{(L_g)^*\,\sigma\,(\xi)\}\,(p) = \theta\,\{\sigma\,(Ad\,(g^{-1})\,\xi\,)\}\,(p) = J(p) \circ Ad\,(g^{-1})\,\xi$$
$$= Ad\,(g^{-1})^* \circ J(p)(\xi)$$

for every $p \in M$ and $\xi \in \mathfrak{g}$. □

9.48 Proposition Let $\pi : T^*M \to M$ be the cotangent bundle of a smooth manifold M, and let $f : M \to M$ be a diffeomorfism. Then we have a diffeomorphism $f^* : T^*M \to T^*M$ defined by

$$(f^*)_p = (f_{*q})^*$$

where $q = f^{-1}(p)$ for $p \in M$, i.e.,

$$f^*(v) = v \circ f_*$$

for $v \in T^*M$, so that (f^*, f^{-1}) is a bundle map and we have a commutative diagram

$$\begin{array}{ccc} T^*M & \xrightarrow{\;f^*\;} & T^*M \\ \Big\downarrow \pi & & \Big\downarrow \pi \\ M & \xleftarrow{\;f\;} & M \end{array}$$

We also have that $(f^*)^*\theta = \theta$, where θ is the canonical 1-form on T^*M.

PROOF : The assertion that (f^*, f^{-1}) is a bundle map follows from Proposition 4.50 applied to the bundle map (f_*^{-1}, f^{-1}) for the tangent bundle. We also have that

$$\begin{aligned} (f^*)^*\theta(v)(w) &= \theta\,(f^*(v))\,((f^*)_*(w)) = f^*(v) \circ \pi_* \circ (f^*)_*(w) \\ &= v \circ (f \circ \pi \circ f^*)_*(w) = v \circ \pi_*(w) = \theta(v)(w) \end{aligned}$$

for $v \in T^*M$ and $w \in T_v(T^*M)$, which shows the last part of the proposition. □

9.49 Proposition Let $\mu : G \times M \to M$ be an action of a Lie group G on a smooth manifold M on the left. Then we have an action $\widehat{\mu} : G \times T^*M \to T^*M$ defined by

$$\widehat{\mu}(g, v) = L^*_{g^{-1}}(v)$$

for $g \in G$ and $v \in T^*M$, which leaves the canonical 1-form θ on T^*M invariant, i.e., the left translation $\hat{L}_g$ of $\widehat{\mu}$ satisfies $(\hat{L}_g)^*\theta = \theta$ for every $g \in G$. The Ad^*-equivariant momentum mapping $J : T^*M \to \mathfrak{g}^*$ for this action defined in Proposition 9.47 is given by

$$J_u = (\sigma_{u*e})^*$$

for $u \in M$, i.e.,

$$\hat{J}(\xi)(v) = v(\,\sigma(\xi)_u)$$

for $\xi \in \mathfrak{g}$, $u \in M$ and $v \in T_u^*M$, where σ is the Lie algebra anti-homomorphism associated with the action μ as defined in Proposition 9.40.

PROOF : Let $\widehat{\sigma}$ be the Lie algebra anti-homomorphism associated with the action $\widehat{\mu}$. As the projection $\pi' : T^*M \to M$ is equivariant with respect to the actions $\widehat{\mu}$ and μ, it follows from Proposition 9.41 that $\widehat{\sigma}(\xi)$ and $\sigma(\xi)$ are π'-related. This implies that

$$\hat{f}(\xi)(v) = \theta(v)(\widehat{\sigma}(\xi)_v) = v \circ \pi'_* \circ \widehat{\sigma}(\xi)(v) = v \circ \sigma(\xi) \circ \pi'(v) = v(\sigma(\xi)_u)$$

for $u \in M$ and $v \in T_u^*M$. □

9.50 Example Let E be an affine space modelled on a finite dimensional vector space V. For each point $q \in E$ we let $\sigma_q : V \to E$ be the affine map defined by $\sigma_q(v) = q + v$ for $v \in V$, having $id_V : V \to V$ as its linear part. If the tangent spaces T_0V and T_qE are canonically identified with V by means of the linear isomorphisms $\omega_0 : T_0V \to V$ and $\omega_q : T_qE \to V$ defined in Lemmas 2.84 and 9.36, we have that

$$\omega_q \circ \sigma_{q\,*\,0} \circ \omega_0^{-1} = id_V .$$

Now let $J : T^*E \to T_0^*V$ be the Ad^*-equivariant momentum mapping for the action of V on E defined in Proposition 9.49. Then we have that $J_q = (\sigma_{q\,*\,0})^*$ so that

$$(\omega_0^*)^{-1} \circ J_q \circ \omega_q^* = id_{V^*} .$$

Identifying the cotangent spaces T_q^*E and T_0^*V with V^* by means of the linear isomorphisms $\widetilde{\omega}_q = (\omega_q^*)^{-1} : T_q^*E \to V^*$ and $(\omega_0^*)^{-1} : T_0^*V \to V^*$, it follows that

$$J(q,p) = p$$

for $(q,p) \in E \times V^*$. This momentum mapping $J : E \times V^* \to V^*$ is called the *linear momentum*.

9.51 Example Let G be a Lie subgroup of $Gl\,(n,\mathbf{R})$, and consider the action $\mu : G \times \mathbf{R}^n \to \mathbf{R}^n$ given by

$$\mu(B,q) = Bq$$

for $B \in G$ and $q \in \mathbf{R}^n$ written as a column vector. For each vector $q \in \mathbf{R}^n$ we let $\sigma_q = \rho_q|_G$, where $\rho_q : \mathfrak{gl}\,(n,\mathbf{R}) \to \mathbf{R}^n$ is the linear map given by $\rho_q(B) = Bq$ for $B \in \mathfrak{gl}\,(n,\mathbf{R})$. Now it follows from Lemma 2.84 that $\sigma_{q\,*\,e} = \rho_q|_{\mathfrak{g}}$ when we canonically identify $T_q\mathbf{R}^n$ with $\mathbf{R}^n$ and $T_e\,\mathfrak{gl}\,(n,\mathbf{R})$ with $\mathfrak{gl}\,(n,\mathbf{R})$ as in Example 8.10 (a). Hence the Ad^*-equivariant momentum mapping $J : \mathbf{R}^n \times (\mathbf{R}^n)^* \to \mathfrak{g}^*$ for the action μ defined in Proposition 9.49 is given by

$$J(q,p)(B) = p(Bq)$$

for $(q,p) \in \mathbf{R}^n \times (\mathbf{R}^n)^*$ and $B \in \mathfrak{g}$.

9.52 Example Let $J_3 : \mathbf{R}^3 \times (\mathbf{R}^3)^* \to \mathfrak{so}(3)^*$ be the Ad^*-equivariant momentum mapping for the action $\mu : SO(3) \times \mathbf{R}^3 \to \mathbf{R}^3$ of $SO(3)$ on the 3-dimensional Euclidean space $\mathbf{R}^3$ as described in Example 9.51. Using the Euclidean metric g_0^3

which is the dot product in $\mathbf{R}^3$ as defined in Definition 7.15, we may identify $(\mathbf{R}^3)^*$ with $\mathbf{R}^3$ by means of the linear isomorphism $e : \mathbf{R}^3 \to (\mathbf{R}^3)^*$ given by

$$e(u)(v) = u \cdot v$$

for $u, v \in \mathbf{R}^3$. By Remark 9.18 we may also identify the Lie algebra $\mathfrak{so}(3)$ with $\mathbf{R}^3$ by means of the Lie algebra isomorphism $\lambda : \mathbf{R}^3 \to \mathfrak{so}(3)$ given by

$$\lambda(u)\, v = u \times v$$

for column vectors $u, v \in \mathbf{R}^3$. With these identifications we obtain a momentum mapping $J = e^{-1} \circ \lambda^* \circ J_3 \circ (id_{\mathbf{R}^3} \times e) : \mathbf{R}^3 \times \mathbf{R}^3 \to \mathbf{R}^3$ given by

$$e \circ J(q,p)(u) = J_3(q, e(p))(\lambda(u)) = e(p)\,(\lambda(u)\, q)$$

$$= p \cdot (u \times q) = (q \times p) \cdot u = e(q \times p)(u)$$

for $q, p, u \in \mathbf{R}^3$ so that

$$J(q,p) = q \times p$$

for $(q,p) \in \mathbf{R}^3 \times \mathbf{R}^3$. This momentum mapping is called the *angular momentum*.

LAGRANGIAN SYSTEMS WITH SYMMETRY

9.53 Noether's theorem Let $\mu : G \times M \to M$ be an action of a Lie group G on a smooth manifold M on the left, and let $(\widetilde{\mu}, \mu)$ be the corresponding action of G on the tangent bundle $\pi : TM \to M$ defined in Proposition 9.4. Let X_E be a Lagrangian vector field for a regular Lagrangian $L : TM \to \mathbf{R}$ which is invariant under the action, i.e.,

$$L(v) = L(gv)$$

for every $v \in TM$ and $g \in G$. Then the action $\widetilde{\mu}$ leaves the Lagrangian 1-form θ_L on TM invariant, i.e., the left translation $\widetilde{L}_g$ of $\widetilde{\mu}$ satisfies $(\widetilde{L}_g)^*\, \theta_L = \theta_L$ for every $g \in G$. The Ad^*-equivariant momentum mapping $J : TM \to \mathfrak{g}^*$ for this action defined in Proposition 9.47 is given by

$$J_u = (\sigma_{u*e})^* \circ F(L)_u$$

for $u \in M$, i.e.,

$$\hat{J}(\xi)(v) = F(L)(v)\,(\sigma(\xi)_u)$$

for $\xi \in \mathfrak{g}$, $u \in M$ and $v \in T_uM$, where σ is the Lie algebra anti-homomorphism associated with the action μ as defined in Proposition 9.40. If $c : I \to TM$ is an integral curve for X_E, then the function $J \circ c$ is constant on I.

PROOF : Let $(\widehat{\mu},\mu)$ be the action of G on the cotangent bundle $\pi' : T^*M \to M$ defined in Proposition 9.49. We first show that the fibre derivative $F(L) : TM \to T^*M$ is equivariant with respect to the actions $\widetilde{\mu}$ and $\widehat{\mu}$ of G on TM and T^*M, i.e., that

$$\hat{L}_g \circ F(L) = F(L) \circ \tilde{L}_g$$

for $g \in G$. By taking the derivative at $v \in T_uM$ on both sides of

$$L_u \circ \tilde{L}_g = L_u \ ,$$

identifying $T_{gv}(T_uM)$ and $T_g(T_uM)$ with T_uM and $T_{L(v)}\mathbf{R}$ with $\mathbf{R}$ in the usual way and using that $\tilde{L}_g$ is linear on T_uM, we obtain

$$F(L)(gv) \circ \tilde{L}_g = F(L)(v) \tag{1}$$

which implies that

$$\hat{L}_g \circ F(L)(v) = F(L)(v) \circ \tilde{L}_{g^{-1}} = F(L) \circ \tilde{L}_g(v)$$

for $v \in TM$ and $g \in G$.

Using this, we now obtain that

$$\begin{aligned}(\tilde{L}_g)^*\,\theta_L &= (\tilde{L}_g)^*\,F(L)^*\,\theta = (F(L) \circ \tilde{L}_g)^*\,\theta \\ &= (\hat{L}_g \circ F(L))^*\,\theta = F(L)^*\,(\hat{L}_g)^*\,\theta = F(L)^*\,\theta = \theta_L\end{aligned}$$

for every $g \in G$. Since $\widetilde{\sigma}(\xi)$ and $\sigma(\xi)$ are π-related, we also have that

$$\begin{aligned}\hat{J}(\xi)(v) &= \theta_L(v)(\widetilde{\sigma}(\xi)_v) = \theta(F(L)(v))\,(F(L)_* \circ \widetilde{\sigma}(\xi)(v)) \\ &= F(L)(v) \circ \pi'_* \circ F(L)_* \circ \widetilde{\sigma}(\xi)(v) = F(L)(v) \circ \pi_* \circ \widetilde{\sigma}(\xi)(v) \\ &= F(L)(v) \circ \sigma(\xi) \circ \pi(v) = F(L)(v)(\sigma(\xi)_u)\end{aligned}$$

for $u \in M$ and $v \in T_uM$. From (1) we see that the action $A : TM \to \mathbf{R}$ and hence also the energy $E = A - L$ is invariant under the operation $\widetilde{\mu}$, since

$$A(v) = F(L)(v)(v) = F(L)(gv)(gv) = A(gv)$$

for every $v \in TM$ and $g \in G$. The last part of the theorem therefore follows from Proposition 9.44. □

9.54 Example Consider a conservative systen consisting of a single particle of mass m moving in $\mathbf{R}^3$, with potential energy function V and Lagrangian L given by

$$L(q,v) = \tfrac{1}{2}\,m\|v\|^{\,2} - V(q)$$

for $(q,v) \in \mathbf{R}^3 \times \mathbf{R}^3$ as described in Example 7.102. Let $\mu : SO(3) \times \mathbf{R}^3 \to \mathbf{R}^3$ be the action of $SO(3)$ on $\mathbf{R}^3$ defined by

$$\mu(B,q) = Bq$$

for $B \in SO(3)$ and $q \in \mathbf{R}^3$ written as a column vector. Using the same notation as in Example 9.52, the Ad^*-equivariant momentum mapping for this action described in Theorem 9.53 is $J = e^{-1} \circ \lambda^* \circ J_3 \circ F(L) : \mathbf{R}^3 \times \mathbf{R}^3 \to \mathbf{R}^3$, which is given by

$$J(q,v) = q \times mv$$

for $(q,v) \in \mathbf{R}^3 \times \mathbf{R}^3$. This momentum mapping is called the *angular momentum* of the system. If the potential energy function V is invariant under the action μ, i.e.,

$$V(q) = V(Bq)$$

for every $q \in \mathbf{R}^3$ and $B \in SO(3)$, then J is a constant of the motion by Noether's theorem.

GRAVITATIONAL CENTRAL FIELDS

9.55 Consider a conservative system consisting of a single particle of mass m moving in $\mathbf{R}^3$ as described in Example 9.54, where the potential energy of the particle only depends on its distance from the center of force at the origin. Then the angular momentum J is a constant of the motion, and the particle is therefore moving in a plane M through the origin having J as its normal vector. Using cylinder coordinates in $\mathbf{R}^3$ with the z-axis along J, so that (r,ϕ) are polar coordinates on M in a coordinate neighbourhood N, the Lagrangian of the system is given by

$$L(v) = \tfrac{1}{2} m(\dot{r}^2 + r^2\dot{\phi}^2) - V(r)$$

for tangent vectors $v \in TN$ with coordinates $(r,\phi,\dot{r},\dot{\phi})$ as described in Example 7.20 and Proposition 7.81. We assume that V is the potential energy of a gravitational force from a mass M at the origin, so that

$$V(r) = -\gamma \frac{Mm}{r}$$

where γ is Newton's gravitational constant.

Since ϕ is a cyclic coordinate, its conjugete momentum $p_\phi \circ F(L) = mr^2\dot{\phi}$ is a constant of the motion, which is equal to the length l of the angular momentum J. The motion of the particle in N is given by an integral curve $\gamma : I \to TN$ for the Lagrangian vector field X_E. From the Lagrange's equations it follows that

$$(r \circ \gamma)'(t) = \dot{r} \circ \gamma(t) \quad \text{and} \quad (\phi \circ \gamma)'(t) = \dot{\phi} \circ \gamma(t)$$

so that

$$m\,\{r \circ \gamma(t)\}^2\,(\phi \circ \gamma)'(t) = l \tag{1}$$

for $t \in I$. Assuming that $J \neq 0$, we see that the function $\phi \circ \gamma$ is a strictly increasing diffeomorphism onto an open interval I'. The orbit of the particle is described by the function $\rho = (r \circ \gamma) \circ (\phi \circ \gamma)^{-1} : I' \to \mathbf{R}$. Since

$$r \circ \gamma(t) = \rho(s) \quad \text{where} \quad s = \phi \circ \gamma(t), \tag{2}$$

it follows from (1) that

$$(r \circ \gamma)'(t) = \rho'(s)\,(\phi \circ \gamma)'(t) = \frac{l}{m}\frac{\rho'(s)}{\rho(s)^2} = -\frac{l}{m}\,u'(s) \tag{3}$$

where we have introduced the function $u = \dfrac{1}{\rho} : I' \to \mathbf{R}$.

By Proposition 7.87 the total energy E is a constant of the motion so that

$$E = \tfrac{1}{2}m(r \circ \gamma)'(t)^2 + U(r \circ \gamma(t)) \tag{4}$$

for $t \in I$, where

$$U(r) = \frac{l^2}{2mr^2} - \gamma\frac{Mm}{r}$$

is called the *effective potential energy*. It has an absolute minimum at the point (r_0, E_0) where

$$r_0 = \frac{l^2}{\gamma M m^2} \quad \text{and} \quad E_0 = -\frac{\gamma M m}{2r_0}. \tag{5}$$

Combining (2), (3) and (4), we now obtain the equation

$$E = \frac{l^2}{2m}u'(s)^2 + \frac{l^2}{2m}u(s)^2 - \gamma M m\, u(s)$$

which can be written in the form

$$(r_0 u'(s))^2 + (r_0 u(s) - 1)^2 = 1 - \frac{E}{E_0}$$

for $s \in I'$. Its solution is given by

$$r_0 u(s) - 1 = \varepsilon \cos(s - s_0)$$

so that

$$\rho(s) = \frac{r_0}{1 + \varepsilon\cos(s - s_0)}$$

for $s \in I'$, where s_0 is an integration constant and $\varepsilon = \left(1 - \dfrac{E}{E_0}\right)^{1/2}$. Hence the particle is moving along a conic section with one focus at the origin and with eccentricity ε. The cases $E = E_0$, $E_0 < E < 0$, $E = 0$ and $E > 0$ correspond to a circle, an ellipse, a parabola and a hyperbola with eccentricities ε satisfying $\varepsilon = 0$, $0 < \varepsilon < 1$, $\varepsilon = 1$ and $\varepsilon > 1$, respectively.

By a rotation of the coordinate system we may assume that $s_0 = 0$, so that the polar equation of the conic section is

$$r = \frac{r_0}{1 + \varepsilon\cos\phi},$$

which can be also written in the form

$$r = r_0 - \varepsilon r \cos\phi. \tag{6}$$

Introducing cartesian coordinates $x = r\cos\phi$ and $y = r\sin\phi$ in the plane M, we see that when $\varepsilon > 0$, the *eccentricity* ε is the ratio of the distances of the particle from the focus at the origin and the line $x = \frac{r_0}{\varepsilon}$, which is called the *directrice*.

In the case $0 \le \varepsilon < 1$ we obtain *Kepler's first law*, saying that the planets are moving in elliptical orbits with the sun at one focus. The other focus is the point $(-2c, 0)$, where c is called the *linear eccentricity*. We consider a circle as a special case of an ellipse where the two foci coincide. The minimal and maximal values of r are

$$r_1 = \frac{r_0}{1-\varepsilon} \quad \text{and} \quad r_2 = \frac{r_0}{1+\varepsilon}.$$

Hence the *semimajor axis* a and the linear eccentricity c are given by

$$a = \tfrac{1}{2}(r_1 + r_2) = \frac{r_0}{1-\varepsilon^2} = \frac{r_0 E_0}{E} \quad \text{and} \quad c = \tfrac{1}{2}(r_2 - r_1) = a\varepsilon, \tag{7}$$

and the *semiminor axis* is

$$b = \sqrt{a^2 - c^2} = a\sqrt{1-\varepsilon^2} = \sqrt{a r_0}. \tag{8}$$

Equation (6) can be written in cartesian form as

$$a\sqrt{x^2+y^2} = b^2 - cx \tag{9}$$

so that

$$\left(2a - \sqrt{x^2+y^2}\,\right)^2 = 4a^2 - 4(b^2 - cx) + (x^2+y^2) = (x+2c)^2 + y^2,$$

showing that the sum of the distances of the point (x, y) from the two foci is $2a$, in agreement with the usual definition of an ellipse. By Equation (9) we also have that

$$a^2(x^2+y^2) = b^2(a^2-c^2) - 2b^2cx + (a^2-b^2)x^2 = a^2(x^2+b^2) - b^2(x+c)^2$$

so that

$$\frac{(x+c)^2}{a^2} + \frac{y^2}{b^2} = 1,$$

which is the cartesian equation of the ellipse. From (5), (7) and (8) we finally obtain

$$E = -\frac{\gamma M m}{2a} \quad \text{and} \quad l^2 = \frac{\gamma M m^2 b^2}{a}.$$

We see that the energy E does not depend on the semiminor axis b. Furthermore, the length l of the angular momentum for a given energy E is largest for a circular orbit.

Chapter 10

FIBRE BUNDLES

INTRODUCTION

10.1 Definition Let M and F be smooth manifolds, and let G be a Lie group which acts on F effectively on the left. A smooth manifold E is called a *fibre bundle* over M with *fibre* F and *structure group* G if the following three conditions are satisfied :

(i) There is a surjective smooth map $\pi : E \to M$ which is called the *projection* of the *total space* E onto the *base space* M. For each point $p \in M$ the inverse image $\pi^{-1}(p)$ is called the *fibre* over p.

(ii) We have a family of G-related local trivializations $\mathscr{D} = \{(t_\alpha, \pi^{-1}(U_\alpha)) | \alpha \in A\}$, in the sense defined below, such that $\{U_\alpha | \alpha \in A\}$ is an open cover of M. By a *local trivialization* we mean a pair $(t, \pi^{-1}(U))$, where U is an open subset of M and $t : \pi^{-1}(U) \to U \times F$ is a diffeomorphism such that the diagram

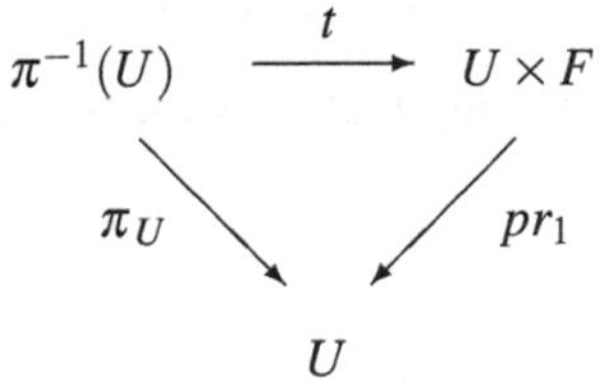

is commutative. Two local trivializations $(t, \pi^{-1}(U))$ and $(s, \pi^{-1}(V))$ are said to be *G-related* if there is a smooth map $\phi : U \cap V \to G$, called the *transition map*, so that

$$s \circ t^{-1}(p, v) = (p, \phi(p)v)$$

for every $p \in U \cap V$ and $v \in F$.

(iii) The family $\mathscr{D}$ is maximal in the sense that if $(t, \pi^{-1}(U))$ is a local trivialization which is G-related to every local trivialization in $\mathscr{D}$, then $(t, \pi^{-1}(U)) \in \mathscr{D}$.

We often use the projection $\pi : E \to M$ instead of the total space E to denote the bundle in order to indicate more of the bundle structure in the notation. F is also called the *typical fibre* to distinguish it from the fibre $\pi^{-1}(p)$ over p.

A family $\mathscr{D}$ satisfying (ii) and (iii) is called a *fibre bundle structure* on E, whereas a family of local trivializations satisfying only (ii) is called a *trivializing cover* of E.

10.2 Proposition If $\mathscr{A}$ is a trivializing cover of a fibre bundle $\pi : E \to M$, then there is a unique fibre bundle structure $\mathscr{D}$ on E containing $\mathscr{A}$.

PROOF: Let $\mathscr{D}$ be the set of all local trivializations in E which are G-related to every local trivialization in $\mathscr{A} = \{(t_\alpha, \pi^{-1}(U_\alpha)) | \alpha \in A\}$. In order to prove that $\mathscr{D}$ is a fibre bundle structure, we must show that any pair of local local trivializations $(t, \pi^{-1}(U))$ and $(s, \pi^{-1}(V))$ in $\mathscr{D}$ are G-related. For each $\alpha \in A$ we have smooth maps $\phi_\alpha : V \cap U_\alpha \to G$ and $\psi_\alpha : U \cap U_\alpha \to G$ so that

$$s \circ t_\alpha^{-1}(p, v) = (p, \phi_\alpha(p)\, v) \quad \text{and} \quad t_\alpha \circ t^{-1}(q, v) = (q, \psi_\alpha(q)\, v)$$

for every $p \in V \cap U_\alpha$, $q \in U \cap U_\alpha$ and $v \in F$. From this it follows that

$$s \circ t^{-1}(p, v) = (p, \phi_\alpha(p)\, \psi_\alpha(p)\, v)$$

for $p \in U \cap V \cap U_\alpha$ and $v \in F$. In particular, we have that

$$\phi_\alpha(p)\, \psi_\alpha(p) = \phi_\beta(p)\, \psi_\beta(p)$$

whenever $p \in U \cap V \cap U_\alpha \cap U_\beta$. This completes the proof that the local trivializations $(t, \pi^{-1}(U))$ and $(s, \pi^{-1}(V))$ are G-related with transition map $\phi : U \cap V \to G$ given by

$$\phi(p) = \phi_\alpha(p)\, \psi_\alpha(p)$$

for $p \in U \cap V \cap U_\alpha$, thus showing that $\mathscr{D}$ is the unique fibre bundle structure on E containing $\mathscr{A}$. □

10.3 Proposition The projection π is a submersion, and the fibre $E_p = \pi^{-1}(p)$ over p is a closed submanifold of E for each point $p \in M$. If $(t, \pi^{-1}(U))$ is a local trivialization around p, then the map $t_p : E_p \to F$ defined by $t_p = pr_2 \circ t|_{E_p}$ is a diffeomorphism. Hence each fibre E_p is diffeomorphic to the typical fibre F.

If $i_p : E_p \to E$ is the inclusion map and $u \in E_p$, then $V_u = i_{p*}(T_u E_p)$ is a subspace of $T_u E$ which is called the *vertical subspace* at u. The vectors in V_u are called *vertical tangent vectors* at u. We have that $V_u = \ker \pi_{*u}$.

PROOF: From the diagram in Definition 10.1 (ii) we see that π is a submersion. Hence E_p is a closed submanifold of E by Proposition 2.41. In the same way we also have that $\{p\} \times F = pr_1{}^{-1}(p)$ is a closed submanifold of $U \times F$.

By Corollary 2.38, the map $s_p : E_p \to \{p\} \times F$ induced by t is therefore a diffeomorphism, and so is the map $pr_2 : \{p\} \times F \to F$, thus showing that $t_p = pr_2 \circ s_p$ is a diffeomorphism.

Since $\pi \circ i_p(u) = p$ for every $u \in E_p$, we have that $\pi_{*u} \circ i_{p*u} = (\pi \circ i_p)_{*u} = 0$ which implies that $V_u \subset \ker \pi_{*u}$. Equality now follows since

$$\dim_u(E_p) = \dim_{t_p(u)}(F) = \dim_u(E) - \dim_p(M) = \dim_u(\ker \pi_{*u}) . \qquad \square$$

10.4 Remark Let $(t, \pi^{-1}(U))$ and $(s, \pi^{-1}(V))$ be two G-related local trivializations in the fibre bundle $\pi : E \to M$ with transition map $\phi : U \cap V \to G$. Then we have that

$$s_p \circ t_p^{-1}(v) = \phi(p)v$$

for every $p \in U \cap V$ and $v \in F$, i.e., the diffeomorphism $s_p \circ t_p^{-1} : F \to F$ coincides with the operation of the group element $\phi(p) \in G$ on the fibre F for every $p \in U \cap V$. Since G acts effectively on F, we know that $\phi(p)$ is uniquely determined by $s_p \circ t_p^{-1}$.

Moreover, we have that

$$s_p(u) = \phi(p)t_p(u)$$

when $p \in U \cap V$ and $u \in E_p$. Indeed, the last formula it obtained from the first by letting $v = t_p(u)$.

10.5 Proposition Let M and F be smooth manifolds, and let G be a Lie group which acts on F effectively on the left. Let $\pi : E \to M$ be a map from a set E, and let $\{U_\alpha | \alpha \in A\}$ be an open cover of M. Suppose that for each $\alpha \in A$, there is a bijection $t_\alpha : \pi^{-1}(U_\alpha) \to U_\alpha \times F$ such that the diagram

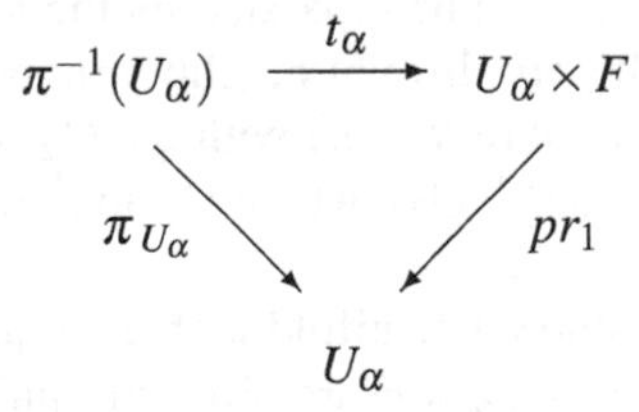

is commutative. Moreover, suppose that for each pair $\alpha, \beta \in A$, there is a smooth map $\phi_{\beta\alpha} : U_\alpha \cap U_\beta \to G$ so that

$$t_\beta \circ t_\alpha^{-1}(p, v) = (p, \phi_{\beta\alpha}(p)v)$$

for every $p \in U_\alpha \cap U_\beta$ and $v \in F$.

Then there is a unique topology, smooth structure and fibre bundle structure $\mathscr{D}$ on E such that $\pi : E \to M$ is a fibre bundle over M with fibre F and structure group G, and such that $(t_\alpha, \pi^{-1}(U_\alpha))$ is a local trivialization for each $\alpha \in A$.

PROOF: We first define a topology τ on E so that the projection π is continuous and the bijections $t_\alpha : \pi^{-1}(U_\alpha) \to U_\alpha \times F$ are homeomorphisms for every $\alpha \in A$. The continuity of π implies that each $\pi^{-1}(U_\alpha)$ must be open in τ, and hence a set O in E must belong to τ if and only if $O \cap \pi^{-1}(U_\alpha) \in \tau$ for every $\alpha \in A$. The requirement that the bijections t_α are homeomorphisms now defines τ uniquely as the collection of sets O in E such that $t_\alpha(O \cap \pi^{-1}(U_\alpha))$ is open in $U_\alpha \times F$ for every $\alpha \in A$. This collection τ is clearly a topology on E, and we must prove that it satisfies the above requirements.

We first show that $\pi^{-1}(U_\alpha)$ is open in τ for each $\alpha \in A$. For every $\beta \in A$ we have that $t_\beta(\pi^{-1}(U_\alpha) \cap \pi^{-1}(U_\beta)) = (U_\alpha \cap U_\beta) \times F$ which is open in $U_\beta \times F$, thereby proving the assertion.

We next show that t_α is a homeomorphism for every $\alpha \in A$. If $O \subset \pi^{-1}(U_\alpha)$ belongs to τ, we have that $t_\alpha(O \cap \pi^{-1}(U_\alpha))$ is open in $U_\alpha \times F$, which proves that t_α is an open map. To show that it is continuous, let O be an open set in $U_\alpha \times F$. For every $\beta \in A$ we have that $t_\beta(t_\alpha^{-1}(O) \cap \pi^{-1}(U_\beta)) = (t_\beta \circ t_\alpha^{-1})(O \cap (U_\alpha \cap U_\beta) \times F)$ which is open in $U_\beta \times F$, and hence $t_\alpha^{-1}(O) \in \tau$ showing that t_α is continuous.

From this it also follows that the projection π is continuous since $\pi_{U_\alpha} = pr_1 \circ t_\alpha$ is the composition of the continuous maps t_α and pr_1 in the above commutative diagram for each $\alpha \in A$.

Now let $\{(y_\gamma, W_\gamma) | \gamma \in C\}$ be an atlas on F, and let $\{(x_\beta, V_\beta) | \beta \in B\}$ be an atlas on M such that $\{V_\beta | \beta \in B\}$ is a refinement of the open cover $\{U_\alpha | \alpha \in A\}$. We then have a map $\rho : B \to A$ such that $V_\beta \subset U_{\rho(\beta)}$ for every $\beta \in B$. By the assumptions in the proposition we now see that the family $\mathscr{B} = \{((x_\beta \times y_\gamma) \circ t_{\rho(\beta)}, t_{\rho(\beta)}^{-1}(V_\beta \times W_\gamma)) | (\beta, \gamma) \in B \times C\}$ is an atlas on E. To show that E is a smooth manifold, we must show that it is Hausdorff and second countable. By Proposition 2.6 we know that M and F have atlases of the above form where B and C are countable, thus showing that E is second countable.

To see that E is Hausdorff, let p_1 and p_2 be two different points in E. If p_1 and p_2 lie in different fibres, they can be separated by the neighbourhoods $\pi^{-1}(W_1)$ and $\pi^{-1}(W_2)$, where W_1 and W_2 are disjoint neighbourhoods of $\pi(p_1)$ and $\pi(p_2)$ in M. If p_1 and p_2 lie in the same fibre $\pi^{-1}(q)$ with $q \in U_\alpha$, they can be separated by the neighbourhoods $t_\alpha^{-1}(O_1)$ and $t_\alpha^{-1}(O_2)$, where O_1 and O_2 are disjoint neighbourhoods of $t_\alpha(p_1)$ and $t_\alpha(p_2)$ in $U_\alpha \times F$.

This shows that E is a smooth manifold with a unique smooth structure such that the bijections $t_\alpha : \pi^{-1}(U_\alpha) \to U_\alpha \times F$ are diffeomorphisms for every $\alpha \in A$. We see that the projection π is smooth since $\pi_{U_\alpha} = pr_1 \circ t_\alpha$ is the composition of the smooth maps t_α and pr_1.

By the assumptions in the proposition the family $\mathscr{A} = \{(t_\alpha, \pi^{-1}(U_\alpha)) | \alpha \in A\}$ is a trivializing cover of E which is contained in a unique fibre bundle structure $\mathscr{D}$ according to Proposition 10.2, and this completes the construction of the fibre bundle $\pi : E \to M$ having $(t_\alpha, \pi^{-1}(U_\alpha))$ as a local trivialization for each $\alpha \in A$. □

10.6 Definition Let $\pi_1 : E_1 \to M_1$ and $\pi_2 : E_2 \to M_2$ be two fibre bundles with the same fibre F and structure group G. A *bundle map* from π_1 to π_2 is a pair $(\tilde{f}, f)$ of smooth maps $\tilde{f} : E_1 \to E_2$ and $f : M_1 \to M_2$ such that the diagram

$$\begin{array}{ccc} E_1 & \xrightarrow{\tilde{f}} & E_2 \\ \pi_1 \downarrow & & \downarrow \pi_2 \\ M_1 & \xrightarrow{f} & M_2 \end{array}$$

is commutative, and for every pair of local trivializations $(t, \pi^{-1}(U))$ and $(s, \pi^{-1}(V))$ in E_1 and E_2, respectively, there is a smooth map $\phi : U \cap f^{-1}(V) \to G$ so that

$$s \circ \tilde{f} \circ t^{-1}(p, v) = (f(p), \phi(p)\, v) \tag{1}$$

for every $p \in U \cap f^{-1}(V)$ and $v \in F$. We let $\tilde{f}_p$ denote the induced smooth map $\tilde{f}_p : \pi_1^{-1}(p) \to \pi_2^{-1}(f(p))$ between the fibres for each $p \in M_1$.

If $\tilde{f}$ and f are diffeomorphisms, $(\tilde{f}, f)$ is called an *equivalence*. In that case it follows from (1) that

$$t \circ \tilde{f}^{-1} \circ s^{-1}(q, v) = (f^{-1}(q), \{\phi \circ f^{-1}(q)\}^{-1} v)$$

for every $q \in V \cap f(U)$ and $v \in F$, so that $(\tilde{f}^{-1}, f^{-1})$ is also a bundle map.

10.7 Definition Let $\pi_1 : E_1 \to M$ and $\pi_2 : E_2 \to M$ be two fibre bundles over the same base space M with the same fibre F and structure group G. A *bundle map* from π_1 to π_2 *over* M is a smooth map $f : E_1 \to E_2$ such that (f, id_M) is a bundle map. If f is a diffeomorphism, it is called an *equivalence over* M.

10.8 Definition Let M be a smooth manifold and G be a Lie Group. By a *system of transition maps* corresponding to an open cover $\{U_\alpha | \alpha \in A\}$ of M with values in G we mean a family $\{\phi_{\beta\,\alpha} | \alpha, \beta \in A\}$ of smooth maps $\phi_{\beta\,\alpha} : U_\alpha \cap U_\beta \to G$ satisfying the consistency condition

$$\phi_{\gamma\beta}(p)\, \phi_{\beta\,\alpha}(p) = \phi_{\gamma\alpha}(p) \tag{1}$$

for every triple of indices $\alpha, \beta, \gamma \in A$ and every $p \in U_\alpha \cap U_\beta \cap U_\gamma$.

10.9 Remark If we put $\alpha = \beta = \gamma$ in (1), we obtain

$$\phi_{\alpha\,\alpha}(p) = e \tag{2}$$

for every $\alpha \in A$ and every $p \in U_\alpha$. Now putting $\alpha = \gamma$ in (1) and using (2), we also have that

$$\phi_{\alpha\,\beta}(p) = \phi_{\beta\,\alpha}(p)^{-1} \tag{3}$$

for every pair of indices $\alpha, \beta \in A$ and every $p \in U_\alpha \cap U_\beta$.

10.10 Proposition Let M and F be smooth manifolds, and let G be a Lie group which acts on F effectively on the left. Given a system $\{\phi_{\beta\,\alpha} | \alpha, \beta \in A\}$ of transition maps corresponding to an open cover $\{U_\alpha | \alpha \in A\}$ of M with values in G, there is a fibre bundle $\pi : E \to M$ with fibre F, structure group G and trivializing cover $\mathscr{A} = \{(t_\alpha, \pi^{-1}(U_\alpha)) | \alpha \in A\}$, where each pair of local trivializations $(t_\alpha, \pi^{-1}(U_\alpha))$ and $(t_\beta, \pi^{-1}(U_\beta))$ in $\mathscr{A}$ are G-related with transition map $\phi_{\beta\,\alpha} : U_\alpha \cap U_\beta \to G$.

PROOF : We first define the set

$$X = \bigcup_{\alpha \in A} U_\alpha \times F \times \{\alpha\},$$

and we say that two elements (p, v, α) and (q, w, β) in X are equivalent, and write $(p, v, \alpha) \sim (q, w, \beta)$, if

$$p = q \quad \text{and} \quad \phi_{\beta\,\alpha}(p)\, v = w\,.$$

It follows from (1) in Definition 10.8 and (2) and (3) in Remark 10.9 that $\sim$ is an equivalence relation, and we denote the equivalence class of (p, v, α) by $[p, v, \alpha]$ and let $E = X/\sim$ be the quotient set. We now have a map $\pi : E \to M$ defined by

$$\pi([p, v, \alpha]) = p\,,$$

and for each $\alpha \in A$ we have a bijection $t_\alpha : \pi^{-1}(U_\alpha) \to U_\alpha \times F$ defined by

$$t_\alpha([p, v, \alpha]) = (p, v)$$

such that the diagram in Proposition 10.5 is commutative. Moreover, we have that

$$t_\beta \circ t_\alpha^{-1}(p, v) = t_\beta([p, v, \alpha]) = t_\beta([p, \phi_{\beta\,\alpha}(p)\, v, \beta]) = (p, \phi_{\beta\,\alpha}(p)\, v)$$

for every $p \in U_\alpha \cap U_\beta$ and $v \in F$, and the proposition therefore follows from Proposition 10.5. □

10.11 Definition Let $\pi : E \to M$ be a fiber bundle, and let V be an open subset of M and $i : V \to M$ be the inclusion mapping. Then a smooth map $s : V \to E$ with $\pi \circ s = i$ is called a *section* of π on V, or a *lifting* of i to E.

If $(t, \pi^{-1}(U))$ is a local trivialization of π with $U \subset V$ and $s : V \to E$ is a map with $\pi \circ s = i$, then it follows from the commutative diagram in Definition 10.1 that $pr_1 \circ t \circ s = id_U$. Hence $s|_U$ is completely determined by the map $h : U \to F$ given by $(t \circ s)(p) = (p, h(p))$ for $p \in U$. h is called the *local representation* of s on U. We see that s is smooth on U if and only if the local representation h is smooth.

INDUCED BUNDLES

10.12 Let $\pi : E \to M$ be a fibre bundle over M with fibre F and structure group G. If $f : N \to M$ is a smooth map from a smooth manifold N, we define the set

$$\widetilde{E} = \{(p, u) \in N \times E \,|\, f(p) = \pi(u)\}$$

and let $\widetilde{\pi} : \widetilde{E} \to N$ and $\widetilde{f} : \widetilde{E} \to E$ be the restrictions to $\widetilde{E}$ of the projections $pr_1 : N \times E \to N$ and $pr_2 : N \times E \to E$ on the first and second factor. Then we have a commutative diagram

$$\begin{array}{ccc} \widetilde{E} & \xrightarrow{\ \widetilde{f}\ } & E \\ {\scriptstyle \widetilde{\pi}}\big\downarrow & & \big\downarrow{\scriptstyle \pi} \\ N & \xrightarrow{\ f\ } & M \end{array}$$

since $f \circ \widetilde{\pi}(p,u) = f(p) = \pi(u) = \pi \circ \widetilde{f}(p,u)$ for $(p,u) \in \widetilde{E}$.

Now let $\mathscr{A} = \{(t_\alpha, \pi^{-1}(U_\alpha)) | \alpha \in A\}$ be a trivializing cover of E. For each $\alpha \in A$ we have an open subset $\widetilde{U}_\alpha = f^{-1}(U_\alpha)$ of N and a bijection $\widetilde{t}_\alpha : \widetilde{\pi}^{-1}(\widetilde{U}_\alpha) \to \widetilde{U}_\alpha \times F$ defined by

$$\widetilde{t}_\alpha(p,u) = (p, t_{\alpha,f(p)}(u)) \tag{1}$$

for $p \in \widetilde{U}_\alpha$ and $u \in E_{f(p)}$. Then $\{\widetilde{U}_\alpha | \alpha \in A\}$ is a open cover of N, and if $\phi_{\beta\,\alpha} : U_\alpha \cap U_\beta \to G$ is the transition map between the local trivializations $(t_\alpha, \pi^{-1}(U_\alpha))$ and $(t_\beta, \pi^{-1}(U_\beta))$, we have that

$$\widetilde{t}_{\beta,p} \circ \widetilde{t}_{\alpha,p}^{-1}(v) = t_{\beta,f(p)} \circ t_{\alpha,f(p)}^{-1}(v) = \phi_{\beta\,\alpha} \circ f(p)\, v \tag{2}$$

so that

$$\widetilde{t}_\beta \circ \widetilde{t}_\alpha^{-1}(p,v) = (p, \phi_{\beta\,\alpha} \circ f(p)\, v)$$

for every $p \in \widetilde{U}_\alpha \cap \widetilde{U}_\beta$ and $v \in F$. By Proposition 10.5 there is therefore a unique topology, smooth structure and fibre bundle structure on $\widetilde{E}$ such that $\widetilde{\pi} : \widetilde{E} \to N$ is a fibre bundle over N with fibre F and structure group G, and such that $(\widetilde{t}_\alpha, \widetilde{\pi}^{-1}(\widetilde{U}_\alpha))$ is a local trivialization for each $\alpha \in A$. It is called the local trivialization in $\widetilde{E}$ associated with the local trivialization $(t_\alpha, \pi^{-1}(U_\alpha))$ in E.

By (1) we see that

$$\widetilde{t}_{\alpha,p} = t_{\alpha,f(p)} \circ \widetilde{f}_p$$

for $p \in \widetilde{U}_\alpha$, so that

$$t_\alpha \circ \widetilde{f} \circ \widetilde{t}_\alpha^{-1}(p,v) = (f(p), v) \tag{3}$$

for $p \in \widetilde{U}_\alpha$ and $v \in F$. Hence it follows from Proposition 10.3 that $(\widetilde{f}, f)$ is a bundle map which induces a diffeomorphism between the fibres $\widetilde{E}_p$ and $E_{f(p)}$ for each $p \in N$.

Since $\widetilde{\pi}$ and $\widetilde{f}$ are the components of the inclusion map $i : \widetilde{E} \to N \times E$, we also have that

$$(id \times t_\alpha) \circ i \circ \widetilde{t}_\alpha^{-1}(p,v) = (p, f(p), v)$$

for $p \in \widetilde{U}_\alpha$ and $v \in F$. Using this, we will show that $\widetilde{E}$ is a closed submanifold of

$N \times E$. If $(p_0, u_0) \in \tilde{\pi}^{-1}(\tilde{U}_\alpha)$ and $t_\alpha(u_0) = (f(p_0), v_0) \in U_\alpha \times F$, we choose local charts $(\tilde{x}, \tilde{V})$, (x, V) and (y, W) around p_0, $f(p_0)$ and v_0, respectively, with $\tilde{V} \subset \tilde{U}_\alpha$ and $f(\tilde{V}) \subset V$. Then we have that

$$(\tilde{x} \times (x \times y) \circ t_\alpha)) \circ i \circ ((\tilde{x} \times y) \circ \tilde{t}_\alpha)^{-1}(a, \xi) = (a, x \circ f \circ \tilde{x}^{-1}(a), \xi)$$

for $a \in \tilde{x}(\tilde{V})$ and $\xi \in y(W)$.

Now let r, n and m be the dimensions of N, M and F, respectively, and let $\phi : \tilde{x}(\tilde{V}) \times \mathbf{R}^n \times \mathbf{R}^m \to \tilde{x}(\tilde{V}) \times \mathbf{R}^m \times \mathbf{R}^n$ be the C^∞-map defined by $\phi(a, b, \xi) = (a, \xi, b - x \circ f \circ \tilde{x}^{-1}(a))$. ϕ is a diffeomorphism with inverse $\psi : \tilde{x}(\tilde{V}) \times \mathbf{R}^m \times \mathbf{R}^n \to \tilde{x}(\tilde{V}) \times \mathbf{R}^n \times \mathbf{R}^m$ given by $\psi(a, \xi, b) = (a, b + x \circ f \circ \tilde{x}^{-1}(a), \xi)$.

If $j : \tilde{x}(\tilde{V}) \times \mathbf{R}^n \times \mathbf{R}^m \to \mathbf{R}^r \times \mathbf{R}^n \times \mathbf{R}^m$ is the inclusion map, we therefore have a local chart $(\tilde{V} \times t_\alpha^{-1}(V \times W), j \circ \phi \circ (\tilde{x} \times (x \times y) \circ t_\alpha))$ around (p_0, u_0) in $N \times E$ having the submanifold property, showing that $\tilde{E}$ is a submanifold of $N \times E$. It is closed as $\tilde{E} = (f \times \pi)^{-1}(\Delta)$, where $\Delta \subset M \times M$ is the diagonal which is closed since M is Hausdorff.

The fibre bundle $\tilde{\pi} : \tilde{E} \to N$ is called the *pull-back* of $\pi : E \to M$ by f or the fibre bundle *induced* from $\pi : E \to M$ by f, and it is denoted by $f^*(E)$ with projection $f^*(\pi)$. The bundle map $(\tilde{f}, f)$ is called the *canonical bundle map* of the induced bundle.

A smooth map $F : N \to E$ with $\pi \circ F = f$ is called a *lifting* of f or a *section* of π *along* f. The set of all liftings of f is denoted by $\Gamma(f; E)$. For each lifting F of f there is a unique section $\tilde{F} : N \to \tilde{E}$ of $\tilde{\pi}$ with $\tilde{f} \circ \tilde{F} = F$, which is given by $\tilde{F}(p) = (p, F(p))$ for $p \in N$. Conversely, if $s : N \to \tilde{E}$ is a section of $\tilde{\pi}$, then $s = \tilde{F}$ where $F = \tilde{f} \circ s$. $\tilde{F}$ is called the section in the induced fibre bundle determined by the lifting F.

If $\pi : E \to M$ is a vector bundle, so is the induced bundle $\tilde{\pi} : \tilde{E} \to N$. Indeed, it follows from (2) that $\tilde{t}_{\beta,p} \circ \tilde{t}_{\alpha,p}^{-1}$ is a linear isomorphism whenever $p \in \tilde{U}_\alpha \cap \tilde{U}_\beta$. Hence there is a well-defined vector space structure on each fibre $\tilde{E}_p$ so that $\tilde{t}_{\alpha,p}$ is a linear isomorphism for any $\alpha \in A$ with $p \in \tilde{U}_\alpha$. From (3) it follows that $(\tilde{f}, f)$ is a bundle map which induces a linear isomorphism between the fibres $\tilde{E}_p$ and $E_{f(p)}$ for each $p \in N$. The vector space structure on $\tilde{E}_p$ is therefore given by

$$a_1(p, v_1) + a_2(p, v_2) = (p, a_1 v_1 + a_2 v_2)$$

for $v_1, v_2 \in E_{f(p)}$ and $a_1, a_2 \in \mathbf{R}$.

If N is an immersed submanifold of M with inclusion map $i : N \to M$, we consider the subset $E|_N = \pi^{-1}(N)$ of E, and let $\pi_N : E|_N \to N$ be the map induced by the projection $\pi : E \to M$. We give $E|_N$ the fibre bundle structure so that the bijection $e : \tilde{E} \to E|_N$ given by $e(p, u) = u$ for $(p, u) \in \tilde{E}$, is an equivalence over N. The fibre bundle $\pi_N : E|_N \to N$ is called the *restriction* of E to N, and $i' = \tilde{i} \circ e^{-1} : E|_N \to E$ is the inclusion map. Given a trivializing cover $\mathscr{A} = \{(t_\alpha, \pi^{-1}(U_\alpha)) | \alpha \in A\}$ of E, we let $\tilde{U}_\alpha = U_\alpha \cap N$ and $t'_\alpha : \pi^{-1}(\tilde{U}_\alpha) \to \tilde{U}_\alpha \times F$ be the map induced by $t_\alpha : \pi^{-1}(U_\alpha) \to U_\alpha \times F$ so that $\tilde{t}_\alpha \circ e^{-1} = t'_\alpha$ for $\alpha \in A$. Then $\mathscr{A}' = \{(t'_\alpha, \pi^{-1}(\tilde{U}_\alpha)) | \alpha \in A\}$ is a trivializing cover of $E|_N$.

10.13 Proposition If $\pi : E \to M$ is a vector bundle and $f : N \to M$ is a smooth map from a smooth manifold N, then the set $\Gamma(f;E)$ of liftings of f is a module over the ring $\mathscr{F}(N)$ of C^∞-functions on N with operations defined by

$$(F_1 + F_2)(p) = F_1(p) + F_2(p) \quad \text{and} \quad (gF)(p) = g(p)F(p)$$

for liftings F_1, F_2 and F and C^∞-functions g on N.

PROOF : Follows from Proposition 2.53 by considering the sections $\widetilde{F}_1$, $\widetilde{F}_2$ and $\widetilde{F}$ in the induced vector bundle $\widetilde{\pi} : \widetilde{E} \to N$ determined by the liftings F_1, F_2 and F, since

$$F_1 + F_2 = \widetilde{f} \circ (\widetilde{F}_1 + \widetilde{F}_2) \quad \text{and} \quad gF = \widetilde{f} \circ (g\widetilde{F})$$

where $(\widetilde{f}, f)$ is the canonical bundle map of $\widetilde{\pi} : \widetilde{E} \to N$. □

10.14 Proposition Let $\pi : E \to M$ be an n-dimensional vector bundle, and let $F_1, \ldots, F_n$ be n liftings of a smooth map $f : N \to M$ from a smooth manifold N which are everywhere linearly independent, i.e., $F_1(p), \ldots, F_n(p) \in \pi^{-1}(f(p))$ are linearly independent for every $p \in N$. Then a map $F : N \to E$ with $\pi \circ F = f$ is a lifting of f if and only if the map $a : N \to \mathbf{R}^n$ defined by

$$F(p) = \sum_{i=1}^{n} a_i(p) F_i(p)$$

for $p \in N$ is smooth.

PROOF : Let $\widetilde{F}_i$ be the section in the induced vector bundle $\widetilde{\pi} : \widetilde{E} \to N$ determined by the lifting F_i for $i = 1, \ldots, n$, and let $\widetilde{F} : N \to \widetilde{E}$ be the map given by $\widetilde{F}(p) = (p, F(p))$ for $p \in N$. Then the proposition follows from Corollary 2.59 since

$$\widetilde{F}(p) = \sum_{i=1}^{n} a_i(p) \widetilde{F}_i(p)$$

for $p \in N$. □

PRINCIPAL FIBRE BUNDLES

10.15 Definition Let M be a smooth manifold and G be a Lie group. Then a *principal fibre bundle* P over M with *structure group* G is a fibre bundle $\pi : P \to M$ where the fibre F coincides with the structure group G which is acting on F by left translation. The principal bundle P is also denoted by $P(M, G)$ and is called a *principal G-bundle* over M.

10.16 If $\pi : P \to M$ is a principal G-bundle, we may define an operation of G on P on the right in the following way. Let $u \in P_p$ and $g \in G$, and choose a local trivialization $(t, \pi^{-1}(U))$ in P with $p \in U$. If $t(u) = (p,h)$, we define $ug \in P_p$ so that $t(ug) = (p,hg)$, i.e.,

$$\pi(ug) = \pi(u) \quad \text{and} \quad t_p(ug) = t_p(u)g \text{ when } p = \pi(u) \in U\,.$$

We must show that the operation of G does not depend on the local trivialization $(t, \pi^{-1}(U))$. So let $(s, \pi^{-1}(V))$ be another local trivialization with $p \in V$ which is G-related to $(t, \pi^{-1}(U))$ with transition map $\phi : U \cap V \to G$. Then we have that

$$s_p(ug) = \phi(p)\,t_p(ug) = \phi(p)\,[t_p(u)\,g] = [\phi(p)t_p(u)]\,g = s_p(u)\,g$$

since $L_{\phi(p)}$ commutes with R_g, showing that we have a well-defined map $\nu : P \times G \to P$ with $\nu(u,g) = ug$ for $u \in P$ and $g \in G$.

Now we have that

$$u\,(g_1 g_2) = (u\,g_1)\,g_2 \quad \text{and} \quad u\,e = u$$

for every $u \in P$ and $g_1, g_2 \in G$, since

$$t_p(u\,(g_1 g_2)) = t_p(u)\,(g_1 g_2) = (t_p(u)\,g_1)\,g_2 = t_p(u\,g_1)\,g_2 = t_p((u\,g_1)\,g_2)$$

and

$$t_p(ue) = t_p(u)\,e = t_p(u)$$

when $u \in P_p$. As the map

$$t \circ \nu \circ (t^{-1} \times id) : U \times G \times G \to U \times G$$

is given by

$$t \circ \nu \circ (t^{-1} \times id)(p,h,g) = (p,hg)$$

for $p \in U$ and $h, g \in G$, it follows that ν is smooth and hence is an operation of G on P on the right.

We see that G acts freely on P, and that the orbits of G are the fibres of P. Indeed, if $ug = u$ for a $u \in P_p$, then $t_p(u)g = t_p(u)$ which implies that $g = e$. Moreover, given $u_1, u_2 \in P_p$ there is a $g \in G$ with $t_p(u_1)\,g = t_p(u_2)$ so that $u_1\,g = u_2$.

Using these results, we can now give an alternative definition of a principal fibre bundle.

10.17 Definition Let M be a smooth manifold and G be a Lie group. A smooth manifold P is called a *principal fibre bundle* over M with *structure group* G if the following three conditions are satisfied :

(i) G acts freely on P on the right.

(ii) There is a surjective smooth map $\pi : P \to M$ which is called the *projection* of the *total space* P onto the *base space* M satisfying

$$\pi(ug) = \pi(u) \text{ for every } u \in P \text{ and } g \in G.$$

For each point $p \in M$ the inverse image $\pi^{-1}(p)$ is called the *fibre* over p.

(iii) For each point $p \in M$ there is an open neighbourhood U around p and a diffeomorphism $t : \pi^{-1}(U) \to U \times G$ of the form $t(u) = (\pi(u), \psi(u))$, where $\psi : \pi^{-1}(U) \to G$ is a map satisfying

$$\psi(ug) = \psi(u)g \text{ for every } u \in \pi^{-1}(U) \text{ and } g \in G.$$

The pair $(t, \pi^{-1}(U))$ is called a *local trivialization* around p.

10.18 Remark It follows from 10.16 that a principal G-bundle $\pi : P \to M$ as defined in Definition 10.15 clearly satisfies Definition 10.17.

Conversely, suppose that $\pi : P \to M$ is a principal G-bundle as defined in Definition 10.17. Then we have a family $\mathscr{A} = \{(t_\alpha, \pi^{-1}(U_\alpha)) | \alpha \in A\}$ of local trivializations such that $\{U_\alpha | \alpha \in A\}$ is an open cover of M, and where each diffeomorphism $t_\alpha : \pi^{-1}(U_\alpha) \to U_\alpha \times G$ is of the form $t_\alpha(u) = (\pi(u), \psi_\alpha(u))$ where $\psi_\alpha : \pi^{-1}(U_\alpha) \to G$ is a map satisfying

$$\psi_\alpha(ug) = \psi_\alpha(u)g \text{ for every } u \in \pi^{-1}(U_\alpha) \text{ and } g \in G.$$

If $u \in \pi^{-1}(U_\alpha \cap U_\beta)$, we have that

$$\psi_\beta(ug)\,\psi_\alpha(ug)^{-1} = \psi_\beta(u)\,\psi_\alpha(u)^{-1}$$

for every $g \in G$, which shows that the map $\psi_{\beta\,\alpha} : \pi^{-1}(U_\alpha \cap U_\beta) \to G$ given by

$$\psi_{\beta\,\alpha}(u) = \psi_\beta(u)\,\psi_\alpha(u)^{-1}$$

is constant on each fibre. Hence it follows that the local trivializations $(t_\alpha, \pi^{-1}(U_\alpha))$ and $(t_\beta, \pi^{-1}(U_\beta))$ are G-related with transition map $\phi_{\beta\,\alpha} : U_\alpha \cap U_\beta \to G$ given by

$$\phi_{\beta\,\alpha}(p) = \psi_{\beta\,\alpha}(t_\alpha^{-1}(p, e))$$

for $p \in U_\alpha \cap U_\beta$. This shows that $\mathscr{A}$ is a trivializing cover of E which is contained in a unique fibre bundle structure $\mathscr{D}$ by Proposition 10.2.

10.19 Definition Let $\pi_i : P_i \to M_i$ be a principal G_i-bundle for $i = 1, 2$. A *homomorphism* from π_1 to π_2 is a triple $(\tilde{f}, f, \psi)$ of smooth maps $\tilde{f} : P_1 \to P_2$ and $f : M_1 \to M_2$ and a Lie group homomorphism $\psi : G_1 \to G_2$ such that

$$\pi_2 \circ \tilde{f} = f \circ \pi_1 \tag{1}$$

and

$$\tilde{f} \circ R_g = R_{\psi(g)} \circ \tilde{f} \tag{2}$$

for every $g \in G_1$. We let $\tilde{f}_p$ denote the induced smooth map $\tilde{f}_p : \pi_1^{-1}(p) \to \pi_2^{-1}(f(p))$ between the fibres for each $p \in M_1$.

If $\tilde{f}$, f and ψ are inclusion maps which are immersions, then $\pi_1 : P_1 \to M_1$ is called a *subbundle* of $\pi_2 : P_2 \to M_2$. If in addition $M_2 = M_1$ and $f = id$, then the homomorphism $(\tilde{f}, f, \psi)$ is called a *reduction of the structure group* and $\pi_1 : P_1 \to M_1$ is called a *reduced subbundle* of $\pi_2 : P_2 \to M_2$. If $\tilde{f}$, f and ψ are diffeomorphisms, then $(\tilde{f}, f, \psi)$ is called an *isomorphism*. An isomorphism of a principal fibre bundle with itself is called an *automorphism*.

10.20 Remark A homomorphism $(\tilde{f}, f, \psi)$ is actually determined by the pair $(\tilde{f}, \psi)$ satisfying (2), since $f(p) = \pi_2 \circ \tilde{f}(u)$ for any $u \in P_1$ with $\pi_1(u) = p$. This is well defined, for if we choose another $u' \in P_1$ with $\pi_1(u') = p$, then $u' = ug$ for a $g \in G$, so that

$$\pi_2 \circ \tilde{f}(u') = \pi_2(\tilde{f}(u)\,\psi(g)) = \pi_2 \circ \tilde{f}(u) \,.$$

If $(t, \pi^{-1}(V))$ is a local trivialization in P and s is the section on V given by $s(p) = t^{-1}(p, e)$ for $p \in V$, then $f|_V = \pi_2 \circ \tilde{f} \circ s$ which shows that f is smooth.

10.21 Example By a *frame* in a vector space V we mean an ordered basis $\mathscr{B} = (v_1, ..., v_n)$ for V. With each frame $\mathscr{B}$ in V we associate a linear isomorphism $F : \mathbf{R}^n \to V$ given by $F(e_i) = v_i$ for $i = 1, ..., n$, where $\mathscr{E} = (e_1, ..., e_n)$ is the standard basis for $\mathbf{R}^n$, so that

$$F(a) = \sum_{i=1}^{n} a^i\, v_i$$

for $a \in \mathbf{R}^n$. Conversely, each linear isomorphism $F : \mathbf{R}^n \to V$ is obtained from the unique frame $(F(e_1), ..., F(e_n))$ in V. We let $\mathscr{L}(V)$ denote the set of all linear isomorphisms $F : \mathbf{R}^n \to V$.

If $\pi : E \to M$ is an n-dimensional vector bundle, we define $\mathscr{L}_p(E) = \mathscr{L}(E_p)$ for $p \in M$. We want to prove that

$$\mathscr{L}(E) = \bigcup_{p \in M} \mathscr{L}_p(E)$$

is a principal fibre bundle over M called the *frame bundle* of E, with structure group $Gl(\mathbf{R}^n)$ and projection $\pi' : \mathscr{L}(E) \to M$ sending each set $\mathscr{L}_p(E)$ to p.

Let $\{(t_\alpha, \pi^{-1}(U_\alpha)) \mid \alpha \in A\}$ be a trivializing cover of E. For each $\alpha \in A$ we have a bijection $\phi_\alpha : \pi'^{-1}(U_\alpha) \to U_\alpha \times Gl(\mathbf{R}^n)$ defined by

$$\phi_\alpha(u) = (p, t_{\alpha,p} \circ u)$$

for $p \in U_\alpha$ and $u \in \mathscr{L}_p(E)$. For each pair of indices $\alpha, \beta \in A$, we have that

$$\phi_\alpha \circ \phi_\beta^{-1}(p, F) = (p, \phi_{\alpha\beta}(p) \circ F)$$

for $(p,F) \in (U_\alpha \cap U_\beta) \times Gl(\mathbf{R}^n)$, where $\phi_{\alpha\beta} : U_\alpha \cap U_\beta \to Gl(\mathbf{R}^n)$ is the differentiable map given by

$$\phi_{\alpha\beta}(p) = t_{\alpha,p} \circ t_{\beta,p}^{-1}$$

for $p \in U_\alpha \cap U_\beta$. By Proposition 10.5 it follows that $\pi' : \mathscr{L}(E) \to M$ is a principal fibre bundle over M with structure group $Gl(\mathbf{R}^n)$ and with $(\phi_\alpha, \pi'^{-1}(U_\alpha))$ as a local trivialization for each $\alpha \in A$.

A map $s : V \to \mathscr{L}(E)$ defined on an open subset V of M is a section of π' if and only if the maps $s_j : V \to E$ defined by $s_j(p) = s(p)(e_j)$ for $p \in V$ and $j = 1, \ldots, n$ are sections of $\pi : E \to M$ forming a local basis for E. Indeed, for each $\alpha \in A$ we have that

$$\psi(t_{\alpha,p} \circ s(p))_{ij} = e^i \circ t_{\alpha,p} \circ s(p)(e_j) = e^i \circ t_{\alpha,p}(s_j(p)) = [\,t_{\alpha,p}(s_j(p))\,]_i$$

for $p \in V \cap U_\alpha$ and $i,j = 1, \ldots, n$, which shows that s is smooth on V if and only if s_j is smooth on V for $j = 1, \ldots, n$. The section s is also written as $s = (s_1, \ldots, s_n)$.

We see that $\mathscr{L}(E)$ is an open submanifold of the total space $\Lambda^1(\mathbf{R}^n; E)$ of the vector bundle $\pi'' : \Lambda^1(\mathbf{R}^n; E) \to M$ defined in 5.27. Hence a section s of π' may be identified with the section $i \circ s$ of π'', where $i : \mathscr{L}(E) \to \Lambda^1(\mathbf{R}^n; E)$ is the inclusion map.

Let $(\hat{f}, f)$ be a bundle map between the n-dimensional vector bundles $\pi_1 : E_1 \to M_1$ and $\pi_2 : E_2 \to M_2$, where $\hat{f} : E_1 \to E_2$ induces a linear isomorphism between the fibres $\pi_1^{-1}(p)$ and $\pi_2^{-1}(f(p))$ for every $p \in M_1$. Then we have a homomorphism $(\tilde{f}, f, id)$ between the frame bundles $\pi_1' : \mathscr{L}(E_1) \to M_1$ and $\pi_2' : \mathscr{L}(E_2) \to M_2$, where $\tilde{f} : \mathscr{L}(E_1) \to \mathscr{L}(E_2)$ is the map given by

$$\tilde{f}(u) = \hat{f}_p \circ u$$

for $p \in M_1$ and $u \in \mathscr{L}_p(E_1)$, and $id : Gl(\mathbf{R}^n) \to Gl(\mathbf{R}^n)$ is the identity map. To see that $\tilde{f}$ is smooth, let $(t_i, \pi_i^{-1}(U_i))$ be a local trivialization in $\pi_i : E_i \to M_i$ and $(\phi_i, \pi_i'^{-1}(U_i))$ be the corresponding local trivialization in $\pi_i' : \mathscr{L}(E_i) \to M_i$ for $i = 1, 2$. Then we have that

$$\phi_2 \circ \tilde{f} \circ \phi_1^{-1}(p, F) = (p, \phi(p) \circ F)$$

for $(p,F) \in (U_1 \cap f^{-1}(U_2)) \times Gl(\mathbf{R}^n)$, where $\phi : U_1 \cap f^{-1}(U_2) \to Gl(\mathbf{R}^n)$ is the smooth map given by

$$\phi(p) = t_{2,f(p)} \circ \hat{f}_p \circ t_{1,p}^{-1}$$

for $p \in U_1 \cap f^{-1}(U_2)$.

10.22 Remark We say that $(\phi, \pi'^{-1}(U))$, where $\phi_p(u) = t_p \circ u$ for $p \in U$ and $u \in \mathscr{L}_p(E)$, is the local trivialization in the frame bundle $\pi' : \mathscr{L}(E) \to M$ corresponding to the local trivialization $(t, \pi^{-1}(U))$ in the vector bundle $\pi : E \to M$.

The frame bundle $\mathscr{L}(TM)$ of the tangent bundle of M is denoted simply by $\mathscr{L}(M)$ and is called the frame bundle of M, and $\mathscr{L}_p(TM)$ is denoted by $\mathscr{L}_p(M)$ for

$p \in M$. If $(t_x, \pi^{-1}(U))$ is the local trivialization in the tangent bundle associated with a local chart (x,U) on M, then the corresponding local trivialization $(\phi, \pi'^{-1}(U))$ in the frame bundle is denoted by $(\phi_x, \pi'^{-1}(U))$. If $X_1, \dots, X_n$ are vector fields on an open subset V of M forming a local basis for TM, then the section $s = (X_1, \dots, X_n)$ in $\pi' : \mathscr{L}(M) \to M$ is called a *frame field* on V.

10.23 Example If V is an n-dimensional vector space, we let $\mathscr{A}(V)$ be the set of invertible affine maps $w : \mathbf{R}^n \to V$ given by $w(b) = u(b) + v$ for $b \in \mathbf{R}^n$, where $u \in \mathscr{L}(V)$ and $v \in V$. We have a bijection $\sigma_V : \mathscr{L}(V) \times V \to \mathscr{A}(V)$ given by

$$\sigma_V(u,v)(b) = u(b) + v$$

for $b \in \mathbf{R}^n$. The Lie group $GA(\mathbf{R}^n)$ defined in Example 9.35 operates on $\mathscr{A}(V)$ on the right by composition, and we have that

$$\begin{aligned}\sigma_V(u,v) \circ \sigma_n(T,a)(b) &= u(T(b)+a)+v = u \circ T(b) + u(a) + v \\ &= \sigma_V(u \circ T, u(a)+v)(b) = \sigma_V((u,v)(T,a))(b)\end{aligned}$$

where $\sigma_n : Gl(\mathbf{R}^n) \times_{id} \mathbf{R}^n \to GA(\mathbf{R}^n)$ is the Lie group isomorphism given by

$$\sigma_n(T,a)(b) = T(b) + a$$

for $b \in \mathbf{R}^n$, and we define an operation of the Lie group $Gl(\mathbf{R}^n) \times_{id} \mathbf{R}^n$ on $\mathscr{L}(V) \times V$ on the right by

$$(u,v)(T,a) = (u \circ T, u(a) + v)$$

for $(u,v) \in \mathscr{L}(V) \times V$ and $(T,a) \in Gl(\mathbf{R}^n) \times_{id} \mathbf{R}^n$.

If $\pi : E \to M$ is an n-dimensional vector bundle, we define $\mathscr{A}_p(E) = \mathscr{A}(E_p)$ for $p \in M$, and let $\sigma_p = \sigma_{E_p} : \mathscr{L}_p(E) \times E_p \to \mathscr{A}_p(E)$ be the bijection defined above. We want to prove that

$$\mathscr{A}(E) = \bigcup_{p \in M} \mathscr{A}_p(E)$$

is a principal fibre bundle over M called the *affine frame bundle* of E, with structure group $GA(\mathbf{R}^n)$ and projection $\pi'' : \mathscr{A}(E) \to M$ sending each set $\mathscr{A}_p(E)$ to p.

Let $\{(t_\alpha, \pi^{-1}(U_\alpha)) \mid \alpha \in A\}$ be a trivializing cover of E where each U_α is the coordinate neighbourhood for a local chart (x_α, U_α) on M. For each $\alpha \in A$ we have a bijection $\phi_\alpha : \pi''^{-1}(U) \to U \times GA(\mathbf{R}^n)$ defined by

$$\phi_\alpha \circ \sigma_p(u,v) = (p, \sigma_n(t_{\alpha,p} \circ u, t_{\alpha,p}(v) + x_\alpha(p)))$$

for $p \in U_\alpha$ and $(u,v) \in \mathscr{L}_p(E) \times E_p$. For each pair of indices $\alpha, \beta \in A$, we have that

$$\begin{aligned}\phi_\alpha \circ \phi_\beta^{-1}(p, \sigma_n(T,a)) &= (p, \sigma_n(t_{\alpha,p} \circ t_{\beta,p}^{-1} \circ T, t_{\alpha,p} \circ t_{\beta,p}^{-1}(a - x_\beta(p)) + x_\alpha(p))) \\ &= (p, \phi_{\alpha\beta}(p) \circ \sigma_n(T,a))\end{aligned}$$

for $p \in U_\alpha \cap U_\beta$ and $(T,a) \in Gl(\mathbf{R}^n) \times_{id} \mathbf{R}^n$, where $\phi_{\alpha\beta} : U_\alpha \cap U_\beta \to GA(\mathbf{R}^n)$ is the smooth map given by

$$\phi_{\alpha\beta}(p) = \sigma_n(t_{\alpha,p} \circ t_{\alpha,p}^{-1}, x_\alpha(p) - t_{\alpha,p} \circ t_{\beta,p}^{-1}(x_\beta(p)))$$

for $p \in U_\alpha \cap U_\beta$. By Proposition 10.5 it follows that $\pi'' : \mathscr{A}(E) \to M$ is a principal fibre bundle over M with structure group $GA(\mathbf{R}^n)$ and with $(\phi_\alpha, \pi''^{-1}(U_\alpha))$ as a local trivialization for each $\alpha \in A$.

The structure group $GA(\mathbf{R}^n)$ operates on $\mathscr{A}(E)$ on the right by composition since

$$\begin{aligned}
&\phi_{\alpha,p}(\sigma_p(u,v)\,\sigma_n(T,a)) = \sigma_n(t_{\alpha,p} \circ u, t_{\alpha,p}(v) + x_\alpha(p)) \circ \sigma_n(T,a) \\
&= \sigma_n((t_{\alpha,p} \circ u) \circ T, t_{\alpha,p} \circ u(a) + t_{\alpha,p}(v) + x_\alpha(p)) \\
&= \sigma_n(t_{\alpha,p} \circ (u \circ T), t_{\alpha,p}(u(a) + v) + x_\alpha(p)) = \phi_{\alpha,p} \circ \sigma_p((u,v)(T,a)) \\
&= \phi_{\alpha,p}(\sigma_p(u,v) \circ \sigma_n(T,a))
\end{aligned}$$

for $(u,v) \in \mathscr{L}_p(E) \times E_p$ and $(T,a) \in Gl(\mathbf{R}^n) \times_{id} \mathbf{R}^n$.

10.24 Remark We say that $(\phi, \pi'^{-1}(U))$, where $\phi_p \circ \sigma_p(u,v) = \sigma_n(t_p \circ u, t_p(v) + x(p))$ for $p \in U$ and $(u,v) \in \mathscr{L}_p(E) \times E_p$, is the local trivialization in the affine frame bundle $\pi' : \mathscr{A}(E) \to M$ corresponding to the local trivialization $(t, \pi^{-1}(U))$ in the vector bundle $\pi : E \to M$, where U is the coordinate neighbourhood for a local chart (x,U) on M.

The affine frame bundle $\mathscr{A}(TM)$ of the tangent bundle of M is denoted simply by $\mathscr{A}(M)$ and is called the affine frame bundle of M, and $\mathscr{A}_p(TM)$ is denoted by $\mathscr{A}_p(M)$ for $p \in M$. If $(t_x, \pi^{-1}(U))$ is the local trivialization in the tangent bundle assosiated with a local chart (x,U) on M, then the corresponding local trivialization $(\phi, \pi'^{-1}(U))$ in the affine frame bundle is denoted by $(\phi_x, \pi'^{-1}(U))$.

10.25 Example Given an n-dimensional vector bundle $\pi : E \to M$, we let $\widetilde{\pi} : \mathscr{E}(E) \to \mathscr{L}(E)$ be the vector bundle induced from $\pi : E \to M$ by the projection $\pi' : \mathscr{L}(E) \to M$ in the frame bundle of E, as shown in the commutative diagram

$$\begin{array}{ccc}
\mathscr{E}(E) & \xrightarrow{\ \widetilde{\pi}'\ } & E \\
\widetilde{\pi} \downarrow & & \downarrow \pi \\
\mathscr{L}(E) & \xrightarrow{\ \pi'\ } & M
\end{array}$$

where

$$\mathscr{E}(E) = \bigcup_{p \in M} \mathscr{L}_p(E) \times E_p\,.$$

If $(t,\pi^{-1}(U))$ is a local trivialization in the vector bundle $\pi : E \to M$, where U is the coordinate neighbourhood for a local chart (x,U) on M, then the corresponding local trivialization $(\widetilde{t},\widetilde{\pi}^{-1}(\widetilde{U}))$ in the vector bundle $\widetilde{\pi} : \mathscr{E}(E) \to \mathscr{L}(E)$ is given by

$$\widetilde{t}(u,v) = (u,t_p(v))$$

for $p \in U$ and $(u,v) \in \mathscr{L}_p(E) \times E_p$, where $\widetilde{U} = \pi'^{-1}(U)$.

We also have that $\widetilde{\pi}' : \mathscr{E}(E) \to E$ is a principal fibre bundle with structure group $Gl(\mathbf{R}^n)$. The corresponding local trivialization $(\widetilde{t},\widetilde{\pi}^{-1}(\widetilde{U}))$ in this bundle is given by

$$\widetilde{t}(u,v) = (v,t_p \circ u)$$

for $p \in U$ and $(u,v) \in \mathscr{L}_p(E) \times E_p$.

We have a smooth map $\psi : \mathscr{E}(E) \to \mathbf{R}^n$ defined by

$$\psi(u,v) = u^{-1}(v)$$

for $p \in M$ and $(u,v) \in \mathscr{L}_p(E) \times E_p$. Indeed, if $(\phi,\pi'^{-1}(U))$ is the local trivialization in the frame bundle $\pi' : \mathscr{L}(E) \to M$ corresponding to the local trivialization $(t,\pi^{-1}(U))$ in $\pi : E \to M$, we have that

$$\psi \circ \widetilde{t}^{-1} \circ (\phi \times id_{\mathbf{R}^n})^{-1}(p,T,a) = \psi(t_p^{-1} \circ T, t_p^{-1}(a)) = T^{-1}(a)$$

for $p \in U$, $T \in Gl(\mathbf{R}^n)$ and $a \in \mathbf{R}^n$.

The map $\sigma : \mathscr{E}(E) \to \mathscr{A}(E)$ given by $\sigma(u,v) = \sigma_p(u,v)$ for $p \in M$ and $(u,v) \in \mathscr{L}_p(E) \times E_p$ is a diffeomorphism, since

$$\phi \circ \sigma \circ \widetilde{t}^{-1} \circ (\phi \times id_{\mathbf{R}^n})^{-1}(p,T,a) = (p,\sigma_n(T,a+x(p)))$$

for $p \in U$, $a \in \mathbf{R}^n$ and $T \in Gl(\mathbf{R}^n)$, so that

$$(x \times \sigma_n^{-1}) \circ \phi_\alpha \circ \sigma \circ \widetilde{t}^{-1} \circ ((x \times id_{Gl(\mathbf{R}^n)}) \circ \phi \times id_{\mathbf{R}^n})^{-1}(b,T,a) = (b,T,a+b)$$

for $a,b \in \mathbf{R}^n$ and $T \in Gl(\mathbf{R}^n)$.

If $\pi : TM \to M$ is the tangent bundle over M, then $\mathscr{E}(TM)$ is denoted simply by $\mathscr{E}(M)$.

10.26 Proposition If $\pi : P \to M$ is a principal G-bundle and $u \in P$, then the map $\sigma_u : G \to P$ defined by $\sigma_u(g) = ug$ for $g \in G$, is an embedding whose image is the fibre P_p where $p = \pi(u)$. Moreover, the image $\sigma_{u*}(\mathfrak{g}) = V_u$ is the vertical subspace at u, and these subspaces form a distribution V on P.

If $X \in \mathfrak{g}$, then the vector field $\sigma(X)$ on P defined by $\sigma(X)_u = \sigma_{u*}(X)$ for $u \in P$, belongs to V and is called the *fundamental vector field* determined by X. If $\{X_1,...,X_m\}$ is a basis for $\mathfrak{g}$, then $\{\sigma(X_1),...,\sigma(X_m)\}$ is a local basis for V on P.

PROOF: From 10.16 we have that $\sigma_u = i_p \circ \rho_u$, where $i_p : P_p \to P$ is the inclusion map and $\rho_u : G \to P_p$ is the diffeomorphism given by

$$\rho_u = t_p^{-1} \circ L_{t_p(u)},$$

i.e.,

$$\rho_u(g) = t_p^{-1}(t_p(u)\,g)$$

for $g \in G$. Hence it follows that σ_u is an embedding whose image is P_p, and we have that

$$\sigma_{u*}(\mathfrak{g}) = i_{p*} \circ \rho_{u*}(\mathfrak{g}) = i_{p*}(T_u P_p) = V_u\,.$$

The last part of the proposition now follows from Proposition 2.62 and 9.37. □

10.27 Remark By Proposition 4.95 and 10.3 we know that the distribution V on P is integrable having the connected components of the fibres of $\pi : P \to M$ as its maximal connected integral manifolds.

ASSOCIATED BUNDLES

10.28 Let $\pi : P \to M$ be a principal G-bundle, and let $\psi : G \to G'$ be a Lie group homomorphism into a Lie group G' which acts on a smooth manifold F effectively on the left. We are going to construct a fibre bundle $\pi' : E \to M$ with fibre F and structure group G', called the *fibre bundle associated with* P. Consider the action of G on the product manifold $X = P \times F$ on the right defined by

$$(u,v)\,g = (u\,g, \psi(g^{-1})\,v)\,,$$

and let $E = P \times_\psi F$ be the set of orbits in X. The orbit of an element $(u,v) \in X$ is denoted by $[u,v] = (u,v)\,G$. We have a canonical projection $p' : X \to E$ given by $p'(u,v) = [u,v]$, and a map $\pi' : E \to M$ defined by $\pi'([u,v]) = \pi(u)$ so that the diagram

$$\begin{array}{ccc} X & \xrightarrow{\;p'\;} & E \\ {\scriptstyle pr_1}\downarrow & & \downarrow{\scriptstyle \pi'} \\ P & \xrightarrow{\;\pi\;} & M \end{array}$$

is commutative, where $pr_1 : X \to P$ is the projection on the first factor.

Now let $\mathscr{A} = \{(t_\alpha, \pi^{-1}(U_\alpha)) \,|\, \alpha \in A\}$ be a trivializing cover of P. For each $\alpha \in A$ we have a surjective map $\tau_\alpha : \pi^{-1}(U_\alpha) \times F \to U_\alpha \times F$ defined by

$$\tau_\alpha(u,v) = (p, \psi \circ t_{\alpha,p}(u)\,v)$$

for $(u,v) \in \pi^{-1}(U_\alpha) \times F$, where $p = \pi(u)$. Since

$$\tau_\alpha(u',v') = \tau_\alpha(u,v) \;\Leftrightarrow\; \pi(u') = p \text{ and } \psi \circ t_{\alpha,p}(u')v' = \psi \circ t_{\alpha,p}(u)v$$
$$\Leftrightarrow\; u' = ug \text{ for a } g \in G \text{ and } \psi \circ t_{\alpha,p}(u')v' = \psi \circ t_{\alpha,p}(u')\,(\psi(g^{-1})v)$$
$$\Leftrightarrow\; u' = ug \text{ and } v' = \psi(g^{-1})\,v \text{ for a } g \in G \;\Leftrightarrow\; [u',v'] = [u,v]\,,$$

we therefore have a well-defined bijection $t'_\alpha : \pi'^{-1}(U_\alpha) \to U_\alpha \times F$ with $t'_\alpha \circ p' = \tau_\alpha$, given by

$$t'_\alpha([u,v]) = (p, \psi \circ t_{\alpha,p}(u)\,v) \tag{1}$$

for $[u,v] \in \pi'^{-1}(U_\alpha)$ where $p = \pi(u)$. If $s_\alpha : U_\alpha \to P$ is the section of π defined by

$$s_\alpha(p) = t_\alpha^{-1}(p,e)$$

for $p \in U_\alpha$, then

$$t'_\alpha([s_\alpha(p),v]) = (p,v) \tag{2}$$

for $(p,v) \in U_\alpha \times F$. Now if $(t_\alpha, \pi^{-1}(U_\alpha))$ and $(t_\beta, \pi^{-1}(U_\beta))$ are two local trivializations in $\mathscr{A}$ which are G-related with transition map $\phi_{\beta\alpha} : U_\alpha \cap U_\beta \to G$, we have that

$$t'_\beta \circ t'^{-1}_\alpha(p,v) = (p, \psi \circ t_{\beta,p}(s_\alpha(p))\,v) = (p, \psi \circ t_{\beta,p} \circ t_{\alpha,p}^{-1}(e)\,v) = (p, \psi \circ \phi_{\beta\alpha}(p)\,v)$$

for every $p \in U_\alpha \cap U_\beta$ and $v \in F$. By Proposition 10.5 there is therefore a unique topology, smooth structure and fibre bundle structure on E such that $\pi' : E \to N$ is a fibre bundle over M with fibre F and structure group G', and such that $(t'_\alpha, \pi'^{-1}(U_\alpha))$ is a local trivialization for each $\alpha \in A$. It is called the local trivialization in E associated with the local trivialization $(t_\alpha, \pi^{-1}(U_\alpha))$ in P.

The canonical projection $p' : X \to E$ is smooth as we see from the commutative diagram

$$
\begin{array}{ccc}
\pi^{-1}(U_\alpha) \times F & \xrightarrow{\;\;p'_{\pi'^{-1}(U_\alpha)}\;\;} & \pi'^{-1}(U_\alpha) \\[2mm]
{\scriptstyle t_\alpha \times id_F}\big\downarrow & \searrow{\scriptstyle \tau_\alpha} & \big\downarrow{\scriptstyle t'_\alpha} \\[2mm]
U_\alpha \times G \times F & \xrightarrow[\;\;id_{U_\alpha} \times \mu\;\;]{} & U_\alpha \times F
\end{array}
$$

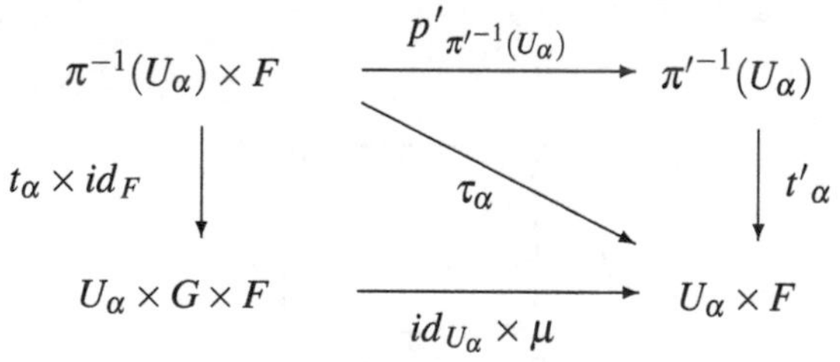

where $\mu : G \times F \to F$ is the group action of G on F defined by $\mu(g,v) = \psi(g)\,v$.

Actually, $p' : X \to E$ is a principal G-bundle over E. To see this, let $\phi : G \times F \to F \times G$ be the diffeomorphism defined by $\phi(g,v) = (\psi(g)\,v,g)$ with inverse $\phi^{-1}(v,g) = (g, \psi(g^{-1})\,v)$.

$$\begin{array}{ccc} G\times F & \xrightarrow{\ \phi\ } & F\times G \\ & \mu \searrow \quad \swarrow pr_1 & \\ & F & \end{array}$$

Then we have local trivializations $(\pi^{-1}(U_\alpha) \times F, \rho_\alpha)$ in X, where $\rho_\alpha : \pi^{-1}(U_\alpha) \times F \to \pi'^{-1}(U_\alpha) \times G$ is the diffeomorphism given by the commutative diagram

$$\begin{array}{ccc} \pi^{-1}(U_\alpha)\times F & \xrightarrow{\quad \rho_\alpha \quad} & \pi'^{-1}(U_\alpha)\times G \\ {\scriptstyle t_\alpha \times id_F}\downarrow & & \downarrow{\scriptstyle t'_\alpha \times id_G} \\ U_\alpha\times G\times F & \xrightarrow{\ id_{U_\alpha}\times\phi\ } & U_\alpha\times F\times G \end{array}$$

so that

$$\rho_\alpha(u,v) = (t'_\alpha \times id_G)^{-1}(\tau_{\alpha,p}(u,v), t_{\alpha,p}(u)) = (p'(u,v), t_{\alpha,p}(u))$$

where $p = \pi(u)$, and

$$\rho_\alpha^{-1}(w,g) = (t_\alpha \times id_F)^{-1}(p, g, \psi(g^{-1})\, t'_{\alpha,p}(w)) = (t_{\alpha,p}^{-1}(g), \psi(g^{-1})\, t'_{\alpha,p}(w))$$

where $p = \pi'(w)$. Using this, we see that

$$\rho_\beta \circ \rho_\alpha^{-1}(w,g) = (w, t_{\beta,p} \circ t_{\alpha,p}^{-1}(g)) = (w, \phi_{\beta\,\alpha}(p)\, g)$$

for every $w \in \pi'^{-1}(U_\alpha \cap U_\beta)$ and $g \in G$.

For each $u \in P$ we have a diffeomorphism $\mu_u : F \to E_p$ from the typical fibre F to the fibre E_p where $p = \pi(u)$, given by

$$\mu_u(v) = [u, v]$$

for $v \in F$. This follows from (1) which implies that

$$t'_{\alpha,p} \circ \mu_u(v) = \psi(t_{\alpha,p}(u))\, v$$

so that

$$\mu_u = t'^{-1}_{\alpha,p} \circ L_{\psi(t_{\alpha,p}(u))}\,,$$

showing that μ_u is the composition of the diffeomorphisms $L_{\psi(t_{\alpha,p}(u))} : F \to F$ and $t'^{-1}_{\alpha,p} : F \to E_p$.

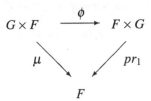

For each $v \in F$ we have a smooth map $\nu_v : P \to E$ given by

$$\nu_v(u) = [u, v]$$

for $u \in P$, so that the diagram

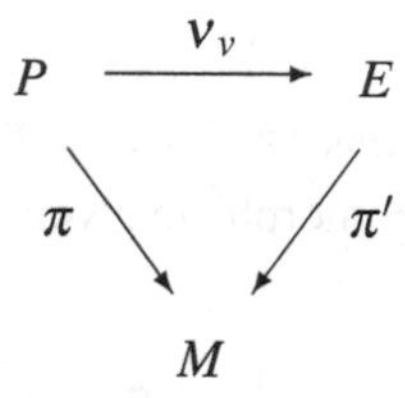

is commutative. This follows from the commutative diagram in the beginning of this section and the relation

$$t_\alpha(u) = (p, t_{\alpha,p}(u))$$

for $u \in \pi^{-1}(U_\alpha)$ where $\pi(u) = p$, which together with (1) implies that

$$t'_\alpha \circ \nu_v \circ t_\alpha^{-1}(p, g) = (p, \psi(g)\, v)$$

for $p \in U_\alpha$ and $g \in G$.

Since

$$[ug, v] = [u, \psi(g)\, v]\,,$$

it follows that

$$\mu_{ug} = \mu_u \circ L_{\psi(g)}$$

and

$$\nu_{\psi(g)\, v} = \nu_v \circ R_g$$

for $u \in P$, $v \in F$ and $g \in G$.

10.29 Examples

(a) Let $\pi : P \to M$ be a principal G-bundle, and let $\pi' : E \to M$ be the fibre bundle associated with P with fibre G obtained from the Lie group homomorphism $id : G \to G$, using that G is acting effectively on itself on the left by left translation. We contend that $\pi' : E \to M$ is equivalent over M to $\pi : P \to M$, with the equivalence $e : E \to P$ given by

$$e([u, g]) = ug$$

for $u \in P$ and $g \in G$. This is clearly well defined, for if $h \in G$, then

$$e([uh, id(h^{-1})\, g]) = uhh^{-1}g = ug = e([u, g])\,.$$

To show that e is an equivalence, let $(t, \pi^{-1}(U))$ be a local trivialization in P, and let $(t', \pi'^{-1}(U))$ be the associated local trivialization in E defined as in 10.28 by

$$t'([u,g]) = (p, t_p(u)g) = (p, t_p(ug)) = t \circ e([u,g])$$

for $u \in \pi^{-1}(p)$ and $g \in G$ where $p \in U$, which shows that

$$t \circ e \circ t'^{-1} = id_{U \times G} .$$

The inverse $e^{-1} : P \to E$ is given by

$$e^{-1}(u) = [u, e]$$

for $u \in P$.

Using the equivalence e, we may therefore identify the fibre bundle E associated with P with fibre G with the principal fibre bundle $\pi : P \to M$ itself, and the diffeomorphism $\mu_u : G \to E_p$ defined in 10.28 is then identified with the diffeomorphism $e_p \circ \mu_u : G \to P_p$ which is induced by the embedding $\sigma_u : G \to P$ defined in Proposition 10.26, since

$$e_p \circ \mu_u(g) = ug = \sigma_u(g)$$

for each $u \in P$ and $g \in G$.

(b) Let $\pi : P \to M$ be a principal G-bundle, and let $\psi : G \to Aut\,(W)$ be a representation of G on an m-dimensional vector space W. Then $Aut\,(W)$ acts effectively on W on the left, and we will show that the fibre bundle $\pi' : E \to M$ with fibre W associated with P is an m-dimensional vector bundle over M. Using the same notation as in 10.28, we consider two local trivializations $(t_\alpha, \pi^{-1}(U_\alpha))$ and $(t_\beta, \pi^{-1}(U_\beta))$ for P which are G-related with transition map $\phi_{\beta\,\alpha} : U_\alpha \cap U_\beta \to G$. Then $t'_\beta \circ t'^{-1}_\alpha$ is a diffeomorphism from $(U_\alpha \cap U_\beta) \times W$ onto itself which is a linear isomorphism on each fibre $\{p\} \times W$ for $p \in U_\alpha \cap U_\beta$, since

$$t'_{\beta,p} \circ t'^{-1}_{\alpha,p} = \psi \circ \phi_{\beta\,\alpha}(p) \in Aut\,(W) .$$

Hence there is a unique vector space structure on each fibre E_p such that $\pi' : E \to M$ is a vector bundle having $((id \times x) \circ t'_\alpha, \pi'^{-1}(U_\alpha))$ as a local trivialization for each $\alpha \in A$, where $x : W \to \mathbf{R}^m$ is any linear isomorphism. Since

$$t'_{\alpha,p}([u,v]) = \psi \circ t_{\alpha,p}(u)(v) ,$$

where $\psi \circ t_{\alpha,p}(u) \in Aut\,(W)$, we have that

$$\begin{aligned} a_1[u,v_1] + a_2[u,v_2] &= t'^{-1}_{\alpha,p}(a_1 t'_{\alpha,p}([u,v_1]) + a_2 t'_{\alpha,p}([u,v_2])) \\ &= [u, a_1 v_1 + a_2 v_2] \end{aligned}$$

for $u \in \pi^{-1}(p)$, $v_1, v_2 \in W$ and $a_1, a_2 \in \mathbf{R}$.

From this it also follows that (p', π) is a bundle map between the vector bundles $pr_1 : P \times W \to P$ and $\pi' : E \to M$ which is a linear isomorphism on each fibre. As usual $pr_1 : P \times W \to P$ is given the vector bundle structure with $(id \times x, pr_1^{-1}(P))$ as a local trivialization, i.e., with product manifold structure and vector space operations

$$a_1(u, v_1) + a_2(u, v_2) = (u, a_1v_1 + a_2v_2)$$

on each fibre $\{u\} \times W$ so that the projection $\rho_u : \{u\} \times W \to W$ on the second factor is a linear isomorphism and

$$t'_{\alpha,p} \circ p' \circ \rho_u^{-1} = \psi \circ t_{\alpha,p}(u) \in Aut\,(W)$$

when $\pi(u) = p \in U_\alpha$, thus showing the above assertions about (p', π). It also follows that the map $\mu_u = p'_u \circ \rho_u^{-1} : W \to E_p$ where $p = \pi(u)$, given by

$$\mu_u(v) = [u, v]$$

for $v \in W$, is a linear isomorphism for each $u \in P$.

(c) Let $\pi : P \to M$ be a principal G-bundle, and let $\pi' : E \to M$ be the associated vector bundle obtained from a representation $\psi : G \to Aut\,(W)$ of G on an m-dimensional vector space W as described in (b). Let $\rho : G \to Aut\,(\mathfrak{gl}\,(W))$ be the representation of the structure group G on the finite dimensional vector space $\mathfrak{gl}\,(W)$ given by

$$\rho(g)(F) = \psi(g) \circ F$$

for $g \in G$ and $F \in \mathfrak{gl}\,(W)$. Then it follows from (b) that the fibre bundle $\pi'' : E' \to M$ with fibre $\mathfrak{gl}\,(W)$ associated with P is a vector bundle which we contend is equivalent over M to the vector bundle $\pi''' : \Lambda^1(W\,; E) \to M$ defined in 5.27, with the equivalence $e : E' \to \Lambda^1(W\,; E)$ given by

$$e([u, F]) = \mu_u \circ F$$

for $u \in P$ and $F \in \mathfrak{gl}\,(W)$. This is clearly well defined, for if $g \in G$, then

$$e([ug, \rho(g^{-1})\,F]) = \mu_{ug} \circ \rho(g^{-1})\,(F) = \mu_u \circ \psi(g) \circ \psi(g^{-1}) \circ F$$
$$= \mu_u \circ F = e([u, F])\,.$$

To show that e is an equivalence, let $(t, \pi^{-1}(U))$ be a local trivialization in P, and let $(t', \pi'^{-1}(U))$ and $(t'', \pi''^{-1}(U))$ be the corresponding local trivializations in E and E', respectively, defined as in 10.28 by

$$t'([u, v]) = (p, \psi(t_p(u))\,(v))$$

and

$$t''([u, F]) = (p, \rho\,(t_p(u))\,(F))$$

for $[u, v] \in \pi'^{-1}(U)$ and $[u, F] \in \pi''^{-1}(U)$ where $p = \pi(u)$. Given a linear isomorphism $x : W \to \mathbf{R}^m$, we obtain an equivalence $e_x : U \times \mathfrak{gl}\,(W) \to U \times \mathfrak{gl}\,(\mathbf{R}^m)$

defined by $e_x(p,F) = (p, x\circ F\circ x^{-1})$ for $(p,F)\in U\times \mathfrak{gl}(W)$. Combining this, we now have that

$$t''([u,F]) = (p, \psi(t_p(u))\circ F) = (p, t'_p\circ \mu_u\circ F) = e_x^{-1}\circ \tau_x\circ e([u,F])$$

where $\tau_x : \pi'''^{-1}(U) \to U\times \mathfrak{gl}(\mathbf{R}^m)$ is the equivalence over U defined in Corollary 5.29, showing that

$$e_x^{-1}\circ \tau_x\circ e\circ t''^{-1} = id_{U\times \mathfrak{gl}(W)}\,.$$

Using the equivalence e, we may therefore identify the fibre bundle E' associated with P with fibre $\mathfrak{gl}(W)$ with the vector bundle $\pi''' : \Lambda^1(W;E) \to M$, and the linear isomorphism $\mu'_u : \mathfrak{gl}(W) \to E'_p$ defined in (b) is then identified with the linear isomorphism $\mu''_u = e_p\circ \mu'_u : \mathfrak{gl}(W) \to \Lambda^1(W;E_p)$ given by

$$\mu''_u(F) = \mu_u\circ F$$

for $u\in P_p$ and $F\in \mathfrak{gl}(W)$ where $p\in M$.
We have a smooth map $\nu = e\circ \nu'_{id_W} : P \to \Lambda^1(W;E)$ given by

$$\nu(u) = \mu_u$$

for $u\in P$. From 10.28 we see that

$$t''\circ \nu'_{id_W}\circ t^{-1} = id_U\times \psi$$

so that

$$e_x^{-1}\circ \tau_x\circ \nu\circ t^{-1} = id_U\times \psi$$

for every local trivialization $(t, \pi^{-1}(U))$ in P.

(d) Let $\pi : P\to M$ be a principal G-bundle, and let $\rho : G\to Aut(W)$ be the representation of G on a finite dimensional vector space W given by

$$\rho(g) = id_W$$

for $g\in G$. Then it follows from (b) that the fibre bundle $\pi' : E\to M$ with fibre W associated with P is a vector bundle which we contend is equivalent over M to the trivial vector bundle $\widetilde{\pi} : E_W \to M$ defined in 5.26, with the equivalence $e : E'\to E_W$ given by

$$e([u,v]) = (\pi(u), v)$$

for $u\in P$ and $v\in W$. This is clearly well defined, for if $g\in G$, then

$$e([ug, \rho(g^{-1})\,v]) = (\pi(ug), v) = (\pi(u), v) = e([u,v])\,.$$

To show that e is an equivalence, let $(t, \pi^{-1}(U))$ be a local trivialization in P, and let $(t', \pi'^{-1}(U))$ be the corresponding local trivialization in E defined as in 10.28 by

$$t'([u,v]) = (p, \rho(t_p(u))(v))$$

for $[u,v] \in \pi'^{-1}(U)$ where $p = \pi(u)$. Combining this, we now have that

$$t'([u,v]) = (p,v) = e([u,v]),$$

showing that

$$id_{U\times W} \circ e \circ t'^{-1} = id_{U\times W}.$$

Using the equivalence e, we may therefore identify the fibre bundle E associated with P with fibre W with the vector bundle $\widetilde{\pi} : E_W \to M$, and the linear isomorphism $\mu_u : W \to E_p$ defined in (b) is then identified with the linear isomorphism $\mu'_u = e_p \circ \mu_u : W \to \{p\} \times W$ given by

$$\mu'_u(v) = (p,v)$$

for $u \in P_p$ and $v \in W$ where $p \in M$.

(e) Let $\pi' : \mathscr{L}(E) \to M$ be the frame bundle of an n-dimensional vector bundle $\pi : E \to M$. Then $id : Gl(\mathbf{R}^n) \to Aut(\mathbf{R}^n)$ is a representation of the structure group $Gl(\mathbf{R}^n)$ on the finite dimensional vector space $\mathbf{R}^n$, so by (b) the fibre bundle $\pi'' : E' \to M$ with fibre $\mathbf{R}^n$ associated with $\mathscr{L}(E)$ is a vector bundle which we contend is equivalent over M to the vector bundle $\pi : E \to M$, with the equivalence $e : E' \to E$ given by

$$e([u,\xi]) = u(\xi)$$

for $u \in \mathscr{L}(E)$ and $\xi \in \mathbf{R}^n$. This is clearly well defined, for if $G \in Gl(\mathbf{R}^n)$, then

$$e([u \circ G, id(G^{-1})\,\xi]) = u \circ G \circ G^{-1}(\xi) = u(\xi) = e([u,\xi]).$$

To show that e is an equivalence, let $(t, \pi^{-1}(U))$ be a local trivialization in the vector bundle $\pi : E \to M$, and let $(\phi, \pi'^{-1}(U))$ and $(t', \pi''^{-1}(U))$ be the corresponding local trivializations in $\mathscr{L}(E)$ and E', respectively, defined as in 10.28 and Example 10.21 by

$$\phi(u) = (p, t_p \circ u)$$

and

$$t'([u,\xi]) = (p, \phi_p(u)\,\xi)$$

for $u \in \mathscr{L}_p(E)$ and $\xi \in \mathbf{R}^n$ where $p \in U$. Combining this, we have that

$$t'([u,\xi]) = (p, t_p \circ u(\xi)) = t \circ e([u,\xi])$$

which shows that

$$t \circ e \circ t'^{-1} = id_{U\times \mathbf{R}^n}.$$

Using the equivalence e, we may therefore identify the fibre bundle E' associated with $\mathscr{L}(E)$ with fibre $\mathbf{R}^n$ with the vector bundle $\pi : E \to M$, and the linear isomorphism $\mu_u : \mathbf{R}^n \to E'_p$ defined in (b) is then identified with $u : \mathbf{R}^n \to E_p$ since

$$e_p \circ \mu_u = u$$

for each $u \in \mathscr{L}(E)$.

(f) Let $\pi' : \mathscr{L}(E) \to M$ be the frame bundle of an n-dimensional vector bundle $\pi : E \to M$. By Proposition 8.13 we have a representation $\rho^k_l : Gl(\mathbf{R}^n) \to Aut(\mathscr{T}^k_l(\mathbf{R}^n))$ of the structure group $Gl(\mathbf{R}^n)$ on the finite dimensional vector space $\mathscr{T}^k_l(\mathbf{R}^n)$ given by

$$\rho^k_l(F) = (F^{-1})^*$$

for $F \in Gl(\mathbf{R}^n)$, i.e.,

$$\rho^k_l(F)(T)(v_1, \dots, v_k, \lambda_1, \dots, \lambda_l) = T(F^{-1}(v_1), \dots, F^{-1}(v_k), \lambda_1 \circ F, \dots, \lambda_l \circ F)$$

for $T \in \mathscr{T}^k_l(\mathbf{R}^n)$, $v_1, \dots, v_k \in \mathbf{R}^n$ and $\lambda_1, \dots, \lambda_l \in (\mathbf{R}^n)^*$. By (b) the fibre bundle $\pi'' : E' \to M$ with fibre $\mathscr{T}^k_l(\mathbf{R}^n)$ associated with $\mathscr{L}(E)$ is therefore a vector bundle which we contend is equivalent over M to the tensor bundle $\pi^k_l : T^k_l(E) \to M$, with the equivalence $e : E' \to T^k_l(E)$ given by

$$e([u, T]) = (u^{-1})^*(T)$$

for $u \in \mathscr{L}(E)$ and $T \in \mathscr{T}^k_l(\mathbf{R}^n)$, i.e.

$$e([u, T])(v_1, \dots, v_k, \lambda_1, \dots, \lambda_l) = T(u^{-1}(v_1), \dots, u^{-1}(v_k), \lambda_1 \circ u, \dots, \lambda_l \circ u)$$

for $u \in \mathscr{L}_p(E)$, $T \in \mathscr{T}^k_l(\mathbf{R}^n)$, $v_1, \dots, v_k \in E_p$ and $\lambda_1, \dots, \lambda_l \in E^*_p$ where $p \in M$. This is clearly well defined, for if $G \in Gl(\mathbf{R}^n)$, then

$$e([u \circ G, \rho^k_l(G^{-1})T]) = (u^{-1})^* \circ (G^{-1})^* \circ G^*(T) = (u^{-1})^*(T) = e([u, T]).$$

To show that e is an equivalence, let $(t, \pi^{-1}(U))$ be a local trivialization in the vector bundle $\pi : E \to M$, and let $(\phi, \pi'^{-1}(U))$ and $(t', \pi''^{-1}(U))$ be the corresponding local trivializations in $\mathscr{L}(E)$ and E', respectively, defined as in 10.28 and Example 10.21 by

$$\phi(u) = (p, t_p \circ u)$$

and

$$t'([u, T]) = (p, \rho^k_l(\phi_p(u))(T))$$

for $u \in \mathscr{L}_p(E)$ and $T \in \mathscr{T}^k_l(\mathbf{R}^n)$ where $p \in U$. Combining this, we have that

$$t'([u, T]) = (p, (t_p^{-1})^* \circ (u^{-1})^*(T)) = \tau \circ e([u, T])$$

where $\tau : (\pi_l^k)^{-1}(U) \to U \times \mathscr{T}_l^k(\mathbf{R}^n)$ is the equivalence over U defined in Proposition 4.53, showing that

$$\tau \circ e \circ t'^{-1} = id_{U \times \mathscr{T}_l^k(\mathbf{R}^n)} \,.$$

Using the equivalence e, we may therefore identify the fibre bundle E' associated with $\mathscr{L}(E)$ with fibre $\mathscr{T}_l^k(\mathbf{R}^n)$ with the tensor bundle $\pi_l^k : T_l^k(E) \to M$, and the linear isomorphism $\mu_u : \mathscr{T}_l^k(\mathbf{R}^n) \to E'_p$ defined in (b) is then identified with $(u^{-1})^* : \mathscr{T}_l^k(\mathbf{R}^n) \to \mathscr{T}_l^k(E_p)$ since

$$e_p \circ \mu_u = (u^{-1})^*$$

for $u \in \mathscr{L}_p(E)$.

(g) Let $\pi' : \mathscr{L}(E) \to M$ be the frame bundle of an n-dimensional vector bundle $\pi : E \to M$. Then $Ad : Gl(\mathbf{R}^n) \to Aut(\mathfrak{gl}(\mathbf{R}^n))$ is a representation of the structure group $Gl(\mathbf{R}^n)$ on the finite dimensional vector space $\mathfrak{gl}(\mathbf{R}^n)$, so by (b) the fibre bundle $\pi'' : E' \to M$ with fibre $\mathfrak{gl}(\mathbf{R}^n)$ associated with $\mathscr{L}(E)$ is a vector bundle which we contend is equivalent over M to the vector bundle $\pi''' : \Lambda^1(E;E) \to M$, with the equivalence $e : E' \to \Lambda^1(E;E)$ given by

$$e([u,F]) = u \circ F \circ u^{-1}$$

for $u \in \mathscr{L}(E)$ and $F \in \mathfrak{gl}(\mathbf{R}^n)$. This is clearly well defined, for if $G \in Gl(\mathbf{R}^n)$, then

$$\begin{aligned} e([u \circ G, Ad(G^{-1})F]) &= (u \circ G) \circ (G^{-1} \circ F \circ G) \circ (G^{-1} \circ u^{-1}) \\ &= u \circ F \circ u^{-1} = e([u,F]). \end{aligned}$$

To show that e is an equivalence, let $(t, \pi^{-1}(U))$ be a local trivialization in the vector bundle $\pi : E \to M$, and let $(\phi, \pi'^{-1}(U))$ and $(t', \pi''^{-1}(U))$ be the corresponding local trivializations in $\mathscr{L}(E)$ and E', respectively, defined as in 10.28 and Example 10.21 by

$$\phi(u) = (p, t_p \circ u)$$

and

$$t'([u,F]) = (p, Ad(\phi_p(u))(F))$$

for $u \in \mathscr{L}_p(E)$ and $F \in \mathfrak{gl}(\mathbf{R}^n)$ where $p \in U$. Combining this, we have that

$$t'([u,F]) = (p, t_p \circ u \circ F \circ u^{-1} \circ t_p^{-1}) = \tau \circ e([u,F])$$

where $\tau : \pi'''^{-1}(U) \to U \times \mathfrak{gl}(\mathbf{R}^n)$ is the equivalence over U defined in Proposition 5.28, showing that

$$\tau \circ e \circ t'^{-1} = id_{U \times \mathfrak{gl}(\mathbf{R}^n)} \,.$$

Using the equivalence e, we may therefore identify the fibre bundle E' associated with $\mathscr{L}(E)$ with fibre $\mathfrak{gl}(\mathbf{R}^n)$ with the vector bundle $\pi''' : \Lambda^1(E;E) \to M$, and the linear isomorphism $\mu_u : \mathfrak{gl}(\mathbf{R}^n) \to E'_p$ defined in (b) is then identified with the linear isomorphism $\mu'_u = e_p \circ \mu_u : \mathfrak{gl}(\mathbf{R}^n) \to \Lambda^1(E_p;E_p)$ given by

$$\mu'_u(F) = u \circ F \circ u^{-1}$$

for $u \in \mathscr{L}_p(E)$ and $F \in \mathfrak{gl}(\mathbf{R}^n)$ where $p \in M$.

(h) Let $\pi' : \mathscr{L}(E) \to M$ be the frame bundle of an n-dimensional vector bundle $\pi : E \to M$, and let $\rho : Gl(\mathbf{R}^n) \to Aut(L(\mathbf{R}^m, \mathbf{R}^n))$ be the representation of the structure group $Gl(\mathbf{R}^n)$ on the finite dimensional vector space $L(\mathbf{R}^m, \mathbf{R}^n)$ given by

$$\rho(G)(F) = G \circ F$$

for $G \in Gl(\mathbf{R}^n)$ and $F \in L(\mathbf{R}^m, \mathbf{R}^n)$. Then it follows from (b) that the fibre bundle $\pi'' : E' \to M$ with fibre $L(\mathbf{R}^m, \mathbf{R}^n)$ associated with $\mathscr{L}(E)$ is a vector bundle which we contend is equivalent over M to the vector bundle $\pi''' : \Lambda^1(\mathbf{R}^m;E) \to M$ defined in 5.27, with the equivalence $e : E' \to \Lambda^1(\mathbf{R}^m;E)$ given by

$$e([u,F]) = u \circ F$$

for $u \in \mathscr{L}(E)$ and $F \in L(\mathbf{R}^m, \mathbf{R}^n)$. This is clearly well defined, for if $G \in Gl(\mathbf{R}^n)$, then

$$e([u \circ G, \rho(G^{-1})F]) = (u \circ G) \circ (G^{-1} \circ F) = u \circ F = e([u,F]).$$

To show that e is an equivalence, let $(t, \pi^{-1}(U))$ be a local trivialization in the vector bundle $\pi : E \to M$, and let $(\phi, \pi'^{-1}(U))$ and $(t', \pi''^{-1}(U))$ be the corresponding local trivializations in $\mathscr{L}(E)$ and E', respectively, defined as in 10.28 and Example 10.21 by

$$\phi(u) = (p, t_p \circ u)$$

and

$$t'([u,F]) = (p, \rho(\phi_p(u))(F))$$

for $u \in \mathscr{L}_p(E)$ and $F \in L(\mathbf{R}^m, \mathbf{R}^n)$ where $p \in U$. Combining this, we have that

$$t'([u,F]) = (p, t_p \circ u \circ F) = \tau \circ e([u,F])$$

where $\tau : \pi'''^{-1}(U) \to U \times L(\mathbf{R}^m, \mathbf{R}^n)$ is the equivalence over U defined in Corollary 5.29, showing that

$$\tau \circ e \circ t'^{-1} = id_{U \times L(\mathbf{R}^m, \mathbf{R}^n)}.$$

Using the equivalence e, we may therefore identify the fibre bundle E' associated with $\mathscr{L}(E)$ with fibre $L(\mathbf{R}^m, \mathbf{R}^n)$ with the vector bundle $\pi''' : \Lambda^1(\mathbf{R}^m;E)$

$\to M$, and the linear isomorphism $\mu_u : L(\mathbf{R}^m, \mathbf{R}^n) \to E'_p$ defined in (b) is then identified with the linear isomorphism $\mu'_u = e_p \circ \mu_u : L(\mathbf{R}^m, \mathbf{R}^n) \to \Lambda^1(\mathbf{R}^m ; E_p)$ given by

$$\mu'_u(F) = u \circ F$$

for $u \in \mathscr{L}_p(E)$ and $F \in L(\mathbf{R}^m, \mathbf{R}^n)$ where $p \in M$.

(i) Let $\pi' : \mathscr{L}(E) \to M$ be the frame bundle of an n-dimensional vector bundle $\pi : E \to M$, and let $\widetilde{\pi} : \widetilde{E} \to M$ be the associated vector bundle obtained from a representation $\psi : Gl(\mathbf{R}^n) \to Aut(W)$ of the structure group $Gl(\mathbf{R}^n)$ on an m-dimensional vector space W. Then we have a representation $\rho_k : Gl(\mathbf{R}^n) \to \mathrm{Aut}(\mathscr{T}^k(\mathbf{R}^n ; W))$ of the structure group $Gl(\mathbf{R}^n)$ on the finite dimensional vector space $\mathscr{T}^k(\mathbf{R}^n ; W)$ given by

$$\rho_k(G)(F) = \psi(G) \circ F \circ \{G^{-1}\}^k$$

for $G \in Gl(\mathbf{R}^n)$ and $F \in \mathscr{T}^k(\mathbf{R}^n ; W)$. By (b) the fibre bundle $\pi'' : E' \to M$ with fibre $\mathscr{T}^k(\mathbf{R}^n ; W)$ associated with $\mathscr{L}(E)$ is a vector bundle which we contend is equivalent over M to the bundle $\pi^k : \mathscr{T}^k(E ; \widetilde{E}) \to M$, with the equivalence $e : E' \to \mathscr{T}^k(E ; \widetilde{E})$ given by

$$e([u, F]) = \widetilde{\mu}_u \circ F \circ \{u^{-1}\}^k$$

for $u \in \mathscr{L}_p(E)$ and $F \in \mathscr{T}^k(\mathbf{R}^n ; W)$ where $p \in M$. This is clearly well defined, for if $G \in Gl(\mathbf{R}^n)$, then

$$e([u \circ G, \rho_k(G^{-1}) F]) = \{\widetilde{\mu}_u \circ \psi(G)\} \circ \{\psi(G)^{-1} \circ F \circ \{G\}^k\} \circ$$
$$\{\{G^{-1}\}^k \circ \{u^{-1}\}^k\} = e([u, F]) .$$

To show that e is an equivalence, let $(t, \pi^{-1}(U))$ be a local trivialization in the vector bundle $\pi : E \to M$, and let $(\phi, \pi'^{-1}(U))$ be the corresponding local trivialization in $\mathscr{L}(E)$ defined as in Example 10.21 by

$$\phi(u) = (p, t_p \circ u)$$

for $u \in \mathscr{L}_p(E)$ where $p \in U$. Given a linear isomorphism $x : W \to \mathbf{R}^m$, we let $z : \mathscr{T}^k(\mathbf{R}^n ; W) \to \mathscr{T}^k(\mathbf{R}^n ; \mathbf{R}^m)$ be the linear isomorphism defined by $z(F) = x \circ F$ for $F \in \mathscr{T}^k(\mathbf{R}^n ; W)$. Furthermore, let $((id \times x) \circ \widetilde{t}, \widetilde{\pi}^{-1}(U))$ and $(t', \pi''^{-1}(U))$ be the local trivializations in $\widetilde{E}$ and E', respectively, defined as in 10.28 by

$$\widetilde{t}([u, v]) = (p, \psi(\phi_p(u))(v))$$

and

$$t'([u, F]) = (p, \rho_k(\phi_p(u))(F))$$

for $u \in \mathscr{L}_p(E)$, $v \in W$ and $F \in \mathscr{T}^k(\mathbf{R}^n ; W)$ where $p \in U$. Combining this, we have that

$$(id \times z) \circ t'([u,F]) = (p, x \circ \psi(\phi_p(u)) \circ F \circ \{\phi_p(u)^{-1}\}^k)$$
$$= (p, x \circ \widetilde{t}_p \circ \widetilde{\mu}_u \circ F \circ \{u^{-1}\}^k \circ \{t_p^{-1}\}^k) = \tau \circ e([u,F])$$

where $\tau : \pi''^{-1}(U) \to U \times \mathscr{T}^k(\mathbf{R}^n;\mathbf{R}^m)$ is the equivalence over U defined in Proposition 5.28, showing that

$$\tau \circ e \circ t'^{-1} \circ (id \times z)^{-1} = id_{U \times \mathscr{T}^k(\mathbf{R}^n;\mathbf{R}^m)} \,.$$

Using the equivalence e, we may therefore identify the fibre bundle E' associated with $\mathscr{L}(M)$ with fibre $\mathscr{T}^k(\mathbf{R}^n;W)$ with the vector bundle $\pi^k : \mathscr{T}^k(E;\widetilde{E}) \to M$, and the linear isomorphism $\mu_u : \mathscr{T}^k(\mathbf{R}^n;W) \to E'_p$ defined in (b) is then identified with the linear isomorphism $\mu'_u = e_p \circ \mu_u : \mathscr{T}^k(\mathbf{R}^n;W) \to \mathscr{T}^k(E_p;\widetilde{E}_p)$ given by

$$\mu'_u(F) = \widetilde{\mu}_u \circ F \circ \{u^{-1}\}^k$$

for $u \in \mathscr{L}_p(M)$ and $F \in \mathscr{T}^k(\mathbf{R}^n;W)$ where $p \in M$.

CONNECTIONS

10.30 Proposition If $\pi : P \to M$ is a principal G-bundle, then

$$(R_g)_* \, \sigma(X) = \sigma(Ad(g^{-1})X)$$

for every $X \in \mathfrak{g}$ and $g \in G$.

PROOF: Since

$$R_g \circ \sigma_u(h) = uhg = ug\, g^{-1}hg = \sigma_{ug} \circ i_{g^{-1}}(h)$$

for $u \in P$ and $h \in G$, we have that

$$((R_g)_* \, \sigma(X))_{ug} = (R_g)_* \, \sigma(X)_u = (R_g)_* \circ \sigma_{u*}(X)$$
$$= \sigma_{ug*} \circ Ad(g^{-1})(X) = \sigma(Ad(g^{-1})X)_{ug} \,.$$ □

10.31 Corollary If $\pi : P \to M$ is a principal G-bundle, then

$$V_{ug} = (R_g)_* V_u$$

for every $u \in P$ and $g \in G$.

PROOF: By Proposition 10.30 we have that

$$(R_g)_* \circ \sigma_{u*} = \sigma_{ug*} \circ Ad\,(g^{-1})$$

so that

$$(R_g)_* V_u = (R_g)_* \circ \sigma_{u*}(\mathfrak{g}) = \sigma_{ug*}(\mathfrak{g}) = V_{ug}\,.$$

□

10.32 Definition A *(principal) connection* in a principal G-bundle $\pi : P \to M$ is a distribution H on P such that

(i) $T_uP = H_u \oplus V_u$ for every $u \in P$,
(ii) $H_{ug} = (R_g)_* H_u$ for every $u \in P$ and $g \in G$.

The subspace H_u of T_uP is called the *horizontal subspace* at u, and the vectors in H_u are called *horizontal tangent vectors* at u.

10.33 If H is a connection in the principal G-bundle $\pi : P \to M$ and $u \in P$, we let $h_u : T_uP \to T_uP$ and $v_u : T_uP \to T_uP$ be the projections on the horizontal and vertical subspace of T_uP uniquely determined by

$$w = h_u(w) + v_u(w) \quad \text{where} \quad h_u(w) \in H_u \quad \text{and} \quad v_u(w) \in V_u$$

for $w \in T_uP$. The vectors $h_u(w)$ and $v_u(w)$ are called the *horizontal* and *vertical* components of w.

Now the vertical subspace V_u is isomorphic to the Lie algebra $\mathfrak{g}$ by Proposition 10.26. If $\tau_u : V_u \to \mathfrak{g}$ is the inverse of the linear isomorphism from $\mathfrak{g}$ to V_u induced by $\sigma_{u*} : TG \to TP$, we therefore have a $\mathfrak{g}$-valued 1-form $\omega_u = \tau_u \circ v_u$ on T_uP so that $\omega_u \circ \sigma_{u*} = id_{\mathscr{G}}$. We will prove that the map $\omega : P \to \Lambda^1(TP\,;\mathfrak{g})$ defined by $\omega(u) = \omega_u$ for $u \in P$, is a $\mathfrak{g}$-valued 1-form on P called the *connection form* of the given connection H. We first need a definition.

10.34 Definition Let $\pi : P \to M$ be a principal G-bundle. Then a $\mathfrak{g}$-valued 1-form ω on P is called a *connection form* if it satisfies

(i) $\omega(u)(\sigma(X)_u) = X$ for every $u \in P$ and $X \in \mathfrak{g}$,
(ii) $R_g^*\,\omega = Ad\,(g^{-1}) \cdot \omega$ for every $g \in G$.

10.35 Proposition If H is a connection in the principal G-bundle $\pi : P \to M$, then the map $\omega : P \to \Lambda^1(TP\,;\mathfrak{g})$ defined in 10.33 is a connection form on P with $H_u = \ker\,\omega(u)$ for $u \in P$.

PROOF: Formula (i) in the definition is just a reformulation of the formula $\omega_u \circ \sigma_{u*} = id_{\mathscr{G}}$ established in 10.33.

We next show that ω is a $\mathfrak{g}$-valued 1-form on P. Let $\mathscr{C} = \{X_1, ..., X_m\}$ be a basis for $\mathfrak{g}$, and let $u \in P$. Choose a local basis $\{Y_{m+1}, ..., Y_{m+n}\}$ for H on an open neighbourhood U of u, and let $Y_j = \sigma(X_j)|_U$ for $j = 1, ..., m$. Then $\{Y_1, ..., Y_{m+n}\}$ is a local basis for TP on U, and we let $\{\omega^1, ..., \omega^{m+n}\}$ be the dual local basis for T^*P on U. In fact, it follows from Corollary 2.59 and Proposition 4.11 that the maps $\omega^j : U \to T^*P$ defined by $\omega^j(w)(Y_{i\,w}) = \delta_{ij}$ for $1 \le i, j \le m+n$ and $w \in U$ are sections in T^*P.

Now the components of ω with respect to the basis $\mathscr{C}$ for $\mathfrak{g}$ are ω^j for $j = 1, ..., m$. Indeed, we have that

$$\omega(u)(Y_{j\,u}) = \begin{cases} X_j & \text{for} \quad 1 \le j \le m \\ 0 & \text{for} \quad m+1 \le j \le m+n \end{cases} .$$

So if $w \in T_uP$ is written as

$$w = \sum_{j=1}^{m+n} \omega^j(u)(w)\, Y_{j\,u} ,$$

then

$$\omega(u)(w) = \sum_{j=1}^{m} \omega^j(u)(w)\, X_j$$

showing the above assertion and completing the proof that ω is a $\mathfrak{g}$-valued 1-form on P.

To prove formula (ii) in the definition, suppose first that $w \in H_u$. Then $(R_g)_*\, w \in H_{ug}$ which implies that

$$R_g^*\, \omega\,(u)\,(w) = \omega(ug)\,((R_g)_*\, w) = 0 = Ad\,(g^{-1})\; \omega(u)(w) .$$

If on the other hand $w \in V_u$, then $w = \sigma\,(X)_u$ for an $X \in \mathfrak{g}$. By Proposition 10.30 we have that

$$(R_g)_*\, w = \sigma\,(Ad\,(g^{-1})\,X)_{ug}$$

so that

$$R_g^*\, \omega\,(u)\,(w) = \omega(ug)\,((R_g)_*\, w) = Ad\,(g^{-1})\,X = Ad\,(g^{-1})\; \omega(u)(w) ,$$

and this completes the proof of formula (ii) since $T_uP = H_u \oplus V_u$. The last part of the proposition follows immediately from the definition of ω. □

10.36 Proposition Let ω be a connection form in the principal G-bundle $\pi : P \to M$. Then there is a unique connection H in P whose connection form is ω, which is given by $H_u = \ker\, \omega(u)$ for $u \in P$.

PROOF: We first show formula (i) in Definition 10.32. By Definition 10.34 (i) and Proposition 10.26 we have that

$$w - \sigma\,(\omega(u)\,(w))_u \in H_u \quad \text{and} \quad \sigma\,(\omega(u)\,(w))_u \in V_u$$

for every $w \in T_uP$, which shows that $T_uP = H_u + V_u$. If $w \in H_u \cap V_u$, then $w =$

$\sigma(X)_u$ for an $X \in \mathfrak{g}$, where $X = \omega(u)(w) = 0$ so that $w = 0$. Hence we have that $T_uP = H_u \oplus V_u$ with projections $h_u : T_uP \to T_uP$ and $v_u : T_uP \to T_uP$ on H_u and V_u given by

$$h_u(w) = w - \sigma(\omega(u)(w))_u \quad \text{and} \quad v_u(w) = \sigma(\omega(u)(w))_u \tag{1}$$

for $w \in T_uP$.

We next show that H is a distribution in P. Let ω^j for $j = 1, ..., m$ be the components of ω with respect to a basis $\mathscr{C} = \{X_1, ..., X_m\}$ for $\mathfrak{g}$, and let $u_0 \in P$. If $\{Y_1, ..., Y_{m+n}\}$ is a local basis for TP on an open neighbourhood U of u_0, then the vector fields $\{Z_1, ..., Z_{m+n}\}$ defined by

$$Z_{iu} = h_u(Y_{iu}) = Y_{iu} - \sigma\Big(\sum_{j=1}^{m} \omega^j(u)(Y_{iu})\, X_j\Big)_u = Y_{iu} - \sum_{j=1}^{m} \omega^j(u)(Y_{iu})\, \sigma(X_j)_u$$

for $i = 1, ..., m+n$ and $u \in U$, span H on U. Hence n of the vector fields Z_i form a local basis for H on an open neighbourhood W of u_0 contained in U, showing that H is a distribution in P.

To show formula (ii) in Definition 10.32, let $w \in H_u$. Then we have that

$$\omega(ug)((R_g)_* w) = R_g^* \omega(u)(w) = Ad(g^{-1})\, \omega(u)(w) = 0$$

so that $(R_g)_* w \in H_{ug}$. Since $(R_g)_*$ is injective and $\dim(H_{ug}) = \dim(H_u)$, it follows that $H_{ug} = (R_g)_* H_u$, which completes the proof that H is a connection in P. From the last formula in (1) and 10.33 we have that

$$\omega(u)(w) = \tau_u \circ v_u(w)$$

for $w \in T_uP$, showing that ω is the connection form of H. □

10.37 Let $\pi : P \to M$ be a principal G-bundle, and let $f : N \to M$ be a smooth map from a smooth manifold N. Then the induced bundle $\widetilde{\pi} : \widetilde{P} \to N$ defined in 10.12 is also a principal G-bundle. The action of G on $\widetilde{P}$ on the right is given by

$$(p, u)\, g = (p, ug) \tag{1}$$

for $(p, u) \in \widetilde{P}$ and $g \in G$, since

$$\widetilde{t}_{\alpha,p}((p,u)\, g) = \widetilde{t}_{\alpha,p}(p,u)\, g = t_{\alpha,f(p)}(u)\, g = t_{\alpha,f(p)}(ug) = \widetilde{t}_{\alpha,p}(p, ug)$$

for $p \in \widetilde{U}_\alpha$, $u \in P_{f(p)}$ and $g \in G$ by 10.12 and 10.16.

By applying the map $\widetilde{f} : \widetilde{P} \to P$ defined in 10.12 on both sides of (1), we obtain

$$\widetilde{f}((p,u)\, g) = ug = \widetilde{f}(p,u)\, g$$

which implies that

$$\widetilde{f} \circ \widetilde{\sigma}_{(p,u)} = \sigma_u \tag{2}$$

where $\widetilde{\sigma}_{(p,u)} : G \to \widetilde{P}$ and $\sigma_u : G \to P$ are the maps defined by

$$\widetilde{\sigma}_{(p,u)}(g) = (p,u)\,g \quad \text{and} \quad \sigma_u(g) = u\,g\,.$$

We also have that

$$\widetilde{f} \circ \widetilde{R}_g = R_g \circ \widetilde{f} \tag{3}$$

where $\widetilde{R}_g$ and R_g are the right translations by g on $\widetilde{P}$ and P, respectively.

Given a connection form ω on P, we will show that $\widetilde{f}^*(\omega)$ is a connection form on $\widetilde{P}$. If $X \in \mathfrak{g}$, and if $\sigma(X)$ and $\widetilde{\sigma}(X)$ are the fundamental vector fields determined by X on P and $\widetilde{P}$, respectively, it follows from (2) that

$$\widetilde{f}_*\, \widetilde{\sigma}(X)_{(p,u)} = \widetilde{f}_*\, \widetilde{\sigma}_{(p,u)*}(X) = \sigma_{u*}(X) = \sigma(X)_u$$

so that

$$\widetilde{f}^*(\omega)(p,u)\,(\widetilde{\sigma}(X)_{(p,u)}) = \omega(u)\,(\sigma(X)_u) = X\,.$$

By (3) and Proposition 5.48 we also have that

$$\widetilde{R}_g^*\, \widetilde{f}^*(\omega) = \widetilde{f}^*(R_g^*\,\omega) = \widetilde{f}^*(Ad\,(g^{-1})\cdot\omega) = Ad\,(g^{-1})\cdot\widetilde{f}^*(\omega)$$

which completes the proof that $\widetilde{f}^*(\omega)$ is a connection form on $\widetilde{P}$.

If H and $\widetilde{H}$ are the connections in P and $\widetilde{P}$, respectively, with connection forms ω and $\widetilde{f}^*(\omega)$, we have that

$$\widetilde{f}^*(\omega)\,(p,u) = \omega(u) \circ \widetilde{f}_{*\,(p,u)}$$

which implies that

$$\widetilde{H}_{(p,u)} = (\widetilde{f}_{*\,(p,u)})^{-1}(H_u)$$

for $p \in N$ and $u \in P_{f(p)}$.

TENSORIAL FORMS

10.38 Definition Let $\pi : P \to M$ be a principal G-bundle and $\rho : G \to \mathrm{Aut}\,(W)$ a representation of G on a finite dimensional vector space W. Then a W-valued k-form $\omega \in \Omega^k(P;W)$ is said to be *pseudo-tensorial* of type (ρ, W) if

$$R_g^*\,\omega = \rho\,(g^{-1})\cdot\omega$$

for every $g \in G$. It is said to be *horizontal* if $\omega(u)\,(w_1, ..., w_k) = 0$ whenever $u \in P$ and at least one $w_i \in T_uP$ is vertical. We say that ω is *tensorial* of type (ρ, W) if it is both horizontal and pseudo-tensorial of type (ρ, W). A vector field X on P is said to be *vertical* if X_u is a vertical vector in T_uP for every $u \in P$.

If H is a connection in P, then a vector-valued k-form ω on P is said to be *vertical* if $\omega(u)(w_1, ..., w_k) = 0$ whenever $u \in P$ and at least one $w_i \in T_uP$ is horizontal. A vector field X on P is said to be *horizontal* if X_u is a horizontal vector in T_uP for every $u \in P$.

10.39 Proposition Let $\pi : P \to M$ be a principal G-bundle, and let $\rho_i : G \to \mathrm{Aut}(W_i)$ be a representation of G on a finite dimensional vector space W_i for $i = 1,2$. If $F : W_1 \to W_2$ is a linear map satisfying

$$F \circ \rho_1(g) = \rho_2(g) \circ F$$

for every $g \in G$, and $\omega \in \Omega^k(P;W_1)$ is a vector-valued k-form on P which is pseudo-tensorial (tensorial) of type (ρ_1, W_1), then the form $F \cdot \omega$ is pseudo-tensorial (tensorial) of type (ρ_2, W_2).

PROOF: If ω is pseudo-tensorial of type (ρ_1, W_1), then it follows from Proposition 5.48 that

$$R_g^*(F \cdot \omega) = F \cdot R_g^*(\omega) = F \cdot (\rho_1(g^{-1}) \cdot \omega) = \rho_2(g^{-1}) \cdot (F \cdot \omega)$$

for every $g \in G$, which shows that $F \cdot \omega$ is pseudo-tensorial of type (ρ_2, W_2). If ω is horizontal, so is $F \cdot \omega$. □

10.40 Proposition Let H be a connection in the principal G-bundle $\pi : P \to M$. Then the maps $h : TP \to TP$ and $v : TP \to TP$, where $h(w)$ and $v(w)$ are the horizontal and vertical components of w for each $w \in TP$, are bundle maps over P called the *horizontal* and *vertical projection* determined by H. We have that

$$(R_g)_* \circ h = h \circ (R_g)_* \quad \text{and} \quad (R_g)_* \circ v = v \circ (R_g)_*$$

for every $g \in G$.

PROOF: By Proposition 2.64 we know that h and v are bundle maps over P. The last part of the proposition follows from the relation

$$(R_g)_*(w) = (R_g)_* \circ h(w) + (R_g)_* \circ v(w)$$

where

$$(R_g)_* \circ h(w) \in H_{ug} \quad \text{and} \quad (R_g)_* \circ v(w) \in V_{ug}$$

for every $w \in T_uP$ by Definition 10.32 and Corollary 10.31. □

10.41 Proposition Let H be a connection in the principal G-bundle $\pi : P \to M$, and let $h : TP \to TP$ and $v : TP \to TP$ be the horizontal and vertical projection determined by H. Then a W-valued k-form $\omega \in \Omega^k(P;W)$ is horizontal if and only if $h^*(\omega) = \omega$ and vertical if and only if $v^*(\omega) = \omega$.

PROOF : Suppose that $w_1, \dots, w_k \in T_uP$, and let $w_{i,1} = h(w_i)$ and $w_{i,2} = v(w_i)$ for $i = 1, \dots, k$. Then we have that

$$\omega(u)(w_1, \dots, w_k) = \sum_{i_1,\dots,i_k=1}^{2} \omega(u)(w_{1,i_1}, \dots, w_{k,i_k}) .$$

If ω is horizontal, it follows that

$$\omega(u)(w_1, \dots, w_k) = \omega(u)(h(w_1), \dots, h(w_k)) = h^* \omega(u)(w_1, \dots, w_k) ,$$

and a similar argument applies when ω is vertical. The converse statements follow directly from the definition. □

COVARIANT DERIVATIVE OF FORMS ON PRINCIPAL BUNDLES

10.42 Definition Let H be a connection in the principal G-bundle $\pi : P \to M$ with horizontal projection $h : TP \to TP$. Then the *exterior covariant derivative* of a vector-valued k-form $\omega \in \Omega^k(P;W)$ is the vector-valued $(k+1)$-form $D\omega \in \Omega^{k+1}(P;W)$ defined by

$$D\omega = h^*(d\omega) ,$$

i.e.,

$$D\omega(u)(w_1, \dots, w_{k+1}) = d\omega(u)(h(w_1), \dots, h(w_{k+1}))$$

for $u \in P$ and $w_1, \dots, w_{k+1} \in T_uP$.

10.43 Proposition Let $\pi : P \to M$ be a principal G-bundle with a connection H, and let $\rho : G \to \mathrm{Aut}(W)$ be a representation of G on a finite dimensional vector space W. If $\omega \in \Omega^k(P;W)$ is a pseudo-tensorial k-form of type (ρ, W), then its exterior covariant derivative $D\omega$ is a tensorial $k+1$-form of type (ρ, W).

PROOF : If $u \in P$ and $w_1, \dots, w_{k+1} \in T_uP$, then

$$\begin{aligned}
R_g^*(D\omega)(u)(w_1, \dots, w_{k+1}) &= (D\omega)(ug)((R_g)_* w_1, \dots, (R_g)_* w_{k+1}) \\
&= (d\omega)(ug)(h \circ (R_g)_* w_1, \dots, h \circ (R_g)_* w_{k+1}) \\
&= (d\omega)(ug)((R_g)_* \circ h(w_1), \dots, (R_g)_* \circ h(w_{k+1})) \\
&= R_g^*(d\omega)(u)(h(w_1), \dots, h(w_{k+1})) = d(R_g^*\omega)(u)(h(w_1), \dots, h(w_{k+1})) \\
&= d(\rho(g^{-1}) \cdot \omega)(u)(h(w_1), \dots, h(w_{k+1})) \\
&= \rho(g^{-1}) \cdot d\omega(u)(h(w_1), \dots, h(w_{k+1})) = \rho(g^{-1}) \cdot D\omega(u)(w_1, \dots, w_{k+1}) ,
\end{aligned}$$

thereby proving that $D\omega$ is pseudo-tensorial of type (ρ, W).

If at least one $w_i \in V_u$, then $h(w_i) = 0$ so that $D\omega(u)(w_1, \ldots, w_{k+1}) = 0$, which shows that $D\omega$ is horizontal. □

THE CURVATURE FORM

10.44 Definition Let $\pi : P \to M$ be a principal G-bundle with a connection H, and let ω be the connection form of H. Then the $\mathfrak{g}$-valued 2-form $\Omega = D\omega$ is called the *curvature form* of H. The connection H is said to be *flat* if $\Omega = 0$.

10.45 Remark From Proposition 10.35 it follows that the connection form ω is a vertical pseudo-tensorial 1-form of type $(Ad, \mathfrak{g})$, and hence the curvature form Ω is a tensorial 2-form of type $(Ad, \mathfrak{g})$.

10.46 Proposition Let $\pi : P \to M$ be a principal G-bundle over a 1-dimensional smooth manifold M. Then every connection H in P is flat.

PROOF: If Ω is the curvature form of H, we must show that $\Omega(u)(w_1, w_2) = 0$ for every $u \in P$ and $w_1, w_2 \in T_uP$. Since M is 1-dimensional, we know that $\dim(H_u) = 1$. If $\{w\}$ is a basis for H_u and $h(w_i) = c_i w$ for $i = 1, 2$, we have that

$$\Omega(u)(w_1, w_2) = c_1 c_2 \, d\omega(u)(w, w) = 0$$

since $d\omega(u)$ is skew symmetric. □

10.47 Proposition Let $\pi : P \to M$ be a principal G-bundle with a connection H, and let ω and Ω be the connection form and the curvature form of H. Then we have that

$$\Omega(X, Y) = d\omega(h \circ X, h \circ Y) = -\,\omega([h \circ X, h \circ Y])$$

for any pair of vector fields X, Y on an open subset U of P.

PROOF: Follows from Remark 5.86 since $\omega(h \circ X) = 0$ and $\omega(h \circ Y) = 0$. □

10.48 Corollary Let H be a connection in the principal G-bundle $\pi : P \to M$, and let Ω be the curvature form of H. Then H is an integrable distribution on P if and only if $\Omega = 0$.

PROOF: If H is integrable, then $[h \circ X, h \circ Y] \in H$ which implies that $\Omega(X, Y) = 0$ for every pair of vector fields X, Y on P.

Conversely, assume that $\Omega = 0$. For any pair of vector fields $X, Y \in H$ defined on some open subset U of P, it follows from Proposition 10.47 that $[X, Y] = [h \circ X, h \circ Y] \in H$, which shows that H is an integrable distribution on P. □

HORIZONTAL LIFTS OF VECTOR FIELDS

10.49 Proposition Let $\pi : P \to M$ be a principal G-bundle with a connection H, and let $i : H \to TP$ be the inclusion map which is a bundle map over P. Then $(\pi_* \circ i, \pi)$ is a bundle map from H to TM such that $\alpha = \pi_* \circ i$ induces a linear isomorphism between the fibres H_u and T_pM for each $u \in P$, where $p = \pi(u)$. If X is a vector field on M, then the pull-back $\alpha^*(X)$ is a section of the bundle $\pi' : H \to P$, and $X^* = i \circ \alpha^*(X)$ is a horizontal vector field on P which is π-related to X and is called the *horizontal lift* of X.

PROOF : If $w \in H_u$ with $\pi_{*u}(w) = 0$, then $w \in H_u \cap V_u = \{0\}$ which shows that α_u is injective. Since $\dim(H_u) = \dim(T_uP) - \dim(V_u) = \dim(T_pM)$ by Definition 10.32 (i) and Proposition 10.3, it follows that α_u is a linear isomorphism. The last part of the propostion now follows from Proposition 2.57. □

10.50 Proposition If $\pi : P \to M$ is a principal G-bundle with a connection H and X^* is the horizontal lift of a vector field X on M, then

$$(R_g)_* X^* = X^*$$

for every $g \in G$. Conversely, if Y is a horizontal vector field on P satisfying $(R_g)_* Y = Y$ for every $g \in G$, then Y is the horizontal lift of a unique vector field X on M.

PROOF : Since $\pi \circ R_g = \pi$, it follows that

$$\pi_* \circ (R_g)_* = \pi_*$$

for every $g \in G$, and by Definition 10.32 and Corollary 2.38 we have a bundle map (r_g, R_g) from the vector bundle $\pi' : H \to P$ to itself, where $r_g : H \to H$ is the map induced by $(R_g)_* : TP \to TP$ so that

$$i \circ r_g = (R_g)_* \circ i$$

for every $g \in G$, where $i : H \to TP$ is the inclusion map.

Combining this, it follows that the bundle map (α, π) from the vector bundle $\pi' : H \to P$ to the tangent bundle $\pi_M : TM \to M$, where $\alpha = \pi_* \circ i$, satisfies

$$\alpha \circ r_g = \pi_* \circ i \circ r_g = \pi_* \circ (R_g)_* \circ i = \pi_* \circ i = \alpha$$

for every $g \in G$, which means that

$$\alpha_{ug} \circ (r_g)_u = \alpha_u$$

and therefore

$$(r_g)_u \circ \alpha_u^{-1} = \alpha_{ug}^{-1}$$

for every $u \in P$ and $g \in G$.

From the definition of the pull-back $\alpha^*(X)$, we now have that

$$\alpha^*(X)_u = \alpha_u^{-1}(X_p) \quad \text{and} \quad \alpha^*(X)_{ug} = \alpha_{ug}^{-1}(X_p)$$

where $p = \pi(u) = \pi(ug)$, so that

$$(r_g)_u \alpha^*(X)_u = \alpha^*(X)_{ug}$$

for every $u \in P$ and $g \in G$. Hence it follows that

$$r_g \circ \alpha^*(X) = \alpha^*(X) \circ R_g$$

which implies that

$$(R_g)_* \circ X^* = (R_g)_* \circ i \circ \alpha^*(X) = i \circ r_g \circ \alpha^*(X) = i \circ \alpha^*(X) \circ R_g = X^* \circ R_g$$

so that

$$(R_g)_* X^* = (R_g)_* \circ X^* \circ R_g^{-1} = X^*$$

for every $g \in G$, showing the first part of the proposition.

To prove that last part, suppose that Y is a horizontal vector field on P satisfying $(R_g)_* Y = Y$ for every $g \in G$. Then we have that

$$\pi_* \circ Y \circ R_g = \pi_* \circ (R_g)_* \circ Y = \pi_* \circ Y$$

for every $g \in G$, which shows that $\pi_* \circ Y$ is constant on each fibre P_p. Hence we have a well-defined map $X : M \to TM$ satisfying $\pi_* \circ Y = X \circ \pi$ given by

$$X(p) = \pi_* \circ Y(u),$$

where u is any point in P with $\pi(u) = p$. If $(t, \pi^{-1}(U))$ is a local trivialization in P, then

$$X(p) = \pi_* \circ Y \circ t^{-1}(p, e)$$

for $p \in U$, and this completes the proof that X is a vector field on M which is π-related to Y. □

10.51 Proposition Let H be a connection in the principal G-bundle $\pi : P \to M$ with horizontal projection $h : TP \to TP$. If X^* and Y^* are the horizontal lifts of the vector fields X and Y on M, then

(1) $(X+Y)^* = X^* + Y^*$,

(2) $(fX)^* = (f \circ \pi) X^*$ for every smooth function $f : M \to \mathbf{R}$,

(3) $h \circ [X^*, Y^*] = [X, Y]^*$.

PROOF: (1) $X^* + Y^*$ is a horizontal vector field on P which is π-related to $X + Y$ since

$$\pi_*(X_u^* + Y_u^*) = \pi_*(X_u^*) + \pi_*(Y_u^*) = X_{\pi(u)} + Y_{\pi(u)}$$

for every $u \in P$.

(2) $(f \circ \pi)X^*$ is a horizontal vector field on P which is π-related to fX since

$$\pi_* \left((f \circ \pi)(u)\, X^*_u \right) = (f \circ \pi)(u)\; \pi_* \left(X^*_u \right) = f(\pi(u))\; X_{\pi(u)}$$

for every $u \in P$.

(3) By Propositions 4.88 and 10.3 we know that $[X^*, Y^*]$ is π-related to $[X, Y]$, and the vertical component of $[X^*, Y^*]_u$ lies in the kernel of π_{*u} for every $u \in P$. Hence we have that

$$\pi_* \left(h\, [X^*, Y^*]_u \right) = \pi_* \left([X^*, Y^*]_u \right) = [X, Y]_{\pi(u)}$$

for $u \in P$, which shows that $h \circ [X^*, Y^*]$ is a horizontal vector field on P which is π-related to $[X, Y]$. □

LOCAL SECTIONS AND TRIVIALIZATIONS

10.52 Proposition Let $s : V \to P$ be a section in a principal G-bundle $\pi : P \to M$ on an open subset V of M. Then the map $\rho : V \times G \to \pi^{-1}(V)$ defined by

$$\rho\,(p, g) = s(p)\, g$$

for $p \in V$ and $g \in G$ is a diffeomorphism, and we have that

$$\rho_* \left(i_{g*}(v) + i_{p*}(w) \right) = (R_g)_* \circ s_*(v) + \sigma\left((L_{g^{-1}})_*\, w \right)_{\rho\,(p,g)} \tag{1}$$

for $p \in V$, $g \in G$, $v \in T_pM$ and $w \in T_gG$, where $i_g : V \to V \times G$ and $i_p : G \to V \times G$ are the embeddings defined by $i_g\,(p') = (p', g)$ and $i_p\,(g') = (p, g')$ for $p' \in V$ and $g' \in G$.

Moreover, $(\rho^{-1}, \pi^{-1}(V))$ is a local trivialization in P, called the local trivialization associated with the section s, and

$$s(p) = \rho\,(p, e)$$

for $p \in V$.

Conversely, any local trivialization $(t, \pi^{-1}(V))$ in P is associated with a unique section s on V which is given by

$$s(p) = t^{-1}(p, e)$$

for $p \in V$.

If $s_k : V_k \to P$ for $k = 1, 2$ are two sections in P with associated local trivializations $(t_k, \pi^{-1}(V_k))$ which are G-related with transition map $\phi : V_1 \cap V_2 \to G$, then

$$s_1(p) = s_2(p)\, \phi(p)$$

for $p \in V_1 \cap V_2$. If $\lambda : V_1 \cap V_2 \to V_2 \times G$ is the map defined by

$$\lambda(p) = (p, \phi(p))$$

for $p \in V_1 \cap V_2$, then

$$\lambda_*(v) = i_{\phi(p)*}(v) + i_{p*} \circ \phi_*(v) \tag{2}$$

and

$$s_{1*}(v) = (R_{\phi(p)})_* \circ s_{2*}(v) + \sigma\left((L_{\phi(p)^{-1}})_* \circ \phi_*(v)\right)_{s_1(p)} \tag{3}$$

for $p \in V_1 \cap V_2$ and $v \in T_pM$.

PROOF: Let $p \in V$ and $g \in G$. Since

$$\rho_g = R_g \circ s \quad \text{and} \quad \rho_p = \sigma_{\rho(p,g)} \circ L_{g^{-1}},$$

we have that

$$\rho_{g*}(v) = (R_g)_* \circ s_*(v) \quad \text{and} \quad \rho_{p*}(w) = \sigma\left((L_{g^{-1}})_* w\right)_{\rho(p,g)}$$

for $v \in T_pM$ and $w \in T_gG$, which proves formula (1).

We next show that ρ is a diffeomorphism, and that $(\rho^{-1}, \pi^{-1}(V))$ is a local trivialization in P. Since $\pi \circ \rho = pr_1$ and G acts freely on P having the fibres of π as orbits, we see that ρ is a bijection. Indeed, if $u \in \pi^{-1}(V)$ and $\pi(u) = p$, then both u and $s(p)$ lie in the fibre $\pi^{-1}(p)$, and there is a unique $g \in G$ with $u = s(p)g = \rho(p,g)$. If we denote this g by $\psi(u)$, we obtain a map $\psi : \pi^{-1}(V) \to G$ so that

$$\rho^{-1}(u) = (\pi(u), \psi(u))$$

for $u \in \pi^{-1}(V)$. As $\pi \circ R_g \circ s = id_V$ for each $g \in G$, we have that

$$v - (R_g)_* \circ s_* \circ \pi_*(v) \in \ker \pi_{*u} = V_u = \sigma_{u*}(\mathfrak{g}) = \sigma_{u*} \circ (L_{g^{-1}})_*(T_gG)$$

for every $v \in T_uP$ where $u = \rho(p,g)$, showing that ρ_{*u} surjective. Since $\dim_u(P) = \dim_p(M) + \dim_g(G)$, it follows from the inverse function theorem that ρ and ρ^{-1} are diffeomorphisms. For every $u = \rho(p,g) \in \pi^{-1}(V)$ and $h \in G$ we have that $uh = \rho(p,gh)$ so that

$$\psi(uh) = gh = \psi(u)h,$$

thus completing the proof that $(\rho^{-1}, \pi^{-1}(V))$ is a local trivialization in P.

Now, if $(t, \pi^{-1}(V))$ is any local trivialization in P, then the map $s : V \to P$ given by

$$s(p) = t^{-1}(p,e)$$

for $p \in V$ is clearly a section on V. From 10.16 it follows that

$$s(p)g = t^{-1}(p,g)$$

for $p \in V$ and $g \in G$, which shows that $(t, \pi^{-1}(V))$ is the local trivialization associated with s.

Finally, if $(t_k, \pi^{-1}(V_k))$ for $k = 1,2$ are two G-related local trivializations in P with transition map $\phi : V_1 \cap V_2 \to G$ associated with the sections s_k, then

$$s_1(p) = t_1^{-1}(p,e) = t_2^{-1} \circ t_2 \circ t_1^{-1}(p,e) = t_2^{-1}(p, \phi(p)\, e) = s_2(p)\, \phi(p)$$

for $p \in V_1 \cap V_2$, and (2) follows from formula (4) in Proposition 2.74. Now we have that

$$s_1(p) = \rho_2 \circ \lambda\,(p)$$

for $p \in V_1 \cap V_2$, where $\rho_2 : V_2 \times G \to \pi^{-1}(V_2)$ is the map defined by $\rho_2(p,g) = s_2\,(p)g$, and formula (3) therefore follows from (1) and (2). □

10.53 Remark For each local chart (x,U) on the manifold M^n, the local trivialization $(\phi_x, \pi'^{-1}(U))$ in the frame bundle $\pi' : \mathscr{L}(M) \to M$ defined in Example 10.21 is associated with the frame field

$$s = \left(\frac{\partial}{\partial x^1}, \ldots, \frac{\partial}{\partial x^n} \right)$$

on U by Remark 2.82.

10.54 Corollary Suppose that $s_1 = (X_1, \ldots, X_n)$ and $s_2 = (X'_1, \ldots, X'_n)$ are frame fields on open subsets V_1 and V_2 of a smooth manifold M, and that the associated local trivializations $(t_1, \pi^{-1}(V_1))$ and $(t_2, \pi^{-1}(V_2))$ in the frame bundle $\pi : \mathscr{L}(M) \to M$ are $Gl\,(\mathbf{R}^n)$-related with transition map $\phi : V_1 \cap V_2 \to Gl\,(\mathbf{R}^n)$. Then we have that

$$X_j(p) = \sum_{i=1}^{n} \phi(p)^i_{\;j}\; X'_i(p) \tag{1}$$

for $p \in V_1 \cap V_2$ and $1 \leq j \leq n$, where the components of $\phi(p)$ are with respect to the basis $\mathscr{D} = \{F^i_j | 1 \leq i,j \leq n\}$ for $\mathfrak{gl}\,(\mathbf{R}^n)$ defined in Example 8.10 (b), where $F^i_j : \mathbf{R}^n \to \mathbf{R}^n$ is the linear map given by $F^i_j(e_k) = \delta_{jk}\, e_i$ for $k = 1, \ldots, n$.

PROOF: By Example 10.21 and Proposition 10.52 we have that

$$X_k(p) = s_1(p)(e_k) = s_2(p) \circ \phi(p)(e_k) = s_2(p) \left(\sum_{ij} \phi(p)^i_{\;j}\; F^i_j(e_k) \right)$$

$$= s_2(p) \left(\sum_i \phi(p)^i_{\;k}\; e_i \right) = \sum_t \phi(p)^i_{\;k}\; X'_i(p)$$

for $p \in V_1 \cap V_2$ and $1 \leq k \leq n$. □

10.55 Proposition Let $s : V \to P$ be a section in a principal G-bundle $\pi : P \to M$ on an open subset V of M, and let $\rho : V \times G \to \pi^{-1}(V)$ be the map defined by

$$\rho\,(p,g) = s(p)\, g$$

for $p \in V$ and $g \in G$. If ω is a tensorial k-form on $\pi^{-1}(V)$ of type (ψ, W), where $\psi : G \to \mathrm{Aut}(W)$ is a representation of G on a finite dimensional vector space W, and if $\widetilde{\omega} = s^*(\omega)$ is the pull-back of ω by s, then

$$\omega(\rho(p,g))(\rho_{g*}(v_1)+\rho_{p*}(w_1),\dots,\rho_{g*}(v_k)+\rho_{p*}(w_k)) = \psi(g^{-1})\,\widetilde{\omega}(p)(v_1,\dots,v_k) \tag{1}$$

for $p \in V$, $g \in G$, $v_1, \dots, v_k \in T_pM$ and $w_1, \dots, w_k \in T_gG$, where $\rho_g = \rho \circ i_g$ and $\rho_p = \rho \circ i_p$.

Conversely, given any W-valued k-form $\widetilde{\omega}$ on V, there is a unique tensorial k-form ω on $\pi^{-1}(V)$ of type (ψ, W) with $\widetilde{\omega} = s^*(\omega)$, which is given by (1).

If $s_k : V_k \to P$ for $k = 1,2$ are two sections in P with associated local trivializations $(t_k, \pi^{-1}(V_k))$ which are G-related with transition map $\phi : V_1 \cap V_2 \to G$, and if $\widetilde{\omega}_k = s_k^*(\omega)$ for a tensorial k-form ω on $\pi^{-1}(V_1 \cap V_2)$ of type (ψ, W), then

$$\widetilde{\omega}_1(p)(v_1, \dots, v_k) = \psi(\phi(p)^{-1})\,\widetilde{\omega}_2(p)(v_1, \dots, v_k) \tag{2}$$

for $p \in V_1 \cap V_2$ and $v_1, \dots, v_k \in T_pM$.

PROOF: By formula (1) in Proposition 10.52 we have that

$$\begin{aligned}\omega(\rho(p,g))(\rho_{g*}(v_1), \dots, \rho_{g*}(v_k)) &= R_g^*\,\omega(s(p))(s_*(v_1), \dots, s_*(v_k)) \\ = \psi(g^{-1})\,\omega(s(p))(s_*(v_1), \dots, s_*(v_k)) &= \psi(g^{-1})\,\widetilde{\omega}(p)(v_1, \dots, v_k)\end{aligned}$$

and

$$\rho_{p*}(w_i) \in V_{\rho(p,g)} \quad \text{for} \quad i = 1, \dots, k$$

which implies (1).

Now let $\widetilde{\omega}$ be an arbitrary W-valued k-form on V, and let $\omega : \pi^{-1}(V) \to \Lambda^k(TP\,; W)$ be the map defined by (1). To show that ω is smooth, we first consider the map $\eta : V \times G \to \Lambda^k(T(V \times G); W)$ given by

$$\eta(p,g)(i_{g*}(v_1)+i_{p*}(w_1), \dots, i_{g*}(v_k)+i_{p*}(w_k)) = \psi(g^{-1})\,\widetilde{\omega}(p)(v_1, \dots, v_k) \tag{3}$$

for $p \in V$, $g \in G$, $v_1, \dots, v_k \in T_pM$ and $w_1, \dots, w_k \in T_gG$. If $\alpha : V \times G \to V$ is the projection on the first factor and $\tilde{\alpha} : V \times G \times W \to V \times W$ is the map given by

$$\tilde{\alpha}(p, g, \xi) = (p, \psi(g)\,\xi)$$

for $p \in V$, $g \in G$ and $\xi \in W$, we have a bundle map $(\tilde{\alpha}, \alpha)$ between the vector bundles $\pi_1 : V \times G \times W \to V \times G$ and $\pi_2 : V \times W \to V$ which is a linear isomorphism on each fibre. Then it follows from Propositions 5.33 that η is a W-valued k-form on $V \times G$ since

$$\eta = \rho_1' \circ (\alpha_*, \tilde{\alpha})^*(\rho_2'^{-1} \circ \widetilde{\omega}),$$

where $\rho_1' : \Lambda^k(T(V \times G); V \times G \times W) \to \Lambda^k(T(V \times G); W)$ and $\rho_2' : \Lambda^k(TV\,; V \times W) \to \Lambda^k(TV\,; W)$ are the equivalences over $V \times G$ and V, respectively, defined in 5.26. Hence it follows that $\omega = (\rho^{-1})^*\,\eta$ is a W-valued k-form on $\pi^{-1}(V)$.

It only remains to show that ω is tensorial of type (ψ, W) and that $\widetilde{\omega} = s^*(\omega)$. Since

$$\sigma_{\rho(p,g)*} \circ (L_{g^{-1}})_* : T_g G \to V_{\rho(p,g)}$$

is a linear isomorphism by Proposition 10.26, it follows from (1) that ω is horizontal. Indeed, if $\rho_{g*}(v_r) + \rho_{p*}(w_r) \in V_{\rho(p,g)}$ for some r, then $v_r = 0$ by formula (1) in Proposition 10.52 so that $\widetilde{\omega}(p)(v_1, \dots, v_k) = 0$. Furthermore, we have that

$$R_h \circ \rho(p,g) = \rho(p,gh)$$

which implies that

$$R_h \circ \rho_g = \rho_{gh} \quad \text{and} \quad R_h \circ \rho_p = \rho_p \circ R_h$$

so that

$$\begin{aligned}
&R_h^* \,\omega(\rho(p,g))\,(\rho_{g*}(v_1) + \rho_{p*}(w_1), \dots, \rho_{g*}(v_k) + \rho_{p*}(w_k)) \\
&= \omega(\rho(p,gh))\,(\rho_{gh*}(v_1) + \rho_{p*} \circ (R_h)_*(w_1), \dots, \rho_{gh*}(v_k) + \rho_{p*} \circ (R_h)_*(w_k)) \\
&= \psi((gh)^{-1})\,\widetilde{\omega}(p)(v_1, \dots, v_k) = \psi(h^{-1})\,\psi(g^{-1})\,\widetilde{\omega}(p)(v_1, \dots, v_k) \\
&= \psi(h^{-1})\,\omega(\rho(p,g))\,(\rho_{g*}(v_1) + \rho_{p*}(w_1), \dots, \rho_{g*}(v_k) + \rho_{p*}(w_k))\,.
\end{aligned}$$

This shows that ω is a tensorial k-form of type (ψ, W) on $\pi^{-1}(V)$, and since

$$s(p) = \rho(p,e) = \rho_e(p)$$

for $p \in V$, we have that

$$s^*(\omega)(p)(v_1, \dots, v_k) = \omega(\rho(p,e))\,(\rho_{e*}(v_1), \dots, \rho_{e*}(v_k)) = \widetilde{\omega}(p)(v_1, \dots, v_k)\,.$$

Finally, if $(t_k, \pi^{-1}(V_k))$ for $k = 1, 2$ are two G-related local trivializations in P with transition map $\phi : V_1 \cap V_2 \to G$ associated with the sections s_k, then

$$s_1(p) = \rho_2 \circ \lambda(p)$$

for $p \in V_1 \cap V_2$, where λ is the map defined in Proposition 10.52 and $\rho_2 : V_2 \times G \to \pi^{-1}(V_2)$ is given by $\rho_2(p,g) = s_2(p)g$. Hence it follows from (1) that

$$\begin{aligned}
&\widetilde{\omega}_1(p)(v_1, \dots, v_k) = \omega(s_1(p))\,(s_{1*}(v_1), \dots, s_{1*}(v_k)) \\
&= \omega(\rho_2(p, \phi(p)))\,(\rho_{2\phi(p)*}(v_1) + \rho_{2p*} \circ \phi_*(v_1), \dots, \rho_{2\phi(p)*}(v_k) + \rho_{2p*} \circ \phi_*(v_k)) \\
&= \psi(\phi(p)^{-1})\,\widetilde{\omega}_2(p)(v_1, \dots, v_k)
\end{aligned}$$

for $p \in V_1 \cap V_2$ and $v_1, \dots, v_k \in T_pM$, which completes the proof of formula (2). □

10.56 Proposition Let $s : V \to P$ be a section in a principal G-bundle $\pi : P \to M$ on an open subset V of M, and let $\rho : V \times G \to \pi^{-1}(V)$ be the map defined by

$$\rho(p,g) = s(p)g$$

for $p \in V$ and $g \in G$. If ω is a connection form on $\pi^{-1}(V)$, and if $\widetilde{\omega} = s^*(\omega)$ is the pull-back of ω by s, then

$$\omega(\rho(p,g))(\rho_{g*}(v) + \rho_{p*}(w)) = Ad(g^{-1})\,\widetilde{\omega}(p)(v) + \theta(g)(w) \tag{1}$$

for $p \in V$, $g \in G$, $v \in T_pM$ and $w \in T_gG$, where $\rho_g = \rho \circ i_g$ and $\rho_p = \rho \circ i_p$, and where θ is the canonical Maurer-Cartan form on G.

Conversely, given any $\mathfrak{g}$-valued 1-form $\widetilde{\omega}$ on V, there is a unique connection form ω on $\pi^{-1}(V)$ with $\widetilde{\omega} = s^*(\omega)$, which is given by (1).

If $s_k : V_k \to P$ for $k = 1,2$ are two sections in P with associated local trivializations $(t_k, \pi^{-1}(V_k))$ which are G-related with transition map $\phi : V_1 \cap V_2 \to G$, and if $\widetilde{\omega}_k = s_k^*(\omega)$ for a connection form ω on $\pi^{-1}(V_1 \cap V_2)$, then

$$\widetilde{\omega}_1(p)(v) = Ad(\phi(p)^{-1})\,\widetilde{\omega}_2(p)(v) + \widetilde{\theta}(p)(v) \tag{2}$$

for $p \in V_1 \cap V_2$ and $v \in T_pM$, where $\widetilde{\theta} = \phi^*(\theta)$ is the pull-back of the canonical Maurer-Cartan form θ on G by ϕ.

PROOF: By formula (1) in Proposition 10.52 we have that

$$\begin{aligned}\omega(\rho(p,g))(\rho_{g*}(v)) &= R_g^*\,\omega(s(p))(s_*(v)) \\ &= Ad(g^{-1})\,\omega(s(p))(s_*(v)) = Ad(g^{-1})\,\widetilde{\omega}(p)(v)\end{aligned}$$

and

$$\omega(\rho(p,g))(\rho_{p*}(w)) = (L_{g^{-1}})_*(w) = \theta(g)(w)$$

which implies (1).

Now let $\widetilde{\omega}$ be an arbitrary $\mathfrak{g}$-valued 1-form on V, and let $\omega : \pi^{-1}(V) \to \Lambda^1(TP\,;\mathfrak{g})$ be the map defined by (1). To show that ω is smooth, we first consider the map $\eta : V \times G \to \Lambda^1(T(V \times G);\mathfrak{g})$ given by

$$\eta(p,g)(i_{g*}(v) + i_{p*}(w)) = Ad(g^{-1})\,\widetilde{\omega}(p)(v) + \theta(g)(w) \tag{3}$$

for $p \in V$, $g \in G$, $v \in T_pM$ and $w \in T_gG$. If $\alpha : V \times G \to V$ and $\gamma : V \times G \to G$ are the projections on the first and second factor, and $\tilde{\alpha} : V \times G \times \mathfrak{g} \to V \times \mathfrak{g}$ is the map given by

$$\tilde{\alpha}(p,g,X) = (p, Ad(g)\,X)$$

for $p \in V$, $g \in G$ and $X \in \mathfrak{g}$, we have a bundle map $(\tilde{\alpha}, \alpha)$ between the vector bundles $\pi_1 : V \times G \times \mathfrak{g} \to V \times G$ and $\pi_2 : V \times \mathfrak{g} \to V$ which is a linear isomorphism on each fibre. Then it follows from Propositions 5.33 and 5.47 that η is a $\mathfrak{g}$-valued 1-form on $V \times G$ since

$$\eta = \rho_1' \circ (\alpha_*, \tilde{\alpha})^*(\rho_2'^{-1} \circ \widetilde{\omega}) + \gamma^*(\theta)\,,$$

where $\rho_1' : \Lambda^k(T(V \times G); V \times G \times \mathfrak{g}) \to \Lambda^k(T(V \times G);\mathfrak{g})$ and $\rho_2' : \Lambda^k(TV\,; V \times \mathfrak{g}) \to \Lambda^k(TV\,;\mathfrak{g})$ are the equivalences over $V \times G$ and V, respectively, defined in 5.26. Hence it follows that $\omega = (\rho^{-1})^*\,\eta$ is a $\mathfrak{g}$-valued 1-form on $\pi^{-1}(V)$.

It only remains to show that ω is a connection form on $\pi^{-1}(V)$ with $\widetilde{\omega} = s^*(\omega)$. If $h \in G$, we have that

$$\sigma_{\rho(p,g)}(h) = \rho(p,gh) = \rho_p \circ L_g(h)$$

which implies that

$$\sigma(X)_{\rho(p,g)} = \sigma_{\rho(p,g)*}(X) = \rho_{p*} \circ (L_g)_*(X)$$

so that

$$\omega(\rho(p,g))(\sigma(X)_{\rho(p,g)}) = \theta(g)((L_g)_*(X)) = X$$

by Definition 8.75. Furthermore, we have that

$$R_h \circ \rho(p,g) = \rho(p,gh)$$

which implies that

$$R_h \circ \rho_g = \rho_{gh} \quad \text{and} \quad R_h \circ \rho_p = \rho_p \circ R_h$$

so that

$$\begin{aligned}
&R_h^* \,\omega(\rho(p,g))(\rho_{g*}(v) + \rho_{p*}(w)) \\
&= \omega(\rho(p,gh))(\rho_{gh*}(v) + \rho_{p*} \circ (R_h)_*(w)) \\
&= Ad((gh)^{-1})\, \widetilde{\omega}(p)(v) + \theta(gh)((R_h)_*(w)) \\
&= Ad((gh)^{-1})\, \widetilde{\omega}(p)(v) + R_h^* \,\theta(g)(w) \\
&= Ad(h^{-1})\, \{Ad(g^{-1})\, \widetilde{\omega}(p)(v) + \theta(g)(w)\} \\
&= Ad(h^{-1})\, \omega(\rho(p,g))(\rho_{g*}(v) + \rho_{p*}(w))
\end{aligned}$$

by Proposition 9.15. This shows that ω is a connection form on $\pi^{-1}(V)$, and since

$$s(p) = \rho(p,e) = \rho_e(p)$$

for $p \in V$, we have that

$$s^*(\omega)(p)(v) = \omega(\rho(p,e))(\rho_{e*}(v)) = \widetilde{\omega}(p)(v)\,.$$

Finally, if $(t_k, \pi^{-1}(V_k))$ for $k = 1,2$ are two G-related local trivializations in P with transition map $\phi : V_1 \cap V_2 \to G$ associated with the sections s_k, then

$$s_1(p) = \rho_2 \circ \lambda(p)$$

for $p \in V_1 \cap V_2$, where λ is the map defined in Proposition 10.52 and $\rho_2 : V_2 \times G \to \pi^{-1}(V_2)$ is given by $\rho_2(p,g) = s_2(p)g$. Hence it follows from (1) that

$$\begin{aligned}\widetilde{\omega}_1(p)(v) &= \omega(s_1(p))(s_{1*}(v)) \\ &= \omega(\rho_2(p,\phi(p)))(\rho_{2\phi(p)*}(v) + \rho_{2p*} \circ \phi_*(v)) \\ &= Ad(\phi(p)^{-1})\,\widetilde{\omega}_2(p)(v) + \theta(\phi(p))(\phi_*(v)) \\ &= Ad(\phi(p)^{-1})\,\widetilde{\omega}_2(p)(v) + \widetilde{\theta}(p)(v)\end{aligned}$$

for $p \in V_1 \cap V_2$ and $v \in T_pM$, which completes the proof of formula (2). □

10.57 Definition If $\pi : P \to M$ is a principal G-bundle with a connection H, then a smooth map $F : N \to P$ from a smooth manifold N is said to be *horizontal* at a point $p \in N$ if $F_*(T_pN) \subset H_{F(p)}$. We say that F is *horizontal* if it is horizontal at every point $p \in N$.

10.58 Remark If ω is the connection form of H, then the map $F : N \to P$ is horizontal at $p \in N$ if and only if $F^*(\omega)(p) = 0$, since $H_{F(p)} = \ker \omega(F(p))$.

If F is a lifting of a smooth map $f : N \to M$, then F is horizontal with respect to ω if and only if the section $\widetilde{F} : N \to \widetilde{E}$ in the induced fibre bundle determined by F is horizontal with respect to $\widetilde{f}^*(\omega)$, since $F^*(\omega) = \widetilde{F}^*(\widetilde{f}^*(\omega))$.

10.59 Proposition Let $\pi : P \to M$ be a principal G-bundle with a connection H, and let ω be the connection form of H. If $s : V \to P$ is a horizontal section of π on an open subset V of M, and $\rho : V \times G \to \pi^{-1}(V)$ is the map defined by

$$\rho(p,g) = s(p)g$$

for $p \in V$ and $g \in G$, then

$$\omega(\rho(p,g))(\rho_{g*}(v) + \rho_{p*}(w)) = \theta(g)(w) \tag{1}$$

for $p \in V$, $g \in G$, $v \in T_pM$ and $w \in T_gG$, where $\rho_g = \rho \circ i_g$ and $\rho_p = \rho \circ i_p$, and where θ is the canonical Maurer-Cartan form on G. Moreover, we have that $H_{\rho(p,g)} = \rho_{g*}(T_pM)$ for $(p,g) \in V \times G$.

If $s_k : V_k \to P$ for $k = 1,2$ are two sections in P with associated local trivializations $(t_k, \pi^{-1}(V_k))$ which are G-related with transition map $\phi : V_1 \cap V_2 \to G$, then

$$s_1(p) = s_2(p)\,\phi(p)$$

for $p \in V_1 \cap V_2$. Assuming that s_2 is horizontal on $V_1 \cap V_2$, we have that s_1 is horizontal on $V_1 \cap V_2$ if and only if ϕ is constant on each connected component of $V_1 \cap V_2$. In particular, if the sections s_1 and s_2 are horizontal on $V_1 \cap V_2$ and $s_1(p_0) = s_2(p_0)$ for a point p_0 in $V_1 \cap V_2$, then s_1 and s_2 coincide on the connected component of p_0 in $V_1 \cap V_2$.

PROOF : Formula (1) follows from Proposition 10.56 since $s^*(\omega) = 0$, and it implies that

$$H_{\rho(p,g)} = \ker \omega\,(\rho\,(p,g)) = \rho_{g*}(T_pM)$$

for $(p,g) \in V \times G$, thus proving the first part of the proposition.

By formula (2) in Proposition 10.56 we have that

$$s_1^*(\omega)(p)(v) = \phi^*(\theta)\,(p)(v) = \theta\,(\phi(p))(\phi_*(v)) = (L_{\phi(p)^{-1}})_* \circ \phi_*(v)$$

for $p \in V_1 \cap V_2$ and $v \in T_pM$, which shows that s_1 is horizontal on $V_1 \cap V_2$ if and only if $\phi_* = 0$. The proposition now follows from Proposition 2.29 and Proposition 13.38 in the appendix. □

HORIZONTAL LIFTS OF CURVES

10.60 Proposition Let $\pi : P \to M$ be a principal G-bundle with a connection H, and let $\beta : I \to P$ be a horizontal lifting of a smooth curve $\gamma : I \to M$ defined on an open interval I. Then a smooth curve $\alpha : I \to P$ is a horizontal lifting of γ if and only if

$$\alpha(t) = \beta(t)\,g$$

for $t \in I$ where $g \in G$.

PROOF : Let $\widetilde{\pi} : \widetilde{P} \to I$ be the principal G-bundle induced by γ with canonical bundle map $(\widetilde{\gamma}, \gamma)$. If $\widetilde{\alpha} : I \to \widetilde{P}$ and $\widetilde{\beta} : I \to \widetilde{P}$ are the sections of $\widetilde{\pi}$ determined by the liftings α and β of γ, respectively, it follows from Remark 10.58 and Proposition 10.59 that α is horizontal if and only if there is a group element $g \in G$ such that

$$\widetilde{\alpha}(t) = \widetilde{\beta}(t)\,g$$

for $t \in I$. By applying $\widetilde{\gamma}$ on both sides and using formula (3) in 10.37, we see that this is equivalent to the assertion that

$$\alpha(t) = \beta(t)\,g$$

for $t \in I$. □

10.61 Proposition Let H be a flat connection in a principal G-bundle $\pi : P \to M^n$, and let $u_0 \in P$ and $p_0 = \pi(u_0)$. Then there is a horizontal section $s : V \to P$ of π defined on some open neighbourhood V of p_0 with $s(p_0) = u_0$.

PROOF : By Corollary 10.48 we know that H is an integrable distribution on P. Let N be an integral manifold of H passing through u_0. If $i : N \to P$ is the inclusion map, it follows from Proposition 10.49 that $\pi \circ i : N \to M$ is an immersion of rank n. By the rank theorem it maps an open neighbourhood U of u_0 in N diffeomorpically onto an open neighbourhood V of p_0 in M. If $\sigma : V \to U$ is the inverse of the map induced by $\pi \circ i$ and $s = i \circ \sigma$, we have that $s(p_0) = u_0$ and $s_*(T_pM) = i_*(T_{\sigma(p)}N) = H_{s(p)}$ for every $p \in V$. □

10.62 Proposition Let $\pi : P \to M$ be a principal G-bundle with a connection form ω, and let $\gamma : I \to M$ be a smooth curve defined on an open interval I. If $t_0 \in I$ and $u_0 \in \pi^{-1}(\gamma(t_0))$, then there is a unique horizontal lifting $\alpha : I \to P$ of γ with $\alpha(t_0) = u_0$.

PROOF: Let $\widetilde{\pi} : \widetilde{P} \to I$ be the principal G-bundle induced by γ with canonical bundle map $(\widetilde{\gamma}, \gamma)$. By 10.37 and Proposition 10.46 we know that $\widetilde{\gamma}^*(\omega)$ is a connection form on $\widetilde{P}$ corresponding to a flat connection since the base manifold I in the induced bundle is 1-dimensional.

Let $\mathscr{J}$ be the set of all open subintervals J' of I containing t_0 such that there exists a horizontal section $s' : J' \to \widetilde{P}$ of $\widetilde{\pi}$ with $s'(t_0) = (t_0, u_0)$, and put $J = \bigcup_{J' \in \mathscr{J}} J'$. Define the horizontal section $s : J \to \widetilde{P}$ as follows. If $t \in J$, choose an interval J' in $\mathscr{J}$ containing t and set $s(t) = s'(t)$. It follows by Propositions 10.61 and 10.59 that $\mathscr{J}$ is nonempty and that s is well defined.

Now we contend that $J = I$. Otherwise, if an endpoint t_1 of J is contained in I, there is by Propositions 10.61 a horizontal section $s'' : J'' \to \widetilde{P}$ of $\widetilde{\pi}$ defined on an open subinterval J'' of I containing t_1. If $t_2 \in J \cap J''$ and $s(t_2) = s''(t_2)g$, we have a horizontal section $s' : J \cup J'' \to \widetilde{P}$ defined by

$$s'(t) = \begin{cases} s(t) & \text{for } t \in J \\ s''(t)g & \text{for } t \in J'' \end{cases}$$

contradicting the assumption that t_1 is an endpoint of J. This shows that there is a unique horizontal section $s : I \to \widetilde{P}$ of $\widetilde{\pi}$ with $s(t_0) = (t_0, u_0)$, and the result now follows from 10.12 and Remark 10.58 with $\alpha = \widetilde{\gamma} \circ s$. □

10.63 Definition Let M be a smooth manifold, and let $p_0, p_1 \in M$. By a *curve* in M from p_0 to p_1 we mean a continuous map $\gamma : [a,b] \to M$ with $\gamma(a) = p_0$ and $\gamma(b) = p_1$. The curve γ is said to be *smooth* if it has an extension to a smooth curve $\gamma_0 : I \to M$ defined on an open interval I containing $[a,b]$. It is called *piecewise smooth* if there is a partition $a = t_0 < \dots < t_r = b$ of $[a,b]$ such that $\gamma|_{[t_i, t_{i+1}]}$ is a smooth curve for each $i = 0, \dots, r-1$.

We define the *inverse* curve $\gamma^{-1} : [a,b] \to M$ from p_1 to p_0 by $\gamma^{-1}(t) = \gamma(a + b - t)$ for $t \in [a,b]$, and we see that γ^{-1} is smooth or piecewise smooth if and only if the same is true for γ.

10.64 Proposition If two points p_0 and p_1 on a smooth manifold M can be joined by a continuous curve $c : [a,b] \to M$, then they can also be joined by a piecewise smooth curve $\gamma : [a,b] \to M$.

PROOF : Let $\mathscr{A} = \{(x_\alpha, U_\alpha) \mid \alpha \in A\}$ be an atlas on M where $x_\alpha(U_\alpha)$ is convex for every $\alpha \in A$, and let ε be the Lebesgue number of the covering $\{c^{-1}(U_\alpha) \mid \alpha \in A\}$ of $[a,b]$. If $a = t_0 < \ldots < t_r = b$ is a partition of $[a,b]$ with mesh less that $2\,\varepsilon$, then $c([t_i, t_{i+1}])$ is contained in a coordinate neighbourhood U_{α_i} for each $i = 0, \ldots, r-1$, and we define

$$\gamma(t) = x_{\alpha_i}^{-1}\left(\frac{t_{i+1}-t}{t_{i+1}-t_i}\; x_{\alpha_i} \circ\, c\,(t_i) + \frac{t-t_i}{t_{i+1}-t_i}\; x_{\alpha_i} \circ\, c\,(t_{i+1})\right)$$

for $t \in [t_i, t_{i+1}]$. □

10.65 Definition Let M be a pseudo-Riemannian manifold. The *arc length* of a piecewise smooth curve $\gamma : [a,b] \to M$ is defined by

$$L(\gamma) = \int_a^b \|\gamma'(t)\|\, dt\,.$$

If $\gamma : I \to M$ is a smooth curve defined on an open or closed interval I with $\|\gamma'(t)\| > 0$ for $t \in I$, we define its *arc length function* $s : I \to \mathbf{R}$ *with initial point* $t_0 \in I$ by

$$s(t) = \int_{t_0}^t \|\gamma'(u)\|\, du$$

for $t \in I$. The curve $\alpha = \gamma \circ s^{-1} : J \to M$ defined on the interval $J = s(I)$ is a reparametrization of γ, called a *parametrization by arc length*, satisfying $\|\alpha'(t)\| = 1$ for $t \in J$.

10.66 Definition Let $\pi : P \to M$ be a principal G-bundle with a connection H. Then a smooth curve $\gamma : [a,b] \to P$ is said to be horizontal if it has an extension to a smooth curve $\gamma_0 : I \to P$ defined on an open interval I containing $[a,b]$ which is horizontal on $[a,b]$. A piecewise smooth curve $\gamma : [a,b] \to P$ is horizontal if there is a partition $a = t_0 < \ldots < t_r = b$ of $[a,b]$ such that $\gamma|_{[t_i,t_{i+1}]}$ is a horizontal smooth curve for each $i = 0, \ldots, r-1$.

10.67 Proposition Let $\pi : P \to M$ be a principal G-bundle with a connection H, and let $\beta : [a,b] \to P$ be a horizontal lifting of a piecewise smooth curve $\gamma : [a,b] \to M$. Then a piecewise smooth curve $\alpha : [a,b] \to P$ is a horizontal lifting of γ if and only if

$$\alpha(t) = \beta(t)\, g \tag{1}$$

for $t \in [a,b]$ where $g \in G$.

PROOF : Let $a = t_0 < \ldots < t_r = b$ be a partition of $[a,b]$ such that $\alpha|_{[t_i,t_{i+1}]}$ and $\beta|_{[t_i,t_{i+1}]}$ are smooth liftings of $\gamma|_{[t_i,t_{i+1}]}$ for $i = 0, \ldots, r-1$. By Proposition 10.60 we know that α is horizontal on $< t_i, t_{i+1} >$ if and only if

$$\alpha(t) = \beta(t)\, g_i \tag{2}$$

for $t \in <t_i, t_{i+1}>$ where $g_i \in G$. If α is horizontal, formula (2) actually holds for $t \in [t_i, t_{i+1}]$ by the continuity of α and β, and it follows by induction that $g_i = g_0$ for $i = 1, \dots, r-1$. Hence formula (1) is true for $t \in [a,b]$ with $g = g_0$.

Conversely, assuming that formula (1) is true we have that α is horizontal on $[t_i, t_{i+1}]$ by the continuity of $\alpha_i^*(\omega)$, where $\alpha_i : I_i \to P$ is an extension of $\alpha|_{[t_i,t_{i+1}]}$ to a smooth curve defined on an open interval I_i containing $[t_i, t_{i+1}]$, and ω is the connection form of H. □

10.68 Proposition Let $\pi : P \to M$ be a principal G-bundle with a connection H, and let $\gamma : [a,b] \to M$ be a piecewise smooth curve in M. If $u_0 \in \pi^{-1}(\gamma(a))$, then there is a unique piecewise smooth horizontal lifting $\alpha : [a,b] \to P$ of γ with $\alpha(a) = u_0$.

PROOF : Let $a = t_0 < \dots < t_r = b$ be a partition of $[a,b]$ such that $\gamma|_{[t_i,t_{i+1}]}$ is a smooth curve for each $i = 0, \dots, r-1$. If $\gamma_i : I_i \to M$ is an extension of $\gamma|_{[t_i,t_{i+1}]}$ to a smooth curve defined on an open interval I_i containing $[t_i, t_{i+1}]$, then it follows from Proposition 10.62 that there are horizontal liftings $\alpha_i : I_i \to P$ of γ_i for $i = 0, \dots, r-1$ with $\alpha_0(t_0) = u_0$ and $\alpha_i(t_i) = \alpha_{i-1}(t_i)$ for $i = 1, \dots, r-1$. Then we have a piecewise smooth horizontal lifting $\alpha : [a,b] \to P$ of γ with $\alpha(a) = u_0$ defined by $\alpha(t) = \alpha_i(t)$ for $t \in [t_i, t_{i+1}]$, thus completing the existence part of the proposition. The uniqueness part follows from Proposition 10.67. □

PARALLEL TRANSPORT

10.69 Definition Let $\pi : P \to M$ be a principal G-bundle with a connection H, and let $\pi' : E \to M$ be the fibre bundle associated with P with fibre F obtained from a Lie group homomorphism $\psi : G \to G'$ into a Lie group G' which acts on the smooth manifold F effectively on the left. For each piecewise smooth curve $\gamma : [a,b] \to M$ in M from p_0 to p_1 we have a map

$$\tau_\gamma : \pi'^{-1}(p_0) \to \pi'^{-1}(p_1),$$

called *parallel transport* or *parallel translation along* γ, defined for each $u \in \pi^{-1}(p_0)$ and $v \in F$ by

$$\tau_\gamma([u,v]) = [\beta(b), v]$$

where $\beta : [a,b] \to P$ is the horizontal lifting of γ with $\beta(a) = u$.

10.70 Remark We must show that τ_γ is well defined, i.e., that

$$\tau_\gamma([ug, \psi(g^{-1})\, v]) = \tau_\gamma([u,v])$$

for every $g \in G$. If $\beta : [a,b] \to P$ is the horizontal lifting of γ with $\beta(a) = u$, then it follows from Proposition 10.67 that $\alpha = R_g \circ \beta$ is the unique horizontal lifting of γ with $\alpha(a) = R_g(u)$ so that

$$\tau_\gamma([ug, \psi(g^{-1})v]) = [\alpha(b), \psi(g^{-1})v] = [\beta(b)g, \psi(g^{-1})v] = [\beta(b), v] = \tau_\gamma([u,v]) .$$

We see that τ_γ is a diffeomorphism since

$$\tau_\gamma = \mu_{\beta(b)} \circ \mu_{\beta(a)}^{-1} ,$$

for any horizontal lift β of γ, where $\mu_{\beta(a)} : F \to E_{p_0}$ and $\mu_{\beta(b)} : F \to E_{p_1}$ are the diffeomorphisms defined in 10.28. As β^{-1} is the horizontal lifting of γ^{-1} with $\beta^{-1}(a) = \beta(b)$ and $\beta^{-1}(b) = \beta(a) = u$, we have that

$$\tau_{\gamma^{-1}} \circ \tau_\gamma([u,v]) = [\beta^{-1}(b), v] = [u,v]$$

so that

$$\tau_\gamma^{-1} = \tau_{\gamma^{-1}} .$$

In the case where $\psi = id_G$ and $F = G$ which is acting on itself by left translation, the fibre bundle $\pi' : E \to M$ may be identified with the principal fibre bundle $\pi : P \to M$ as in Example 10.29 (a). The parallel transport $\tau_\gamma : \pi'^{-1}(p_0) \to \pi'^{-1}(p_1)$ along γ in E is then identified with the parallel transport

$$\widetilde{\tau}_\gamma = e_{p_1} \circ \tau_\gamma \circ e_{p_0}^{-1} : \pi^{-1}(p_0) \to \pi^{-1}(p_1)$$

along γ in P which is given by

$$\widetilde{\tau}_\gamma(u) = \beta(b) ,$$

where $\beta : [a,b] \to P$ is the horizontal lifting of γ with $\beta(a) = u$.

10.71 Definition Let $\pi : P \to M$ be a principal G-bundle with a connection H, and let $\pi' : E \to M$ be the fibre bundle associated with P with fibre F obtained from a Lie group homomorphism $\psi : G \to G'$ into a Lie group G' which acts on the smooth manifold F effectively on the left. For each smooth curve $\gamma : I \to M$ in M defined on an open interval I and each $t_0, t_1 \in I$ we have a map

$$\tau_{t_1}^{t_0} : \pi'^{-1}(\gamma(t_0)) \to \pi'^{-1}(\gamma(t_1)) ,$$

called *parallel transport* or *parallel translation along γ from t_0 to t_1*, defined for each $u \in \pi^{-1}(\gamma(t_0))$ and $v \in F$ by

$$\tau_{t_1}^{t_0}([u,v]) = [\beta(t_1), v]$$

where $\beta : I \to P$ is the horizontal lifting of γ with $\beta(t_0) = u$.

10.72 Remark By Propositions 10.60 and 10.62 it follows in the same way as in Remark 10.70 that $\tau_{t_1}^{t_0}$ is a well-defined diffeomorphism with inverse

$$(\tau_{t_1}^{t_0})^{-1} = \tau_{t_0}^{t_1} ,$$

and we have that

$$\tau_{t_1}^{t_0} = \mu_{\beta(t_1)} \circ \mu_{\beta(t_0)}^{-1}$$

for any horizontal lift β of γ, where $\mu_{\beta(t_0)} : F \to E_{\gamma(t_0)}$ and $\mu_{\beta(t_1)} : F \to E_{\gamma(t_1)}$ are the diffeomorphisms defined in 10.28. It also follows from the definition that

$$\tau_{t_2}^{t_1} \circ \tau_{t_1}^{t_0} = \tau_{t_2}^{t_0}$$

for all $t_0, t_1, t_2 \in I$.

In the case where $\psi = id_G$ and $F = G$ which is acting on itself by left translation, the fibre bundle $\pi' : E \to M$ may be identified with the principal fibre bundle $\pi : P \to M$ as in Example 10.29 (a). The parallel transport $\tau_{t_1}^{t_0} : \pi'^{-1}(\gamma(t_0)) \to \pi'^{-1}(\gamma(t_1))$ along γ in E from t_0 to t_1 is then identified with the parallel transport

$$\widetilde{\tau}_{t_1}^{t_0} = e_{\gamma(t_1)} \circ \tau_{t_1}^{t_0} \circ e_{\gamma(t_0)}^{-1} : \pi^{-1}(\gamma(t_0)) \to \pi^{-1}(\gamma(t_1))$$

along γ in P from t_0 to t_1 which is given by

$$\widetilde{\tau}_{t_1}^{t_0}(u) = \beta(t_1) \,,$$

where $\beta : I \to P$ is the horizontal lifting of γ with $\beta(t_0) = u$.

FORMS IN ASSOCIATED BUNDLES

10.73 Proposition Let $\pi : P \to M$ be a principal G-bundle, and let $\pi' : E \to M$ be the associated vector bundle obtained from a representation $\psi : G \to Aut\,(W)$ of G on a finite dimensional vector space W as defined in Example 10.29 (b). For each $u \in P$, let $\mu_u : W \to E_p$ where $\pi(u) = p$ be the linear isomorphism given by

$$\mu_u(v) = [u, v]$$

for $v \in W$. If ω is a bundle-valued k-form on M with values in E, then we have a vector-valued k-form $\overline{\omega}$ on P with values in W given by

$$\overline{\omega}(u)(w_1, \dots, w_k) = \mu_u^{-1} \circ \omega(\pi(u))(\pi_*(w_1), \dots, \pi_*(w_k)) \qquad (1)$$

for $u \in P$ and $w_1, \dots, w_k \in T_uP$, which is tensorial of type (ψ, W). Conversely, given a tensorial W-valued k-form η on P of type (ψ, W), there is a unique E-valued k-form ω on M with $\eta = \overline{\omega}$, which is given by

$$\omega(p)(v_1, \dots, v_k) = \mu_u \circ \eta(u)(w_1, \dots, w_k) \qquad (2)$$

for $p \in M$ and $v_1, \dots, v_k \in T_pM$, where $u \in P$ and $w_1, \dots, w_k \in T_uP$ are chosen so that $\pi(u) = p$ and $\pi_*(w_i) = v_i$ for $i = 1, \dots, k$. If $s : V \to P$ is a section of $\pi : P \to M$ on an open subset V of M, and $\widetilde{\eta} = s^*(\eta)$ is the pull-back of η by s, then

$$\omega(p) = \mu_{s(p)} \circ \widetilde{\eta}(p)$$

for $p \in V$.

PROOF : Using the bundle map (p', π) between the vector bundles $pr_1 : P \times W \to P$ and $\pi' : E \to M$ defined in Example 10.29 (b), which is a linear isomorphism on each fibre, it follows from Proposition 5.33 that the pull-back $(\pi_*, p')^*(\omega)$ is a bundle-valued k-form on P with values in $P \times W$. Now we have that

$$\overline{\omega} = \rho' \circ (\pi_*, p')^*(\omega) ,$$

where ρ' is the equivalence over P between the vector bundles $\Lambda^k(TP; P \times W) \to P$ and $\Lambda^k(TP; W) \to P$ defined as in 5.26 by

$$\rho'(\alpha) = \rho_u \circ \alpha$$

for $\alpha \in \Lambda^k(T_uP; \{u\} \times W)$, where $\rho_u : \{u\} \times W \to W$ is the projection on the second factor so that

$$\mu_u^{-1} = \rho_u \circ p_u'^{-1}$$

for each $u \in P$. Hence it follows that $\overline{\omega}$ is a vector-valued k-form on P with values in W which is clearly horizontal since $V_u = \ker \pi_{*u}$ by Proposition 10.3.

To show that it is tensorial of type (ψ, W), we use that

$$\mu_u(v) = [u, v] = [ug, \psi(g^{-1})\, v] = \mu_{ug} \circ \psi(g^{-1})(v)$$

for $u \in P$, $v \in W$ and $g \in G$ so that

$$\mu_u = \mu_{ug} \circ \psi(g^{-1}). \tag{3}$$

Therefore we have that

$$\mu_{ug}^{-1} = \psi(g^{-1}) \circ \mu_u^{-1}$$

which implies that

$$R_g^*\, \overline{\omega}(u)\,(w_1, \ldots, w_k) = \overline{\omega}(ug)\,((R_g)_*\, w_1, \ldots, (R_g)_*\, w_k)$$

$$\mu_{ug}^{-1} \circ \omega(\pi(u))(\pi_*(w_1)), \ldots, \pi_*(w_k)) = \psi(g^{-1}) \cdot \overline{\omega}(u)\,(w_1, \ldots, w_k)$$

for $u \in P$ and $w_1, \ldots, w_k \in T_uP$.

Conversely, assume that η is a W-valued k-form on P which is tensorial of type (ψ, W), and let $\omega : M \to \Lambda^k(TM; E)$ be the map defined by (2). We must first show that ω is well defined and does not depend on the choice of u and $w_1, \ldots, w_k$. If w_i is replaced by another element $w_i' \in T_uP$ with $\pi_*(w_i') = v_i$, then $w_i - w_i' \in \ker \pi_{*u} = V_u$ which shows that $\omega(p)(v_1, \ldots, v_k)$ is unchanged since η is horizontal. Moreover, if u is replaced by another element $u' \in \pi^{-1}(p)$, then $u' = ug$ for a $g \in G$. Using (3) we therefore have that

$$\mu_u \circ \eta(u)(w_1, \ldots, w_k) = \mu_{ug} \circ \psi(g^{-1}) \circ \eta(u)(w_1, \ldots, w_k)$$

$$= \mu_{ug} \circ R_g^*\, \eta(u)\,(w_1, \ldots, w_k) = \mu_{ug} \circ \eta\,(ug)\,((R_g)_*\, w_1, \ldots, (R_g)_*\, w_k)$$

$$= \mu_{u'} \circ \eta(u')(w_1, \ldots, w_k)$$

since η is tensorial of type (ψ, W), thus completing the proof that ω is well defined.

To show that ω is an E-valued k-form on M, let $s : V \to P$ be a section in P defined on an open subset V of M with associated local trivialization $(t, \pi^{-1}(V))$, and let $(t', \pi'^{-1}(V))$ be the associated local trivialization in E. Then t' is an equivalence over V between the vector bundles $\pi'_V : \pi'^{-1}(V) \to V$ and $pr_1 : V \times W \to V$, and it follows from formula (2) in 10.28 that

$$\omega(p)(v_1, \dots, v_k) = [s(p), \eta(s(p))(s_*(v_1), \dots, s_*(v_k))]$$
$$= [s(p), s^*(\eta)(p)(v_1, \dots, v_k)] = t'^{-1}(p, s^*(\eta)(p)(v_1, \dots, v_k))$$

for $p \in V$ and $v_1, \dots, v_k \in T_pM$. Hence we have that

$$\omega|_V = (id_{V*}, t')^*(\rho'^{-1} \circ s^*(\eta))$$

where $\rho' : \Lambda^k(TM; V \times W) \to \Lambda^k(TM; W)$ is the equivalence over V defined in 5.26, and the result therefore follows by Proposition 5.33. □

10.74 Proposition Let $\pi : P \to M$ be a principal G-bundle, and let $\pi' : E \to M$ be the associated vector bundle obtained from a representation $\psi : G \to Aut\,(W)$ of G on a finite dimensional vector space W. If ω is a bundle-valued k-form on M with values in E and X_i^* is the horizontal lift of the vector field X_i on M for $i = 1, \dots, k$, then we have that

$$\overline{\omega}\,(X_1^*, \dots, X_k^*) = \overline{\omega\,(X_1, \dots, X_k)}\,.$$

PROOF: For every $u \in P$ we have that

$$\overline{\omega}\,(X_1^*, \dots, X_k^*)(u) = \overline{\omega}\,(u)(X_{1,u}^*, \dots, X_{k,u}^*)$$
$$= \mu_u^{-1} \circ \omega(\pi(u))(\pi_*(X_{1,u}^*), \dots, \pi_*(X_{k,u}^*))$$
$$= \mu_u^{-1} \circ \omega(\pi(u))\,(X_{1,\pi(u)}, \dots, X_{k,\pi(u)})$$
$$= \mu_u^{-1} \circ \omega\,(X_1, \dots, X_k)(\pi(u)) = \overline{\omega\,(X_1, \dots, X_k)}(u)\,.$$

□

10.75 Definition Let $\pi : P \to M$ be a principal G-bundle with a connection H, and let $\pi' : E \to M$ be the associated vector bundle obtained from a representation $\psi : G \to Aut\,(W)$ of G on a finite dimensional vector space W. Then the *exterior covariant derivative* of a bundle-valued k-form $\omega \in \Omega^k\,(M; E)$ is the bundle-valued $(k+1)$-form $D\omega \in \Omega^{k+1}\,(M; E)$ which is obtained from the vector-valued $k+1$-form $D\overline{\omega} \in \Omega^{k+1}\,(P; W)$ as in Proposition 10.73, i.e.,

$$\overline{D\omega} = D\overline{\omega}$$

or

$$D\omega(p)(v_1, \dots, v_{k+1}) = \mu_u \circ d\overline{\omega}(u)(w_1, \dots, w_{k+1})$$

for $p \in M$ and $v_1, \dots, v_{k+1} \in T_pM$, where $u \in P$ and $w_1, \dots, w_{k+1} \in H_u$ are chosen so that $\pi(u) = p$ and $\pi_*(w_i) = v_i$ for $i = 1, \dots, k+1$.

10.76 Let $\pi : P \to M$ be a principal G-bundle, and let $\pi'_i : E_i \to M$ be the associated vector bundle obtained from a representation $\psi_i : G \to Aut\,(W_i)$ of G on a finite dimensional vector space W_i for $i = 1, 2, 3$. Suppose that $\nu : W_1 \times W_2 \to W_3$ is a bilinear map such that

$$\psi_3(g) \circ \nu = \nu \circ (\psi_1(g) \times \psi_2(g))$$

for every $g \in G$. Then we have a bilinear map $\nu_p : E_{1,p} \times E_{2,p} \to E_{3,p}$ for each $p \in M$ defined by

$$\nu_p\,([u, v_1], [u, v_2]) = [u, \nu\,(v_1, v_2)]$$

for $u \in \pi^{-1}(p)$, $v_1 \in W_1$ and $v_2 \in W_2$. It is well defined since

$$\nu_p\,([u\,g, \psi_1(g^{-1})\,v_1], [u\,g, \psi_2(g^{-1})\,v_2]) = [u\,g, \nu\,(\psi_1(g^{-1})\,v_1, \psi_2(g^{-1})\,v_2)]$$

$$= [u\,g, \psi_3(g^{-1})\,\nu(v_1, v_2)] = [u, \nu\,(v_1, v_2)]$$

for every $g \in G$. If $\mu_{i,u} : W_i \to E_{i,p}$ where $\pi(u) = p$ is the linear isomorphism defined by $\mu_{i,u}(v) = [u, v]$ for $v \in W_i$, we have that

$$\mu_{3,u} \circ \nu = \nu_p \circ (\mu_{1,u} \times \mu_{2,u}) \tag{1}$$

for each $u \in P$.

Now if $\omega \in \Omega^k(V; E_1)$ and $\eta \in \Omega^l(V; E_2)$ are bundle-valued differential forms on an open subset V of M, we have a bundle-valued differential form $\omega \wedge_\nu \eta \in \Omega^{k+l}(V; E_3)$, called the *wedge product* of ω and η with respect to ν, defined by

$$\overline{\omega \wedge_\nu \eta} = \overline{\omega} \wedge_\nu \overline{\eta}$$

or

$$(\omega \wedge_\nu \eta)(p)(v_1, \dots, v_{k+l}) = \mu_{3,u} \circ\, (\overline{\omega} \wedge_\nu \overline{\eta})(u)(w_1, \dots, w_{k+1}) \tag{2}$$

for $p \in V$ and $v_1, \dots, v_{k+l} \in T_pM$, where $u \in P$ and $w_1, \dots, w_{k+l} \in T_uP$ are chosen so that $\pi(u) = p$ and $\pi_*(w_i) = v_i$ for $i = 1, \dots, k+1$. Using (1) we have that

$$\mu_{3,u} \circ\, (\overline{\omega}\,(u) \wedge_\nu \overline{\eta}\,(u)) = \mu_{1,u} \circ \overline{\omega}\,(u) \wedge_{\nu_p} \mu_{2,u} \circ \overline{\eta}\,(u)$$

which together with (2) shows that

$$(\omega \wedge_\nu \eta)(p) = \omega\,(p) \wedge_{\nu_p} \eta\,(p)$$

for $p \in V$. Using Proposition 5.69 and Definition 10.75 of the exterior covariant derivative, we have that

$$D(\omega \wedge_\nu \eta) = (D\,\omega) \wedge_\nu \eta + (-1)^k\,\omega \wedge_\nu (D\eta)$$

We will often omit the reference to ν in the notation if it is clear from the context.

COVARIANT DERIVATIVE OF SECTIONS IN ASSOCIATED VECTOR BUNDLES

10.77 Definition Let $\pi : P \to M$ be a principal G-bundle with a connection H, and let $\pi' : E \to M$ be the associated vector bundle obtained from a representation $\psi : G \to Aut\,(W)$ of G on a finite dimensional vector space W. If $s \in \Gamma(M;E)$ is a section of π', then the vector-valued function $\bar{s} \in \mathscr{F}(P\,;W)$ is given by

$$\bar{s}\,(u) = \mu_u^{-1} \circ s(\pi(u))$$

for $u \in P$. The *exterior covariant derivative* of s is the bundle-valued 1-form $Ds \in \Omega^1(M;E)$ given by

$$Ds\,(p)(v) = \mu_u \circ d\bar{s}\,(u)(w) = \mu_u \circ w(\bar{s}\,)$$

for $p \in M$ and $v \in T_pM$, where $u \in P$ and $w \in H_u$ are chosen so that $\pi(u) = p$ and $\pi_*(w) = v$, and w is considered to act componentwise as a local derivation at u as defined in 5.49. Ds is usually denoted by ∇s.

The element $Ds\,(p)(v) \in E_p$, also denoted by $\nabla_v s$, is called the *covariant derivative of s at p with respect to the tangent vector* $v \in T_pM$. If X is a vector field on M, then the section $Ds\,(X) \in \Gamma(M;E)$, which is denoted by $\nabla_X s$, is called the *covariant derivative of s with respect to X*. It is given by

$$\nabla_X s\,(p) = \nabla_{X_p} s = \mu_u \circ X_u^*\,(\bar{s}\,)$$

for $p \in M$, where $u \in P$ is chosen so that $\pi(u) = p$, and X^* is the horizontal lifting of X.

10.78 Proposition Let $\pi : P \to M$ be a principal G-bundle with a connection H, and let $\pi' : E \to M$ be the associated vector bundle obtained from a representation $\psi : G \to Aut\,(W)$ of G on a finite dimensional vector space W. Suppose that $v \in T_pM$ is the tangent vector to a smooth curve $\gamma : I \to M$ at p, where I is an open interval containing t_0 and $\gamma(t_0) = p$. If $s \in \Gamma(M;E)$ is a section of π', then the covariant derivative of s at p with respect to v is given by

$$\nabla_v s = \lim_{h\to 0} \frac{1}{h}\,[\tau_{t_0}^{t_0+h}\, s(\gamma(t_0+h)) - s(p)] = \frac{d}{dt}\Big|_{t_0} \tau_{t_0}^{t}\, s(\gamma(t))\,,$$

where $\tau_{t_0}^{t} : \pi'^{-1}(\gamma(t)) \to \pi'^{-1}(p)$ is the parallel transport in E along γ from t to t_0, and the limit is taken with respect to the vector space topology on E_p defined in Proposition 13.117 in the appendix.

PROOF: Choose a $u \in \pi^{-1}(p)$ and a $w \in H_u$ with $\pi_*(w) = v$, and let $\alpha : I \to P$ be the horizontal lift of γ with $\alpha(t_0) = u$. Then we also have that $\alpha'(t_0) = w$, since $\alpha'(t_0) \in H_u$ and $\pi_*(\alpha'(t_0)) = \gamma'(t_0) = v$. By Remark 10.72 we now have that

$$\tau^t_{t_0} = (\tau^{t_0}_t)^{-1} = \mu_u \circ \mu^{-1}_{\alpha(t)}$$

so that

$$\tau^t_{t_0} s(\gamma(t)) = \mu_u \circ \bar{s}\,(\alpha(t))$$

for $t \in I$. Hence it follows from 5.49 and Lemma 2.84 that

$$\frac{d}{dt}\bigg|_{t_0} \tau^t_{t_0} s(\gamma(t)) = \mu_u \circ \bar{s}_*(\alpha'(t_0)) = \mu_u \circ \bar{s}_*(w) = \mu_u \circ w(\bar{s}) = \nabla_v s$$

which completes the proof of the proposition. □

10.79 Remark Let $\pi : P \to M$ be a principal G-bundle with a connection H, and let $\pi' : E \to M$ be the associated vector bundle obtained from a representation $\psi : G \to Aut\,(W)$ of G on a finite dimensional vector space W. Suppose that $v \in T_pM$ is the tangent vector to a smooth curve $\gamma : I \to M$ at p, where I is an open interval containing 0 and $\gamma(0) = p$, and let $s \in \Gamma(M;E)$ be a section of π'.

If $\mathscr{B} = \{v_1, \ldots, v_n\}$ is a basis for E_p, we let $v_i(t) = \tau^0_t(v_i)$ for $i = 1, \ldots, n$ and $t \in I$ be the vectors obtained by parallel transport in E along the curve γ from 0 to t. Then $\mathscr{B}_t = \{v_1(t), \ldots, v_n(t)\}$ is a basis for $E_{\gamma(t)}$, since $\tau^0_t : E_p \to E_{\gamma(t)}$ is a linear isomorphism. If

$$s(\gamma(t)) = \sum_{i=1}^{n} c_i(t)\, v_i(t)\,,$$

then

$$\tau^t_0 s(\gamma(t)) = \sum_{i=1}^{n} c_i(t)\, v_i$$

so that c_i are the components with respect to $\mathscr{B}$ of the smooth curve $c : I \to E_p$ given by $c(t) = \tau^t_0 s(\gamma(t)) = \mu_u \circ \bar{s}\,(\alpha(t))$. Hence we have that

$$\nabla_v s = \sum_{i=1}^{n} c_i'(0)\, v_i$$

by Lemma 2.84.

10.80 Proposition Let $\pi : P \to M$ be a principal G-bundle with a connection H, and let $\pi' : E \to M$ be the associated vector bundle obtained from a representation $\psi : G \to Aut\,(W)$ of G on a finite dimensional vector space W. Then we have that

(1) $\nabla_{X_1+X_2} s = \nabla_{X_1} s + \nabla_{X_2} s$

(2) $\nabla_X (s_1 + s_2) = \nabla_X s_1 + \nabla_X s_2$

(3) $\nabla_{fX} s = f\, \nabla_X s$

(4) $\nabla_X(fs) = X(f)\,s + f\,\nabla_X s$

for all vector fields $X, X_1, X_2 \in \mathscr{T}_1(M)$, sections $s, s_1, s_2 \in \Gamma(M;E)$ and smooth functions $f \in \mathscr{F}(M)$.

PROOF: We see that (1), (2) and (3) follow immediately from Definition 10.77 and Proposition 10.51. To prove (4), we use that

$$\overline{fs}\,(u) = \mu_u^{-1}(f(\pi(u))\,s(\pi(u))) = f(\pi(u))\,\mu_u^{-1}(s(\pi(u))) = f \circ \pi\,(u)\,\bar{s}\,(u)$$

for $u \in P$, which implies that

$$\overline{fs} = (f \circ \pi)\,\bar{s}\,.$$

Since

$$X_u^*(f \circ \pi) = \pi_*(X_u^*)(f) = X_p(f) = X(f)(p)$$

where $\pi(u) = p$, it follows from Proposition 5.50 that

$$X_u^*(\overline{fs}\,) = X(f)(p)\,\bar{s}(u) + f(p)\,X_u^*(\bar{s}\,)\,.$$

By applying μ_u on both sides, we now obtain

$$\nabla_X(fs)(p) = X(f)(p)\,s(p) + f(p)\,\nabla_X s\,(p)$$

which completes the proof of (4). □

10.81 Proposition Let $\pi : P \to M$ be a principal G-bundle with a connection H, and let W a finite dimensional vector space. Consider the associated vector bundle obtained from the representation $\rho : G \to Aut\,(W)$ given by $\rho(g) = id_W$ for $g \in G$, which may be identified with the trivial vector bundle $\widetilde{\pi} : E_W \to M$ as described in Example 10.29 (d). Then each vector-valued smooth function $f : M \to W$ on M corresponds to a section $s = (id_M, f)$ of $\widetilde{\pi}$, and we have that

$$\nabla_v s = (p, v(f))$$

for every $p \in M$ and $v \in T_pM$.

PROOF: Since

$$\bar{s}\,(u) = \mu_u^{-1} \circ s\,(\pi(u)) = \mu_u^{-1}(\pi(u), f \circ \pi(u)) = f \circ \pi\,(u)$$

for $u \in P$, we have that

$$\bar{s} = f \circ \pi\,.$$

Hence it follows from 5.51 and Proposition 5.63 that

$$\nabla_v s = \mu_u \circ d\bar{s}\,(u)(w) = \mu_u \circ \pi^*(df)(u)(w) = \mu_u \circ df(p)(v) = (p, v(f))\,,$$

where $u \in P$ and $w \in H_u$ are chosen so that $\pi(u) = p$ and $\pi_*(w) = v$. □

COVARIANT DERIVATIVE OF TENSOR FIELDS

10.82 Remark If $\pi : \mathscr{L}(M) \to M$ is the frame bundle over a smooth manifold M^n with a connection H, then the associated vector bundle obtained from the representation $\rho^k_l : Gl(\mathbf{R}^n) \to Aut(\mathscr{T}^k_l(\mathbf{R}^n))$ may be identified with the tensor bundle $\pi^k_l : T^k_l(M) \to M$ as in Example 10.29 (f). Suppose that $v \in T_pM$ is the tangent vector to a smooth curve $\gamma : I \to M$ at p, where I is an open interval containing 0 and $\gamma(0) = p$, and let $T \in \mathscr{T}^k_l(M)$ be a tensor field of type $\binom{k}{l}$ on M. Choose a $u \in \mathscr{L}_p(M)$, and let $\alpha : I \to \mathscr{L}(M)$ be the horizontal lift of γ with $\alpha(0) = u$.

Now let $\mathscr{B} = \{v_1, \ldots, v_n\}$ be a basis for T_pM with dual basis $\mathscr{B}^* = \{v^1, \ldots, v^n\}$, and let $v_i(t) = \tau^0_t(v_i) = \alpha(t) \circ u^{-1}(v_i)$ and $v^i(t) = \tau^0_t(v^i) = v^i \circ u \circ \alpha(t)^{-1}$ for $i = 1, \ldots, n$ and $t \in I$ be the vectors and covectors obtained by parallel transport in TM and T^*M along the curve γ from 0 to t. Then $\mathscr{B}_t = \{v_1(t), \ldots, v_n(t)\}$ is a basis for $T_{\gamma(t)}M$ with dual basis $\mathscr{B}^*_t = \{v^1(t), \ldots, v^n(t)\}$, since $\tau^0_t : T_pM \to T_{\gamma(t)}M$ is a linear isomorphism.

Since $\tau^0_t = (u \circ \alpha(t)^{-1})^*$ in the tensor bundle $T^k_l(M)$, it follows by induction from Remark 4.54 that

$$v^{i_1}(t) \otimes \cdots \otimes v^{i_k}(t) \otimes v_{j_1}(t) \otimes \cdots \otimes v_{j_l}(t) = \tau^0_t(v^{i_1} \otimes \cdots \otimes v^{i_k} \otimes v_{j_1} \otimes \cdots \otimes v_{j_l})$$

for $(i_1, \ldots, i_k, j_1, \ldots, j_l) \in I^{k+l}_n$ and $t \in I$, which shows that the parallel transport in $T^k_l(M)$ of the basis vectors in $\mathscr{T}^k_l(\mathscr{B})$ are obtained by taking the tensor product of the parallel transported basis vectors from $\mathscr{B}$ and $\mathscr{B}^*$. If

$$T(\gamma(t)) = \sum_{\substack{i_1, \ldots, i_k \\ j_1, \ldots, j_l}} C_p{}^{j_1, \ldots, j_l}_{i_1, \ldots, i_k}(t)\, v^{i_1}(t) \otimes \cdots \otimes v^{i_k}(t) \otimes v_{j_1}(t) \otimes \cdots \otimes v_{j_l}(t),$$

then

$$\tau^t_0 T(\gamma(t)) = \sum_{\substack{i_1, \ldots, i_k \\ j_1, \ldots, j_l}} C_p{}^{j_1, \ldots, j_l}_{i_1, \ldots, i_k}(t)\, v^{i_1} \otimes \cdots \otimes v^{i_k} \otimes v_{j_1} \otimes \cdots \otimes v_{j_l}$$

so that $C_p{}^{j_1, \ldots, j_l}_{i_1, \ldots, i_k}$ are the components with respect to $\mathscr{T}^k_l(\mathscr{B})$ of the smooth curve $c : I \to \mathscr{T}^k_l(T_pM)$ given by $c(t) = \tau^t_0 T(\gamma(t)) = (u^{-1})^* \circ \overline{T}(\alpha(t))$. Hence we have that

$$\nabla_v T = \sum_{\substack{i_1, \ldots, i_k \\ j_1, \ldots, j_l}} \frac{d}{dt}\bigg|_0 C_p{}^{j_1, \ldots, j_l}_{i_1, \ldots, i_k}\, v^{i_1} \otimes \cdots \otimes v^{i_k} \otimes v_{j_1} \otimes \cdots \otimes v_{j_l}$$

by Lemma 2.84.

10.83 Proposition Let X be a vector field on the smooth manifold M.

(1) If f is a smooth function on M, then $\nabla_X f = X(f)$.

(2) If $T \in \mathscr{T}^{k_1}_{l_1}(M)$ and $S \in \mathscr{T}^{k_2}_{l_2}(M)$, then $\nabla_X(T \otimes S) = (\nabla_X T) \otimes S + T \otimes (\nabla_X S)$.

(3) $\nabla_X \delta = 0$, where $\delta \in \mathscr{T}^1_1(M)$ is the Kronecker delta tensor.

(4) ∇_X commutes with contractions, i.e., if $T \in \mathscr{T}^k_l(M)$, and if $1 \le r \le k$ and $1 \le s \le l$, then $\nabla_X(C^r_s T) = C^r_s(\nabla_X T)$.

(5) If $T \in \mathscr{T}^k_l(M)$, and if $X_1, \dots, X_k \in \mathscr{T}_1(M)$ and $\lambda_1, \dots, \lambda_l \in \mathscr{T}^1(M)$, then

$$\begin{aligned}\nabla_X\left[T(X_1,\dots,X_k,\lambda_1,\dots,\lambda_l)\right] &= (\nabla_X T)(X_1,\dots,X_k,\lambda_1,\dots,\lambda_l)\\ &+\sum_{i=1}^{k} T(X_1,\dots,\nabla_X X_i,\dots,X_k,\lambda_1,\dots,\lambda_l)\\ &+\sum_{i=1}^{l} T(X_1,\dots,X_k,\lambda_1,\dots,\nabla_X\lambda_i,\dots,\lambda_l)\,.\end{aligned}$$

PROOF: Let f be a smooth function on M which can be considered to be a tensor field of type $\binom{0}{0}$. If $p \in M$ and $\mathscr{B}$ is a basis for T_pM, we have that $\mathscr{T}^0_0(T_pM) = \mathbf{R}$ with basis $\mathscr{T}^0_0(\mathscr{B}) = \{1\}$. Suppose that X_p is the tangent vector to a smooth curve $\gamma : I \to M$ at p, where I is an open interval containing 0 and $\gamma(0) = p$. If $f(\gamma(t)) = c(t)$ for $t \in I$, then it follows from 2.77 and Remark 10.82 that $\nabla_{X_p} f = c'(0) = X_p(f)$, which completes the proof of (1).

Point (2) to (4) follow directly from Remark 10.82. Point (5) follows from (2) and (4) since $T(X_1,\dots,X_k,\lambda_1,\dots,\lambda_l)$ can be obtained from $T \otimes X_1 \otimes \cdots \otimes X_k \otimes \lambda_1 \otimes \cdots \otimes \lambda_l$ by applying contractions repeatedly. □

COVARIANT DERIVATIVE OF SECTIONS ALONG SMOOTH MAPS

10.84 Let $\pi : P \to M$ be a principal G-bundle, and let $\pi' : E \to M$ be the fibre bundle associated with P with fibre F obtained from a Lie group homomorphism $\psi : G \to G'$ into a Lie group G' which acts on the smooth manifold F effectively on the left. Consider a smooth map $f : N \to M$ from a smooth manifold N, and let $\widetilde{\pi} : \widetilde{P} \to N$ and $\widetilde{\pi}' : \widetilde{E} \to N$ be fibre bundles induced from $\pi : P \to M$ and $\pi' : E \to M$ by f. We contend that the fibre bundle $\pi'' : E' \to N$ with fibre F associated with $\widetilde{\pi} : \widetilde{P} \to N$ is equivalent over N to the induced fibre bundle $\widetilde{\pi}' : \widetilde{E} \to N$, with the equivalence $e : E' \to \widetilde{E}$ given by

$$e([(p,u),v]) = (p,[u,v])$$

for $p \in N$, $u \in P$ and $v \in F$ where $f(p) = \pi(u)$. This is clearly well defined, for if $g \in G$, then

$$e([(p,u)g,\psi(g^{-1})v]) = [(p,ug),\psi(g^{-1})v] = (p,[ug,\psi(g^{-1})v])$$
$$= (p,[u,v]) = e([(p,u),v]).$$

To show that e is an equivalence, let $(t',\pi'^{-1}(U))$ be the local trivialization in E associated with a local trivialization $(t,\pi^{-1}(U))$ in P, defined as in 10.28 by

$$t'([u,v]) = (q,\psi\circ t_q(u)v)$$

for $q\in U$, $u\in P_q$ and $v\in F$. Let $(\widetilde{t},\widetilde{\pi}^{-1}(\widetilde{U}))$ and $(\widetilde{t}\,',\widetilde{\pi}'^{-1}(\widetilde{U}))$, where $\widetilde{U}=f^{-1}(U)$, be the corresponding local trivializations in $\widetilde{P}$ and $\widetilde{E}$, respectively, defined as in 10.12 by

$$\widetilde{t}(p,u) = (p,t_{f(p)}(u))$$

and

$$\widetilde{t}\,'(p,[u,v]) = (p,t'_{f(p)}([u,v])) = (p,\psi\circ t_{f(p)}(u)v)$$

for $p\in\widetilde{U}$, $u\in P_{f(p)}$ and $v\in F$. Now if $(t'',\pi''^{-1}(U))$ is the local trivialization in E' associated with $(\widetilde{t},\widetilde{\pi}^{-1}(\widetilde{U}))$, we have that

$$t''([(p,u),v]) = (p,\psi\circ\widetilde{t}_p(p,u)v) = (p,\psi\circ t_{f(p)}(u)v)$$

for $p\in\widetilde{U}$, $u\in P_{f(p)}$ and $v\in F$, which shows that

$$\widetilde{t}\,'\circ e\circ t''^{-1} = id_{\widetilde{U}\times F}\,.$$

Using the equivalence e, we may therefore identify the fibre bundle E' associated with $\widetilde{P}$ with fibre F with the fibre bundle $\widetilde{E}$ induced from E by f, and the diffeomorphism $\mu'_{(p,u)}:F\to E'_p$ defined in 10.28 is then identified with the diffeomorphism $\widetilde{\mu}_{(p,u)} = e_p\circ\mu'_{(p,u)}:F\to\widetilde{E}_p$ given by

$$\widetilde{\mu}_{(p,u)}(v) = (p,\mu_u(v))$$

for $p\in N$, $u\in P_{f(p)}$ and $v\in F$, where $\mu_u:F\to E_{f(p)}$ is the corresponding diffeomorphism for the fibre bundle E. If $(\widetilde{f}\,',f)$ is the canonical bundle map of the induced bundle $\widetilde{\pi}':\widetilde{E}\to N$, we have that

$$\widetilde{f}'_p\circ\widetilde{\mu}_{(p,u)} = \mu_u$$

for $p\in N$ and $u\in P_{f(p)}$.

10.85 Let $\pi:P\to M$ be a principal G-bundle with a connection H, and let $\pi':E\to M$ be the associated vector bundle obtained from a representation $\psi:G\to Aut\,(W)$ of G on a finite dimensional vector space W. Consider a smooth map $f:N\to M$ from a smooth manifold N, and let $\widetilde{\pi}:\widetilde{P}\to N$ and $\widetilde{\pi}':\widetilde{E}\to N$ be the fibre bundles induced from $\pi:P\to M$ and $\pi':E\to M$ by f, respectively, with canonical bundle maps $(\widetilde{f},f)$ and $(\widetilde{f}\,',f)$.

If $F : N \to E$ is a lifting of f and $v \in T_pN$ is a tangent vector at a point $p \in N$, we define the *covariant derivative of F at p with respect to v* by

$$\nabla_v F = \widetilde{f}'(\nabla_v \widetilde{F}) ,$$

where $\widetilde{F}$ is the section determined by F in the induced vector bundle $\widetilde{\pi}' : \widetilde{E} \to N$, which is given by $\widetilde{F}(p) = (p, F(p))$ for $p \in N$. If X is a vector field on N, we define

$$\nabla_X F(p) = \nabla_{X_p} F$$

for $p \in N$, so that

$$\nabla_X F = \widetilde{f}' \circ \nabla_X \widetilde{F} .$$

Hence $\nabla_X F : N \to E$ is the lifting of f corresponding to the section $\nabla_X \widetilde{F}$ in the induced vector bundle $\widetilde{\pi}' : \widetilde{E} \to N$.

Now let $F \in \mathscr{F}(\widetilde{P}; W)$ be the vector-valued function given by

$$\overline{F}(p,u) = \mu_u^{-1} \circ F(p)$$

for $p \in N$ and $u \in P_{f(p)}$. As $\mu_u = \widetilde{f}_p' \circ \widetilde{\mu}_{(p,u)}$ and $F = \widetilde{f}' \circ \widetilde{F}$, this means that

$$\overline{F}(p,u) = \widetilde{\mu}_{(p,u)}^{-1} \circ \widetilde{F}(\widetilde{\pi}'(p,u))$$

for $(p,u) \in \widetilde{P}$. Using Definition 10.77 we therefore have that

$$\nabla_v \widetilde{F} = \widetilde{\mu}_{(p,u)} \circ w(\overline{F})$$

so that

$$\nabla_v F = \mu_u \circ w(\overline{F})$$

for $p \in N$ and $v \in T_pN$, where $u \in P_{f(p)}$ and $w \in \widetilde{H}_{(p,u)}$ is chosen so that $\widetilde{\pi}_*(w) = v$, and where w is considered to act componentwise as a local derivation at (p,u) as defined in 5.49. We recall from 10.37 that

$$\widetilde{H}_{(p,u)} = (\widetilde{f}_{*\,(p,u)})^{-1}(H_u) .$$

If X^* is the horizontal lifting of X, we have that

$$\nabla_X F(p) = \mu_u \circ X_u^*(\overline{F})$$

for $p \in N$ and $u \in P_{f(p)}$.

10.86 Remark If $F : I \to E$ is a lifting of a smooth curve $\gamma : I \to M$ on M defined on an open interval I, and we use the standard local chart (r, I) on I where $r : I \to \mathbf{R}$ is the inclusion map, then $\nabla_{\frac{d}{dr}} F$ is also denoted by $\frac{DF}{dr}$ and is called the *covariant derivative of F along* γ.

10.87 Proposition Let $\pi : P \to M$ be a principal G-bundle with a connection H, and let $\pi' : E \to M$ be the associated vector bundle obtained from a representation $\psi : G \to Aut(W)$ of G on a finite dimensional vector space W. If $f : N \to M$ is a smooth map from a smooth manifold N and $s : M \to E$ is a section of π' on M, then we have that

$$\nabla_v(s \circ f) = \nabla_{f_*(v)}\, s$$

for every $v \in TN$.

PROOF : Let $\widetilde{\pi} : \widetilde{P} \to N$ be the fibre bundle induced from $\pi : P \to M$ by f with canonical bundle map $(\widetilde{f}, f)$. If $F = s \circ f$, then the map $\overline{F} \in \mathscr{F}(\widetilde{P}\,; W)$ is given by

$$\overline{F}\,(p,u) = \mu_u^{-1} \circ s \circ f\,(p) = \mu_u^{-1} \circ s \circ \pi\,(u) = \bar{s}\,(u) = \bar{s} \circ \widetilde{f}\,(p,u)$$

for $(p,u) \in \widetilde{P}$. If $w \in \widetilde{H}_{(p,u)}$ is chosen so that $\widetilde{\pi}_*(w) = v$, then $\widetilde{f}_*(w) \in H_u$ and

$$\pi_* \circ \widetilde{f}_*(w) = f_* \circ \widetilde{\pi}_*(w) = f_*(v)\,.$$

Hence we have that

$$\nabla_{f_*(v)}\, s = \mu_u \circ \widetilde{f}_*(w)(\bar{s}) = \mu_u \circ w(\bar{s} \circ \widetilde{f}) = \nabla_v(s \circ f)$$

which completes the proof of the proposition. □

10.88 Proposition Let $\pi : P \to M$ be a principal G-bundle with a connection H, and let $\pi' : E \to M$ be the associated vector bundle obtained from a representation $\psi : G \to Aut(W)$ of G on a finite dimensional vector space W. If $f : N \to M$ is a smooth map from a smooth manifold N, then we have that

(1) $\nabla_{X_1+X_2} F = \nabla_{X_1} F + \nabla_{X_2} F$

(2) $\nabla_X(F_1 + F_2) = \nabla_X F_1 + \nabla_X F_2$

(3) $\nabla_{gX} F = g\, \nabla_X F$

(4) $\nabla_X(gF) = X(g)\, F + g\, \nabla_X F$

for all vector fields $X, X_1, X_2 \in \mathscr{T}_1(N)$, liftings $F, F_1, F_2 \in \Gamma(f\,; E)$ of f and smooth functions $g \in \mathscr{F}(N)$.

PROOF : Follows from 10.85 and Proposition 10.80 applied to the sections $\widetilde{F}$, $\widetilde{F}_1$ and $\widetilde{F}_2$ in the vector bundle $\widetilde{\pi}' : \widetilde{E} \to N$ induced from $\pi' : E \to M$ by f. □

10.89 Lemma Let $\pi : P \to M$ be a principal G-bundle with a connection H, and let $\pi' : E \to M$ be the associated vector bundle obtained from a representation $\psi : G \to Aut(W)$ of G on a finite dimensional vector space W. Suppose that $f : N \to M$ is a smooth map from a smooth manifold N, and that $X \in \mathscr{T}_1(N)$ is a vector field on N. If $F_1, F_2 \in \Gamma(f\,; E)$ are liftings of f with $F_1|_V = F_2|_V$ for an open subset V of N, then we have that $(\nabla_X F_1)|_V = (\nabla_X F_2)|_V$.

PROOF: Considering the difference $F_2 - F_1$, it is clearly enough to prove that $(\nabla_X F)|_V = 0$ for each lifting $F \in \Gamma(f;E)$ of f with $F|_V = 0$. To show this, let $p \in V$ and choose a smooth function $h \in \mathscr{F}(N)$ with $h = 1$ on $N - V$ and $h(p) = 0$. Such a function h exists by Corollary 2.24. Then we have that

$$\nabla_X F(p) = \nabla_X (hF)(p) = X(h)(p)\, F(p) + h(p)\, \nabla_X F(p) = 0$$

which completes the proof since p was an arbitrary point in V. □

10.90 Proposition Let $\pi : P \to M$ be a principal G-bundle with a connection H, and let $\pi' : E \to M$ be the associated vector bundle obtained from a representation $\psi : G \to Aut\,(W)$ of G on a finite dimensional vector space W. Suppose that $F : I \to E$ is a lifting of the smooth curve $\gamma : I \to M$ defined on an open interval I. Using the standard local chart (r, I) on I where $r : I \to \mathbf{R}$ is the inclusion map, the covariant derivative of F along γ at $t_0 \in I$ is given by

$$\frac{DF}{dr}(t_0) = \lim_{h \to 0} \frac{1}{h} \left[\tau_{t_0}^{t_0+h} F(t_0 + h) - F(t_0) \right] = \frac{d}{dt}\bigg|_{t_0} \tau_{t_0}^{t} F(t)\,,$$

where $\tau_{t_0}^{t} : \pi'^{-1}(\gamma(t)) \to \pi'^{-1}(\gamma(t_0))$ is the parallel transport in E along γ from t to t_0, and the limit is taken with respect to the vector space topology on $E_{\gamma(t_0)}$ defined in Proposition 13.117 in the appendix.

PROOF: Let $\widetilde{\pi} : \widetilde{P} \to I$ and $\widetilde{\pi}' : \widetilde{E} \to I$ be the fibre bundles induced from $\pi : P \to M$ and $\pi' : E \to M$ by γ, and let $(\widetilde{\gamma}', \gamma)$ be the canonical bundle map of $\widetilde{\pi}' : \widetilde{E} \to I$. If $\widetilde{F}$ is the section in $\widetilde{\pi}' : \widetilde{E} \to I$ determined by F which is a lifting of the identity map $\beta = id_I : I \to I$, it follows from Proposition 10.78 that

$$\nabla_{\frac{d}{dr}\big|_{t_0}} \widetilde{F} = \frac{d}{dt}\bigg|_{t_0} \widetilde{\tau}_{t_0}^{t} \widetilde{F}(\beta(t)), \tag{1}$$

where $\widetilde{\tau}_{t_0}^{t} : \widetilde{\pi}'^{-1}(\beta(t)) \to \widetilde{\pi}'^{-1}(\beta(t_0))$ is the parallel transport in $\widetilde{E}$ along β from t to t_0.

Now let $\alpha : I \to P$ be a horizontal lift of γ. Then the section $\widetilde{\alpha} : I \to \widetilde{P}$ in $\widetilde{\pi} : \widetilde{P} \to I$ determined by α, which is given by $\widetilde{\alpha}(t) = (t, \alpha(t))$ for $t \in I$, is a horizontal lift of β by Remark 10.58. The proposition therefore follows by applying $\widetilde{\gamma}'$ on both sides of (1), since

$$\begin{aligned} \widetilde{\gamma}' \circ \widetilde{\tau}_{t_0}^{t} \circ \widetilde{F}(\beta(t)) &= \widetilde{\gamma}' \circ \widetilde{\mu}_{(t_0, \alpha(t_0))} \circ \widetilde{\mu}_{(t, \alpha(t))}^{-1} \circ \widetilde{F}(t) \\ &= \mu_{\alpha(t_0)} \circ \mu_{\alpha(t)}^{-1} \circ F(t) = \tau_{t_0}^{t} \circ F(t) \end{aligned}$$

for $t \in I$ by 10.85 and Remark 10.72. □

10.91 Definition Let $\pi : P \to M$ be a principal G-bundle with a connection H, and let $\pi' : E \to M$ be the associated vector bundle obtained from a representation $\psi : G \to Aut\,(W)$ of G on a finite dimensional vector space W. Consider a smooth curve $\gamma : I \to M$ on M defined on an open interval I, and let (r, I) be the standard local chart on I where $r : I \to \mathbf{R}$ is the inclusion map. Then a lifting $F : I \to E$ of γ is said to be *parallel along* γ if the covariant derivative $DF/dr = 0$ on I.

10.92 Proposition Let $\pi : P \to M$ be a principal G-bundle with a connection H, and let $\pi' : E \to M$ be the associated vector bundle obtained from a representation $\psi : G \to Aut\,(W)$ of G on a finite dimensional vector space W. Then a lifting $F : I \to E$ of a smooth curve $\gamma : I \to M$ defined on an open interval I is parallel along γ if and only if the following assertion is satisfied :

For each pair of points $t_0, t_1 \in I$ we have that

$$\tau_{t_1}^{t_0} F(t_0) = F(t_1)\,,$$

where $\tau_{t_1}^{t_0} : \pi'^{-1}(\gamma(t_0)) \to \pi'^{-1}(\gamma(t_1))$ is the parallel transport in E along γ from t_0 to t_1.

PROOF : If $c : I \to E_{\gamma(t_1)}$ is the smooth curve given by $c(t) = \tau_{t_1}^{t} F(t)$ for $t \in I$, then it follows from Proposition 10.90 that

$$c'(t) = \frac{d}{ds}\bigg|_t \tau_{t_1}^{s} F(s) = \tau_{t_1}^{t} \frac{d}{ds}\bigg|_t \tau_t^{s} F(s) = \tau_{t_1}^{t} \frac{DF}{dr}(t)$$

for $t \in I$. Hence the curve c is constant if and only if $DF/dr = 0$ on I. □

LINEAR CONNECTIONS

10.93 Proposition Let $\pi : \mathscr{L}(M) \to M$ be the frame bundle over a smooth manifold M^n. Then we have an $\mathbf{R}^n$-valued 1-form θ on $\mathscr{L}(M)$ defined by

$$\theta(u) = u^{-1} \circ \pi_{*u}$$

for $u \in \mathscr{L}(M)$, called the *canonical form* or the *dual form* on $\mathscr{L}(M)$, which is tensorial of type $(id, \mathbf{R}^n)$ with $\ker\theta(u) = V_u$ for every $u \in \mathscr{L}(M)$. We have that $\theta = \overline{\varepsilon}$, where $\varepsilon \in \Omega^1(M; TM)$ is the bundle-valued 1-form on M defined in Example 5.45.

PROOF : Follows from Propositions 10.73 and 10.3. □

10.94 Remark If $s = (X_1, \dots, X_n)$ is a frame field on an open subset V of M, then the 1-forms $s^*\theta^i$ for $i = 1, \dots, n$ form the dual local basis for $\Lambda^1(TM)$ on V. Indeed, we have that

$$s^*\theta(p)(v) = \theta(s(p))(s_*(v)) = s(p)^{-1}(v)$$

for $p \in V$ and $v \in T_pM$, showing that

$$s^*\theta^i(p)(X_j(p)) = e^i \circ s^*\theta(p) \circ s(p)(e_j) = \delta_{ij}$$

for $p \in V$ and $i, j = 1, \dots, n$.

10.95 Definition A connection H in the frame bundle $\pi : \mathscr{L}(M) \to M$ over a smooth manifold M is called a *linear connection* on M.

10.96 Proposition Let H be a linear connection on a smooth manifold M^n, and let $\xi \in \mathbf{R}^n$. Then we have a horizontal vector field $B(\xi)$ on $\mathscr{L}(M)$, called the *basic vector field* determined by ξ, where $B(\xi)_u$ is the unique vector in H_u such that

$$\pi_*(B(\xi)_u) = u(\xi) \tag{1}$$

for $u \in \mathscr{L}(M)$, i.e.,

$$B(\xi)_u = \alpha_u^{-1} \circ u(\xi) \tag{2}$$

where (α, π) is the bundle map from H to TM defined in Proposition 10.49.

If θ and ω is the dual form and the connection form of H, respectively, then the map $B : \mathbf{R}^n \to \mathscr{T}_1\mathscr{L}(M)$ is uniquely determined by the fact that

$$\theta(u)\,(B(\xi)_u) = \xi \quad \text{and} \quad \omega(u)\,(B(\xi)_u) = 0 \tag{3}$$

for every $u \in \mathscr{L}(M)$ and $\xi \in \mathbf{R}^n$.

PROOF: The last formula in (3) is equivalent to the assertion that $B(\xi)_u \in H_u$, and the first formula in (3) is obtained from (1) by applying the linear isomorphism u^{-1} on both sides.

It only remains to prove that $B(\xi)$ is a vector field on $\mathscr{L}(M)$. If $\{Y_1, \dots, Y_n\}$ is a local basis for H on an open subset V of $\mathscr{L}(M)$, we must show that the map $a : V \to \mathbf{R}^n$ defined by

$$B(\xi)_u = \sum_{i=1}^{n} a^i(u)\,Y_{iu}$$

for $u \in V$, is smooth. By applying $\theta(u)$ on both sides, it follows from (3) that

$$\xi = \sum_{i=1}^{n} a^i(u)\,h_i(u)$$

where

$$h_i(u) = \theta(u)(Y_{iu}) = u^{-1} \circ \alpha_u(Y_{iu})$$

for $i = 1, \dots, n$. Since u and α_u are linear isomorphisms, these vectors form a basis for $\mathbf{R}^n$. We therefore have a smooth map $h : V \to \mathrm{Gl}(n, \mathbf{R})$, where $h(u)$ is the matrix with $h_1(u), \dots, h_n(u)$ as column vectors for $u \in V$. Writing the vectors ξ and $a(u)$ as column vectors as well, the above equation can be written as

$$\xi = h(u)\,a(u)$$

so that

$$a(u) = h(u)^{-1}\,\xi$$

for $u \in V$. This shows that a is smooth and hence that $B(\xi)$ is a vector field on $\mathscr{L}(M)$. □

10.97 Proposition Given a linear connection H on a smooth manifold M^n, then

$$(R_F)_*\,B(\xi) = B(F^{-1}(\xi)) \tag{1}$$

for every $\xi \in \mathbf{R}^n$ and $F \in Gl(\mathbf{R}^n)$. Moreover, $B(\xi)$ is nowhere zero when $\xi \neq 0$, and if $\{\xi_1, \dots, \xi_n\}$ is a basis for $\mathbf{R}^n$, then $\{B(\xi_1), \dots, B(\xi_n)\}$ is a local basis for H on $\mathscr{L}(M)$.

PROOF: For each $u \in \mathscr{L}(M)$ it follows from Definition 10.32 (ii) that $(R_F)_*\,B(\xi)_u$ and $B(F^{-1}(\xi))_{u \circ F}$ are vectors in $H_{u \circ F}$ such that

$$\pi_*((R_F)_*\,B(\xi)_u) = \pi_*(B(\xi)_u) = u(\xi) = u \circ F(F^{-1}(\xi)) = \pi_*(B(F^{-1}(\xi))_{u \circ F})$$

since $\pi \circ R_F = \pi$. Hence

$$(R_F)_* \circ B(\xi) = B(F^{-1}(\xi)) \circ R_F$$

which completes the proof of (1).

The last part of the Proposition follows from formula (2) in Proposition 10.96 since $\alpha_u^{-1} \circ u$ is a linear isomorphism for every $u \in \mathscr{L}(M)$. □

10.98 Definition Let H be a linear connection on a smooth manifold M^n, and let θ be the dual form on $\mathscr{L}(M)$. Then the $\mathbf{R}^n$-valued 2-form $\Theta = D\theta$ is called the *torsion form* of H on $\mathscr{L}(M)$.

10.99 Remark From Proposition 10.43 and 10.93 it follows that the torsion form Θ is a tensorial 2-form of type $(id, \mathbf{R}^n)$.

10.100 Definition Let H be a linear connection on a smooth manifold M^n, and let $\pi' : TM \to M$ be the tangent bundle. Then we have a bundle-valued 2-form $T \in \Omega^2(M; TM)$, called the *torsion form* on M, defined by $\overline{T} = \Theta$, where Θ is the torsion form of H on $\mathscr{L}(M)$, i.e.,

$$T(p)(v_1, v_2) = u \circ \Theta(u)(w_1, w_2)$$

for $p \in M$ and $v_1, v_2 \in T_pM$, where $u \in \mathscr{L}(M)$ and $w_1, w_2 \in T_u\mathscr{L}(M)$ are chosen so that $\pi(u) = p$ and $\pi_*(w_i) = v_i$ for $i = 1,2$. By Proposition 10.93 we have that $T = D\varepsilon$, where $\varepsilon \in \Omega^1(M;TM)$ is the bundle-valued 1-form on M defined in Example 5.45.

10.101 Proposition Suppose that H is a linear connection on a smooth manifold M^n, and that $s = (X_1, \dots, X_n)$ is a frame field on an open subset V of M and $\mathscr{E} = \{e_1, \dots, e_n\}$ is the standard basis for $\mathbf{R}^n$. Let Θ and T be the torsion forms of H on $\mathscr{L}(M)$ and M, respectively, and $\widetilde{\Theta} = s^*(\Theta)$ be the pull-back of Θ by s. Then we have that

$$T(X_j, X_k) = \sum_{i=1}^{n} T_{jk}^{i} X_i ,$$

where the smooth functions $T_{jk}^{i} : V \to \mathbf{R}$, called the *components* of T with respect to the frame field $s = (X_1, \dots, X_n)$, are defined by

$$T_{jk}^{i} = \widetilde{\Theta}^{i}(X_j, X_k)$$

for $1 \leq i, j, k \leq n$, where $\widetilde{\Theta}^{i}$ are the components of $\widetilde{\Theta}$ with respect to the basis $\mathscr{E}$. If $\widetilde{\theta}^{i} = s^*\theta^{i}$ for $i = 1, \dots, n$ are the pull-backs of the components of the dual form θ by s, which form a dual local basis on V to the frame field $s = (X_1, \dots, X_n)$ as described in Remark 10.94, then we have that

$$\widetilde{\Theta}^{i} = \sum_{j<k} T_{jk}^{i} \widetilde{\theta}^{j} \wedge \widetilde{\theta}^{k}$$

for $1 \leq i \leq n$.

PROOF: By Proposition 10.73 and Example 10.29 (e) we have that

$$T(X_j, X_k) = s \cdot \widetilde{\Theta}(X_j, X_k) = \sum_{i=1}^{n} \widetilde{\Theta}^{i}(X_j, X_k)\, X_i$$

for $1 \leq j, k \leq n$. □

10.102 Proposition If T is the torsion form of a linear connection H on a smooth manifold M, then we have that

$$T(X, Y) = \nabla_X Y - \nabla_Y X - [X, Y]$$

for every vector field X and Y on M.

PROOF: Since $\theta = \overline{\varepsilon}$ and $X = \varepsilon(X)$, where $\varepsilon \in \Omega^1(M;TM)$ is the bundle-valued 1-form on M defined in Example 5.45, it follows from Proposition 10.74 that $\overline{X} = \theta(X^*)$ where X^* is the horizontal lift of X, and similarly $\overline{Y} = \theta(Y^*)$. Hence we have that

$$\nabla_X Y(p) = u \circ X_u^*(\theta(Y^*))$$

for $p \in M$, where $u \in P$ is chosen so that $\pi(u) = p$. By Remark 5.86 it therefore follows that

$$\begin{aligned} T(X,Y)(p) &= T(p)(X_p,Y_p) = u \circ d\theta(u)(X_u^*,Y_u^*) \\ &= u \circ X_u^*(\theta(Y^*)) - u \circ Y_u^*(\theta(X^*)) - u \circ \theta(u)([X^*,Y^*]_u) \\ &= \nabla_X Y(p) - \nabla_Y X(p) - [X,Y]_p , \end{aligned}$$

since

$$u \circ \theta(u)([X^*,Y^*]_u) = \pi_*([X^*,Y^*]_u) = [X,Y]_p$$

by Proposition 4.88. □

10.103 Definition Let $\pi : P \to M$ be a principal G-bundle with a connection H, and let $\pi' : P \times_{Ad} \mathfrak{g} \to M$ be the associated vector bundle obtained from the adjoint representation $Ad : G \to Aut(\mathfrak{g})$ of G on the Lie algebra $\mathfrak{g}$. Then we have a bundle-valued 2-form $R \in \Omega^2(M;P \times_{Ad} \mathfrak{g})$, called the *curvature form* on M, defined by $\overline{R} = \Omega$, where Ω is the curvature form of H on P. i.e.,

$$R(p)(v_1,v_2) = \mu_u \circ \Omega(u)(w_1,w_2)$$

for $p \in M$ and $v_1, v_2 \in T_pM$, where $u \in P$ and $w_1, w_2 \in T_uP$ are chosen so that $\pi(u) = p$ and $\pi_*(w_i) = v_i$ for $i = 1,2$.

10.104 Remark If $\pi' : \mathscr{L}(E) \to M$ is the frame bundle of an n-dimensional vector bundle $\pi : E \to M$ with a connection H, then the associated vector bundle obtained from the adjoint representation $Ad : Gl(\mathbf{R}^n) \to Aut(\mathfrak{gl}(\mathbf{R}^n))$ may be identified with the vector bundle $\pi'' : \Lambda^1(E;E) \to M$ as in Example 10.29 (g). We may therefore assume that the curvature form $R \in \Omega^2(M;\Lambda^1(E;E))$, and we have that

$$R(p)(v_1,v_2) = u \circ \Omega(u)(w_1,w_2) \circ u^{-1}$$

for $p \in M$ and $v_1, v_2 \in T_pM$, where $u \in \mathscr{L}_p(E)$ and $w_1, w_2 \in T_u\mathscr{L}(E)$ are chosen so that $\pi'_*(w_i) = v_i$ for $i = 1,2$. If X and Y are vector fields on M and Z is a section of π on M, then $R(X,Y)$ is a section of π'' on M, and hence $R(X,Y)Z$ is again a section of π on M.

10.105 Proposition Let $\pi' : \mathscr{L}(E) \to M$ be the frame bundle of an n-dimensional vector bundle $\pi : E \to M$, and consider the associated vector bundle obtained from the adjoint representation $Ad : Gl(\mathbf{R}^n) \to Aut(\mathfrak{gl}(\mathbf{R}^n))$ which may be identified with the vector bundle $\pi'' : \Lambda^1(E;E) \to M$ as in Example 10.29 (g). Suppose that ω is a bundle-valued k-form on M with values in $\Lambda^1(E;E)$, and that $\eta = \overline{\omega}$. Let $\mathscr{E} = \{e_1, \dots, e_n\}$ be the standard basis for $\mathbf{R}^n$ and $\mathscr{D} = \{F_j^i | 1 \le i,j \le n\}$ be the basis for $\mathfrak{gl}(\mathbf{R}^n)$ defined in Example 8.10 (b), where $F_j^i : \mathbf{R}^n \to \mathbf{R}^n$ is the linear map given by

$F_j^i(e_l) = \delta_{jl}\, e_i$ for $l = 1, \dots, n$. If $s = (s_1, \dots, s_n)$ is a section of $\pi' : \mathscr{L}(E) \to M$ on an open subset V of M, and $\widetilde{\eta} = s^*(\eta)$ is the pull-back of η by s, then we have that

$$\omega(p)(v_1, \dots, v_k)\, s_j(p) = \sum_{i=1}^{n} \widetilde{\eta}\,^i_j(p)(v_1, \dots, v_k)\, s_i(p)$$

for $p \in V$, $v_1, \dots, v_k \in T_pM$ and $j = 1, \dots, n$, where $\widetilde{\eta}\,^i_j$ are the components of $\widetilde{\eta}$ with respect to the basis $\mathscr{D}$.

PROOF : By Proposition 10.73 and Example 10.29 (g) we know that

$$\omega(p)(v_1, \dots, v_k) \circ s(p) = s(p) \circ \widetilde{\eta}(p)(v_1, \dots, v_k)$$

for $p \in V$ and $v_1, \dots, v_k \in T_pM$. Hence it follows that

$$\begin{aligned}
&\omega(p)(v_1, \dots, v_k)\, s_l(p) = \omega(p)(v_1, \dots, v_k) \circ s(p)(e_l) \\
&= s(p) \circ \widetilde{\eta}(p)(v_1, \dots, v_k)(e_l) = \sum_{ij} \widetilde{\eta}\,^i_j(p)(v_1, \dots, v_k)\, s(p) \circ F_j^i(e_l) \\
&= \sum_{i=1}^{n} \widetilde{\eta}\,^i_l(p)(v_1, \dots, v_k)\, s(p)(e_i) = \sum_{i=1}^{n} \widetilde{\eta}\,^i_l(p)(v_1, \dots, v_k)\, s_i(p)
\end{aligned}$$

for $p \in V$, $v_1, \dots, v_k \in T_pM$ and $l = 1, \dots, n$. □

10.106 Corollary Suppose that H is a linear connection on a smooth manifold M^n, and that $s = (X_1, \dots, X_n)$ is a frame field on an open subset V of M. Let $\mathscr{E} = \{e_1, \dots, e_n\}$ be the standard basis for $\mathbf{R}^n$ and $\mathscr{D} = \{F_j^i \mid 1 \le i, j \le n\}$ be the basis for $\mathfrak{gl}(\mathbf{R}^n)$ defined in Example 8.10 (b), where $F_j^i : \mathbf{R}^n \to \mathbf{R}^n$ is the linear map given by $F_j^i(e_k) = \delta_{jk}\, e_i$ for $k = 1, \dots, n$. If Ω and R are the curvature forms of H on $\mathscr{L}(M)$ and M, respectively, and $\widetilde{\Omega} = s^*(\Omega)$ is the pull-back of Ω by s, then we have that

$$R(X_k, X_l)\, X_j = \sum_{i=1}^{n} R^i_{jkl}\, X_i\,,$$

where the smooth functions $R^i_{jkl} : V \to \mathbf{R}$, called the *components* of R with respect to the frame field $s = (X_1, \dots, X_n)$, are defined by

$$R^i_{jkl} = \widetilde{\Omega}^i_j(X_k, X_l)$$

for $1 \le i, j, k, l \le n$, where $\widetilde{\Omega}^i_j$ are the components of $\widetilde{\Omega}$ with respect to the basis $\mathscr{D}$. If $\widetilde{\theta}\,^i = s^* \theta\,^i$ for $i = 1, \dots, n$ are the pull-backs of the components of the dual form θ by s, which form a dual local basis on V to the frame field $s = (X_1, \dots, X_n)$ as described in Remark 10.94, then we have that

$$\widetilde{\Omega}^i_j = \sum_{k<l} R^i_{jkl}\, \widetilde{\theta}\,^k \wedge \widetilde{\theta}\,^l$$

for $1 \le i, j \le n$.

10.107 Proposition If R is the curvature form defined in Remark 10.104, then we have that

$$R(X,Y)Z = \nabla_X \nabla_Y Z - \nabla_Y \nabla_X Z - \nabla_{[X,Y]} Z$$

for every vector field X and Y on M, and every section Z of π on M.

PROOF : Choose $u \in \mathscr{L}_p(E)$ for a $p \in M$, and let X^* and Y^* be the horizontal lifts of X and Y, respectively. By Proposition 10.47 we have that

$$\Omega(u)(X^*_u, Y^*_u) = -\,\omega(u)([X^*, Y^*]_u) = -\,\omega(u)(v([X^*, Y^*]_u)),$$

and it follows from Proposition 10.26 that the vertical component

$$v([X^*, Y^*]_u) = \sigma(F)_u$$

for a unique $F \in \mathfrak{gl}(\mathbf{R}^n)$. Hence we have that

$$\Omega(u)(X^*_u, Y^*_u) = -F$$

so that

$$R(X,Y)Z(p) = u \circ \Omega(u)(X^*_u, Y^*_u) \circ \overline{Z}(u) = -\,u \circ F \circ \overline{Z}(u). \tag{1}$$

On the other hand we have that

$$\overline{\nabla_Y Z} = \overline{DZ(Y)} = \overline{DZ}(Y^*) = d\overline{Z}(Y^*) = Y^*(\overline{Z})$$

so that

$$\begin{aligned}
&\nabla_X \nabla_Y Z(p) - \nabla_Y \nabla_X Z(p) - \nabla_{[X,Y]} Z(p) \\
&= u \circ X^*_u(Y^*(\overline{Z})) - u \circ Y^*_u(X^*(\overline{Z})) - u \circ h([X^*, Y^*]_u)(\overline{Z}) \qquad (2) \\
&= u \circ v([X^*, Y^*]_u)(\overline{Z}) = u \circ \sigma(F)_u(\overline{Z}).
\end{aligned}$$

By Proposition 9.39 the integral curve $\gamma_u : \mathbf{R} \to \mathscr{L}(E)$ for the fundamental vector field $\sigma(F)$ with initial condition u is given by

$$\gamma_u(t) = u \circ \exp(tF)$$

for $t \in \mathbf{R}$. Using that $\overline{Z}$ is tensorial, we define a curve $c : \mathbf{R} \to \mathbf{R}^n$ by

$$c(t) = \overline{Z} \circ \gamma_u(t) = \overline{Z}(u \circ \exp(tF)) = \exp(-tF) \circ \overline{Z}(u) = \Lambda_u \circ \phi_{-F}(t)$$

for $t \in \mathbf{R}$, where $\phi_{-F} : \mathbf{R} \to Gl(\mathbf{R}^n)$ is the one-parameter subgroup defined in Proposition 8.29, and $\Lambda_u : \mathfrak{gl}(\mathbf{R}^n) \to \mathbf{R}^n$ is the linear map given by

$$\Lambda_u(G) = G \circ \overline{Z}(u)$$

for $G \in \mathfrak{gl}(\mathbf{R}^n)$. By 5.49 and 2.70 it follows that

$$\sigma(F)_u(\overline{Z}) = c'(0) = \Lambda_u \circ \phi'_{-F}(0) = -F \circ \overline{Z}(u) \tag{3}$$

if we canonically identify $T_{c(0)}(T_u \mathscr{L}(E))$ with $T_u \mathscr{L}(E)$ as in Lemma 2.84. The proposition now follows from formula (1), (2) and (3). □

10.108 Let R be the curvature form of a linear connection H on a smooth manifold M. By 5.25, Proposition 5.31 and Remark 4.54 we have an equivalence $\phi : \mathscr{T}^2(TM;\Lambda^1(TM;TM)) \to T^3_1(M)$ over M defined by

$$\phi(F)(v_1,v_2,v_3,\lambda) = \lambda \circ F(v_2,v_3)(v_1)$$

for $p \in M$, $F \in \mathscr{T}^2(T_pM;\Lambda^1(T_pM;T_pM))$, $v_1,v_2,v_3 \in T_pM$ and $\lambda \in T_p^*M$. The contraction $\mathrm{Ric} = C^2_1(\phi \circ R) \in \mathscr{T}^2(M)$ is called the *Ricci curvature tensor* of H. We see that

$$\mathrm{Ric}(p)(v_1,v_2) = \mathrm{tr}(G)$$

for $p \in M$ and $v_1,v_2 \in T_pM$, where $G : T_pM \to T_pM$ is the linear map given by

$$G(v) = R(p)(v,v_2)(v_1)$$

for $v \in T_pM$. If $s = (X_1, \dots ,X_n)$ is a frame field on an open subset V of M, it follows from Corollary 10.106 that

$$\mathrm{Ric}(X_j,X_l) = \sum_{k=1}^{n} R^k_{jkl}\,.$$

10.109 Theorem Let $\pi : P \to M$ be a principal G-bundle with a connection H, and let $\pi' : E \to M$ be the associated vector bundle obtained from a representation $\psi : G \to Aut\,(W)$ of G on a finite dimensional vector space W. Let $\nu : P \to \Lambda^1(W\,;E)$ be the smooth map defined in Example 10.29 (c), which is given by

$$\nu(u)(v) = \mu_u(v) = [u,v]$$

for $u \in P$ and $v \in W$. If s is a section of π on an open subset V of M, and $\widetilde{\omega} = s^*(\omega)$ is the pull-back by s of the connection form ω of H, then we have that

$$\nabla(\nu \circ s)(p)(v) = (\nu \circ s)(p) \circ (\psi_* \cdot \widetilde{\omega})(p)(v)$$

for $p \in V$ and $v \in T_pM$.

PROOF: If $\rho : V \times G \to \pi^{-1}(V)$ is the diffeomorphism defined by

$$\rho(p,g) = s(p)\,g$$

for $p \in V$ and $g \in G$, then

$$\overline{\nu \circ s} \circ \rho(p,g) = \mu^{-1}_{s(p)g} \circ \nu(s(p)) = \psi(g^{-1}) \circ \mu^{-1}_{s(p)} \circ \mu_{s(p)} = \psi(g^{-1})\,.$$

Now let $i_e : V \to V \times G$ and $i_p : G \to V \times G$ be the embeddings defined by $i_e(p') = (p',e)$ and $i_p(g') = (p,g')$ for $p' \in V$ and $g' \in G$, and let $\rho_p = \rho \circ i_p$. For each $p \in V$ and $v \in T_pM$ there is by Proposition 10.26 a unique $X \in \mathfrak{g}$ with $s_*(v) + \rho_{p*}(X) \in H_{s(p)}$ so that

$$0 = \omega(s(p))(s_*(v) + \rho_{p*}(X)) = \widetilde{\omega}(p)(v) + X\,.$$

Hence it follows from 5.51, Proposition 5.63 and Corollary 8.33 that

$$d\,\overline{v\circ s}\,(s(p))\,(s_*(v)+\rho_{p*}(X)) = \rho^* d\,\overline{v\circ s}\,(p,e)\,(i_{e*}(v)+i_{p*}(X))$$
$$= d\,(\overline{v\circ s}\circ\rho)\,(p,e)\,(i_{e*}(v)+i_{p*}(X)) = -\,\psi_*(X) = (\psi_*\cdot\widetilde{\omega})\,(p)(v)$$

so that

$$\nabla(v\circ s)(p)(v) = \mu_{s(p)}\circ d\overline{v\circ s}(s(p))(s_*(v)+\rho_{p*}(X))$$
$$= (v\circ s)(p)\circ(\psi_*\cdot\widetilde{\omega})(p)(v).$$

□

10.110 Proposition Suppose that $\pi' : \mathscr{L}(E)\to M$ is the frame bundle of an n-dimensional vector bundle $\pi : E\to M$ with a connection H, and that $s=(s_1,\dots,s_n)$ is a section of π' on an open subset V of M. Let $\mathscr{E}=\{e_1,\dots,e_n\}$ be the standard basis for $\mathbf{R}^n$ and $\mathscr{D}=\{F_j^i | 1\le i,j\le n\}$ be the basis for $\mathfrak{gl}(\mathbf{R}^n)$ defined in Example 8.10 (b), where $F_j^i : \mathbf{R}^n\to\mathbf{R}^n$ is the linear map given by $F_j^i(e_k)=\delta_{jk}\,e_i$ for $k=1,\dots,n$. If ω is the connection form of H and $\widetilde{\omega}=s^*(\omega)$ is the pull-back of ω by s, then we have that

$$\nabla s_j(p)(v) = \sum_{i=1}^{n} \widetilde{\omega}_j^i(p)(v)\, s_i(p)$$

for $p\in V$, $v\in T_pM$ and $j=1,\dots,n$, where $\widetilde{\omega}_j^i$ are the components of $\widetilde{\omega}$ with respect to the basis $\mathscr{D}$.

PROOF: For each k we have that $\bar{s}_k=\Lambda_k\cdot\bar{s}$, where $\Lambda_k : \mathfrak{gl}(\mathbf{R}^n)\to\mathbf{R}^n$ is the linear map given by $\Lambda_k(F)=F(e_k)$ for $F\in\mathfrak{gl}(\mathbf{R}^n)$. Hence it follows by Proposition 5.62 and Theorem 10.109 that

$$\nabla s_k(p)(v) = \nabla s(p)(v)(e_k) = s(p)\circ\widetilde{\omega}(p)(v)(e_k) = \sum_{ij}\widetilde{\omega}_j^i(p)(v)s(p)\circ F_j^i(e_k)$$
$$= \sum_{i=1}^{n}\widetilde{\omega}_k^i(p)(v)s(p)(e_i) = \sum_{i=1}^{n}\widetilde{\omega}_k^i(p)(v)s_i(p)$$

for $p\in V$, $v\in T_pM$ and $k=1,\dots,n$. □

10.111 Corollary Suppose that H is a linear connection on a smooth manifold M^n, and that $s=(X_1,\dots,X_n)$ is a frame field on an open subset V of M. Let $\mathscr{E}=\{e_1,\dots,e_n\}$ be the standard basis for $\mathbf{R}^n$ and $\mathscr{D}=\{F_j^i | 1\le i,j\le n\}$ be the basis for $\mathfrak{gl}(\mathbf{R}^n)$ defined in Example 8.10 (b), where $F_j^i : \mathbf{R}^n\to\mathbf{R}^n$ is the linear map given by $F_j^i(e_k)=\delta_{jk}\,e_i$ for $k=1,\dots,n$. If ω is the connection form of H and $\widetilde{\omega}=s^*(\omega)$ is the pull-back of ω by s, then we have that

$$\nabla_{X_k}X_j = \sum_{i=1}^{n}\Gamma_{kj}^i\, X_i\,,$$

where the smooth functions $\Gamma^i_{kj} : V \to \mathbf{R}$, called the *connection components* or the *connection coefficients* of H with respect to the frame field $s = (X_1, \dots, X_n)$, are defined by

$$\Gamma^i_{kj} = \widetilde{\omega}^i_j(X_k)$$

for $1 \le i, j, k \le n$, where $\widetilde{\omega}^i_j$ are the components of $\widetilde{\omega}$ with respect to the basis $\mathscr{D}$. If $\widetilde{\theta}^{\,i} = s^*\theta^{\,i}$ for $i = 1, \dots, n$ are the pull-backs of the components of the dual form θ by s, which form a dual local basis on V to the frame field $s = (X_1, \dots, X_n)$ as described in Remark 10.94, then we have that

$$\widetilde{\omega}^i_j = \sum_{k=1}^{n} \Gamma^i_{kj}\, \widetilde{\theta}^{\,k}$$

for $1 \le i, j \le n$.

10.112 Remark Let M^n be a smooth manifold with a linear connection H. By the *connection components* or the *connection coefficients* Γ^i_{kj} of H with respect to a local chart (x, U) on M we mean the connection coefficients with respect to the coordinate frame field

$$s = \left(\frac{\partial}{\partial x^1}, \dots, \frac{\partial}{\partial x^n} \right).$$

In the same way we define the components T^i_{jk} and R^i_{jkl} of the torsion and curvature forms T and R of H with respect to the local chart (x, U).

10.113 Proposition Let H be a linear connection on a smooth manifold M^n, and let Γ^i_{kj} be the connection coefficients of H with respect to a frame field $s = (X_1, \dots, X_n)$ on an open subset V of M. If X and Y are vector fields on V with

$$X = \sum_{i=1}^{n} a^i X_i \quad \text{and} \quad Y = \sum_{i=1}^{n} b^i X_i\,,$$

then we have that

$$\nabla_X Y = \sum_{i=1}^{n} \left(\sum_{j=1}^{n} a^j\, b^i{}_{;j} \right) X_i$$

where

$$b^i{}_{;j} = X_j(b^i) + \sum_{k=1}^{n} b^k\, \Gamma^i_{jk}$$

for $1 \le i, j \le n$.

PROOF: By Proposition 10.80 and Corollary 10.111 we have that

$$\nabla_X Y = \sum_{ij} a^j\, \nabla_{X_j}(b^i X_i) = \sum_{ij} a^j\, \{X_j(b^i)\, X_i + b^i\, \nabla_{X_j} X_i\}$$

$$= \sum_{ij} a^j\, X_j(b^i)\, X_i + \sum_{kj} a^j\, b^k\, \nabla_{X_j} X_k = \sum_{ij} a^j \left(X_j(b^i) + \sum_k b^k\, \Gamma^i_{jk} \right) X_i\,.$$

□

10.114 Corollary Let H be a linear connection on a smooth manifold M^n, and let Γ^i_{kj} be the connection coefficients of H with respect to a local chart (x,U) on M. If X and Y are vector fields on U with

$$X = \sum_{i=1}^{n} a^i \frac{\partial}{\partial x^i} \quad \text{and} \quad Y = \sum_{i=1}^{n} b^i \frac{\partial}{\partial x^i},$$

then we have that

$$\nabla_X Y = \sum_{i=1}^{n} \left(\sum_{j=1}^{n} a^j\, b^i{}_{;j} \right) \frac{\partial}{\partial x^i}$$

where

$$b^i{}_{;j} = \frac{\partial b^i}{\partial x^j} + \sum_{k=1}^{n} b^k\, \Gamma^i_{jk}$$

for $1 \leq i, j \leq n$.

10.115 Proposition Suppose that $\pi' : \mathscr{L}(E) \to M$ is the frame bundle of an n-dimensional vector bundle $\pi : E \to M$ with a connection H, and that $s = (s_1, \ldots, s_n)$ is a section of π' on an open subset V of M with a dual local basis $(\alpha^1, \ldots, \alpha^n)$ in the dual bundle $\pi^* : E^* \to M$. Let $\mathscr{E} = \{e_1, \ldots, e_n\}$ be the standard basis for $\mathbf{R}^n$ with dual basis $\mathscr{E}^* = \{e^1, \ldots, e^n\}$, and let $\mathscr{D} = \{F_j^{\,i} | 1 \leq i, j \leq n\}$ be the basis for $\mathfrak{gl}(\mathbf{R}^n)$ defined in Example 8.10 (b), where $F_j^{\,i} : \mathbf{R}^n \to \mathbf{R}^n$ is the linear map given by $F_j^{\,i}(e_k) = \delta_{jk}\, e_i$ for $k = 1, \ldots, n$. If ω is the connection form of H and $\widetilde{\omega} = s^*(\omega)$ is the pull-back of ω by s, then we have that

$$\nabla \alpha^i(p)(v) = -\sum_{j=1}^{n} \widetilde{\omega}_j^i(p)(v)\, \alpha^j(p)$$

for $p \in V$, $v \in T_pM$ and $i = 1, \ldots, n$, where $\widetilde{\omega}_j^i$ are the components of $\widetilde{\omega}$ with respect to the basis $\mathscr{D}$.

PROOF: From Example 10.29 (f) we know that the dual bundle $\pi^* : E^* \to M$ is associated to the frame bundle $\mathscr{L}(E)$ by means of the representation $\psi : Gl(\mathbf{R}^n) \to Aut((\mathbf{R}^n)^*)$ given by

$$\psi(F) = (F^{-1})^*$$

for $F \in Gl(\mathbf{R}^n)$, i.e.,

$$\psi(F)(\lambda) = \lambda \circ F^{-1}$$

for $\lambda \in (\mathbf{R}^n)^*$. By Corollary 8.33 it follows that $\psi_* : \mathfrak{gl}(\mathbf{R}^n) \to End((\mathbf{R}^n)^*)$ is given by

$$\psi_*(F) = -F^*$$

for $F \in \mathfrak{gl}(\mathbf{R}^n)$, so that

$$\psi_*(F_j^{\,i})(e^k) = -\, e^k \circ F_j^{\,i} = -\sum_{l=1}^{n} e^k \circ F_j^{\,i}(e_l)\, e^l = -\sum_{l=1}^{n} \delta_{jl}\, e^k(e_i)\, e^l = -\,\delta_{ik}\, e^j$$

for $k = 1, \ldots, n$. We also have a smooth map $\nu : \mathscr{L}(E) \to \Lambda^1((\mathbf{R}^n)^* ; E^*)$ given by

$$\nu(u) = (u^{-1})^*$$

for $u \in \mathscr{L}(E)$.

For each k we now have that $\overline{\alpha^k} = \Lambda_k \cdot \overline{\nu \circ s}$, where $\Lambda_k : \mathfrak{gl}((\mathbf{R}^n)^*) \to (\mathbf{R}^n)^*$ is the linear map given by $\Lambda_k(F) = F(e^k)$ for $F \in \mathfrak{gl}((\mathbf{R}^n)^*)$. Hence it follows by Proposition 5.62 and Theorem 10.109 that

$$\nabla\alpha^k(p)(v) = \nabla(\nu \circ s)(p)(v)(e^k) = (\nu \circ s)(p) \circ (\psi_* \cdot \widetilde{\omega})(p)(v)(e^k)$$

$$= \sum_{ij} \widetilde{\omega}_j^i(p)(v)\, (\nu \circ s)(p) \circ \psi_*(F_j^i)(e^k) = -\sum_{j=1}^n \widetilde{\omega}_j^k(p)(v)\, (\nu \circ s)(p)(e^j)$$

$$= -\sum_{i=1}^n \widetilde{\omega}_j^k(p)(v)\, \alpha^j(p)$$

for $p \in V$, $v \in T_pM$ and $k = 1, \ldots, n$. □

10.116 Corollary Suppose that H is a linear connection on a smooth manifold M^n, and that $s = (X_1, \ldots, X_n)$ is a frame field on an open subset V of M with a dual local basis $(\alpha^1, \ldots, \alpha^n)$ in the cotangent bundle $\pi^* : T^*M \to M$. Let $\mathscr{E} = \{e_1, \ldots, e_n\}$ be the standard basis for $\mathbf{R}^n$ and $\mathscr{D} = \{F_j^i | 1 \le i, j \le n\}$ be the basis for $\mathfrak{gl}(\mathbf{R}^n)$ defined in Example 8.10 (b), where $F_j^i : \mathbf{R}^n \to \mathbf{R}^n$ is the linear map given by $F_j^i(e_k) = \delta_{jk}\, e_i$ for $k = 1, \ldots, n$. If ω is the connection form of H and $\widetilde{\omega} = s^*(\omega)$ is the pull-back of ω by s, then we have that

$$\nabla_{X_k}\alpha^i = -\sum_{j=1}^n \Gamma_{kj}^i\, \alpha^j ,$$

where Γ_{kj}^i are the connection coefficients of H with respect to the frame field $s = (X_1, \ldots, X_n)$ given by

$$\Gamma_{kj}^i = \widetilde{\omega}_j^i(X_k)$$

for $1 \le i, j, k \le n$, where $\widetilde{\omega}_j^i$ are the components of $\widetilde{\omega}$ with respect to the basis $\mathscr{D}$.

10.117 Proposition Let H be a linear connection on a smooth manifold M^n, and let Γ_{kj}^i be the connection coefficients of H with respect to a frame field $s = (X_1, \ldots, X_n)$ on an open subset V of M. Let $(\alpha^1, \ldots, \alpha^n)$ be the dual local basis in the cotangent bundle $\pi^* : T^*M \to M$. If X is a vector field and α a covector field on V with

$$X = \sum_{i=1}^n a^i X_i \quad \text{and} \quad \alpha = \sum_{i=1}^n b_i\, \alpha^i ,$$

then we have that

$$\nabla_X \alpha = \sum_{i=1}^{n} \left(\sum_{j=1}^{n} a^j \, b_{i;j} \right) \alpha^i$$

where

$$b_{i;j} = X_j(b_i) - \sum_{k=1}^{n} b_k \Gamma^k_{ji}$$

for $1 \leq i, j \leq n$.

PROOF : By Proposition 10.80 and Corollary 10.116 we have that

$$\nabla_X \alpha = \sum_{ij} a^j \, \nabla_{X_j}(b_i \alpha^i) = \sum_{ij} a^j \, \{X_j(b_i) \, \alpha^i + b_i \nabla_{X_j} \alpha^i\}$$

$$= \sum_{ij} a^j \, X_j(b_i) \, \alpha^i + \sum_{kj} a^j \, b_k \nabla_{X_j} \alpha^k = \sum_{ij} a^j \left(X_j(b_i) - \sum_k b_k \Gamma^k_{ji} \right) \alpha^i \, .$$

□

10.118 Corollary Let H be a linear connection on a smooth manifold M^n, and let Γ^i_{kj} be the connection coefficients of H with respect to a local chart (x, U) on M. If X is a vector field and α a covector field on U with

$$X = \sum_{i=1}^{n} a^i \frac{\partial}{\partial x^i} \quad \text{and} \quad \alpha = \sum_{i=1}^{n} b_i dx^i \, ,$$

then we have that

$$\nabla_X \alpha = \sum_{i=1}^{n} \left(\sum_{j=1}^{n} a^j \, b_{i;j} \right) dx^i$$

where

$$b_{i;j} = \frac{\partial b_i}{\partial x^j} - \sum_{k=1}^{n} b_k \Gamma^k_{ji}$$

for $1 \leq i, j \leq n$.

10.119 Proposition Let H be a linear connection on a smooth manifold M^n, and let Γ^i_{kj} be the connection coefficients of H with respect to a frame field $s = (X_1, \dots, X_n)$ on an open subset V of M. Let $(\alpha^1, \dots, \alpha^n)$ be the dual local basis in the cotangent bundle $\pi^* : T^*M \to M$. If X is a vector field and T a tensor field of type $\binom{k}{l}$ on V with

$$X = \sum_{i=1}^{n} a^i X_i$$

and

$$T = \sum_{\substack{i_1, \dots, i_k \\ j_1, \dots, j_l}} B^{j_1, \dots, j_l}_{i_1, \dots, i_k} \, \alpha^{i_1} \otimes \cdots \otimes \alpha^{i_k} \otimes X_{j_1} \otimes \cdots \otimes X_{j_l} \, ,$$

then we have that

$$\nabla_X T = \sum_{\substack{i_1,\dots,i_k \\ j_1,\dots,j_l}} \left(\sum_{h=1}^{n} a^h B^{j_1,\dots,j_l}_{i_1,\dots,i_k;h} \right) \alpha^{i_1} \otimes \cdots \otimes \alpha^{i_k} \otimes X_{j_1} \otimes \cdots \otimes X_{j_l}$$

where

$$B^{j_1,\dots,j_l}_{i_1,\dots,i_k;h} = X_h(B^{j_1,\dots,j_l}_{i_1,\dots,i_k}) + \sum_{m=1}^{l} \sum_{j=1}^{n} B^{j_1,\dots,j_{m-1},j,j_{m+1},\dots,j_l}_{i_1,\dots,i_k} \Gamma^{j_m}_{hj}$$

$$- \sum_{m=1}^{k} \sum_{i=1}^{n} B^{j_1,\dots,j_l}_{i_1,\dots,i_{m-1},i,i_{m+1},\dots,i_k} \Gamma^{i}_{h\,i_m} .$$

PROOF: Using that

$$B^{j_1,\dots,j_l}_{i_1,\dots,i_k;h} = \nabla_{X_h} T \left(X_{i_1}, \dots, X_{i_k}, \alpha^{j_1}, \dots, \alpha^{j_l} \right) ,$$

the result follows from Proposition 10.83 (5) and Corollaries 10.111 and 10.116. □

10.120 Corollary Let H be a linear connection on a smooth manifold M^n, and let Γ^i_{kj} be the connection coefficients of H with respect to a local chart (x,U) on M. If X is a vector field and T a tensor field of type $\binom{k}{l}$ on U with

$$X = \sum_{i=1}^{n} a^i \frac{\partial}{\partial x^i}$$

and

$$T = \sum_{\substack{i_1,\dots,i_k \\ j_1,\dots,j_l}} B^{j_1,\dots,j_l}_{i_1,\dots,i_k} dx^{i_1} \otimes \cdots \otimes dx^{i_k} \otimes \frac{\partial}{\partial x^{j_1}} \otimes \cdots \otimes \frac{\partial}{\partial x^{j_l}} ,$$

then we have that

$$\nabla_X T = \sum_{\substack{i_1,\dots,i_k \\ j_1,\dots,j_l}} \left(\sum_{h=1}^{n} a^h B^{j_1,\dots,j_l}_{i_1,\dots,i_k;h} \right) dx^{i_1} \otimes \cdots \otimes dx^{i_k} \otimes \frac{\partial}{\partial x^{j_1}} \otimes \cdots \otimes \frac{\partial}{\partial x^{j_l}}$$

where

$$B^{j_1,\dots,j_l}_{i_1,\dots,i_k;h} = \frac{\partial B^{j_1,\dots,j_l}_{i_1,\dots,i_k}}{\partial x^h} + \sum_{m=1}^{l} \sum_{j=1}^{n} B^{j_1,\dots,j_{m-1},j,j_{m+1},\dots,j_l}_{i_1,\dots,i_k} \Gamma^{j_m}_{hj}$$

$$- \sum_{m=1}^{k} \sum_{i=1}^{n} B^{j_1,\dots,j_l}_{i_1,\dots,i_{m-1},i,i_{m+1},\dots,i_k} \Gamma^{i}_{h\,i_m} .$$

10.121 Proposition Let $\pi : P \to M$ be a principal G-bundle with a connection H, and let $\pi' : E \to M$ be the associated vector bundle obtained from a representation $\psi : G \to Aut\,(W)$ of G on a finite dimensional vector space W. Let Γ^i_{jk} be the components of H with respect to a frame field $s = (X_1, \ldots, X_n)$ on an open subset V of M. If $F : I \to TM$ is a lifting of a smooth curve $\gamma : I \to M$ lying in V, with

$$F = \sum_{i=1}^{n} a^i \{X_i \circ \gamma\} \quad \text{and} \quad \gamma' = \sum_{i=1}^{n} b^i \{X_i \circ \gamma\}$$

for smooth functions $a^i : I \to \mathbf{R}$ and $b^i : I \to \mathbf{R}$, and we use the standard local chart (r, I) on I where $r : I \to \mathbf{R}$ is the inclusion map, then we have that

$$\frac{DF}{dr} = \sum_{i=1}^{n} \left(a^{i\,\prime} + \sum_{j,k=1}^{n} \{\Gamma^i_{jk} \circ \gamma\}\, b^j\, a^k \right) \{X_i \circ \gamma\}\,.$$

PROOF : Let $F_i : I \to TM$ be the lifting of γ given by $F_i = X_i \circ \gamma$ for $i = 1, \ldots, n$. Then it follows from Remark 2.83 and Propositions 10.87 and 10.88 that

$$\frac{DF_k}{dr}(t) = \nabla_{\gamma'(t)} X_k = \sum_{j=1}^{n} b^j(t)\, \nabla_{F_j(t)} X_k = \sum_{i=1}^{n} \left(\sum_{j=1}^{n} b^j(t)\, \Gamma^i_{jk}(\gamma(t)) \right) X_i(\gamma(t))$$

for $t \in I$ and $k = 1, \ldots, n$, which implies that

$$\frac{DF}{dr} = \sum_{k=1}^{n} \{a^{k\,\prime}\, F_k + a^k\, \frac{DF_k}{dr}\} = \sum_{i=1}^{n} \left(a^{i\,\prime} + \sum_{j,k=1}^{n} \{\Gamma^i_{jk} \circ \gamma\}\, b^j\, a^k \right) \{X_i \circ \gamma\}\,.$$

□

10.122 Proposition Let H be a linear connection on a smooth manifold M^n, and let Γ^i_{kj} be the connection coefficients of H with respect to a local chart (x, U) on M. Then the components T^i_{jk} and R^i_{jkl} of the torsion and curvature forms of H with respect to (x, U) are given by

(1) $T^i_{jk} = \Gamma^i_{jk} - \Gamma^i_{kj}$

(2) $R^i_{jkl} = \dfrac{\partial \Gamma^i_{lj}}{\partial x^k} - \dfrac{\partial \Gamma^i_{kj}}{\partial x^l} + \sum_{h=1}^{n} (\Gamma^h_{lj}\, \Gamma^i_{kh} - \Gamma^h_{kj}\, \Gamma^i_{lh})$

PROOF: (1) By Proposition 10.102 we have that

$$T\left(\frac{\partial}{\partial x^j}, \frac{\partial}{\partial x^k}\right) = \nabla_{\frac{\partial}{\partial x^j}} \frac{\partial}{\partial x^k} - \nabla_{\frac{\partial}{\partial x^k}} \frac{\partial}{\partial x^j} = \sum_{i=1}^{n} (\Gamma^i_{jk} - \Gamma^i_{kj})\, \frac{\partial}{\partial x^i}$$

(2) By Propositions 10.107 and 10.80 we have that

$$R\left(\frac{\partial}{\partial x^k},\frac{\partial}{\partial x^l}\right)\frac{\partial}{\partial x^j}=\nabla_{\frac{\partial}{\partial x^k}}\left(\nabla_{\frac{\partial}{\partial x^l}}\frac{\partial}{\partial x^j}\right)-\nabla_{\frac{\partial}{\partial x^l}}\left(\nabla_{\frac{\partial}{\partial x^k}}\frac{\partial}{\partial x^j}\right)$$

$$=\nabla_{\frac{\partial}{\partial x^k}}\left(\sum_{h=1}^{n}\Gamma_{lj}^h\frac{\partial}{\partial x^h}\right)-\nabla_{\frac{\partial}{\partial x^l}}\left(\sum_{h=1}^{n}\Gamma_{kj}^h\frac{\partial}{\partial x^h}\right)$$

$$=\sum_{i=1}^{n}\left\{\left(\frac{\partial\Gamma_{lj}^i}{\partial x^k}+\sum_{h=1}^{n}\Gamma_{lj}^h\,\Gamma_{kh}^i\right)-\left(\frac{\partial\Gamma_{kj}^i}{\partial x^l}+\sum_{h=1}^{n}\Gamma_{kj}^h\,\Gamma_{lh}^i\right)\right\}\frac{\partial}{\partial x^i}$$

□

10.123 Proposition Suppose that H is a linear connection on a smooth manifold M^n, and that $s_k : V_k \to \mathscr{L}(M)$ for $k=1,2$ are two sections in the frame bundle $\pi : \mathscr{L}(M) \to M$ with associated local trivializations $(t_k, \pi^{-1}(V_k))$ which are $Gl(\mathbf{R}^n)$-related with transition map $\phi : V_1 \cap V_2 \to Gl(\mathbf{R}^n)$. If $\Gamma_{kj}^i : V_1 \to \mathbf{R}$ and $\Gamma'^{\,i}_{kj} : V_2 \to \mathbf{R}$ are the connection coefficients of H with respect to the frame fields $s_1 = (X_1, \ldots, X_n)$ and $s_2 = (X_1', \ldots, X_n')$, respectively, then

$$\Gamma_{kj}^i(p)=\sum_{rst}\Gamma'^{\,r}_{ts}(p)\;\phi(p)^t_k\;\phi(p)^s_j\;\{\phi(p)^{-1}\}^i_r+\sum_r X_k(\phi^{\,r}_j)(p)\;\{\phi(p)^{-1}\}^i_r \quad (1)$$

for $p \in V_1 \cap V_2$ and $1 \le i,j,k \le n$, where the components in $\mathfrak{gl}(\mathbf{R}^n)$ are with respect to the basis $\mathscr{D} = \{F_j^i \mid 1 \le i,j \le n\}$ defined in Example 8.10 (b), where $F_j^i : \mathbf{R}^n \to \mathbf{R}^n$ is the linear map given by $F_j^i(e_k) = \delta_{jk}\, e_i$ for $k = 1, \ldots, n$.

Conversely, assume that we have a family of frame fields defined on sets which form an open cover of M, and for each frame field a family $\{\Gamma_{kj}^i,\ 1 \le i,j,k \le n\}$ of smooth functions satisfying the transformation rule given in formula (1). Then there is a unique linear connection H on M having these Γ_{kj}^i as connection coefficients.

PROOF: If ω is the connection form of H and $\widetilde{\omega}_k = s_k^*(\omega)$ for $k = 1,2$, then it follows from 5.51, Lemma 2.84 and Example 9.8 (c) that formula (2) in Proposition 10.56 is equivalent to

$$\widetilde{\omega}_1(p)(v)=\phi(p)^{-1}\circ\widetilde{\omega}_2(p)(v)\circ\phi(p)+\phi(p)^{-1}\circ d\phi(p)(v) \quad (2)$$

for $p \in V_1 \cap V_2$ and $v \in T_pM$. So if we let $v = X_k(p)$ and use that $F_j^i \circ F_s^r = \delta_{jr}\, F_s^i$, we obtain the equivalent formula

$$(\widetilde{\omega}_1)^i_j(X_k)(p)=\sum_{rs}\{\phi(p)^{-1}\}^i_r\;(\widetilde{\omega}_2)^r_s(X_k)(p)\;\phi(p)^s_j$$
$$+\sum_r\{\phi(p)^{-1}\}^i_r\;X_k(\phi^{\,r}_j)(p)$$

for $p \in V_1 \cap V_2$ and $1 \le i,j,k \le n$. Since we have that

$$X_k(p)=\sum_t\phi(p)^t_k\;X_t'(p)$$

for $p \in V_1 \cap V_2$ and $1 \leq k \leq n$ by Corollary 10.54, we therefore see that formula (1) in our proposition is equivalent to formula (2) in Proposition 10.56. □

10.124 Corollary Let H be a linear connection on a smooth manifold M^n. If $\Gamma^i_{kj} : U \to \mathbf{R}$ and $\Gamma'^i_{kj} : U' \to \mathbf{R}$ are the connection coefficients of H with respect to the local charts (x,U) and (x',U') on M, then

$$\Gamma^i_{kj} = \sum_{rst} \Gamma'^r_{ts} \frac{\partial x'^t}{\partial x^k} \frac{\partial x'^s}{\partial x^j} \frac{\partial x^i}{\partial x'^r} + \sum_r \frac{\partial^2 x'^r}{\partial x^k \partial x^j} \frac{\partial x^i}{\partial x'^r} \tag{1}$$

on $U \cap U'$ for $1 \leq i,j,k \leq n$.

Conversely, assume that we for each local chart in an atlas on M have a family $\{\Gamma^i_{kj},\ 1 \leq i,j,k \leq n\}$ of smooth functions satisfying the transformation rule given in formula (1). Then there is a unique linear connection H on M having these Γ^i_{kj} as connection coefficients.

PROOF: To obtain formula (1), apply Proposition 10.123 to the coordinate frame fields

$$s_1 = \left(\frac{\partial}{\partial x^1}, \dots, \frac{\partial}{\partial x^n} \right) \quad \text{and} \quad s_2 = \left(\frac{\partial}{\partial x'^1}, \dots, \frac{\partial}{\partial x'^n} \right)$$

where $\phi(p) = t_{x',p} \circ t_{x,p}^{-1} = D(x' \circ x^{-1})(x(p))$ for $p \in U \cap U'$. □

KOSZUL CONNECTIONS

10.125 Definition A *Koszul connection* on a smooth manifold M is a map $\nabla : \mathscr{T}_1(M) \times \mathscr{T}_1(M) \to \mathscr{T}_1(M)$, where $\nabla(X,Y)$ is denoted by $\nabla_X Y$, satisfying

(1) $\nabla_{X_1+X_2} Y = \nabla_{X_1} Y + \nabla_{X_2} Y$

(2) $\nabla_X (Y_1 + Y_2) = \nabla_X Y_1 + \nabla_X Y_2$

(3) $\nabla_{fX} Y = f\, \nabla_X Y$

(4) $\nabla_X (fY) = X(f)\, Y + f\, \nabla_X Y$

for all vector fields $X, X_1, X_2, Y, Y_1, Y_2 \in \mathscr{T}_1(M)$ and smooth functions $f \in \mathscr{F}(M)$.

10.126 Remark Given a Koszul connection ∇ and a vector field Y on a smooth manifold M, the map $F : \mathscr{T}_1(M) \to \mathscr{T}_1(M)$ defined by $F(X) = \nabla_X Y$ for $X \in \mathscr{T}_1(M)$ is linear over $\mathscr{F}(M)$. By Proposition 5.36 there is therefore a unique bundle-valued 1-form $\nabla Y \in \Omega^1(M;TM)$ with $F = (\nabla Y)_M$. Denoting $\nabla Y(p)(v)$ by $\nabla_v Y$ for $p \in M$ and $v \in T_pM$, we therefore have that

$$(\nabla_X Y)_p = \nabla_{X_p} Y$$

for $p \in M$. Hence the value of $\nabla_X Y$ at a point $p \in M$ only depends on the value of X at p and, as we shall see in the next lemma, on the local behavior of Y around p.

10.127 Lemma Let ∇ be a Koszul connection on a smooth manifold M, and let X be a vector field on M. If Y_1, Y_2 are vector fields on M with $Y_1|_V = Y_2|_V$ for an open subset V of M, then we have that $(\nabla_X Y_1)|_V = (\nabla_X Y_2)|_V$.

PROOF: Considering the difference $Y_2 - Y_1$, it is clearly enough to prove that $(\nabla_X Y)|_V = 0$ for each vector field Y on M with $Y|_V = 0$. To show this, let $p \in V$ and choose a smooth function $h \in \mathscr{F}(M)$ with $h = 1$ on $M - V$ and $h(p) = 0$. Such a function h exists by Corollary 2.24. Then we have that

$$\nabla_X Y(p) = \nabla_X (hY)(p) = X(h)(p)\, Y(p) + h(p)\, \nabla_X Y(p) = 0$$

which completes the proof since p was an arbitrary point in V. □

10.128 Proposition A Koszul connection ∇ on a smooth manifold M induces a unique Koszul connection ∇_V on each open subset V of M so that

$$(\nabla_X Y)|_V = (\nabla_V)_{(X|_V)}(Y|_V)$$

for every vector field X and Y on M. The connection ∇_V is usually denoted simply by ∇.

PROOF: Let $\widetilde{X}$ and $\widetilde{Y}$ be vector fields on V. Given a point $p \in V$, we choose vector fields X and Y on M which coincide with $\widetilde{X}$ and $\widetilde{Y}$, respectively, on an open neighbourhood U of p with $\overline{U} \subset V$ using Proposition 2.56 (1), and we define

$$(\nabla_V)_{\widetilde{X}} \widetilde{Y}(p) = \nabla_X Y(p) \,.$$

By Remark 10.126 and Lemma 10.127 we see that ∇_V is well defined, being independent of the choice of the vector fields X and Y. Using the same choice for every point p in the open set U defined above, we also see that ∇_V is a Koszul connection. □

10.129 Definition Let ∇ be a Koszul connection on a smooth manifold M, and let $s = (X_1, \dots, X_n)$ be a frame field on an open subset V of M. Then the smooth functions $\Gamma^i_{kj} : V \to \mathbf{R}$ defined by

$$\nabla_{X_k} X_j = \sum_{i=1}^{n} \Gamma^i_{kj}\, X_i$$

for $1 \leq j, k \leq n$, are called the *components* or the *coefficients* of ∇ with respect to the frame field $s = (X_1, \dots, X_n)$.

10.130 Proposition Let ∇ be a Koszul connection on a smooth manifold M, and let Γ^i_{kj} be the components of ∇ with respect to a frame field $s = (X_1, \dots, X_n)$ on an open subset V of M. If X and Y are vector fields on V with

$$X = \sum_{i=1}^{n} a^i X_i \quad \text{and} \quad Y = \sum_{i=1}^{n} b^i X_i \,,$$

then we have that

$$\nabla_X Y = \sum_{i=1}^{n} \left(\sum_{j=1}^{n} a^j \, b^i{}_{;j} \right) X_i$$

where

$$b^i{}_{;j} = X_j(b^i) + \sum_{k=1}^{n} b^k \, \Gamma^i_{jk}$$

for $1 \leq i, j \leq n$.

PROOF : By Definitions 10.125 and 10.129 we have that

$$\nabla_X Y = \sum_{ij} a^j \, \nabla_{X_j}(b^i X_i) = \sum_{ij} a^j \, \{X_j(b^i)\, X_i + b^i \, \nabla_{X_j} X_i\}$$

$$= \sum_{ij} a^j \, X_j(b^i)\, X_i + \sum_{kj} a^j \, b^k \, \nabla_{X_j} X_k = \sum_{ij} a^j \left(X_j(b^i) + \sum_k b^k \, \Gamma^i_{jk} \right) X_i \, .$$

□

10.131 Proposition Suppose that ∇ is a Koszul connection on a smooth manifold M^n, and that $s_k : V_k \to \mathscr{L}(M)$ for $k = 1,2$ are two sections in the frame bundle $\pi : \mathscr{L}(M) \to M$ with associated local trivializations $(t_k, \pi^{-1}(V_k))$ which are $Gl\,(\mathbf{R}^n)$-related with transition map $\phi : V_1 \cap V_2 \to Gl\,(\mathbf{R}^n)$. If $\Gamma^i_{kj} : V_1 \to \mathbf{R}$ and $\Gamma'^{\,i}_{kj} : V_2 \to \mathbf{R}$ are the components of ∇ with respect to the frame fields $s_1 = (X_1, \dots, X_n)$ and $s_2 = (X'_1, \dots, X'_n)$, respectively, then

$$\Gamma^i_{kj}(p) = \sum_{rst} \Gamma'^{\,r}_{ts}(p) \; \phi(p)^t_k \; \phi(p)^s_j \; \{\phi(p)^{-1}\}^i_r + \sum_r X_k(\phi^r_j)(p) \; \{\phi(p)^{-1}\}^i_r \quad (1)$$

for $p \in V_1 \cap V_2$ and $1 \leq i, j, k \leq n$, where the components in $\mathfrak{gl}\,(\mathbf{R}^n)$ are with respect to the basis $\mathscr{D} = \{F^i_j | 1 \leq i, j \leq n\}$ defined in Example 8.10 (b), where $F^i_j : \mathbf{R}^n \to \mathbf{R}^n$ is the linear map given by $F^i_j(e_k) = \delta_{jk} \, e_i$ for $k = 1, \dots, n$.

PROOF : By Corollary 10.54 we have that

$$X_j(p) = \sum_{s=1}^{n} \phi(p)^s_j \; X'_s(p) \quad \text{and} \quad X_k(p) = \sum_{t=1}^{n} \phi(p)^t_k \; X'_t(p)$$

for $p \in V_1 \cap V_2$ and $1 \leq j, k \leq n$, so that

$$\nabla_{X_k} X_j(p) = \sum_{rt} \phi(p)^t_k \left(X'_t(\phi^r_j)(p) + \sum_s \phi(p)^s_j \; \Gamma'^{\,r}_{ts}(p) \right) X'_r(p)$$

$$= \sum_r \left(\sum_{st} \Gamma'^{\,r}_{ts}(p) \; \phi(p)^t_k \; \phi(p)^s_j + X_k(\phi^r_j)(p) \right) X'_r(p)$$

by Proposition 10.130. Combining this with

$$X'_r(p) = \sum_{i=1}^{n} \{\phi(p)^{-1}\}^i_r \, X_i(p) \,,$$

we obtain formula (1). □

10.132 Corollary The covariant derivative ∇ of a linear connection H on a smooth manifold M is a Koszul connection on M. Conversely, given a Koszul connection ∇ on M, there is a unique linear connection H on M having ∇ as its covariant derivative.

PROOF: The first assertion follows from Proposition 10.80, and the second assertion from Propositions 10.123 and 10.131. □

STRUCTURE EQUATIONS

10.133 Lemma Let H be a linear connection on a smooth manifold M^n. If $F \in \mathfrak{gl}(\mathbf{R}^n)$ and $\xi \in \mathbf{R}^n$, we have that

$$[\,\sigma(F), B(\xi)\,] = B(F(\xi))\,.$$

PROOF: According to Proposition 4.86 we have that $[\,\sigma(F), B(\xi)\,] = L_{\sigma(F)} B(\xi)$, and the flow $\gamma : \mathbf{R} \times \mathscr{L}(M) \to \mathscr{L}(M)$ of the fundamental vector field $\sigma(F)$ is given by

$$\gamma(t,u) = u \circ \, exp\,(tF)$$

for $t \in \mathbf{R}$ and $u \in \mathscr{L}(M)$ by Proposition 9.39.

Now fix $u \in \mathscr{L}(M)$, and let $c : \mathbf{R} \to T_u\mathscr{L}(M)$ be the curve defined by

$$\begin{aligned} c(t) &= [\,\gamma_t^* B(\xi)\,]_u = [\,R^*_{exp\,(tF)} B(\xi)\,]_u = [\,(R_{exp\,(-tF)})_* B(\xi)\,]_u \\ &= B(exp\,(tF)\,\xi)_u = \Lambda_u \circ \phi_F(t) \end{aligned}$$

for $t \in \mathbf{R}$, where $\phi_F : \mathbf{R} \to Gl\,(\mathbf{R}^n)$ is the one-parameter subgroup defined in Proposition 8.29, and $\Lambda_u : \mathfrak{gl}(\mathbf{R}^n) \to T_u\mathscr{L}(M)$ is the linear map given by

$$\Lambda_u(G) = B(G(\xi))_u$$

for $G \in \mathfrak{gl}(\mathbf{R}^n)$. Using Proposition 4.76 it follows that

$$[L_{\sigma(F)} B(\xi)]_u = c'(0) = \Lambda_u \circ \phi'_F(0) = B(F(\xi))_u$$

if we canonically identify $T_{c(0)}(T_u\mathscr{L}(M))$ with $T_u\mathscr{L}(M)$ as in Lemma 2.84. □

10.134 Lemma Let $\pi : P \to M$ be a principal G-bundle with a connection H. If $X \in \mathfrak{g}$ and Y is a horizontal vector field on P, then the vector field $[\,\sigma(X), Y\,]$ is also horizontal.

PROOF : According to Proposition 4.86 we have that $[\,\sigma(X), Y\,] = L_{\sigma(X)} Y$, and the flow $\gamma : \mathbf{R} \times P \to P$ of the fundamental vector field $\sigma(X)$ is given by

$$\gamma(t,u) = u\ exp\ (tX)$$

for $t \in \mathbf{R}$ and $u \in P$ by Proposition 9.39.

Now fix $u \in P$, and let $c : \mathbf{R} \to T_u P$ be the curve defined by

$$c(t) = [\,\gamma_t^* Y\,]_u = [R^*_{exp\ (tX)} Y\,]_u = (R_{exp\ (-tX)})_* Y_{u\ exp\ (tX)}$$

for $t \in \mathbf{R}$. By Definition 10.32 (ii) we know that c actually is a curve which lies in H_u, and therefore

$$[L_{\sigma(X)} Y\,]_u = c'(0) \in H_u$$

by Proposition 4.76 if we canonically identify $T_{c(0)}(T_u P)$ with $T_u P$ as in Lemma 2.84. □

10.135 Theorem (The first structure equation) Let H be a linear connection on a smooth manifold M^n, and let ω, θ and Θ be the connection form, the dual form and the torsion form of H. Then we have that

$$d\theta(u)(w_1,w_2) = -\{\omega(u)(w_1)(\theta(u)(w_2)) - \omega(u)(w_2)(\theta(u)(w_1))\} + \Theta(u)(w_1,w_2)$$

for every $u \in \mathscr{L}(M)$ and $w_1, w_2 \in T_u \mathscr{L}(M)$.

PROOF : If $w_1, w_2 \in H_u$, then $\omega(u)(w_i) = 0$ and $h(w_i) = w_i$ for $i = 1,2$. The structure equation therefore follows from the definition of the torsion form Θ in this case.

Suppose next that $w_1, w_2 \in V_u$. Then the right-hand side of the structure equation is 0 since $h(w_i) = 0$ and $\theta(u)(w_i) = 0$ for $i = 1,2$. To prove that the left-hand side is also 0, let $F_i = \omega(u)(w_i)$ so that $\sigma(F_i)_u = w_i$ for $i = 1,2$. Since $\sigma(F)$ is a vertical vector field on $\mathscr{L}(M)$ so that $\theta(\sigma(F))(u) = 0$ for every $u \in \mathscr{L}(M)$ and $F \in \mathfrak{gl}(\mathbf{R}^n)$, and $\sigma : \mathfrak{gl}(\mathbf{R}^n) \to \mathscr{T}_1 \mathscr{L}(M)$ is a Lie algebra homomorphism, it follows from Remark 5.86 and Lemma 2.78 (2) that

$$d\,\theta(u)(w_1,w_2) = d\,\theta\,(\sigma(F_1), \sigma(F_2))(u)$$

$$= \sigma(F_1)(\,\theta\,(\sigma(F_2)))(u) - \sigma(F_2)(\,\theta\,(\sigma(F_1)))(u) - \theta\,([\sigma(F_1), \sigma(F_2)])(u) = 0$$

from which we obtain the structure equation.

Finally, suppose that $w_1 \in V_u$ and $w_2 \in H_u$, and let $F = \omega(u)(w_1)$ and $\xi = \theta(u)(w_2)$ so that $\sigma(F)_u = w_1$ and $B(\xi)_u = w_2$. Since $\theta\,(B(\xi))$ and $\theta\,(\sigma(F))$ are constant functions on $\mathscr{L}(M)$ with values ξ and 0, respectively, it follows from Remark 5.86 and Lemmas 2.78 (2) and 10.133 that

$$\begin{aligned}
d\theta(u)(w_1,w_2) &= d\theta(\sigma(F),B(\xi))(u) \\
&= \sigma(F)(\theta(B(\xi)))(u) - B(\xi)(\theta(\sigma(F)))(u) - \theta([\sigma(F),B(\xi)])(u) \\
&= -\,\theta(u)\,(B(F(\xi))_u) = -F(\xi) = -\,\omega(u)(w_1)\,(\theta(u)(w_2))\,.
\end{aligned}$$

From this we obtain the structure equation also in this case since $\omega(u)(w_2)=0$, $\theta(u)(w_1)=0$ and $h(w_1)=0$ so that $\Theta(u)(w_1,w_2)=0$. Since both sides of the structure equation are bilinear and skew symmetric as a function of w_1 and w_2, the theorem follows from these three cases. □

10.136 Remark The first structure equation can also be written in the form

$$d\theta = -\,\omega\wedge\theta+\Theta$$

where the wedge product is with respect to the bilinear map

$$\nu : \mathfrak{gl}(\mathbf{R}^n)\times\mathbf{R}^n\to\mathbf{R}^n$$

given by $\nu(F,\xi)=F(\xi)$.

10.137 Corollary Let H be a linear connection on a smooth manifold M^n, and let ω, θ and Θ be the connection form, the dual form and the torsion form of H. Then we have that

$$d\theta^{\,i} = -\sum_{j=1}^{n}\ \omega^i_j\wedge\theta^{\,j}+\Theta^{\,i}$$

for $i=1,\dots,n$, where $\theta^{\,i}$ and $\Theta^{\,i}$ are the components of θ and Θ with respect to the standard basis $\mathscr{E}=\{e_1,\dots,e_n\}$ for $\mathbf{R}^n$, and ω^i_j are the components of ω with respect to the basis $\mathscr{D}=\{F^{\,i}_j\,|\,1\le i,j\le n\}$ for $\mathfrak{gl}(\mathbf{R}^n)$ defined in Example 8.10 (b), where $F^{\,i}_j:\mathbf{R}^n\to\mathbf{R}^n$ is the linear map given by $F^{\,i}_j(e_k)=\delta_{jk}\,e_i$ for $k=1,\dots,n$.

PROOF: Since $\nu\,(F^{\,i}_j,e_k)=\delta_{jk}\,e_i$ for $1\le i,j,k\le n$, it follows from Proposition 5.68 that

$$(\omega\wedge\theta)(u)(w_1,w_2)=\sum_{ij}\ (\omega^i_j\wedge\theta^{\,j})(u)(w_1,w_2)\ e_i$$

for every $u\in\mathscr{L}(M)$ and $w_1,w_2\in T_u\mathscr{L}(M)$. □

10.138 Theorem (The second structure equation) Let $\pi:P\to M$ be a principal G-bundle with a connection H, and let ω and Ω be the connection form and the curvature form of H. Then we have that

$$d\omega(u)(w_1,w_2) = -[\,\omega(u)(w_1),\ \omega(u)(w_2)\,]+\Omega(u)(w_1,w_2)$$

for every $u\in P$ and $w_1,w_2\in T_uP$.

PROOF : If $w_1, w_2 \in H_u$, then $\omega(u)(w_i) = 0$ and $h(w_i) = w_i$ for $i = 1,2$. The structure equation therefore follows from the definition of the curvature form Ω in this case.

Suppose next that $w_1, w_2 \in V_u$, and let $X_i = \omega(u)(w_i)$ so that $\sigma(X_i)_u = w_i$ for $i = 1,2$. Since $\omega(\sigma(X_i))$ is a constant function on P with value X_i for $i = 1,2$, and $\sigma : \mathfrak{g} \to \mathscr{T}_1 P$ is a Lie algebra homomorphism, it follows from Remark 5.86 and Lemma 2.78 (2) that

$$\begin{aligned} &d\,\omega(u)(w_1,w_2) = d\,\omega(\sigma(X_1),\sigma(X_2))(u) \\ &= \sigma(X_1)(\omega(\sigma(X_2)))(u) - \sigma(X_2)(\omega(\sigma(X_1)))(u) - \omega([\sigma(X_1),\sigma(X_2)])(u) \\ &= -\,\omega(u)(\sigma([X_1,X_2])_u) = -[X_1,X_2] = -[\omega(u)(w_1),\,\omega(u)(w_2)]\,. \end{aligned}$$

From this we obtain the structure equation since $h(w_i) = 0$ for $i = 1,2$ so that $\Omega(u)(w_1,w_2) = 0$.

Finally, suppose that $w_1 \in V_u$ and $w_2 \in H_u$. Then the right-hand side of the structure equation is 0 since $h(w_1) = 0$ and $\omega(u)(w_2) = 0$. To prove that the left-hand side is also 0, let $X = \omega(u)(w_1)$ so that $\sigma(X)_u = w_1$, and choose a horizontal vector field Y on P with $Y_u = w_2$ using Proposition 2.56 (2). Since $\omega(\sigma(X))$ is a constant function on P with value X and the vector field $[\sigma(X),Y]$ is horizontal by Lemma 10.134, it follows from Remark 5.86 and Lemma 2.78 (2) that

$$\begin{aligned} &d\,\omega(u)(w_1,w_2) = d\,\omega(\sigma(X),Y)(u) \\ &= \sigma(X)(\omega(Y))(u) - Y(\omega(\sigma(X)))(u) - \omega([\sigma(X),Y])(u) = 0 \end{aligned}$$

which proves the structure equation also in this case. Since both sides of the structure equation are bilinear and skew symmetric as a function of w_1 and w_2, the theorem follows from these three cases. □

10.139 Remark The formula in Theorem 10.138 is also called the *structure equation of Elie Cartan*, and it can be written in the form

$$d\omega = -\frac{1}{2}\ \omega \wedge \omega + \Omega$$

where the wedge product is with respect to the bilinear map

$$\nu : \mathfrak{g} \times \mathfrak{g} \to \mathfrak{g}$$

given by $\nu(X,Y) = [X,Y]$.

10.140 Corollary Let H be a linear connection on a smooth manifold M^n, and let ω and Ω be the connection form and the curvatore form of H. Then we have that

$$d\omega^i_j = -\sum_{k=1}^{n}\ \omega^i_k \wedge \omega^k_j + \Omega^i_j$$

for $i,j = 1,...,n$, where ω^i_j and Ω^i_j are the components of ω and Ω with respect to the basis $\mathscr{D} = \{F_j^i | 1 \le i, j \le n\}$ for $\mathfrak{gl}(\mathbf{R}^n)$ defined in Example 8.10 (b), where $F_j^i : \mathbf{R}^n \to \mathbf{R}^n$ is the linear map given by $F_j^i(e_k) = \delta_{jk}\, e_i$ for $k = 1, \ldots, n$.

PROOF: Since

$$\nu\,(F_j^i, F_l^k) = \delta_{jk}\, F_l^i - \delta_{li}\, F_j^k$$

for $1 \le i, j, k, l \le n$ by Example 8.10 (b), it follows from Proposition 5.68 that

$$(\omega \wedge \omega)(u)(w_1, w_2) = \sum_{ikl} (\omega_k^i \wedge \omega_l^k)(u)(w_1, w_2)\; F_l^i$$

$$- \sum_{jkl} (\omega_j^l \wedge \omega_l^k)(u)(w_1, w_2)\; F_j^k = 2 \sum_{ijk} (\omega_k^i \wedge \omega_j^k)(u)(w_1, w_2)\; F_j^i$$

for every $u \in \mathscr{L}(M)$ and $w_1, w_2 \in T_u\mathscr{L}(M)$. □

10.141 Theorem (Bianchi's 1st identity) Let H be a linear connection on a smooth manifold M^n, and let Ω, θ and Θ be the curvature form, the dual form and the torsion form of H. Then we have that

$$D\Theta = \Omega \wedge \theta\,.$$

PROOF: If we apply d on both sides of the first structure equation

$$d\theta = -\;\omega \wedge \theta + \Theta$$

and use Propositions 5.64 (3) and 5.69, we obtain

$$0 = -(d\,\omega) \wedge \theta + \omega \wedge (d\,\theta) + d\Theta\,.$$

If $h : T\mathscr{L}(M) \to T\mathscr{L}(M)$ is the horizontal projection determined by H, we therefore have that

$$D\Theta = h^*(d\Theta) = h^*(d\,\omega) \wedge h^*(\theta) - h^*(\omega) \wedge h^*(d\,\theta) = \Omega \wedge \theta$$

by 5.65 and Proposition 10.41, since $h^*(d\,\omega) = D\omega = \Omega$ by the definition of Ω, and $h^*(\theta) = \theta$ and $h^*(\omega) = 0$ as θ is horizontal and ω is vertical. □

10.142 Theorem (Bianchi's 2nd identity) Let H be a connection in the principal G-bundle $\pi : P \to M$, and let Ω be the curvature form of H. Then we have that

$$D\Omega = 0\,.$$

PROOF: If we apply d on both sides of the second structure equation

$$d\omega = -\tfrac{1}{2}\;\omega \wedge \omega + \Omega$$

and use Propositions 5.64 (3) and 5.69, we obtain

$$0 = -\tfrac{1}{2}\ (d\,\omega) \wedge \omega + \tfrac{1}{2}\ \omega \wedge (d\,\omega) + d\Omega\,.$$

If $h : TP \to TP$ is the horizontal projection determined by H, we therefore have that

$$D\Omega = h^*(d\Omega) = \tfrac{1}{2}\ h^*(d\,\omega) \wedge h^*(\omega) - \tfrac{1}{2}\ h^*(\omega) \wedge h^*(d\,\omega) = 0$$

by 5.65 since $h^*(\omega) = 0$ as ω is vertical. □

10.143 Proposition Let $\pi : \mathscr{L}(M) \to M$ be the frame bundle over a smooth manifold M^n with a connection H, and let $\pi' : E \to M$ be the associated vector bundle obtained from a representation $\psi : Gl\,(\mathbf{R}^n) \to Aut\,(W)$ of the structure group $Gl\,(\mathbf{R}^n)$ on a finite dimensional vector space W.

If $\omega \in \Omega^k(M;E)$ is a bundle-valued k-form on M with values in E, we let $\eta \in \Omega^k(\mathscr{L}(M);W)$ be the corresponding vector-valued k-form on $\mathscr{L}(M)$ with values in W. Considering ω as a section in the vector bundle $\pi^k : \mathscr{T}^k(TM;E) \to M$, we have that

$$\nabla_v\,\omega(v_1, \ldots, v_k) = \mu_{u_0} \circ w(\eta(B(\xi_1), \ldots, B(\xi_k)))$$

for $p \in M$ and $v, v_1, \ldots, v_k \in T_pM$, where $u_0 \in \mathscr{L}(M)$, $w \in H_{u_0}$ and $\xi_1, \ldots, \xi_k \in \mathbf{R}^n$ are chosen so that $\pi(u_0) = p$, $\pi_*(w) = v$ and $u_0(\xi_i) = v_i$ for $i = 1, \ldots, k$.

PROOF : By Definition 10.77 and Example 10.29 (i) the section ω has a corresponding vector-valued function $\widetilde{\eta}$ on $\mathscr{L}(M)$ with values in $\mathscr{T}^k(\mathbf{R}^n;W)$ which is given by

$$\widetilde{\eta}(u) = \mu_u^{-1} \circ \omega(\pi(u)) \circ u^k$$

for $u \in \mathscr{L}(M)$. If $\Lambda : \mathscr{T}^k(\mathbf{R}^n;W) \to W$ is the linear map defined by

$$\Lambda(F) = F(\xi_1, \ldots, \xi_k)$$

for $F \in \mathscr{T}^k(\mathbf{R}^n;W)$, we therefore have that

$$\begin{aligned}(\Lambda \cdot \widetilde{\eta})(u) &= \widetilde{\eta}(u)(\xi_1, \ldots, \xi_k) = \mu_u^{-1} \circ \omega(\pi(u))(u(\xi_1), \ldots, u(\xi_k)) \\ &= \eta(u)(B(\xi_1)_u, \ldots, B(\xi_k)_u) = \eta(B(\xi_1), \ldots, B(\xi_k))\,(u)\end{aligned}$$

for $u \in \mathscr{L}(M)$. Now we have that

$$\nabla_v\,\omega = \mu_{u_0} \circ d\widetilde{\eta}(u_0)(w) \circ \{u_0^{-1}\}^k\,,$$

so using Proposition 5.62 we obtain

$$\begin{aligned}\nabla_v\,\omega(v_1, \ldots, v_k) &= \mu_{u_0} \circ d\widetilde{\eta}(u_0)(w)\,(\xi_1, \ldots, \xi_k) = \mu_{u_0} \circ (\Lambda \cdot d\widetilde{\eta})(u_0)(w) \\ &= \mu_{u_0} \circ d(\Lambda \cdot \widetilde{\eta})(u_0)(w) = \mu_{u_0} \circ w(\Lambda \cdot \widetilde{\eta}) = \mu_{u_0} \circ w(\eta(B(\xi_1), \ldots, B(\xi_k)))\,.\end{aligned}$$

□

10.144 Proposition Let $\pi : \mathscr{L}(M) \to M$ be the frame bundle over a smooth manifold M^n with a connection H, and let T be the torsion form of H on M. Then we have that

$$T(p)(v_1,v_2) = -\pi_*([B(\xi_1),B(\xi_2)]_u)$$

for $p \in M$ and $v_1, v_2 \in T_pM$, where $u \in \mathscr{L}(M)$ and $\xi_1, \xi_2 \in \mathbf{R}^n$ are chosen so that $\pi(u) = p$ and $u(\xi_i) = v_i$ for $i = 1,2$.

PROOF : Using Remark 5.86, Lemma 2.78 (2) and Proposition 10.93, we have that

$$\begin{aligned}
T(p)(v_1,v_2) &= u \circ d\theta(B(\xi_1),B(\xi_2))(u) \\
&= u \circ B(\xi_1)_u(\theta(B(\xi_2))) - u \circ B(\xi_2)_u(\theta(B(\xi_1))) - u \circ \theta(u)([B(\xi_1),B(\xi_2)]_u) \\
&= -\pi_*([B(\xi_1),B(\xi_2)]_u),
\end{aligned}$$

since $\theta(B(\xi_i))(u) = \xi_i$ for $u \in \mathscr{L}(M)$ and $i = 1,2$ by Proposition 10.96. □

10.145 Proposition Let $\pi : \mathscr{L}(M) \to M$ be the frame bundle over a smooth manifold M^n with a connection H, and let $\pi' : E \to M$ be the associated vector bundle obtained from a representation $\psi : Gl(\mathbf{R}^n) \to Aut(W)$ of the structure group $Gl(\mathbf{R}^n)$ on a finite dimensional vector space W. Let T be the torsion form of H on M.

If $\omega \in \Omega^2(M;E)$ is a bundle-valued 2-form on M with values in E, we let $\eta \in \Omega^2(\mathscr{L}(M);W)$ be the corresponding vector-valued 2-form on $\mathscr{L}(M)$ with values in W. Considering ω as a section in the vector bundle $\pi^2 : \mathscr{T}^2(TM;E) \to M$, we have that

$$\mu_u \circ D\eta(u)(w_1,w_2,w_3) =$$

$$\sum_{\sigma \in A_3} \{(\nabla_{v_{\sigma(1)}}\omega)(v_{\sigma(2)},v_{\sigma(3)}) + \omega(p)(T(p)(v_{\sigma(1)},v_{\sigma(2)}),v_{\sigma(3)})\}$$

for $p \in M$ and $v_1, v_2, v_3 \in T_pM$, where $u \in \mathscr{L}(M)$ and $w_1, w_2, w_3 \in H_u$ are chosen so that $\pi(u) = p$ and $\pi_*(w_i) = v_i$ for $i = 1,2,3$, and A_3 denotes the alternating group of three elements.

PROOF : Choose $\xi_1, \xi_2, \xi_3 \in \mathbf{R}^n$ so that $u(\xi_i) = v_i$ for $i = 1,2,3$. Using Remark 5.85 and Propositions 10.143 and 10.144, we then have that

$$\mu_u \circ D\eta\,(u)(w_1,w_2,w_3) = \mu_u \circ d\eta\,(B(\xi_1),B(\xi_2),B(\xi_3))(u)$$

$$= \sum_{\sigma\in A_3} \{\mu_u \circ B(\xi_{\sigma(1)})_u(\eta\,(B(\xi_{\sigma(2)}),B(\xi_{\sigma(3)})))$$

$$-\ \mu_u \circ \eta(u)([B(\xi_{\sigma(1)}),B(\xi_{\sigma(2)})]_u,B(\xi_{\sigma(3)})_u)\}$$

$$= \sum_{\sigma\in A_3} \{(\nabla_{v_{\sigma(1)}}\omega)(v_{\sigma(2)},v_{\sigma(3)}) + \omega(p)(T(p)(v_{\sigma(1)},v_{\sigma(2)}),v_{\sigma(3)})\}\,.$$

□

10.146 Corollary Let R and T be the curvature form and the torsion form of a linear connection H on a smooth manifold M. Then Bianchi's 1st and 2nd identity can be written in the form

(1) $$\sum_{\sigma\in A_3} R(p)(v_{\sigma(1)},v_{\sigma(2)})\,v_{\sigma(3)} =$$

$$\sum_{\sigma\in A_3} \{(\nabla_{v_{\sigma(1)}}T)(v_{\sigma(2)},v_{\sigma(3)}) + T(p)(T(p)(v_{\sigma(1)},v_{\sigma(2)}),v_{\sigma(3)})\}\,,\ \text{and}$$

(2) $$\sum_{\sigma\in A_3} \{(\nabla_{v_{\sigma(1)}}R)(v_{\sigma(2)},v_{\sigma(3)}) + R(p)(T(p)(v_{\sigma(1)},v_{\sigma(2)}),v_{\sigma(3)})\} = 0$$

respectively, for every $p \in M$ and $v_1,v_2,v_3 \in T_pM$, where A_3 denotes the alternating group of three elements.

PROOF: Choose $u \in \mathscr{L}(M)$ and $w_1,w_2,w_3 \in H_u$ so that $\pi(u) = p$ and $\pi_*(w_i) = v_i$ for $i = 1,2,3$. Then we have that

$$u \circ (\Omega\wedge\theta)(u)(w_1,w_2,w_3) = \frac{(2+1)!}{2!\,1!}\ u \circ \mathscr{A}(\Omega\otimes\theta)(u)(w_1,w_2,w_3)$$

$$= \frac{1}{2!\,1!} \sum_{\sigma\in S_3} \varepsilon(\sigma)\{u \circ\ \Omega(u)(w_{\sigma(1)},w_{\sigma(2)}) \circ u^{-1}\}\{u \circ\ \theta(u)(w_{\sigma(3)})\}$$

$$= \sum_{\sigma\in A_3} \{u \circ\ \Omega(u)(w_{\sigma(1)},w_{\sigma(2)}) \circ u^{-1}\}\,\pi_*(w_{\sigma(3)}) = \sum_{\sigma\in A_3} R(p)(v_{\sigma(1)},v_{\sigma(2)})\,v_{\sigma(3)}$$

The corollary now follows from Theorems 10.141 and 10.142 and Proposition 10.145. □

GEODESICS

10.147 Definition Let M be a smooth manifold with a linear connection H. Then a smooth curve $\gamma : I \to M$ defined on an open interval I is called a *geodesic* on M if γ' is parallel along γ. The geodesic γ is called *maximal* if it has no extension to a geodesic on any larger open interval.

10.148 Proposition Let $\pi : \mathscr{L}(M) \to M$ be the frame bundle over a smooth manifold M^n with a connection H, and let $\gamma : I \to M$ be a smooth curve defined on an open interval I containing t_0 with a horizontal lifting $\beta : I \to \mathscr{L}(M)$. Then γ is a geodesic if and only if β is an integral curve for the basic vector field $B(\xi)$ determined by the vector $\xi \in \mathbf{R}^n$ with $\beta(t_0)(\xi) = \gamma'(t_0)$.

PROOF : The proposition follows from Definition 10.71 and Proposition 10.96 which implies that

$$\pi_*(\beta'(t)) = \gamma'(t) \quad \text{and} \quad \pi_*(B(\xi)_{\beta(t)}) = \beta(t)(\xi) = \tau_t^{t_0}(\gamma'(t_0))$$

for every $t \in I$. □

10.149 Proposition Let M^n be a smooth manifold with a linear connection H, and let $v \in T_pM$ for a $p \in M$. Then there is a unique maximal geodesic $\gamma : I \to M$ with $\gamma(0) = p$ and $\gamma'(0) = v$. Denoting this geodesic by $\gamma_v : I(v) \to M$, we have that $I(tv) = t^{-1}I(v)$ when $t \in \mathbf{R} - \{0\}$ and $I(0) = \mathbf{R}$, and $\gamma_{tv}(s) = \gamma_v(ts)$ for $s \in I(tv)$.

PROOF : Let $\pi : \mathscr{L}(M) \to M$ be the frame bundle over M, choose a $u \in \pi^{-1}(p)$ and let $\xi \in \mathbf{R}^n$ be the vector with $u(\xi) = v$. If $\beta : I \to \mathscr{L}(M)$ is the unique maximal integral curve for the basic vector field $B(\xi)$ with initial condition u, then it follows from Proposition 10.148 that $\gamma = \pi \circ \beta$ is the unique maximal geodesic on M with $\gamma(0) = p$ and $\gamma'(0) = v$.

To prove the second part of the proposition, let $\mu_t : \mathbf{R} \to \mathbf{R}$ be the smooth map defined by $\mu_t(s) = ts$ for $s \in \mathbf{R}$. Then the horizontal lift $\beta \circ \mu_t$ of $\gamma_v \circ \mu_t$ is the maximal integral curve for the basic vector field $B(t\xi)$ with initial condition u, since

$$(\beta \circ \mu_t)'(s) = t\beta'(ts) = t\,B(\xi)(\beta(ts)) = B(t\xi)(\beta \circ \mu_t(s))$$

by Definition 2.69. As $u(t\xi) = tv$, we conclude that $\gamma_{tv} = \gamma_v \circ \mu_t$. □

10.150 Using the above notation, we let $\mathscr{D}$ be the subset of TM given by $\mathscr{D} = \{v \in TM \,|\, 1 \in I(v)\}$, and we define the *exponential map* $exp : \mathscr{D} \to M$ by

$$exp(v) = \gamma_v(1)$$

for $v \in \mathscr{D}$. For each $p \in M$, we let $\mathscr{D}_p = \mathscr{D} \cap T_pM$ and call the restriction $exp_p = exp \mid_{\mathscr{D}_p} : \mathscr{D}_p \to M$ the *exponential map at p*.

10.151 Remark By Proposition 10.149 the maximal geodesic $\gamma_v : I(v) \to M$ is given by

$$\gamma_v(t) = \gamma_{tv}(1) = exp\,(tv)$$

for $t \in I(v)$, which is equivalent to the conditions $1 \in I(tv)$ or $tv \in \mathscr{D}$.

10.152 Let $\tilde{\pi} : \mathscr{E}(M) \to \mathscr{L}(M)$ be the vector bundle induced from the tangent bundle $\pi : TM \to M$ by the projection $\pi' : \mathscr{L}(M) \to M$ in the frame bundle, as shown in the commutative diagram in Example 10.25, and let $\widetilde{\mathscr{D}} = \tilde{\pi}'^{-1}(\mathscr{D})$. Then we have a map $E : \widetilde{\mathscr{D}} \to \mathscr{L}(M)$ defined by

$$E(u,v) = \beta_{u,v}(1)$$

for $(u,v) \in \widetilde{\mathscr{D}}$, where $\beta_{u,v}$ is the unique horizontal lifting of the maximal geodesic γ_v with $\beta_{u,v}(0) = u$. From Proposition 10.148 we know that $\beta_{u,v}$ is an integral curve for the basic vector field $B(u^{-1}(v))$, and we have a commutative diagram

$$\begin{array}{ccc} \widetilde{\mathscr{D}} & \xrightarrow{\ \tilde{\pi}'\ } & \mathscr{D} \\ {\scriptstyle E}\downarrow & & \downarrow{\scriptstyle exp} \\ \mathscr{L}(M) & \xrightarrow{\ \pi'\ } & M \end{array}$$

10.153 Proposition Let M^n be a smooth manifold with a linear connection H. Then $E : \widetilde{\mathscr{D}} \to \mathscr{L}(M)$ and $exp : \mathscr{D} \to M$ are smooth maps defined on the open subsets $\widetilde{\mathscr{D}}$ and $\mathscr{D}$ of $\mathscr{E}(M)$ and TM, respectively, and $\mathscr{D}_p$ is an open star- shaped neighbourhood of $0 \in T_pM$ for each $p \in M$.

PROOF : We first show that $\widetilde{\mathscr{D}}$ is an open subset of $\mathscr{E}(M)$ and that E is smooth. Let Y be the vector field on $\mathscr{L}(M) \times \mathscr{E}(M)$ defined by

$$Y(w,u,v) = (B(u^{-1}(v))_w, 0) = (\alpha_w^{-1} \circ w \circ u^{-1}(v), 0)$$

for $w \in \mathscr{L}(M)$ and $(u,v) \in \mathscr{E}(M)$, where (α, π) is the bundle map from H to TM defined in Proposition 10.49. The flow $\gamma : \mathscr{D}(Y) \to \mathscr{L}(M) \times \mathscr{E}(M)$ for Y is given by

$$\gamma(t,w,u,v) = (\beta\,(u^{-1}(v))(t,w), u, v)\,,$$

where $\beta(\xi)$ is the flow for the basic vector field $B(\xi)$ for each $\xi \in \mathbf{R}^n$. Hence we have that

$$\widetilde{\mathscr{D}} = (\widetilde{\pi}, id_{\mathscr{E}(M)})^{-1}(\mathscr{D}_1(Y))$$

and

$$E(u,v) = \pi_1 \circ \gamma_1(u,u,v)$$

for $(u,v) \in \widetilde{\mathscr{D}}$, where $\pi_1 : \mathscr{L}(M) \times \mathscr{E}(M) \to \mathscr{L}(M)$ is the projection on the first factor, thus completing the proof that E is smooth with an open domain $\widetilde{\mathscr{D}}$.

As the projection $\widetilde{\pi}' : \mathscr{E}(M) \to TM$ is an open mapping onto TM by Propositions 10.3 and 2.43, it follows that $\mathscr{D} = \widetilde{\pi}'(\widetilde{\mathscr{D}})$ is open in TM. For each frame field $s : V \to \mathscr{L}(M)$ defined on an open subset V of M we have that

$$exp\,(v) = \pi' \circ E(s \circ \pi(v), v)$$

for $v \in \mathscr{D} \cap \pi^{-1}(V)$, which shows that exp is smooth. □

10.154 Proposition For each $p \in M$ we have that $exp_{\,p*0} : T_pM \to T_pM$ is the identity map. Hence $exp_{\,p}$ maps an open star-shaped neighbourhood U_p of $0 \in T_pM$ diffeomorphically onto an open neighbourhood V_p of $p \in M$.

We let $log_{\,p} : V_p \to U_p$ denote the inverse of the diffeomorphism induced by $exp_{\,p}$.

PROOF: Using the curves γ_v and ψ_v on M and T_pM, respectively, defined in Proposition 10.149 and Lemma 8.36, we have that

$$exp_{\,p*}(v) = (exp \circ \psi_v)'(0) = \gamma_v'(0) = v$$

for every $v \in T_pM$ by 2.70 and Proposition 10.151, showing that $exp_{\,p*0} = id$. The last part of the proposition now follows from the inverse function theorem. □

10.155 Remark If $x : T_pM \to \mathbf{R}^n$ is a linear isomorphism and $y = x \circ log_{\,p}$, then (y, V_p) is a local chart around p on M, called a *normal chart* or a *normal coordinate system* on M. Let $\mathscr{E} = \{e_1, ..., e_n\}$ be the standard basis for $\mathbf{R}^n$ and $\mathscr{C} = \{v_1, ..., v_n\}$ be a basis for T_pM so that $x(v_i) = e_i$ for $i = 1, ..., n$. Then the coordinate map y is given by

$$y(exp_{\,p}(\sum_{i=1}^{n} a_i v_i)) = a$$

for $a \in x(U_p)$.

10.156 By 10.12 we know that the bundle map $(\widetilde{\pi}', \pi')$ between the vector bundles $\widetilde{\pi} : \mathscr{E}(M) \to \mathscr{L}(M)$ and $\pi : TM \to M$ induces a linear isomorphism $\widetilde{\pi}'_u : \mathscr{E}_u(M) \to T_pM$ for each $p \in M$ and $u \in \mathscr{L}_p(M)$. We have that $s_u = E \circ \widetilde{\pi}'^{-1}_u \circ log_{\,p} : V_p \to \mathscr{L}(M)$ is a frame field where $s_u(q)$ is obtained from u by parallel transport along the geodesic $\gamma_{log_p(q)}$ for each $q \in V_p$. Indeed, if $v = log_p(q)$ and $\beta_{u,v}$ is the unique horizontal lifting of the maximal geodesic γ_v with $\beta_{u,v}(0) = u$, then

$$s_u(q) = E \circ \tilde{\pi}_u'^{-1}(v) = E(u,v) = \beta_{u,v}(1) = \tau_1^0(u)$$

where τ_1^0 denotes parallel transport in $\mathscr{L}(M)$ along the geodesic γ_v from 0 to 1, and

$$\pi' \circ s_u(q) = \pi' \circ \beta_{u,v}(1) = \gamma_v(1) = exp_p(v) = q\,.$$

10.157 Proposition Let M be a smooth manifold with a linear connection H. Then a smooth curve $\gamma : I \to M$ defined on an open interval I is a geodesic on M if and only if for each local chart (x,U) on M, we have that

$$(x^i \circ \gamma)''(t) + \sum_{j,k=1}^{n} \Gamma^i_{jk}(\gamma(t))\,(x^j \circ \gamma)'(t)\,(x^k \circ \gamma)'(t) = 0$$

for $t \in I \cap \gamma^{-1}(U)$ and $i = 1, \ldots, n$, where Γ^i_{jk} are the components of H with respect to the coordinate frame field $s = \left(\frac{\partial}{\partial x^1}, \ldots, \frac{\partial}{\partial x^n}\right)$. These are called the *geodesic equations.*

PROOF : Follows from Proposition 10.121 since

$$\gamma'(t) = \sum_{i=1}^{n} (x^i \circ \gamma)'(t)\, \frac{\partial}{\partial x^i}\bigg|_{\gamma(t)}\,.$$

□

10.158 Proposition Let M^n be a smooth manifold with a linear connection H, and let Γ^i_{jk} be the components of H with respect to a normal coordinate system (y, V_p) around a point p on M as defined in Remark 10.155. Then we have that

$$\Gamma^i_{jk}(p) + \Gamma^i_{kj}(p) = 0$$

for $1 \le i,j,k \le n$. If the torsion $T(p) = 0$, then

$$\Gamma^i_{jk}(p) = 0$$

for $1 \le i,j,k \le n$.

PROOF : Assume that $y = x \circ log_p$ where $x : T_pM \to \mathbf{R}^n$ is a linear isomorphism, and that $x(v) = a$ for a tangent vector $v \in T_pM$. Then

$$y \circ \gamma_v(t) = y \circ exp_p(tv) = ta$$

for $t \in I(v)$, so that

$$\sum_{j,k=1}^{n} \Gamma^i_{jk}(p)\, a^j a^k = 0$$

for $i = 1, \ldots, n$ by Proposition 10.157. By choosing $a = e_j$ we see that $\Gamma^i_{jj}(p) = 0$ for $1 \le i,j \le n$. The formula in the proposition is now obtained by choosing $a = e_j + e_k$. □

METRICAL CONNECTIONS

10.159 Definition Let $\pi : P \to M$ be a principal G-bundle with a connection H, and let g be a fibre metric in the associated vector bundle $\pi' : E \to M$ obtained from a representation $\psi : G \to Aut\,(W)$ of G on a finite dimensional vector space W. Then H is said to be a *metrical* connection or a connection *compatible* with g if parallel transport in E preserves the fibre metric, i.e., if the parallel transport $\tau_{t_1}^{t_0} : \pi'^{-1}(\gamma(t_0)) \to \pi'^{-1}(\gamma(t_1))$ in E from t_0 to t_1 along any smooth curve $\gamma : I \to M$ on M, where t_0 and t_1 are arbitrary points in the open interval I, is a linear isometry.

By Propositions 2.56 (2) and 10.92 this is equivalent to saying that the function $c : I \to \mathbf{R}$ defined by $c(t) = g(\gamma(t))\,(F_1(t), F_2(t))$ for $t \in I$, is constant for every pair of parallel liftings F_1 and F_2 along γ. The function c is usually denoted by $<F_1, F_2>$.

10.160 Proposition Let $\pi : P \to M$ be a principal G-bundle with a connection H, and let g be a fibre metric in the associated vector bundle $\pi' : E \to M$ obtained from a representation $\psi : G \to Aut\,(W)$ of G on a finite dimensional vector space W. Then the connection H is compatible with g if and only if the following condition is satisfied:

If $F_1 : I \to E$ and $F_2 : I \to E$ are liftings of an arbitrary smooth curve $\gamma : I \to M$ on M defined on an open interval I, and if (r, I) is the standard local chart on I where $r : I \to \mathbf{R}$ is the inclusion map, then

$$\frac{d}{dr} <F_1, F_2> = \left\langle \frac{DF_1}{dr}, F_2 \right\rangle + \left\langle F_1, \frac{DF_2}{dr} \right\rangle. \tag{1}$$

PROOF : If F_1 and F_2 are parallel along γ so that $DF_1/dr = 0$ and $DF_2/dr = 0$ on I, then it follows from (1) that $<F_1, F_2>$ is a constant function on I. Hence (1) implies that H is compatible with g, which completes the proof of the if part of the proposition.

Suppose conversely that the connection H is compatible with g, and consider a smooth curve $\gamma : I \to M$ on M defined on an open interval I. Choose an orthonormal basis $\mathscr{B}_{t_0} = \{v_1, \ldots, v_n\}$ in the fibre $E_{\gamma(t_0)}$ with

$$<v_i, v_i> = \varepsilon_i = \begin{cases} -1 & \text{for } \ 1 \le i \le r \\ \ \ 1 & \text{for } \ r+1 \le i \le n \end{cases}$$

for a point $t_0 \in I$, and let $V_i : I \to E$ be the lifting of γ obtained from v_i by parallel transport along γ for $i = 1, \ldots, n$. We know that V_i is smooth since

$$V_i(t) = \tau_t^{t_0}(v_i) = \mu_{\alpha(t)} \circ \mu_{\alpha(t_0)}^{-1}(v_i) = \nu_{w_i} \circ \alpha(t)$$

for $t \in I$, where $\alpha : I \to P$ is a horizontal lifting of γ, and $\nu_{w_i} : P \to E$ with $w_i = \mu_{\alpha(t_0)}^{-1}(v_i) \in W$ is the smooth map defined in 10.28. Since $\tau_t^{t_0}$ is an isometry, we have that $\mathscr{B}_t = \{V_1(t),...,V_n(t)\}$ is an orthonormal basis in the fibre $E_{\gamma(t)}$ with $< V_i(t), V_i(t) > = \varepsilon_i$ for $i = 1, \dots, n$ and $t \in I$.

Now if $F_1 : I \to E$ and $F_2 : I \to E$ are two liftings of γ, then there are smooth maps $a : I \to \mathbf{R}^n$ and $b : I \to \mathbf{R}^n$ so that

$$F_1 = \sum_{i=1}^{n} a_i V_i \quad \text{and} \quad F_2 = \sum_{i=1}^{n} b_i V_i$$

by Proposition 10.14. As $DV_i/dr = 0$ on I for $i = 1, \dots, n$, it follows from Proposition 10.88 that

$$\frac{DF_1}{dr} = \sum_{i=1}^{n} \frac{da_i}{dr} V_i \quad \text{and} \quad \frac{DF_2}{dr} = \sum_{i=1}^{n} \frac{db_i}{dr} V_i \,.$$

Using that the liftings $V_1, \dots, V_n$ are everywhere orthonormal, we therefore have that

$$\frac{d}{dr} < F_1, F_2 > = \frac{d}{dr} \sum_{i=1}^{n} \varepsilon_i a_i b_i = \sum_{i=1}^{n} \varepsilon_i \left(\frac{da_i}{dr} b_i + a_i \frac{db_i}{dr} \right)$$

$$= \left\langle \frac{DF_1}{dr}, F_2 \right\rangle + \left\langle F_1, \frac{DF_2}{dr} \right\rangle$$

which completes the proof of the only if part of the proposition. □

10.161 Lemma Let $\pi : P \to M$ be a principal G-bundle with a connection H, and let g be a fibre metric in the associated vector bundle $\pi' : E \to M$ obtained from a representation $\psi : G \to Aut\ (W)$ of G on a finite dimensional vector space W. Suppose that

$$X < s_1, s_2 > \; = \; < \nabla_X s_1, s_2 > + < s_1, \nabla_X s_2 > \tag{1}$$

for all vector fields $X \in \mathscr{T}_1(M)$ and sections $s_1, s_2 \in \Gamma(M;E)$. If $f : N \to M$ is a smooth map from a smooth manifold N, then we have that

$$Y < F_1, F_2 > \; = \; < \nabla_Y F_1, F_2 > + < F_1, \nabla_Y F_2 > \tag{2}$$

for all vector fields $Y \in \mathscr{T}_1(N)$ and liftings F_1, F_2 of f.

PROOF: Fix a point $p \in N$, and let $\sigma_1, \dots, \sigma_n$ be a local basis for E on an open neighbourhood U of $f(p)$. If $F_1 : N \to E$ and $F_2 : N \to E$ are liftings of f, then there are smooth maps $\alpha : f^{-1}(U) \to \mathbf{R}^n$ and $\beta : f^{-1}(U) \to \mathbf{R}^n$ so that

$$F_1(q) = \sum_{i=1}^{n} \alpha_i(q)\, \sigma_i \circ f(q) \quad \text{and} \quad F_2(q) = \sum_{j=1}^{n} \beta_j(q)\, \sigma_j \circ f(q)$$

for $q \in f^{-1}(U)$ by Proposition 10.14.

Choose an open neighbourhood V of $f(p)$ with $\overline{V} \subset U$. Then it follows by Proposition 2.56 (1) and Corollary 2.25 that there are sections $s_1, \dots, s_n$ of π' on M and smooth maps $a : N \to \mathbf{R}^n$ and $b : N \to \mathbf{R}^n$ which coincide with $\sigma_1, \dots, \sigma_n$ and α, β

on $\overline{V}$ and $f^{-1}(\overline{V})$, respectively. Using Propositions 10.87 and 10.88 we therefore have that

$$\nabla_{Y_p} F_1 = \sum_{i=1}^{n} \{ Y_p(a_i)\, s_i(f(p)) + a_i(p)\, \nabla_{f_*(Y_p)}\, s_i \},$$

and we have a similar formula for $\nabla_{Y_p} F_2$. Combining these and using (1) it follows that

$$\begin{aligned}
&<\nabla_{Y_p} F_1, F_2(p)> + <F_1(p), \nabla_{Y_p} F_2> \\
&= \sum_{i,j} \{ Y_p(a_i)\, b_j(p) + a_i(p)\, Y_p(b_j) \} <s_i(f(p)), s_j(f(p))> \\
&+ \sum_{i,j} a_i(p)\, b_j(p)\, \{ <\nabla_{f_*(Y_p)}\, s_i, s_j(f(p))> + <s_i(f(p)), \nabla_{f_*(Y_p)}\, s_j> \} \\
&= \sum_{i,j} \{ Y_p(a_i b_j) <s_i(f(p)), s_j(f(p))> + a_i(p)\, b_j(p)\, f_*(Y_p) <s_i, s_j> \} \\
&= \sum_{i,j} Y_p\, (a_i b_j <s_i \circ f, s_j \circ f>) = Y_p <F_1, F_2>
\end{aligned}$$

which completes the proof of (2), since the point $p \in N$ was arbtrary. □

10.162 Proposition Let $\pi : P \to M$ be a principal G-bundle with a connection H, and let g be a fibre metric in the associated vector bundle $\pi' : E \to M$ obtained from a representation $\psi : G \to Aut\,(W)$ of G on a finite dimensional vector space W. Then the connection H is compatible with g if and only if

$$X <s_1, s_2> = <\nabla_X s_1, s_2> + <s_1, \nabla_X s_2> \tag{1}$$

for all vector fields $X \in \mathscr{T}_1(M)$ and sections $s_1, s_2 \in \Gamma(M; E)$.

PROOF : Suppose that H is compatible with g, and let $p \in M$ and $\gamma : I \to M$ be a smooth curve defined on an open interval I containing 0 with $\gamma(0) = p$ and $\gamma'(0) = X_p$. If (r, I) is the standard local chart on I where $r : I \to \mathbf{R}$ is the inclusion map, then it follows from 2.77 and Proposition 10.87 that

$$X_p <s_1, s_2> = \tfrac{d}{dr}\Big|_0 <s_1 \circ \gamma, s_2 \circ \gamma>$$

and

$$\nabla_{X_p} s_i = \nabla_{\gamma_*\left(\frac{d}{dr}\big|_0\right)} s_i = \nabla_{\frac{d}{dr}\big|_0} (s_i \circ \gamma) = \frac{D(s_i \circ \gamma)}{dr}(0)$$

for $i = 1, 2$. Formula (1) hence follows at the arbitrary point $p \in M$ by applying Proposition 10.160 to the liftings $s_1 \circ \gamma$ and $s_2 \circ \gamma$ of γ.

Conversely, assuming that (1) is satisfied, it follows from Proposition 10.160 and Lemma 10.161, applied to an arbitrary smooth curve $\gamma : I \to M$ defined on an open interval I and the vector field $\frac{d}{dr}$ on I, that H is compatible with g. □

10.163 Proposition Let X and Y be vector fields and α a 1-form on a pseudo-Riemannian manifold M with a metric g. Then we have that

$$(\nabla_X Y)^\flat = \nabla_X Y^\flat \quad \text{and} \quad (\nabla_X \alpha)^\sharp = \nabla_X \alpha^\sharp$$

when ∇ is the covariant derivative corresponding to a linear connection H on M compatible with g.

PROOF: By Propositions 10.162 and 10.83 we have that

$$\begin{aligned}(\nabla_X Y)^\flat(Z) &= <\nabla_X Y, Z> = X<Y,Z> - <Y, \nabla_X Z> \\ &= \nabla_X(Y^\flat(Z)) - Y^\flat(\nabla_X Z) = (\nabla_X Y^\flat)(Z)\end{aligned}$$

for every vector field Z on M, which completes the proof of the first formula. The second formula is obtained from the first with $Y = \alpha^\sharp$ by applying $g^\sharp$ on both sides. □

10.164 Corollary Let X be a vector field and T a tensor field of type $\binom{k}{l}$ on a pseudo-Riemannian manifold M with a metric g. Then we have that

$$L^r_s(\nabla_X T) = \nabla_X L^r_s(T) \quad \text{and} \quad R^r_s(\nabla_X T) = \nabla_X R^r_s(T)$$

when ∇ is the covariant derivative corresponding to a linear connection H on M compatible with g.

PROOF: Follows from Propositions 10.163 and 10.83. □

10.165 Theorem Let M be a pseudo-Riemannian manifold with a metric g, and let $T \in \Omega^2(M;TM)$ be a bundle-valued 2-form on M with values in the tangent bundle. Then there is a unique linear connection H on M with torsion form T which is compatible with g. It is given by

$$\begin{aligned}<\nabla_X Y, Z> = \frac{1}{2}\{&X<Z,Y> + Y<Z,X> - Z<X,Y> \\ &+ <X,[Z,Y]> + <Y,[Z,X]> + <Z,[X,Y]> \\ &+ <X,T(Z,Y)> + <Y,T(Z,X)> + <Z,T(X,Y)>\}\end{aligned} \tag{1}$$

for every vector field $X,Y,Z \in \mathscr{T}_1(M)$. In the case when $T = 0$, H is called the *Levi–Civita connection* on M.

PROOF: If H is compatible with g, then it follows from Proposition 10.162 that

$$Z<X,Y> = <\nabla_Z X, Y> + <X, \nabla_Z Y> \tag{2}$$

for each vector field $X, Y, Z \in \mathscr{T}_1(M)$. By a permutation of the vector fields X, Y and Z, we also have that

$$X <Z,Y> = <\nabla_X Z,Y> + <Z,\nabla_X Y> \tag{3}$$

and

$$Y <Z,X> = <\nabla_Y Z,X> + <Z,\nabla_Y X>, \tag{4}$$

from which we obtain

$$2 <\nabla_X Y,Z> - X <Z,Y> - Y <Z,X> + Z <X,Y>$$

$$= <X,\nabla_Z Y - \nabla_Y Z> + <Y,\nabla_Z X - \nabla_X Z> + <Z,\nabla_X Y - \nabla_Y X> .$$

Using Proposition 10.102 we therefore obtain formula (1) which shows the uniqueness of ∇ and therefore of H.

To show the existence of H, we fix vector fields X and Y on M, and let $\omega : \mathscr{T}_1(M) \to \mathscr{F}(M)$ be the map where $\omega(Z)$ is the right-hand side of (1) for every vector field $Z \in \mathscr{T}_1(M)$. Then ω is linear over $\mathscr{F}(M)$. Indeed, if $f \in \mathscr{F}(M)$, we have that

$$\omega(fZ) = \frac{1}{2} \{ X <fZ,Y> + Y <fZ,X> - fZ <X,Y>$$

$$+ <X,[fZ,Y]> + <Y,[fZ,X]> + <fZ,[X,Y]>$$

$$+ <X,T(fZ,Y)> + <Y,T(fZ,X)> + <fZ,T(X,Y)> \} \tag{5}$$

$$= f\,\omega(Z) + \frac{1}{2} \{ X(f) <Z,Y> + Y(f) <Z,X>$$

$$- Y(f) <X,Z> - X(f) <Y,Z> \} = f\,\omega(Z).$$

Hence ω may be considered to be a 1-form on M, and by Proposition 7.18 there is a unique vector field $\nabla_X Y$ on M so that $\omega(Z) = <\nabla_X Y,Z>$ for every $Z \in \mathscr{T}_1(M)$.

Next we need to show that ∇ satisfies the conditions (1) – (4) for a Koszul connection given in Definition 10.125. The first two conditions follow immediately from formula (1). To prove the last two conditions, let $f \in \mathscr{F}(M)$. Then we have that

$$<\nabla_{fX} Y,Z> = \frac{1}{2} \{ fX <Z,Y> + Y <Z,fX> - Z <fX,Y>$$

$$+ <fX,[Z,Y]> + <Y,[Z,fX]> + <Z,[fX,Y]>$$

$$+ <fX,T(Z,Y)> + <Y,T(Z,fX)> + <Z,T(fX,Y)> \} \tag{6}$$

$$= <f\,\nabla_X Y,Z> + \frac{1}{2} \{ Y(f) <Z,X> - Z(f) <X,Y>$$

$$+ Z(f) <Y,X> - Y(f) <Z,X> \} = <f\,\nabla_X Y,Z>$$

and

$$<\nabla_X(fY),Z> = \frac{1}{2}\{X<Z,fY> + fY<Z,X> - Z<X,fY>$$

$$+ <X,[Z,fY]> + <fY,[Z,X]> + <Z,[X,fY]>$$

$$+ <X,T(Z,fY)> + <fY,T(Z,X)> + <Z,T(X,fY)>\} \qquad (7)$$

$$= <f\nabla_X Y,Z> + \frac{1}{2}\{X(f)<Z,Y> - Z(f)<X,Y>$$

$$+ Z(f)<X,Y> + X(f)<Z,Y>\} = <X(f)Y + f\nabla_X Y,Z>$$

for every $Z \in \mathscr{T}_1(M)$, which completes the proof that ∇ is a Koszul connection and hence gives rize to a unique connection H by Corollary 10.132.

Finally, we must show that H has torsion form T and is compatible with the metric g. This follows from (1) which implies that

$$<\nabla_X Y - \nabla_Y X - [X,Y],Z> = <T(X,Y),Z>$$

and

$$<\nabla_X Y,Z> + <Y,\nabla_X Z> = X<Y,Z>$$

for every $X,Y,Z \in \mathscr{T}_1(M)$. □

10.166 Proposition Let M^n be a pseudo-Riemannian manifold with a metric g, and let (x,U) be a local chart on M. Then the components of the Levi-Civita connection with respect to the coordinate frame field

$$s = \left(\frac{\partial}{\partial x^1}, \dots, \frac{\partial}{\partial x^n}\right)$$

are given by

$$\Gamma^i_{jk} = \sum_{l=1}^{n} g^{il}\,[jk,l]$$

where

$$[jk,l] = \frac{1}{2}\left\{\frac{\partial g_{lk}}{\partial x^j} + \frac{\partial g_{lj}}{\partial x^k} - \frac{\partial g_{jk}}{\partial x^l}\right\}.$$

The terms $[jk,l]$ and Γ^i_{jk} are called *Christoffel symbols* of the first and second kind, respectively.

PROOF: Inserting the vector fields $X = \frac{\partial}{\partial x^j}$, $Y = \frac{\partial}{\partial x^k}$ and $Z = \frac{\partial}{\partial x^l}$ into formula (1) in Theorem 10.165 and using Corollary 10.111, we obtain

$$[jk,l] = \sum_{i=1}^{n} g_{il}\,\Gamma^i_{jk} = \frac{1}{2}\left\{\frac{\partial g_{lk}}{\partial x^j} + \frac{\partial g_{lj}}{\partial x^k} - \frac{\partial g_{jk}}{\partial x^l}\right\}.$$

□

10.167 Proposition Let $\pi : TM \to M$ be the tangent bundle of a pseudo-Riemannian manifold M^n with a metric g, and let $L : TM \to \mathbf{R}$ be the hyperregular Lagrangian defined by

$$L(v) = \tfrac{1}{2}\, g(u)(v,v)$$

for $u \in M$ and $v \in T_uM$ as described in Proposition 7.101. Then a smooth curve $\gamma : I \to M$ is a geodesic on M if and only if the curve $\gamma' : I \to TM$ is an integral curve for the Lagrangian vector field X_E of L.

PROOF: Let (x,U) be a local chart on M, and let (y,V) be the corresponding local chart on TM defined in Proposition 7.81. We denote the two component maps of $y : V \to \mathbf{R}^n \times \mathbf{R}^n$ by q and $\dot{q}$, so that $q = x \circ \pi$. Since

$$L|_V = \tfrac{1}{2} \sum_{i,j=1}^{n} (g_{ij} \circ \pi)\, \dot{q}^i \dot{q}^j \,,$$

the Lagrange's equations for the curve $\alpha = \gamma'$ take the form

$$(x^i \circ \gamma)'(t) = (\dot{q}^i \circ \alpha)(t)$$

and

$$\Big\{ \sum_{j=1}^{n} (g_{ij} \circ \gamma)(\dot{q}^j \circ \alpha) \Big\}'(t) = \frac{1}{2} \sum_{j,k=1}^{n} \frac{\partial g_{jk}}{\partial x^i}(\gamma(t))\, (\dot{q}^j \circ \alpha)(t)\, (\dot{q}^k \circ \alpha)(t)$$

for $t \in I \cap \gamma^{-1}(U)$ and $i = 1, \ldots, n$. By Definition 2.69 the first equation is obviously true. The second equation can also be written in the form

$$\sum_{j=1}^{n} g_{ij}(\gamma(t))\, (x^j \circ \gamma)''(t)$$

$$= -\sum_{j,k=1}^{n} \Big\{ \frac{\partial g_{ij}}{\partial x^k}(\gamma(t)) - \frac{1}{2} \frac{\partial g_{jk}}{\partial x^i}(\gamma(t)) \Big\} (x^j \circ \gamma)'(t)\, (x^k \circ \gamma)'(t)$$

$$= -\sum_{j,k=1}^{n} \frac{1}{2} \Big\{ \frac{\partial g_{ik}}{\partial x^j}(\gamma(t)) + \frac{\partial g_{ij}}{\partial x^k}(\gamma(t)) - \frac{\partial g_{jk}}{\partial x^i}(\gamma(t)) \Big\} (x^j \circ \gamma)'(t)\, (x^k \circ \gamma)'(t)\,,$$

which is equivalent to the geodesic equation

$$(x^i \circ \gamma)''(t) + \sum_{j,k=1}^{n} \Gamma^i_{jk}(\gamma(t))\, (x^j \circ \gamma)'(t)\, (x^k \circ \gamma)'(t) = 0$$

for $t \in I \cap \gamma^{-1}(U)$ and $i = 1, \ldots, n$. □

10.168 Proposition Let $\pi_i : P_i \to M_i$ be a principle G_i-bundle for $i = 1, 2$, and let

$(\tilde{f}, f, \psi)$ be a homomorphism from π_1 to π_2 where $f : M_1 \to M_2$ is a diffeomorphism. If H_1 is a connection in P_1, then there is a unique connection H_2 in P_2 so that

$$\tilde{f}_*(H_{1,u}) = H_{2,\tilde{f}(u)}$$

for every $u \in P_1$. If ω_i and Ω_i are the connection form and the curvature form of H_i on P_i, respectively, for $i = 1, 2$, then

$$\tilde{f}^*\omega_2 = \psi_{*}\cdot\, \omega_1 \quad \text{and} \quad \tilde{f}^*\Omega_2 = \psi_{*}\cdot\, \Omega_1 \,.$$

PROOF : We first prove the uniqueness of the connection H_2. Let $v \in P_2$, and choose a $u \in \pi_1^{-1}(f^{-1} \circ \pi_2(v))$. Since $\pi_2(\tilde{f}(u)) = f(\pi_1(u)) = \pi_2(v)$, there is a $g \in G_2$ with $v = \tilde{f}(u)\, g$. Hence we have that

$$H_{2,v} = (R_g)_* H_{2,\tilde{f}(u)} = (R_g)_* \circ \tilde{f}_*(H_{1,u})$$

which completes the proof of the uniqueness of H_2.

To prove existence of the connection H_2, let $s_1 : V_1 \to P_1$ be a section in $\pi_1 : P_1 \to M_1$ defined on an open subset V_1 of M_1. Then $s_2 = \tilde{f} \circ s_1 \circ f^{-1}$ is a section in $\pi_2 : P_2 \to M_2$ on the open subset $V_2 = f(V_1)$ of M_2. Let $\widetilde{\omega}_1 = s_1^*(\omega_1)$ be the pull-back of ω_1 by s_1, and let $\widetilde{\omega}_2$ be the $\mathfrak{g}_2$-valued 1-form on V_2 given by $\widetilde{\omega}_2 = \psi_*\cdot (f^{-1})^*\,\widetilde{\omega}_1$ so that $f^*\,\widetilde{\omega}_2 = \psi_* \cdot \widetilde{\omega}_1$ by Proposition 5.48. By Proposition 10.56 there is a unique connection form η_2 on $\pi_2{}^{-1}(V_2)$ with $\widetilde{\omega}_2 = s_2^*(\eta_2)$. We want to show that $\tilde{f}^*\,\eta_2 = \psi_* \cdot\, \eta_1$, where η_1 is the restriction of ω_1 to $\pi_1{}^{-1}(V_1)$.

If $\rho_r : V_r \times G_r \to \pi_r{}^{-1}(V_r)$ is the map defined by

$$\rho_r(p,g) = s_r(p)\, g$$

for $p \in V_r$ and $g \in G_r$, and θ_r is the canonical Maurer-Cartan form on G_r for $r = 1, 2$, then we have that

$$\eta_r(\rho_r(p,g))\,(\rho_{r,g\,*}(v) + \rho_{r,p\,*}(w)) = Ad\,(g^{-1})\ \widetilde{\omega}_r(p)(v) + \theta_r(g)(w)$$

for $p \in V_r$, $g \in G_r$, $v \in T_pM_r$ and $w \in T_gG_r$. Furthermore, we have that

$$\tilde{f} \circ \rho_1(p,g) = \tilde{f}(s_1(p)\, g) = \tilde{f} \circ s_1(p)\ \psi(g) = s_2 \circ f(p)\ \psi(g) = \rho_2(f(p), \psi(g))$$

so that

$$\tilde{f} \circ \rho_{1,g} = \rho_{2,\psi(g)} \circ f \quad \text{and} \quad \tilde{f} \circ \rho_{1,p} = \rho_{2,f(p)} \circ \psi$$

for $p \in V_1$ and $g \in G_1$, and we also have that $\psi^*\,\theta_2 = \psi_* \cdot \theta_1$ since

$$\begin{aligned} \psi^*\,\theta_2(g)(w) &= \theta_2(\psi(g))(\psi_*(w)) = (L_{\psi(g)^{-1}})_* \circ\ \psi_*(w) \\ &= (L_{\psi(g)^{-1}} \circ\ \psi)_*(w) = (\psi \circ L_{g^{-1}})_*(w) = \psi_*\ \theta_1(g)(w) \end{aligned}$$

for $g \in G_1$ and $w \in T_gG_1$. Using this and Proposition 9.14, it follows that

$$
\begin{aligned}
&\tilde{f}^*\eta_2(\rho_1(p,g))(\rho_{1,g*}(v)+\rho_{1,p*}(w))\\
&=\eta_2(\tilde{f}\circ\rho_1(p,g))(\tilde{f}_*\circ\rho_{1,g*}(v)+\tilde{f}_*\circ\rho_{1,p*}(w))\\
&=\eta_2(\rho_2(f(p),\psi(g)))(\rho_{2,\psi(g)*}\circ f_*(v)+\rho_{2,f(p)*}\circ\psi_*(w))\\
&=Ad(\psi(g)^{-1})\,\widetilde{\omega}_2(f(p))(f_*(v))+\theta_2(\psi(g))(\psi_*(w))\\
&=Ad(\psi(g)^{-1})\,f^*\widetilde{\omega}_2(p)(v)+\psi^*\theta_2(g)(w)\\
&=Ad(\psi(g)^{-1})\,\psi_*\widetilde{\omega}_1(p)(v)+\psi_*\theta_1(g)(w)\\
&=\psi_*\{Ad(g^{-1})\,\widetilde{\omega}_1(p)(v)+\theta_1(g)(w)\}\\
&=\psi_*\eta_1(\rho_1(p,g))(\rho_{1,g*}(v)+\rho_{1,p*}(w))
\end{aligned}
$$

for $p\in V_1$, $g\in G_1$, $v\in T_pM_1$ and $w\in T_gG_1$, which completes the proof that $\tilde{f}^*\eta_2=\psi_*\cdot\eta_1$.

Now let $\widetilde{H}_r$ be the connection on $\pi_r{}^{-1}(V_r)$ with connection form η_r for $r=1,2$. Then we have that $\tilde{f}_*(\widetilde{H}_{1,u})\subset\widetilde{H}_{2,\tilde{f}(u)}$ for $u\in\pi_1{}^{-1}(V_1)$. If $i_r:\widetilde{H}_r\to TP_r$ is the inclusion map and $\alpha_r=\pi_{r*}\circ i_r$, then it follows from Proposition 10.49 that $\tilde{f}_{*u}$ induces a linear isomorphism $\alpha^{-1}_{2,\tilde{f}(u)}\circ f_{*\pi_1(u)}\circ\alpha_{1,u}$ between $\widetilde{H}_{1,u}$ and $\widetilde{H}_{2,\tilde{f}(u)}$ so that in fact $\tilde{f}_*(\widetilde{H}_{1,u})=\widetilde{H}_{2,\tilde{f}(u)}$ for each $u\in\pi_1{}^{-1}(V_1)$.

Suppose that $\widetilde{H}_2'$ is the connection on $\pi_2{}^{-1}(V_2')$ obtained in the same way as above from another section $s_1':V_1'\to P_1$ in $\pi_1:P_1\to M_1$ where $V_2'=f(V_1')$. Then it follows from the uniqueness part of the proposition that $\widetilde{H}_2$ and $\widetilde{H}_2'$ coincide on $\pi_2{}^{-1}(V_2\cap V_2')$. Hence there is a connection H_2 in P_2 with a connection form ω_2 satisfying the properties in the proposition.

Using Propositions 5.62 and 5.63, we also have that $\tilde{f}^*(d\,\omega_2)=\psi_*\cdot d\,\omega_1$. Hence it follows by the second structure equation and the fact that ψ_* is a Lie algebra homomorphism that

$$
\begin{aligned}
&\tilde{f}^*\Omega_2(u)(w_1,w_2)=\Omega_2(\tilde{f}(u))(\tilde{f}_*(w_1),\tilde{f}_*(w_2))\\
&=d\,\omega_2(\tilde{f}(u))(\tilde{f}_*(w_1),\tilde{f}_*(w_2))+[\,\omega_2(\tilde{f}(u))(\tilde{f}_*(w_1)),\,\omega_2(\tilde{f}(u))(\tilde{f}_*(w_2))\,]\\
&=\tilde{f}^*(d\,\omega_2)(u)(w_1,w_2)+[\,\tilde{f}^*\omega_2(u)(w_1),\,\tilde{f}^*\omega_2(u)(w_2)\,]\\
&=\psi_*\{d\,\omega_1(u)(w_1,w_2)+[\,\omega_1(u)(w_1),\,\omega_1(u)(w_2)\,]\}=\psi_*\Omega_1(u)(w_1,w_2)
\end{aligned}
$$

for $u\in P_1$ and $w_1,w_2\in T_uP_1$, which shows that $\tilde{f}^*\Omega_2=\psi_*\cdot\Omega_1$ and completes the proof of the proposition. □

10.169 Definition Let $\pi_1:P_1\to M$ be a reduced subbundle of a principal fibre bundle $\pi_2:P_2\to M$. Then a connection H_2 in P_2 is said to be *reducible* to a connection H_1 in P_1 if H_2 is obtained from H_1 in the way described in Proposition 10.168.

10.170 Proposition Let $\pi : E \to M$ be an n-dimensional vector bundle with a fibre metric g of index k, and let ε be a signature compatible with g. Then we have a reduced subbundle $\pi'' : \mathscr{O}_\varepsilon(E) \to M$ of the frame bundle $\pi' : \mathscr{L}(E) \to M$, called the *orthonormal frame bundle of E of signature* ε, with fibre $\mathscr{O}_{\varepsilon,p}(E) = \{u \in \mathscr{L}_p(E) | u^* g(p) = g^n_\varepsilon\}$ for $p \in M$ and structure group $O_\varepsilon(\mathbf{R}^n)$. If an ε is not specified, we will assume that ε is the canonical signature for g, and the bundle $\mathscr{O}_\varepsilon(E)$ is denoted simply by $\mathscr{O}(E)$ and is called the *orthonormal frame bundle* of E.

PROOF: Let $p_0 \in M$, and choose an orthonormal basis $\mathscr{B} = \{v_1, ..., v_n\}$ for E_{p_0} with signature ε. If $(t, \pi^{-1}(U))$ is a local trivialization in the vector bundle $\pi : E \to M$, where U is an open neighbourhood of p_0 on M, then we have a section $s' = (s'_1, \ldots, s'_n)$ of $\pi' : \mathscr{L}(E) \to M$ on U with $s'_i(p_0) = v_i$ for $i = 1, \ldots, n$, given by

$$s'_i(p) = t_{x,p}^{-1} \circ t_{x,p_0}(v_i)$$

for $p \in U$ and $i = 1, \ldots, n$. Let V be an open connected neighbourhood of p_0 contained in U so that the $i \times i$-matrix $\left(g(p)(s'_\alpha(p), s'_\beta(p)) \right)$, where $1 \leq \alpha, \beta \leq i$, is non-singular for each $p \in V$ and $1 \leq i \leq n$. Then the subspace $V_i(p)$ of E_p spanned by $\{s'_1(p), \ldots, s'_i(p)\}$ is non-degenerate for $p \in V$ and $1 \leq i \leq n$. Using this, we are going to constuct a section $s = (s_1, \ldots, s_n)$ of $\pi'' : \mathscr{O}_\varepsilon(E) \to M$ on V with $s_i(p_0) = v_i$ for $i = 1, \ldots, n$.

By applying the Gram-Schmidt process to $\{s'_1(p), \ldots, s'_n(p)\}$, we obtain an orthogonal basis $\{s''_1(p), \ldots, s''_n(p)\}$ for E_p with $s''_i(p_0) = v_i$ for $i = 1, \ldots, n$, given by

$$s''_1(p) = s'_1(p)$$

and

$$s''_i(p) = s'_i(p) - \sum_{j=1}^{i-1} \frac{g(p)\,(s'_i(p), s''_j(p))}{g(p)\,(s''_j(p), s''_j(p))}\, s''_j(p)$$

for $p \in V$ and $i = 2, \ldots, n$. Indeed, assuming inductively that $\{s''_1(p), \ldots, s''_{i-1}(p)\}$ is an orthogonal basis for $V_{i-1}(p)$, we have that $s''_j(p) \notin V_{i-1}(p)^\perp$ so that $g(p)(s''_j(p), s''_j(p)) \neq 0$ for $j = 1, \ldots, i-1$ since $V_{i-1}(p)$ is a non-degenerate subspace of E_p, thus showing that $s''_i(p)$ is well defined. Furthermore, $g(p)(s''_i(p), s''_r(p)) = 0$ for $r = 1, \ldots, i-1$ which shows that $\{s''_1(p), \ldots, s''_i(p)\}$ is an orthogonal basis for $V_i(p)$. Hence we have an orthonormal basis $\{s_1(p), \ldots, s_n(p)\}$ for E_p with $s_i(p_0) = v_i$ for $i = 1, \ldots, n$, given by

$$s_i(p) = s''_i(p) \;/\; \|s''_i(p)\|$$

for $p \in V$ and $i = 1, \ldots, n$.

Since the map $f_i : V \to \mathbf{R}$ given by $f_i(p) = g(p)(s_i(p), s_i(p))$ for $p \in V$ is smooth and V is connected, we have that $f_i(p) = g(p_0)(v_i, v_i) = \varepsilon_i$ for $p \in V$ and $i = 1, \ldots, n$. This implies that

$$\begin{aligned} s(p)^* g(p)(a,b) &= g(p)(s(p)\,a, s(p)\,b) \\ &= g(p)\left(\sum_{i=1}^n a^i\, s_i(p), \sum_{j=1}^n b^j\, s_j(p) \right) = g^n_\varepsilon(a,b) \end{aligned}$$

for every $a,b \in \mathbf{R}^n$, which shows that $s(p) \in \mathscr{O}_{\varepsilon,p}(E)$ for $p \in V$. Hence the local trivialization $(\rho^{-1}, \pi'^{-1}(V))$ in the frame bundle $\pi' : \mathscr{L}(E) \to M$, where $\rho : V \times Gl(\mathbf{R}^n) \to \pi'^{-1}(V)$ is the diffeomorphism given by

$$\rho(p,G) = s(p) \circ G$$

for $p \in V$ and $G \in Gl(\mathbf{R}^n)$, induces a bijection $t_s : \pi''^{-1}(V) \to V \times O_\varepsilon(\mathbf{R}^n)$. Indeed, since $\rho(p,G)^* g(p) = G^* g^n_\varepsilon$ for $p \in V$, we have that $\rho(p,G) \in \pi''^{-1}(V)$ for $p \in V$ if and only if $G \in O_\varepsilon(\mathbf{R}^n)$. If $\tilde{s} = (\tilde{s}_1, \dots, \tilde{s}_n)$ is another section of $\pi'' : \mathscr{O}_\varepsilon(E) \to M$ of index ε on some open set $\widetilde{V}$, and if $\phi : V \cap \widetilde{V} \to Gl(\mathbf{R}^n)$ is the transition map between the corresponding local trivializations in $\mathscr{L}(E)$, then $\phi(p) \in O_\varepsilon(\mathbf{R}^n)$ for $p \in V \cap \widetilde{V}$, since $s(p) = \tilde{s}(p) \circ \phi(p)$ by Proposition 10.52 which implies that $\phi(p)^* g^n_\varepsilon = \phi(p)^* \tilde{s}(p)^* g(p) = s(p)^* g(p) = g^n_\varepsilon$ for $p \in V \cap \widetilde{V}$. Hence ϕ induces a smooth map $\psi : V \cap \widetilde{V} \to O_\varepsilon(\mathbf{R}^n)$ so that

$$t_{\tilde{s}} \circ t_s^{-1}(p,G) = (p, \psi(p) \circ G)$$

for every $p \in V \cap \widetilde{V}$ and $G \in O_\varepsilon(\mathbf{R}^n)$. By Proposition 10.5 there is therefore a unique topology, smooth structure and fibre bundle structure on $\mathscr{O}_\varepsilon(E)$ such that $\pi'' : \mathscr{O}_\varepsilon(E) \to M$ is a principal fibre bundle over M with structure group $O_\varepsilon(\mathbf{R}^n)$, and such that $(t_s, \pi''^{-1}(V))$ is a local trivialization for each section $s = (s_1, \dots, s_n)$ of $\pi'' : \mathscr{O}_\varepsilon(E) \to M$ of index ε on some open set V. We see that the inclusion map $i : \mathscr{O}_\varepsilon(E) \to \mathscr{L}(E)$ is an immersion, and hence $\pi'' : \mathscr{O}_\varepsilon(E) \to M$ is a reduced subbundle of the frame bundle $\pi' : \mathscr{L}(E) \to M$. □

10.171 Remark If M^n is a pseudo-Riemannian manifold and ε is a signature compatible with its metric g, then the orthonormal frame bundle $\mathscr{O}_\varepsilon(TM)$ of the tangent bundle of M is denoted simply by $\mathscr{O}_\varepsilon(M)$ and is called the *orthonormal frame bundle of M of signature* ε. It is a reduced subbundle of the frame bundle $\mathscr{L}(M)$ of M. If ε is the canonical signature for g, the bundle $\mathscr{O}_\varepsilon(M)$ is denoted simply by $\mathscr{O}(M)$ and is called the *orthonormal frame bundle* of M. The fibres of these orthonormal frame bundles are denoted by $\mathscr{O}_{\varepsilon,p}(M)$ and $\mathscr{O}_p(M)$, respectively, for $p \in M$.

10.172 Proposition Let $\pi : E \to M$ be an n-dimensional vector bundle with a fibre metric g of index k, and let ε be a signature compatible with g. Then a connection H in $\mathscr{L}(E)$ is a metrical connection if and only if it is reducible to a connection H' in the orthonormal frame bundle $\mathscr{O}_\varepsilon(E)$ of E of signature ε.

PROOF: Suppose that H is reducible to a connection H' in $\mathscr{O}_\varepsilon(E)$, and let $\gamma : I \to M$ be a smooth curve on M defined on an open interval I. If $t_0, t_1 \in I$ and $u \in \mathscr{O}_{\varepsilon,\gamma(t_0)}(E)$, then the horizontal lifting $\beta : I \to \mathscr{O}_\varepsilon(E)$ of γ in $\mathscr{O}_\varepsilon(E)$ with $\beta(t_0) = u$ is also horizontal in $\mathscr{L}(E)$. Hence if $v_0, v_1 \in E_{\gamma(t_0)}$ and $\xi_i = u^{-1}(v_i)$ for $i = 1,2$, we have that

$$g(\gamma(t_0))(v_0,v_1) = u^*g(\gamma(t_0))(\xi_0,\xi_1) = g^n_\varepsilon(\xi_0,\xi_1)$$
$$= \beta(t_1)^*g(\gamma(t_1))(\xi_0,\xi_1) = g(\gamma(t_1))(\tau^{t_0}_{t_1}(v_0),\tau^{t_0}_{t_1}(v_1))$$

which shows that the parallel transport $\tau^{t_0}_{t_1}$ is an isometry.

Now assume conversely that H is a metrical connection, and let $v \in H_u$ for a $u \in \mathscr{O}_{\varepsilon,p}(E)$. Choose a smooth curve $\gamma : I \to M$ defined on an open interval I containing t_0 with $\gamma(t_0) = p$ and $\gamma'(t_0) = \pi_*(v)$, where $\pi : \mathscr{L}(E) \to M$ is the projection in the frame bundle, and let $\beta : I \to \mathscr{L}(E)$ be the horizontal lifting of γ with $\beta(t_0) = u$. Then we have that $\beta'(t_0) = v$, and

$$\beta(t)^*g(\gamma(t))(\xi_0,\xi_1) = \beta(t_0)^*g(\gamma(t_0))(\xi_0,\xi_1) = g^n_\varepsilon(\xi_0,\xi_1)$$

for $\xi_0,\xi_1 \in \mathbf{R}^n$, which shows that $\beta(t) \in \mathscr{O}_\varepsilon(E)$ for all $t \in I$. Hence we have that $H_u \subset T_u\mathscr{O}_\varepsilon(E)$ for every $u \in \mathscr{O}_\varepsilon(E)$, showing that H is reducible to a connection in $\mathscr{O}_\varepsilon(E)$. □

10.173 Proposition Let M^n be a pseudo-Riemannian manifold with a metric g of index r, and let $s : V \to \mathscr{O}(M)$ be a section in the orthonormal frame bundle $\pi : \mathscr{O}(M) \to M$. Let $\widetilde{\theta} = s^*\theta$ be the pull-back of the dual form θ by s, and let $\widetilde{\Theta}$ be an $\mathbf{R}^n$-valued 2-form on V. Then there is a unique $\mathfrak{o}_r(\mathbf{R}^n)$-valued 1-form $\widetilde{\omega}$ on V which satisfies the structure equation

$$d\widetilde{\theta} = -\,\widetilde{\omega}\wedge\widetilde{\theta} + \widetilde{\Theta}. \tag{1}$$

PROOF: Let $\widetilde{\theta}^{\,i}$ and $\widetilde{\Theta}^{\,i}$ be the components of $\widetilde{\theta}$ and $\widetilde{\Theta}$ with respect to the standard basis $\mathscr{E} = \{e_1, \dots, e_n\}$ for $\mathbf{R}^n$, and let $\widetilde{\omega}^i_j$ be the components of $\widetilde{\omega}$ with respect to the basis $\mathscr{D} = \{F^i_j | 1 \le i,j \le n\}$ for $\mathfrak{gl}(\mathbf{R}^n)$ defined in Example 8.10 (b), where $F^i_j : \mathbf{R}^n \to \mathbf{R}^n$ is the linear map given by $F^i_j(e_k) = \delta_{jk}\, e_i$ for $k = 1, \dots, n$. By Remark 10.94 we know that $\widetilde{\theta}^{\,i}$ for $i = 1, \dots, n$ form a dual local basis on V to the orthonormal frame field $s = (X_1, \dots, X_n)$. Hence there are unique smooth functions a^i_{jk}, b^i_{jk} and c^i_{jk} so that

$$\widetilde{\omega}^i_j = \sum_k a^i_{jk}\,\widetilde{\theta}^{\,k} \tag{2}$$

for $i,j = 1,\dots,n$, and

$$d\widetilde{\theta}^{\,i} = \tfrac{1}{2}\sum_{jk} b^i_{jk}\,\widetilde{\theta}^{\,j}\wedge\widetilde{\theta}^{\,k} \quad \text{and} \quad \widetilde{\Theta}^{\,i} = \tfrac{1}{2}\sum_{jk} c^i_{jk}\,\widetilde{\theta}^{\,j}\wedge\widetilde{\theta}^{\,k}$$

for $i = 1,\dots,n$, where

$$b^i_{jk} = -b^i_{kj} \quad \text{and} \quad c^i_{jk} = -c^i_{kj}. \tag{3}$$

Since the form $\widetilde{\omega}$ is $\mathfrak{o}_r(\mathbf{R}^n)$-valued, we have that

$$s_{ij}\,a^j_{ik} = -a^i_{jk}, \tag{4}$$

where $s_{ij} = sgn\,(i-r-1/2)(j-r-1/2)$. We have that

$$s_{ij}\, s_{jk}\, s_{ki} = 1 \tag{5}$$

for every triple of indices i, j and k. Indeed, we can partition the index set $\{1,...,n\}$ in two disjoint classes $\{1,...,r\}$ and $\{r+1,...,n\}$. Then two indices i and j belong to the same class if and only if $s_{ij} = 1$. Now there are two possibilities. Either all three indices i, j and k belong to the same class, in which case all the factors in the product are 1, or two indices can belong to one class and the third to the other class, in which case two of the factors in the product are -1 and the third is 1. Using the structure equation, we now obtain

$$\tfrac{1}{2}\sum_{jk} b^i_{jk}\,\widetilde{\theta}^{\,j}\wedge\widetilde{\theta}^{\,k} = d\widetilde{\theta}^{\,i} = -\sum_j \widetilde{\omega}^i_j\wedge\widetilde{\theta}^{\,j} + \widetilde{\Theta}^i = \sum_{jk} a^i_{jk}\,\widetilde{\theta}^{\,j}\wedge\widetilde{\theta}^{\,k} + \tfrac{1}{2}\sum_{jk} c^i_{jk}\,\widetilde{\theta}^{\,j}\wedge\widetilde{\theta}^{\,k}$$

so that

$$a^i_{jk} - a^i_{kj} = b^i_{jk} - c^i_{jk} \tag{6}$$

which together with (4) implies that

$$a^i_{jk} + s_{ki}\, a^k_{ij} = b^i_{jk} - c^i_{jk}. \tag{7}$$

By interchanging i, j and k cyclically, and multiplying the next equation by s_{ij} and the last by $-s_{ki}$ using (5), we obtain

$$s_{ij}\, a^j_{ki} + a^i_{jk} = s_{ij}\,(b^j_{ki} - c^j_{ki}) \tag{8}$$

and

$$-s_{ki}\, a^k_{ij} - s_{ij}\; a^j_{ki} = -s_{ki}\,(b^k_{ij} - c^k_{ij}). \tag{9}$$

Adding these three equations and dividing by 2, we have that

$$a^i_{jk} = \tfrac{1}{2}\,(b^i_{jk} + s_{ij}\, b^j_{ki} - s_{ki}\, b^k_{ij}) \;-\; \tfrac{1}{2}\,(c^i_{jk} + s_{ij}\, c^j_{ki} - s_{ki}\, c^k_{ij}) \tag{10}$$

which shows the uniqueness of the form $\widetilde{\omega}$.

To prove existence, suppose that $\widetilde{\omega}$ is the form whose components are defined by (2), where the coefficients a^i_{jk} are the smooth functions defined by (10). Using (3) and (5) we have that

$$s_{ij}\, a^j_{ik} = \tfrac{1}{2}\,(s_{ij}\, b^j_{ik} + b^i_{kj} - s_{ki}\, b^k_{ji}) \;-\; \tfrac{1}{2}\,(s_{ij}\, c^j_{ik} + c^i_{kj} - s_{ki}\, c^k_{ji}) = -a^i_{jk}$$

and

$$a^i_{kj} = \tfrac{1}{2}\,(b^i_{kj} + s_{ki}\, b^k_{ji} - s_{ij}\, b^j_{ik}) \;-\; \tfrac{1}{2}\,(c^i_{kj} + s_{ki}\, c^k_{ji} - s_{ij}\, c^j_{ik}) = a^i_{jk} - (b^i_{jk} - c^i_{jk})$$

which shows that formula (4) and (6) are satisfied. Hence $\widetilde{\omega}$ is $\mathfrak{o}_r(\mathbf{R}^n)$-valued, and we have that

$$-\sum_{j} \widetilde{\omega}_j^i \wedge \widetilde{\theta}^j = \sum_{jk} a_{jk}^i \widetilde{\theta}^j \wedge \widetilde{\theta}^k = \sum_{j<k} (a_{jk}^i - a_{kj}^i) \widetilde{\theta}^j \wedge \widetilde{\theta}^k$$

$$= \sum_{j<k} b_{jk}^i \widetilde{\theta}^j \wedge \widetilde{\theta}^k - \sum_{j<k} c_{jk}^i \widetilde{\theta}^j \wedge \widetilde{\theta}^k = d\widetilde{\theta}^i - \widetilde{\Theta}^i$$

for $i = 1, ..., n$, which shows that $\widetilde{\omega}$ satisfies the structure Equation (1). □

10.174 Proposition Let M^n be a pseudo-Riemannian manifold with a metric g of index r, and let $s_k : V_k \to \mathscr{O}(M)$ for $k = 1, 2$ be two sections in the orthonormal frame bundle $\pi : \mathscr{O}(M) \to M$ with associated local trivializations $(t_k, \pi^{-1}(V_k))$ which are $O_r(\mathbf{R}^n)$-related with transition map $\phi : V_1 \cap V_2 \to O_r(\mathbf{R}^n)$. Let $\widetilde{\theta} = \widetilde{i}^*(\theta)$ be the pullback of the dual form θ on $\mathscr{L}(M)$, where $\widetilde{i} : \mathscr{O}(M) \to \mathscr{L}(M)$ is the inclusion map, and let Θ be an $\mathbf{R}^n$-valued 2-form on $\mathscr{O}(M)$ which is tensorial of type $(id, \mathbf{R}^n)$. Then the unique $\mathfrak{o}_r(\mathbf{R}^n)$-valued 1-forms $\widetilde{\omega}_k$ on V_k which satisfy the structure equation

$$d\widetilde{\theta}_k = -\, \widetilde{\omega}_k \wedge \widetilde{\theta}_k + \widetilde{\Theta}_k \tag{1}$$

for $k = 1, 2$, where $\widetilde{\theta}_k = s_k^*(\widetilde{\theta})$ and $\widetilde{\Theta}_k = s_k^*(\Theta)$, are related by the relation

$$\widetilde{\omega}_1(p)(v) = \phi(p)^{-1} \circ \widetilde{\omega}_2(p)(v) \circ \phi(p) + \phi(p)^{-1} \circ d(i \circ \phi)(p)(v) \tag{2}$$

for $p \in V_1 \cap V_2$ and $v \in T_pM$, where $i : O_r(\mathbf{R}^n) \to \mathfrak{gl}(\mathbf{R}^n)$ is the inclusion map.

PROOF: Because of the uniqueness, it is enough to prove that the form $\widetilde{\omega}_1$ defined by formula (2) satisfies the structure Equation (1) for $p \in V_1 \cap V_2$ and $v \in T_pM$, supposing that this is true for $\widetilde{\omega}_2$.

If $\alpha : O_r(\mathbf{R}^n) \to O_r(\mathbf{R}^n)$ is the map given by $\alpha(g) = g^{-1}$ and $\psi = i \circ \alpha \circ \phi$, it follows by Proposition 10.55 that

$$\widetilde{\theta}_1 = \psi \wedge \widetilde{\theta}_2 \quad \text{and} \quad \widetilde{\Theta}_1 = \psi \wedge \widetilde{\Theta}_2 .$$

Using Proposition 5.69 we therefore have that

$$d\widetilde{\theta}_1 = d\psi \wedge \widetilde{\theta}_2 + \psi \wedge d\widetilde{\theta}_2 = d\psi \wedge \widetilde{\theta}_2 - \psi \wedge (\widetilde{\omega}_2 \wedge \widetilde{\theta}_2) + \psi \wedge \widetilde{\Theta}_2 = -\, \widetilde{\omega}_1 \wedge \widetilde{\theta}_1 + \widetilde{\Theta}_1$$

since

$$d\psi(p)(v_1) = \psi_{*p}(v_1) = \alpha_{*\phi(p)} \circ d(i \circ \phi)(p)(v_1)$$
$$= -\, \phi(p)^{-1} \circ d(i \circ \phi)(p)(v_1) \circ \phi(p)^{-1}$$

by 5.51 and Corollary 8.33, which implies that

$$d\psi(p)(v_1)\, (\widetilde{\theta}_2(p)(v_2)) - \psi(p) \circ \widetilde{\omega}_2(p)(v_1)\, (\widetilde{\theta}_2(p)(v_2))$$
$$= -\, \phi(p)^{-1} \circ d(i \circ \phi)(p)(v_1) \circ \phi(p)^{-1} (\widetilde{\theta}_2(p)(v_2))$$
$$-\, \phi(p)^{-1} \circ \widetilde{\omega}_2(p)(v_1) \circ \phi(p) \circ \phi(p)^{-1} (\widetilde{\theta}_2(p)(v_2))$$
$$= -\, \widetilde{\omega}_1(p)(v_1)\, (\widetilde{\theta}_1(p)(v_2))$$

for $p \in V_1 \cap V_2$ and $v_1, v_2 \in T_pM$. □

10.175 Let M^n be a pseudo-Riemannian manifold with a metric g, and let R be the curvature form of the Levi-Civita connection on M. Using the equivalence $\phi : \mathscr{T}^2(TM;\Lambda^1(TM;TM)) \to T^3_1(M)$ over M defined in 10.108 and lowering the contravariant index, we obtain a covariant tensor field $\mathscr{R} = L^1_1(\phi \circ R) \in \mathscr{T}^4(M)$ on M of degree 4, called the *covariant curvature tensor* on M, which is given by

$$\mathscr{R}(p)(v_1,v_2,v_3,v_4) = \phi \circ R(p)(v_2,v_3,v_4,g(p)^\flat(v_1)) = <R(p)(v_3,v_4)\; v_2,v_1>$$

for $p \in M$ and $v_1,v_2,v_3,v_4 \in T_pM$. If R^i_{jkl} are the components of the curvature form R with respect to a frame field $s = (X_1, \dots, X_n)$ on an open subset V of M, and if g_{ij} and $\mathscr{R}_{ijkl}$ are the components of the local representations of g and $\mathscr{R}$ on V, then

$$\mathscr{R}_{ijkl} = \sum_{r=1}^{n} g_{ir} R^r_{jkl}$$

for $1 \le i,j,k,l \le n$.

10.176 Proposition Let $T \in \mathscr{T}^4(V)$ be a covariant tensor of degree 4 on a vector space V satisfying the relations

(1) $T(v_1,v_2,v_3,v_4) = -T(v_2,v_1,v_3,v_4)$,

(2) $T(v_1,v_2,v_3,v_4) = -T(v_1,v_2,v_4,v_3)$ and

(3) $T(v_1,v_2,v_3,v_4) + T(v_1,v_3,v_4,v_2) + T(v_1,v_4,v_2,v_3) = 0$

for every $v_1,v_2,v_3,v_4 \in V$. Then T also satisfies the relation

(4) $T(v_1,v_2,v_3,v_4) = T(v_3,v_4,v_1,v_2)$

for $v_1,v_2,v_3,v_4 \in V$.

A tensor T with these properties is called *curvaturelike.*

PROOF : Using the first three relations we have that

$$\begin{aligned}
2\,T(v_1,v_2,v_3,v_4) &= T(v_1,v_2,v_3,v_4) - T(v_2,v_1,v_3,v_4) \\
&= -\,\{T(v_1,v_3,v_4,v_2) + T(v_1,v_4,v_2,v_3)\} \\
&\quad + \{T(v_2,v_3,v_4,v_1) + T(v_2,v_4,v_1,v_3)\} \\
&= -\,\{T(v_3,v_1,v_2,v_4) + T(v_3,v_2,v_4,v_1)\} \\
&\quad + \{T(v_4,v_1,v_2,v_3) + T(v_4,v_2,v_3,v_1)\} \\
&= T(v_3,v_4,v_1,v_2) - T(v_4,v_3,v_1,v_2) = 2\,T(v_3,v_4,v_1,v_2)
\end{aligned}$$

for $v_1,v_2,v_3,v_4 \in V$. □

10.177 Proposition Let $T \in \mathscr{T}^4(V)$ be a curvaturelike tensor on a vector space V satisfying the relation

$$T(v_1, v_2, v_1, v_2) = 0$$

for every $v_1, v_2 \in V$. Then we have that $T = 0$.

PROOF: Combining the above relation for T with relation (4) in Proposition 10.176, we have that

$$\begin{aligned} 0 &= T(v_1 + v_3, v_2, v_1 + v_3, v_2) \\ &= 0 + T(v_1, v_2, v_3, v_2) + T(v_3, v_2, v_1, v_2) + 0 \\ &= 2\,T(v_1, v_2, v_3, v_2) \end{aligned}$$

which shows that

$$T(v_1, v_2, v_3, v_2) = 0$$

for every $v_1, v_2, v_3 \in V$. Now combining this with relation (2) in Proposition 10.176, we see that

$$\begin{aligned} 0 &= T(v_1, v_2 + v_4, v_3, v_2 + v_4) \\ &= 0 + T(v_1, v_2, v_3, v_4) + T(v_1, v_4, v_3, v_2) + 0 \\ &= T(v_1, v_2, v_3, v_4) - T(v_1, v_4, v_2, v_3) \end{aligned}$$

which implies that

$$T(v_1, v_2, v_3, v_4) = T(v_1, v_4, v_2, v_3)$$

for $v_1, v_2, v_3, v_4 \in V$. This shows that $T(v_1, v_2, v_3, v_4)$ is unchanged by a cyclic permutation of the vectors v_2, v_3 and v_4. Using relation (3) in Proposition 10.176 we therefore conclude that

$$T(v_1, v_2, v_3, v_4) = 0$$

for every $v_1, v_2, v_3, v_4 \in V$. □

10.178 Proposition Let $T \in \mathscr{T}^4(V)$ be a curvaturelike tensor on a vector space V, and let $\{v_1, v_2\}$ and $\{w_1, w_2\}$ are two pairs of vectors in V with $w_j = \sum_{i=1}^{2} a_{ij}\, v_i$ for $j = 1, 2$. Then we have that

$$T(w_1, w_2, w_1, w_2) = \det(a_{ij})^2\, T(v_1, v_2, v_1, v_2)\,.$$

PROOF: Follows from relation (1) and (2) in Proposition 10.176. □

10.179 Example Let V be a finite dimensional vector space with a metric g. Then the tensor $T_g \in \mathscr{T}^4(V)$ defined by

$$T_g(v_1, v_2, v_3, v_4) = g(v_1, v_3)\, g(v_2, v_4) - g(v_1, v_4)\, g(v_2, v_3)$$

for $v_1, v_2, v_3, v_4 \in V$ is curvaturelike. We see immediately that it satisfies the first two relations in Proposition 10.176, and relation (3) follows since

$$\begin{aligned}
&T_g(v_1,v_2,v_3,v_4)+T_g(v_1,v_3,v_4,v_2)+T_g(v_1,v_4,v_2,v_3)\\
&= g(v_1,v_3)\, g(v_2,v_4) - g(v_1,v_4)\, g(v_2,v_3)\\
&+ g(v_1,v_4)\, g(v_3,v_2) - g(v_1,v_2)\, g(v_3,v_4)\\
&+ g(v_1,v_2)\, g(v_4,v_3) - g(v_1,v_3)\, g(v_4,v_2) = 0
\end{aligned}$$

for every $v_1,v_2,v_3,v_4 \in V$. We let $Q_g : V\times V \to \mathbf{R}$ be the function defined by

$$Q_g(v_1,v_2) = T_g(v_1,v_2,v_1,v_2) = g(v_1,v_1)\, g(v_2,v_2) - g(v_1,v_2)^2$$

for $v_1,v_2 \in V$. From 7.2 it follows that a 2-dimensional subspace W of V with a basis $\{v_1,v_2\}$ is non-degenerate if and only if $Q_g(v_1,v_2) \neq 0$.

10.180 Proposition Let M be a pseudo-Riemannian manifold with a metric g, and let $\mathscr{R}$ be the covariant curvature tensor on M. Then $\mathscr{R}(p)$ is a curvaturelike tensor on T_pM for every $p \in M$.

PROOF : We must show that $\mathscr{R}(p)$ satisfies relation (1) – (3) in Proposition 10.176 for every $p \in M$.

(1) Let ε be a signature compatible with g. By Proposition 10.172 we know that the Levi-Civita connection H is reducible to a connection H' in the orthonormal frame bundle $\pi' : \mathscr{O}_\varepsilon(M) \to M$. If Ω and Ω' are the curvature forms of H and H' on $\mathscr{L}(M)$ and $\mathscr{O}_\varepsilon(M)$, respectively, it follows from Proposition 10.168 that $i^*\Omega = \psi_*\cdot\, \Omega'$, where $i : \mathscr{O}_\varepsilon(M) \to \mathscr{L}(M)$ and $\psi : O_\varepsilon(\mathbf{R}^n) \to Gl(\mathbf{R}^n)$ are the inclusion maps. Choosing $u \in \mathscr{O}_{\varepsilon,p}(M)$ and $w_3,w_4 \in T_u\mathscr{O}_\varepsilon(M)$ so that $\pi'_*(w_r) = v_r$ for $r = 3,4$, we now have that

$$\begin{aligned}
&<R(p)(v_3,v_4)\; v_2,v_1> = g(p)(u\circ i^*\Omega(u)(w_3,w_4)\circ u^{-1}(v_2),v_1)\\
&= g^n_\varepsilon(\,\Omega'(u)(w_3,w_4)\circ u^{-1}(v_2),u^{-1}(v_1))\\
&= -g^n_\varepsilon(\,\Omega'(u)(w_3,w_4)\circ u^{-1}(v_1),u^{-1}(v_2))\\
&= -g(p)(u\circ i^*\Omega(u)(w_3,w_4)\circ u^{-1}(v_1),v_2) = -<R(p)(v_3,v_4)\; v_1,v_2>
\end{aligned}$$

by Remark 10.104 and Example 8.58 (b).

(2) Follows since the curvature form R on M satisfies the relation

$$R(p)(v_3,v_4) = -R(p)(v_4,v_3)$$

for $p \in M$ and $v_3,v_4 \in T_pM$.

(3) Follows from Corollary 10.146 (1). □

10.181 Definition Let M be a pseudo-Riemannian manifold with a metric g, and let $\mathscr{R}$ be the covariant curvature tensor on M. A 2-dimensional subspace of the tangent space T_pM at a point $p \in M$ is called a *tangent plane* to M at p. For each non-degenerate tangent plane W at p, the number

$$K(W) = \frac{\mathscr{R}(p)(v_1, v_2, v_1, v_2)}{Q_{g(p)}(v_1, v_2)} ,$$

where $\{v_1, v_2\}$ is a basis for W and $Q_{g(p)} : T_pM \times T_pM \to \mathbf{R}$ is the function defined in Example 10.179, is called the *sectional curvature* of W. If $\dim_p(M) = 2$, then $K(T_pM)$ is called the *Gaussian curvature* at p.

10.182 Remark It follows from Proposition 10.178 and Example 10.179 that the sectional curature $K(W)$ is well defined, being independent of the choice of basis $\{v_1, v_2\}$ for W, and with $Q_{g(p)}(v_1, v_2) \neq 0$ when W is non-degenerate. In the next proposition we will show that the covariant curvature tensor $\mathscr{R}$ is completely determined by the sectional curvature K on M. We first need a lemma.

10.183 Lemma Let V be a finite dimensional vector space with a metric g. Then the set

$$B = \{(v_1, v_2) \in V \times V \,|\, Q_g(v_1, v_2) \neq 0\}$$

is dense in $V \times V$.

PROOF : Consider two vectors $u_1, u_2 \in V$, and let O_1 and O_2 be open neighbourhoods around u_1 and u_2, respectively. We must show that $(O_1 \times O_2) \cap B \neq \emptyset$.

We first show that there are vectors $v_1 \in O_1$ and $v_2 \in O_2$ which are linearly independent. Let $\{e_1, e_2\}$ be a basis for a 2-dimensional subspace W of V containing u_1 and u_2, and suppose that $u_j = \sum_{i=1}^{2} a_{ij}\, e_i$ for $j = 1, 2$. Then there is a real number t so that $v_i = u_i + te_i \in O_i$ for $i = 1, 2$, and

$$\begin{vmatrix} a_{11}+t & a_{12} \\ a_{21} & a_{22}+t \end{vmatrix} = \begin{vmatrix} a_{11} & a_{12} \\ a_{21} & a_{22} \end{vmatrix} + (a_{11}+a_{22})\,t + t^2 \neq 0$$

which implies that v_1 and v_2 are linearly independent.

If $Q_g(v_1, v_2) \neq 0$, then $(v_1, v_2) \in (O_1 \times O_2) \cap B$. On the other hand, if $Q_g(v_1, v_2) = 0$, the subspace W is degenerate, so the metric g must be indefinite. We contend that there exists a vector $v_3 \in V$ with $Q_g(v_1, v_3) < 0$. Indeed, if $g(v_1, v_1) = 0$ we choose a $v_3 \in V$ with $g(v_1, v_3) \neq 0$, and if $g(v_1, v_1) \neq 0$ we choose a $v_3 \in V$ so that $g(v_3, v_3)$ and $g(v_1, v_1)$ have opposite sign. Now there is a real number t so that $v_2 + tv_3 \in O_2$ and

$$Q_g(v_1, v_2 + tv_3) = 2t\, T_g(v_1, v_2, v_1, v_3) + t^2\, Q_g(v_1, v_3) \neq 0 ,$$

which implies that $(v_1, v_2 + tv_3) \in (O_1 \times O_2) \cap B$. □

10.184 Proposition Let M be a pseudo-Riemannian manifold with a metric g, and let $\mathscr{R}$ and K be the covariant curvature tensor and the sectional curvature on M. Let $T \in \mathscr{T}^4(T_pM)$ be a curvaturelike tensor in the tangent space T_pM, and suppose that

$$K(W) = \frac{T(v_1,v_2,v_1,v_2)}{Q_{g(p)}(v_1,v_2)}$$

for each non-degenerate tangent plane W at p, where $\{v_1,v_2\}$ is a basis for W. Then we have that $T = \mathscr{R}(p)$.

PROOF : Follows from Proposition 10.177 and Lemma 10.183 applied to the curvaturelike tensor $T - \mathscr{R}(p)$. □

10.185 Let M^n be a pseudo-Riemannian manifold with a metric g, and let Ric be the Ricci curvature tensor of the Levi-Civita connection on M. If $p \in M$ and $\mathscr{B} = \{w_1,...,w_n\}$ is an orthonormal basis for T_pM with signature ε, we have that

$$\mathrm{Ric}\,(p)(v_1,v_2) = \sum_{i=1}^{n} \varepsilon_i < R(p)(w_i,v_2)\, v_1, w_i > = \sum_{i=1}^{n} \varepsilon_i \mathscr{R}(p)(w_i,v_1,w_i,v_2)$$

for $v_1,v_2 \in T_pM$. Using Proposition 10.180 (4) we see that

$$\mathrm{Ric}\,(p)(v_1,v_2) = \mathrm{Ric}\,(p)(v_2,v_1)$$

for $p \in M$ and $v_1,v_2 \in T_pM$, showing that the Ricci tensor of the Levi-Civita connection is symmetric.

Taking the contraction of Ric after raising the first index, we obtain a smooth function $S = C_1^1(R_1^{\,1}(\mathrm{Ric})) \in \mathscr{F}(M)$ on M, called the *scalar curvature* on M. If R^i_{jkl} are the components of the curvature form R with respect to a frame field $s = (X_1, \dots ,X_n)$ on an open subset V of M, and if g_{ij} and Ric_{ij} are the components of the local representations of g and Ric on V, then

$$S = \sum_{i,j} g^{ij}\,\mathrm{Ric}_{ij} = \sum_{i,j,k} g^{ij} R^k_{ikj}\ .$$

10.186 Proposition Let M^n be a pseudo-Riemannian manifold with a metric g, and let R be the curvature form of the Levi-Civita connection H on M. If R^i_{jkl} are the components of R with respect to a frame field $s = (X_1, \dots ,X_n)$ on an open subset V of M, then Bianchi's 2nd identity can be written in the form

$$R^i_{jkl;m} + R^i_{jlm;k} + R^i_{jmk;l} = 0$$

for $1 \leq i,j,k,l,m \leq n$.

PROOF : Let $(\alpha^1, \dots ,\alpha^n)$ be the dual local basis of the frame field s. Using the equivalence $\phi : \mathscr{T}^2(TM;\Lambda^1(TM;TM)) \to T^{\,3}_{\,1}(M)$ over M defined in 10.108, we have that

$$\phi \circ R = \sum_{i,j,k,l} R^i_{jkl}\ \alpha^j \otimes \alpha^k \otimes \alpha^l \otimes X_i\ ,$$

so that

$$\nabla_{X_m}(\phi \circ R) = \sum_{i,j,k,l} R^i_{jkl;m}\, \alpha^j \otimes \alpha^k \otimes \alpha^l \otimes X_i$$

for $1 \leq m \leq n$ by Proposition 10.119. To see that

$$\nabla_{X_m}(\phi \circ R) = \phi \circ \nabla_{X_m} R\,,$$

we consider the frame bundle $\pi : \mathscr{L}(M) \to M$ over M and the vector-valued functions $\overline{\phi \circ R}$ and $\overline{R}$ on $\mathscr{L}(M)$. For each $p \in M$ and $u \in \mathscr{L}_p(M)$ we have a commutative diagram

$$\begin{array}{ccc} \mathscr{T}^2(\mathbf{R}^n; \mathfrak{gl}(\mathbf{R}^n)) & \xrightarrow{\ \psi\ } & \mathscr{T}^3_1(\mathbf{R}^n) \\ \widehat{\mu}_u \big\downarrow & & \big\downarrow \widetilde{\mu}_u \\ \mathscr{T}^2(T_pM; \Lambda^1(T_pM; T_pM)) & \xrightarrow{\ \phi_p\ } & \mathscr{T}^3_1(T_pM) \end{array}$$

where $\psi : \mathscr{T}^2(\mathbf{R}^n; \mathfrak{gl}(\mathbf{R}^n)) \to \mathscr{T}^3_1(\mathbf{R}^n)$ is the linear isomorphism defined by

$$\psi(G)(a_1, a_2, a_3, \lambda) = \lambda \circ G(a_2, a_3)(a_1)$$

for $G \in \mathscr{T}^2(\mathbf{R}^n; \mathfrak{gl}(\mathbf{R}^n))$, $a_1, a_2, a_3 \in \mathbf{R}^n$ and $\lambda \in (\mathbf{R}^n)^*$, and $\widehat{\mu}_u$ and $\widetilde{\mu}_u$ are the linear isomorphisms defined in Example 10.29 (i) and (f). By Definition 10.77 it follows that

$$\overline{\phi \circ R}\,(u) = \widetilde{\mu}_u^{-1} \circ \phi \circ R(\pi(u)) = \psi \circ \widehat{\mu}_u^{-1} \circ R(\pi(u)) = \psi \circ \overline{R}\,(u)$$

for $u \in \mathscr{L}(M)$, so that

$$\begin{aligned} \nabla_v(\phi \circ R) &= \widetilde{\mu}_u \circ d\,\overline{\phi \circ R}\,(u)(w) = \widetilde{\mu}_u \circ \psi \circ d\overline{R}\,(u)(w) \\ &= \phi \circ \widehat{\mu}_u \circ d\overline{R}\,(u)(w) = \phi\,(\nabla_v R) \end{aligned}$$

for $p \in M$ and $v \in T_pM$, where $u \in \mathscr{L}(M)$ and $w \in H_u$ are chosen so that $\pi(u) = p$ and $\pi_*(w) = v$. The proposition now follows from Corollary 10.146 (2). □

10.187 Remark Contracting Bianchi's 2nd identity with respect to the indices i and m leads to

$$\sum_{m,r} g^{mr} \mathscr{R}_{rjkl;m} - \mathrm{Ric}_{jl;k} + \mathrm{Ric}_{jk;l} = \sum_m R^m_{jkl;m} + \sum_m R^m_{jlm;k} + \sum_m R^m_{jmk;l} = 0$$

for $1 \leq j,k,l \leq n$. Raising the index j and contracting with l then gives

$$\sum_m \mathrm{Ric}^m_{k;m} - S_{;k} + \sum_l \mathrm{Ric}^l_{k;l} = -\sum_{m,r,l} g^{mr} R^l_{rkl;m} - S_{;k} + \sum_l \mathrm{Ric}^l_{k;l} = 0$$

which implies that

$$\sum_l (\mathrm{Ric}^l_k - \tfrac{1}{2} S\, \delta^l_k)_{;l} = 0$$

for $1 \leq k \leq n$. Lowering the index l gives the covariant form

$$\sum_l (\mathrm{Ric}_{kl} - \tfrac{1}{2} S\, g_{kl})_{;l} = 0$$

for $1 \leq k \leq n$. The tensor $G = \mathrm{Ric} - \frac{1}{2} S g$ is called the *Einstein curvature tensor.*

Using this we can now write down the fundamental equation of general relativity, the *Einstein field equation*

$$G + \Lambda g = \kappa T\,,$$

where Λ is the *cosmological constant* and T is the *energy-momentum tensor.* The fundamental constant κ, called the *Einstein constant,* is given by

$$\kappa = \frac{8\pi\gamma}{c^4}\,,$$

where $\gamma = 6.673 \cdot 10^{-11}\,\mathrm{N\,m^2/\,kg^2}$ is *Newton's gravitational constant* and $c = 2.998 \cdot 10^8\,\mathrm{m/\,s}$ is the *speed of light* in vacuum.

In both special and general relativity one often replaces the SI units with *natural units* where $c = 1$, so that time is measured in meters where $1\,\mathrm{s} = 2.998 \cdot 10^8\,\mathrm{m}$. In general relativity one also uses *geometrized units* where in addition $\gamma = 1$, so that $\kappa = 8\pi$. Since $\gamma/c^2 = 7.425 \cdot 10^{-28}\,\mathrm{m/\,kg}$, this means that we are also measuring mass in meters where $1\,\mathrm{kg} = 7.425 \cdot 10^{-28}\,\mathrm{m}$.

The fundamental idea of general relativity is that, contrary to Newton's theory of gravity, there are no gravitational forces. Instead the energy-momentum tensor generates a curvature in spacetime, where free particles move along geodesics.

The *spacetime* of *events* is a 4-dimensional Lorentz manifold, and the path of a particle through spacetime is called its *world line*. The world lines of material particles are timelike, while photons follow null world lines. We imagine that each material particle is equipped with a clock measuring its *proper time*, which in natural units is the arc length of its world line from some specified starting point. The world line is usually parametrized by proper time.

The energy-momentum tensor for a *perfect fluid* is given by

$$T = (\rho + p)\, u^\flat \otimes u^\flat + p g\,,$$

where ρ and p are the *energy density* and *pressure* measured by an observer co-moving with the fluid. They are functions on the spacetime manifold M, and u is a timelike unit vector field on M called the *flow vector field* of the fluid.

10.188 Example By Proposition 2.41 we know that the sphere

$$\mathbf{S}_a^2 = \{(x,y,z) \in \mathbf{R}^3 | x^2+y^2+z^2 = a^2\}$$

with radius $a > 0$ is a closed submanifold of the Euclidean space $\mathbf{R}^3$. Let $i_a : \mathbf{S}_a^2 \to \mathbf{R}^3$ be the inclusion map and $p : \mathbf{R} \times \mathbf{R}^2 \to \mathbf{R}^2$ be the projection on the last factor. If (ψ, V) is the spherical coordinate system on $\mathbf{R}^3$ defined in Example 7.21 with coordinate functions r, θ and ϕ, and if $V_a = V \cap \mathbf{S}_a^2$ and $\psi_a = p \circ \psi \circ i_a$, then (ψ_a, V_a) is a local chart on $\mathbf{S}_a^2$ with coordinate functions $\theta \circ i_a$ and $\phi \circ i_a$ which we also denote simply by θ and ϕ. Let $g_a = i_a^* g$ be the pull-back to $\mathbf{S}_a^2$ of the standard metric g on $\mathbf{R}^3$. Then we have that

$$g_a|_{V_a} = a^2\, d\theta^2 + a^2 \sin^2\theta\, d\phi^2 .$$

We see that

$$X_1 = \frac{1}{a}\frac{\partial}{\partial\theta} \quad \text{and} \quad X_2 = \frac{1}{a\sin\theta}\frac{\partial}{\partial\phi}$$

form an orthonormal frame field on V_a with dual local basis

$$\widetilde{\theta}^{\,1} = a\, d\theta \quad \text{and} \quad \widetilde{\theta}^{\,2} = a \sin\theta\, d\phi .$$

Since

$$d\widetilde{\theta}^{\,1} = 0$$

$$d\widetilde{\theta}^{\,2} = a\cos\theta\, d\theta \wedge d\phi = \frac{1}{a}\cot\theta\; \widetilde{\theta}^{\,1} \wedge \widetilde{\theta}^{\,2} ,$$

the non-zero coefficients b^i_{jk} and a^i_{jk} are

$$b^2_{12} = -\, b^2_{21} = \frac{1}{a}\cot\theta$$

and

$$a^1_{22} = -\, a^2_{12} = \frac{1}{2}\,(b^1_{22} + b^2_{21} - b^2_{12}) = -\frac{1}{a}\cot\theta .$$

This shows that

$$\widetilde{\omega}^1_2 = -\,\widetilde{\omega}^2_1 = -\frac{1}{a}\cot\theta\; \widetilde{\theta}^{\,2} = -\cos\theta\, d\phi$$

so that

$$\widetilde{\Omega}^1_2 = -\,\widetilde{\Omega}^2_1 = d\widetilde{\omega}^1_2 + \widetilde{\omega}^1_1 \wedge \widetilde{\omega}^1_2 + \widetilde{\omega}^1_2 \wedge \widetilde{\omega}^2_2 = \sin\theta\, d\theta \wedge d\phi = \frac{1}{a^2}\; \widetilde{\theta}^{\,1} \wedge \widetilde{\theta}^{\,2} .$$

We have a diffeomorphism $\sigma_a : \mathbf{R}^3 - \{0\} \to (0,\infty) \times \mathbf{S}_a^2$ defined by

$$\sigma_a(v) = (\,\|v\|, av/\|v\|\,)$$

for $v \in \mathbf{R}^3 - \{0\}$, with inverse $\rho_a : (0,\infty) \times \mathbf{S}_a^2 \to \mathbf{R}^3 - \{0\}$ given by

$$\rho_a(r,u) = (r/a)\,u$$

for $(r,u) \in (0,\infty) \times \mathbf{S}_a^2$, so that

$$\sigma_a(V) = (0,\infty) \times V_a \quad \text{and} \quad p \circ \psi = \psi_a \circ \sigma_{a,2} .$$

THE SCHWARZSCHILD – DE SITTER SPACETIME

10.189 We will now find the vacuum solution of Einstein's field equation with cosmological constant Λ for a spacetime which is static and spherically symmetric. Let M be a Lorentz manifold with a metric g so that

$$g|_N = -e^{2\alpha}\,dt^2 + e^{2\beta}\,dr^2 + r^2(d\theta^2 + \sin^2\theta\, d\phi^2)$$

in natural units for a local chart (ψ, N) on M with coordinate functions t, r, θ and ϕ, where $\alpha = \tilde{\alpha}\circ r$ and $\beta = \tilde{\beta}\circ r$ for real functions $\tilde{\alpha}$ and $\tilde{\beta}$. The partial derivatives $\frac{\partial\alpha}{\partial r}$ and $\frac{\partial\beta}{\partial r}$ are denoted by α' and β', respectively, and are given by $\alpha' = \tilde{\alpha}'\circ r$ and $\beta' = \tilde{\beta}'\circ r$. We see that

$$X_1 = e^{-\alpha}\frac{\partial}{\partial t}\ ,\quad X_2 = e^{-\beta}\frac{\partial}{\partial r}\ ,\quad X_3 = \frac{1}{r}\frac{\partial}{\partial\theta}\quad \text{and}\quad X_4 = \frac{1}{r\sin\theta}\frac{\partial}{\partial\phi}$$

form an orthonormal frame field on N with dual local basis

$$\widetilde{\theta}^{\,1} = e^{\alpha}\,dt\ ,\quad \widetilde{\theta}^{\,2} = e^{\beta}\,dr\ ,\quad \widetilde{\theta}^{\,3} = r\,d\theta\quad \text{and}\quad \widetilde{\theta}^{\,4} = r\sin\theta\, d\phi\ .$$

Since

$$\begin{aligned}
d\widetilde{\theta}^{\,1} &= -\,\alpha' e^{-\beta}\,\widetilde{\theta}^{\,1}\wedge\widetilde{\theta}^{\,2}\\
d\widetilde{\theta}^{\,2} &= 0\\
d\widetilde{\theta}^{\,3} &= \tfrac{1}{r}\,e^{-\beta}\,\widetilde{\theta}^{\,2}\wedge\widetilde{\theta}^{\,3}\\
d\widetilde{\theta}^{\,4} &= \tfrac{1}{r}\,e^{-\beta}\,\widetilde{\theta}^{\,2}\wedge\widetilde{\theta}^{\,4} + \tfrac{1}{r}\cot\theta\,\widetilde{\theta}^{\,3}\wedge\widetilde{\theta}^{\,4},
\end{aligned}$$

the non-zero coefficients b^i_{jk} and a^i_{jk} are

$$\begin{aligned}
b^1_{12} &= -\,b^1_{21} = -\,\alpha' e^{-\beta}\\
b^i_{2i} &= -\,b^i_{i2} = \tfrac{1}{r}\,e^{-\beta} \qquad \text{for}\quad i = 3,4\\
b^4_{34} &= -\,b^4_{43} = \tfrac{1}{r}\cot\theta
\end{aligned}$$

and

$$\begin{aligned}
a^1_{21} = a^2_{11} &= \tfrac{1}{2}\,(b^1_{21} - b^2_{11} - b^1_{12}) = \alpha' e^{-\beta}\\
a^2_{ii} = -\,a^i_{2i} &= \tfrac{1}{2}\,(b^2_{ii} + b^i_{i2} - b^i_{2i}) = -\tfrac{1}{r}\,e^{-\beta} \qquad \text{for}\quad i = 3,4\\
a^3_{44} = -\,a^4_{34} &= \tfrac{1}{2}\,(b^3_{44} + b^4_{43} - b^4_{34}) = -\tfrac{1}{r}\cot\theta
\end{aligned}$$

This shows that the non-zero components of $\widetilde{\omega}$ are

$$\widetilde{\omega}^1_2 = \widetilde{\omega}^2_1 = \alpha' \, e^{-\beta} \; \widetilde{\theta}^{\,1} = \alpha' \, e^{\alpha-\beta} \; dt$$

$$\widetilde{\omega}^2_3 = -\, \widetilde{\omega}^3_2 = -\tfrac{1}{r} \, e^{-\beta} \; \widetilde{\theta}^{\,3} = -\, e^{-\beta} \; d\theta$$

$$\widetilde{\omega}^2_4 = -\, \widetilde{\omega}^4_2 = -\tfrac{1}{r} \, e^{-\beta} \; \widetilde{\theta}^{\,4} = -\, e^{-\beta} \sin\theta \; d\phi$$

$$\widetilde{\omega}^3_4 = -\, \widetilde{\omega}^4_3 = -\tfrac{1}{r} \, \cot\theta \; \widetilde{\theta}^{\,4} = -\, \cos\theta \; d\phi$$

so that

$$\begin{aligned}
\widetilde{\Omega}^1_2 &= \widetilde{\Omega}^2_1 = d\widetilde{\omega}^1_2 = -\, e^{\alpha-\beta} \, (\alpha'' + \alpha'^{\,2} - \alpha'\beta') \; dt \wedge dr \\
&= -\, e^{-2\beta} \, (\alpha'' + \alpha'^{\,2} - \alpha'\beta') \; \widetilde{\theta}^{\,1} \wedge \widetilde{\theta}^{\,2} \\
\widetilde{\Omega}^1_3 &= \widetilde{\Omega}^3_1 = \widetilde{\omega}^1_2 \wedge \widetilde{\omega}^2_3 = -\tfrac{1}{r} \, \alpha' \, e^{-2\beta} \; \widetilde{\theta}^{\,1} \wedge \widetilde{\theta}^{\,3} \\
\widetilde{\Omega}^1_4 &= \widetilde{\Omega}^4_1 = \widetilde{\omega}^1_2 \wedge \widetilde{\omega}^2_4 = -\tfrac{1}{r} \, \alpha' \, e^{-2\beta} \; \widetilde{\theta}^{\,1} \wedge \widetilde{\theta}^{\,4} \\
\widetilde{\Omega}^2_3 &= -\, \widetilde{\Omega}^3_2 = d\widetilde{\omega}^2_3 + \widetilde{\omega}^2_4 \wedge \widetilde{\omega}^4_3 = \beta' \, e^{-\beta} \; dr \wedge d\theta = \tfrac{1}{r} \, \beta' \, e^{-2\beta} \; \widetilde{\theta}^{\,2} \wedge \widetilde{\theta}^{\,3} \\
\widetilde{\Omega}^2_4 &= -\, \widetilde{\Omega}^4_2 = d\widetilde{\omega}^2_4 + \widetilde{\omega}^2_3 \wedge \widetilde{\omega}^3_4 = \beta' \, e^{-\beta} \sin\theta \; dr \wedge d\phi \\
&\quad -\, e^{-\beta} \cos\theta \; d\theta \wedge d\phi + e^{-\beta} \cos\theta \; d\theta \wedge d\phi = \tfrac{1}{r} \, \beta' \, e^{-2\beta} \; \widetilde{\theta}^{\,2} \wedge \widetilde{\theta}^{\,4} \\
\widetilde{\Omega}^3_4 &= -\, \widetilde{\Omega}^4_3 = d\widetilde{\omega}^3_4 + \widetilde{\omega}^3_2 \wedge \widetilde{\omega}^2_4 = \sin\theta \; d\theta \wedge d\phi - e^{-2\beta} \sin\theta \; d\theta \wedge d\phi \\
&= \tfrac{1}{r^2} \, (1 - e^{-2\beta}) \; \widetilde{\theta}^{\,3} \wedge \widetilde{\theta}^{\,4}
\end{aligned}$$

By Corollary 10.106 the non-zero components of the curvature form R are

$$R^1_{212} = -R^1_{221} = R^2_{112} = -R^2_{121} = -\, e^{-2\beta} \, (\alpha'' + \alpha'^{\,2} - \alpha'\beta')$$

$$R^1_{i1i} = -R^1_{ii1} = R^i_{11i} = -R^i_{1i1} = -\tfrac{1}{r} \, \alpha' \, e^{-2\beta} \quad \text{for} \quad i = 3, 4$$

$$R^2_{i2i} = -R^2_{ii2} = -R^i_{22i} = R^i_{2i2} = \tfrac{1}{r} \, \beta' \, e^{-2\beta} \quad \text{for} \quad i = 3, 4$$

$$R^3_{434} = -R^3_{443} = -R^4_{334} = R^4_{343} = \tfrac{1}{r^2} \, (1 - e^{-2\beta})$$

Using the components of R with equal contravariant and second covariant index, we find the non-zero components of the Ricci tensor

$$\mathrm{Ric}_{11} = R^2_{121} + R^3_{131} + R^4_{141} = e^{-2\beta} \, (\alpha'' + \alpha'^{\,2} - \alpha'\beta') + \tfrac{2}{r} \, \alpha' \, e^{-2\beta}$$

$$\mathrm{Ric}_{22} = R^1_{212} + R^3_{232} + R^4_{242} = -\, e^{-2\beta} (\alpha'' + \alpha'^{\,2} - \alpha'\beta') + \tfrac{2}{r} \, \beta' \, e^{-2\beta}$$

$$\mathrm{Ric}_{33} = R^1_{313} + R^2_{323} + R^4_{343} = \tfrac{1}{r} \, (\beta' - \alpha') \, e^{-2\beta} + \tfrac{1}{r^2} \, (1 - e^{-2\beta})$$

$$\mathrm{Ric}_{44} = R^1_{414} + R^2_{424} + R^3_{434} = \tfrac{1}{r} \, (\beta' - \alpha') \, e^{-2\beta} + \tfrac{1}{r^2} \, (1 - e^{-2\beta})$$

and the scalar curvature

$$S = -\mathrm{Ric}_{11} + \mathrm{Ric}_{22} + \mathrm{Ric}_{33} + \mathrm{Ric}_{44}$$
$$= -2\,e^{-2\beta}\,(\alpha'' + \alpha'^{\,2} - \alpha'\beta') + \tfrac{4}{r}\,(\beta' - \alpha')\,e^{-2\beta} + \tfrac{2}{r^2}\,(1 - e^{-2\beta})$$

From these we obtain the non-zero components of the Einstein tensor

$$G_{11} = \mathrm{Ric}_{11} + \tfrac{1}{2}\,S = \tfrac{2}{r}\,\beta' e^{-2\beta} + \tfrac{1}{r^2}\,(1 - e^{-2\beta}) = \tfrac{1}{r^2}\,\{r(1 - e^{-2\beta})\}'$$
$$G_{22} = \mathrm{Ric}_{22} - \tfrac{1}{2}\,S = \tfrac{2}{r}\,\alpha' e^{-2\beta} - \tfrac{1}{r^2}\,(1 - e^{-2\beta})$$
$$= \tfrac{2}{r}\,(\alpha + \beta)' e^{-2\beta} - \tfrac{1}{r^2}\,\{r(1 - e^{-2\beta})\}'$$
$$G_{ii} = \mathrm{Ric}_{ii} - \tfrac{1}{2}\,S = e^{-2\beta}\,(\alpha'' + \alpha'^{\,2} - \alpha'\beta') - \tfrac{1}{r}\,(\beta' - \alpha')\,e^{-2\beta}$$
$$= e^{-2\beta}\left\{(\alpha+\beta)'' + (\alpha+\beta)'\left(\alpha' - 2\beta' + \tfrac{1}{r}\right)\right\} - \left\{e^{-2\beta}(\beta'' - 2\beta'^2) + \tfrac{2}{r}\beta' e^{-2\beta}\right\}$$
$$= e^{-2\beta}\left\{(\alpha+\beta)'' + (\alpha+\beta)'\left(\alpha' - 2\beta' + \tfrac{1}{r}\right)\right\} - \frac{1}{2r}\{r(1 - e^{-2\beta})\}'' \quad \text{for } i = 3,4$$

We see that the vacuum field equation $G + \Lambda g = 0$ is equivalent to the conditions

$$\alpha + \beta = C \quad \text{and} \quad r(1 - e^{-2\beta}) = R_s + \frac{\Lambda}{3}\,r^3$$

for constants C and R_s. Changing the constant C simply corresponds to a rescaling of the time coordinate t. Hence we can choose $C = 0$ so that

$$e^{2\alpha} = e^{-2\beta} = 1 - \frac{R_s}{r} - \frac{\Lambda}{3}\,r^2$$

and

$$g|_N = -\left(1 - \tfrac{R_s}{r} - \tfrac{\Lambda}{3}r^2\right)dt^2 + \left(1 - \tfrac{R_s}{r} - \tfrac{\Lambda}{3}r^2\right)^{-1}dr^2 + r^2(d\theta^2 + \sin^2\theta\,d\phi^2), \quad (1)$$

where g is called the *Schwarzschild–de Sitter metric*. The value of the constant R_s, called the *Schwarzschild radius*, will be determined in Section 10.190. If $\Lambda = 0$, the metric g is called the *Schwarzschild metric*, and t, r, θ and ϕ are called *Schwarzschild coordinates*. If $R_s = 0$, the metric g is called the *de Sitter metric* when $\Lambda > 0$ and the *anti de Sitter metric* when $\Lambda < 0$, and t, r, θ and ϕ are called *Eddington coordinates*.

The coordinate expression

$$g|_N = -\left(1 - \tfrac{R_s}{r}\right)dt^2 + \left(1 - \tfrac{R_s}{r}\right)^{-1}dr^2 + r^2(d\theta^2 + \sin^2\theta\;d\phi^2), \quad (2)$$

of the Schwarzschild metric is not defined for $r = 0$ and $r = R_s$. Hence the Schwarzschild spacetime M has two connected components $M_1 = \mathbf{R} \times (R_s, \infty) \times \mathbf{S}^2$ and $M_2 = \mathbf{R} \times (0, R_s) \times \mathbf{S}^2$. The metric on M is given by

$$g = -\left(1 - \tfrac{R_s}{r}\right)dt^2 + \left(1 - \tfrac{R_s}{r}\right)^{-1}dr^2 + r^2 g_1,$$

where g_1 is the metric on the 2-sphere $\mathbf{S}^2$ of radius 1 described in Example 10.188. The coordinate neighbourhood N has the connected components $N_1 = \mathbf{R} \times (R_s, \infty) \times V_1$ and $N_2 = \mathbf{R} \times (0, R_s) \times V_1$, where (ψ_1, V_1) is the local chart on $\mathbf{S}^2$ with coordinate functions θ and ϕ.

The value $r = 0$ corresponds to a physical singularity in the Schwarzschild spacetime where the curvature tensors diverge. On the other hand, at $r = R_s$ the curvature tensors are finite and well behaved. This value of r is therefore only a coordinate singularity which can be removed by choosing new coordinates. Physically it corresponds to a horizon in the spacetime, separating the components M_1 and M_2 which are called the exterior Schwarzschild spacetime and the black hole, as will be explained in Section 10.191.

10.190 Timelike geodesics in the Schwarzschild spacetime The Lagrangian of a test particle of unit mass moving in the coordinate neighbourhood N in the Schwarzschild spacetime M is given by

$$L(v) = \tfrac{1}{2}\left\{-\left(1-\tfrac{R_s}{r}\right)\dot{t}^2 + \left(1-\tfrac{R_s}{r}\right)^{-1}\dot{r}^2 + r^2\dot{\theta}^2 + r^2\sin^2\theta\,\dot{\phi}^2\right\} \qquad (1)$$

for tangent vectors $v \in TN$ with coordinates $(t, r, \theta, \phi, \dot{t}, \dot{r}, \dot{\theta}, \dot{\phi})$ as described in 10.189 and Proposition 7.81. Since t and ϕ are cyclic coordinates, their conjugete momenta

$$p_t \circ F(L) = -\left(1-\tfrac{R_s}{r}\right)\dot{t} \quad \text{and} \quad p_\phi \circ F(L) = r^2\sin^2\theta\,\dot{\phi}$$

are constants of the motion. If $\gamma : I \to TN$ is an integral curve for the Lagrangian vector field X_E of L, we denote the constant fuctions $p_t \circ F(L) \circ \gamma$ and $p_\phi \circ F(L) \circ \gamma$ simply by p_t and p_ϕ. As well as being real numbers and coordinate functions of the local chart (ψ, N), we also let t, r, θ, ϕ, $\dot{t}$, $\dot{r}$, $\dot{\theta}$ and $\dot{\phi}$ denote the components of the map $(\psi \times id) \circ t_\psi \circ \gamma$ as functions of the proper time τ. Then it follows from Lagrange's equations that

$$\tfrac{d}{d\tau}(r^2\dot{\theta}) = \tfrac{d}{d\tau}\left(\frac{\partial L}{\partial \dot{\theta}} \circ \gamma\right) = \frac{\partial L}{\partial \theta} \circ \gamma = r^2 \sin\theta\,\cos\theta\,\dot{\phi}^2 = \frac{p_\phi^2\cot\theta}{r^2\sin^2\theta}\,.$$

Multiplying by $2\,r^2\dot{\theta}$ we therefore have that

$$\tfrac{d}{d\tau}\{(r^2\dot{\theta})^2 + (p_\phi\cot\theta)^2\} = 0. \qquad (2)$$

By the spherical symmertry there is no loss of generality assuming that $\theta(\tau_0) = \pi/2$ and $\dot{\theta}(\tau_0) = 0$ at some time τ_0. Then it follows from (2) that $\theta(\tau) = \pi/2$ and $\dot{\theta}(\tau) = 0$ for every $\tau \in I$, showing that the particle is moving in the equatorial plane.

As the path of the test particle is parameterized by proper time, it now follows from (1) that

$$-\left(1-\tfrac{R_s}{r}\right)^{-1}p_t^2 + \left(1-\tfrac{R_s}{r}\right)^{-1}\dot{r}^2 + \frac{p_\phi^2}{r^2} = -1$$

so that

$$\dot{r}^2 + \left(1 - \frac{R_s}{r}\right)\left(1 + \frac{p_\phi^2}{r^2}\right) = p_t^2 .$$

This implies that

$$\tfrac{1}{2}\dot{r}^2 + V(r) = E \tag{3}$$

where

$$V(r) = -\frac{R_s}{2r} + \frac{p_\phi^2}{2r^2} - \frac{R_s p_\phi^2}{2r^3} \quad \text{and} \quad E = \tfrac{1}{2}(p_t^2 - 1) .$$

The first two terms in the formula for $V(r)$ constitute the effective Newtonian potential for the motion of a test particle of unit mass in a gravitational central field from a mass M at the origin as described in 9.55, where

$$R_s = 2\gamma M .$$

The third term is a relativstic correction, and Equation (3) can be solved by numerical methods. It was a major breakthrough in the general theory of relativity when this relativistic term was used to predict a perihelion precession of 43 arc seconds per century for the orbit of the planet Mercury, which was exactly the part of the observed precession which could not be accounted for by Newton's theory.

10.191 Eddington–Finkelstein coordinates In this section we will see how the coordinate singularity $r = R_s$ in the local expression for the Schwarzschild metric given in (2) in 10.189 can be removed by introducing new coordinates. We first define a new radial coordinate function

$$\widetilde{r} = r + R_s \log\left|\frac{r}{R_s} - 1\right| ,$$

which is called a *tortoise coordinate* since $\widetilde{r} \to -\infty$ when $r \to R_s$. As

$$d\widetilde{r} = \left(1 - \frac{R_s}{r}\right)^{-1} dr ,$$

the Schwarzschild metric then takes the form

$$g|_{N_j} = \left(1 - \frac{R_s}{r}\right)(-dt^2 + d\widetilde{r}^2) + r^2(d\theta^2 + \sin^2\theta\, d\phi^2) \tag{1}$$

for $j = 1, 2$. Let $\gamma : I \to N_1$ be the geodesic for a radially moving light ray in the exterior Schwarzschild spacetime. Then we have that

$$(t + \widetilde{r}) \circ \gamma = \text{const}$$

for ingoing light and

$$(t - \widetilde{r}) \circ \gamma = \text{const}$$

for outgoing light.

Introducing a new coordinate function $v = t + \widetilde{r}$ we obtain a local chart (ψ_{ie}, N)

on M with coordinate functions v, r, θ and ϕ, called *ingoing Eddington–Finkelstein coordinates*, satisfying

$$-dt^2 + d\widetilde{r}^2 = -dv\,(dv - 2\,d\widetilde{r})$$

so that

$$g|_N = -\left(1 - \frac{R_s}{r}\right) dv^2 + 2\,dv\,dr + r^2(d\theta^2 + \sin^2\theta\; d\phi^2). \qquad (2)$$

Since $\det(g_{ij})|_N = -r^4 \sin^2\theta$, this coordinate expression of the Schwarzschild metric is well defined for all $r > 0$. Hence the coordinate singularity at $r = R_s$ has been removed, and the Lorentz manifold M and the coordinate neighbourhood N can be replaced by $E = \mathbf{R} \times (0,\infty) \times \mathbf{S}^2$ and $V = \mathbf{R} \times (0,\infty) \times V_1$, with the metric

$$g_{ie} = -\left(1 - \frac{R_s}{r}\right) dv^2 + 2\,dv\,dr + r^2 g_1\,.$$

Now we have that

$$v \circ \gamma = \text{const}$$

for ingoing light and

$$(v - 2\widetilde{r}) \circ \gamma = \text{const}$$

for outgoing light. This shows that the surface $r = R_s$ acts as a one-way membrane, letting the light rays cross only from the outside to the inside. Since the light cannot escape from the region $r < R_s$, it is called a *black hole*, and the surface $r = R_s$ is called a *horizon*.

We can also introduce a coordinate function $u = t - \widetilde{r}$ and obtain a local chart (ψ_{oe}, N) on M with coordinate functions u, r, θ and ϕ, called *outgoing Eddington–Finkelstein coordinates*, satisfying

$$-dt^2 + d\widetilde{r}^2 = -du\,(du + 2\,d\widetilde{r})$$

so that

$$g|_N = -\left(1 - \frac{R_s}{r}\right) du^2 - 2\,du\,dr + r^2(d\theta^2 + \sin^2\theta\; d\phi^2). \qquad (3)$$

The Lorentz manifold M and the coordinate neighbourhood N can again be replaced by E and V, but this time with the metric

$$g_{oe} = -\left(1 - \frac{R_s}{r}\right) du^2 - 2\,du\,dr + r^2 g_1\,.$$

Now we have that

$$(u + 2\widetilde{r}) \circ \gamma = \text{const}$$

for ingoing light and

$$u \circ \gamma = \text{const}$$

for outgoing light. Hence the light rays can cross the surface $r = R_s$ only from the inside to the outside, and the region $r < R_s$ is called a *white hole*.

The Lorentz manifolds $(M_1, g|_{M_1})$, $(M_1, g_{ie}|_{M_1})$ and $(M_1, g_{oe}|_{M_1})$ are isometric isomorphic, as are the Lorentz manifolds $(M_2, g|_{M_2})$, $(M_2, g_{ie}|_{M_2})$ and $(M_2, g_{oe}|_{M_2})$. But the Lorentz manifolds (E, g_{ie}) and (E, g_{oe}) are not isometric isomorphic, in agreement with the fact that a black hole and a white hole have different physical interpretations. However, in the next section we will see that both (E, g_{ie}) and (E, g_{oe}) can be isometrically embedded in a larger Lorentz manifold.

10.192 Kruskal-Szekeres coordinates Using the coordinate functions

$$v = t + \widetilde{r} \quad , \quad u = t - \widetilde{r}$$

described in 10.191, and fixing functions $\varepsilon_1 : M \to \{-1, 1\}$ and $\varepsilon_2 : M \to \{-1, 1\}$ which are constant on M_1 and M_2 with

$$\varepsilon_1 \, \varepsilon_2 = \operatorname{sgn}\Big(\frac{r}{R_s} - 1 \Big) \,,$$

we can define new coordinate functions

$$\widetilde{v} = \varepsilon_1 \exp\Big(\frac{v}{2R_s} \Big) \quad , \quad \widetilde{u} = -\, \varepsilon_2 \exp\Big(-\frac{u}{2R_s} \Big)$$

and

$$T = \tfrac{1}{2} \, (\widetilde{v} + \widetilde{u}) \quad , \quad R = \tfrac{1}{2} \, (\widetilde{v} - \widetilde{u})$$

so that

$$R^2 - T^2 = -\, \widetilde{v} \, \widetilde{u} = \operatorname{sgn}\Big(\frac{r}{R_s} - 1 \Big) \exp\Big(\frac{\widetilde{r}}{R_s} \Big) = \Big(\frac{r}{R_s} - 1 \Big) \exp\Big(\frac{r}{R_s} \Big). \tag{1}$$

The transformation from the Schwarzschild spacetime M is given by

$$T = \varepsilon_1 \Big(\frac{r}{R_s} - 1 \Big)^{1/2} \exp\Big(\frac{r}{2R_s} \Big) \sinh\Big(\frac{t}{2R_s} \Big)$$

$$R = \varepsilon_1 \Big(\frac{r}{R_s} - 1 \Big)^{1/2} \exp\Big(\frac{r}{2R_s} \Big) \cosh\Big(\frac{t}{2R_s} \Big)$$

when $r > R_s$, mapping M_1 onto

$$K_{2-\varepsilon_1} = \{ (T, R) \in \mathbf{R}^2 \, | \, |T| < \varepsilon_1 R \, \} \times \mathbf{S}^2 \,,$$

and by

$$T = \varepsilon_1 \Big(1 - \frac{r}{R_s} \Big)^{1/2} \exp\Big(\frac{r}{2R_s} \Big) \cosh\Big(\frac{t}{2R_s} \Big)$$

$$R = \varepsilon_1 \Big(1 - \frac{r}{R_s} \Big)^{1/2} \exp\Big(\frac{r}{2R_s} \Big) \sinh\Big(\frac{t}{2R_s} \Big)$$

when $0 < r < R_s$, mapping M_2 onto

$$K_{3-\varepsilon_1} = \{ (T, R) \in \mathbf{R}^2 \, | \, |R| < \varepsilon_1 T < \sqrt{R^2 + 1} \, \} \times \mathbf{S}^2 \,.$$

The inverse transformation is given by (1) and by

$$\tanh\left(\frac{t}{2R_s}\right) = \frac{T}{R}$$

when $r > R_s$, and

$$\tanh\left(\frac{t}{2R_s}\right) = \frac{R}{T}$$

when $0 < r < R_s$. Now using that

$$\begin{aligned} -dT^2 + dR^2 &= -\,d\widetilde{v}\,d\widetilde{u} = \frac{1}{4R_s^2}\,(-\,\widetilde{v}\,\widetilde{u})\,(-\,dv\,du) \\ &= \frac{r}{4R_s^3}\left(1 - \frac{R_s}{r}\right)\exp\left(\frac{r}{R_s}\right)(-dt^2 + d\widetilde{r}^2), \end{aligned}$$

we have that

$$g|_N = \frac{4R_s^3}{r}\exp\left(-\frac{r}{R_s}\right)(-dT^2 + dR^2) + r^2(d\theta^2 + \sin^2\theta\, d\phi^2) \qquad (2)$$

where r is given implicitely by (1). As this coordinate expression of the Schwarzschild metric is well defined for all $r > 0$, and using all possible alternatives for ε_1, we obtain a new Lorentz manifold

$$K = \{(T,R) \in \mathbf{R}^2 \mid T^2 - R^2 < 1\} \times \mathbf{S}^2$$

and a local chart (ψ_k, W) on K with coordinate neighbourhood

$$W = \{(T,R) \in \mathbf{R}^2 \mid T^2 - R^2 < 1\} \times V_1$$

and coordinate functions T, R, θ and ϕ, called *Kruskal–Szekeres coordinates*. The metric g_k on K is

$$g_k = \frac{4R_s^3}{r}\exp\left(-\frac{r}{R_s}\right)(-dT^2 + dR^2) + r^2 g_1\,.$$

To find the transformations from the Eddington–Finkelstein extensions, we use that

$$\widetilde{u} = -\,\varepsilon_2\exp\left(\frac{\widetilde{r}}{R_s}\right)\exp\left(-\frac{v}{2R_s}\right) = -\,\varepsilon_1\left(\frac{r}{R_s} - 1\right)\exp\left(\frac{r}{R_s}\right)\exp\left(-\frac{v}{2R_s}\right)$$

and

$$\widetilde{v} = \varepsilon_1\exp\left(\frac{\widetilde{r}}{R_s}\right)\exp\left(\frac{u}{2R_s}\right) = \varepsilon_2\left(\frac{r}{R_s} - 1\right)\exp\left(\frac{r}{R_s}\right)\exp\left(\frac{u}{2R_s}\right),$$

which imply that

$$\begin{aligned} T &= \frac{\varepsilon_1}{2}\left\{\exp\left(\frac{v}{2R_s}\right) - \left(\frac{r}{R_s} - 1\right)\exp\left(\frac{r}{R_s}\right)\exp\left(-\frac{v}{2R_s}\right)\right\} \\ R &= \frac{\varepsilon_1}{2}\left\{\exp\left(\frac{v}{2R_s}\right) + \left(\frac{r}{R_s} - 1\right)\exp\left(\frac{r}{R_s}\right)\exp\left(-\frac{v}{2R_s}\right)\right\} \end{aligned}$$

and

$$T = \frac{\varepsilon_2}{2}\left\{\left(\frac{r}{R_s} - 1\right)\exp\left(\frac{r}{R_s}\right)\exp\left(\frac{u}{2R_s}\right) - \exp\left(-\frac{u}{2R_s}\right)\right\}$$
$$R = \frac{\varepsilon_2}{2}\left\{\left(\frac{r}{R_s} - 1\right)\exp\left(\frac{r}{R_s}\right)\exp\left(\frac{u}{2R_s}\right) + \exp\left(-\frac{u}{2R_s}\right)\right\}.$$

The inverse transformations are given by (1) and by

$$\varepsilon_1 \exp\left(\frac{v}{2R_s}\right) = R + T \quad , \quad \varepsilon_2 \exp\left(-\frac{u}{2R_s}\right) = R - T\,.$$

Choosing $\varepsilon_1 = 1$ and $\varepsilon_2 = 1$ in these formulae, we see that the regions

$$K_{12} = \{((T,R),\mathbf{u}) \in K \,|\, R > -T\} \quad \text{and} \quad K_{14} = \{((T,R),\mathbf{u}) \in K \,|\, R > T\}$$

are isometric to the ingoing and outgoing Eddington–Finkelstein extension of K_1, respectively. Hence K_2 is isometric to the black hole, and K_4 is isometric to the white hole. The regions K_1 and K_3 are both isometric to the external Schwarzschild spacetime. But these two regions cannot be connected by any timelike or null curve. An object entering the black hole K_2 cannot avoid the singularity at $T^2 - R^2 = 1$ where $r = 0$.

10.193 Outgoing and ingoing coordinates in the de Sitter spacetime The coordinate expression

$$g|_N = -\left(1 - \frac{\Lambda}{3}r^2\right)dt^2 + \left(1 - \frac{\Lambda}{3}r^2\right)^{-1}dr^2 + r^2(d\theta^2 + \sin^2\theta\, d\phi^2), \quad (1)$$

of the de Sitter metric, obtained from (1) in 10.189 with $R_s = 0$ and $\Lambda > 0$, is not defined for $r = R_d = (3/\Lambda)^{1/2}$. Hence the de Sitter spacetime M has two connected components $M_1 = \mathbf{R} \times (0,R_d) \times \mathbf{S}^2$ and $M_2 = \mathbf{R} \times (R_d,\infty) \times \mathbf{S}^2$. The metric on M is given by

$$g = -\left(1 - \frac{\Lambda}{3}r^2\right)dt^2 + \left(1 - \frac{\Lambda}{3}r^2\right)^{-1}dr^2 + r^2 g_1\,,$$

where g_1 is the metric on the 2-sphere $\mathbf{S}^2$ of radius 1 described in Example 10.188. The coordinate neighbourhood N has the connected components $N_1 = \mathbf{R} \times (0,R_d) \times V_1$ and $N_2 = \mathbf{R} \times (R_d,\infty) \times V_1$, where (ψ_1, V_1) is the local chart on $\mathbf{S}^2$ with coordinate functions θ and ϕ.

In this section we will see how the coordinate singularity $r = R_d$ in the local expression for the de Sitter metric given in (1) can be removed by introducing new coordinates. We first define a new radial coordinate function

$$\tilde{r} = \frac{R_d}{2}\log\left|\frac{R_d + r}{R_d - r}\right|\,,$$

which is called a *tortoise coordinate* since $\tilde{r} \to \infty$ when $r \to R_d$. As

$$d\tilde{r} = \left(1 - \frac{\Lambda}{3}r^2\right)^{-1}dr\,,$$

the de Sitter metric then takes the form

$$g|_{N_j} = \left(1 - \frac{\Lambda}{3} r^2\right)(-dt^2 + d\widetilde{r}^2) + r^2(d\theta^2 + \sin^2\theta\; d\phi^2) \qquad (2)$$

for $j = 1,2$. Let $\gamma : I \to N_1$ be the geodesic for a radially moving light ray in the interior de Sitter spacetime. Then we have that

$$(t - \widetilde{r}) \circ \gamma = \text{const}$$

for outgoing light and

$$(t + \widetilde{r}) \circ \gamma = \text{const}$$

for ingoing light.

Introducing a new coordinate function $v = t - \widetilde{r}$ we obtain a local chart (ψ_{oe}, N) on M with coordinate functions v, r, θ and ϕ, called *outgoing coordinates*, satisfying

$$-dt^2 + d\widetilde{r}^2 = -dv\,(dv + 2\,d\widetilde{r})$$

so that

$$g|_N = -\left(1 - \frac{\Lambda}{3} r^2\right) dv^2 - 2\,dv\,dr + r^2(d\theta^2 + \sin^2\theta\; d\phi^2). \qquad (3)$$

Since $\det(g_{ij})|_N = -r^4 \sin^2\theta$, this coordinate expression of the de Sitter metric is well defined for all $r > 0$. Hence the coordinate singularity at $r = R_d$ has been removed, and the Lorentz manifold M and the coordinate neighbourhood N can be replaced by $E = \mathbf{R} \times (0,\infty) \times \mathbf{S}^2$ and $V = \mathbf{R} \times (0,\infty) \times V_1$, with the metric

$$g_{oe} = -\left(1 - \frac{\Lambda}{3} r^2\right) dv^2 - 2\,dv\,dr + r^2 g_1\,.$$

Now we have that

$$v \circ \gamma = \text{const}$$

for outgoing light and

$$(v + 2\widetilde{r}) \circ \gamma = \text{const}$$

for ingoing light. This shows that the surface $r = R_d$ acts as a one-way membrane, letting the light rays cross only from the inside to the outside.

We can also introduce a coordinate function $u = t + \widetilde{r}$ and obtain a local chart (ψ_{ie}, N) on M with coordinate functions u, r, θ and ϕ, called *ingoing coordinates*, satisfying

$$-dt^2 + d\widetilde{r}^2 = -du\,(du - 2\,d\widetilde{r})$$

so that

$$g|_N = -\left(1 - \frac{\Lambda}{3} r^2\right) du^2 + 2\,du\,dr + r^2(d\theta^2 + \sin^2\theta\; d\phi^2). \qquad (4)$$

The Lorentz manifold M and the coordinate neighbourhood N can again be replaced by E and V, but this time with the metric

$$g_{ie} = -\left(1 - \frac{\Lambda}{3} r^2\right) du^2 + 2\,du\,dr + r^2 g_1\,.$$

Now we have that

$$(u-2\widetilde{r})\circ\gamma = \text{const}$$

for outgoing light and

$$u\circ\gamma = \text{const}$$

for ingoing light. Hence the light rays can cross the surface $r = R_d$ only from the outside to the inside.

The Lorentz manifolds $(M_1, g|_{M_1})$, $(M_1, g_{oe}|_{M_1})$ and $(M_1, g_{ie}|_{M_1})$ are isometric isomorphic, as are the Lorentz manifolds $(M_2, g|_{M_2})$, $(M_2, g_{oe}|_{M_2})$ and $(M_2, g_{ie}|_{M_2})$. But the Lorentz manifolds (E, g_{oe}) and (E, g_{ie}) are not isometric isomorphic. However, in the next section we will see that both (E, g_{oe}) and (E, g_{ie}) can be isometrically embedded in a larger Lorentz manifold which is analogous to the Kruskal–Szekeres extension of the Schwarzschild spacetime.

10.194 Blau–Guendelman–Guth coordinates Using the coordinate functions

$$v = t-\widetilde{r} \quad , \quad u = t+\widetilde{r}$$

described in 10.193, and fixing functions $\varepsilon_1 : M \to \{-1,1\}$ and $\varepsilon_2 : M \to \{-1,1\}$ which are constant on M_1 and M_2 with

$$\varepsilon_1\varepsilon_2 = \operatorname{sgn}(R_d - r)\,,$$

we can define new coordinate functions

$$\widetilde{v} = \varepsilon_1 \exp\Big(\frac{v}{R_d}\Big) \quad , \quad \widetilde{u} = -\,\varepsilon_2 \exp\Big(-\frac{u}{R_d}\Big)$$

and

$$T = \tfrac{1}{2}\,(\widetilde{v}+\widetilde{u}) \quad , \quad R = \tfrac{1}{2}\,(\widetilde{v}-\widetilde{u})$$

so that

$$R^2 - T^2 = -\,\widetilde{v}\,\widetilde{u} = \operatorname{sgn}(R_d - r)\,\exp\Big(-\frac{2\widetilde{r}}{R_d}\Big) = \frac{R_d - r}{R_d + r}. \tag{1}$$

The transformation from the de Sitter spacetime M is given by

$$T = \varepsilon_1\left(\frac{R_d - r}{R_d + r}\right)^{1/2} \sinh\Big(\frac{t}{R_d}\Big)$$

$$R = \varepsilon_1\left(\frac{R_d - r}{R_d + r}\right)^{1/2} \cosh\Big(\frac{t}{R_d}\Big)$$

when $0 < r < R_d$, mapping M_1 onto

$$B_{2-\varepsilon_1} = \{(T,R) \in \mathbf{R}^2 \,|\, |T| < \varepsilon_1 R < \sqrt{T^2+1}\,\} \times \mathbf{S}^2\,,$$

and by

$$T = \varepsilon_1 \left(\frac{r - R_d}{r + R_d}\right)^{1/2} \cosh\left(\frac{t}{R_d}\right)$$

$$R = \varepsilon_1 \left(\frac{r - R_d}{r + R_d}\right)^{1/2} \sinh\left(\frac{t}{R_d}\right)$$

when $r > R_d$, mapping M_2 onto

$$B_{3-\varepsilon_1} = \{(T,R) \in \mathbf{R}^2 | |R| < \varepsilon_1 T < \sqrt{R^2+1}\} \times \mathbf{S}^2 .$$

The inverse transformation is given by

$$\frac{r}{R_d} = \frac{1-(R^2-T^2)}{1+(R^2-T^2)}, \tag{2}$$

and by

$$\tanh\left(\frac{t}{R_d}\right) = \frac{T}{R}$$

whcn $0 < r < R_d$, and

$$\tanh\left(\frac{t}{R_d}\right) = \frac{R}{T}$$

when $r > R_d$. Now using that

$$-dT^2 + dR^2 = -\,d\tilde{v}\,d\tilde{u} = \frac{1}{R_d^2}\,(-\,\tilde{v}\,\tilde{u})\,(-\,dv\,du) = \frac{1}{R_d^2}\left(\frac{R_d - r}{R_d + r}\right)(-dt^2 + d\tilde{r}^2)\,,$$

we have that

$$g|_N = (R_d + r)^2(-dT^2 + dR^2) + r^2(d\theta^2 + \sin^2\theta\, d\phi^2) \tag{3}$$

where r is given by (2). As this coordinate expression of the de Sitter metric is well defined for all $r > 0$, and using all possible alternatives for ε_1, we obtain a new Lorentz manifold

$$B = \{(T,R) \in \mathbf{R}^2 \mid |R^2 - T^2| < 1\} \times \mathbf{S}^2$$

and a local chart (ψ_b, W) on B with coordinate neighbourhood

$$W = \{(T,R) \in \mathbf{R}^2 \mid |R^2 - T^2| < 1\} \times V_1$$

and coordinate functions T, R, θ and ϕ, called *Blau–Guendelman–Guth* (BGG) *coordinates.* The metric g_b on B is

$$g_b = (R_d + r)^2(-dT^2 + dR^2) + r^2 g_1\,.$$

To find the transformations from the extensions (E, g_{oe}) and (E, g_{ie}) of (M, g), we use that

$$\tilde{u} = -\,\varepsilon_2 \exp\left(-\frac{2\tilde{r}}{R_d}\right)\exp\left(-\frac{v}{R_d}\right) = -\,\varepsilon_1\left(\frac{R_d - r}{R_d + r}\right)\exp\left(-\frac{v}{R_d}\right)$$

and

$$\widetilde{v} = \varepsilon_1 \exp\left(-\frac{2\widetilde{r}}{R_d}\right) \exp\left(\frac{u}{R_d}\right) = \varepsilon_2 \left(\frac{R_d - r}{R_d + r}\right) \exp\left(\frac{u}{R_d}\right),$$

which imply that

$$T = \frac{\varepsilon_1}{2} \left\{ \exp\left(\frac{v}{R_d}\right) - \left(\frac{R_d - r}{R_d + r}\right) \exp\left(-\frac{v}{R_d}\right) \right\}$$

$$R = \frac{\varepsilon_1}{2} \left\{ \exp\left(\frac{v}{R_d}\right) + \left(\frac{R_d - r}{R_d + r}\right) \exp\left(-\frac{v}{R_d}\right) \right\}$$

and

$$T = \frac{\varepsilon_2}{2} \left\{ \left(\frac{R_d - r}{R_d + r}\right) \exp\left(\frac{u}{R_d}\right) - \exp\left(-\frac{u}{R_d}\right) \right\}$$

$$R = \frac{\varepsilon_2}{2} \left\{ \left(\frac{R_d - r}{R_d + r}\right) \exp\left(\frac{u}{R_d}\right) + \exp\left(-\frac{u}{R_d}\right) \right\}.$$

The inverse transformations are given by (2) and by

$$\varepsilon_1 \exp\left(\frac{v}{R_d}\right) = R + T \quad , \quad \varepsilon_2 \exp\left(-\frac{u}{R_d}\right) = R - T .$$

Choosing $\varepsilon_1 = 1$ and $\varepsilon_2 = 1$ in these formulae, we see that the regions

$$B_{12} = \{((T,R),\mathbf{u}) \in B \,|\, R > -T\} \quad \text{and} \quad B_{14} = \{((T,R),\mathbf{u}) \in B \,|\, R > T\}$$

are isometric to the outgoing and ingoing extension of B_1, respectively. The regions B_1 and B_3 are both isometric to the internal de Sitter spacetime, but these two regions cannot be connected by any timelike or null curve.

By (2) we also have that

$$R_d + r = \frac{2R_d}{1 + (R^2 - T^2)}$$

so that

$$\begin{aligned}
&(R_d + r)^2 \, (-dT^2 + dR^2) \\
&= d\,(R_d + r)\,[1 + (R^2 - T^2)]\, dr + (R_d + r)^2 \, (dR^2 - dT^2) \\
&= [1 + (R^2 - T^2)]\, dr^2 + (R_d + r)\, d(R^2 - T^2)\, dr + (R_d + r)^2 \, (dR^2 - dT^2) \\
&= -\{T\,dr + (R_d + r)\, dT\}^2 + \{R\,dr + (R_d + r)\, dR\}^2 + dr^2 \\
&= -\{d\,(R_d + r)\,T\}^2 + \{d\,(R_d + r)\,R\}^2 + dr^2
\end{aligned}$$

and

$$-\{(R_d+r)T\}^2+\{(R_d+r)R\}^2+r^2$$
$$=R_d^2\left\{\frac{4(R^2-T^2)}{[1+(R^2-T^2)]^2}+\frac{[1-(R^2-T^2)]^2}{[1+(R^2-T^2)]^2}\right\}=R_d^2\,.$$

Hence there is an isometric embedding $\alpha : B \to \mathbf{R}_1^5$ of B as a 4-pseudosphere

$$\mathbf{S}_{R_d}^4=\{(v,w,x,y,z)\in\mathbf{R}^5\,|\,-v^2+w^2+x^2+y^2+z^2=R_d^2\}$$

with radius R_d in the pseudo-Euclidean space $\mathbf{R}_1^5$ given by

$$\alpha((T,R),\mathbf{u})=R_d\left(\frac{2}{[1+(R^2-T^2)]^2}\,(T,R)\,,\,\frac{[1-(R^2-T^2)]^2}{[1+(R^2-T^2)]^2}\,\mathbf{u}\right) \tag{4}$$

for $((T,R),\mathbf{u})\in B$. On the coordinate domain W we have that

$$\begin{aligned}
v\circ\alpha|_W &= (R_d+r)T\\
w\circ\alpha|_W &= (R_d+r)R\\
x\circ\alpha|_W &= r\sin\theta\,\cos\phi\\
y\circ\alpha|_W &= r\sin\theta\,\sin\phi\\
z\circ\alpha|_W &= r\cos\theta
\end{aligned} \tag{5}$$

where r is given by (2), so that

$$v\circ\alpha|_W=\frac{2R_dT}{1+(R^2-T^2)}$$
$$w\circ\alpha|_W=\frac{2R_dR}{1+(R^2-T^2)}$$
$$x\circ\alpha|_W=\frac{1-(R^2-T^2)}{1+(R^2-T^2)}\,R_d\sin\theta\,\cos\phi$$
$$y\circ\alpha|_W=\frac{1-(R^2-T^2)}{1+(R^2-T^2)}\,R_d\sin\theta\,\sin\phi$$
$$z\circ\alpha|_W=\frac{1-(R^2-T^2)}{1+(R^2-T^2)}\,R_d\cos\theta$$

in BGG coordinates. On the open submanifold

$$U=\{(T,R)\in\mathbf{R}^2\,|\,0<|R^2-T^2|<1\}\times V_1$$

we also have that

$$v \circ \alpha|_U = \begin{cases} \varepsilon_1 (R_d^2 - r^2)^{1/2} \sinh(t/R_d) & \text{when} \quad 0 < r < R_d \\ \varepsilon_1 (r^2 - R_d^2)^{1/2} \cosh(t/R_d) & \text{when} \quad r > R_d \end{cases}$$

$$w \circ \alpha|_U = \begin{cases} \varepsilon_1 (R_d^2 - r^2)^{1/2} \cosh(t/R_d) & \text{when} \quad 0 < r < R_d \\ \varepsilon_1 (r^2 - R_d^2)^{1/2} \sinh(t/R_d) & \text{when} \quad r > R_d \end{cases}$$

$$x \circ \alpha|_U = r \sin\theta \cos\phi$$
$$y \circ \alpha|_U = r \sin\theta \sin\phi$$
$$z \circ \alpha|_U = r \cos\theta$$

in Eddington coordinates.

AFFINE TRANSFORMATIONS AND KILLING VECTOR FIELDS

10.195 Definition Let M and N be smooth manifolds with linear connections H and $\widetilde{H}$, respectively. Then a smooth map $f : M \to N$ is called *affine* if

$$\widetilde{\tau}_{t_1}^{t_0} \circ f_{*\gamma(t_0)} = f_{*\gamma(t_1)} \circ \tau_{t_1}^{t_0}$$

for every smooth curve $\gamma : I \to M$ defined on an open interval I and for each pair of points $t_0, t_1 \in I$, where $\tau_{t_1}^{t_0}$ and $\widetilde{\tau}_{t_1}^{t_0}$ are the parallel translations from t_0 to t_1 in TM and TN along γ and $f \circ \gamma$, respectively.

10.196 Remark If $f : M \to N$ is an affine map and $F : I \to TM$ is a lifting of a smooth curve $\gamma : I \to M$ which is parallel along γ, then it follows from Proposition 10.92 that the lifting $f_* \circ F$ of $f \circ \gamma$ is parallel along $f \circ \gamma$. Applying this to the lifting γ' of γ, it follows that $f \circ \gamma$ is a geodesic on N whenever γ is a geodesic on M. It also follows that

$$exp \circ f_* |_{\mathscr{D}} = f \circ exp \, .$$

Indeed, let $v \in \mathscr{D}_p$ and $\gamma_v : I(v) \to M$ be the unique maximal geodesic on M with $\gamma(0) = p$ and $\gamma'(0) = v$. Then $f \circ \gamma_v$ is a geodesic on N with $f \circ \gamma_v(0) = f(p)$ and $(f \circ \gamma_v)'(0) = f_*(v)$, and we have that

$$exp \circ f_*(v) = f \circ \gamma_v(1) = f \circ exp\,(v) \, .$$

10.197 Proposition Let M and N be smooth manifolds with linear connections H and H', respectively, and let $f : M \to N$ be an affine map. If X, Y and Z are vector fields on M which are f-related to the vector fields X', Y' and Z' on N, then we have that

(1) $\nabla_X Y$ and $\nabla'_{X'} Y'$ are f-related, where ∇ and ∇' are the covariant derivatives of H and H' on M and N, respectively.

(2) $T(X,Y)$ and $T'(X',Y')$ are f-related, where T and T' are the torsion tensors of H and H' on M and N, respectively.

(3) $R(X,Y)Z$ and $R'(X',Y')Z'$ are f-related, where R and R' are the curvature tensors of H and H' on M and N, respectively.

PROOF: (1) Let $\gamma : I \to M$ be a smooth curve defined on an open interval I containing t_0 with $\gamma(t_0) = p$ and $\gamma'(t_0) = X_p$, and let $\tau^t_{t_0}$ and $\tilde{\tau}^t_{t_0}$ be the parallel translations from t to t_0 in TM and TN along γ and $f \circ \gamma$, respectively. Then we have that

$$f_{*p} \circ \tau^t_{t_0} Y(\gamma(t)) = \tilde{\tau}^t_{t_0} \circ f_{*\gamma(t)} Y(\gamma(t)) = \tilde{\tau}^t_{t_0} Y'(f \circ \gamma(t))$$

for $t \in I$. Since $(f \circ \gamma)'(t_0) = f_*(X_p) = X'_{f(p)}$, it follows from Proposition 10.78 that

$$f_*(\nabla_{X_p} Y) = \nabla'_{X'_{f(p)}} Y',$$

showing that $\nabla_X Y$ and $\nabla'_{X'} Y'$ are f-related.

(2) Follows from (1) and Propositions 4.88 and 10.102.

(3) Follows from (1) and Propositions 4.88 and 10.107. □

10.198 Let $f : M \to N$ be a diffeomorphism between the n-dimensional smooth manifolds M and N. Then we have an isomorphism $(\tilde{f}, f, id)$ between the frame bundles $\pi : \mathscr{L}(M) \to M$ and $\pi' : \mathscr{L}(N) \to N$ obtained from the equivalence (f_*, f) between the tangent bundles $\tilde{\pi} : TM \to M$ and $\tilde{\pi}' : TN \to N$ as described in Example 10.21, where the map $\tilde{f} : \mathscr{L}(M) \to \mathscr{L}(N)$ is said to be *induced* by f and is given by

$$\tilde{f}(u) = f_{*p} \circ u$$

for $p \in M$ and $u \in \mathscr{L}_p(M)$.

If θ and θ' are the dual forms on $\mathscr{L}(M)$ and $\mathscr{L}(N)$, we have that

$$\tilde{f}^*(\theta') = \theta \tag{1}$$

since

$$\tilde{f}^*(\theta')(u) = \theta'(\tilde{f}(u)) \circ \tilde{f}_{*u} = \tilde{f}(u)^{-1} \circ \pi'_{*\tilde{f}(u)} \circ \tilde{f}_{*u}$$

$$= (f_{*p} \circ u)^{-1} \circ (f \circ \pi)_{*u} = u^{-1} \circ \pi_{*u} = \theta(u)$$

for $p \in M$ and $u \in \mathscr{L}_p(M)$. Moreover, $\tilde{f} : \mathscr{L}(M) \to \mathscr{L}(N)$ is the unique diffeomorphism with $\pi \circ \tilde{f} = f \circ \pi$ satisfying (1). Indeed, if $\tilde{g} : \mathscr{L}(M) \to \mathscr{L}(N)$ is a diffeomorphism with $\pi \circ \tilde{g} = f \circ \pi$ and $\tilde{g}^*(\theta') = \theta$, then

$$u^{-1} \circ \pi_{*u} = \theta(u) = \tilde{g}^*(\theta')(u) = \theta'(\tilde{g}(u)) \circ \tilde{g}_{*u}$$
$$= \tilde{g}(u)^{-1} \circ \pi'_{*\tilde{g}(u)} \circ \tilde{g}_{*u} = \tilde{g}(u)^{-1} \circ f_{*p} \circ \pi_{*u}$$

which implies that

$$\tilde{g}(u) = f_{*p} \circ u = \tilde{f}(u)$$

for $p \in M$ and $u \in \mathscr{L}_p(M)$.

The fundamental vector fields $\sigma(X)$ and $\sigma'(X)$ on $\mathscr{L}(M)$ and $\mathscr{L}(N)$ are $\tilde{f}$-related for each $X \in \mathfrak{gl}(\mathbf{R}^n)$, since

$$\sigma'_{\tilde{f}(u)}(F) = \tilde{f}(u) \circ F = R_F \circ \tilde{f}(u) = \tilde{f} \circ R_F(u) = \tilde{f}(u \circ F) = \tilde{f} \circ \sigma_u(F)$$

for $u \in \mathscr{L}(M)$ and $F \in Gl(\mathbf{R}^n)$, which implies that

$$\sigma'(X)_{\tilde{f}(u)} = \sigma'_{\tilde{f}(u)*}(X) = \tilde{f}_* \circ \sigma_{u*}(X) = \tilde{f}_*(\sigma(X)_u)$$

for $u \in \mathscr{L}(M)$ and $X \in \mathfrak{gl}(\mathbf{R}^n)$.

10.199 Proposition Let $f : M \to N$ be a diffeomorphism between the n-dimensional smooth manifolds M and N with linear connections H and H', respectively, and let $\tilde{f} : \mathscr{L}(M) \to \mathscr{L}(N)$ be the map induced by f. Then the following assertions are equivalent :

(1) f is affine.

(2) $\tilde{f}_*(H_u) = H'_{\tilde{f}(u)}$ for every $u \in \mathscr{L}(M)$.

(3) $\tilde{f}^*(\omega') = \omega$, where ω and ω' are the connection forms of H and H'.

(4) If X and Y are vector fields on M which are f-related to the vector fields X' and Y' on N, and if ∇ and ∇' are the covariant derivatives of H and H' on M and N, respectively, then $\nabla_X Y$ and $\nabla'_{X'} Y'$ are f-related.

(5) The basic vector fields $B(\xi)$ and $B'(\xi)$ on $\mathscr{L}(M)$ and $\mathscr{L}(N)$ are $\tilde{f}$-related for each $\xi \in \mathbf{R}^n$.

PROOF : We first show that (3) implies (2). Assuming that $\tilde{f}^*(\omega') = \omega$, we have that $\omega'(\tilde{f}(u)) \circ \tilde{f}_{*u} = \omega(u)$ which implies that $\tilde{f}_*(H_u) \subset H'_{\tilde{f}(u)}$ for every $u \in \mathscr{L}(M)$. Now since H_u and $H'_{\tilde{f}(u)}$ both have dimension n by Proposition 10.49, we thus obtain (2).

We next show that (2) implies (1). Assuming that $\beta : I \to \mathscr{L}(M)$ is a horizontal lift of a smooth curve $\gamma : I \to M$ in M defined on an open interval I, it follows from (2) that $\tilde{f} \circ \beta$ is a horizontal lift of $f \circ \gamma$. Hence if $\tau_{t_1}^{t_0}$ and $\widetilde{\tau}_{t_1}^{t_0}$ are the parallel translations from t_0 to t_1 in TM and TN along γ and $f \circ \gamma$, we have that

$$\widetilde{\tau}_{t_1}^{t_0} = (\tilde{f} \circ \beta)(t_1) \circ (\tilde{f} \circ \beta)(t_0)^{-1} = \{f_{*\gamma(t_1)} \circ \beta(t_1)\} \circ \{f_{*\gamma(t_0)} \circ \beta(t_0)\}^{-1}$$
$$= f_{*\gamma(t_1)} \circ \tau_{t_1}^{t_0} \circ f_{*\gamma(t_0)}^{-1},$$

showing that (1) is true.

From Proposition 10.197 we know that (1) implies (4), so it only remains to show that (4) implies (3) in order to prove that the first four assertions are equivalent. By Proposition 10.168 there is a linear connection H'' on N with a connection form ω'' satisfying $\tilde{f}^*(\omega'') = \omega$. From the first part of the proof we know that the corresponding covariant derivative ∇'' satisfies (4) with H' and ∇' replaced with H'' and ∇''. Assuming that (4) is also true for H' and ∇', it follows that ∇' and ∇'' are two coinciding Koszul connections on N. By Corollary 10.132 we therefore have that $\omega' = \omega''$, which shows that (3) is true.

Finally, we will show that (5) is equivalent to (2). We first note that

$$\pi'_* \circ \tilde{f}_*(B(\xi)_u) = f_* \circ \pi_*(B(\xi)_u) = f_* \circ u(\xi) = \tilde{f}(u)(\xi)$$

for $u \in \mathscr{L}(M)$ and $\xi \in \mathbf{R}^n$. Assuming (2), we also know that $B(\xi)_u \in H_u$ implies that $\tilde{f}_*(B(\xi)_u) \in H'_{\tilde{f}(u)}$. Hence it follows from Proposition 10.96 that

$$\tilde{f}_*(B(\xi)_u) = B'(\xi)_{\tilde{f}(u)}$$

for every $u \in \mathscr{L}(M)$ and $\xi \in \mathbf{R}^n$, which shows that (5) is true.

Conversely, it follows from the last equality that (5) implies (2), since

$$H_u = \{B(\xi)_u | \xi \in \mathbf{R}^n\} \quad \text{and} \quad H'_{\tilde{f}(u)} = \{B'(\xi)_{\tilde{f}(u)} | \xi \in \mathbf{R}^n\}$$

by formula (2) in Proposition 10.96. □

10.200 Proposition Let $\pi : \mathscr{L}(M) \to M$ be the frame bundle of a smooth manifold M, and let X be a vector field on M with global flow $\gamma : \mathscr{D}(X) \to M$. Then we have a map $\widetilde{\gamma} : \mathscr{B} \to \mathscr{L}(M)$ defined on the open subset $\mathscr{B} = (id \times \pi)^{-1}(\mathscr{D}(X))$ of $\mathbf{R} \times \mathscr{L}(M)$ which is given by

$$\widetilde{\gamma}(t,u) = \widetilde{\gamma}_t(u)$$

for $(t,u) \in \mathscr{B}$, where $\widetilde{\gamma}_t : \pi^{-1}(\mathscr{D}_t(X)) \to \pi^{-1}(\mathscr{D}_{-t}(X))$ is the map induced by the diffeomorphism $\gamma_t : \mathscr{D}_t(X) \to \mathscr{D}_{-t}(X)$ as described in 10.198.

The curve $\widetilde{\gamma}_u : I(p) \to \mathscr{L}(M)$ defined by $\widetilde{\gamma}_u(t) = \widetilde{\gamma}(t,u)$ for $t \in I(p)$ is smooth for each $p \in M$ and $u \in \mathscr{L}_p(M)$, and we have a vector field $\widetilde{X}$ on $\mathscr{L}(M)$ with global flow $\widetilde{\gamma}$ which is given by $\widetilde{X}(u) = \widetilde{\gamma}'_u(0)$ for $u \in \mathscr{L}(M)$. $\widetilde{X}$ is called the *natural lift* of X, and it is the unique vector field on $\mathscr{L}(M)$ satisfying

(1) $(R_F)_*\widetilde{X} = \widetilde{X}$ for every $F \in Gl(\mathbf{R}^n)$.

(2) $L_{\widetilde{X}}\,\theta = 0$.

(3) $\widetilde{X}$ and X are π-related.

Conversely, given a vector field $\widetilde{X}$ on $\mathscr{L}(M)$ satisfying assertion (1) and (2), there is a unique vector field X on M which satisfies assertion (3).

PROOF : We first show that $\widetilde{X}$ is smooth on $\mathscr{L}(M)$. Given a point p on M, we let (y,V) be a local chart around p, and choose an open neighbourhood U of p and an open interval I containing 0 so that $\gamma(I\times U)\subset V$. Now let $x=y|_U$, and let $(\phi_x,\pi^{-1}(U))$ and $(\phi_y,\pi^{-1}(V))$ be the corresponding local trivializations in the frame bundle $\mathscr{L}(M)$ as defined in Remark 10.22. Furthermore, let $(\beta, Gl(\mathbf{R}^n))$ be the local chart on $Gl(\mathbf{R}^n)$ given by

$$\beta^{\,b(i,j)}(G)=e^j\circ G(e_i)$$

for $G\in Gl(\mathbf{R}^n)$ and $i,j=1,\ldots,n$, where $b: I_n\times I_n\to I_{n^2}$ is the bijection defined in Remark 4.27 and $\mathscr{E}=\{e_1,\ldots,e_n\}$ is the standard basis for $\mathbf{R}^n$ with dual basis $\mathscr{E}^*=\{e^1,\ldots,e^n\}$. Then we obtain local charts $(\widetilde{x},\pi^{-1}(U))$ and $(\widetilde{y},\pi^{-1}(V))$ on $\mathscr{L}(M)$ where $\widetilde{x}=(x\times\beta)\circ\phi_x$ and $\widetilde{y}=(y\times\beta)\circ\phi_y$, and we have that

$$\widetilde{X}(u)=[\widetilde{y},(\widetilde{y}\circ\widetilde{\gamma}_u)'(0)]_u$$

so that

$$t_{\widetilde{y}}\circ\widetilde{X}(u)=(u,h(u))$$

where $h(u)=(\widetilde{y}\circ\widetilde{\gamma}_u)'(0)$ for $u\in\pi^{-1}(U)$. Now we have that

$$\begin{aligned}\phi_y\circ\widetilde{\gamma}_u(t)&=(\gamma_t(p),t_{y,\gamma_t(p)}\circ\gamma_{t*p}\circ t_{x,p}^{-1}\circ t_{x,p}\circ u)\\&=(\gamma_t(p),D(y\circ\gamma_t\circ x^{-1})(x(p))\circ\phi_{x,p}(u))\\&=(\gamma(t,p),D_2(y\circ\gamma\circ(id\times x)^{-1})(t,x(p))\circ\phi_{x,p}(u))\end{aligned}$$

for $(t,u)\in I\times\pi^{-1}(U)$ where $\pi(u)=p$. This implies that $\widetilde{\gamma}_u$ is smooth in I for $u\in\pi^{-1}(U)$, and that

$$\begin{aligned}h\circ\widetilde{x}^{-1}(a,b)=(&D_1(y\circ\gamma\circ(id\times x)^{-1})(0,a),\\&\beta(D_1D_2(y\circ\gamma\circ(id\times x)^{-1})(0,a)\circ\beta^{-1}(b)))\end{aligned}$$

for $(a,b)\in\widetilde{x}(\pi^{-1}(U))\subset\mathbf{R}^n\times\mathbf{R}^{n^2}$, thus showing that $\widetilde{X}$ is smooth.

Using that $(\widetilde{\gamma}_t,\gamma_t,id)$ is a homomorphism, we see that $\widetilde{X}$ satisfies (3) since

$$\pi\circ\widetilde{\gamma}_u(t)=\pi\circ\widetilde{\gamma}_t(u)=\gamma_t\circ\pi(u)=\gamma_{\pi(u)}(t)$$

when $t\in I(\pi(u))$, which implies that

$$\pi_*\circ\widetilde{X}(u)=(\pi\circ\widetilde{\gamma}_u)'(0)=\dot{\gamma}_{\pi(u)}(0)=X\circ\pi(u)$$

for $u\in\mathscr{L}(M)$.

We next prove that $\widetilde{\gamma}$ is the global flow for $\widetilde{X}$. From Proposition 3.42 (2) we know that

$$\gamma_s\circ\gamma_t(p)=\gamma_{s+t}(p)$$

for $p\in\mathscr{D}_t(X)\cap\gamma_t^{-1}(\mathscr{D}_s(X))$, which implies that

$$\widetilde{\gamma}_{\widetilde{\gamma}_u(t)}(s)=\widetilde{\gamma}(s,\widetilde{\gamma}(t,u))=\widetilde{\gamma}_s\circ\widetilde{\gamma}_t(u)=\widetilde{\gamma}_{s+t}(u)=\widetilde{\gamma}(s+t,u)=\widetilde{\gamma}_u(s+t)$$

for $u \in \pi^{-1}(\mathscr{D}_t(X)) \cap \widetilde{\gamma}_t^{-1}(\pi^{-1}(\mathscr{D}_s(X)))$ when $s \in I(p) - t$. By taking the derivative with respect to s at $s = 0$ we see that

$$\widetilde{X}(\widetilde{\gamma}_u(t)) = \widetilde{\gamma}_u'(t)$$

when $t \in I(p)$ for each $p \in M$ and $u \in \mathscr{L}_p(M)$. Hence it follows from Proposition 3.48 that each $\widetilde{\gamma}_u$ is a maximal integral curve for $\widetilde{X}$ with initial condition u, thus showing that $\widetilde{\gamma}$ is the global flow for $\widetilde{X}$.

Since $(\widetilde{\gamma}_t, \gamma_t, id)$ is a homomorphism, it follows from Corollary 3.50 that $\widetilde{X}$ satisfies (1) since

$$R_F \circ \widetilde{\gamma}_t = \widetilde{\gamma}_t \circ R_F$$

for every $t \in \mathbf{R}$ and $F \in Gl(\mathbf{R}^n)$. $\widetilde{X}$ also satisfies (2) since

$$\widetilde{\gamma}_t^*(\theta)(u) = \theta(u)$$

for every $u \in \mathscr{L}(M)$ and $t \in I(\pi(u))$ by formula (1) in 10.198.

To show the uniqueness of $\widetilde{X}$, let $\widehat{X}$ be a vector field on $\mathscr{L}(M)$ which is π-related to X and satisfies $L_{\widehat{X}}\,\theta = 0$ and $(R_F)_*\widehat{X} = \widehat{X}$ for every $F \in Gl(\mathbf{R}^n)$. Let $\widehat{\gamma} : \mathscr{D}(\widehat{X}) \to \mathscr{L}(M)$ be its global flow, and let $\bar{\gamma} : \mathscr{A} \to M$ be the restriction of γ to $\mathscr{A} = (id \times \pi)(\mathscr{D}(\widehat{X}))$. By Proposition 3.49 the condition $(R_F)_*\widehat{X} = \widehat{X}$ implies that

$$\mathscr{D}(\widehat{X}) = (id \times R_F)(\mathscr{D}(\widehat{X}))$$

for every $F \in Gl(\mathbf{R}^n)$, which shows that $\mathscr{D}(\widehat{X}) = (id \times \pi)^{-1}(\mathscr{A})$. As $\widehat{X}$ and X are π-related, it follows from Proposition 3.48 that $\pi \circ \widehat{\gamma}_u$ is an integral curve for X with initial condition $\pi(u)$, so that $\pi \circ \widehat{\gamma}_u(t) = \bar{\gamma}_{\pi(u)}(t)$ for $u \in \mathscr{L}(M)$ and $t \in \widehat{I}(u)$. This implies that

$$\pi \circ \widehat{\gamma}_t = \bar{\gamma}_t \circ \pi$$

for $t \in \mathbf{R}$. Finally, the condition $L_{\widehat{X}}\,\theta = 0$ implies that

$$\widehat{\gamma}_t^*(\theta)(u) = \theta(u)$$

for every $t \in \mathbf{R}$ and $u \in \mathscr{D}_t(\widehat{X})$ by Proposition 4.90. Now it follows from 10.198 that $\widehat{\gamma} = \widetilde{\gamma}|_{\mathscr{D}(\widehat{X})}$ which shows that $\widehat{X} = \widetilde{X}$.

To prove the last part of the proposition, consider a vector field $\widetilde{X}$ on $\mathscr{L}(M)$ satisfying assertion (1) and (2), and let

$$X(p) = \pi_* \circ \widetilde{X}(u)$$

for $p \in M$, where $u \in \mathscr{L}(M)$ is chosen so that $\pi(u) = p$. To see that X is well defined, let $u_1, u_2 \in \mathscr{L}(M)$ with $\pi(u_1) = \pi(u_2)$, and choose $F \in Gl(\mathbf{R}^n)$ with $R_F(u_1) = u_2$. Then it follows from (1) that

$$\begin{aligned}\pi_* \circ \widetilde{X}(u_2) &= \pi_* \circ \widetilde{X} \circ R_F(u_1) = \pi_* \circ (R_F)_* \circ \widetilde{X}(u_1)\\ &= (\pi \circ R_F)_* \circ \widetilde{X}(u_1) = \pi_* \circ \widetilde{X}(u_1)\,.\end{aligned}$$

If $s : U \to \mathscr{L}(M)$ is a frame field on an open subset U of M, we have that $X|_U = \pi_* \circ \widetilde{X} \circ s$ which shows that X is smooth and therefore is a vector field on M satisfying $\pi_* \circ \widetilde{X} = X \circ \pi$. □

10.201 Definition Let M be a smooth manifold with a linear connection H. Then a vector field X on M with global flow $\gamma : \mathscr{D}(X) \to M$ is called an *infinitesimal affine transformation* of M if the diffeomorphism $\gamma_t : \mathscr{D}_t(X) \to \mathscr{D}_{-t}(X)$ is affine for each $t \in \mathbf{R}$.

10.202 Proposition Let X be a vector field on the smooth manifold M^n with a linear connection H, and let $\widetilde{X}$ be the natural lift of X. Then the following assertions are equivalent :

(1) X is an infinitesimal affine transformation of M.

(2) $L_{\widetilde{X}}\, \omega = 0$, where ω is the connection form of H.

(3) $[\widetilde{X}, B(\xi)] = 0$ for every $\xi \in \mathbf{R}^n$, where $B(\xi)$ is the basic vector field determined by ξ.

PROOF : Let $\widetilde{\gamma} : \mathscr{D}(\widetilde{X}) \to \mathscr{L}(M)$ be the global flow for $\widetilde{X}$. Then we have that (1) and (2) are equivalent, since they are both equivalent to the assertion that

$$\widetilde{\gamma}_t^*(\omega)(u) = \omega(u)$$

for every $u \in \mathscr{L}(M)$ and $t \in I(u)$ by Propositions 4.90 and 10.199. We also have that (1) and (3) are equivalent, since they are both equivalent to the assertion that

$$\widetilde{\gamma}_t^*(B(\xi))_u = B(\xi)_u$$

for every $\xi \in \mathbf{R}^n$, $u \in \mathscr{L}(M)$ and $t \in I(u)$ by Propositions 4.86, 4.90 and 10.199. □

10.203 Let $f : M \to N$ be a diffeomorphism between the n-dimensional pseudo-Riemannian manifolds M and N with metrics g and g' of index k. If $\tilde{f} : \mathscr{L}(M) \to \mathscr{L}(N)$ is the map induced by f, we have that

$$\tilde{f}(u)^* g'(f(p)) = u^*(f^* g')(p)$$

for every $p \in M$ and $u \in \mathscr{L}_p(M)$. Hence f is an isometry if and only if

$$\tilde{f}(\mathscr{O}_\varepsilon(M)) \subset \mathscr{O}_\varepsilon(N)$$

where ε is any signature compatible with g.

10.204 Proposition Let $f : M \to N$ be an isometry between the pseudo-Riemannian manifolds M and N with Levi-Civita connections H and H'. Then the map f is affine.

PROOF : Let H'' be the connection on N obtained from H in the way described in Proposition 10.168, so that

$$\tilde{f}_*(H_u) = H'_{\tilde{f}(u)}$$

for every $u \in \mathscr{L}(M)$. Then it follows from Propositions 10.199 and 10.197 that f is affine with respect to the connections H and H'', and that H'' has no torsion. By Proposition 10.172 we know that H is reducible to a connection in $\mathscr{O}_\varepsilon(M)$. As f is an isometry, the same must be true for H'' by 10.203. This implies that H'' must coincide with the Levi–Civita connection H' on N. □

10.205 Definition Let M be a pseudo-Riemannian manifold with a metric g. Then a vector field X on M with global flow $\gamma\colon \mathscr{D}(X) \to M$ is called an *infinitesimal isometry* or a *Killing vector field* on M if the diffeomorphism $\gamma_t : \mathscr{D}_t(X) \to \mathscr{D}_{-t}(X)$ is an isometry for each $t \in \mathbf{R}$. By Proposition 10.204 every infinitesimal isometry is an infinitesimal affine transformation of M with respect to the Levi–Civita connection.

10.206 Proposition Let X be a vector field on the pseudo-Riemannian manifold M^n with a metric g, and let $\widetilde{X}$ be the natural lift of X. Then the following assertions are equivalent:

(1) X is an infinitesimal isometry.

(2) $\widetilde{X}|_{\mathscr{O}_\varepsilon(M)}$ belongs to $T\mathscr{O}_\varepsilon(M)$.

(3) $L_X\, g = 0$.

(4) $X(g(Y,Z)) = g([X,Y],Z) + g(Y,[X,Z])$ for every vector field Y and Z on M.

(5) $g(\nabla_Y X, Z) + g(\nabla_Z X, Y) = 0$ for every vector field Y and Z on M.

PROOF : Let $\gamma\colon \mathscr{D}(X) \to M$ be the global flow for X. Then we have that (1) and (2) are equivalent, since they are both equivalent to the assertion that

$$\tilde{\gamma}_t\,(\mathscr{O}_\varepsilon(\mathscr{D}_t(X))) \subset \mathscr{O}_\varepsilon(\mathscr{D}_{-t}(X))$$

for every $t \in \mathbf{R}$ by 10.203.

We also have that (1) and (3) are equivalent, since they are both equivalent to the assertion that

$$\gamma_t^*(g)(p) = g(p)$$

for every $p \in M$ and $t \in I(p)$ by Proposition 4.90.

Furthermore, we have that (3) and (4) are equivalent, since

$$\begin{aligned} X(g(Y,Z)) &= L_X(g(Y,Z)) = (L_X g)(Y,Z) + g(L_X Y, Z) + g(Y, L_X Z) \\ &= (L_X g)(Y,Z) + g([X,Y],Z) + g(Y,[X,Z]) \end{aligned}$$

by Remark 4.77 and Propositions 4.79 (5) and 4.86.

Using that the Levi–Civita connection has no torsion so that

$$T(X,Y) = \nabla_X Y - \nabla_Y X - [X,Y] = 0$$

for every vector field Y, we finally see that (4) and (5) are equivalent since

$$\begin{aligned} g(\nabla_Y X, Z) + g(\nabla_Z X, Y) &= g(\nabla_X Y - [X,Y], Z) + g(Y, \nabla_X Z - [X,Z]) \\ &= \{g(\nabla_X Y, Z) + g(Y, \nabla_X Z)\} - \{g([X,Y],Z) + g(Y,[X,Z])\} \\ &= X(g(Y,Z)) - \{g([X,Y],Z) + g(Y,[X,Z])\} \end{aligned}$$

by Proposition 10.162. □

10.207 Let X be a vector field on the pseudo-Riemannian manifold M^n with a metric g, and let (x,U) be a local chart on M. If

$$X\,|_U = \sum_{i=1}^{n} \xi^i \frac{\partial}{\partial x^i}$$

and

$$g\,|_U = \sum_{ij} g_{ij}\, dx^i \otimes dx^j\,,$$

then it follows from Propositions 4.76 and 10.166 and Corollary 10.118 that

$$L_X g\,|_U = \sum_{ij} B_{ij}\, dx^i \otimes dx^j$$

where

$$\begin{aligned} B_{ij} &= \sum_{k=1}^{n} \left\{ \xi^k \frac{\partial g_{ij}}{\partial x^k} + g_{kj} \frac{\partial \xi^k}{\partial x^i} + g_{ki} \frac{\partial \xi^k}{\partial x^j} \right\} \\ &= \sum_{k=1}^{n} \left\{ \frac{\partial (g_{kj}\xi^k)}{\partial x^i} + \frac{\partial (g_{ki}\xi^k)}{\partial x^j} - \xi^k \left(\frac{\partial g_{kj}}{\partial x^i} + \frac{\partial g_{ki}}{\partial x^j} - \frac{\partial g_{ij}}{\partial x^k} \right) \right\} \\ &= \sum_{k=1}^{n} \left\{ \frac{\partial (g_{kj}\xi^k)}{\partial x^i} + \frac{\partial (g_{ki}\xi^k)}{\partial x^j} - 2\,\xi^k \sum_{l=1}^{n} g_{kl}\,\Gamma^l_{ij} \right\} \\ &= \frac{\partial \xi_j}{\partial x^i} + \frac{\partial \xi_i}{\partial x^j} - 2 \sum_{l=1}^{n} \xi_l\, \Gamma^l_{ij} = \xi_{j;i} + \xi_{i;j}\,. \end{aligned}$$

Hence X is a Killing vector field if and only if

$$\xi_{j;i} + \xi_{i;j} = 0$$

for every local chart (x,U) and $1 \le i, j \le n$.

CONFORMAL TRANSFORMATIONS

10.208 Definition A diffeomorphism $f : M \to N$ between two pseudo-Riemannian manifolds M and N with metrics g and g' is said to be *conformal* if

$$f^*(g') = e^{2\sigma} g$$

for a smooth function $\sigma \in \mathscr{F}(M)$. Two metrics g and $\widetilde{g}$ on a smooth manifold M are said to be *conformally related* if the identity map $id : M \to M$ is conformal with respect to g and $\widetilde{g}$, i.e., if

$$\widetilde{g} = e^{2\sigma} g$$

for a smooth function $\sigma \in \mathscr{F}(M)$.

10.209 Proposition Let g and $\widetilde{g}$ be two metrics on a smooth manifold M which are conformally related so that

$$\widetilde{g} = e^{2\sigma} g$$

for a smooth function $\sigma \in \mathscr{F}(M)$, and let R and $\widetilde{R}$ be the corresponding curvature forms on M. If

$$U = \operatorname{grad} \sigma$$

and $B \in \Omega^1(M;TM)$ is the bundle-valued 1-form on M defined by

$$B(X) = -X(\sigma)\,U + \nabla_X U + \tfrac{1}{2}\,U(\sigma)\,X$$

for vector fields X on M, then we have that

$$\begin{aligned}\widetilde{R}(X,Y)\,Z = R(X,Y)\,Z &+ \{g(X,Z)\,B(Y) - g(Y,Z)\,B(X)\} \\ &+ \{g(B(X),Z)\,Y - g(B(Y),Z)\,X\}\end{aligned} \tag{1}$$

for every vector field X, Y and Z on M.

PROOF: Let H and $\widetilde{H}$ be the Levi–Civita connections of g and $\widetilde{g}$ with covariant derivatives ∇ and $\widetilde{\nabla}$, respectively, and let $K \in \mathscr{T}^2(M;TM)$ be the bundle-valued covariant tensor field of degree 2 on M defined by

$$K(X,Y) = \widetilde{\nabla}_X Y - \nabla_X Y$$

for vector fields X and Y on M. We will first show that

$$K(X,Y) = X(\sigma)\,Y + Y(\sigma)\,X - g(X,Y)\,U. \tag{2}$$

As the connections H and $\widetilde{H}$ have no torsion, the tensor field K is symmetric by Proposition 10.102. Since H and $\widetilde{H}$ are metrical connections compatible with g and $\widetilde{g}$, respectively, it follows from Proposition 10.162 that

$$X(\widetilde{g}(Y,Z)) = \widetilde{g}(\widetilde{\nabla}_X Y,Z) + \widetilde{g}(Y,\widetilde{\nabla}_X Z) = e^{2\sigma}\{g(\widetilde{\nabla}_X Y,Z) + g(Y,\widetilde{\nabla}_X Z)\}$$

and

$$X(\widetilde{g}(Y,Z)) = X(e^{2\sigma}g(Y,Z)) = 2X(\sigma)\, e^{2\sigma}g(Y,Z)$$
$$+\, e^{2\sigma}\{g(\nabla_X Y,Z)+g(Y,\nabla_X Z)\}$$

which implies that

$$g(K(X,Y),Z)+g(Y,K(X,Z)) = 2X(\sigma)\, g(Y,Z) \tag{3}$$

for each vector field $X,Y,Z \in \mathscr{T}_1(M)$. By a permutation of the vector fields X, Y and Z, we also have that

$$g(K(Y,X),Z)+g(X,K(Y,Z)) = 2\, Y(\sigma)\, g(X,Z) \tag{4}$$

and

$$g(K(Z,X),Y)+g(X,K(Z,Y)) = 2Z(\sigma)\, g(X,Y) \tag{5}$$

from which we obtain

$$g(K(X,Y),Z) = X(\sigma)\, g(Y,Z)+Y(\sigma)\, g(X,Z)-Z(\sigma)\, g(X,Y). \tag{6}$$

Now using that

$$Z(\sigma) = d\sigma(Z) = g(U,Z)\,,$$

formula (6) implies that

$$g(K(X,Y)-X(\sigma)\,Y-Y(\sigma)\,X+g(X,Y)\,U,Z)=0$$

for every vector field Z, which completes the proof of formula (2). By Proposition 10.107 we have that

$$\widetilde{R}(X,Y)Z = \widetilde{\nabla}_X\widetilde{\nabla}_Y Z - \widetilde{\nabla}_Y\widetilde{\nabla}_X Z - \widetilde{\nabla}_{[X,Y]} Z \tag{7}$$

where

$$\widetilde{\nabla}_X\widetilde{\nabla}_Y Z = \nabla_X(\nabla_Y Z+K(Y,Z))+K(X,\nabla_Y Z+K(Y,Z)) \tag{8}$$

and

$$\widetilde{\nabla}_{[X,Y]} Z = \nabla_{[X,Y]} Z+K([X,Y],Z). \tag{9}$$

Formula (2) implies that

$$\begin{aligned}
\nabla_X K(Y,Z)+K(X,\nabla_Y Z) &= X(Y(\sigma))\,Z+Y(\sigma)\,\nabla_X Z\\
&+\, X(Z(\sigma))\,Y+Z(\sigma)\,\nabla_X Y-X(g(Y,Z))\,U-g(Y,Z)\,\nabla_X U\\
&+\, X(\sigma)\,\nabla_Y Z+(\nabla_Y Z)(\sigma)\,X-g(X,\nabla_Y Z)\,U\\
&= \{X(\sigma)\,\nabla_Y Z+Y(\sigma)\,\nabla_X Z\}-\{g(X,\nabla_Y Z)+g(Y,\nabla_X Z)\}\,U\\
&+\, \{g(U,\nabla_X Z)\,Y+g(U,\nabla_Y Z)\,X\}+\{X(Y(\sigma))\,Z+Z(\sigma)\,\nabla_X Y\\
&-\, g(\nabla_X Y,Z)\,U\}-g(Y,Z)\,B_1(X)+g(B_1(X),Z)\,Y
\end{aligned}$$

where

$$B_1(X) = \nabla_X U\ ,$$

so that

$$\begin{aligned}\{\nabla_X K(Y,Z) + K(X,\nabla_Y Z)\} - \{\nabla_Y K(X,Z) + K(Y,\nabla_X Z)\} \\ - \ K([X,Y],Z) = \{g(X,Z)\,B_1(Y) - g(Y,Z)\,B_1(X)\} \\ + \ \{g(B_1(X),Z)\,Y - g(B_1(Y),Z)\,X\}.\end{aligned} \tag{10}$$

We also have that

$$\begin{aligned}K(X,K(Y,Z)) &= X(\sigma)\,\{Y(\sigma)\,Z + Z(\sigma)\,Y - g(Y,Z)\,U\} \\ &+ \ \{Y(\sigma)\,Z(\sigma) + Z(\sigma)\,Y(\sigma) - g(Y,Z)\,U(\sigma)\}\,X \\ &- \ g(X, Y(\sigma)\,Z + Z(\sigma)\,Y - g(Y,Z)\,U)\,U \\ &= X(\sigma)\,Y(\sigma)\,Z + Z(\sigma)\,\{X(\sigma)\,Y + Y(\sigma)\,X\} \\ &- \ \{X(\sigma)\,g(Y,Z) + Y(\sigma)\,g(X,Z)\}\,U - Z(\sigma)\,g(X,Y)\,U \\ &- \ g(Y,Z)\,B_2(X) - g(B_2(Y),Z)\,X\end{aligned}$$

where

$$B_2(X) = -X(\sigma)\,U + \tfrac{1}{2}\,U(\sigma)\,X\ ,$$

so that

$$\begin{aligned}K(X,K(Y,Z)) - K(Y,K(X,Z)) = \{g(X,Z)\,B_2(Y) - g(Y,Z)\,B_2(X)\} \\ + \ \{g(B_2(X),Z)\,Y - g(B_2(Y),Z)\,X\}.\end{aligned} \tag{11}$$

Since $B = B_1 + B_2$, formula (1) in the proposition now follows from formula (7) – (11). □

Chapter 11

ISOMETRIC IMMERSIONS AND THE SECOND FUNDAMENTAL FORM

CONNECTIONS IN REDUCED SUBBUNDLES

11.1 Proposition Let $\pi : P \to M$ be a principal G-bundle with a connection form ω, and let H be a Lie subgroup of G so that

$$\mathfrak{g} = \mathfrak{h} \oplus W$$

for a subspace W of $\mathfrak{g}$ with $Ad(H)(W) \subset W$. If $\pi' : Q \to M$ is a reduced subbundle of $\pi : P \to M$ with structure group H, and if $i : Q \to P$ is the inclusion map and $\rho : \mathfrak{g} \to \mathfrak{h}$ is the projection on $\mathfrak{h}$, then $\rho \cdot i^*(\omega)$ is a connection form on Q.

Suppose that $\eta \in \Omega^k(P;\mathfrak{g})$ is a $\mathfrak{g}$-valued k-form on P which is tensorial of type $(Ad, \mathfrak{g})$, and let $\rho' : \mathfrak{g} \to W$ be the projection on W and $\psi : H \to \mathrm{Aut}(W)$ be the representation of H on W obtained from Ad. Then $\rho' \cdot i^*(\eta)$ is a W-valued k-form on Q which is tensorial of type (ψ, W).

PROOF : Let $\alpha : H \to G$ be the inclusion map, and let $u \in Q$. Then we have that

$$i \circ \sigma_u(h) = i \circ R_h(u) = R_{\alpha(h)} \circ i(u) = \sigma_{i(u)} \circ \alpha(h)$$

for $h \in H$ by Definition 10.19, which implies that

$$i_* \, \sigma(X)_u = i_* \, \sigma_{u*}(X) = \sigma_{i(u)*} \, \alpha_*(X) = \sigma(\alpha_*(X))_{i(u)}$$

for $X \in \mathfrak{h}$. Hence we have that

$$\rho \cdot i^*(\omega)(u)(\sigma(X)_u) = \rho \circ \omega(i(u))(\sigma(\alpha_*(X))_{i(u)}) = \rho \circ \alpha_*(X) = X$$

which shows (i) in Definition 10.34. Using Proposition 5.48 we also have that

$$R_h^*(\rho \cdot i^*(\omega)) = \rho \cdot R_h^* \, i^*(\omega) = \rho \cdot i^* \, R_{\alpha(h)}^*(\omega)$$

$$= (\rho \circ Ad(\alpha(h)^{-1})) \cdot i^*(\omega) = Ad(h^{-1}) \cdot (\rho \cdot i^*(\omega)) \,,$$

since $Ad(\alpha(h^{-1}))(W) \subset W$ and

$$Ad(\alpha(h^{-1})) \circ \alpha_*(X) = \alpha_* \circ Ad(h^{-1})(X)$$

for $X \in \mathfrak{h}$ by Proposition 9.14, thereby showing (ii) in Definition 10.34.

In the same way we see that

$$\begin{aligned} R_h^*(\rho' \cdot i^*(\eta)) &= \rho' \cdot R_h^* \, i^*(\eta) = \rho' \cdot i^* \, R_{\alpha(h)}^*(\eta) \\ &= (\rho' \circ Ad(\alpha(h)^{-1})) \cdot i^*(\eta) = \psi(h^{-1}) \cdot (\rho' \cdot i^*(\eta)), \end{aligned}$$

which shows that $\rho' \cdot i^*(\eta)$ is pseudo-tensorial of type (ψ, W). It is also horizontal, since $\pi \circ i = \pi'$ implies that $i_*(\ker \pi'_{*u}) \subset \ker \pi_{*i(u)}$ for $u \in Q$. Hence i_* maps vertical vectors in T_uQ into vertical vectors in $T_{i(u)}P$ by Proposition 10.3. □

THE NORMAL BUNDLE AND THE BUNDLE OF ADAPTED ORTHONORMAL FRAMES

11.2 Let N^m and M^n be pseudo-Riemannian manifolds with metrics g and g' of index r and k, respectively, and let $f : M \to N$ be an isometric immersion. Then it follows from Proposition 7.9 that $f_*(T_pM)$ and $f_*(T_pM)^{\perp}$ are non-degenerate subspaces of $T_{f(p)}N$ with

$$T_{f(p)}N = f_*(T_pM) \oplus f_*(T_pM)^{\perp}$$

for every $p \in M$. Let $\widetilde{\pi}_M : f^*(TN) \to M$ be the pullback of the tangent bundle $\widetilde{\pi} : TN \to N$ by f with canonical bundle map $(\widetilde{f}, f)$. The pullback $\widetilde{g} = \widetilde{f}^*(g)$ of the metric g is a fibre metric in $f^*(TN)$ of index r. There is also a bundle map $\widetilde{i}_1 : TM \to f^*(TN)$ over M from the tangent bundle $\widetilde{\pi}_1 : TM \to M$ so that $\widetilde{f} \circ \widetilde{i}_1 = f_*$, which is given by $\widetilde{i}_1(v) = (\widetilde{\pi}_1, f_*)(v) = (p, f_*(v))$ for $p \in M$ and $v \in T_pM$.

We want to define a subbundle $\widetilde{\pi}_2 : \mathcal{N}(M) \to M$ of $\widetilde{\pi}_M : f^*(TN) \to M$ of dimension $d = m - n$, called the *normal bundle* of M in N, with fibre

$$\mathcal{N}_p(M) = \widetilde{i}_1(T_pM)^{\perp}$$

for $p \in M$. If $\widetilde{i}_2 : \mathcal{N}(M) \to f^*(TN)$ is the inclusion map, then $g'' = \widetilde{i}_2^*(\widetilde{g})$ is a fibre metric in $\mathcal{N}(M)$ of index $l = r - k$. Let $(\widetilde{f}_1, f)$ and $(\widetilde{f}_2, f)$ be the bundle maps from $\widetilde{\pi}_1 : TM \to M$ and $\widetilde{\pi}_2 : \mathcal{N}(M) \to M$ to $\widetilde{\pi} : TN \to N$, respectively, where $\widetilde{f}_r = \widetilde{f} \circ \widetilde{i}_r$ for $r = 1, 2$ so that $\widetilde{f}_1 = f_*$ and $\widetilde{f}_2 = \widetilde{f}|_{\mathcal{N}(M)}$. We also have bundle maps $\eta_1 : f^*(TN) \to TM$ and $\eta_2 : f^*(TN) \to \mathcal{N}(M)$ over M so that

$$v = \widetilde{i}_1 \circ \eta_1(v) + \widetilde{i}_2 \circ \eta_2(v)$$

for $v \in f^*(TN)$. For each $p \in M$ it follows that

$$v = \widetilde{f}_1 \circ \widetilde{\eta}_{1,p}(v) + \widetilde{f}_2 \circ \widetilde{\eta}_{2,p}(v)$$

for $v \in T_{f(p)}N$, where $\widetilde{\eta}_{1,p} : T_{f(p)}N \to T_pM$ and $\widetilde{\eta}_{2,p} : T_{f(p)}N \to \mathscr{N}_p(M)$ are the linear maps defined by $\widetilde{\eta}_{r,p} = \eta_{r,p} \circ \widetilde{f}_p^{-1}$ for $r = 1, 2$.

Now let $\pi : \mathscr{O}_\varepsilon(N) \to N$ and $\pi_M : \mathscr{O}_\varepsilon(f^*(TN)) \to M$ be the orthonormal frame bundles of N and $f^*(TN)$, respectively, with signature ε given by

$$\varepsilon_i = \begin{cases} -1 & \text{for} \quad 1 \leq i \leq k \quad \text{and} \quad n+1 \leq i \leq n+l \\ 1 & \text{for} \quad k+1 \leq i \leq n \quad \text{and} \quad n+l+1 \leq i \leq m \end{cases}.$$

We have a bundle map (f', f) from $\pi_M : \mathscr{O}_\varepsilon(f^*(TN)) \to M$ to $\pi : \mathscr{O}_\varepsilon(N) \to N$, where $f'(u) = \widetilde{f}_p \circ u$ for $p \in M$ and $u \in \mathscr{O}_{\varepsilon,p}(f^*(TN))$.

The orthonormal frame bundle $\pi_M : \mathscr{O}_\varepsilon(f^*(TN)) \to M$ may also be identified with the pullback bundle $\pi'_M : f^*(\mathscr{O}_\varepsilon(N)) \to M$ of $\pi : \mathscr{O}_\varepsilon(N) \to N$ by f by means of the equivalence $e : f^*(\mathscr{O}_\varepsilon(N)) \to \mathscr{O}_\varepsilon(f^*(TN))$ over M given by

$$e(p,u)(a) = (p, u(a))$$

for $p \in M$, $u \in \mathscr{O}_{\varepsilon,f(p)}(N)$ and $a \in \mathbf{R}^m$, since

$$e(p,u)^*\widetilde{g}(p) = (\widetilde{f}_p \circ e(p,u))^* g(f(p)) = u^* g(f(p))$$

for $p \in M$ and $u \in \mathscr{O}_{\varepsilon,f(p)}(N)$. We have that $(f' \circ e, f)$ is the canonical bundle map of $\pi'_M : f^*(\mathscr{O}_\varepsilon(N)) \to M$.

We want to define a reduced subbundle $\pi' : \mathscr{O}(N,M) \to M$ of $\pi_M : \mathscr{O}_\varepsilon(f^*(TN)) \to M$, called the bundle of *adapted orthonormal frames*, with fibre

$$\mathscr{O}_p(N,M) = \{u \in \mathscr{O}_{\varepsilon,p}(f^*(TN)) | u(\mathbf{R}^n \times \{0\}) = \widetilde{i}_1(T_pM)\}$$

for $p \in M$ and structure group $O_k(\mathbf{R}^n) \times O_l(\mathbf{R}^d)$. Let (i', id_M, ϕ) be the corresponding homomorphism, where $i' : \mathscr{O}(N,M) \to \mathscr{O}_\varepsilon(f^*(TN))$ and $\phi : O_k(\mathbf{R}^n) \times O_l(\mathbf{R}^d) \to O_\varepsilon(\mathbf{R}^m)$ are the inclusion maps. Furthermore, let $j_1 : \mathbf{R}^n \to \mathbf{R}^n \times \mathbf{R}^d$ and $j_2 : \mathbf{R}^d \to \mathbf{R}^n \times \mathbf{R}^d$ be the linear maps given by $j_1(a) = (a, 0)$ and $j_2(b) = (0, b)$ for $a \in \mathbf{R}^n$ and $b \in \mathbf{R}^d$, and let $k_1 : \mathbf{R}^n \times \mathbf{R}^d \to \mathbf{R}^n$ and $k_2 : \mathbf{R}^n \times \mathbf{R}^d \to \mathbf{R}^d$ be the projections on the first and second factor.

11.3 Proposition The normal bundle $\widetilde{\pi}_2 : \mathscr{N}(M) \to M$ is a subbundle of $\widetilde{\pi}_M : f^*(TN) \to M$, and the bundle $\pi' : \mathscr{O}(N,M) \to M$ of adapted orthonormal frames is a reduced subbundle of $\pi_M : \mathscr{O}_\varepsilon(f^*(TN)) \to M$.

PROOF : Let $p_0 \in M$, and choose an orthonormal basis $\mathscr{C} = \{v_1, \ldots, v_m\}$ for $T_{f(p_0)}N$ with signature ε so that $v_i = f_*(w_i)$ for $i = 1, \ldots, n$, where $\mathscr{B} = \{w_1, \ldots, w_n\}$ is an orthonormal basis for $T_{p_0}M$. By Proposition 2.34 there are local charts (x, U) and (y, V) around p_0 and $f(p_0)$ on M and N, respectively, such that $f(U) = \{q \in V | y^{n+1}(q) = \ldots = y^m(q) = 0\}$ and $x^1(q) = y^1 \circ f(q)$, . . . , $x^n(q) = y^n \circ f(q)$ for $q \in U$. Hence $y \circ f = j_1 \circ x$, which implies that

$$t_{y,f(p)} \circ f_{*p} = j_1 \circ t_{x,p} \quad \text{and} \quad f_{*p} \circ t_{x,p}^{-1} = t_{y,f(p)}^{-1} \circ j_1$$

for $p \in U$ by the commutative diagram in 2.70, where $(t_x, \widetilde{\pi}_1^{-1}(U))$ and $(t_y, \widetilde{\pi}^{-1}(V))$ are the local trivializations in the tangent bundles $\widetilde{\pi}_1 : TM \to M$ and $\widetilde{\pi} : TN \to N$ associated with the local charts (x,U) and (y,V), respectively.

We have an adapted frame field $(Y_1, \dots, Y_m)$ on U with $Y_i(p_0) = (p_0, v_i)$ for $i = 1, \dots, m$, given by

$$Y_i(p) = \widetilde{f}_p^{\,-1} \circ t_{y,f(p)}^{-1} \circ t_{y,f(p_0)}(v_i)$$

for $p \in U$ and $i = 1, \dots, m$, so that

$$Y_i(p) = \widetilde{\imath}_1 \circ t_{x,p}^{-1} \circ t_{x,p_0}(w_i) \in \widetilde{\imath}_1(T_pM)$$

for $p \in U$ and $i = 1, \dots, n$. Let W be an open connected neighbourhood of p_0 contained in U so that the $i \times i$-matrix $\left(\widetilde{g}(p)(Y_\alpha(p), Y_\beta(p))\right)$, where $1 \le \alpha, \beta \le i$, is non-singular for each $p \in W$ and $1 \le i \le m$. Then the subspace $V_i(p)$ of $f^*(TN)_p$ spanned by $\{ Y_1(p), \dots, Y_i(p) \}$ is non-degenerate for $p \in W$ and $1 \le i \le m$. Using this, we are going to constuct an adapted orthonormal frame field $(X_1, \dots, X_m)$ on W with $X_i(p_0) = (p_0, v_i)$ for $i = 1, \dots, m$.

By applying the Gram-Schmidt process to $\{ Y_1(p), \dots, Y_m(p) \}$, we obtain an orthogonal basis $\{ Z_1(p), \dots, Z_m(p) \}$ for $f^*(TN)_p$ with $Z_i(p_0) = (p_0, v_i)$ for $i = 1, \dots, m$, given by

$$Z_1(p) = Y_1(p)$$

and

$$Z_i(p) = Y_i(p) - \sum_{j=1}^{i-1} \frac{\widetilde{g}(p)\,(Y_i(p), Z_j(p))}{\widetilde{g}(p)\,(Z_j(p), Z_j(p))}\, Z_j(p)$$

for $p \in W$ and $i = 2, \dots, m$. Indeed, assuming inductively that $\{ Z_1(p), \dots, Z_{i-1}(p) \}$ is an orthogonal basis for $V_{i-1}(p)$, we have that $Z_j(p) \notin V_{i-1}(p)^\perp$ so that $\widetilde{g}(p)(Z_j(p), Z_j(p)) \neq 0$ for $j = 1, \dots, i-1$ since $V_{i-1}(p)$ is a non-degenerate subspace of $f^*(TN)_p$, thus showing that $Z_i(p)$ is well defined. Furthermore, $\widetilde{g}(p)(Z_i(p), Z_r(p)) = 0$ for $r = 1, \dots, i-1$ which shows that $\{ Z_1(p), \dots, Z_i(p) \}$ is an orthogonal basis for $V_i(p)$. Hence we have an adapted orthonormal frame field $s = (X_1, \dots, X_m)$ on W with $X_i(p_0) = (p_0, v_i)$ for $i = 1, \dots, m$, given by

$$X_i(p) = Z_i(p) \,/\, \|Z_i(p)\|$$

for $p \in W$ and $i = 1, \dots, m$. As $\{X_{n+1}(p), \dots, X_m(p)\}$ is a basis for $\mathscr{N}_p(M)$ for $p \in W$, it follows from Proposition 2.62 that the normal bundle $\widetilde{\pi}_2 : \mathscr{N}(M) \to M$ is a subbundle of $\widetilde{\pi}_M : f^*(TN) \to M$.

Since the map $h_i : W \to \mathbf{R}$ given by $h_i(p) = \widetilde{g}(p)(X_i(p), X_i(p))$ for $p \in W$ is smooth and W is connected, we have that $h_i(p) = g(f(p_0))(v_i, v_i) = \varepsilon_i$ for $p \in W$ and $i = 1, \dots, m$. This implies that

$$s(p)^*\widetilde{g}(p)(a,b) = \widetilde{g}(p)(s(p)\,a, s(p)\,b)$$
$$= \widetilde{g}(p)\left(\sum_{i=1}^m a^i\, X_i(p), \sum_{j=1}^m b^j\, X_j(p) \right) = g_\varepsilon^m(a,b)$$

for every $a,b \in \mathbf{R}^m$, which shows that $s(p) \in \mathscr{O}_p(N,M)$ for $p \in W$. Hence the local trivialization $(\rho^{-1}, \pi_M^{-1}(W))$ in the bundle $\pi_M : \mathscr{O}_\varepsilon(f^*(TN)) \to M$, where $\rho : W \times O_\varepsilon(\mathbf{R}^m) \to \pi_M^{-1}(W)$ is the diffeomorphism given by

$$\rho(p,G) = s(p) \circ G$$

for $p \in W$ and $G \in O_\varepsilon(\mathbf{R}^m)$, induces a bijection $t_s : \pi'^{-1}(W) \to W \times O_k(\mathbf{R}^n) \times O_l(\mathbf{R}^d)$. Indeed, since $\rho(p,G)^{-1}(\tilde{i}_1(T_pM)) = G^{-1}(\mathbf{R}^n \times \{0\})$ for $p \in W$, we have that $\rho(p,G) \in \pi'^{-1}(W)$ for $p \in W$ if and only if $G \in O_k(\mathbf{R}^n) \times O_l(\mathbf{R}^d)$. If $s' = (X_1', \dots, X_m')$ is another adapted orthonormal frame field on some open set W', and if $\phi : W \cap W' \to O_\varepsilon(\mathbf{R}^m)$ is the transition map between the corresponding local trivializations in $\mathscr{O}_\varepsilon(f^*(TN))$, then $\phi(p) \in O_k(\mathbf{R}^n) \times O_l(\mathbf{R}^d)$ for $p \in W \cap W'$, since $s(p) = s'(p) \circ \phi(p)$ by Proposition 10.52 which implies that $\phi(p)(\mathbf{R}^n \times \{0\}) = \phi(p) \circ s(p)^{-1}(\tilde{i}_1(T_pM)) = s'(p)^{-1}(\tilde{i}_1(T_pM)) = \mathbf{R}^n \times \{0\}$ for $p \in W \cap W'$. Hence ϕ induces a smooth map $\psi : W \cap W' \to O_k(\mathbf{R}^n) \times O_l(\mathbf{R}^d)$ so that

$$t_{s'} \circ t_s^{-1}(p,G) = (p, \psi(p) \circ G)$$

for every $p \in W \cap W'$ and $G \in O_k(\mathbf{R}^n) \times O_l(\mathbf{R}^d)$. By Proposition 10.5 there therefore is a unique topology, smooth structure and fibre bundle structure on $\mathscr{O}(N,M)$ such that $\pi' : \mathscr{O}(N,M) \to M$ is a principal fibre bundle over M with structure group $O_k(\mathbf{R}^n) \times O_l(\mathbf{R}^d)$, and such that $(t_s, \pi'^{-1}(W))$ is a local trivialization for each adapted orthonormal frame field $s = (X_1, \dots, X_m)$ on some open set W. We see that the inclusion map $i' : \mathscr{O}(N,M) \to \mathscr{O}_\varepsilon(f^*(TN))$ is an immersion, and hence $\pi' : \mathscr{O}(N,M) \to M$ is a reduced subbundle of the bundle $\pi_M : \mathscr{O}_\varepsilon(f^*(TN)) \to M$. □

11.4 Proposition There are homomorphisms $(\lambda_1, id_M, \phi_1)$ and $(\lambda_2, id_M, \phi_2)$ from the bundle $\pi' : \mathscr{O}(N,M) \to M$ of adapted orthonormal frames to the orthonormal frame bundles $\pi_1 : \mathscr{O}(M) \to M$ and $\pi_2 : \mathscr{O}(\mathscr{N}(M)) \to M$, where $\lambda_1 : \mathscr{O}(N,M) \to \mathscr{O}(M)$ and $\lambda_2 : \mathscr{O}(N,M) \to \mathscr{O}(\mathscr{N}(M))$ are the maps defined by

$$\tilde{i}_{r,p} \circ \lambda_r(u) = u \circ j_r$$

for $p \in M$, $u \in \mathscr{O}_p(N,M)$ and $r = 1,2$, and where $\phi_1 : O_k(\mathbf{R}^n) \times O_l(\mathbf{R}^d) \to O_k(\mathbf{R}^n)$ and $\phi_2 : O_k(\mathbf{R}^n) \times O_l(\mathbf{R}^d) \to O_l(\mathbf{R}^d)$ are the projections on the first and second factor.

PROOF: Using that

$$j_r \circ \phi_r(G) = G \circ j_r$$

for $G \in O_k(\mathbf{R}^n) \times O_l(\mathbf{R}^d)$ and $r = 1,2$, we see that

$$\tilde{i}_{r,p} \circ \lambda_r(u) \circ \phi_r(G) = u \circ j_r \circ \phi_r(G) = u \circ G \circ j_r$$

which implies that

$$\lambda_r(u \circ G) = \lambda_r(u) \circ \phi_r(G) \tag{1}$$

for $p \in M$, $u \in \mathscr{O}_p(N,M)$ and $G \in O_k(\mathbf{R}^n) \times O_l(\mathbf{R}^d)$. Hence we have that

$$\pi_r \circ \lambda_r = id_M \circ \pi'$$

and

$$\lambda_r \circ R_G = R_{\phi_r(G)} \circ \lambda_r$$

for $G \in O_k(\mathbf{R}^n) \times O_l(\mathbf{R}^d)$, and it only remains to prove that λ_r is differentible.

Let $s = (X_1, \ldots, X_m)$ be an adapted orthonormal frame field on an open set W in M. By Proposition 3.14 there is a unique orthonormal frame field $s_1 = (X_1', \ldots, X_n')$ on W in $\mathscr{O}(M)$ with $\widetilde{i}_1 \circ X_j' = X_j$ for $j = 1, \ldots, n$, and we let $s_2 = (X_{n+1}, \ldots, X_m)$ so that $s_r = \lambda_r \circ s$ for $r = 1, 2$. If $(t_s, \pi'^{-1}(W))$, $(t_{s_1}, \pi_1^{-1}(W))$ and $(t_{s_2}, \pi_2^{-1}(W))$ are the local trivializations in $\mathscr{O}(N,M)$, $\mathscr{O}(M)$ and $\mathscr{O}(\mathscr{N}(M))$ associated with the sections s, s_1 and s_2, it follows by replacing u with $s(p)$ in (1) that

$$\lambda_r \circ t_s^{-1}(p, G) = t_{s_r}^{-1}(p, \phi_r(G))$$

for $p \in W$ and $G \in O_k(\mathbf{R}^n) \times O_l(\mathbf{R}^d)$. Hence we have that

$$t_{s_r} \circ \lambda_r \circ t_s^{-1} = id_W \times \phi_r ,$$

showing that λ_r is smooth and completing the proof that $(\lambda_r, id_M, \phi_r)$ is a homomorphism for $r = 1, 2$. □

11.5 Remark The principal fibre bundles and homorphisms described in 11.2 and in Propositions 11.3 and 11.4 can be summarized in the commutative diagrams

$$\begin{array}{ccccc}
\mathscr{O}(N,M) & \xrightarrow{i'} & \mathscr{O}_\varepsilon(f^*(TN)) & \xrightarrow{f'} & \mathscr{O}_\varepsilon(N) \\
\pi' \downarrow & & \pi_M \downarrow & & \downarrow \pi \\
M & \xrightarrow{id_M} & M & \xrightarrow{f} & N
\end{array}$$

and

$$\begin{array}{ccccc}
\mathscr{O}(M) & \xleftarrow{\lambda_1} & \mathscr{O}(N,M) & \xrightarrow{\lambda_2} & \mathscr{O}(\mathscr{N}(M)) \\
\pi_1 \downarrow & & \pi' \downarrow & & \downarrow \pi_2 \\
M & \xleftarrow{id_M} & M & \xrightarrow{id_M} & M
\end{array}$$

11.6 Let N^m and M^n be pseudo-Riemannian manifolds, and let $f : M \to N$ be an isometric immersion. Let $exp : \mathscr{D} \to N$ be the exponential map in the tangent bundle $\widetilde{\pi} : TN \to N$, and consider the smooth map $\phi = exp \circ \widetilde{f}_2 : \widetilde{\mathscr{D}} \to N$ defined on the open set $\widetilde{\mathscr{D}} = \widetilde{f}_2^{-1}(\mathscr{D})$ in $\mathscr{N}(M)$ containing $\zeta(M)$, where $\zeta : M \to \mathscr{N}(M)$ is the zero section in the normal bundle $\widetilde{\pi}_2 : \mathscr{N}(M) \to M$ and $(\widetilde{f}_2, f)$ is the bundle map from $\widetilde{\pi}_2 : \mathscr{N}(M) \to M$ to $\widetilde{\pi} : TN \to N$ defined in 11.2. Using that

$$\phi \circ \zeta = f\,,$$

we see that $\phi_{*\,\zeta(p)} : T_{\zeta(p)}\mathscr{N}(M) \to T_{f(p)}N$ is the linear isomorphism given by

$$\phi_{*\,\zeta(p)}(\zeta_*(v_1) + i_{p*}(v_2)) = f_*(v_1) + \widetilde{f}_2 \circ \omega_{\zeta(p)}(v_2)$$

for $p \in M$, $v_1 \in T_pM$ and $v_2 \in T_{\zeta(p)}\mathscr{N}_p(M)$, where $i_p : \mathscr{N}_p(M) \to \mathscr{N}(M)$ is the inclusion map and $\omega_{\zeta(p)} : T_{\zeta(p)}\mathscr{N}_p(M) \to \mathscr{N}_p(M)$ is the canonical identification defined in Lemma 2.84.

11.7 Theorem Let N^m and M^n be pseudo-Riemannian manifolds, and let $f : M \to N$ be an isometric embedding. Then the map ϕ defined in 11.6 induces a diffeomorphism from an open subset W of $\mathscr{N}(M)$ containing $\zeta(M)$ to an open subset U of N containing $f(M)$.

This diffeomorphism is called a *tubular map*, and the sets W and U are called *tubular neighbourhoods* for f in $\mathscr{N}(M)$ and N, respectively. When f is the inclusion map, they are called *tubular neighbourhoods* of M in $\mathscr{N}(M)$ and N.

PROOF: By the inverse function theorem ϕ induces a diffeomorphism $\phi_p : W_p \to V_p''$ from an open neighbourhood W_p of $\zeta(p)$ in $\mathscr{N}(M)$ to an open neighbourhood V_p'' of $f(p)$ in N for each $p \in M$, and we may assume that

$$\phi_p(W_p \cap \zeta(M)) = V_p'' \cap f(M). \tag{1}$$

Indeed, since $\phi_p(W_p \cap \zeta(M)) \subset V_p'' \cap f(M)$ is an open set in $f(M)$ in the subspace topology, we have that $\phi_p(W_p \cap \zeta(M)) = f(M) \cap O$ for an open set O in N. Replacing V_p'' and W_p by $V_p'' \cap O$ and $\phi_p^{-1}(V_p'' \cap O)$, respectively, we see that condition (1) is satisfied.

Let $\{V_p' | p \in M\}$ and $\{V_p | p \in M\}$ be locally finite open covers of the open submanifold $V'' = \bigcup_{p\in M} V_p''$ of N with $\overline{V_p} \subset V_p' \subset V_p''$ for $p \in M$, and let

$$V = \{u \in V'' | \phi_p^{-1}(u) = \phi_q^{-1}(u) \text{ whenever } u \in \overline{V_p} \cap \overline{V_q} \text{ for any } p, q \in M\}\,.$$

For each $p \in M$ there is an open neighbourhood U_p' of $f(p)$ in N and a finite set of points $p_1, \dots, p_r$ in M so that $U_p' \cap V_q = \emptyset$ for every $q \in M - \{p_1, \dots, p_r\}$. If

$$f(p) \in \bigcap_{i=1}^{k} \overline{V_{p_i}} \cap \bigcap_{i=k+1}^{r} \overline{V_{p_i}}^{\,c}\,,$$

then it follows from (1) that

$$\zeta(p) \in \bigcap_{i=1}^{k} \phi_{p_i}^{-1}(V'_{p_i})$$

which shows that

$$U_p = \phi\left(\bigcap_{i=1}^{k} \phi_{p_i}^{-1}(V'_{p_i})\right) \cap \bigcap_{i=k+1}^{r} \overline{V}_{p_i}^{\,c} \cap U'_p$$

is an open neighbourhood of $f(p)$ contained in V. Thus $U = \bigcup_{p\in M} U_p$ and $W = \bigcup_{p\in M} \phi_p^{-1}(U_p)$ satisfy the conditions in the theorem, U being an open subset of N with $f(M) \subset U \subset V$. □

THE SECOND FUNDAMENTAL FORM

11.8 We now want to apply Proposition 11.1 to the bundle $\pi' : \mathscr{O}(N,M) \to M$ of adapted orthonormal frames, which is a reduced subbundle of $\pi_M : \mathscr{O}_\varepsilon(f^*(TN)) \to M$, in order to obtain a connection form on $\mathscr{O}(N,M)$ from the Levi–Civita connection on $\mathscr{O}_\varepsilon(N)$. We have that $O_k(\mathbf{R}^n) \times O_l(\mathbf{R}^d)$ is a Lie subgroup of $O_\varepsilon(\mathbf{R}^m)$ so that

$$\mathfrak{o}_\varepsilon(\mathbf{R}^m) = i_{1*}\,\mathfrak{o}_k(\mathbf{R}^n) \oplus i_{2*}\,\mathfrak{o}_l(\mathbf{R}^d) \oplus \mathfrak{k}\,,$$

where $i_1 : O_k(\mathbf{R}^n) \to O_k(\mathbf{R}^n) \times O_l(\mathbf{R}^d)$ and $i_2 : O_l(\mathbf{R}^d) \to O_k(\mathbf{R}^n) \times O_l(\mathbf{R}^d)$ are the Lie group homomorphisms given by $i_1(F) = F \times id_{\mathbf{R}^d}$ and $i_2(G) = id_{\mathbf{R}^n} \times G$ for $F \in O_k(\mathbf{R}^n)$ and $G \in O_l(\mathbf{R}^d)$, and

$$\mathfrak{k} = \{F \in \mathfrak{o}_\varepsilon(\mathbf{R}^m) | F(\mathbf{R}^n \times \{0\}) \subset \{0\} \times \mathbf{R}^d \text{ and } F(\{0\} \times \mathbf{R}^d) \subset \mathbf{R}^n \times \{0\}\}$$

is a subspace of $\mathfrak{o}_\varepsilon(\mathbf{R}^m)$ which satisfies

$$Ad\,(O_k(\mathbf{R}^n) \times O_l(\mathbf{R}^d))(\mathfrak{k}) \subset \mathfrak{k}\,.$$

Let $\rho_1 : \mathfrak{o}_\varepsilon(\mathbf{R}^m) \to i_{1*}\,\mathfrak{o}_k(\mathbf{R}^n) + i_{2*}\,\mathfrak{o}_l(\mathbf{R}^d)$, $\rho_2 : \mathfrak{o}_\varepsilon(\mathbf{R}^m) \to i_{1*}\,\mathfrak{o}_k(\mathbf{R}^n) + i_{2*}\,\mathfrak{o}_l(\mathbf{R}^d)$ and $\rho_3 : \mathfrak{o}_\varepsilon(\mathbf{R}^m) \to \mathfrak{k}$ be the projections on $i_{1*}\,\mathfrak{o}_k(\mathbf{R}^n)$, $i_{2*}\,\mathfrak{o}_l(\mathbf{R}^d)$ and $\mathfrak{k}$, respectively, and let $\rho = \rho_1 + \rho_2$. We have linear maps $\rho_{21} : \mathfrak{k} \to L(\mathbf{R}^n,\mathbf{R}^d)$ and $\rho_{12} : \mathfrak{k} \to L(\mathbf{R}^d,\mathbf{R}^n)$ given by $\rho_{21}(F) = k_2 \circ F \circ j_1$ and $\rho_{12}(F) = k_1 \circ F \circ j_2$ for $F \in \mathfrak{k}$, and we let $j_3 : \mathfrak{k} \to \mathfrak{o}_\varepsilon(\mathbf{R}^m)$ be the inclusion map and $\psi : O_k(\mathbf{R}^n) \times O_l(\mathbf{R}^d) \to \mathrm{Aut}(\mathfrak{k})$ be the representation obtained from Ad.

If θ_1 and θ are the dual forms on $\mathscr{O}(M)$ and $\mathscr{O}_\varepsilon(N)$, respectively, we have that

$$j_1 \cdot \lambda_1^*(\theta_1) = (f' \circ i')^*(\theta)\,,$$

since

$$j_1 \circ \lambda_1^*(\theta_1)(u)(v) = j_1 \circ \theta_1(\lambda_1(u))(\lambda_{1*}(v)) = j_1 \circ \lambda_1(u)^{-1} \circ (\pi_1 \circ \lambda_1)_*(v)$$
$$= u^{-1} \circ \widetilde{i}_1 \circ \pi'_*(v) = (\widetilde{f}_p \circ u)^{-1} \circ f_* \circ \pi'_*(v) = (f' \circ i')(u)^{-1} \circ \pi_* \circ (f' \circ i')_*(v)$$
$$= (f' \circ i')^*(\theta)(u)(v)$$

for $p \in M$, $u \in \mathscr{O}_p(N,M)$ and $v \in T_u\mathscr{O}(N,M)$. It also follows that

$$(f' \circ i')^*(\theta)(u) = u^{-1} \circ \widetilde{i}_1 \circ \pi'_{*u}$$

for $u \in \mathscr{O}(N,M)$.

Let ω be the Levi–Civita connection form on $\mathscr{O}_\varepsilon(N)$ satisfying the first structure equation

$$d\theta = -\ \omega \wedge \theta\ .$$

If $\omega_M = f'^*(\omega)$ is the pull-back of ω to $\mathscr{O}_\varepsilon(f^*(TN))$, and $\theta' = (f' \circ i')^*(\theta)$ and $\omega' = (f' \circ i')^*(\omega)$ are the pull-backs of θ and ω to $\mathscr{O}(N,M)$, it follows from Propositions 5.63 and 5.70 that

$$d\theta' = -\ \omega' \wedge \theta'\ . \tag{1}$$

By 10.37 and Proposition 11.1 we know that $\rho \cdot \omega'$ is a connection form on $\mathscr{O}(N,M)$, and we let ω_0, ω_1 and ω_2 be the connection forms on $\mathscr{O}_\varepsilon(f^*(TN))$, $\mathscr{O}(M)$ and $\mathscr{O}(\mathscr{N}(M))$ obtained from $\rho \cdot \omega'$ in the way described in Proposition 10.168 so that

$$i'^*(\omega_0) = (\phi_* \circ \rho) \cdot \omega' \quad \text{and} \quad \lambda_r^*(\omega_r) = (\phi_{r*} \circ \rho) \cdot \omega'\ ,$$

and hence

$$i_{r*} \cdot \lambda_r^*(\omega_r) = \rho_r \cdot \omega' \tag{2}$$

for $r = 1,2$. Since

$$\rho_2 \cdot \omega'(u)(\mathbf{R}^n \times \{0\}) = \{0\} \times \{0\} \quad \text{and} \quad \rho_3 \cdot \omega'(u)(\mathbf{R}^n \times \{0\}) \subset \{0\} \times \mathbf{R}^d$$

for $u \in \mathscr{O}(N,M)$, it follows from (1) that

$$d\theta_1 = -\ \omega_1 \wedge \theta_1$$

showing that ω_1 is the Levi–Civita connection form on $\mathscr{O}(M)$, and

$$\tau \wedge \theta' = 0 \tag{3}$$

where $\tau = \rho_3 \cdot \omega'$ is a $\mathfrak{k}$-valued 1-form on $\mathscr{O}(N,M)$, called the *second fundamental form* on $\mathscr{O}(N,M)$, which is tensorial of type $(\psi, \mathfrak{k})$. This follows from Proposition 11.1 since

$$j_3 \cdot \tau = \omega' - (\phi_* \circ \rho) \cdot \omega' = i'^*(\tau_M)\ ,$$

where $\tau_M = \omega_M - \omega_0$ is the difference of two connection forms on $\mathscr{O}_\varepsilon(f^*(TN))$ and therefore is tensorial of type $(Ad, \mathfrak{o}_\varepsilon(\mathbf{R}^m))$.

11.9 By Example 10.29 (b) the fibre bundle $\pi'' : E \to M$ with fibre $\mathfrak{k}$ associated with $\mathscr{O}(N,M)$ is a vector bundle which we contend is equivalent over M to the vector bundle $\pi''' : \mathscr{Q} \to M$ with fibre

$$\mathscr{Q}_p = \{F \in \mathfrak{o}_\varepsilon(f^*(TN)_p) | F(\widetilde{i}_1(T_pM)) \subset \mathscr{N}_p(M) \text{ and}$$
$$F(\mathscr{N}_p(M)) \subset \widetilde{i}_1(T_pM)\}$$

for $p \in M$, with the equivalence $e : E \to Q$ given by

$$e([u,F]) = u \circ F \circ u^{-1}$$

for $u \in \mathscr{O}(N,M)$ and $F \in \mathfrak{k}$. This is clearly well defined, for if $G \in O_k(\mathbf{R}^n) \times O_l(\mathbf{R}^d)$, then

$$e([u\,G, \psi(G^{-1})\,F]) = (u \circ G) \circ (G^{-1} \circ F \circ G) \circ (G^{-1} \circ u^{-1})$$
$$= u \circ F \circ u^{-1} = e([u,F]) .$$

To show that e is an equivalence, let $(t_s, \pi'^{-1}(W))$ be the local trivialization in $\mathscr{O}(N,M)$ assosiated with an adapted orthonormal frame field s on some open set W in M, and let $(t'_s, \pi''^{-1}(W))$ be the corresponding local trivialization in E defined as in 10.28 by

$$t'_s([u,F]) = (p, \psi(t_{s,p}(u))(F))$$

for $u \in \mathscr{O}_p(N,M)$ and $F \in \mathfrak{k}$ where $p \in W$. By (2) in 10.28 we also have a local trivialization $(\widetilde{t}_s, \widetilde{\pi}_M^{-1}(W))$ in the vector bundle $\widetilde{\pi}_M : f^*(TN) \to M$ given by

$$\widetilde{t}_s(s(p)(a)) = (p,a)$$

for $p \in W$ and $a \in \mathbf{R}^m$. Combining this, we have that

$$t'_s([u,F]) = (p, s(p)^{-1} \circ u \circ F \circ u^{-1} \circ s(p)) = \tau_s \circ e([u,F])$$

where $\tau_s : \pi'''^{-1}(W) \to W \times \mathfrak{k}$ is the equivalence over W induced by the equivalence $\tau'_s : \pi_M'^{-1}(W) \to W \times \mathfrak{gl}(\mathbf{R}^m)$ defined in Proposition 5.28 for the vector bundle $\pi'_M : \Lambda^1(f^*(TN)\,;f^*(TN)) \to M$ using the local trivialization $(\widetilde{t}_s, \widetilde{\pi}_M^{-1}(W))$ in the vector bundle $\widetilde{\pi}_M : f^*(TN) \to M$. Hence we have that

$$\tau_s \circ e \circ t_s'^{-1} = id_{W \times \mathfrak{k}} .$$

Using the equivalence e, we may therefore identify the fibre bundle E associated with $\mathscr{O}(N,M)$ with fibre $\mathfrak{k}$ with the vector bundle $\pi''' : \mathscr{Q} \to M$, and the linear isomorphism $\mu_u : \mathfrak{k} \to E_p$ defined in Example 10.29 (b) is then identified with the linear isomorphism $\mu'_u = e_p \circ \mu_u : \mathfrak{k} \to \mathscr{Q}_p$ given by

$$\mu'_u(F) = u \circ F \circ u^{-1}$$

for $u \in \mathscr{O}_p(N,M)$ and $F \in \mathfrak{k}$ where $p \in M$.

Now we have a unique bundle-valued 1-form $\alpha \in \Omega^1(M;\mathscr{Q})$ with $\tau = \overline{\alpha}$, called the *second fundamental form* on M, defined by

$$\alpha(p)(v) = u \circ \tau(u)(w) \circ u^{-1}$$

for $p \in M$ and $v \in T_pM$, where $u \in \mathscr{O}_p(N,M)$ and $w \in T_u\mathscr{O}(N,M)$ is chosen so that $\pi'_*(w) = v$. We also have that

$$\alpha(p)(v_1)(\widetilde{i}_1(v_2)) = u \circ \tau(u)(w_1)\,(\theta'(u)(w_2))$$

for $p \in M$ and $v_1, v_2 \in T_pM$, where $u \in \mathscr{O}_p(N,M)$ and $w_r \in T_u\mathscr{O}(N,M)$ is chosen so that $\pi'_*(w_r) = v_r$ for $r = 1,2$. Hence it follows from (3) in 11.8 that

$$\alpha(p)(v_1)(\widetilde{i}_1(v_2)) = \alpha(p)(v_2)(\widetilde{i}_1(v_1))$$

since

$$\alpha(p)(v_1)(\widetilde{i}_1(v_2)) - \alpha(p)(v_2)(\widetilde{i}_1(v_1)) = u \circ (\tau \wedge \theta')(u)(w_1, w_2) = 0\,.$$

11.10 Let $\mathscr{E} = \{e_1, \dots, e_m\}$ be the standard basis for $\mathbf{R}^m$ and $\mathscr{D} = \{F_j^i \mid 1 \le i, j \le m\}$ be the basis for $\mathfrak{gl}(\mathbf{R}^m)$ defined in Example 8.10 (b), where $F_j^i : \mathbf{R}^m \to \mathbf{R}^m$ is the linear map given by $F_j^i(e_r) = \delta_{jr}\, e_i$ for $r = 1, \dots, m$. If $s = (Z_1, \dots, Z_m)$ is an adapted orthonormal frame field on an open set V in M, and $\widetilde{\tau} = s^*(\tau)$ is the pull-back of τ by s, it follows from Proposition 10.105 that

$$\alpha(p)(v)\, Z_i(p) = \sum_{r=n+1}^{m} \widetilde{\tau}_i^{\,r}(p)(v)\, Z_r(p)$$

for $p \in V$, $v \in T_pM$ and $i = 1, \dots, n$, and

$$\alpha(p)(v)\, Z_r(p) = \sum_{i=1}^{n} \widetilde{\tau}_r^{\,i}(p)(v)\, Z_i(p)$$

for $p \in V$, $v \in T_pM$ and $r = n+1, \dots, m$, where $\widetilde{\tau}_j^{\,i}$ are the components of $\widetilde{\tau}$ with respect to the basis $\mathscr{D}$. Since the form $\widetilde{\tau}$ is $\mathfrak{o}_\varepsilon(\mathbf{R}^m)$-valued, we have that

$$s_{ir}\, \widetilde{\tau}_r^{\,i} = -\widetilde{\tau}_i^{\,r}$$

for $i = 1, \dots, n$ and $r = n+1, \dots, m$, where $s_{ir} = sgn\,(i - k - 1/2)(r - n - l - 1/2)$.

By Proposition 3.14 there is a unique orthonormal frame field $s_1 = (Y_1, \dots, Y_n)$ on V in $\mathscr{O}(M)$ with $\widetilde{i}_1 \circ Y_j = Z_j$ for $j = 1, \dots, n$, so that $s_1 = \lambda_1 \circ s$. Let $\widetilde{\theta}^{\,j} = s^*(\theta'^{\,j}) = (f' \circ i' \circ s)^*(\theta^{\,j}) = s_1^*(\theta_1^j)$ for $j = 1, \dots, n$ be the pull-backs of the non-zero components of the 1-form θ' by s, which form a dual local basis on V to the frame field s_1 as described in Remark 10.94. Then we have that

$$\widetilde{\tau}_j^{\,r} = \sum_{i=1}^{n} A_{ij}^r\, \widetilde{\theta}^{\,i}$$

for $1 \leq j \leq n$ and $n+1 \leq r \leq m$, where the smooth functions $A^r_{ij} : V \to \mathbf{R}$ are defined by

$$A^r_{ij} = \widetilde{\tau}^r_j(Y_i)$$

for $1 \leq i, j \leq n$ and $n+1 \leq r \leq m$. Using formula (3) in 11.8 we have that

$$\widetilde{\tau} \wedge \widetilde{\theta} = 0$$

so that

$$0 = \sum_{j=1}^{n} \widetilde{\tau}^r_j \wedge \widetilde{\theta}^j = \sum_{i<j} (A^r_{ij} - A^r_{ji})\, \widetilde{\theta}^i \wedge \widetilde{\theta}^j$$

for $n+1 \leq r \leq m$. This implies that

$$A^r_{ij} = A^r_{ji}$$

for $1 \leq i, j \leq n$ and $n+1 \leq r \leq m$.

THE SHAPE TENSOR

11.11 Consider the representation $\psi_h : O_k(\mathbf{R}^n) \times O_l(\mathbf{R}^d) \to \mathrm{Aut}(\mathscr{T}^h(\mathbf{R}^n; \mathbf{R}^d))$ of the structure group $O_k(\mathbf{R}^n) \times O_l(\mathbf{R}^d)$ in the bundle $\mathscr{O}(N,M)$ of adapted orthonormal frames on the finite dimensional vector space $\mathscr{T}^h(\mathbf{R}^n; \mathbf{R}^d)$ given by

$$\psi_h(G)(T) = \phi_2(G) \circ T \circ \{\phi_1(G)^{-1}\}^h$$

for $G \in O_k(\mathbf{R}^n) \times O_l(\mathbf{R}^d)$ and $T \in \mathscr{T}^h(\mathbf{R}^n; \mathbf{R}^d)$. By Example 10.29 (b) the fibre bundle $\pi'' : E \to M$ with fibre $\mathscr{T}^h(\mathbf{R}^n; \mathbf{R}^d)$ associated with $\mathscr{O}(N,M)$ is a vector bundle which we contend is equivalent over M to the bundle $\pi^h : \mathscr{T}^h(TM; \mathscr{N}(M)) \to M$, with the equivalence $e : E \to \mathscr{T}^h(TM; \mathscr{N}(M))$ given by

$$e([u,T]) = \lambda_2(u) \circ T \circ \{\lambda_1(u)^{-1}\}^h$$

for $u \in \mathscr{O}_p(N,M)$ and $T \in \mathscr{T}^h(\mathbf{R}^n; \mathbf{R}^d)$ where $p \in M$. This is clearly well defined, for if $G \in O_k(\mathbf{R}^n) \times O_l(\mathbf{R}^d)$, then

$$e([u \circ G, \psi_h(G^{-1})\, T]) = \{\lambda_2(u) \circ \phi_2(G)\} \circ \{\phi_2(G)^{-1} \circ T \circ \{\phi_1(G)\}^h\} \circ$$
$$\{\{\phi_1(G)^{-1}\}^h \circ \{\lambda_1(u)^{-1}\}^h\} = e([u,T])\,.$$

To show that e is an equivalence, let $(t_s, \pi'^{-1}(W))$ be the local trivialization in $\mathscr{O}(N,M)$ assosiated with an adapted orthonormal frame field s on some open set W in M, and let $(t'_s, \pi''^{-1}(W))$ be the corresponding local trivialization in E defined as in 10.28 by

$$t'_s([u,T]) = (p, \psi_h(t_{s,p}(u))(T))$$

for $u \in \mathscr{O}_p(N,M)$ and $T \in \mathscr{T}^h(\mathbf{R}^n;\mathbf{R}^d)$ where $p \in W$. Since

$$\lambda_r(u) = \lambda_r(s(p)) \circ \phi_r(s(p)^{-1} \circ u)$$

for $u \in \mathscr{O}_p(N,M)$ and $r = 1,2$, it follows that

$$\phi_r(s(p)^{-1} \circ u) = s_r(p)^{-1} \circ \lambda_r(u)$$

where $s_r = \lambda_r \circ s$. Combining this, we have that

$$t'_s([u,T]) = (p, s_2(p)^{-1} \circ \lambda_2(u) \circ T \circ \{\lambda_1(u)^{-1}\}^h \circ \{s_1(p)\}^h) = \tau_s \circ e([u,T]),$$

where $\tau_s : (\pi^h)^{-1}(W) \to W \times \mathscr{T}^h(\mathbf{R}^n;\mathbf{R}^d)$ is the equivalence over W defined in Proposition 5.28, using the local trivializations $(\widetilde{t}_{s_1}, \widetilde{\pi}_1^{-1}(W))$ and $(\widetilde{t}_{s_2}, \widetilde{\pi}_2^{-1}(W))$ in the vector bundles $\widetilde{\pi}_1 : TM \to M$ and $\widetilde{\pi}_2 : \mathscr{N}(M) \to M$ given as in formula (2) in 10.28 by

$$\widetilde{t}_{s_r}(s_r(p)(a_r)) = (p, a_r)$$

for $p \in W$, $a_1 \in \mathbf{R}^n$ and $a_2 \in \mathbf{R}^d$. Hence we have that

$$\tau_s \circ e \circ t'^{-1}_s = id_{W \times \mathscr{T}^h(\mathbf{R}^n;\mathbf{R}^d)}.$$

Using the equivalence e, we may therefore identify the fibre bundle E associated with $\mathscr{O}(N,M)$ with fibre $\mathscr{T}^h(\mathbf{R}^n;\mathbf{R}^d)$ with the vector bundle $\pi^h : \mathscr{T}^h(TM;\mathscr{N}(M)) \to M$, and the linear isomorphism $\mu_u : \mathscr{T}^h(\mathbf{R}^n;\mathbf{R}^d) \to E_p$ defined in Example 10.29 (b) is then identified with the linear isomorphism $\mu'_u = e_p \circ \mu_u : \mathscr{T}^h(\mathbf{R}^n;\mathbf{R}^d) \to \mathscr{T}^h(T_pM;\mathscr{N}_p(M))$ given by

$$\mu'_u(T) = \lambda_2(u) \circ T \circ \{\lambda_1(u)^{-1}\}^h$$

for $u \in \mathscr{O}_p(N,M)$ and $T \in \mathscr{T}^h(\mathbf{R}^n;\mathbf{R}^d)$ where $p \in M$.

If $\phi_p : \mathscr{T}^h(T_pM;\mathscr{N}_p(M)) \to \mathscr{T}^h_1(T_pM,\mathscr{N}_p(M))$ and $\phi : \mathscr{T}^h(\mathbf{R}^n;\mathbf{R}^d) \to \mathscr{T}^h_1(\mathbf{R}^n, \mathbf{R}^d)$ are the linear isomorphisms defined in 5.15, we have that

$$(\phi_p \circ \mu'_u)(T)(v_1, \ldots, v_h, \lambda) = \phi(T)(\lambda_1(u)^{-1}(v_1), \ldots, \lambda_1(u)^{-1}(v_h), \lambda \circ \lambda_2(u))$$

for $v_1, \ldots, v_h \in T_pM$, $\lambda \in \mathscr{N}_p{}^*(M)$, $u \in \mathscr{O}_p(N,M)$ and $T \in \mathscr{T}^h(\mathbf{R}^n;\mathbf{R}^d)$.

Suppose that $v \in T_pM$ is the tangent vector to a smooth curve $\gamma : I \to M$ at p, where I is an open interval containing t_0 and $\gamma(t_0) = p$. If T is a section in the bundle $\pi^h : \mathscr{T}^h(TM;\mathscr{N}(M)) \to M$, then the covariant derivative of T at p with respect to v is given by

$$\nabla_v T = \lim_{h \to 0} \frac{1}{h} [(\tau_2)_{t_0}^{t_0+h} \circ T(\gamma(t_0+h)) \circ \{(\tau_1)_{t_0+h}^{t_0}\}^h - T(p)]$$

$$= \frac{d}{dt}\bigg|_{t_0} (\tau_2)_{t_0}^{t} \circ T(\gamma(t)) \circ \{(\tau_1)_t^{t_0}\}^h,$$

where $(\tau_1)_t^{t_0} : \pi_1{}^{-1}(p) \to \pi_1{}^{-1}(\gamma(t))$ is the parallel transport in TM along γ from t_0 to t, and $(\tau_2)_{t_0}^t : \pi_2{}^{-1}(\gamma(t)) \to \pi_2{}^{-1}(p)$ is the parallel transport in $\mathscr{N}(M)$ along γ from t to t_0. The limit is taken with respect to the vector space topology on $\mathscr{T}^h(T_pM;\mathscr{N}_p(M))$ defined in Proposition 13.117 in the appendix.

11.12 Since the representation $\psi_1 : O_k(\mathbf{R}^n) \times O_l(\mathbf{R}^d) \to \mathrm{Aut}(L(\mathbf{R}^n,\mathbf{R}^d))$ defined in 11.11 satisfies

$$\rho_{21} \circ \psi(G) = \psi_1(G) \circ \rho_{21}$$

for every $G \in O_k(\mathbf{R}^n) \times O_l(\mathbf{R}^d)$, it follows from Proposition 10.39 that the vector-valued 1-form $\rho_{21} \cdot \tau$ on $\mathscr{O}(N,M)$ is tensorial of type $(\psi_1, L(\mathbf{R}^n,\mathbf{R}^d))$. Hence we have a unique bundle-valued 1-form $\alpha_{21} \in \Omega^1(M;\Lambda^1(TM;\mathscr{N}(M)))$ with $\rho_{21} \cdot \tau = \overline{\alpha_{21}}$, which is given by

$$\alpha_{21}(p)(v) = \lambda_2(u) \circ k_2 \circ \tau(u)(w) \circ j_1 \circ \lambda_1(u)^{-1} = \eta_2 \circ \alpha(p)(v) \circ \tilde{\imath}_1$$

for $p \in M$ and $v \in T_pM$, where $u \in \mathscr{O}_p(N,M)$ and $w \in T_u\mathscr{O}(N,M)$ is chosen so that $\pi'_*(w) = v$.

Using Proposition 5.31, we also have a symmetric covariant tensor field $\mathrm{II} \in \mathscr{T}^2(M;\mathscr{N}(M))$ of degree 2 on M with values in the normal bundle $\widetilde{\pi}_2 : \mathscr{N}(M) \to M$, called the *shape tensor*, which is defined by

$$\mathrm{II}(p)(v_1,v_2) = \alpha_{21}(p)(v_1)(v_2)$$

for $p \in M$ and $v_1, v_2 \in T_pM$.

11.13 Suppose that $v \in T_pM$ is the tangent vector to a smooth curve $\gamma : I \to M$ at p, where I is an open interval containing 0 and $\gamma(0) = p$, and let T be a section in the bundle $\pi^2 : \mathscr{T}^2(TM;\mathscr{N}(M)) \to M$. Choose a $u \in \mathscr{O}_p(N,M)$, and let $\alpha : I \to \mathscr{O}(N,M)$ be the horizontal lift of γ with $\alpha(0) = u$.

Now let $\mathscr{B} = \{v_1,\ldots,v_n\}$ and $\mathscr{C} = \{w_1,\ldots,w_d\}$ be bases for T_pM and $\mathscr{N}_p(M)$, respectively, with dual bases $\mathscr{B}^* = \{v^1,\ldots,v^n\}$ and $\mathscr{C}^* = \{w^1,\ldots,w^d\}$. We let $v_i(t) = (\tau_1)_t^0(v_i) = \lambda_1(\alpha(t)) \circ \lambda_1(u)^{-1}(v_i)$ and $v^i(t) = (\tau_1)_t^0(v^i) = v^i \circ \lambda_1(u) \circ \lambda_1(\alpha(t))^{-1}$ for $i = 1,\ldots,n$ and $t \in I$ be the vectors and covectors obtained by parallel transport in TM and T^*M along the curve γ from 0 to t. Similarly, we let $w_i(t) = (\tau_2)_t^0(w_i) = \lambda_2(\alpha(t)) \circ \lambda_2(u)^{-1}(w_i)$ and $w^i(t) = (\tau_2)_t^0(w^i) = w^i \circ \lambda_2(u) \circ \lambda_2(\alpha(t))^{-1}$ for $i = 1,\ldots,d$ and $t \in I$ be the vectors and covectors obtained by parallel transport in $\mathscr{N}(M)$ and $\mathscr{N}^*(M)$. Then $\mathscr{B}_t = \{v_1(t),\ldots,v_n(t)\}$ is a basis for $T_{\gamma(t)}M$ with dual basis $\mathscr{B}_t^* = \{v^1(t),\ldots,v^n(t)\}$, since $(\tau_1)_t^0 : T_pM \to T_{\gamma(t)}M$ is a linear isomorphism. Similarly, $\mathscr{C}_t = \{w_1(t),\ldots,w_d(t)\}$ is a basis for $\mathscr{N}_{\gamma(t)}(M)$ with dual basis $\mathscr{C}_t^* = \{w^1(t),\ldots,w^d(t)\}$.

If τ_t^0 denotes parallel transport along γ from 0 to t in $\mathscr{T}_1^2(TM,\mathscr{N}(M))$, it follows from 11.11 that

$$v^{i_1}(t) \otimes v^{i_2}(t) \otimes w_j(t) = \tau_t^0(v^{i_1} \otimes v^{i_2} \otimes w_j)$$

for $(i_1,i_2,j) \in I_n^2 \times I_d$ and $t \in I$, which shows that the parallel transport in $\mathscr{T}_1^2(TM,\mathscr{N}(M))$ of the basis vectors in $\mathscr{T}_1^2(\mathscr{B},\mathscr{C})$ are obtained by taking the tensor product of the parallel transported basis vectors from $\mathscr{B}$ and $\mathscr{C}^*$. If

$$\phi \circ T(\gamma(t)) = \sum_{i_1,i_2,j} C_{p\,i_1,i_2}^{\;j}(t)\, v^{i_1}(t) \otimes v^{i_2}(t) \otimes w_j(t)\,,$$

then

$$\phi \circ \tau_0^t T(\gamma(t)) = \sum_{i_1,i_2,j} C_{p\,i_1,i_2}^{\;j}(t)\, v^{i_1} \otimes v^{i_2} \otimes w_j$$

where τ_0^t now denotes parallel transport in $\mathscr{T}^2(TM;\mathscr{N}(M))$, so that $C_{p\,i_1,i_2}^{\;j}$ are the components with respect to $\mathscr{T}_1^2(\mathscr{B},\mathscr{C})$ of the smooth curve $c : I \to \mathscr{T}_1^2(T_pM, \mathscr{N}_p(M))$ given by $c(t) = \phi \circ \tau_0^t T(\gamma(t))$. Hence we have that

$$\phi \circ \nabla_v T = \sum_{i_1,i_2,j} \frac{d}{dt}\bigg|_0 C_{p\,i_1,i_2}^{\;j}\, v^{i_1} \otimes v^{i_2} \otimes w_j \tag{1}$$

by Lemma 2.84.

If X, Y and Z are vector fields on M, we have the relation

$$\nabla_{X_p}^2\,[T(Y,Z)] = (\nabla_{X_p} T)(Y_p,Z_p) + T(p)(\nabla_{X_p}^1 Y, Z_p) + T(p)(Y_p, \nabla_{X_p}^1 Z). \tag{2}$$

for $p \in M$. Indeed, writing

$$Y_{\gamma(t)} = \sum_k A_p^k(t)\, v_k(t) \quad \text{and} \quad Z_{\gamma(t)} = \sum_k B_p^k(t)\, v_k(t)\,,$$

we obtain

$$T(\gamma(t))(Y_{\gamma(t)}, Z_{\gamma(t)}) = \sum_j \left(\sum_{i_1,i_2} C_{p\,i_1,i_2}^{\;j}(t)\, A_p^{i_1}(t)\, B_p^{i_2}(t) \right) w_j(t)\,.$$

Hence formula (2) follows from (1) and Remark 10.79.

THE SHAPE OPERATOR

11.14 Consider the representation $\psi'_1 : O_k(\mathbf{R}^n) \times O_l(\mathbf{R}^d) \to \mathrm{Aut}(L(\mathbf{R}^d,\mathbf{R}^n))$ of the structure group $O_k(\mathbf{R}^n) \times O_l(\mathbf{R}^d)$ in the bundle $\mathscr{O}(N,M)$ of adapted orthonormal frames on the finite dimensional vector space $L(\mathbf{R}^d,\mathbf{R}^n)$ given by

$$\psi'_1(G)(F) = \phi_1(G) \circ F \circ \phi_2(G)^{-1}$$

for $G \in O_k(\mathbf{R}^n) \times O_l(\mathbf{R}^d)$ and $F \in L(\mathbf{R}^d,\mathbf{R}^n)$. By Example 10.29 (b) the fibre bundle $\pi'' : E \to M$ with fibre $L(\mathbf{R}^d,\mathbf{R}^n)$ associated with $\mathscr{O}(N,M)$ is a vector bundle which we contend is equivalent over M to the bundle $\widehat{\pi}^1 : \Lambda^1(\mathscr{N}(M);TM) \to M$, with the equivalence $e : E \to \Lambda^1(\mathscr{N}(M);TM)$ given by

$$e([u,F]) = \lambda_1(u) \circ F \circ \lambda_2(u)^{-1}$$

for $u \in \mathscr{O}_p(N,M)$ and $F \in L(\mathbf{R}^d,\mathbf{R}^n)$ where $p \in M$. This is clearly well defined, for if $G \in O_k(\mathbf{R}^n) \times O_l(\mathbf{R}^d)$, then

$$e([u \circ G, \psi_1'(G^{-1})\, F]) = \{\lambda_1(u) \circ \phi_1(G)\} \circ \{\phi_1(G)^{-1} \circ F \circ \phi_2(G)\} \circ \{\phi_2(G)^{-1} \circ \lambda_2(u)^{-1}\} = e([u,F])\,.$$

To show that e is an equivalence, let $(t_s, \pi'^{-1}(W))$ be the local trivialization in $\mathscr{O}(N,M)$ assosiated with an adapted orthonormal frame field s on some open set W in M, and let $(t_s', \pi''^{-1}(W))$ be the corresponding local trivialization in E defined as in 10.28 by

$$t_s'([u,F]) = (p, \psi_1'(t_{s,p}(u))(F))$$

for $u \in \mathscr{O}_p(N,M)$ and $F \in L(\mathbf{R}^d,\mathbf{R}^n)$ where $p \in W$. Since

$$\lambda_r(u) = \lambda_r(s(p)) \circ \phi_r(s(p)^{-1} \circ u)$$

for $u \in \mathscr{O}_p(N,M)$ and $r = 1,2$, it follows that

$$\phi_r(s(p)^{-1} \circ u) = s_r(p)^{-1} \circ \lambda_r(u)$$

where $s_r = \lambda_r \circ s$. Combining this, we have that

$$t_s'([u,F]) = (p, s_1(p)^{-1} \circ \lambda_1(u) \circ F \circ \lambda_2(u)^{-1} \circ s_2(p)) = \tau_s \circ e([u,F])\,,$$

where $\tau_s : (\widehat{\pi}^1)^{-1}(W) \to W \times L(\mathbf{R}^d,\mathbf{R}^n)$ is the equivalence over W defined in Proposition 5.28, using the local trivializations $(\widetilde{t}_{s_1}, \widetilde{\pi}_1^{-1}(W))$ and $(\widetilde{t}_{s_2}, \widetilde{\pi}_2^{-1}(W))$ in the vector bundles $\widetilde{\pi}_1 : TM \to M$ and $\widetilde{\pi}_2 : \mathscr{N}(M) \to M$ given as in formula (2) in 10.28 by

$$\widetilde{t}_{s_r}(s_r(p)(a_r)) = (p, a_r)$$

for $p \in W$, $a_1 \in \mathbf{R}^n$ and $a_2 \in \mathbf{R}^d$. Hence we have that

$$\tau_s \circ e \circ t_s'^{-1} = id_{W \times L(\mathbf{R}^d,\mathbf{R}^n)}\,.$$

Using the equivalence e, we may therefore identify the fibre bundle E associated with $\mathscr{O}(N,M)$ with fibre $L(\mathbf{R}^d,\mathbf{R}^n)$ with the vector bundle $\widehat{\pi}^1 : \Lambda^1(\mathscr{N}(M);TM) \to M$, and the linear isomorphism $\mu_u : L(\mathbf{R}^d,\mathbf{R}^n) \to E_p$ defined in Example 10.29 (b) is then identified with the linear isomorphism $\mu_u' = e_p \circ \mu_u : L(\mathbf{R}^d,\mathbf{R}^n) \to \Lambda^1(\mathscr{N}_p(M);T_pM)$ given by

$$\mu_u'(F) = \lambda_1(u) \circ F \circ \lambda_2(u)^{-1}$$

for $u \in \mathscr{O}_p(N,M)$ and $F \in L(\mathbf{R}^d,\mathbf{R}^n)$ where $p \in M$.

11.15 Since the representation $\psi_1': O_k(\mathbf{R}^n) \times O_l(\mathbf{R}^d) \to \mathrm{Aut}(L(\mathbf{R}^d, \mathbf{R}^n))$ defined in 11.14 satisfies

$$\rho_{12} \circ \psi(G) = \psi_1'(G) \circ \rho_{12}$$

for every $G \in O_k(\mathbf{R}^n) \times O_l(\mathbf{R}^d)$, it follows from Proposition 10.39 that the vector-valued 1-form $\rho_{12} \cdot \tau$ on $\mathscr{O}(N,M)$ is tensorial of type $(\psi_1', L(\mathbf{R}^d, \mathbf{R}^n))$. Hence we have a unique bundle-valued 1-form $\alpha_{12} \in \Omega^1(M; \Lambda^1(\mathscr{N}(M); TM))$ with $\rho_{12} \cdot \tau = \overline{\alpha_{12}}$, which is given by

$$\alpha_{12}(p)(v) = \lambda_1(u) \circ k_1 \circ \tau(u)(w) \circ j_2 \circ \lambda_2(u)^{-1} = \eta_1 \circ \alpha(p)(v) \circ \widetilde{i}_2$$

for $p \in M$ and $v \in T_pM$, where $u \in \mathscr{O}_p(N,M)$ and $w \in T_u\mathscr{O}(N,M)$ is chosen so that $\pi'_*(w) = v$.

Using Proposition 5.32, we also have a section S in the vector bundle $\widehat{\pi}: \Lambda^1(\mathscr{N}(M); \Lambda^1(TM; TM)) \to M$ defined by

$$S(p)(v_1)(v_2) = -\alpha_{12}(p)(v_2)(v_1)$$

for $p \in M$, $v_1 \in \mathscr{N}_p(M)$ and $v_2 \in T_pM$. By Example 8.58 (b) we have the relation

$$g'(p)(S(p)(v_1)(v_2), v_3) = g''(p)(\mathrm{II}(p)(v_2, v_3), v_1)$$

for $p \in M$, $v_1 \in \mathscr{N}_p(M)$ and $v_2, v_3 \in T_pM$. Since the shape tensor II is symmetric, it follows that $S(p)(v)$ is a self-adjoint linear operator on T_pM for every $p \in M$ and $v \in \mathscr{N}_p(M)$. If Z is a section in the normal bundle $\widetilde{\pi}_2 : \mathscr{N}(M) \to M$ on an open subset V of M, then $S(Z) \in \Omega^1(V; TM))$ is a bundle-valued 1-form on V with values in TM. It is also denoted by S_Z and is called the *shape operator* determined by Z.

THE FORMULAE OF GAUSS AND WEINGARTEN

11.16 Let ∇' and ∇ be the covariant derivatives corresponding to the connection forms ω_M and ω on $\mathscr{O}_\varepsilon(f^*(TN))$ and $\mathscr{O}_\varepsilon(N)$, respectively, and let Z be a section in $\widetilde{\pi}_M : f^*(TN) \to M$. Then we have that

$$\nabla_v(\widetilde{f} \circ Z) = \widetilde{f}(\nabla'_v Z)$$

when $v \in T_pM$. If Y_1 and Y_2 are sections in $\widetilde{\pi}_1 : TM \to M$ and $\widetilde{\pi}_2 : \mathscr{N}(M) \to M$, respectively, then

$$\widetilde{\eta}_{r,p}(\nabla_v(\widetilde{f}_r \circ Y_r)) = \eta_r(\nabla'_v(\widetilde{i}_r \circ Y_r))$$

for $r = 1, 2$.

Suppose that Z is the pull-back of a vector field Z' on N as defined in Proposition 2.57. Then it follows from Proposition 10.87 that

$$\nabla_{f_*(v)} Z' = \widetilde{f}(\nabla'_v Z)$$

when $v \in T_pM$, and if X and X' are vector fields on M and N, respectively, which are f-related, then

$$\nabla_{X'} Z' \circ f = \widetilde{f} \circ \nabla'_X Z .$$

11.17 Theorem Let ∇' and ∇^0 be the covariant derivatives corresponding to the connection forms ω_M and ω_0 on $\mathscr{O}_\varepsilon(f^*(TN))$, and let Z be a section in $\widetilde{\pi}_M : f^*(TN) \to M$. If $v \in T_pM$, then

$$\nabla'_v Z = \nabla^0_v Z + \alpha(p)(v)\, Z_p .$$

PROOF: By Definition 10.77 we know that

$$\nabla'_v Z - \nabla^0_v Z = u \circ (w - w_0)(\overline{Z}) ,$$

where $u \in \mathscr{O}_{\varepsilon,p}(f^*(TN))$, $w \in \ker \omega_M(u)$ and $w_0 \in \ker \omega_0(u)$, and where $\overline{Z} : \mathscr{O}_\varepsilon(f^*(TN)) \to \mathbf{R}^m$ is the smooth map defined by

$$\overline{Z}(u) = u^{-1} \circ Z(\pi_M(u))$$

for $u \in \mathscr{O}_\varepsilon(f^*(TN))$. If $\sigma_u : O_\varepsilon(\mathbf{R}^m) \to \mathscr{O}_\varepsilon(f^*(TN))$ is defined as in Proposition 10.26 by $\sigma_u(G) = u \circ G$ for $G \in O_\varepsilon(\mathbf{R}^m)$, we have that

$$\overline{Z} \circ \sigma_u(G) = \overline{Z}(u \circ G) = G^{-1} \circ \overline{Z}(u) = \Lambda_u \circ \phi\,(G)$$

for $G \in O_\varepsilon(\mathbf{R}^m)$, where $\phi : O_\varepsilon(\mathbf{R}^m) \to O_\varepsilon(\mathbf{R}^m)$ is the smooth map given by $\phi\,(G) = G^{-1}$ for $G \in O_\varepsilon(\mathbf{R}^m)$, and $\Lambda_u : \mathfrak{gl}\,(\mathbf{R}^m) \to \mathbf{R}^m$ is the linear map given by

$$\Lambda_u(F) = F \circ \overline{Z}(u)$$

for $F \in \mathfrak{gl}\,(\mathbf{R}^m)$. As $w - w_0 \in V_u$, we also have that

$$\begin{aligned} w - w_0 &= \sigma_{u*} \circ \omega_M(u)(w - w_0) = -\sigma_{u*} \circ \omega_M(u)(w_0) \\ &= -\sigma_{u*} \circ (\omega_M - \omega_0)(u)(w_0) = -\sigma_{u*} \circ \tau_M(u)(w_0) \end{aligned}$$

so that

$$\begin{aligned} (w - w_0)(\overline{Z}) &= -\tau_M(u)(w_0)(\overline{Z} \circ \sigma_u) = -\Lambda_u \circ \phi_* \circ \tau_M(u)(w_0) \\ &= \Lambda_u \circ \tau_M(u)(w_0) = \tau_M(u)(w_0) \circ \overline{Z}(u) \end{aligned}$$

by Corollary 8.33. This implies that

$$\nabla'_v Z - \nabla^0_v Z = u \circ \tau_M(u)(w_0) \circ u^{-1} \circ Z(p) = \alpha(p)(v)\, Z_p ,$$

where the last equality follows by choosing u in $\mathscr{O}_p(N,M)$. □

11.18 Proposition Let ∇^0, ∇^1 and ∇^2 be the covariant derivatives corresponding to the connection forms $\rho \cdot \omega'$, ω_1 and ω_2, and let Y_1 and Y_2 be sections in $\widetilde{\pi}_1 : TM \to M$ and $\widetilde{\pi}_2 : \mathscr{N}(M) \to M$, respectively. If $v \in T_pM$, then

$$\nabla^0_v (\widetilde{i}_r \circ Y_r) = \widetilde{i}_r (\nabla^r_v Y_r)$$

for $r = 1,2$, and ∇^0 is also the covariant derivative corresponding to the connection form ω_0 on $\mathscr{O}_\varepsilon(f^*(TN))$, as defined in Theorem 11.17.

PROOF: If $Z_r = \widetilde{i}_r \circ Y_r$, then $\overline{Z}_r = j_r \circ \overline{Y}_r \circ \lambda_r$ since

$$\overline{Z}_r(u) = u^{-1} \circ \widetilde{i}_r \circ Y_r(\pi'(u)) = j_r \circ \lambda_r(u)^{-1} \circ Y_r(\pi'(u)) = j_r \circ \overline{Y}_r \circ \lambda_r(u)$$

for $u \in \mathscr{O}(N,M)$ and $r = 1,2$. Now choose a $u \in \mathscr{O}_p(N,M)$ and a $w \in \ker \rho \cdot \omega'(u)$, and let $u_r = \lambda_r(u)$ and $w_r = \lambda_{r*}(w)$ for $r = 1,2$. Then we have that

$$\omega_r(u_r)(w_r) = \lambda_r^*(\omega_r)(u)(w) = (\phi_{r*} \circ \rho) \cdot \omega'(u)(w) = 0$$

so that

$$\nabla^0_v Z_r = u \circ d\overline{Z}_r(u)(w) = u \circ d(j_r \circ \overline{Y}_r \circ \lambda_r)(u)(w)$$

$$= u \circ j_r \circ \lambda_r^*(d\overline{Y}_r)(u)(w) = \widetilde{i}_r \circ u_r \circ d\overline{Y}_r(u_r)(w_r) = \widetilde{i}_r (\nabla^r_v Y_r),$$

which completes the proof of the proposition. □

11.19 Corollary Let ∇, ∇^1 and ∇^2 be the covariant derivatives corresponding to the metrical connections without torsion in TN, TM and $\mathscr{N}(M)$, and let Y_1 and Y_2 be sections in $\widetilde{\pi}_1 : TM \to M$ and $\widetilde{\pi}_2 : \mathscr{N}(M) \to M$, respectively. If $v \in T_pM$, then

$$\widetilde{\eta}_{r,p} (\nabla_v(\widetilde{f}_r \circ Y_r)) = \nabla^r_v Y_r$$

and

$$\widetilde{i}_{3-r} \circ \widetilde{\eta}_{3-r,p} (\nabla_v(\widetilde{f}_r \circ Y_r)) = \alpha(p)(v)(\widetilde{i}_r \circ Y_r(p))$$

for $r = 1,2$.

11.20 Remark Writing $\nabla_v(\widetilde{f}_r \circ Y_r)$ as a sum of the two orthogonal components given in the corollary is called the *Gauss' formula* when $r = 1$ and the *Weingarten formula* when $r = 2$.

11.21 Let N^m and M^1 be pseudo-Riemannian manifolds with metrics g and g' of index r and k, respectively, and let $f : M \to N$ be an isometric immersion and $\widetilde{u}$ be a unit vector field on M. The index k of g' can be either 0 or 1, and we have that

$$<\widetilde{u}(p), \widetilde{u}(p)> = \varepsilon$$

for $p \in M$, where $\varepsilon = 1 - 2k$.

Let ∇' and ∇^0 be the covariant derivatives corresponding to the connection forms ω_M and ω_0 on $\mathscr{O}_\varepsilon(f^*(TN))$, and let Z be a section in $\widetilde{\pi}_M : f^*(TN) \to M$. We have that $Z = Z_1 + Z_2$ where $Z_n = \widetilde{i}_n \circ \eta_n \circ Z$ for $n = 1, 2$, and we let $u = \widetilde{i}_1 \circ \widetilde{u}$. Since

$$\alpha(p)(v)(Z_1(p)) = \widetilde{i}_2 \circ \eta_2(\nabla'_v Z_1) = \widetilde{i}_2 \circ \eta_2(\nabla'_v(\varepsilon <Z,u> u))$$
$$= \varepsilon <Z(p),u(p)> \nabla'_v u$$

and

$$\alpha(p)(v)(Z_2(p)) = \widetilde{i}_1 \circ \eta_1(\nabla'_v Z_2) = \varepsilon <\nabla'_v Z_2, u(p)> u(p)$$
$$= -\varepsilon <Z(p), \nabla'_v u> u(p)\,,$$

it follows from Theorem 11.17 that

$$\nabla^0_v Z = \nabla'_v Z + \varepsilon <Z(p), \nabla'_v u> u(p) - \varepsilon <Z(p), u(p)> \nabla'_v u$$

for $p \in M$ and $v \in T_pM$. When N is a Lorenz manifold, $\nabla^0_v Z$ is called the *Fermi derivative* of Z at p with respect to v. The corresponding parallel transport is called *Fermi–Walker transport*.

STRAIN AND VORTICITY

11.22 Proposition Let X be a vector field on a pseudo-Riemannian manifold M^n, and let F be a vector field along an integral curve $c : I \to M$ for X obtained by Lie transport. Then F satisfies *Jacobi's equation*

$$\frac{DF}{dr}(t) = \nabla_{F(t)} X$$

for $t \in I$, using the standard local chart (r, I) on I where $r : I \to \mathbf{R}$ is the inclusion map.

PROOF: Let (x, U) be a local chart around a point $c(t_0) = p_0$ on the curve c, and suppose that

$$F(t) = \sum_{i=1}^{n} a^i(t) \left.\frac{\partial}{\partial x^i}\right|_{c(t)}$$

for $t \in c^{-1}(U)$ and

$$X(p) = \sum_{i=1}^{n} b^i(p) \left.\frac{\partial}{\partial x^i}\right|_{p}$$

for $p \in U$. If $\gamma : \mathscr{D}(X) \to M$ is the global flow for X, we have that $c(t) = \gamma_{p_0}(t - t_0)$ and $F(t) = \gamma_{t-t_0\,*}F(t_0)$ for $t \in I$. This implies that

$$a^i(t) = F(t)(x^i) = F(t_0)(x^i \circ \gamma_{t-t_0}) = \sum_{j=1}^{n} a^j(t_0)\, \frac{\partial (x^i \circ \gamma_{t-t_0})}{\partial x^j}\,(p_0)$$

for $t \in c^{-1}(U)$, so that

$$a^{i\,\prime}(t_0) = \sum_{j=1}^{n} a^j(t_0)\, \frac{\partial^2 (x^i \circ \gamma)}{\partial x^j \partial t}\,(0, p_0) = \sum_{j=1}^{n} a^j(t_0)\, \frac{\partial b^i}{\partial x^j}\,(p_0)\,.$$

By Proposition 10.121 and Corollary 10.114 we therefore have that

$$\frac{DF}{dr}(t_0) = \sum_{i=1}^{n} \left(a^{i\,\prime}(t_0) + \sum_{j,k=1}^{n} \Gamma^i_{jk}(p_0)\, b^j(p_0)\, a^k(t_0) \right) \frac{\partial}{\partial x^i}\bigg|_{p_0}$$

$$= \sum_{i=1}^{n} \left(\sum_{j=1}^{n} a^j(t_0) \left(\frac{\partial b^i}{\partial x^j}(p_0) + \sum_{k=1}^{n} b^k(p_0)\, \Gamma^i_{jk}(p_0) \right) \right) \frac{\partial}{\partial x^i}\bigg|_{p_0} = \nabla_{F(t_0)} X\,.$$

□

11.23 Let N^m and the open interval I be pseudo-Riemannian manifolds with metrics g and g' of index 1, and let $c : I \to N$ be an isometric embedding which is an integral curve for a unit vector field u on N. Let $\zeta : I \to \mathscr{N}(I)$ be the zero section in the normal bundle $\widetilde{\pi}_2 : \mathscr{N}(I) \to I$. Then the map ϕ defined in 11.6 induces a diffeomorphism, also denoted by ϕ, between tubular neighbourhoods W and U for c in $\mathscr{N}(I)$ and N as described in Theorem 11.7. Disregarding the connected components of W and U not containing $\zeta(I)$ and $c(I)$, respectively, we may assume that W and U are connected.

Suppose that $h = \|\widetilde{\pi}_{2*} \circ \phi_*^{-1} \circ u\|$ is a non-zero function on U, and let $\tilde{u} = (1/h)\, u|_U$ and $\hat{u} = \phi^* \tilde{u}$. While the integral curves of u are parametrized by arc length, we will show that $\tilde{u}$ corresponds to a reparametrization of these curves so that a vector field F along c obtained by Lie transport with respect to $\tilde{u}$ and being orthogonal to c at a $t_0 \in I$, will remain orthogonal to c at all $t \in I$. By taking the Fermi derivative of the section obtained from F in the normal bundle, we will derive expressions for the rate of stretching (strain) and the rate of rotation (vorticity) in the space orthogonal to u at each point on c with respect to a Fermi–Walker transported basis.

If the dimension of N is $m = 4$, the flow of a timelike unit vector field u can be thought of as a family of world lines of reference particles or observers constituting a *reference frame* R. The vector field u is called the *four-velocity field* of these observers, and the 3-dimensional subspace of the tangent space T_pN at a point $p \in N$ orthogonal to $u(p)$ is called the *simultaneity space* or the *rest space* of the observer at p.

Using the standard local chart (r, I) on I, where $r : I \to \mathbf{R}$ is the inclusion map, we see that $\frac{d}{dr}$ is a unit vector field on I since it is c-related to u by means of the isometric embedding c. As we also have that $\|\widetilde{\pi}_{2*}\circ\hat{u}\|=1$ on W, this implies that

$$\widetilde{\pi}_{2*}\circ\hat{u}=a\,\frac{d}{dr}\circ\widetilde{\pi}_2$$

where $a:W\to\mathbf{R}$ is a smooth function with $|a|=1$. Furthermore, $h\circ c=1$ so that

$$\widetilde{\pi}_{2*}\circ\hat{u}\circ\zeta=c_*^{-1}\circ u\circ c=\frac{d}{dr}\,.$$

Since W is connected it follows that $a=1$, showing that the vector field $\hat{u}$ on W is $\widetilde{\pi}_2$-related to the vector field $\frac{d}{dr}$ on I.

If $\gamma:\mathscr{D}(\hat{u})\to W$ is the global flow for $\hat{u}$, then it follows from Proposition 3.48 that $\widetilde{\pi}_2\circ\gamma_p$ is an integral curve for $\frac{d}{dr}$ with initial condition $\widetilde{\pi}_2(p)$, i.e.,

$$\widetilde{\pi}_2\circ\gamma_p(t)=\widetilde{\pi}_2(p)+t$$

when $p\in W$ and $t\in I(p)$. Hence

$$\widetilde{\pi}_2\circ\gamma_{t-t_0}=T_{t-t_0}\circ\widetilde{\pi}_2$$

which implies that

$$\widetilde{\pi}_{2*}\circ\gamma_{t-t_0*}=T_{t-t_0*}\circ\widetilde{\pi}_{2*}\tag{1}$$

for $t,t_0\in I$, where $T_{t-t_0}:I-t+t_0\to I$ is translation by $t-t_0$.

Now let F be a vector field along c obtained by Lie transport with respect to the vector field $\tilde{u}$, and with $F(t_0)\in c_*(T_{t_0}I)^\perp$ for a $t_0\in I$. Then $\widehat{F}=\phi_*^{-1}\circ F$ is a vector field along ζ obtained by Lie transport with respect to the vector field $\hat{u}$ so that $\widehat{F}(t)=\gamma_{t-t_0*}\widehat{F}(t_0)$ for $t\in I$, and $\widehat{F}(t_0)\in V_{\zeta(t_0)}=\ker\widetilde{\pi}_{2*\zeta(t_0)}$ is a vertical tangent vector. From (1) it follows that $\widehat{F}(t)\in V_{\zeta(t)}$ is also a vertical tangent vector so that $F(t)\in c_*(T_tI)^\perp$ for every $t\in I$. Hence $F=\widetilde{c}_2\circ Y$ for a section Y in the normal bundle $\widetilde{\pi}_2:\mathscr{N}(I)\to I$, and the Fermi derivative of Y at t with respect to $v=\left.\frac{d}{dr}\right|_t$ is given by

$$\begin{aligned}\nabla_v^2 Y &= \widetilde{\eta}_{2,t}(\nabla_v F)=\widetilde{\eta}_{2,t}(\nabla_{F(t)}\widetilde{u})\\ &=\widetilde{\eta}_{2,t}\left(F(t)\left(\frac{1}{h}\right)u+\frac{1}{h\circ c(t)}\nabla_{F(t)}u\right)=\widetilde{\eta}_{2,t}(\nabla_{F(t)}u)\end{aligned}$$

for $t\in I$, where ∇ is the covariant derivative corresponding to the Levi–Civita connection on N and $\widetilde{\eta}_{2,t}:T_{c(t)}N\to\mathscr{N}_t(I)$ is the linear map defined in 11.2.

Let $\nu:TN\to TN$ be the bundle map over N given by

$$\nu_p=\tilde{c}_{2,t}\circ\widetilde{\eta}_{2,t}$$

for $p=c(t)\in N$, where c is an integral curve for u through p as described above. We see that $\nu_p:T_pN\to T_pN$ is the orthogonal projection on the fibre

$$\ker u^\flat(p)=u(p)^\perp=c_*(T_tI)^\perp=\tilde{c}_2(\mathscr{N}_t(I))$$

in the distribution $\ker u^\flat$ on N. If (x,O) is a local chart on N and

$$u|_O = \sum_{i=1}^m u^i \frac{\partial}{\partial x^i}\ , \tag{2}$$

we have that

$$\nu_p\left(\left.\frac{\partial}{\partial x^i}\right|_p\right) = \left.\frac{\partial}{\partial x^i}\right|_p + \left\langle \left.\frac{\partial}{\partial x^i}\right|_p , u(p)\right\rangle u(p) = \sum_{k=1}^m \nu_i^k(p) \left.\frac{\partial}{\partial x^k}\right|_p$$

for $p \in O$, where

$$\nu_i^k = \delta_i^k + u^k u_i$$

for $i,k = 1, \ldots ,m$.

By composing the bundle-valued 1-form $\nabla u \in \Omega^1(N;TN)$ with the equivalence $\phi : \mathscr{T}^1(TN;TN) \to \mathscr{T}^1_1(TN,TN)$ over N defined in 5.25 and lowering the contravariant index, we obtain a covariant tensor field $\widetilde{\tau} = L^1_1(\phi \circ \nabla u) \in \mathscr{T}^2(N)$ on N of degree 2, which is given by

$$\widetilde{\tau}(p)(v_1,v_2) = \phi \circ \nabla u(p)(v_2, g(p)^\flat(v_1)) = < v_1, \nabla_{v_2} u > = \nabla_{v_2} u^\flat(v_1)$$

for $p \in N$ and $v_1, v_2 \in T_pN$. The last equality follows from Proposition 10.163. Using Corollary 10.118 we therefore have that

$$\widetilde{\tau}|_O = \sum_{i,j=1}^m u_{i;j}\, dx^i \otimes dx^j\ .$$

The pull-back tensor field $\tau = \nu^*\widetilde{\tau}$ is given by

$$\tau(p)(v_1,v_2) = < \nu(p)(v_1), \nabla_{\nu(p)(v_2)} u > = < v_1, \nabla_{\nu(p)(v_2)} u >$$

for $p \in N$ and $v_1, v_2 \in T_pN$.

Now let $\widetilde{\theta}$ and $\widetilde{\omega}$ be the symmetric and antisymmetric parts of $\widetilde{\tau}$, and let $\theta = \nu^*\widetilde{\theta}$ and $\omega = \nu^*\widetilde{\omega}$ be their pull-backs by ν. The symmetric tensor field θ is called the *rate of strain*, and the 2-form ω the *vorticity* or the *rate of rotation*. We have that

$$\tau|_O = \sum_{i,j=1}^m \tau_{ij}\, dx^i \otimes dx^j$$

where

$$\tau_{ij} = \sum_{k,l=1}^m u_{k;l}\, \nu_i^k \nu_j^l = \sum_{l=1}^m u_{i;l}\, \nu_j^l$$

for $i,j = 1, \ldots ,m$. The corresponding components of $\theta|_O$ and $\omega|_O$ are

$$\theta_{ij} = \tfrac{1}{2} \sum_{k,l=1}^m \{u_{k;l} + u_{l;k}\}\, \nu_i^k \nu_j^l \tag{3}$$

and

$$\omega_{ij} = \tfrac{1}{2} \sum_{k,l=1}^{m} \{u_{k,l} - u_{l,k}\}\, v_i^k v_j^l \tag{4}$$

for $i, j = 1, \dots, m$, where $u_{k,l} = \dfrac{\partial u_k}{\partial x^l}$. From (2) we also have that

$$u^\flat|_O = \sum_{i=1}^{m} u_i\, dx^i$$

so that

$$du^\flat|_O = \sum_{i,j=1}^{m} u_{i,j}\; dx^j \wedge dx^i = -\sum_{i,j=1}^{m} (u_{i,j} - u_{j,i})\; dx^i \otimes dx^j = -2\, \tilde{\omega}|_O\,.$$

This implies that

$$j^* du^\flat = -2\, j^* \omega$$

where $j : \ker(u^\flat) \to TN$ is the inclusion map, since $v \circ j = j$. By Corollary 5.88 it follows that the distribution $\ker(u^\flat)$ is integrable if and only if the vorticity ω vanishes.

Given an orientation on N, we have an induced orientation on $\ker(u^\flat)$ so that an ordered basis $(v_2, \dots, v_m)$ in $\ker(u^\flat(p))$ is positively oriented if and only if $(u(p), v_2, \dots, v_m)$ is a positively oriented basis in T_pN for each $p \in N$. Then the metric volume element in $\ker(u^\flat)$ is $\eta = i_u \tilde{\eta}$, where $\tilde{\eta}$ is the metric volume element in N. Indeed, if the positively oriented basis $(v_2, \dots, v_m)$ in $\ker(u^\flat(p))$ is orthonormal, then $\tilde{\eta}(p) = u(p)^* \wedge v_2^* \wedge \cdots \wedge v_m^*$ and $\eta(p) = v_2^* \wedge \cdots \wedge v_m^*$.

If the dimension of N is $m = 4$, the *vorticity vector field* Ω on N is defined by

$$\Omega(p) = -(* j^* \omega(p))^\sharp$$

so that

$$i_{\Omega(p)}\, \eta(p) = -j^* \omega(p) = \tfrac{1}{2}\, j^* du^\flat(p)$$

for $p \in N$. If

$$\Omega(p) = \sum_{i=2}^{4} \Omega^i v_i \quad , \quad j^* \omega(p) = \sum_{j,k=2}^{4} \omega_{jk}\, v_j^* \otimes v_k^*$$

and

$$\eta(p) = \sum_{i,j,k=2}^{4} \eta_{ijk}\, v_i^* \otimes v_j^* \otimes v_k^*$$

for a basis $\{v_2, \dots, v_4\}$ in $\ker(u^\flat(p))$ with dual basis $\{v_2^*, \dots, v_4^*\}$, then

$$\sum_{i=2}^{4} \Omega^i\, \eta_{ijk} = \omega_{kj} \tag{5}$$

for $2 \le j, k \le 4$.

11.24 Let N be a Lorentz manifold with a stationary metric g so that

$$g|_O = -e^{2\alpha}\left(dx^1 - \sum_{i=2}^{4} w_i\,dx^i\right)^2 + \sum_{i,j=2}^{4} h_{ij}\,dx^i dx^j$$

in natural units for a local chart (x,O) on N, where $\alpha = \tilde{\alpha}\circ(x^2,x^3,x^4)$, $w_i = \tilde{w}_i\circ(x^2,x^3,x^4)$ and $h_{ij} = \tilde{h}_{ij}\circ(x^2,x^3,x^4)$ for real-valued functions $\tilde{\alpha}$, $\tilde{w}_i$ and $\tilde{h}_{ij}$ on $\mathbf{R}^3$. Let u be a unit vector field on N so that

$$u|_O = e^{-\alpha}\frac{\partial}{\partial x^1}\,,$$

which means that the coordinate system (x,O) is *comoving* in the reference frame R defined by u. Hence each reference particle in R has a world line c which is an integral curve for u satisfying $x_i\circ c = \text{constant}$ for $i = 2,3,4$. Since

$$u^1 = e^{-\alpha} \quad \text{and} \quad u^i = 0 \quad \text{for} \quad i = 2,3,4$$

and

$$u_1 = -e^{\alpha} \quad \text{and} \quad u_i = e^{\alpha}w_i \quad \text{for} \quad i = 2,3,4,$$

the non-zero components of v are

$$v_i^k = \delta_1^k w_i + \delta_i^k \quad \text{for} \quad i = 2,3,4 \quad \text{and} \quad k = 1,2,3,4.$$

From (4) in 11.23 it follows that the non-zero components of ω are given by

$$\begin{aligned}2\,\omega_{ij} &= (u_{1,1} - u_{1,1})\,w_i w_j + (u_{1,j} - u_{j,1})\,w_i + (u_{i,1} - u_{1,i})\,w_j + (u_{i,j} - u_{j,i})\\ &= -e^{\alpha}(\alpha_{,j}w_i - \alpha_{,i}w_j) + (e^{\alpha}w_i)_{,j} - (e^{\alpha}w_j)_{,i} = e^{\alpha}(w_{i,j} - w_{j,i})\end{aligned}$$

for $i,j = 2,3,4$.

Now fix an arbitrary point $p\in O$, and introduce a local chart $(\tilde{x},O)$ on N where

$$\tilde{x}^1 = e^{\alpha(p)}\left(x^1 - \sum_{i=2}^{4} w_i(p)\,x^i\right) \quad \text{and} \quad \tilde{x}^i = x^i \quad \text{for} \quad i = 2,3,4,$$

so that

$$g(p) = -\,d\tilde{x}^1(p)^2 + \sum_{i,j=2}^{4} h_{ij}(p)\,d\tilde{x}^i(p)\,d\tilde{x}^j(p)\,.$$

Assuming that N has an orientation such that the local chart (x,O) is positively oriented, it follows from 7.26 that the metric volume element in $\ker(u^\flat(p))$ is given by

$$\eta(p) = \sum_{i,j,k=2}^{4} \eta_{ijk}(p)\;d\tilde{x}^i(p)\otimes d\tilde{x}^j(p)\otimes d\tilde{x}^k(p)$$

where

$$\eta_{ijk}(p) = \det(h_{rs}(p))^{1/2}\,\varepsilon_{ijk}$$

for $i, j, k = 2,3,4$. If the vorticity vector $\Omega(p)$ is given by

$$\Omega(p) = \sum_{i=2}^{4} \Omega^i(p) \left.\frac{\partial}{\partial \tilde{x}^i}\right|_p = \sum_{i=2}^{4} \Omega^i(p) \left(\left.\frac{\partial}{\partial x^i}\right|_p + w_i(p) \left.\frac{\partial}{\partial x^1}\right|_p \right),$$

then it follows from (5) in 11.23 that

$$\det(h_{rs}(p))^{1/2} \sum_{i=2}^{4} \Omega^i(p)\, \varepsilon_{ijk} = \omega_{kj}(p)$$

for every $p \in O$, so that

$$\det(h_{rs})^{1/2} \sum_{i=2}^{4} \Omega^i\, \varepsilon_{ijk} = \omega_{kj}$$

for $2 \leq j,k \leq 4$. Multiplying by $\frac{1}{2}\, \eta^{ijk} = \frac{1}{2}\, \det(h_{rs})^{-1/2}\, \varepsilon^{ijk}$ and summing over $j,k = 2,3,4$, we obtain

$$\Omega^i = \tfrac{1}{2}\, \det(h_{rs})^{-1/2} \sum_{j,k=2}^{4} \varepsilon^{ijk}\, \omega_{kj} = \tfrac{1}{2}\, e^{\alpha}\, \det(h_{rs})^{-1/2} \sum_{j,k=2}^{4} \varepsilon^{ijk}\, w_{k,j}$$

for $i = 2,3,4$, and

$$\|\Omega|_O\| = \Big(\sum_{i,j=2}^{4} h_{ij}\, \Omega^i\, \Omega^j \Big)^{1/2}.$$

11.25 Example Let N be an oriented Lorentz manifold with a stationary and cylindrically symmetric metric g so that

$$\begin{aligned} g|_O &= -f\, dt^2 + 2k\, dt\, d\phi + l\, d\phi^2 + e^{\mu}(dr^2 + dz^2) \\ &= -f \left(dt - \frac{k}{f}\, d\phi \right)^2 + \frac{D^2}{f}\, d\phi^2 + e^{\mu}(dr^2 + dz^2) \end{aligned}$$

in natural units for a positively oriented local chart (ψ, O) on N with coordinate functions t, r, ϕ and z, where $f = \tilde{f} \circ r$, $k = \tilde{k} \circ r$, $l = \tilde{l} \circ r$ and $\mu = \tilde{\mu} \circ r$ for real functions $\tilde{f}$, $\tilde{k}$, $\tilde{l}$ and $\tilde{\mu}$, and where $D^2 = fl + k^2 > 0$ and $f > 0$. Then we have that

$$\Omega|_O = \tfrac{1}{2}\, \sqrt{f} \left(e^{2\mu}\, \frac{D^2}{f} \right)^{-1/2} \left(\frac{k}{f} \right)_{,r} \frac{\partial}{\partial z} = \frac{e^{-\mu} f}{2D} \left(\frac{k}{f} \right)_{,r} \frac{\partial}{\partial z}$$

and

$$\|\Omega|_O\| = \frac{e^{-\mu/2} f}{2D} \left| \left(\frac{k}{f} \right)_{,r} \right|.$$

THE EQUATIONS OF GAUSS, RICCI AND CODAZZI

11.26 We now consider the second structure equation

$$d\omega = -\tfrac{1}{2}\ \omega \wedge \omega + \Omega\,,$$

where Ω is the curvature form on $\mathscr{O}_\varepsilon(N)$, and the wedge product is with respect to the bilinear map

$$\nu : \mathfrak{o}_\varepsilon(\mathbf{R}^m) \times \mathfrak{o}_\varepsilon(\mathbf{R}^m) \to \mathfrak{o}_\varepsilon(\mathbf{R}^m)$$

given by $\nu(X,Y) = [X,Y]$. If $\Omega' = (f' \circ i')^*(\Omega)$ is the pull-back of Ω to $\mathscr{O}(N,M)$, it follows from Propositions 5.63 and 5.70 that

$$d\omega' = -\tfrac{1}{2}\ \omega' \wedge \omega' + \Omega'. \tag{1}$$

For each $u \in \mathscr{O}(N,M)$ we have that

$$(\rho_1 \cdot \omega') \wedge (\rho_2 \cdot \omega')(u) = 0,$$
$$(\rho \cdot \omega') \wedge \tau(u) \subset \mathfrak{k},$$
$$(\rho_1 \cdot \omega') \wedge (\rho_1 \cdot \omega')(u) \subset i_{1*}\, \mathfrak{o}_k(\mathbf{R}^n),$$
$$(\rho_2 \cdot \omega') \wedge (\rho_2 \cdot \omega')(u) \subset i_{2*}\, \mathfrak{o}_l(\mathbf{R}^d) \qquad \text{and}$$
$$\tau \wedge \tau(u) \subset i_{1*}\, \mathfrak{o}_k(\mathbf{R}^n) + i_{2*}\, \mathfrak{o}_l(\mathbf{R}^d).$$

Hence (1) implies that

$$d(\rho_r \cdot \omega') = -\tfrac{1}{2}(\rho_r \cdot \omega') \wedge (\rho_r \cdot \omega') - \tfrac{1}{2}\rho_r \cdot (\tau \wedge \tau) + \rho_r \cdot \Omega' \tag{2}$$

for $r = 1,2$, and

$$d\tau = -(\rho \cdot \omega') \wedge \tau + \rho_3 \cdot \Omega'. \tag{3}$$

If Ω_1 and Ω_2 are the curvature forms of ω_1 and ω_2 on $\mathscr{O}(M)$ and $\mathscr{O}(\mathscr{N}(M))$, respectively, we have that

$$d\omega_r = -\tfrac{1}{2}\ \omega_r \wedge \omega_r + \Omega_r$$

so that

$$d(\rho_r \cdot \omega') = -\tfrac{1}{2}\ (\rho_r \cdot \omega') \wedge (\rho_r \cdot \omega') + i_{r*} \cdot \lambda_r^*(\Omega_r)$$

for $r = 1,2$ by (2) in 11.8. Combining this equation with (2) we find that

$$i_{r*} \cdot \lambda_r^*(\Omega_r) = -\tfrac{1}{2}\ \rho_r \cdot (\tau \wedge \tau) + \rho_r \cdot \Omega' \tag{4}$$

for $r = 1,2$.

11.27 Consider the representations $\rho_1 : O_l(\mathbf{R}^d) \to \mathrm{Aut}(\mathfrak{gl}(\mathbf{R}^d))$ and $\rho_2, \rho_3 : O_l(\mathbf{R}^d) \to \mathrm{Aut}(L(\mathbf{R}^n, \mathbf{R}^d))$ of the structure group $O_l(\mathbf{R}^d)$ in the orthonormal frame bundle $\pi_2 : \mathscr{O}(\mathscr{N}(M)) \to M$ given by

$$\rho_1(G)(F_1) = G \circ F_1 \quad , \quad \rho_2(G)(F_2) = F_2 \quad \text{and} \quad \rho_3(G)(F_3) = G \circ F_3$$

for $G \in O_l(\mathbf{R}^d)$, $F_1 \in \mathfrak{gl}(\mathbf{R}^d)$ and $F_2, F_3 \in L(\mathbf{R}^n, \mathbf{R}^d)$. By Example 10.29 (d) and (h) the corresponding associated vector bundles may be identified with $\pi_{22} : \Lambda^1(\mathbf{R}^d ; \mathscr{N}(M)) \to M$, $\tilde{\pi}_{21} : E_{L(\mathbf{R}^n, \mathbf{R}^d)} \to M$ and $\pi_{21} : \Lambda^1(\mathbf{R}^n ; \mathscr{N}(M)) \to M$. Let

$$\kappa : \mathfrak{gl}(\mathbf{R}^d) \times L(\mathbf{R}^n, \mathbf{R}^d) \to L(\mathbf{R}^n, \mathbf{R}^d)$$

be the bilinear map given by

$$\kappa(F_1, F_2) = F_1 \circ F_2$$

for $F_1 \in \mathfrak{gl}(\mathbf{R}^d)$ and $F_2 \in L(\mathbf{R}^n, \mathbf{R}^d)$, which satisfies

$$\rho_3(G) \circ \kappa = \kappa \circ (\rho_1(G) \times \rho_2(G))$$

for every $G \in O_l(\mathbf{R}^d)$, and let ∇^{22} and ∇^{21} be the covariant derivatives in $\pi_{22} : \Lambda^1(\mathbf{R}^d ; \mathscr{N}(M)) \to M$ and $\pi_{21} : \Lambda^1(\mathbf{R}^n ; \mathscr{N}(M)) \to M$ corresponding to the connection form ω_2 on $\mathscr{O}(\mathscr{N}(M))$. Given a section $s \in \Gamma(M; \Lambda^1(\mathbf{R}^d ; \mathscr{N}(M)))$ and a vector-valued function $f \in \mathscr{F}(M; L(\mathbf{R}^n, \mathbf{R}^d))$, it follows from 10.76 and Proposition 10.81 that

$$\nabla^{21}_v (s \wedge_\kappa (id_M, f)) = (\nabla^{22}_v s) \circ f(p) + s(p) \circ v(f)$$

for every $p \in M$ and $v \in T_pM$.

We also have a representation $\rho : O_k(\mathbf{R}^n) \to \mathrm{Aut}(\mathfrak{gl}(\mathbf{R}^n))$ of the structure group $O_k(\mathbf{R}^n)$ in the orthonormal frame bundle $\pi_1 : \mathscr{O}(M) \to M$ given by

$$\rho(G)(F) = G \circ F$$

for $G \in O_k(\mathbf{R}^n)$ and $F \in \mathfrak{gl}(\mathbf{R}^n)$. By Example 10.29 (h) the corresponding associated vector bundle may be identified with $\pi_{11} : \Lambda^1(\mathbf{R}^n ; TM) \to M$, and we let ∇^{11} be the covariant derivative in this bundle corresponding to the connection form ω_1 on $\mathscr{O}(M)$. The smooth maps $v_1 : \mathscr{O}(M) \to \Lambda^1(\mathbf{R}^n ; TM)$ and $v_2 : \mathscr{O}(\mathscr{N}(M)) \to \Lambda^1(\mathbf{R}^d ; \mathscr{N}(M))$ defined in Example 10.29 (c) are the inclusion maps.

11.28 Proposition Let R' and R be the curvature tensors in $f^*(TN)$ and N, respectively, and let $(\widehat{f}, f)$ be the bundle map from $\widehat{\pi}' : \Lambda^1(f^*(TN); f^*(TN)) \to M$ to $\widehat{\pi} : \Lambda^1(TN; TN) \to N$ given by

$$\widehat{f}(F) = \widetilde{f}_p \circ F \circ \widetilde{f}_p^{-1}$$

for $p \in M$ and $F \in \Lambda^1(f^*(TN)_p ; f^*(TN)_p)$. Then $R' = (f_*, \widehat{f})^*(R)$ is the pull-back of R by $(f_*, \widehat{f})$ as defined in Proposition 5.33, i.e.,

$$R'(p)(v_1, v_2) = \widetilde{f}_p^{-1} \circ R(f(p))(f_*(v_1), f_*(v_2)) \circ \widetilde{f}_p$$

for $p \in M$ and $v_1, v_2 \in T_pM$.

PROOF : Choose $u \in \mathscr{O}_p(N,M)$ and vectors $w_1, w_2 \in T_u\mathscr{O}(N,M)$ so that $\pi'_*(w_i) = v_i$ for $i = 1,2$. Then we have that

$$\begin{aligned}
R'(p)(v_1,v_2) &= u \circ \Omega'(u)(w_1,w_2) \circ u^{-1} \\
&= \widetilde{f}_p^{-1} \circ (\widetilde{f}_p \circ u) \circ \Omega'(u)(w_1,w_2) \circ (\widetilde{f}_p \circ u)^{-1} \circ \widetilde{f}_p \\
&= \widetilde{f}_p^{-1} \circ (f' \circ i')(u) \circ \Omega(f' \circ i'(u))\,((f' \circ i')_*(w_1), (f' \circ i')_*(w_2)) \\
&\quad \circ (f' \circ i')(u)^{-1} \circ \widetilde{f}_p \\
&= \widetilde{f}_p \circ R(f(p))(f_*(v_1), f_*(v_2)) \circ \widetilde{f}_p,
\end{aligned}$$

since $\pi(f' \circ i'(u)) = f(\pi'(u)) = f(p)$ and $\pi_*((f' \circ i')_*(w_i)) = f_*(\pi'_*(w_i)) = f_*(v_i)$ for $i = 1,2$. □

11.29 Theorem Let N and M be pseudo-Riemannian manifolds, and let $f : M \to N$ be an isometric immersion. If R, R_1 and R_2 are the curvature tensors in N, M and $\mathscr{N}(M)$, respectively, and α is the second fundamental form, then we have that

$$\begin{aligned}
\widetilde{\eta}_{r,p} \circ R(f(p))(f_*(v_1), f_*(v_2)) \circ \widetilde{f}_r &= R_r(p)(v_1,v_2) \\
&+ \eta_{r,p} \circ \alpha(p)(v_1) \circ \alpha(p)(v_2) \circ \widetilde{i}_r - \eta_{r,p} \circ \alpha(p)(v_2) \circ \alpha(p)(v_1) \circ \widetilde{i}_r
\end{aligned}$$

for every $p \in M$ and $v_1, v_2 \in T_pM$.

FIRST PROOF : The theorem follows from formula (4) in 11.26 and Proposition 11.28. Choose $u \in \mathscr{O}_p(N,M)$ and vectors $w_1, w_2 \in T_u\mathscr{O}(N,M)$ so that $\pi'_*(w_i) = v_i$ for $i = 1,2$. Then we have that

$$\begin{aligned}
&\eta_{r,p} \circ u \circ i_{r*} \cdot \lambda_r^*(\Omega_r)(u)(w_1,w_2) \circ u^{-1} \circ \widetilde{i}_r \\
&= \lambda_r(u) \circ k_r \circ i_{r*} \cdot \lambda_r^*(\Omega_r)(u)(w_1,w_2) \circ j_r \circ \lambda_r(u)^{-1} \\
&= \lambda_r(u) \circ \Omega_r(\lambda_r(u))(\lambda_{r*}(w_1), \lambda_{r*}(w_2)) \circ \lambda_r(u)^{-1} \\
&= R_r(p)(v_1,v_2)
\end{aligned}$$

since $\pi_r(\lambda_r(u)) = \pi'(u) = p$ and $\pi_{r*}(\lambda_{r*}(w_i)) = \pi'_*(w_i) = v_i$ for $i = 1,2$. Furthermore, we have that

$$\begin{aligned}
\eta_{r,p} \circ u \circ \rho_r \cdot \Omega'(u)(w_1,w_2) \circ u^{-1} \circ \widetilde{i}_r &= \eta_{r,p} \circ R'(p)(v_1,v_2) \circ \widetilde{i}_r \\
&= \widetilde{\eta}_{r,p} \circ R(f(p))(f_*(v_1), f_*(v_2)) \circ \widetilde{f}_r
\end{aligned}$$

and

$$\begin{aligned}
&\tfrac{1}{2}\, \eta_{r,p} \circ u \circ \rho_r \cdot (\tau \wedge \tau)(u)(w_1,w_2) \circ u^{-1} \circ \widetilde{i}_r \\
&= \eta_{r,p} \circ u \circ [\,\tau(u)(w_1),\, \tau(u)(w_2)\,] \circ u^{-1} \circ \widetilde{i}_r \\
&= \eta_{r,p} \circ \{u \circ \tau(u)(w_1) \circ u^{-1}\} \circ \{u \circ \tau(u)(w_2) \circ u^{-1}\} \circ \widetilde{i}_r \\
&\quad - \eta_{r,p} \circ \{u \circ \tau(u)(w_2) \circ u^{-1}\} \circ \{u \circ \tau(u)(w_1) \circ u^{-1}\} \circ \widetilde{i}_r \\
&= \eta_{r,p} \circ \alpha(p)(v_1) \circ \alpha(p)(v_2) \circ \widetilde{i}_r - \eta_{r,p} \circ \alpha(p)(v_2) \circ \alpha(p)(v_1) \circ \widetilde{i}_r
\end{aligned}$$

which completes the proof of the theorem. □

SECOND PROOF : Let X and Y be vector fields on M with $X_p = v_1$ and $Y_p = v_2$, and let $Z_r = \eta_r \circ Z'$ for the section $Z' = \widetilde{f}^*(Z)$ in $\widetilde{\pi}_M : f^*(TN) \to M$ which is the pull-back of a vector field Z on N as defined in Proposition 2.57. Then it follows from Propositions 10.107, 11.28 and Corollary 11.19 that

$$\begin{aligned}
&\widetilde{\eta}_{r,p} \circ R(f(p))(f_*(X_p), f_*(Y_p)) \circ \widetilde{f}_r(Z_{r,p}) \\
&\quad = \eta_r \circ R'(p)(v_1, v_2) \circ \widetilde{i}_r(Z_{r,p}) \\
&\quad = \{ \nabla^r_{X_p} \nabla^r_Y Z_r + \eta_r \circ \alpha(p)(X_p) \circ \alpha(p)(Y_p) \circ \widetilde{i}_r(Z_{r,p}) \} \\
&\qquad - \{ \nabla^r_{Y_p} \nabla^r_X Z_r + \eta_r \circ \alpha(p)(Y_p) \circ \alpha(p)(X_p) \circ \widetilde{i}_r(Z_{r,p}) \} - \nabla^r_{[X,Y](p)} Z_r \\
&\quad = R_r(p)(X_p, Y_p)(Z_{r,p}) + \eta_r \circ \alpha(p)(X_p) \circ \alpha(p)(Y_p) \circ \widetilde{i}_r(Z_{r,p}) \\
&\qquad - \eta_r \circ \alpha(p)(Y_p) \circ \alpha(p)(X_p) \circ \widetilde{i}_r(Z_{r,p}). \qquad \square
\end{aligned}$$

11.30 Corollary Let N and M be pseudo-Riemannian manifolds, and let $f : M \to N$ be an isometric immersion. If R, R_1 and R_2 are the curvature tensors in N, M and $\mathcal{N}(M)$, respectively, and α is the second fundamental form, then we have that

$$\begin{aligned}
<R(f(p))(f_*(v_1), f_*(v_2))(\widetilde{f}_r(v_3)), \widetilde{f}_r(v_4)> &= <R_r(p)(v_1, v_2)(v_3), v_4> \\
&+ <\alpha(p)(v_1)(\widetilde{i}_r(v_3)), \alpha(p)(v_2)(\widetilde{i}_r(v_4))> \\
&- <\alpha(p)(v_2)(\widetilde{i}_r(v_3)), \alpha(p)(v_1)(\widetilde{i}_r(v_4))>
\end{aligned}$$

for every $p \in M$, $v_1, v_2 \in T_pM$ and $v_3, v_4 \in \widetilde{\eta}_{r,p}(T_{f(p)}N)$.

PROOF : By Theorem 11.29 we have that

$$\begin{aligned}
&<R(f(p))(f_*(v_1), f_*(v_2))(\widetilde{f}_r(v_3)), \widetilde{f}_r(v_4)> \\
&\quad = <\widetilde{\eta}_{r,p} \circ R(f(p))(f_*(v_1), f_*(v_2))(\widetilde{f}_r(v_3)), v_4> \\
&\quad = <R_r(p)(v_1, v_2)(v_3), v_4> \\
&\qquad + <\alpha(p)(v_1) \circ \alpha(p)(v_2)(\widetilde{i}_r(v_3)), \widetilde{i}_r(v_4)> \\
&\qquad - <\alpha(p)(v_2) \circ \alpha(p)(v_1)(\widetilde{i}_r(v_3)), \widetilde{i}_r(v_4)> \\
&\quad = <R_r(p)(v_1, v_2)(v_3), v_4> \\
&\qquad + <\alpha(p)(v_1)(\widetilde{i}_r(v_3)), \alpha(p)(v_2)(\widetilde{i}_r(v_4))> \\
&\qquad - <\alpha(p)(v_2)(\widetilde{i}_r(v_3)), \alpha(p)(v_1)(\widetilde{i}_r(v_4))>
\end{aligned}$$

where the last equality follows from formula (1) in Example 8.58 (b). □

11.31 Remark The formula in Theorem 11.29 or its corollary is called *Gauss' equation* when $r=1$ and the *Ricci equation* when $r=2$. Gauss' equation can also be written in the following two forms:

11.32 Corollary Let N and M be pseudo-Riemannian manifolds, and let $f: M \to N$ be an isometric immersion. If R and R_1 are the curvature tensors in N and M, respectively, and II is the shape tensor, then we have that

$$\begin{aligned} &< R(f(p))(f_*(v_1), f_*(v_2))\,(f_*(v_3)), f_*(v_4) > \;=\; < R_1(p)(v_1,v_2)(v_3), v_4 > \\ &\quad + < \mathrm{II}(p)(v_1,v_3), \mathrm{II}(p)(v_2,v_4) > \;-\; < \mathrm{II}(p)(v_2,v_3), \mathrm{II}(p)(v_1,v_4) > \end{aligned}$$

for every $p \in M$ and $v_1, v_2, v_3, v_4 \in T_pM$.

11.33 Corollary Let N and M be pseudo-Riemannian manifolds, and let $f: M \to N$ be an isometric immersion. If K and K_1 are the sectional curvatures in N and M, respectively, and II is the shape tensor, then we have that

$$\begin{aligned} &K(f_*(W)) = K_1(W) \\ &\quad - \frac{< \mathrm{II}(p)(v_1,v_1), \mathrm{II}(p)(v_2,v_2) > \;-\; < \mathrm{II}(p)(v_1,v_2), \mathrm{II}(p)(v_1,v_2) >}{< v_1, v_1 > < v_2, v_2 > \;-\; < v_1, v_2 >^2} \end{aligned}$$

for every $p \in M$ and every non-degenerate tangent plane W to M at p, where $\{v_1, v_2\}$ is a basis for W.

11.34 Theorem (The Codazzi Equation) Let N and M be pseudo-Riemannian manifolds, and let $f: M \to N$ be an isometric immersion. If R is the curvature tensor in N and II is the shape tensor, we have that

$$\widetilde{\eta}_{2,p} \circ R(f(p))(f_*(v_1), f_*(v_2))\,(f_*(v_3)) = (\nabla_{v_1}\mathrm{II})\,(v_2, v_3) - (\nabla_{v_2}\mathrm{II})\,(v_1, v_3)$$

for every $p \in M$ and $v_1, v_2, v_3 \in T_pM$.

FIRST PROOF : Let s be an adapted orthonormal frame field on M, and let $s_r = \lambda_r \circ s$ for $r=1,2$. Choose $a_3 \in \mathbf{R}^n$ so that $s_1(p)(a_3) = v_3$, and let X, Y and Z be vector fields on M with $X_p = v_1$, $Y_p = v_2$ and $Z(q) = s_1(q)(a_3)$ for $q \in M$. By formula (3) in 11.26 we have that

$$\begin{aligned} &\widetilde{\eta}_{2,p} \circ R(f(p))(f_*(X_p), f_*(Y_p)) \circ f_*(v_3) \\ &= \eta_2 \circ R'(p)(X_p, Y_p) \circ \widetilde{i}_1(v_3) \\ &= \eta_2 \circ s(p) \circ (\rho_3 \cdot \widetilde{\Omega}')(p)(X_p, Y_p) \circ s(p)^{-1} \circ \widetilde{i}_1(v_3) \\ &= s_2(p) \circ k_2 \circ (\rho_3 \cdot \widetilde{\Omega}')(p)(X_p, Y_p) \circ j_1(a_3) \\ &= s_2(p) \circ k_2 \circ d\widetilde{\tau}(X,Y)(p) \circ j_1(a_3) \\ &\quad + \; s_2(p) \circ k_2 \circ \{(\rho \cdot \widetilde{\omega}') \wedge \widetilde{\tau}\}(p)(X_p, Y_p) \circ j_1(a_3), \end{aligned}$$

where

$$d\widetilde{\tau}(X,Y)(p) = X_p(\widetilde{\tau}(Y)) - Y_p(\widetilde{\tau}(X)) - \widetilde{\tau}([X,Y])(p)$$
$$= \{X_p(\widetilde{\tau}(Y)) - \widetilde{\tau}(p)(\nabla^1_{X_p} Y)\} - \{Y_p(\widetilde{\tau}(X)) - \widetilde{\tau}(p)(\nabla^1_{Y_p} X)\}$$

and

$$k_2 \circ \{(\rho \cdot \widetilde{\omega}') \wedge \widetilde{\tau}\}(p)(X_p,Y_p) \circ j_1$$
$$= k_2 \circ \{(\rho_2 \cdot \widetilde{\omega}')(p)(X_p) \circ \widetilde{\tau}(p)(Y_p) - \widetilde{\tau}(p)(Y_p) \circ (\rho_1 \cdot \widetilde{\omega}')(p)(X_p)\} \circ j_1$$
$$- \; k_2 \circ \{(\rho_2 \cdot \widetilde{\omega}')(p)(Y_p) \circ \widetilde{\tau}(p)(X_p) - \widetilde{\tau}(p)(X_p) \circ (\rho_1 \cdot \widetilde{\omega}')(p)(Y_p)\} \circ j_1.$$

The second part after the final equality sign in each of the last two formulas follows from the first part by interchanging X and Y. Concerning the first parts, we have that

$$s_2(p) \circ k_2 \circ \{(\rho_2 \cdot \widetilde{\omega}')(p)(X_p) \circ \widetilde{\tau}(p)(Y_p) + X_p(\widetilde{\tau}(Y))\} \circ j_1(a_3)$$
$$= \{s_2(p) \circ \widetilde{\omega}_2(p)(X_p) \circ (\rho_{21} \cdot \widetilde{\tau})(Y)(p) + s_2(p) \circ X_p((\rho_{21} \cdot \widetilde{\tau})(Y))\}(a_3)$$
$$= \{\nabla^{22}_{X_p}(\nu_2 \circ s_2) \circ (\rho_{21} \cdot \widetilde{\tau})(Y)(p) + (\nu_2 \circ s_2)(p) \circ X_p((\rho_{21} \cdot \widetilde{\tau})(Y))\}(a_3)$$
$$= \nabla^{21}_{X_p}\{(\nu_2 \circ s_2) \wedge_\kappa (id_M, (\rho_{21} \cdot \widetilde{\tau})(Y))\}(a_3) = \nabla^2_{X_p}(\mathrm{II}(Y,Z)),$$

since

$$\{(\nu_2 \circ s_2) \wedge_\kappa (id_M, (\rho_{21} \cdot \widetilde{\tau})(Y))\}(q)(a_3) = s_2(q) \circ (\rho_{21} \cdot \widetilde{\tau})(Y)(q)(a_3)$$
$$= \eta_2 \circ s(q) \circ \widetilde{\tau}(Y)(q) \circ j_1(a_3) = \eta_2 \circ \alpha(Y)(q) \circ \widetilde{i}_1 \circ s_1(q)(a_3) = \mathrm{II}(Y,Z)(q)$$

for $q \in M$. Furthermore, we have that

$$s_2(p) \circ k_2 \circ \widetilde{\tau}(p)(\nabla^1_{X_p} Y) \circ j_1(a_3) = \eta_2 \circ s(p) \circ \widetilde{\tau}(p)(\nabla^1_{X_p} Y) \circ j_1(a_3)$$
$$= \eta_2 \circ \alpha(p)(\nabla^1_{X_p} Y) \circ \widetilde{i}_1 \circ s_1(p)(a_3) = \mathrm{II}(p)(\nabla^1_{X_p} Y, Z_p)$$

and

$$s_2(p) \circ k_2 \circ \widetilde{\tau}(p)(Y_p) \circ (\rho_1 \cdot \widetilde{\omega}')(p)(X_p) \circ j_1(a_3)$$
$$= \eta_2 \circ s(p) \circ \widetilde{\tau}(p)(Y_p) \circ s(p)^{-1} \circ \widetilde{i}_1 \circ s_1(p) \circ \widetilde{\omega}_1(p)(X_p)(a_3)$$
$$= \eta_2 \circ \alpha(p)(Y_p) \circ \widetilde{i}_1 \circ \nabla^{11}_{X_p}(\nu_1 \circ s_1)(a_3) = \mathrm{II}(p)(Y_p, \nabla^1_{X_p} Z).$$

The theorem now follows from formula (2) in 11.13, as the final term in the last three formulas combine to $(\nabla_{X_p} \mathrm{II})(Y_p, Z_p)$. □

SECOND PROOF : Let X, Y and Z be vector fields on M with $X_p = v_1$, $Y_p = v_2$ and

$Z_p = v_3$. Then it follows from Proposition 10.107, Corollary 11.19 and formula (2) in 11.13 that

$$\begin{aligned}
&\widetilde{\eta}_{2,p} \circ R(f(p))(f_*(X_p), f_*(Y_p)) \circ f_*(Z_p) \\
&= \eta_2 \circ R'(p)(X_p, Y_p) \circ \widetilde{i}_1(Z_p) \\
&= \{\nabla^2_{X_p}[\,\alpha(Y)(\widetilde{i}_1 \circ Z)\,] + \eta_2 \circ \alpha(p)(X_p)(\widetilde{i}_1(\nabla^1_{Y_p} Z))\} \\
&\quad - \{\nabla^2_{Y_p}[\,\alpha(X)(\widetilde{i}_1 \circ Z)\,] + \eta_2 \circ \alpha(p)(Y_p)(\widetilde{i}_1(\nabla^1_{X_p} Z))\} \\
&\quad - \eta_2 \circ \alpha(p)(\nabla^1_{X_p} Y - \nabla^1_{Y_p} X)(\widetilde{i}_1(Z_p)) \\
&= \{\nabla^2_{X_p}[\mathrm{II}(Y,Z)] - \mathrm{II}(p)(\nabla^1_{X_p} Y, Z_p) - \mathrm{II}(p)(Y_p, \nabla^1_{X_p} Z)\} \\
&\quad - \{\nabla^2_{Y_p}[\mathrm{II}(X,Z)] - \mathrm{II}(p)(\nabla^1_{Y_p} X, Z_p) - \mathrm{II}(p)(X_p, \nabla^1_{Y_p} Z)\} \\
&= (\nabla_{X_p} \mathrm{II})(Y_p, Z_p) - (\nabla_{Y_p} \mathrm{II})(X_p, Z_p).
\end{aligned}$$

□

PSEUDO-RIEMANNIAN HYPERSURFACES

11.35 Let M^n be a pseudo-Riemannian hypersurface of a pseudo-Riemannian manifold N^{n+1}, with metrics g' and g of index k and r, respectively. Then the normal bundle $\widetilde{\pi}_2 : \mathscr{N}(M) \to M$ is 1-dimensional with index $l = r - k$ which can be 0 or 1. The number $\varepsilon = 1 - 2\,l$ is called the *sign* of the hypersurface M. A section s of $\widetilde{\pi}_2$ on an open subset V of M is called a *normal field* to M on V. It is called a *unit normal field* if $\|s(p)\| = 1$ for every $p \in V$.

11.36 Proposition Let $f : N \to \mathbf{R}$ be a smooth function on a pseudo-Riemannian manifold N^{n+1} with a metric g, and let $a \in \mathbf{R}$ be a point in the image of f. If $<\operatorname{grad} f, \operatorname{grad} f>$ is either positive or negative in a neighbourhood of $f^{-1}(a)$, then $M = f^{-1}(a)$ is a closed pseudo-Riemannian hypersurface of N with sign equal to the sign of $<\operatorname{grad} f, \operatorname{grad} f>$, and with $U = (\operatorname{grad} f / \|\operatorname{grad} f\|)|_M$ as a unit normal field.

PROOF : Since

$$df = (\operatorname{grad} f)^\flat \neq 0$$

in a neighbourhood of $f^{-1}(a)$, it follows from Proposition 2.41 that M is a closed hypersurface of N. If $i : M \to N$ is the inclusion map, then

$$<\operatorname{grad} f(p), i_*(v)> = df(p)(i_*(v)) = i^*(df)(p)(v) = d(f \circ i)(p)(v) = 0$$

for $p \in M$ and $v \in T_pM$, which shows that $\operatorname{grad} f(p) \notin i_*(T_pM)$. Hence it follows that i^*g is a metric on M, and that $(\operatorname{grad} f)|_M$ is a normal field to M. □

11.37 Let V be a vector space of dimension n with a metric g. Then we have a fibre metric h in TV defined by

$$h(p) = \omega_p^* g$$

for $p \in V$, where $\omega_p : T_pV \to V$ is the canonical identification given by $\omega_p = x^{-1} \circ t_{x,p}$ for any linear isomorphism $x : V \to \mathbf{R}^n$, as described in Lemma 2.84. Let $\mathscr{E} = \{e_1, ..., e_n\}$ be the standard basis for $\mathbf{R}^n$ and $\mathscr{B} = \{v_1, ..., v_n\}$ be a basis for V so that $x(v_i) = e_i$ for $i = 1, ..., n$. Then $\mathscr{B}^* = \{x^1, ..., x^n\}$ is the dual basis of $\mathscr{B}$. If

$$g = \sum_{i,j} g_{ij}\, x^i \otimes x^j \,,$$

then

$$h = \sum_{i,j} g_{ij}\, dx^i \otimes dx^j$$

since

$$h(p) \left(\left. \frac{\partial}{\partial x^i} \right|_p , \left. \frac{\partial}{\partial x^j} \right|_p \right) = g(v_i, v_j) = g_{ij}$$

for $p \in V$. By Proposition 10.166 it follows that all the Christoffel symbols are zero, so that the Levi–Civita connection on V is flat.

Now let $f : V \to \mathbf{R}$ be the smooth function defined by

$$f(p) = g(p, p)$$

for $p \in V$. Then we have that

$$f = <P, P>$$

where

$$P = \sum_{i=1}^{n} x^i \frac{\partial}{\partial x^i}$$

is the position vector field on V defined in Example 3.4, since

$$f(p) = g(p, p) = h(p)(\omega_p^{-1}(p), \omega_p^{-1}(p)) = h(p)(P(p), P(p))$$

for $p \in V$. By Corollary 10.114 and Proposition 10.166 we have that

$$\nabla_X P = X$$

so that

$$<\operatorname{grad} f, X> = X(f) = X<P, P> = 2<P, \nabla_X P> = 2<P, X>$$

for every vector field X on V. This shows that

$$\operatorname{grad} f = 2P$$

and

$$<\operatorname{grad} f, \operatorname{grad} f> = 4f\ .$$

If $a > 0$ and $\varepsilon = \pm 1$, it follows from Proposition 11.36 that $Q_{a,\varepsilon} = f^{-1}(\varepsilon a^2)$ is a closed pseudo-Riemannian hypersurface of V with sign ε and unit normal field $U = P'/a$ where $P' = P|_{Q_{a,\varepsilon}}$. The shape tensor is given by

$$\mathrm{II}(X,Y) = -\,\varepsilon <X,Y> U/a$$

for vector fields X and Y on $Q_{a,\varepsilon}$, since

$$<\mathrm{II}(X,Y),U> = <\eta_2 \circ \nabla'_X(\widetilde{i}_1 \circ Y),U> = <\nabla'_X(\widetilde{i}_1 \circ Y),P'>/a$$

$$= -<\widetilde{i}_1 \circ Y, \nabla'_X P'>/a = -<\widetilde{i}_1 \circ Y, \widetilde{i}_1 \circ X>/a = -<X,Y>/a$$

by 11.16 and Corollaries 11.19. Since V is flat, it follows from Corollary 11.33 that the sectional curvature K of $Q_{a,\varepsilon}$ is given by

$$K(W) = \varepsilon/a^2$$

for every $p \in Q_{a,\varepsilon}$ and every non-degenerate tangent plane W at p when $n \geq 3$.

11.38 Definition Let $f_k : \mathbf{R}^{n+1}_k \to \mathbf{R}$ be the smooth function on the pseudo-Euclidean space $\mathbf{R}^{n+1}_k$ defined by

$$f_k(p) = g^{n+1}_k(p,p)$$

for $p \in \mathbf{R}^{n+1}_k$, and let $a > 0$. Then the set

$$\mathbf{S}^n_k(a) = f_k^{-1}(a^2) = \{p \in \mathbf{R}^{n+1}_k \mid g^{n+1}_k(p,p) = a^2\}$$

is called the *pseudosphere* with radius a of dimension n and index k. It is a closed pseudo-Riemannian hypersurface of $\mathbf{R}^{n+1}_k$ with sign 1. The set

$$\mathbf{H}^n_k(a) = f_{k+1}^{-1}(-a^2) = \{p \in \mathbf{R}^{n+1}_{k+1} \mid g^{n+1}_{k+1}(p,p) = -a^2\}$$

is called the *pseudohyperboloid* with radius a of dimension n and index k. It is a closed pseudo-Riemannian hypersurface of $\mathbf{R}^{n+1}_{k+1}$ with sign -1. The sets $\mathbf{S}^n_0(a)$ and $\mathbf{H}^n_0(a)$ will also be denoted by $\mathbf{S}^n_a$ and $\mathbf{H}^n_a$, respectively.

11.39 Consider a vector space V of dimension n with a metric g, and let M be a pseudo-Riemannian hypersurface in V with sign ε. Let $Q_{1,\varepsilon}$ be the hypersurface in V defined in 11.37, and let $\widetilde{i} : M \to V$ and $\widehat{i} : Q_{1,\varepsilon} \to V$ be the inclusion maps. By Lemma 2.84 we have an equivalence $\omega : TV \to V \times V$ over V obtained from the canonical identification of the tangent space T_pV with V for each $p \in V$. It is given by $\omega = (id_V \times x^{-1}) \circ t_x$ for any linear isomorphism $x : V \to \mathbf{R}^n$. If Z is a unit normal field on an open subset U of M, we have a smooth map $\phi : U \to Q_{1,\varepsilon}$, called the *Gauss map* determined by Z, defined by

$$\widehat{i} \circ \phi = pr_2 \circ \omega \circ Z$$

where $pr_2 : V \times V \to V$ is the projection on the second factor. Since

$$\omega_{\phi(p)}^{-1} \circ \omega_p(Z_p) = \omega_{\phi(p)}^{-1}(\phi(p)) = P_{\phi(p)}$$

which is a unit normal to $Q_{1,\varepsilon}$ at $\phi(p)$, it follows that $\omega_{\phi(p)}^{-1} \circ \omega_p : T_pV \to T_{\phi(p)}V$ induces a linear isomorphism $\tau_p : T_pM \to T_{\phi(p)}\, Q_{1,\varepsilon}$ so that

$$\widehat{i}_{*\phi(p)} \circ \tau_p = \omega_{\phi(p)}^{-1} \circ \omega_p \circ \widetilde{i}_{*p}$$

for $p \in U$. Now let

$$v = \sum_{i=1}^{n} a^i \frac{\partial}{\partial x^i}\bigg|_p$$

be a vector in T_pM. Using that

$$Z = \sum_{i=1}^{n} (x^i \circ \phi) \frac{\partial}{\partial x^i},$$

it follows from Corollaries 11.19 and 10.114 that

$$-S_Z(p)(v) = \eta_1(\nabla'_v Z) = \eta_1\left(\sum_{i=1}^{n}\left(\sum_{j=1}^{n} a^j \frac{\partial(x^i \circ \phi)}{\partial x^j}(p)\right)\frac{\partial}{\partial x^i}\bigg|_p\right)$$

$$= \eta_1 \circ \omega_p^{-1} \circ \omega_{\phi(p)} \circ (\widehat{i} \circ \phi)_{*p}(v) = \tau_p^{-1} \circ \phi_{*p}(v),$$

which shows that

$$-S_Z(p) = \tau_p^{-1} \circ \phi_{*p}$$

for $p \in U$.

11.40 Consider a vector space V of dimension n with a Riemannian metric g, and let M be a hypersurface in V, and S_Z be the shape operator determined by a unit normal field Z on an open subset U of M. By 11.15 we know that $S_Z(p)$ is a self-adjoint linear operator on T_pM for each point $p \in U$. Hence there is an orthonormal basis $\{v_1, \ldots, v_{n-1}\}$ in T_pM consisting of eigenvectors for $S_Z(p)$ with eigenvalues $\lambda_1, \ldots, \lambda_{n-1}$, respectively. By Corollary 11.33 it follows that the sectional curvature of the tangent plane W to M at p, spanned by v_i and v_j where $i \neq j$, is given by

$$K(W) = <\mathrm{II}(p)(v_i,v_i), \mathrm{II}(p)(v_j,v_j)> - <\mathrm{II}(p)(v_i,v_j), \mathrm{II}(p)(v_i,v_j)>$$

$$= <S_Z(p)(v_i), v_i><S_Z(p)(v_j), v_j> - <S_Z(p)(v_i), v_j>^2 = \lambda_i \lambda_j.$$

The eigenvalues $\lambda_1, \ldots, \lambda_{n-1}$ are called the *principal curvatures* at p. If they are different, the corresponding eigenvectors define the *principal directions* at p. When $V = \mathbf{R}^3$ with the standard metric g, then $K(T_pM)$ is the Gaussian curvature at p defined in Definition 10.181, in agreement with our Definition in 1.8.

THE ROBERTSON–WALKER SPACETIME

11.41 Let M be a Lorentz manifold with a metric g so that

$$g|_V = -dt^2 + a^2(d\chi^2 + r^2(d\theta^2 + \sin^2\theta\, d\phi^2))$$

for a local chart (ψ, V) on M with coordinate functions t, χ, θ and ϕ, where $a = \tilde{a} \circ t$ and $r = \tilde{r} \circ \chi$ for real functions $\tilde{a}$ and $\tilde{r}$. This is the general form of a metric in a homogeneous and isotropic universe model using natural units. The local chart is assumed to be a *comoving coordinate system* so that each test particle has constant spatial coordinates. The coordinate t, called the *cosmic time*, is the proper time for each test particle. The partial derivatives $\frac{\partial a}{\partial t}$ and $\frac{\partial r}{\partial \chi}$ are denoted by $\dot{a}$ and r', respectively, and are given by $\dot{a} = \tilde{a}' \circ t$ and $r' = \tilde{r}' \circ \chi$. We see that

$$X_1 = \frac{\partial}{\partial t}\ , \quad X_2 = \frac{1}{a}\frac{\partial}{\partial \chi}\ , \quad X_3 = \frac{1}{ar}\frac{\partial}{\partial \theta} \quad \text{and} \quad X_4 = \frac{1}{ar\sin\theta}\frac{\partial}{\partial \phi}$$

form an orthonormal frame field on V with dual local basis

$$\tilde{\theta}^1 = dt\ , \quad \tilde{\theta}^2 = a\, d\chi\ , \quad \tilde{\theta}^3 = ar\, d\theta \quad \text{and} \quad \tilde{\theta}^4 = ar\sin\theta\, d\phi\ .$$

Since

$$\begin{aligned}
d\tilde{\theta}^1 &= 0\\
d\tilde{\theta}^2 &= \tfrac{\dot{a}}{a}\ \tilde{\theta}^1 \wedge \tilde{\theta}^2\\
d\tilde{\theta}^3 &= \tfrac{\dot{a}}{a}\ \tilde{\theta}^1 \wedge \tilde{\theta}^3 + \tfrac{r'}{ar}\ \tilde{\theta}^2 \wedge \tilde{\theta}^3\\
d\tilde{\theta}^4 &= \tfrac{\dot{a}}{a}\ \tilde{\theta}^1 \wedge \tilde{\theta}^4 + \tfrac{r'}{ar}\ \tilde{\theta}^2 \wedge \tilde{\theta}^4 + \tfrac{1}{ar}\cot\theta\ \tilde{\theta}^3 \wedge \tilde{\theta}^4\ ,
\end{aligned}$$

the non-zero coefficients b^i_{jk} and a^i_{jk} are

$$\begin{aligned}
b^i_{1i} &= -\, b^i_{i1} = \tfrac{\dot{a}}{a} \quad &&\text{for } i = 2,3,4\\
b^i_{2i} &= -\, b^i_{i2} = \tfrac{r'}{ar} \quad &&\text{for } i = 3,4\\
b^4_{34} &= -\, b^4_{43} = \tfrac{1}{ar}\cot\theta
\end{aligned}$$

and

$$\begin{aligned}
a^1_{ii} &= a^i_{1i} = \tfrac{1}{2}\,(b^1_{ii} - b^i_{i1} + b^i_{1i}) = \tfrac{\dot{a}}{a} \qquad &&\text{for } i = 2,3,4\\
a^2_{ii} &= -\, a^i_{2i} = \tfrac{1}{2}\,(b^2_{ii} + b^i_{i2} - b^i_{2i}) = -\tfrac{r'}{ar} \quad &&\text{for } i = 3,4\\
a^3_{44} &= -\, a^4_{34} = \tfrac{1}{2}\,(b^3_{44} + b^4_{43} - b^4_{34}) = -\tfrac{1}{ar}\cot\theta
\end{aligned}$$

This shows that

$$\widetilde{\omega}^1_2 = \widetilde{\omega}^2_1 = \frac{\dot{a}}{a}\ \widetilde{\theta}^{\,2} = \dot{a}\, d\chi$$

$$\widetilde{\omega}^1_3 = \widetilde{\omega}^3_1 = \frac{\dot{a}}{a}\ \widetilde{\theta}^{\,3} = \dot{a} r\, d\theta$$

$$\widetilde{\omega}^1_4 = \widetilde{\omega}^4_1 = \frac{\dot{a}}{a}\ \widetilde{\theta}^{\,4} = \dot{a} r \sin\theta\, d\phi$$

$$\widetilde{\omega}^2_3 = -\,\widetilde{\omega}^3_2 = -\frac{r'}{ar}\ \widetilde{\theta}^{\,3} = -\,r'\, d\theta$$

$$\widetilde{\omega}^2_4 = -\,\widetilde{\omega}^4_2 = -\frac{r'}{ar}\ \widetilde{\theta}^{\,4} = -\,r' \sin\theta\, d\phi$$

$$\widetilde{\omega}^3_4 = -\,\widetilde{\omega}^4_3 = -\frac{1}{ar}\cot\theta\ \widetilde{\theta}^{\,4} = -\,\cos\theta\, d\phi$$

so that

$$\begin{aligned}\widetilde{\Omega}^1_2 = \widetilde{\Omega}^2_1 &= d\widetilde{\omega}^1_2 + \widetilde{\omega}^1_3 \wedge \widetilde{\omega}^3_2 + \widetilde{\omega}^1_4 \wedge \widetilde{\omega}^4_2\\ &= \ddot{a}\, dt \wedge d\chi = \frac{\ddot{a}}{a}\ \widetilde{\theta}^{\,1} \wedge \widetilde{\theta}^{\,2}\end{aligned}$$

$$\begin{aligned}\widetilde{\Omega}^1_3 = \widetilde{\Omega}^3_1 &= d\widetilde{\omega}^1_3 + \widetilde{\omega}^1_2 \wedge \widetilde{\omega}^2_3 + \widetilde{\omega}^1_4 \wedge \widetilde{\omega}^4_3\\ &= \ddot{a} r\, dt \wedge d\theta + \dot{a} r'\, d\chi \wedge d\theta - \dot{a} r'\, d\chi \wedge d\theta = \frac{\ddot{a}}{a}\ \widetilde{\theta}^{\,1} \wedge \widetilde{\theta}^{\,3}\end{aligned}$$

$$\begin{aligned}\widetilde{\Omega}^1_4 = \widetilde{\Omega}^4_1 &= d\widetilde{\omega}^1_4 + \widetilde{\omega}^1_2 \wedge \widetilde{\omega}^2_4 + \widetilde{\omega}^1_3 \wedge \widetilde{\omega}^3_4\\ &= \ddot{a} r \sin\theta\, dt \wedge d\phi + \dot{a} r' \sin\theta\, d\chi \wedge d\phi + \dot{a} r \cos\theta\, d\theta \wedge d\phi\\ &\quad - \dot{a} r' \sin\theta\, d\chi \wedge d\phi - \dot{a} r \cos\theta\, d\theta \wedge d\phi = \frac{\ddot{a}}{a}\ \widetilde{\theta}^{\,1} \wedge \widetilde{\theta}^{\,4}\end{aligned}$$

$$\begin{aligned}\widetilde{\Omega}^2_3 = -\,\widetilde{\Omega}^3_2 &= d\widetilde{\omega}^2_3 + \widetilde{\omega}^2_1 \wedge \widetilde{\omega}^1_3 + \widetilde{\omega}^2_4 \wedge \widetilde{\omega}^4_3\\ &= -\,r''\, d\chi \wedge d\theta + \dot{a}^2 r\, d\chi \wedge d\theta = \left(\frac{\dot{a}^2}{a^2} - \frac{r''}{ra^2}\right) \widetilde{\theta}^{\,2} \wedge \widetilde{\theta}^{\,3}\end{aligned}$$

$$\begin{aligned}\widetilde{\Omega}^2_4 = -\,\widetilde{\Omega}^4_2 &= d\widetilde{\omega}^2_4 + \widetilde{\omega}^2_1 \wedge \widetilde{\omega}^1_4 + \widetilde{\omega}^2_3 \wedge \widetilde{\omega}^3_4\\ &= -\,r'' \sin\theta\, d\chi \wedge d\phi - r' \cos\theta\, d\theta \wedge d\phi + \dot{a}^2 r \sin\theta\, d\chi \wedge d\phi\\ &\quad + r' \cos\theta\, d\theta \wedge d\phi = \left(\frac{\dot{a}^2}{a^2} - \frac{r''}{ra^2}\right) \widetilde{\theta}^{\,2} \wedge \widetilde{\theta}^{\,4}\end{aligned}$$

$$
\begin{aligned}
\widetilde{\Omega}_4^3 = -\widetilde{\Omega}_3^4 &= d\widetilde{\omega}_4^3 + \widetilde{\omega}_1^3 \wedge \widetilde{\omega}_4^1 + \widetilde{\omega}_2^3 \wedge \widetilde{\omega}_4^2 \\
&= \sin\theta\; d\theta \wedge d\phi + \dot{a}^2 r^2 \sin\theta\; d\theta \wedge d\phi - (r')^2 \sin\theta\; d\theta \wedge d\phi \\
&= \left(\frac{\dot{a}^2}{a^2} + \frac{1}{r^2 a^2} - \frac{(r')^2}{r^2 a^2}\right) \widetilde{\theta}^{\,3} \wedge \widetilde{\theta}^{\,4} \\
&= \left\{\left(\frac{\dot{a}^2}{a^2} - \frac{r''}{r a^2}\right) + \frac{1}{a^2}\left\{\frac{1}{r^2} + \left(\frac{r'}{r}\right)'\right\}\right\} \widetilde{\theta}^{\,3} \wedge \widetilde{\theta}^{\,4} \\
&= \left\{\left(\frac{\dot{a}^2}{a^2} - \frac{r''}{r a^2}\right) - \frac{1}{2a^2}\left(\frac{r}{r'}\right)\left\{\frac{1}{r^2} - \left(\frac{r'}{r}\right)^2\right\}'\right\} \widetilde{\theta}^{\,3} \wedge \widetilde{\theta}^{\,4}
\end{aligned}
$$

Because of the isotropy we must have that

$$
\frac{1}{r^2} - \left(\frac{r'}{r}\right)^2 = k
$$

for a constant k, so that

$$
r'^{\,2} = 1 - kr^2. \tag{1}
$$

Introducing the local chart (ρ, V) on M with coordinate functions t, r, θ and ϕ, we now havc that

$$
g|_V = -dt^2 + a^2\left(\frac{dr^2}{1-kr^2} + r^2(d\theta^2 + \sin^2\theta\; d\phi^2)\right)
$$

which is the *Robertson–Walker* form of the metric. Rescaling r and a by the factors $|k|^{1/2}$ and $|k|^{-1/2}$, respectively, when $k \neq 0$, we may assume that k has the values 1, 0 or -1, and the solution of the differential Equation (1) with initial conditions $\tilde{r}(0) = 0$ and $\tilde{r}'(0) \geq 0$ is given by

$$
r = \begin{cases} \sin\chi & \text{for } k = 1 \quad , \quad 0 < \chi < \pi \\ \chi & \text{for } k = 0 \quad , \quad 0 < \chi < \infty \\ \sinh\chi & \text{for } k = -1 \quad , \quad 0 < \chi < \infty \end{cases} .
$$

The induced metric $\widetilde{g}$ on the simultaneity space $V_{t_0} = \psi^{-1}(\{t_0\} \times \mathbf{R}^3)$ at the time $t = t_0$ is given by

$$
\widetilde{g} = \widetilde{a}(t_0)^2(d\chi^2 + r^2(d\theta^2 + \sin^2\theta\; d\phi^2))
$$

When $k = 0$ we see from Example 7.21 that this is the Euclidean metric on $\mathbf{R}^3$ after a rescaling of the radial coordinate χ by the factor $\widetilde{a}(t_0)$. When $k = 1$ and $k = -1$ it is the metric of the 3-sphere $\mathbf{S}^3_{\widetilde{a}(t_0)}$ with radius $\widetilde{a}(t_0)$ in the Euclidean space $\mathbf{R}^4$, and of the upper sheet of the 3-hyperboloid $\mathbf{H}^3_{\widetilde{a}(t_0)}$ in the Minkowski space $\mathbf{R}^4_1$, respectively, as we see from Examples 7.22 and 7.23. By 11.37 it follows that the sectional curvature of V_{t_0}, called the *spatial curvature* at the time t_0, is $K = k/\widetilde{a}(t_0)^2$. The three cases $k = 1$, $k = 0$ and $k = -1$ with positive, zero and negative spatial curvatures are called *closed*, *flat* and *open* universe models, respectively.

THE FRIEDMANN COSMOLOGICAL MODELS

11.42 The dynamics of the Robertson–Walker spacetime is determined by the scale factor $\widetilde{a}(t)$ which can be found by solving Einstein's field equation. By Corollary 10.106 the non-zero components of the curvature form R are

$$R^1_{i1i} = -R^1_{ii1} = R^i_{11i} = -R^i_{1i1} = \frac{\ddot{a}}{a} \quad \text{for} \quad i = 2,3,4$$

$$R^j_{iji} = -R^j_{iij} = \frac{\dot{a}^2+k}{a^2} \quad \text{for} \quad i \neq j \in \{2,3,4\}$$

Using the components of R with equal contravariant and second covariant index, we find the non-zero components of the Ricci tensor

$$\mathrm{Ric}_{11} = -\frac{3\ddot{a}}{a}$$

$$\mathrm{Ric}_{ii} = \frac{\ddot{a}}{a} + 2\,\frac{\dot{a}^2+k}{a^2} \quad \text{for} \quad i = 2,3,4$$

and the scalar curvature

$$S = -\mathrm{Ric}_{11} + \mathrm{Ric}_{22} + \mathrm{Ric}_{33} + \mathrm{Ric}_{44} = \frac{6\ddot{a}}{a} + 6\,\frac{\dot{a}^2+k}{a^2}$$

From these we obtain the non-zero components of the Einstein tensor

$$G_{11} = \mathrm{Ric}_{11} + \frac{1}{2}\,S = 3\,\frac{\dot{a}^2+k}{a^2}$$

$$G_{ii} = \mathrm{Ric}_{ii} - \frac{1}{2}\,S = -2\,\frac{\ddot{a}}{a} - \frac{\dot{a}^2+k}{a^2} \quad \text{for} \quad i = 2,3,4$$

We assume that the universe is a perfect fluid with energy-momentum tensor given by

$$T = (\rho + p)\, u^\flat \otimes u^\flat + p\,g\,,$$

as described in Remark 10.187. Since the universe is homogeneous and isotropic, the energy density ρ and the pressure p can only be functions of time, and the spatial components of the flow vector field u must be zero. This means that $\rho = \tilde{\rho} \circ t$ and $p = \tilde{p} \circ t$ for real functions $\tilde{\rho}$ and $\tilde{p}$, and $u = X_1$ so that $u^\flat = -\widetilde{\theta}^{\,1}$. Hence the non-zero components of the energy-momentum tensor are

$$T_{11} = \rho \quad \text{and} \quad T_{ii} = p \quad \text{for} \quad i = 2,3,4.$$

From the Einstein field equation we now obtain the *Friedmann equations*

$$3\,\frac{\dot{a}^2+k}{a^2} = \kappa\rho + \Lambda \quad \text{and} \quad -2\,\frac{\ddot{a}}{a} - \frac{\dot{a}^2+k}{a^2} = \kappa p - \Lambda\,.$$

Inserting the first Friedmann equation into the second we obtain

$$\frac{\ddot{a}}{a} = -\frac{\kappa}{6}\,(\rho + 3\,p) + \frac{\Lambda}{3}. \tag{1}$$

Multipying the first Friedmann equation with $a^2/\,3$, taking the derivative with respect to t and dividing by $2a\dot{a}$ yields

$$\frac{\ddot{a}}{a} = \frac{\kappa}{6a}\,\frac{d}{da}(\rho\,a^2) + \frac{\Lambda}{3}. \tag{2}$$

Now subtracting Equation (1) from (2) and multipying with $6a^2/\,\kappa$, we have that

$$a\,\frac{d}{da}(\rho\,a^2) + (\rho + 3\,p)\,a^2 = 0$$

which implies that

$$\frac{\partial}{\partial t}(\rho\,a^3) + p\,\frac{\partial}{\partial t}(a^3) = 0. \tag{3}$$

Since a^3 and $\rho\,a^3$ are proportional to the volume V and the energy U of any fluid element, this equation means that

$$dU + p\,dV = 0$$

which is the first law of thermodynamics for an adiabatic process. This is in agreement with the homogeneity and isotropy of our universe model which does not allow any temperature gradient or heat flow.

We now assume that the perfect fluid obeys the barotropic equation of state

$$p = w\,\rho$$

where w is a constant. For dust, radiation and Lorentz invariant vacuum energy the values of w are 0, $1/3$ and -1, respectively. By dust we mean a set of collisionless, nonrelativistic particles with vanishing pressure, and by radiation we mean either electromagnetic radiation or massive particles with relative velocities close to the speed of light. Multiplying Equation (3) with $(a^3)^w$ we have that

$$\frac{\partial}{\partial t}\{\rho\,a^{3\,(w+1)}\} = 0\,,$$

which implies that

$$\rho = \rho_0\left(\frac{a}{a_0}\right)^{-3\,(w+1)} \tag{4}$$

for constants ρ_0 and a_0. We see that ρ_0 is the energy density of the fluid at the time t_0 when $\tilde{a}(t_0) = a_0$. We let t_0 be the present time, and ρ_0 and a_0 be the present values of the energy density and the scale factor. If ρ_{m0}, ρ_{r0} and ρ_{v0} are the present energy densities of non-interacting dust, radiation and Lorentz invariant vacuum energy, we have that

$$\rho = \rho_{m0}\left(\frac{a}{a_0}\right)^{-3} + \rho_{r0}\left(\frac{a}{a_0}\right)^{-4} + \rho_{v0}. \tag{5}$$

The energy density of dust falls off as a^{-3} due to the decrease in number density of particles as the universe expands. The energy density of radiation falls off faster as a^{-4} because of the additional loss of energy by a factor a^{-1} due to the redshift of each photon. This is also the case for massive relativistic particles which behave like a gas of photons, as their rest energy can be considered to be negligible compaired to their total energy. From the Friedmann equations we see that the cosmological constant Λ can also be interpreted as Lorentz invariant vacuum energy, and can therefore be removed from the equations by including the term Λ/κ in ρ_{v0}.

To find the time evolution of the scale factor a we now use the first Friedmann equation with $\Lambda = 0$, which implies that

$$\left(\frac{\dot{a}}{a}\right)^2 = \frac{\kappa}{3}\rho - \frac{k}{a^2}. \tag{6}$$

If ρ satisfies Equation (4) with $w > -1/3$, the fate of the universe is determined by k. When $k = 0$ or $k = -1$, it will expand forever, but when $k = 1$ it will stop expanding and recollapse to a Big Crunch. If $w < -1/3$, the universe will expand forever for all values of k.

Using the metric g we see that the instantaneous physical distance l from an observer at $\chi = 0$ to a galaxy with radial coordinate $\chi = \chi_0$ is given by $l = a\,\chi_0$. The velocity of the galaxy relative to the observer due to the expansion or contraction of the universe is

$$\frac{dl}{dt} = \frac{\dot{a}}{a}\, a\chi_0 = Hl, \tag{7}$$

where $H = \dot{a}/a$ is called the *Hubble parameter*. Its present value, called the *Hubble constant*, is approximately $H_0 = 20\,(\mathrm{km/s})/\mathrm{Mly}$. Equation (7) is called *Hubble's law*, stating that the velocity of a galaxy is proportional to its distance. The time $t_H = 1/H$ is called the *Hubble age* of the universe. It coincides with the age t of the universe if $\dot{a}$ is constant, which means that the velocity of each fluid particle comoving with the expansion of the universe relative to an observer at $\chi = 0$ is constant.

The energy density ρ_c in a flat universe model, called the *critical energy density*, is given by

$$H^2 = \frac{\kappa}{3}\rho_c$$

which is obtained from Equation (6) with $k = 0$. Its present value is approximately $\rho_{c0} = 2 \cdot 10^{-26}\,\mathrm{kg/m^3}$. Introducing the *relative energy density* $\Omega = \rho/\rho_c$ of a universe model, we see from Equation (6) that its spatial curvature is

$$\frac{k}{a^2} = H^2(\Omega - 1).$$

Hence the universe model is closed if $\Omega > 1$, flat if $\Omega = 1$ and open if $\Omega < 1$. The curvature radius is

$$a = \frac{1}{H}\sqrt{\frac{k}{\Omega - 1}}$$

when $\Omega \neq 1$. We denote the present relative energy densities of non-interacting dust, radiation and vacuum by Ω_{m0}, Ω_{r0} and Ω_{v0}, respectively.

11.43 Universe models with dust and radiation We introduce a new time coordinate η, called the *parametric time*, so that $\eta = \tilde{\eta} \circ t$ for a real function $\tilde{\eta}$ with $\tilde{\eta}' = 1/\tilde{a}$ and $\tilde{\eta}(0) = 0$. Then we have that

$$\frac{\partial a}{\partial \eta} = a\dot{a}\,.$$

In a universe model with dust and radiation it follows from (5) and (6) in 11.42 that

$$\left(\frac{\partial a}{\partial \eta}\right)^2 = \frac{\kappa}{3}\, a_0^3\,(\rho_{m0}\, a + \rho_{r0}\, a_0) - k a^2\,,$$

which can be written as

$$\left(\frac{\partial a}{\partial \eta}\right)^2 = 2\,\alpha\, a + \beta^2 - k a^2 \tag{1}$$

where

$$\alpha = \frac{\kappa}{6}\,\rho_{m0}\, a_0^3 = \frac{1}{2}\,\Omega_{m0}\, H_0^2\, a_0^3 \quad \text{and} \quad \beta = \left(\frac{\kappa}{3}\,\rho_{r0}\right)^{1/2} a_0^2 = \Omega_{r0}^{1/2}\, H_0\, a_0^2\,.$$

When $k \neq 0$, Equation (1) can also be written as

$$\left(\frac{\partial a}{\partial \eta}\right)^2 = \gamma - k\left(a - \frac{\alpha}{k}\right)^2 \tag{2}$$

where $\gamma = \dfrac{\alpha^2}{k} + \beta^2$. Now let $\varepsilon = |\gamma|$ when $\gamma \neq 0$, and let $\varepsilon = 1$ when $\gamma = 0$. Introducing the function $b = \varepsilon^{-1/2}\left(a - \dfrac{\alpha}{k}\right)$, we have that

$$\left(\frac{\partial b}{\partial \eta}\right)^2 = \sigma - k b^2 \tag{3}$$

where $\sigma = \gamma/\,\varepsilon$. When $k = 1$, the solution of this differential equation is

$$b = \sin(\eta + c)$$

so that

$$a = \alpha + \varepsilon^{1/2}\sin(\eta + c)\,,$$

where c is an integration constant. The initial conditions $\widetilde{a}(0) = 0$ and $\widetilde{a}'(0) \geq 0$ imply that

$$\sin c = -\,\varepsilon^{-1/2}\,\alpha \quad \text{and} \quad \cos c = \varepsilon^{-1/2}\,\beta\,,$$

which give the solution

$$a = \alpha(1 - \cos\eta) + \beta \sin\eta$$
$$t = \alpha(\eta - \sin\eta) + \beta(1 - \cos\eta)$$

When $k = -1$, the solution of the differential Equation (3) is

$$b = \frac{1}{2}\left(e^{\eta+c} - \sigma e^{-(\eta+c)}\right)$$

so that

$$a = -\alpha + \varepsilon^{1/2}\left\{\frac{1}{2}\left(e^{c} - \sigma e^{-c}\right)\cosh\eta + \frac{1}{2}\left(e^{c} + \sigma e^{-c}\right)\sinh\eta\right\},$$

where c is an integration constant. The initial conditions $\widetilde{a}(0) = 0$ and $\widetilde{a}'(0) \geq 0$ imply that

$$\frac{1}{2}\left(e^{c} - \sigma e^{-c}\right) = \varepsilon^{-1/2}\alpha \quad \text{and} \quad \frac{1}{2}\left(e^{c} + \sigma e^{-c}\right) = \varepsilon^{-1/2}\beta,$$

which give the solution

$$\begin{aligned} a &= \alpha(\cosh\eta - 1) + \beta\sinh\eta \\ t &= \alpha(\sinh\eta - \eta) + \beta(\cosh\eta - 1) \end{aligned}$$

When $k = 0$, the solution of Equation (1) with initial condition $\widetilde{a}(0) = 0$ is

$$\begin{aligned} a &= \frac{1}{2}\alpha\eta^2 + \beta\eta \\ t &= \frac{1}{6}\alpha\eta^3 + \frac{1}{2}\beta\eta^2 \end{aligned}$$

11.44 Vacuum dominated universe models We now consider universe models with Lorentz invariant vacuum energy corresponding to a cosmological constant $\Lambda > 0$. It follows from (5) and (6) in 11.42 that

$$\dot{a}^2 = \frac{\kappa}{3}\rho_{v0}\,a^2 - k,$$

which can be written as

$$\dot{a}^2 = \omega^2 a^2 - k \tag{1}$$

where

$$\omega = \left(\frac{\kappa}{3}\rho_{v0}\right)^{1/2} = \Omega_{v0}^{1/2} H_0 = (\Lambda/3)^{1/2} = 1/R_d. \tag{2}$$

A solution in this case is

$$a = \begin{cases} \frac{1}{\omega}\cosh\omega t & \text{for} \quad k = 1 \quad , \quad -\infty < t < \infty \\ \frac{1}{\omega}e^{\omega t} & \text{for} \quad k = 0 \quad , \quad -\infty < t < \infty \\ \frac{1}{\omega}\sinh\omega t & \text{for} \quad k = -1 \quad , \quad 0 < t < \infty \end{cases}.$$

Note that the initial condition $\widetilde{a}(0) = 0$ can only be applied when $k = -1$. When $k = 0$ and $k = 1$ the universe is infinitely old without a Big Bang singularity.

Using the comoving coordinate system (ψ, V) described in 11.41 with coordinate functions t, χ, θ and ϕ we have that

$$g|_V = -dt^2 + a^2(d\chi^2 + r^2(d\theta^2 + \sin^2\theta\, d\phi^2)), \tag{3}$$

where r is the function on V given by

$$r = \begin{cases} \sin\chi & \text{for } k=1 \quad , \quad 0<\chi<\pi \\ \chi & \text{for } k=0 \quad , \quad 0<\chi<\infty \\ \sinh\chi & \text{for } k=-1 \quad , \quad 0<\chi<\infty \end{cases}$$

satisfying the relation

$$r'^2 = 1 - kr^2. \tag{4}$$

For each k we are going to find an isometric embedding $\alpha : M \to \mathbf{R}^5_1$ of M as either the entire or a part of the 4-pseudosphere

$$\mathbf{S}^4_1(R_d) = \{(v,w,x,y,z) \in \mathbf{R}^5 | -v^2 + w^2 + x^2 + y^2 + z^2 = R_d^2\}$$

with radius R_d in the pseudo-Euclidean space $\mathbf{R}^5_1$. This also shows that the spacetime M actually is the same as either the entire or a part of the extended de Sitter spacetime described in 10.189 and 10.194.

We first consider the cases where $k \neq 0$. Using (1), (4) and the relations $\ddot{a} = \omega^2 a$ and $r'' = -kr$ we have that

$$\begin{aligned} -dt^2 + a^2 d\chi^2 &= \frac{1}{k}\,[\,\dot{a}^2(r'^2 + kr^2) - \omega^2 a^2]\, dt^2 + a^2(r'^2 + kr^2)\, d\chi^2 \\ &= \frac{1}{k}\,(\dot{a}r'dt - kard\chi)^2 - \frac{1}{k}\,\omega^2 a^2 dt^2 + (\dot{a}rdt + ar'd\chi)^2 \\ &= \frac{1}{k\,\omega^2}\,[\,d(\omega ar')^2 - d\dot{a}^2] + d(ar)^2 \end{aligned}$$

and

$$\begin{aligned} &\frac{1}{k\,\omega^2}\,[\,(\omega ar')^2 - \dot{a}^2] + (ar)^2 \\ &= \frac{1}{k\,\omega^2}\,[\,\omega^2 a^2(1-kr^2) - (\omega^2 a^2 - k)] + (ar)^2 = \frac{1}{\omega^2} = R_d^2\,. \end{aligned}$$

Hence there is an isometric embedding $\alpha : M \to \mathbf{R}^5_1$ given by

$$\begin{aligned} v\circ\alpha|_V &= \begin{cases} \dot{a}/\omega & \text{when } k=1 \\ ar' & \text{when } k=-1 \end{cases} \\ w\circ\alpha|_V &= \begin{cases} ar' & \text{when } k=1 \\ \dot{a}/\omega & \text{when } k=-1 \end{cases} \\ x\circ\alpha|_V &= ar\sin\theta\,\cos\phi \\ y\circ\alpha|_V &= ar\sin\theta\,\sin\phi \\ z\circ\alpha|_V &= ar\cos\theta \end{aligned} \tag{5}$$

on the coordinate domain V.

When $k = 1$ we have that $\Omega_{v0} > 1$ so that $R_d = (3/\Lambda)^{1/2} < 1/H_0$ according to (2). The isometric embedding $\alpha : M \to \mathbf{R}^5_1$ is given by

$$\alpha(t,\chi,\mathbf{u}) = R_d\,(\sinh(t/R_d),\, \cosh(t/R_d)\,\cos\chi,\, \cosh(t/R_d)\,\sin\chi\;\mathbf{u}) \qquad (6)$$

for $(t,\chi,\mathbf{u}) \in \mathbf{R} \times [0,\pi] \times \mathbf{S}^2$, mapping M onto the entire pseudosphere $\mathbf{S}^4_1(R_d)$. On the coordinate domain V we have that

$$\begin{aligned}
v \circ \alpha|_V &= R_d \sinh(t/R_d)\\
w \circ \alpha|_V &= R_d \cosh(t/R_d)\,\cos\chi\\
x \circ \alpha|_V &= R_d \cosh(t/R_d)\,\sin\chi\,\sin\theta\,\cos\phi\\
y \circ \alpha|_V &= R_d \cosh(t/R_d)\,\sin\chi\,\sin\theta\,\sin\phi\\
z \circ \alpha|_V &= R_d \cosh(t/R_d)\,\sin\chi\,\cos\theta
\end{aligned}$$

where $-\infty < t < \infty$, $0 < \chi < \pi$, $0 < \theta < \pi$ and $-\pi < \phi < \pi$. We can also verify directly that α is an isometry in this case. Using Proposition 4.18 we have that

$$\begin{aligned}
\alpha^* dv|_V &= \cosh(t/R_d)\,dt\\
\alpha^* dw|_V &= \sinh(t/R_d)\cos\chi\,dt - R_d\cosh(t/R_d)\sin\chi\,d\chi\\
\alpha^* dx|_V &= \sinh(t/R_d)\sin\chi\,\sin\theta\,\cos\phi\,dt + R_d\cosh(t/R_d)\cos\chi\,\sin\theta\,\cos\phi\,d\chi\\
&\quad + R_d\cosh(t/R_d)\sin\chi\,\cos\theta\,\cos\phi\,d\theta - R_d\cosh(t/R_d)\sin\chi\,\sin\theta\,\sin\phi\,d\phi\\
\alpha^* dy|_V &= \sinh(t/R_d)\sin\chi\,\sin\theta\,\sin\phi\,dt + R_d\cosh(t/R_d)\cos\chi\,\sin\theta\,\sin\phi\,d\chi\\
&\quad + R_d\cosh(t/R_d)\sin\chi\,\cos\theta\,\sin\phi\,d\theta + R_d\cosh(t/R_d)\sin\chi\,\sin\theta\,\cos\phi\,d\phi\\
\alpha^* dz|_V &= \sinh(t/R_d)\sin\chi\,\cos\theta\,dt + R_d\cosh(t/R_d)\cos\chi\,\cos\theta\,d\chi\\
&\quad - R_d\cosh(t/R_d)\sin\chi\,\sin\theta\,d\theta.
\end{aligned}$$

The Lorentz metric on $\mathbf{R}^5_1$ is given by

$$g^5_1 = -\,dv^2 + dw^2 + dx^2 + dy^2 + dz^2$$

in cartesian coordinates. Hence we have that

$$\alpha^* g^5_1|_V = -\,dt^2 + R_d^2\cosh^2(t/R_d)\,[\,d\chi^2 + \sin^2\chi\,(d\theta^2 + \sin^2\theta\;d\phi^2)\,]\,.$$

The simultaneity space V_{t_0} at the time $t = t_0$ is the intersection between the pseudosphere $\mathbf{S}^4_1(R_d)$ and the hyperplane $v = R_d\sinh(t_0/R_d)$. This intersection is a 3-sphere $\mathbf{S}^3_{a_0}$ with radius $a_0 = R_d\cosh(t_0/R_d)$ in the Euclidean space $\mathbf{R}^4$ as described in 11.41 and Example 7.22. The radius initially shrinks until it reaches a minimal value R_d at $t_0 = 0$, and then it increases, describing a "bouncing universe" without a Big Bang.

When $k = -1$ we have that $\Omega_{v0} < 1$ so that $R_d = (3/\Lambda)^{1/2} > 1/H_0$ according to (2). The isometric embedding $\alpha : M \to \mathbf{R}^5_1$ is given by

$$\alpha(t,\chi,\mathbf{u}) = R_d\,(\sinh(t/R_d)\,\cosh\chi,\, \cosh(t/R_d),\, \sinh(t/R_d)\,\sinh\chi\;\mathbf{u}) \qquad (7)$$

for $(t,\chi,\mathbf{u}) \in (0,\infty)\times[0,\infty)\times\mathbf{S}^2$, mapping M onto the open submanifold

$$\{(v,w,x,y,z)\in\mathbf{S}^4_1(R_d)\mid v>0, w>R_d\}$$

of the pseudosphere. On the coordinate domain V we have that

$$\begin{aligned}
v\circ\alpha|_V &= R_d\sinh(t/R_d)\cosh\chi\\
w\circ\alpha|_V &= R_d\cosh(t/R_d)\\
x\circ\alpha|_V &= R_d\sinh(t/R_d)\sinh\chi\,\sin\theta\,\cos\phi\\
y\circ\alpha|_V &= R_d\sinh(t/R_d)\sinh\chi\,\sin\theta\,\sin\phi\\
z\circ\alpha|_V &= R_d\sinh(t/R_d)\sinh\chi\,\cos\theta
\end{aligned}$$

where $0<t<\infty, 0<\chi<\infty, 0<\theta<\pi$ and $-\pi<\phi<\pi$. In this case we can also verify directly that α is an isometry since

$$\begin{aligned}
\alpha^*dv|_V &= \cosh(t/R_d)\cosh\chi\,dt+R_d\sinh(t/R_d)\sinh\chi\,d\chi\\
\alpha^*dw|_V &= \sinh(t/R_d)\,dt\\
\alpha^*dx|_V &= \cosh(t/R_d)\sinh\chi\,\sin\theta\,\cos\phi\,dt+R_d\sinh(t/R_d)\cosh\chi\,\sin\theta\,\cos\phi\,d\chi\\
&\quad+R_d\sinh(t/R_d)\sinh\chi\,\cos\theta\,\cos\phi\,d\theta-R_d\sinh(t/R_d)\sinh\chi\,\sin\theta\,\sin\phi\,d\phi\\
\alpha^*dy|_V &= \cosh(t/R_d)\sinh\chi\,\sin\theta\,\sin\phi\,dt+R_d\sinh(t/R_d)\cosh\chi\,\sin\theta\,\sin\phi\,d\chi\\
&\quad+R_d\sinh(t/R_d)\sinh\chi\,\cos\theta\,\sin\phi\,d\theta+R_d\sinh(t/R_d)\sinh\chi\,\sin\theta\,\cos\phi\,d\phi\\
\alpha^*dz|_V &= \cosh(t/R_d)\sinh\chi\,\cos\theta\,dt+R_d\sinh(t/R_d)\cosh\chi\,\cos\theta\,d\chi\\
&\quad-R_d\sinh(t/R_d)\sinh\chi\,\sin\theta\,d\theta.
\end{aligned}$$

Hence we have that

$$\alpha^*g^5_1|_V = -dt^2+R_d^2\sinh^2(t/R_d)\,[\,d\chi^2+\sinh^2\chi\,(d\theta^2+\sin^2\theta\;d\phi^2)\,]\,.$$

The simultaneity space V_{t_0} at the time $t=t_0$ is the intersection between the pseudosphere $\mathbf{S}^4_1(R_d)$ and the hyperplane $w=R_d\cosh(t_0/R_d)$. This intersection is the upper sheet of the 3-hyperboloid $\mathbf{H}^3_{a_0}$ with radius $a_0=R_d\sinh(t_0/R_d)$ in the Minkowski space $\mathbf{R}^4_1$ as described in 11.41 and Example 7.23.

We finally consider the case where $k=0$. Using the relations $\dot a=\omega a$ and $r'=1$ we have that

$$\begin{aligned}
&-dt^2+a^2d\chi^2\\
&=\dot a\,dt\,(-\dot a^{-2}\cdot\dot a\,dt-\dot a r^2\,dt-2ar\,d\chi)+(\dot a^2r^2\,dt^2+2a\dot a ar\,dt\,d\chi+a^2\,d\chi^2)\\
&=\frac{1}{\omega^2}\,d\dot a\,d(\dot a^{-1}-\dot a r^2)+(\dot a r\,dt+a\,d\chi)^2\\
&=\frac{1}{4\,\omega^2}\,\{-d[\dot a-(\dot a^{-1}-\dot a r^2)]^2+d[\dot a+(\dot a^{-1}-\dot a r^2)]^2\}+d(ar)^2
\end{aligned}$$

and

$$\frac{1}{4\,\omega^2}\{-[\dot a-(\dot a^{-1}-\dot a r^2)]^2+[\dot a+(\dot a^{-1}-\dot a r^2)]^2\}+(ar)^2$$
$$=\frac{1}{\omega^2}\,\dot a(\dot a^{-1}-\dot a r^2)+(ar)^2=\frac{1}{\omega^2}=R_d^2\,.$$

Hence there is an isometric embedding $\alpha : M\to \mathbf{R}^5_1$ given by

$$\begin{aligned} v\circ\alpha|_V &= \frac{1}{2\,\omega}\,[\dot a-(\dot a^{-1}-\dot a r^2)]\\ w\circ\alpha|_V &= \frac{1}{2\,\omega}\,[\dot a+(\dot a^{-1}-\dot a r^2)]\\ x\circ\alpha|_V &= ar\sin\theta\,\cos\phi\\ y\circ\alpha|_V &= ar\sin\theta\,\sin\phi\\ z\circ\alpha|_V &= ar\cos\theta \end{aligned} \tag{8}$$

which also implies that

$$\begin{aligned} (v+w)\circ\alpha|_V &= \dot a/\omega\\ (v-w)\circ\alpha|_V &= (\dot a r^2-\dot a^{-1})/\omega \end{aligned}$$

on the coordinate domain V.

When $k=0$ we have that $\Omega_{v0}=1$ so that $R_d=(3/\Lambda)^{1/2}=1/H_0$ according to (2). The isometric embedding $\alpha : M\to \mathbf{R}^5_1$ is given by

$$\alpha(t,\chi,\mathbf{u})=R_d\,(\sinh(t/R_d)+\tfrac{1}{2}\,\chi^2\,e^{t/R_d}\,,\cosh(t/R_d)-\tfrac{1}{2}\,\chi^2\,e^{t/R_d}\,,e^{t/R_d}\,\chi\,\mathbf{u}) \tag{9}$$

for $(t,\chi,\mathbf{u})\in\mathbf{R}\times[0,\infty)\times\mathbf{S}^2$, mapping M onto the open submanifold

$$\{(v,w,x,y,z)\in\mathbf{S}^4_1(R_d)\mid v+w>0\}$$

of the pseudosphere. On the coordinate domain V we have that

$$\begin{aligned} v\circ\alpha|_V &= R_d\sinh(t/R_d)\,+\tfrac{1}{2}\,R_d\,\chi^2\,e^{t/R_d}\\ w\circ\alpha|_V &= R_d\cosh(t/R_d)\,-\tfrac{1}{2}\,R_d\,\chi^2\,e^{t/R_d}\\ x\circ\alpha|_V &= R_d\,e^{t/R_d}\,\chi\,\sin\theta\,\cos\phi\\ y\circ\alpha|_V &= R_d\,e^{t/R_d}\,\chi\,\sin\theta\,\sin\phi\\ z\circ\alpha|_V &= R_d\,e^{t/R_d}\,\chi\,\cos\theta \end{aligned}$$

so that

$$\begin{aligned} (v+w)\circ\alpha|_V &= R_d\,e^{t/R_d}\\ (v-w)\circ\alpha|_V &= R_d\,\chi^2\,e^{t/R_d}-R_d\,e^{-t/R_d} \end{aligned}$$

where $-\infty < t < \infty$, $0 < \chi < \infty$, $0 < \theta < \pi$ and $-\pi < \phi < \pi$. We can also verify directly that α is an isometry. Using Proposition 4.18 we see that

$$\alpha^* dv|_V = [\cosh(t/R_d) + \tfrac{1}{2}\chi^2 e^{t/R_d}]\, dt + R_d\, \chi\, e^{t/R_d}\, d\chi$$

$$\alpha^* dw|_V = [\sinh(t/R_d) - \tfrac{1}{2}\chi^2 e^{t/R_d}]\, dt - R_d\, \chi\, e^{t/R_d}\, d\chi$$

$$\begin{aligned}\alpha^* dx|_V &= e^{t/R_d}\,\chi\, \sin\theta\, \cos\phi\; dt + R_d\, e^{t/R_d} \sin\theta\, \cos\phi\; d\chi \\ &\quad + R_d\, e^{t/R_d}\,\chi\, \cos\theta\, \cos\phi\; d\theta - R_d\, e^{t/R_d}\,\chi\, \sin\theta\, \sin\phi\; d\phi\end{aligned}$$

$$\begin{aligned}\alpha^* dy|_V &= e^{t/R_d}\,\chi\, \sin\theta\, \sin\phi\; dt + R_d\, e^{t/R_d} \sin\theta\, \sin\phi\; d\chi \\ &\quad + R_d\, e^{t/R_d}\,\chi\, \cos\theta\, \sin\phi\; d\theta + R_d\, e^{t/R_d}\,\chi\, \sin\theta\, \cos\phi\; d\phi\end{aligned}$$

$$\alpha^* dz|_V = e^{t/R_d}\,\chi\, \cos\theta\; dt + R_d\, e^{t/R_d} \cos\theta\; d\chi - R_d\, e^{t/R_d}\,\chi\, \sin\theta\; d\theta.$$

Hence we have that

$$\alpha^* g_1^5|_V = -dt^2 + R_d^2\, e^{2t/R_d}\, [d\chi^2 + \chi^2 (d\theta^2 + \sin^2\theta\; d\phi^2)]\,.$$

The simultaneity space V_{t_0} at the time $t = t_0$ is the intersection between the pseudo-sphere $\mathbf{S}_1^4(R_d)$ and the hyperplane $v + w = R_d\, e^{t_0/R_d}$.

11.45 Flat universe models with vacuum energy and either dust or radiation We now consider flat universe models with energy density

$$\rho = \rho_{v0} + (\rho_{c0} - \rho_{v0}) \left(\frac{a}{a_0}\right)^{-s},$$

consisting of Lorentz invariant vacuum energy, and of dust when $s = 3$ or radiation when $s = 4$. The constants ρ_{c0}, ρ_{v0} and a_0 are the present values of the critical energy density, the vacuum energy density and the scale factor. We see that the present value of the total energy density is ρ_{c0} in agreement with the assumption that the universe is flat. As usual we denote the present relative energy density of vacuum by Ω_{v0}. Now it follows from (6) in 11.42 that

$$a^{s-2}\dot{a}^2 = \frac{\kappa}{3} \{\rho_{v0}\, a^s + (\rho_{c0} - \rho_{v0})\, a_0^s\}\,,$$

which can be written as

$$\left(\frac{s}{2}\right)^2 a^{s-2}\dot{a}^2 = \omega^2 (a^s + A^s) \tag{1}$$

where

$$\omega = \frac{s}{2}\left(\frac{\kappa}{3}\rho_{v0}\right)^{1/2} = \frac{s}{2}\,\Omega_{v0}^{1/2} H_0 \qquad \text{and}$$

$$A = \left(\frac{\rho_{c0} - \rho_{v0}}{\rho_{v0}}\right)^{1/s} a_0 = \left(\frac{1 - \Omega_{v0}}{\Omega_{v0}}\right)^{1/s} a_0\,.$$

Introducing the function $b = \left(\frac{a}{A}\right)^{s/2}$, Equation (1) takes the form

$$\dot{b}^2 = \omega^2 (b^2 + 1) \tag{2}$$

having the solution

$$b = \sinh(\omega t + c) \, ,$$

where c is an integration constant. Hence the solution of Equation (1) with initial condition $\tilde{a}(0) = 0$ is

$$a = A \sinh^{2/s}(\omega t) \, .$$

The Hubble parameter and the critical energy density are

$$H = \frac{\dot{a}}{a} = \frac{2}{s} \, \omega \coth(\omega t)$$

and

$$\rho_c = \frac{3}{\kappa} H^2 = \rho_{v0} \coth^2(\omega t) \, ,$$

and the relative energy density of vacuum is therefore

$$\Omega_v = \tanh^2(\omega t) \, .$$

The ratio of the age t of the universe to the Hubble age t_H is given by

$$t/t_H = Ht = \frac{2}{s} \, \omega t \coth(\omega t) = \frac{2 \operatorname{arctanh} \sqrt{\Omega_v}}{s \sqrt{\Omega_v}} \, .$$

At the present time $t = t_0$ we obtain

$$\omega t_0 = \operatorname{arctanh} \sqrt{\Omega_{v0}} = \operatorname{arcsinh} \left(\frac{a_0}{A}\right)^{s/2} .$$

Using the observed value $\Omega_{v0} = 0.7$ and $s = 3$, we therefore have that

$$A/a_0 = 0.75 \quad \text{and} \quad \omega t_0 = 1.2 \, .$$

Hence

$$a/a_0 = 0.75 \, \sinh^{2/3}(1.2 \, t/t_0)$$

is a realistic expression for the expansion of the universe, and $H_0 t_0$ is approximately equal to 1. This means that the present age t_0 of the universe is approximately equal to its Hubble age t_{H0} as a result of a period of decelerated expansion followed by a period of accelerated expansion.

Chapter 12

JET BUNDLES

BUNDLES

12.1 Definition Let M and F be smooth manifolds. A smooth manifold E is called a *bundle* over M with *fibre* F if the following two conditions are satisfied:

(i) There is a surjective smooth map $\pi : E \to M$ which is called the *projection* of the *total space* E onto the *base space* M. For each point $p \in M$ the inverse image $\pi^{-1}(p)$ is called the *fibre* over p.

(ii) For each point $p \in M$ there is an open neighbourhood U around p and a diffeomorphism $t : \pi^{-1}(U) \to U \times F$ such that the diagram

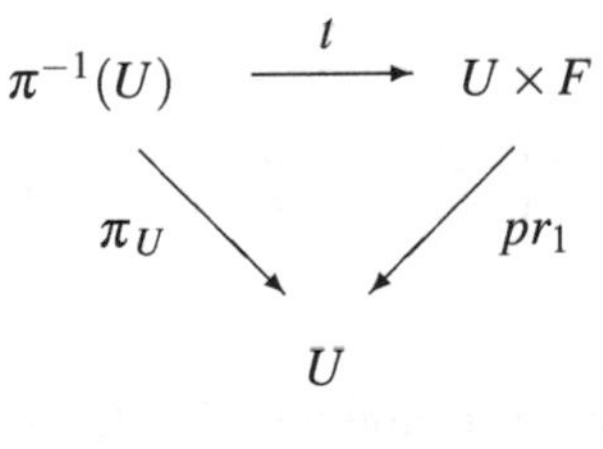

is commutative. The pair $(t, \pi^{-1}(U))$ is called a *local trivialization* around p.

We often use the projection $\pi : E \to M$ instead of the total space E to denote the bundle in order to indicate more of the bundle structure in the notation. F is also called the *typical fibre* to distinguish it from the fibre $\pi^{-1}(p)$ over p.

A family of local trivializations $\{(t_\alpha, \pi^{-1}(U_\alpha)) | \alpha \in A\}$, where $\{U_\alpha | \alpha \in A\}$ is an open cover of M, is called a *trivializing cover* of E.

12.2 Proposition If $\pi : E \to M$ is a bundle with fibre F and p is a point in M, then there is a local trivialization $(t, \pi^{-1}(U))$ around p where U is the coordinate neighbourhood of a local chart (x, U) on M.

PROOF: If $(t_1, \pi^{-1}(U_1))$ is a local trivialization and (x_2, U_2) a local chart around p, let $U = U_1 \cap U_2$ and $x = x_2|_U$, and let $t : \pi^{-1}(U) \to U \times F$ be the diffeomorphism induced by t_1. □

12.3 Proposition The projection π is a submersion, and the fibre $E_p = \pi^{-1}(p)$ over p is a closed submanifold of E for each point $p \in M$. If $(t, \pi^{-1}(U))$ is a local trivialization around p, then the map $t_p : E_p \to F$ defined by $t_p = pr_2 \circ t|_{E_p}$ is a diffeomorphism. Hence each fibre E_p is diffeomorphic to the typical fibre F.

If $i_p : E_p \to E$ is the inclusion map and $u \in E_p$, then $V_u = i_{p*}(T_u E_p)$ is a subspace of $T_u E$ which is called the *vertical subspace* at u. The vectors in V_u are called *vertical tangent vectors* at u. We have that $V_u = \ker \pi_{*u}$ for each $u \in E$, and these subspaces form a distribution V on E. The subbundle $\pi' : V \to E$ of $\pi'' : TE \to E$ is called the *vertical tangent bundle* of $\pi : E \to M$.

Suppose that $(t, \pi^{-1}(U))$ is a local trivialization around a point u_0 in E with $t(u_0) = (p_0, v_0)$, and that (x, O) and (y, Q) are local charts on M^n and F^m around p_0 and v_0, respectively, with $O \subset U$. Then (z, W), where $W = t^{-1}(O \times Q)$ and $z = (x \times y) \circ t$, is a local chart around u_0 on E, and

$$\left\{ \frac{\partial}{\partial z^{n+1}}, \dots, \frac{\partial}{\partial z^{n+m}} \right\}$$

is a local basis for V on W.

Let $pr_2 : \mathbf{R}^n \times \mathbf{R}^m \to \mathbf{R}^m$ be the projection on the second factor. Then we have a local trivialization $(t', \pi'^{-1}(W))$ in the vertical tangent bundle $\pi' : V \to E$, where

$$t' = (id_W \times pr_2) \circ t_z|_{\pi'^{-1}(W)}$$

so that

$$t' \left(\sum_{k=1}^{m} a_k \left. \frac{\partial}{\partial z^{n+k}} \right|_u \right) = (u, a)$$

for $u \in W$ and $a \in \mathbf{R}^m$.

PROOF: From the diagram in Definition 12.1 (ii) we see that π is a submersion. Hence E_p is a closed submanifold of E by Proposition 2.41. In the same way we also have that $\{p\} \times F = pr_1{}^{-1}(p)$ is a closed submanifold of $U \times F$.

By Corollary 2.38, the map $s_p : E_p \to \{p\} \times F$ induced by t is therefore a diffeomorphism, and so is the map $pr_2 : \{p\} \times F \to F$, thus showing that $t_p = pr_2 \circ s_p$ is a diffeomorphism.

Since $\pi \circ i_p\,(u) = p$ for every $u \in E_p$, we have that $\pi_{*u} \circ i_{p*u} = (\pi \circ i_p)_{*u} = 0$ which implies that $V_u \subset \ker \pi_{*u}$. Equality now follows since

$$\dim_u(E_p) = \dim_{t_p(u)}(F) = \dim_u(E) - \dim_p(M) = \dim_u(\ker \pi_{*u}) .$$

To prove the last assertions in the proposition, let $j_p : Q \to O \times Q$ be the embedding defined by $j_p(v) = (p, v)$ for $p \in O$ and $v \in Q$. Then we have that

$$t_* \left(\left. \frac{\partial}{\partial z^{n+k}} \right|_u \right) = j_{p*} \left(\left. \frac{\partial}{\partial y^k} \right|_v \right)$$

for $k = 1, \ldots, m$, where $t(u) = (p,v)$. Indeed, we have that

$$t_*\left(\left.\frac{\partial}{\partial z^{n+k}}\right|_u\right)(f) = \frac{\partial(f\circ t)}{\partial z^{n+k}}(u) = D_{n+k}(f\circ t\circ z^{-1})(z(u))$$

$$= D_{n+k}(f\circ(x\times y)^{-1})(x(p),y(v)) = D_k(f\circ j_p\circ y^{-1})(y(v))$$

$$= \frac{\partial(f\circ j_p)}{\partial y^k}(v) = j_{p*}\left(\left.\frac{\partial}{\partial y^k}\right|_v\right)(f)$$

for every smooth function f defined on an open neighbourhood of (p,v) in $O\times Q$, completing the proof of the last part of the proposition. □

12.4 Remark The coordinate functions z^i for $i = 1,2,\ldots,n$ and $z^{n+\alpha}$ for $\alpha = 1,2,\ldots,m$ on E defined in the proposition are also denoted by x^i and y^α. It will be clear from the context whether x^i and y^α denote coordinate functions on E or on M and F, respectively.

12.5 Remark Let $\pi : TM \to M$ be the tangent bundle of a smooth manifold M^n. For each local chart (x,U) on M we obtain an induced local chart (z,W) on TM where $W = \pi^{-1}(U)$ and $z = (x\times id)\circ t_x$, so that

$$z\left(\sum_{i=1}^{n} a_i \left.\frac{\partial}{\partial x^i}\right|_u\right) = (x(u),a)$$

for $u\in U$ and $a\in\mathbf{R}^n$. The coordinate functions z^i and z^{n+i} for $i = 1,2,\ldots,n$ on TM are also denoted by x^i and $\dot{x}^i$. It will be clear from the context whether x^i denote coordinate functions on TM or on M.

12.6 Proposition Let M and F be smooth manifolds, and let $\pi : E\to M$ be a map from a set E. Let $\{U_\alpha \mid \alpha\in A\}$ be an open cover of M. Suppose that for each $\alpha\in A$, there is a bijection $t_\alpha : \pi^{-1}(U_\alpha)\to U_\alpha\times F$ such that the diagram

$$\begin{array}{ccc} \pi^{-1}(U_\alpha) & \xrightarrow{\;t_\alpha\;} & U_\alpha\times F \\ & {\scriptstyle \pi_{U_\alpha}}\searrow \quad \swarrow{\scriptstyle pr_1} & \\ & U_\alpha & \end{array}$$

is commutative. Moreover, suppose that for each pair $\alpha,\beta\in A$, the map $t_\beta\circ t_\alpha^{-1}$ is a diffeomorphism from $(U_\alpha\cap U_\beta)\times F$ onto itself.

Then there is a unique topology and smooth structure on E such that $\pi : E \to M$ is a bundle over M with fibre F, and such that $(t_\alpha, \pi^{-1}(U_\alpha))$ is a local trivialization for each $\alpha \in A$.

PROOF : We first define a topology τ on E so that the projection π is continuous and the bijections $t_\alpha : \pi^{-1}(U_\alpha) \to U_\alpha \times F$ are homeomorphisms for every $\alpha \in A$. The continuity of π implies that each $\pi^{-1}(U_\alpha)$ must be open in τ, and hence a set O in E must belong to τ if and only if $O \cap \pi^{-1}(U_\alpha) \in \tau$ for every $\alpha \in A$. The requirement that the bijections t_α are homeomorphisms now defines τ uniquely as the collection of sets O in E such that $t_\alpha(O \cap \pi^{-1}(U_\alpha))$ is open in $U_\alpha \times F$ for every $\alpha \in A$. This collection τ is clearly a topology on E, and we must prove that it satisfies the above requirements.

We first show that $\pi^{-1}(U_\alpha)$ is open in τ for each $\alpha \in A$. For every $\beta \in A$ we have that $t_\beta(\pi^{-1}(U_\alpha) \cap \pi^{-1}(U_\beta)) = (U_\alpha \cap U_\beta) \times F$ which is open in $U_\beta \times F$, thereby proving the assertion.

We next show that t_α is a homeomorphism for every $\alpha \in A$. If $O \subset \pi^{-1}(U_\alpha)$ belongs to τ, we have that $t_\alpha(O \cap \pi^{-1}(U_\alpha))$ is open in $U_\alpha \times F$, which proves that t_α is an open map. To show that it is continuous, let O be an open set in $U_\alpha \times F$. For every $\beta \in A$ we have that $t_\beta(t_\alpha^{-1}(O) \cap \pi^{-1}(U_\beta)) = (t_\beta \circ t_\alpha^{-1})(O \cap (U_\alpha \cap U_\beta) \times F)$ which is open in $U_\beta \times F$, and hence $t_\alpha^{-1}(O) \in \tau$ showing that t_α is continuous.

From this it also follows that the projection π is continuous since $\pi_{U_\alpha} = pr_1 \circ t_\alpha$ is the composition of the continuos maps t_α and pr_1 in the above commutative diagram for each $\alpha \in A$.

Now let $\{(y_\gamma, W_\gamma) | \gamma \in C\}$ be an atlas on F, and let $\{(x_\beta, V_\beta) | \beta \in B\}$ be an atlas on M such that $\{V_\beta | \beta \in B\}$ is a refinement of the open cover $\{U_\alpha | \alpha \in A\}$. We then have a map $\rho : B \to A$ such that $V_\beta \subset U_{\rho(\beta)}$ for every $\beta \in B$. By the assumptions in the proposition we now see that the family $\mathscr{B} = \{((x_\beta \times y_\gamma) \circ t_{\rho(\beta)}, t_{\rho(\beta)}^{-1}(V_\beta \times W_\gamma)) | (\beta, \gamma) \in B \times C\}$ is an atlas on E. To show that E is a smooth manifold, we must show that it is Hausdorff and second countable. By Proposition 2.6 we know that M and F have atlases of the above form where B and C are countable, thus showing that E is second countable.

To see that the E is Hausdorff, let p_1 and p_2 be two different points in E. If p_1 and p_2 lie in different fibres, they can be separated by the neighbourhoods $\pi^{-1}(W_1)$ and $\pi^{-1}(W_2)$, where W_1 and W_2 are disjoint neighbourhoods of $\pi(p_1)$ and $\pi(p_2)$ in M. If p_1 and p_2 lie in the same fibre $\pi^{-1}(q)$ with $q \in U_\alpha$, they can be separated by the neighbourhoods $t_\alpha^{-1}(O_1)$ and $t_\alpha^{-1}(O_2)$, where O_1 and O_2 are disjoint neighbourhoods of $t_\alpha(p_1)$ and $t_\alpha(p_2)$ in $U_\alpha \times F$.

This shows that E is a smooth manifold with a unique smooth structure such that the bijections $t_\alpha : \pi^{-1}(U_\alpha) \to U_\alpha \times F$ are diffeomorphisms for every $\alpha \in A$. We see that the projection π is smooth since $\pi_{U_\alpha} = pr_1 \circ t_\alpha$ is the composition of the smooth maps t_α and pr_1. This completes the construction of the bundle $\pi : E \to M$ over M with fibre F, having $(t_\alpha, \pi^{-1}(U_\alpha))$ as a local trivialization for each $\alpha \in A$. □

12.7 Definition Let $\pi_1 : E_1 \to M_1$ and $\pi_2 : E_2 \to M_2$ be two bundles. A *bundle map* from π_1 to π_2 is a pair $(\tilde{f}, f)$ of smooth maps $\tilde{f} : E_1 \to E_2$ and $f : M_1 \to M_2$

such that the diagram

$$\begin{array}{ccc} E_1 & \xrightarrow{\tilde{f}} & E_2 \\ \pi_1 \downarrow & & \downarrow \pi_2 \\ M_1 & \xrightarrow{f} & M_2 \end{array}$$

is commutative. We let $\tilde{f}_p$ denote the induced smooth map $\tilde{f}_p : \pi_1^{-1}(p) \to \pi_2^{-1}(f(p))$ between the fibres for each $p \in M_1$. If $\tilde{f}$ and f are diffeomorphisms, $(\tilde{f}, f)$ is called an *equivalence*.

12.8 Definition Let $\pi_1 : E_1 \to M$ and $\pi_2 : E_2 \to M$ be two bundles over the same base space M. A *bundle map* from π_1 to π_2 *over* M is a smooth map $f : E_1 \to E_2$ such that (f, id_M) is a bundle map. If f is a diffeomorphism, it is called an *equivalence over* M.

12.9 Definition A bundle $\pi : E \to M$ with fibre F is said to be *trivial* if it is equivalent over M with the bundle $pr_1 : M \times F \to M$, where pr_1 is the projection on the first factor.

12.10 Definition Let $\pi : E \to M$ be a bundle over M with fibre F, and let V be an open subset of M and $i : V \to M$ be the inclusion mapping. Then a smooth map $s : V \to E$ with $\pi \circ s = i$ is called a *section* of π on V, or a *lifting* of i to E. The set of all sections of π on V is denoted by $\Gamma(V;E)$ or $\Gamma_V(E)$. Given a point p on M, we let $\Gamma_p(E)$ denote the set of all sections $s : V \to E$, each defined on some open neighbourhood V of p.

If $(t, \pi^{-1}(U))$ is a local trivialization of π with $U \subset V$ and $s : V \to E$ is a map with $\pi \circ s = i$, then it follows from the commutative diagram in Definition 12.1 that $pr_1 \circ t \circ s = id_U$. Hence $s|_U$ is completely determined by the map $h : U \to F$ given by $(t \circ s)(p) = (p, h(p))$ for $p \in U$. h is called the *local representation* of s on U. We see that s is smooth on U if and only if the local representation h is smooth.

12.11 Let $\pi : E \to M$ be a bundle over M with fibre F. If $f : N \to M$ is a smooth map from a smooth manifold N, we define the set

$$\widetilde{E} = \{(p,u) \in N \times E \,|\, f(p) = \pi(u)\}$$

and let $\widetilde{\pi} : \widetilde{E} \to N$ and $\widetilde{f} : \widetilde{E} \to E$ be the restrictions to $\widetilde{E}$ of the projections $pr_1 : N \times E \to N$ and $pr_2 : N \times E \to E$ on the first and second factor. Then we have a commutative diagram

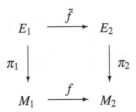

$$\begin{array}{ccc} \widetilde{E} & \xrightarrow{\widetilde{f}} & E \\ \widetilde{\pi}\downarrow & & \downarrow\pi \\ N & \xrightarrow{f} & M \end{array}$$

since $f \circ \widetilde{\pi}(p,u) = f(p) = \pi(u) = \pi \circ \widetilde{f}(p,u)$ for $(p,u) \in \widetilde{E}$.

Now let $\mathscr{A} = \{(t_\alpha, \pi^{-1}(U_\alpha)) | \alpha \in A\}$ be a trivializing cover of E. For each $\alpha \in A$ we have an open subset $\widetilde{U}_\alpha = f^{-1}(U_\alpha)$ of N and a bijection $\widetilde{t}_\alpha : \widetilde{\pi}^{-1}(\widetilde{U}_\alpha) \to \widetilde{U}_\alpha \times F$ defined by

$$\widetilde{t}_\alpha(p,u) = (p, t_{\alpha,f(p)}(u)) \tag{1}$$

for $p \in \widetilde{U}_\alpha$ and $u \in E_{f(p)}$. Then $\{\widetilde{U}_\alpha | \alpha \in A\}$ is a open cover of N, and we will show that the map $\widetilde{t}_\beta \circ \widetilde{t}_\alpha^{-1}$ is a diffeomorphism from $(\widetilde{U}_\alpha \cap \widetilde{U}_\beta) \times F$ onto itself for each pair of indices $\alpha, \beta \in A$. Let $\phi_{\beta\,\alpha} : (U_\alpha \cap U_\beta) \times F \to F$ be the second component of the smooth map $t_\beta \circ t_\alpha^{-1}$ so that

$$t_\beta \circ t_\alpha^{-1}(q,v) = (q, \phi_{\beta\,\alpha}(q,v))$$

for $q \in U_\alpha \cap U_\beta$ and $v \in F$. Then we have that

$$\widetilde{t}_{\beta,p} \circ \widetilde{t}_{\alpha,p}^{-1}(v) = t_{\beta,f(p)} \circ t_{\alpha,f(p)}^{-1}(v) = \phi_{\beta\,\alpha}(f(p),v) \tag{2}$$

so that

$$\widetilde{t}_\beta \circ \widetilde{t}_\alpha^{-1}(p,v) = (p, \phi_{\beta\,\alpha} \circ (f \times id_F)(p,v))$$

for every $p \in \widetilde{U}_\alpha \cap \widetilde{U}_\beta$ and $v \in F$. Since this is also true with the indices α and β interchanged, we have shown the above assertion for $\widetilde{t}_\beta \circ \widetilde{t}_\alpha^{-1}$. By Proposition 12.6 there is therefore a unique topology and smooth structure on $\widetilde{E}$ such that $\widetilde{\pi} : \widetilde{E} \to N$ is a bundle over N with fibre F, and such that $(\widetilde{t}_\alpha, \widetilde{\pi}^{-1}(\widetilde{U}_\alpha))$ is a local trivialization for each $\alpha \in A$. It is called the local trivialization in $\widetilde{E}$ associated with the local trivialization $(t_\alpha, \pi^{-1}(U_\alpha))$ in E.

By (1) we see that

$$\widetilde{t}_{\alpha,p} = t_{\alpha,f(p)} \circ \widetilde{f}_p$$

for $p \in \widetilde{U}_\alpha$, so that

$$t_\alpha \circ \widetilde{f} \circ \widetilde{t}_\alpha^{-1}(p,v) = (f(p),v) \tag{3}$$

for $p \in \widetilde{U}_\alpha$ and $v \in F$. Hence it follows from Proposition 12.3 that $(\widetilde{f}, f)$ is a bundle map which induces a diffeomorphism between the fibres $\widetilde{E}_p$ and $E_{f(p)}$ for each $p \in N$.

Since $\widetilde{\pi}$ and $\widetilde{f}$ are the components of the inclusion map $i : \widetilde{E} \to N \times E$, we also have that

$$(id \times t_\alpha) \circ i \circ \widetilde{t}_\alpha^{-1}(p,v) = (p, f(p), v)$$

for $p \in \widetilde{U}_\alpha$ and $v \in F$. Using this, we will show that $\widetilde{E}$ is a closed submanifold of $N \times E$. If $(p_0, u_0) \in \widetilde{\pi}^{-1}(\widetilde{U}_\alpha)$ and $t_\alpha(u_0) = (f(p_0), v_0) \in U_\alpha \times F$, we choose local charts $(\widetilde{x}, \widetilde{V})$, (x, V) and (y, W) around p_0, $f(p_0)$ and v_0, respectively, with $\widetilde{V} \subset \widetilde{U}_\alpha$ and $f(\widetilde{V}) \subset V$. Then we have that

$$(\widetilde{x} \times (x \times y) \circ t_\alpha)) \circ i \circ ((\widetilde{x} \times y) \circ \widetilde{t}_\alpha)^{-1}(a, \xi) = (a, x \circ f \circ \widetilde{x}^{-1}(a), \xi)$$

for $a \in \widetilde{x}(\widetilde{V})$ and $\xi \in y(W)$.

Now let r, n and m be the dimensions of N, M and F, respectively, and let $\phi : \widetilde{x}(\widetilde{V}) \times \mathbf{R}^n \times \mathbf{R}^m \to \widetilde{x}(\widetilde{V}) \times \mathbf{R}^m \times \mathbf{R}^n$ be the C^∞-map defined by $\phi(a, b, \xi) = (a, \xi, b - x \circ f \circ \widetilde{x}^{-1}(a))$. ϕ is a diffeomorphism with inverse $\psi : \widetilde{x}(\widetilde{V}) \times \mathbf{R}^m \times \mathbf{R}^n \to \widetilde{x}(\widetilde{V}) \times \mathbf{R}^n \times \mathbf{R}^m$ given by $\psi(a, \xi, b) = (a, b + x \circ f \circ \widetilde{x}^{-1}(a), \xi)$.

If $j : \widetilde{x}(\widetilde{V}) \times \mathbf{R}^n \times \mathbf{R}^m \to \mathbf{R}^r \times \mathbf{R}^n \times \mathbf{R}^m$ is the inclusion map, we therefore have a local chart $(\widetilde{V} \times t_\alpha^{-1}(V \times W), j \circ \phi \circ (\widetilde{x} \times (x \times y) \circ t_\alpha))$ around (p_0, u_0) in $N \times E$ having the submanifold property, showing that $\widetilde{E}$ is a submanifold of $N \times E$. It is closed as $\widetilde{E} = (f \times \pi)^{-1}(\Delta)$, where $\Delta \subset M \times M$ is the diagonal which is closed since M is Hausdorff.

The bundle $\widetilde{\pi} : \widetilde{E} \to N$ is called the *pull-back* of $\pi : E \to M$ by f or the bundle *induced* from $\pi : E \to M$ by f, and it is denoted by $f^*(E)$ with projection $f^*(\pi)$.

A smooth map $F : N \to E$ with $\pi \circ F = f$ is called a *lifting* of f. The set of all liftings of f is denoted by $\Gamma(f; E)$. For each lifting F of f there is a unique section $\widetilde{F} : N \to \widetilde{E}$ of $\widetilde{\pi}$ with $\widetilde{f} \circ \widetilde{F} = F$, which is given by $\widetilde{F}(p) = (p, F(p))$ for $p \in N$. Conversely, if $s : N \to \widetilde{E}$ is a section of $\widetilde{\pi}$, then $s = \widetilde{F}$ where $F = \widetilde{f} \circ s$. $\widetilde{F}$ is called the section in the induced bundle determined by the lifting F.

If N is an immersed submanifold of M with inclusion map $i : N \to M$, we consider the subset $E|_N = \pi^{-1}(N)$ of E, and let $\pi_N : E|_N \to N$ be the map induced by the projection $\pi : E \to M$. We give $E|_N$ the topology and manifold structure so that the bijection $e : \widetilde{E} \to E|_N$ given by $e(p, u) = u$ for $(p, u) \in \widetilde{E}$, is an equivalence over N. The bundle $\pi_N : E|_N \to N$ is called the *restriction* of E to N, and $i' = \widetilde{i} \circ e^{-1} : E|_N \to E$ is the inclusion map. Given a trivializing cover $\mathscr{A} = \{(t_\alpha, \pi^{-1}(U_\alpha)) | \alpha \in A\}$ of E, we let $\widetilde{U}_\alpha = U_\alpha \cap N$ and $t'_\alpha : \pi^{-1}(\widetilde{U}_\alpha) \to \widetilde{U}_\alpha \times F$ be the map induced by $t_\alpha : \pi^{-1}(U_\alpha) \to U_\alpha \times F$ so that $\widetilde{t}_\alpha \circ e^{-1} = t'_\alpha$ for $\alpha \in A$. Then $\mathscr{A}' = \{(t'_\alpha, \pi^{-1}(\widetilde{U}_\alpha)) | \alpha \in A\}$ is a trivializing cover of $E|_N$.

12.12 Definition Let $f : N \to M$ be a smooth map and $\pi : TM \to M$ be the tangent bundle of M. Then a smooth map $X : N \to TM$ which is a lifting of f is called a *vector field along* f.

Let $\widetilde{\pi} : f^*(TM) \to N$ be the pullback of the tangent bundle by f. Then a form $\omega \in \Omega^k(N; f^*(TM))$ is called a *bundle-valued k-form along* f.

12.13 Let $\pi : E \to M$ be a bundle over M with fibre F, and let $\pi'' : TE \to E$ be the tangent bundle of E and $\widetilde{\pi}' : \pi^*(TM) \to E$ be the pullback of the tangent bundle $\pi' : TM \to M$ of M. From the commutative diagram

$$\begin{array}{ccc} TE & \xrightarrow{\pi_*} & TM \\ {\scriptstyle \pi''}\downarrow & & \downarrow{\scriptstyle \pi'} \\ E & \xrightarrow{\pi} & M \end{array}$$

we see that the map $(\pi'', \pi_*) : TE \to E \times TM$ induces a bundle map $\rho : TE \to \pi^*(TM)$ over E.

12.14 Proposition Let $\pi : E \to M$ be a bundle over M^n with fibre F^m. Then the pullback $\widetilde{\pi}' : \pi^*(\Lambda^k(TM)) \to E$ of the bundle $\pi' : \Lambda^k(TM) \to M$ can be identified with a subbundle of $\pi'' : \Lambda^k(TE) \to E$ by means of the bundle map $\alpha : \pi^*(\Lambda^k(TM)) \to \Lambda^k(TE)$ over E defined by

$$\alpha(u, \omega) = \omega \circ (\pi_{*u})^k$$

for $p \in M$, $u \in E_p$ and $\omega \in \Lambda^k(T_pM)$, which is an embedding. We have that

$$\alpha_u = \rho_u^*$$

for $u \in E$, where ρ is the bundle map over E defined in 12.13. If (z, W) is a local chart around a point u_0 on E as defined in Proposition 12.3, then

$$\{dz^{i_1} \wedge \cdots \wedge dz^{i_k} \mid 1 \le i_1 < \ldots < i_k \le n\}$$

is a local basis for $\alpha(\pi^*(\Lambda^k(TM)))$ on W. A k-form ω on E belongs to $\alpha(\pi^*(\Lambda^k(TM)))$ if and only if $\omega(u)(w_1, \ldots, w_k) = 0$ whenever $u \in E$ and at least one $w_i \in T_uE$ is vertical. A form on E satisfying this condition is said to be *horizontal* over M. The set of all horizontal k-forms on E is denoted by $\Omega_0^k(E)$.

PROOF : We have that

$$\alpha(u, dx^{i_1}(\pi(u)) \wedge \cdots \wedge dx^{i_k}(\pi(u))) = \pi^*(dx^{i_1} \wedge \cdots \wedge dx^{i_k})(u)$$
$$= d(x^{i_1} \circ \pi)(u) \wedge \cdots \wedge d(x^{i_k} \circ \pi)(u) = dz^{i_1}(u) \wedge \cdots \wedge dz^{i_k}(u)$$

for $u \in W$ and $1 \le i_1 < \ldots < i_k \le n$, which shows the first part of the proposition. It only remains to prove that every horizontal k-form ω on E belongs to $\alpha(\pi^*(\Lambda^k(TM)))$. Let $u \in E_p$ and define $\eta \in \Lambda^k(T_pM)$ by

$$\eta(v_1, \ldots, v_n) = \omega(u)(w_1, \ldots, w_k)$$

for $v_1, \ldots, v_k \in T_pM$, where $w_1, \ldots, w_k \in T_uE$ are chosen so that $\pi_*(w_i) = v_i$ for $i = 1, \ldots, k$. We must show that η is well defined and does not depend on the choice of $w_1, \ldots, w_k$. If w_i is replaced by another element $w_i' \in T_uE$ with $\pi_*(w_i') = v_i$, then $w_i - w_i' \in \ker \pi_{*u} = V_u$ which shows that $\eta(v_1, \ldots, v_k)$ is unchanged since ω is horizontal. As $\omega(u) = \alpha(u, \eta)$, it follows that ω belongs to $\alpha(\pi^*(\Lambda^k(TM)))$. □

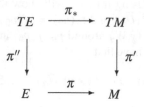

12.15 Remark In the case $k = 1$, the pullback $\tilde{\pi}' : \pi^*(T^*M) \to E$ of the cotangent bundle $\pi' : T^*M \to M$ of M is called the *cotangent bundle of E horizontal over M*. We have that

$$\alpha(\pi^*(T^*M)_u) = A(V_u)$$

for $u \in E$.

AFFINE BUNDLES

12.16 Definition A bundle $\pi' : A \to M$ with fibre $\mathbf{R}^n$ is said to be *affine* if the following two conditions are satisfied :

(i) There is a vector bundle $\pi : E \to M$ so that A_p is an affine space modelled on E_p for every $p \in M$.

(ii) For each point $p \in M$ there are local trivializations $(t', \pi'^{-1}(U))$ and $(t, \pi^{-1}(U))$ around p in A and E, respectively, so that t'_q is an affine map with linear part t_q for every $q \in U$.

We say that $\pi' : A \to M$ is modelled on the vector bundle $\pi : E \to M$, and that $(t', \pi'^{-1}(U))$ is modelled on the local trivialsation $(t, \pi^{-1}(U))$ in E.

12.17 Definition Let $\pi'_1 : A_1 \to M_1$ and $\pi'_2 : A_2 \to M_2$ be two affine bundles modelled on the vector bundles $\pi_1 : E_1 \to M_1$ and $\pi_2 : E_2 \to M_2$, respectively. Then a bundle map $(\tilde{f}, f)$ from π'_1 to π'_2 is said to be *affine* if there is a bundle map $(\tilde{g}, f)$ from π_1 to π_1 so that $\tilde{f}_p$ is an affine map with linear part $\tilde{g}_p$ for every $p \in M$. We say that $(\tilde{g}, f)$ is the *linear part* of $(\tilde{f}, f)$. If $\tilde{f}$ and f are diffeomorphisms, it is called an *affine equivalence*.

In particular, if $\pi'_1 : A_1 \to M$ and $\pi'_2 : A_2 \to M$ are two affine bundles over the same base space M, then an *affine bundle map* from π'_1 to π'_2 *over* M is a smooth map $f : A_1 \to A_2$ such that (id_M, f) is an affine bundle map. If f is a diffeomorphism, it is called an *affine equivalence over M*.

12.18 Proposition Let $\pi' : A \to M$ be an affine bundle modelled on the vector bundle $\pi : E \to M$, and let $s : M \to A$ be a section of π'. Then we have an affine equivalence $S : E \to A$ over M with $id_E : E \to E$ as its linear part, given by

$$S(v) = s \circ \pi(v) + v$$

for $v \in E$. Its inverse is given by

$$S^{-1}(u) = u - s \circ \pi'(u)$$

for $u \in A$. There is a unique vector bundle structure on A so that S is an equivalence of vector bundles, and we have that s is its zero section.

PROOF : Let $(t', \pi'^{-1}(U))$ be a local trivialization in A is modelled on a local trivialization $(t, \pi^{-1}(U))$ in E, and let $h : U \to \mathbf{R}^n$ be the local representation of s on U so that

$$t' \circ s(p) = (p, h(p))$$

for $p \in U$. Then we have that

$$t' \circ S \circ t^{-1}(p,a) = (p, h(p) + a)$$

for $p \in U$ and $a \in \mathbf{R}^n$, showing the first part of the proposition.

As in Example 9.34 there is now a unique vector space structure on each fibre A_p so that $S_p : E_p \to A_p$ is a linear isomorphism. It is given by

$$au + bv = s(p) + a(u - s(p)) + b(v - s(p))$$

for $a, b \in \mathbf{R}$ and $u, v \in A_p$. This shows that $\pi' : A \to M$ is a vector bundle satisfying the assertions in the proposition, with $(t \circ S^{-1}, \pi'^{-1}(U))$ as a local trivialization for each local trivialization $(t, \pi^{-1}(U))$ in E. □

12.19 Proposition Let $\pi' : A \to M$ be an affine bundle modelled on the m-dimensional vector bundle $\pi : E \to M$ over M^n, and let $\pi'' : V \to A$ be the vertical tangent bundle of $\pi' : A \to M$ and $\widetilde{\pi} : \widetilde{E} \to A$ be the bundle induced from $\pi : E \to M$ by π'. Then we have an equivalence $\eta : V \to \widetilde{E}$ over A given by

$$\eta \circ i_{p*}(w) = (u, \omega_u(w))$$

for $p \in M$, $u \in A_p$ and $w \in T_u(A_p)$, where $i_p : A_p \to A$ is the inclusion map and $\omega_u : T_u(A_p) \to E_p$ is the canonical identification defined in Lemma 9.36. If $(t', \pi'^{-1}(U))$ is a local trivialization around p in A modelled on the local trivialization $(t, \pi^{-1}(U))$ in E, we have that

$$\omega_u = t_p^{-1} \circ t_{t'_p, u} \,.$$

We also have a canonical bundle map $(\widetilde{\eta}, \pi')$ from $\pi'' : V \to A$ to $\pi : E \to M$ where $\widetilde{\eta} = \widetilde{\pi}' \circ \eta$, which induces a linear isomorphism between the fibres V_u and $E_{\pi'(u)}$ for each $u \in A$.

PROOF : Let $(\widetilde{t}, \widetilde{\pi}^{-1}(\widetilde{U}))$ and $(t'', \pi''^{-1}(\widetilde{U}))$, where $\widetilde{U} = \pi'^{-1}(U)$, be the local trivializations in $\widetilde{E}$ and V, respectively, obtained from the local trivializations $(t, \pi^{-1}(U))$ and $(t', \pi'^{-1}(U))$ in E and A as described in 12.11 and Proposition 12.3. Then we have that

$$\widetilde{t} \circ \eta = t''$$

since

$$\widetilde{t}_u \circ \eta_u \circ i_{p*}(w) = t_p \circ \omega_u(w) = t_{t'_p, u}(w) = t''_u \circ i_{p*}(w)$$

for $p \in U$, $u \in A_p$ and $w \in T_u(A_p)$, showing that η is an equivalence over A. The last part of the proposition follows from 10.12. □

DERIVATIONS AND THE FRÖLICHER–NIJENHUIS BRACKET

12.20 Definition Given a linear map $\theta \in \Lambda^1(V;W)$ between the finite dimensional vector spaces V and W, we define the *interior product* of a vector-valued r-form $\alpha \in \Lambda^r(V;W)$ on V with values in W and a k-form $\omega \in \Lambda^k(W)$ on W to be the $(r+k-1)$-form $i_{\alpha,\theta}\ \omega \in \Lambda^{r+k-1}(V)$ on V given by

$$i_{\alpha,\theta}\ \omega(v_1,\ldots,v_{r+k-1}) = \frac{1}{r!\,(k-1)!}\sum_{\sigma\in S_{r+k-1}} \varepsilon(\sigma)\,\omega(\alpha(v_{\sigma(1)},\ldots,v_{\sigma(r)}),\theta(v_{\sigma(r+1)}),\ldots,\theta(v_{\sigma(r+k-1)})) \tag{1}$$

for $v_1,\ldots,v_{r+k-1} \in V$. We let $i_\alpha\,\omega = 0$ when $k = 0$. Note that

$$i_{\alpha,\theta} = \frac{(r+k-1)!}{r!\,(k-1)!}\,\mathscr{A}\circ I_{\alpha,\theta}$$

where $\mathscr{A}$ is the alternation operator in $\mathscr{T}^{r+k-1}(V)$ and $I_{\alpha,\theta} : \mathscr{T}^k(W) \to \mathscr{T}^{r+k-1}(V)$ is the linear map given by

$$I_{\alpha,\theta}\ \omega = \omega\circ(\alpha\times\theta^{k-1})$$

for $\omega \in \mathscr{T}^k(W)$, i.e.,

$$\begin{aligned} I_{\alpha,\theta}\ \omega(v_1,\ldots,v_{r+k-1}) &= \omega(\alpha(v_1,\ldots,v_r),\theta(v_{r+1}),\ldots,\theta(v_{r+k-1})) \\ &= \theta^*(i_{\alpha(v_1,\ldots,v_r)}\,\omega)(v_{r+1},\ldots,v_{r+k-1}) \end{aligned} \tag{2}$$

for $\omega \in \mathscr{T}^k(W)$ and $v_1,\ldots,v_{r+k-1} \in V$.

12.21 Proposition Let $(\tilde{f},f)$ be a bundle map between the vector bundles $\pi' : E' \to N$ and $\pi : E \to M$ of dimensions m and n, respectively, and let $\widetilde{\pi} : \widetilde{E} \to N$ be the pullback of $\pi : E \to M$ by f. Let $s : M \to \mathscr{T}^k(E)$ be a section in the vector bundle $\pi^k : \mathscr{T}^k(E) \to M$. Then we have a bundle map

$$\psi : \mathscr{T}^r(E';\widetilde{E}) \to \mathscr{T}^{r+k-1}(E')$$

over N between the vector bundles $\pi'' : \mathscr{T}^r(E';\widetilde{E}) \to N$ and $\pi'^{\,r+k-1} : \mathscr{T}^{r+k-1}(E') \to N$, which is given by

$$\psi(\alpha) = I_{\alpha,\tilde{f}_p}\ s(f(p))$$

for $p \in N$ and $\alpha \in \mathscr{T}^r(E'_p;\widetilde{E}_p)$.

PROOF: Let $(t',\pi^{-1}(V))$ and $(t,\pi^{-1}(U))$ be local trivializations in the vector bundles $\pi' : E' \to N$ and $\pi : E \to M$, respectively, with $f(V) \subset U$, and let $\tau : (\pi^k)^{-1}(U) \to U \times \mathscr{T}^k(\mathbf{R}^n)$ and $\tau' : (\pi'^{\,r+k-1})^{-1}(V) \to V \times \mathscr{T}^{r+k-1}(\mathbf{R}^m)$ be the corresponding equivalences over U and V defined in Proposition 4.53. Let $\tau'' : (\pi'')^{-1}(V) \to V \times \mathscr{T}^r(\mathbf{R}^m;\mathbf{R}^n)$ be the equivalence over V defined in Proposition 5.28. Then we have that

$$\tau \circ s(q) = (q, h(q))$$

for $q \in U$, where $h : U \to \mathscr{T}^k(\mathbf{R}^n)$ is the smooth map given by $h(q) = (t_q^{-1})^* s(q)$. We also have that

$$t \circ \tilde{f} \circ t'^{-1}(p,v) = (f(p), F(p)(v))$$

for $(p,v) \in V \times \mathbf{R}^m$, where $F : V \to \mathscr{T}^1(\mathbf{R}^m;\mathbf{R}^n)$ is the smooth map given by $F(p) = t_{f(p)} \circ \tilde{f}_p \circ t_p'^{-1}$ for $p \in V$. From this it follows that

$$\tau' \circ \psi \circ \tau''^{-1}(p,S) = (p, h(f(p)) \circ \{S \times F(p)^{k-1}\})$$

for $(p,S) \in V \times \mathscr{T}^r(\mathbf{R}^m;\mathbf{R}^n)$, since

$$\begin{aligned}
&\tau' \circ \psi \circ \tau''^{-1}(p,S)(a_1, \dots, a_{r+k-1}) \\
&= (t_p'^{-1})^* \{ I_{t_{f(p)}^{-1} \circ S \circ (t_p')^r, \tilde{f}_p} \; s(f(p)) \}(a_1, \dots, a_{r+k-1}) \\
&= (t_{f(p)}^{-1})^* \{ s(f(p)) \} (S(a_1, \dots, a_r), \\
&\qquad t_{f(p)} \circ \tilde{f}_p \circ t_p'^{-1}(a_{r+1}), \dots, t_{f(p)} \circ \tilde{f}_p \circ t_p'^{-1}(a_{r+k-1})) \\
&= h(f(p))(S(a_1, \dots, a_r), F(p)(a_{r+1}), \dots, F(p)(a_{r+k-1}))
\end{aligned}$$

for $a_1, \dots, a_{r+k-1} \in \mathbf{R}^m$. □

12.22 Remark Let $\omega : M \to \Lambda^k(E)$ be a section in the vector bundle $\pi' : \Lambda^k(E) \to M$. Then the bundle map ψ obtained from ω as described in Proposition 12.21 also gives rise to a bundle map

$$\phi : \Lambda^r(E'; \widetilde{E}) \to \Lambda^{r+k-1}(E')$$

over N given by

$$\phi(\alpha) = i_{\alpha, \tilde{f}_p} \; \omega(f(p))$$

for $p \in N$ and $\alpha \in \Lambda^r(E'_p; \widetilde{E}_p)$.

12.23 Proposition If $\theta \in \Lambda^1(V;W)$, $\alpha \in \Lambda^r(V;W)$ and $\lambda_1, \dots, \lambda_k \in W^*$, we have that

$$i_{\alpha,\theta}(\lambda_1 \wedge \cdots \wedge \lambda_k) = \sum_{i=1}^{k} (-1)^{i+1} i_{\alpha,\theta}(\lambda_i) \wedge \theta^*(\lambda_1 \wedge \cdots \wedge \lambda_{i-1} \wedge \lambda_{i+1} \wedge \cdots \wedge \lambda_k) .$$

PROOF : If $v_1, \ldots, v_{r+k-1} \in V$, it follows from Proposition 5.13 (1) that

$$I_{\alpha,\theta}(\lambda_1 \wedge \cdots \wedge \lambda_k)(v_1, \ldots, v_{r+k-1})$$

$$= (\lambda_1 \wedge \cdots \wedge \lambda_k)(\alpha(v_1, \ldots, v_r), \theta(v_{r+1}), \ldots, \theta(v_{r+k-1}))$$

$$= \sum_{\sigma \in S_k} \varepsilon(\sigma)\, \lambda_{\sigma(1)}(\alpha(v_1, \ldots, v_r))\, \lambda_{\sigma(2)}(\theta(v_{r+1})) \cdots \lambda_{\sigma(k)}(\theta(v_{r+k-1}))$$

$$= \begin{vmatrix} \lambda_1(\alpha(v_1, \ldots, v_r)) & \lambda_1(\theta(v_{r+1})) & \cdots & \lambda_1(\theta(v_{r+k-1})) \\ \lambda_2(\alpha(v_1, \ldots, v_r)) & \lambda_2(\theta(v_{r+1})) & \cdots & \lambda_2(\theta(v_{r+k-1})) \\ \vdots & \vdots & & \vdots \\ \lambda_k(\alpha(v_1, \ldots, v_r)) & \lambda_k(\theta(v_{r+1})) & \cdots & \lambda_k(\theta(v_{r+k-1})) \end{vmatrix}$$

$$= \sum_{i=1}^{k} (-1)^{i+1} \lambda_i(\alpha(v_1, \ldots, v_r))$$

$$(\lambda_1 \wedge \cdots \wedge \lambda_{i-1} \wedge \lambda_{i+1} \wedge \cdots \wedge \lambda_k)(\theta(v_{r+1}), \ldots, \theta(v_{r+k-1}))$$

$$= \sum_{i=1}^{k} (-1)^{i+1} I_{\alpha,\theta}(\lambda_i) \otimes \theta^*(\lambda_1 \wedge \cdots \wedge \lambda_{i-1} \wedge \lambda_{i+1} \wedge \cdots \wedge \lambda_k)(v_1, \ldots, v_{r+k-1})$$

where we have expanded the determinant along the first column. The result now follows by applying the alternation operator on both sides and using Definition 5.11. □

12.24 Proposition If $\theta \in \Lambda^1(V;W)$, $\alpha \in \Lambda^r(V;W)$, $\omega \in \Lambda^k(W)$ and $\eta \in \Lambda^l(W)$, we have that

$$i_{\alpha,\theta}(\omega \wedge \eta) = i_{\alpha,\theta}(\omega) \wedge \theta^*(\eta) + (-1)^{k(r-1)}\, \theta^*(\omega) \wedge i_{\alpha,\theta}(\eta)\,.$$

PROOF : If $\omega = \omega_1 \wedge \cdots \wedge \omega_k$ and $\eta = \eta_1 \wedge \cdots \wedge \eta_l$, where $\omega_1, \ldots, \omega_k, \eta_1, \ldots, \eta_l \in W^*$, it follows from Proposition 12.23 that

$$i_{\alpha,\theta}(\omega \wedge \eta) = \sum_{i=1}^{k} (-1)^{i+1} i_{\alpha,\theta}(\omega_i) \wedge \theta^*(\omega_1 \wedge \cdots \wedge \omega_{i-1} \wedge \omega_{i+1} \wedge \cdots \wedge \omega_k \wedge \eta)$$

$$+ \sum_{j=1}^{l} (-1)^{k+j+1} i_{\alpha,\theta}(\eta_j) \wedge \theta^*(\omega \wedge \eta_1 \wedge \cdots \wedge \eta_{j-1} \wedge \eta_{j+1} \wedge \cdots \wedge \eta_l)$$

$$= i_{\alpha,\theta}(\omega) \wedge \theta^*(\eta) + (-1)^{k(r-1)}\, \theta^*(\omega) \wedge i_{\alpha,\theta}(\eta)\,.$$

The general result now follows since every form $\omega \in \Lambda^k(W)$ and $\eta \in \Lambda^l(W)$ can be expressed as a linear combination of forms of the above type. □

12.25 Proposition Let

$$\begin{array}{ccc} V_1 & \xrightarrow{F_1} & V_2 \\ {\scriptstyle \theta_1}\downarrow & & \downarrow{\scriptstyle \theta_2} \\ W_1 & \xrightarrow[F_2]{} & W_2 \end{array}$$

be a commutative diagram of vector spaces and linear maps, and suppose that $\alpha_1 \in \Lambda^r(V_1;W_1)$ and $\alpha_2 \in \Lambda^r(V_2;W_2)$ are vector-valued r-forms satisfying

$$F_2 \circ \alpha_1 = \alpha_2 \circ {F_1}^r .$$

Then it follows that

$$i_{\alpha_1,\theta_1} F_2^*(\omega) = F_1^*(i_{\alpha_2,\theta_2}\, \omega)$$

for $\omega \in \Lambda^k(W_2)$.

PROOF: We have that

$$\begin{aligned} I_{\alpha_1,\theta_1} F_2^*(\omega) &= I_{\alpha_1,\theta_1}(\omega \circ {F_2}^k) = \omega \circ {F_2}^k \circ (\alpha_1 \times {\theta_1}^{k-1}) \\ &= \omega \circ (\alpha_2 \times {\theta_2}^{k-1}) \circ {F_1}^{r+k-1} = F_1^*(\omega \circ (\alpha_2 \times {\theta_2}^{k-1})) = F_1^*(I_{\alpha_2,\theta_2}\, \omega) , \end{aligned}$$

and the result now follows from Proposition 5.9. □

12.26 Proposition Let $f : N \to M$ be a smooth map. If $\alpha \in \Omega^r(N; f^*(TM))$ is a bundle-valued r-form along f and ω a k-form on M, we have an $(r+k-1)$-form $i_\alpha\, \omega$ on N, called the *interior product* of α and ω, defined by

$$(i_\alpha\, \omega)(p) = i_{\alpha(p), f_{*p}}\, \omega(f(p))$$

for each $p \in N$.

PROOF: We have that $i_\alpha\, \omega = \phi \circ \alpha$, where $\phi : \Lambda^r(TN; f^*(TM)) \to \Lambda^{r+k-1}(TN)$ is the bundle map over N obtained from ω as described in Remark 12.22. □

12.27 Let V and W be vector spaces, and let $\mathscr{B} = \{u_1, \ldots, u_n\}$ be a basis for V with dual basis $\mathscr{B}^* = \{u_1^*, \ldots, u_n^*\}$. Then we have a linear isomorphism

$$\widetilde{\phi} : L(W^*, \Lambda^k(V)) \to \Lambda_1^k(V, W)$$

defined by

$$\widetilde{\phi}(G)(v_1, \ldots, v_k, \lambda) = G(\lambda)(v_1, \ldots, v_k)$$

for $G \in L(W^*, \Lambda^k(V))$, $v_1, ..., v_k \in V$ and $\lambda \in W^*$, with inverse

$$\widetilde{\psi} : \Lambda_1^k(V,W) \to L(W^*, \Lambda^k(V))$$

given by

$$\widetilde{\psi}(\omega)(\lambda) = \sum_{i_1<...<i_k} \omega(u_{i_1}, ..., u_{i_k}, \lambda)\; u_{i_1}^* \wedge \cdots \wedge u_{i_k}^*$$

for $\omega \in \Lambda_1^k(V,W)$ and $\lambda \in W^*$. Indeed, we have that

$$\widetilde{\phi}(\widetilde{\psi}(\omega))(u_{j_1}, ..., u_{j_k}, \lambda) = \widetilde{\psi}(\omega)(\lambda)(u_{j_1}, ..., u_{j_k}) = \omega(u_{j_1}, ..., u_{j_k}, \lambda)$$

for $1 \leq j_1 < ... < j_k \leq n$ and

$$\begin{aligned}\widetilde{\psi}(\widetilde{\phi}(G))(\lambda) &= \sum_{i_1<...<i_k} \widetilde{\phi}(G)(u_{i_1}, ..., u_{i_k}, \lambda)\; u_{i_1}^* \wedge \cdots \wedge u_{i_k}^* \\ &= \sum_{i_1<...<i_k} G(\lambda)(u_{i_1}, ..., u_{i_k})\; u_{i_1}^* \wedge \cdots \wedge u_{i_k}^* = G(\lambda)\end{aligned}$$

which shows that $\widetilde{\psi} = \widetilde{\phi}^{-1}$. In particular, it follows that $\widetilde{\psi}$ does not depend on the choice of the basis $\mathscr{B}$.

If the space W is also finite dimensional, then it follows from 5.15 that we have a linear isomorphism

$$\mu_* : \Lambda^k(V) \otimes W \to \Lambda_1^k(V,W)$$

induced by the bilinear map

$$\mu : \Lambda^k(V) \times W \to \Lambda_1^k(V,W)$$

defined by

$$\mu\,(\omega, w) = \omega \otimes w$$

for $\omega \in \Lambda^k(V)$ and $w \in W$, i.e.,

$$\mu\,(\omega, w)(v_1, ..., v_k, \lambda) = \omega(v_1, ..., v_k)\, \lambda\,(w)$$

for $v_1, ..., v_k \in V$ and $\lambda \in W^*$. The composed map

$$\widetilde{\psi} \circ \mu : \Lambda^k(V) \times W \to L(W^*, \Lambda^k(V))$$

is given by

$$\widetilde{\psi} \circ \mu\,(\omega, w)(\lambda) = \lambda(w)\, \omega$$

for $\omega \in \Lambda^k(V)$, $w \in W$ and $\lambda \in W^*$.

Now let $\overline{\phi} : \Lambda^k(V;W) \to \Lambda_1^k(V,W)$ and $\overline{\psi} : \Lambda_1^k(V,W) \to \Lambda^k(V;W)$ be the linear isomorphisms induced by the isomorphisms $\phi : \mathscr{T}^k(V;W) \to \mathscr{T}_1^k(V,W)$ and $\psi : \mathscr{T}_1^k(V,W) \to \mathscr{T}^k(V;W)$ defined in 5.15. Then we have a linear isomorphism

$$\widetilde{\psi} \circ \overline{\phi} : \Lambda^k(V;W) \to L(W^*, \Lambda^k(V))$$

which is given by

$$\widetilde{\psi} \circ \overline{\phi}\,(\omega)(\lambda) = \lambda \circ \omega$$

for $\omega \in \Lambda^k(V;W)$ and $\lambda \in W^*$. Its inverse is the isomorphism

$$\overline{\psi} \circ \widetilde{\phi} : L(W^*, \Lambda^k(V)) \to \Lambda^k(V;W)$$

given by

$$\overline{\psi} \circ \widetilde{\phi}\,(G)(v_1, \ldots, v_k) = \sum_{i=1}^{m} G(w_i^*)(v_1, \ldots, v_k)\, w_i$$

for $G \in L(W^*, \Lambda^k(V))$ and $v_1, \ldots, v_k \in V$, where $\mathscr{C} = \{w_1, \ldots, w_m\}$ is an arbitrary basis for W with dual basis $\mathscr{C}^* = \{w_1^*, \ldots, w_m^*\}$.

12.28 Definition Let $f : N \to M$ be a smooth map between the smooth manifolds N and M. An $\mathbf{R}$-linear map $D : \Omega(M) \to \Omega(N)$ is called a *derivation along f of degree r* if it satisfies the following two properties:

(i) If $\omega \in \Omega^k(M)$, then $D\omega \in \Omega^{k+r}(N)$,
(ii) $D(\omega \wedge \eta) = (D\omega) \wedge f^*(\eta) + (-1)^{kr} f^*(\omega) \wedge (D\eta)$ whenever $\omega \in \Omega^k(M)$ and $\eta \in \Omega^l(M)$ (Leibniz' rule).

The derivation D is said to be of *type i_** if $Df = 0$ for every $f \in \Omega^0(M)$, and of *type d_** if

$$D \circ d = (-1)^{kr}\, d \circ D\,.$$

12.29 Proposition Let $f : N \to M$ be a smooth map between the smooth manifolds N and M, and let $\alpha \in \Omega^r(N; f^*(\,TM))$ be a bundle-valued r-form along f. Then we have a derivation $i_\alpha : \Omega(M) \to \Omega(N)$ along f of type i_* and degree $(r-1)$, where $i_\alpha\, \omega$ is the interior product of α and ω for each k-form ω on M.

Conversely, for every derivation $D : \Omega(M) \to \Omega(N)$ along f of type i_* and degree $(r-1)$, there is a unique bundle-valued r-form α along f so that $D = i_\alpha$. The map $D_1 : \Omega^1(M) \to \Omega^r(N)$ induced by D is linear over $\mathscr{F}(M)$, and for each such map D_1 there is a unique bundle-valued r-form α along f so that $D_1(\omega) = i_\alpha\, \omega$ for $\omega \in \Omega^1(M)$.

PROOF : The first part of the proposition follows from Propositions 12.24 and 12.26. To prove the last part, we first note that

$$D_1(g\,\omega) = (Dg) \wedge f^*(\omega) + f^*(g)\, D_1\, \omega = f^*(g)\, D_1\, \omega$$

for every $g \in \Omega^0(M)$ and $\omega \in \Omega^1(M)$, showing that D_1 is linear over $\mathscr{F}(M)$.

From this it follows that given a point $p \in N$, $D_1\, \omega(p)$ only depends on the value of ω at $f(p)$, i.e., $D_1\, \omega_1(p) = D_1\, \omega_2(p)$ whenever $\omega_1(f(p)) = \omega_2(f(p))$. Considering the difference $\omega_2 - \omega_1$ it is clearly enough to prove that $D_1\, \omega\,(p) = 0$ if

$\omega(f(p)) = 0$. To show this, we choose a local chart (x,U) around $f(p)$ and a smooth function $h : M \to \mathbf{R}$ with $h(f(p)) = 1$ and $\operatorname{supp}(h) \subset U$. If

$$\omega|_U = \sum_{i=1}^{n} b^i dx^i ,$$

we extend the functions $h\, b^i$ and 1-forms $h\, dx^i$ on U to smooth functions g_i and 1-forms θ_i on M, respectively, by defining them to be zero outside U. Then we have that

$$h^2 \omega = \sum_{i=1}^{n} g^i \theta_i$$

so that

$$(h \circ f)^2 D_1 \omega = D_1 (h^2 \omega) = \sum_{i=1}^{n} (g^i \circ f) D_1 \theta_i .$$

Evaluating at p and using that $h(f(p)) = 1$ and $g^i(f(p)) = 0$ for $i = 1, \ldots, n$, we conclude that $D_1\, \omega\, (p) = 0$ in the case when $\omega(f(p)) = 0$.

Now assume that $D = i_\alpha$ for a bundle-valued r-form α along f. Then we have that

$$D_1\, \omega(p) = i_\alpha\, \omega(p) = \omega(f(p)) \circ \alpha(p)$$

for $\omega \in \Omega^1(M)$ and $p \in N$. Let $\mathscr{C}_p = \{w_1, \ldots, w_n\}$ be a basis for $T_{f(p)}M$ with dual basis $\mathscr{C}_p^* = \{w_1^*, \ldots, w_n^*\}$, and choose 1-forms $\omega^1, \ldots, \omega^n$ on M with $\omega^i(f(p)) = w_i^*$ for $i = 1, \ldots, n$ using Proposition 2.56 (2). Then

$$\alpha(p)(v_1, \ldots, v_r) = \sum_{i=1}^{n} D_1\, \omega^i(p)(v_1, \ldots, v_r)\, w_i \tag{1}$$

for $v_1, \ldots, v_r \in T_pN$, which shows the uniqueness of α.

To show existence, we now define α by formula (1) for each $p \in N$. From 12.27 and the above arguments we know that it is well defined. It only remains to show that α is smooth on N. Fix $p \in N$, and choose a local chart (x,U) around $f(p)$ on M and an open neighbourhood V of $f(p)$ with $\overline{V} \subset U$. By Proposition 2.56 (1) there are 1-forms ω^i on M which coincide with dx^i on $\overline{V}$ for $i = 1, \ldots, n$. Then we have that

$$\alpha(q)(v_1, \ldots, v_r) = \sum_{i=1}^{n} D_1\, \omega^i(q)(v_1, \ldots, v_r) \left. \frac{\partial}{\partial x^i} \right|_{f(q)}$$

for $q \in f^{-1}(V)$ and $v_1, \ldots, v_r \in T_qN$, showing that α is smooth at p by Proposition 5.44 (2). As p was an arbitrary point in N, this completes the proof that α is smooth. Now since D and i_α coincide on $\Omega^0(M)$ and $\Omega^1(M)$, they coincide on $\Omega(M)$ by Leibniz' rule. □

12.30 Proposition Let $f : N \to M$ be a smooth map between the smooth manifolds N and M, and let $\alpha \in \Omega^r(N; f^*(\,TM))$ be a bundle-valued r-form along f.

Then we have a derivation $d_\alpha : \Omega(M) \to \Omega(N)$ along f of type d_* and degree r defined by

$$d_\alpha = i_\alpha \circ d + (-1)^r d \circ i_\alpha$$

Conversely, for every derivation $D : \Omega(M) \to \Omega(N)$ along f of type d_* and degree r, there is a unique bundle-valued r-form α along f so that $D = d_\alpha$.

PROOF : We see that d_α satisfies Leibniz' rule since

$$\begin{aligned}
d_\alpha(\omega \wedge \eta) &= i_\alpha\{(d\omega)\wedge\eta + (-1)^k \omega \wedge (d\eta)\}\\
&+ (-1)^r d\{(i_\alpha\omega)\wedge f^*(\eta) + (-1)^{k(r-1)} f^*(\omega)\wedge(i_\alpha\eta)\}\\
&= i_\alpha(d\omega)\wedge f^*(\eta) + (-1)^{(k+1)(r-1)} f^*(d\omega)\wedge(i_\alpha\eta)\\
&+ (-1)^k\{(i_\alpha\omega)\wedge f^*(d\eta) + (-1)^{k(r-1)} f^*(\omega)\wedge i_\alpha(d\eta)\}\\
&+ (-1)^r\{d(i_\alpha\omega)\wedge f^*(\eta) + (-1)^{r+k-1}(i_\alpha\omega)\wedge f^*(d\eta)\\
&+ (-1)^{k(r-1)}\{f^*(d\omega)\wedge(i_\alpha\eta) + (-1)^k f^*(\omega)\wedge d(i_\alpha\eta)\}\}\\
&= \{i_\alpha(d\omega) + (-1)^r d(i_\alpha\omega)\}\wedge f^*(\eta)\\
&+ (-1)^{kr} f^*(\omega)\wedge\{i_\alpha(d\eta) + (-1)^r d(i_\alpha\eta)\}\\
&+ \{(-1)^{(k+1)(r-1)} + (-1)^r(-1)^{k(r-1)}\} f^*(d\omega)\wedge(i_\alpha\eta)\\
&+ \{(-1)^k + (-1)^r(-1)^{r+k-1}\}(i_\alpha\omega)\wedge f^*(d\eta)\\
&= (d_\alpha\omega)\wedge f^*(\eta) + (-1)^{kr} f^*(\omega)\wedge(d_\alpha\eta)
\end{aligned}$$

whenever $\omega \in \Omega^k(M)$ and $\eta \in \Omega^l(M)$. Since d_α is clearly **R**-linear, it is a derivation along f of degree r which is of type d_* since

$$d_\alpha \circ d = (-1)^r d \circ i_\alpha \circ d = (-1)^r d \circ d_\alpha\,,$$

thus proving the first part of the proposition.

To prove the last part, let $D : \Omega(M) \to \Omega(N)$ be a derivation along f of type d_* and degree r. Given vector fields $X_1, \ldots, X_r \in \mathscr{T}_1(N)$, we have a map

$$F : \mathscr{F}(M) \to \mathscr{F}(N)$$

defined by

$$F(g) = (Dg)_N(X_1, \ldots, X_r)$$

for $g \in \mathscr{F}(M)$, satisfying

$$F(g_1 g_2) = f^*(g_2)F(g_1) + f^*(g_1)F(g_2)$$

for $g_1, g_2 \in \mathscr{F}(M)$. Then $F = X_N$ for a vector field X along f, and we obtain a map

$$G\colon \underbrace{\mathscr{T}_1(N) \times \cdots \times \mathscr{T}_1(N)}_{r} \to \Gamma(N; f^*(TM))$$

which is multilinear and skew symmetric over $\mathscr{F}(N)$. Hence there is a unique bundle-valued r-form $\alpha \in \Omega^r(N; f^*(TM))$ along f such that $G = \alpha_N$. We have that

$$d_\alpha g(p) = i_\alpha(dg)(p) = dg(f(p)) \circ \alpha(p)$$

for $p \in N$, so that

$$(d_\alpha g)_N(X_1, \dots, X_r)(p) = G(X_1, \dots, X_r)(p)(g) = (Dg)_N(X_1, \dots, X_r)(p)$$

for $g \in \mathscr{F}(M)$, $X_1, \dots, X_r \in \mathscr{T}_1(N)$ and $p \in N$, showing that $D = d_\alpha$. □

12.31 Proposition Let $f : N \to M$ be a smooth map between the smooth manifolds N and M. Then every derivation $D : \Omega(M) \to \Omega(N)$ along f can be written as a sum

$$D = i_D + d_D\,,$$

where i_D and d_D are derivations along f of type i_* and type d_*, respectively.

PROOF : Let d_D be the derivation along f of type d_* so that $d_D g = Dg$ for $g \in \Omega^0(M)$, and let $i_D = D - d_D$. □

12.32 Consider the commutative diagram

$$\begin{array}{ccc} N_1 & \xrightarrow{\;g_1\;} & N_2 \\ {\scriptstyle f_1}\big\downarrow & & \big\downarrow{\scriptstyle f_2} \\ M_1 & \xrightarrow[\;g_2\;]{} & M_2 \end{array}$$

of smooth manifolds and mappings. Two bundle-valued r-forms $\alpha_1 \in \Omega^r(N_1; f_1^*(TM_1))$ and $\alpha_2 \in \Omega^r(N_2; f_2^*(TM_2))$ along f_1 and f_2, respectively, are said to be (g_1, g_2)-related if

$$g_{2*f_1(p)} \circ \alpha_1(p) = \alpha_2(g_1(p)) \circ (g_{1*p})^r$$

for $p \in N_1$, i.e.,

$$g_{2*} \circ \alpha_1(p)(v_1, \dots, v_r) = \alpha_2(g_1(p))(g_{1*}(v_1), \dots, g_{1*}(v_r))$$

for $p \in N_1$ and $v_1, \dots, v_r \in T_pN_1$. By Proposition 12.25 this implies that

$$i_{\alpha_1}\, g_2^*(\omega) = g_1^*(i_{\alpha_2}\, \omega)$$

for $\omega \in \Omega^k(M_2)$, since

$$(i_{\alpha_1}\, g_2^*(\omega))(p) = i_{\alpha_1(p),f_{1*p}}\; g_2^*(\omega)\,(f_1(p)) = i_{\alpha_1(p),f_{1*p}}\; g_{2*f_1(p)}^*(\omega(g_2(f_1(p))))$$
$$= g_{1*p}^*(i_{\alpha_2(g_1(p)),f_{2*g_1(p)}}\;\omega(f_2(g_1(p)))) = g_{1*p}^*(i_{\alpha_2}\,\omega(g_1(p))) = g_1^*(i_{\alpha_2}\,\omega)(p)$$

for every $p \in N_1$. It also follows that

$$d_{\alpha_1}\, g_2^*(\omega) = g_1^*(d_{\alpha_2}\,\omega)$$

for $\omega \in \Omega^k(M_2)$. Note that α_2 is uniquely determined by α_1 when g_1 is a surjective submersion.

Assuming that f_1 and g_1 are surjective submersions, let $\alpha_1 \in \Omega^r(N_1; f_1^*(TM_1))$ and $\alpha_2 \in \Omega^r(N_2; f_2^*(TM_2))$ be (g_1,g_2)-related bundle-valued r-forms along f_1 and f_2, respectively, and let $\beta_1 \in \Omega^s(N_1; g_1^*(TN_2))$ and $\beta_2 \in \Omega^s(M_1; g_2^*(TM_2))$ be (f_1,f_2)-related bundle-valued s-forms along g_1 and g_2. Consider the map $D : \Omega(M_2) \to \Omega(N_1)$ given by

$$D = d_{\alpha_1} \circ d_{\beta_2} - (-1)^{rs}\, d_{\beta_1} \circ d_{\alpha_2}\,.$$

We see that D satisfies Leibniz' rule since

$$D(\omega \wedge \eta) = d_{\alpha_1}\{(d_{\beta_2}\,\omega) \wedge g_2^*(\eta) + (-1)^{ks} g_2^*(\omega) \wedge (d_{\beta_2}\,\eta)\}$$
$$-\,(-1)^{rs} d_{\beta_1}\{(d_{\alpha_2}\,\omega) \wedge f_2^*(\eta) + (-1)^{kr} f_2^*(\omega) \wedge (d_{\alpha_2}\,\eta)\}$$
$$= d_{\alpha_1}(d_{\beta_2}\,\omega) \wedge (g_2 \circ f_1)^*(\eta) + (-1)^{(k+s)r} f_1^*(d_{\beta_2}\,\omega) \wedge d_{\alpha_1}(g_2^*(\eta))$$
$$+\,(-1)^{ks}\{d_{\alpha_1}(g_2^*(\omega)) \wedge f_1^*(d_{\beta_2}\,\eta) + (-1)^{kr}(g_2 \circ f_1)^*(\omega) \wedge d_{\alpha_1}(d_{\beta_2}\,\eta)\}$$
$$-\,(-1)^{rs}\{d_{\beta_1}(d_{\alpha_2}\,\omega) \wedge (f_2 \circ g_1)^*(\eta) + (-1)^{(k+r)s} g_1^*(d_{\alpha_2}\,\omega) \wedge d_{\beta_1}(f_2^*(\eta))$$
$$+\,(-1)^{kr}\{d_{\beta_1}(f_2^*(\omega)) \wedge g_1^*(d_{\alpha_2}\,\eta) + (-1)^{ks}(f_2 \circ g_1)^*(\omega) \wedge d_{\beta_1}(d_{\alpha_2}\,\eta)\}\}$$
$$= \{d_{\alpha_1}(d_{\beta_2}\,\omega) - (-1)^{rs} d_{\beta_1}(d_{\alpha_2}\,\omega)\} \wedge (g_2 \circ f_1)^*(\eta)$$
$$+\,(-1)^{k(r+s)}(g_2 \circ f_1)^*(\omega) \wedge \{d_{\alpha_1}(d_{\beta_2}\,\eta) - (-1)^{rs} d_{\beta_1}(d_{\alpha_2}\,\eta)\}$$
$$+\,\{(-1)^{(k+s)r} - (-1)^{rs}(-1)^{kr}\} f_1^*(d_{\beta_2}\,\omega) \wedge g_1^*(d_{\alpha_2}\,\eta)$$
$$+\,\{(-1)^{ks} - (-1)^{rs}(-1)^{(k+r)s}\}\; g_1^*(d_{\alpha_2}\,\omega) \wedge f_1^*(d_{\beta_2}\,\eta)$$
$$= (D\omega) \wedge (g_2 \circ f_1)^*(\eta) + (-1)^{k(r+s)}(g_2 \circ f_1)^*(\omega) \wedge (D\eta)$$

whenever $\omega \in \Omega^k(M_2)$ and $\eta \in \Omega^l(M_2)$. Since D is clearly **R**-linear, it is a derivation along $g_2 \circ f_1$ of degree $r+s$ which is of type d_*.

By Proposition 12.30 there is now a unique bundle-valued $(r+s)$-form $[\,\alpha_1, \beta_1\,] \in \Omega^{r+s}(N_1;\, (g_2 \circ f_1)^*(TM_2))$ along $g_2 \circ f_1$, called the *Frölicher–Nijenhuis bracket* of

α_1 and β_1, so that

$$D = d_{[\alpha_1,\beta_1]} .$$

12.33 Proposition Let $\alpha,\beta \in \Omega^1(M;TM)$ be bundle-valued 1-forms on M. Then we have that

$$\begin{aligned}[\alpha,\beta](X,Y) = \beta(\alpha([X,Y])) + [\alpha(X),\beta(Y)] - \beta([\alpha(X),Y]) - \beta([X,\alpha(Y)])\\ + \alpha(\beta([X,Y])) + [\beta(X),\alpha(Y)] - \alpha([\beta(X),Y]) - \alpha([X,\beta(Y)])\end{aligned}$$

for every pair of vector fields X and Y on M.

PROOF : For each $f \in \mathscr{F}(M)$ we have that

$$\begin{aligned}&[\alpha,\beta](X,Y)(f) = df([\alpha,\beta](X,Y)) = i_{[\alpha,\beta]}(df)(X,Y) = d_{[\alpha,\beta]}(f)(X,Y)\\ &= d_\alpha \circ d_\beta(f)(X,Y) + d_\beta \circ d_\alpha(f)(X,Y) ,\end{aligned}$$

and Remark 5.86 implies that

$$\begin{aligned}&d_\alpha \circ d_\beta(f)(X,Y) = (i_\alpha \circ d - d \circ i_\alpha) \circ i_\beta(df)(X,Y)\\ &= d(df \cdot \beta)(\alpha(X),Y) + d(df \cdot \beta)(X,\alpha(Y)) - d(df \cdot \beta \cdot \alpha)(X,Y)\\ &= \{\alpha(X)(df \cdot \beta(Y)) - Y(df \cdot \beta \cdot \alpha(X)) - df \cdot \beta([\alpha(X),Y])\}\\ &+ \{X(df \cdot \beta \cdot \alpha(Y)) - \alpha(Y)(df \cdot \beta(X)) - df \cdot \beta([X,\alpha(Y)])\}\\ &- \{X(df \cdot \beta \cdot \alpha(Y)) - Y(df \cdot \beta \cdot \alpha(X)) - df \cdot \beta \cdot \alpha([X,Y])\}\\ &= \beta(\alpha([X,Y]))(f) + \alpha(X)\,\beta(Y)(f) - \alpha(Y)\,\beta(X)(f)\\ &- \beta([\alpha(X),Y])(f) - \beta([X,\alpha(Y)])(f)\end{aligned}$$

Combining this we obtain the formula in the proposition. □

12.34 Corollary Let $\alpha \in \Omega^1(M;TM)$ be a bundle-valued 1-form on M. Then we have that

$$\tfrac{1}{2}\,[\alpha,\alpha](X,Y) = \alpha(\alpha([X,Y])) + [\alpha(X),\alpha(Y)] - \alpha([\alpha(X),Y]) - \alpha([X,\alpha(Y)])$$

for every pair of vector fields X and Y on M.

FIRST ORDER JET BUNDLES

12.35 Definition Let $\pi : E \to M$ be a bundle over M with fibre F, and let p be a point on M. We say that two sections $\phi, \psi \in \Gamma_p(E)$ have *contact of order* 0 at p if $\phi(p) = \psi(p)$. They are said to have *kth order contact* at p for $k \geq 1$ if ϕ_* and ψ_* have $(k-1)$st order contact at every point in T_pM. This defines an equivalence relation in $\Gamma_p(E)$ for each $k \geq 0$, and the equivalence class containing ϕ is called the *k-jet of* ϕ *at* p and is denoted by $j_p^k\phi$. The points p and $\phi(p)$ are called the *source* and *target* of $j_p^k\phi$, respectively, and the set of equivalence classes in $\Gamma_p(E)$ is called the *k-jet space of E at p* and is denoted by J_p^kE. We define the *k-jet manifold of E* to be the disjoint union

$$J^kE = \bigcup_{p \in M} J_p^kE\,.$$

The 0-jet manifold J^0E is identified with E. The maps $\pi_k : J^kE \to M$ and $\pi_{k,0} : J^kE \to E$ defined by $\pi_k(j_p^k\phi) = p$ and $\pi_{k,0}(j_p^k\phi) = \phi(p)$, respectively, for $p \in M$ and $\phi \in \Gamma_p(E)$, are called the *source* and *target projections*. If $0 \leq l \leq k$, the l-jet projection $\pi_{k,l} : J^kE \to J^lE$ is the map defined by $\pi_{k,l}(j_p^k\phi) = j_p^l\phi$. We have that $\pi_{k,m} = \pi_{l,m} \circ \pi_{k,l}$ when $0 \leq m \leq l \leq k$, and $\pi_k = \pi \circ \pi_{k,0}$ when $k \geq 0$.

12.36 Proposition Let $\pi : E \to M$ be a bundle over M^n with fibre F^m, and let $\widetilde{\pi}' : \pi^*(TM) \to E$ be the pullback of the tangent bundle $\pi' : TM \to M$ by π, and $\pi'' : V \to E$ be the vertical bundle of π. Then $\pi_{1,0} : J^1E \to E$ is an affine bundle over E modelled on the vector bundle $\pi''' : \Lambda^1(\pi^*(TM);V) \to E$. A section of $\pi_{1,0}$ is called a *jet field* on E.

PROOF: Consider a local chart (z,W) around a point u_0 on E as described in Proposition 12.3, and let $z_1 : W \to \mathbf{R}^n$ and $z_2 : W \to \mathbf{R}^m$ be the component maps of $z : W \to \mathbf{R}^n \times \mathbf{R}^m$. Then we have a bijection $\tau_z : \pi_{1,0}^{-1}(W) \to W \times L(\mathbf{R}^n, \mathbf{R}^m)$ which is given by

$$\tau_z(j_p^1\phi) = (\phi(p), D(z_2 \circ \phi \circ x^{-1})(x(p)))$$

for $p \in O$ and $\phi \in \Gamma_p(E)$ with $\phi(p) = u \in W$. We see that τ_z is well defined since

$$D(z_2 \circ \phi \circ x^{-1})(x(p)) = pr_2 \circ t_{z,u} \circ \phi_{*p} \circ t_{x,p}^{-1}\,,$$

where $pr_2 : \mathbf{R}^n \times \mathbf{R}^m \to \mathbf{R}^m$ is the projection on the second factor.

Suppose that $(\widetilde{z}, \widetilde{W})$ is another local chart on E obtained in the same way from a local trivialization $(\widetilde{t}, \pi^{-1}(\widetilde{U}))$ in E and local charts $(\widetilde{x}, \widetilde{O})$ and $(\widetilde{y}, \widetilde{Q})$ on M and F, respectively, with $\widetilde{O} \subset \widetilde{U}$. Then we have that

$$\begin{gathered}\tau_{\widetilde{z}} \circ \tau_z^{-1}(u,G) = \\ (u, \{D_1(\widetilde{z}_2 \circ z^{-1})(z(u)) + D_2(\widetilde{z}_2 \circ z^{-1})(z(u)) \circ G\} \circ D(x \circ \widetilde{x}^{-1})(\widetilde{z}_1(u)))\end{gathered} \tag{1}$$

for $u \in W \cap \widetilde{W}$ and $G \in L(\mathbf{R}^n, \mathbf{R}^m)$. Hence it follows from Proposition 12.6 that $\pi_{1,0} : J^1E \to E$ is a bundle over E with fibre $\mathbf{R}^{nm}$, and with $((id_W \times \beta) \circ \tau_z, \pi_{1,0}^{-1}(W))$ as a local trivialization for each local chart (z, W) on E of the above form, using the linear isomorphism $\beta : L(\mathbf{R}^n, \mathbf{R}^m) \to \mathbf{R}^{nm}$ defined in Corollary 5.29.

By Proposition 5.28 we have a local trivialization $((id_W \times \beta) \circ \tau_z', \pi'''^{-1}(W))$ in the vector bundle $\pi''' : \Lambda^1(\pi^*(TM); V) \to E$ associated with the local chart (z, W) on E, where $\tau_z' : \pi'''^{-1}(W) \to W \times L(\mathbf{R}^n, \mathbf{R}^m)$ is the equivalence over W defined by

$$\tau_z'(F) = (u, pr_2 \circ t_{z,u} \circ F \circ t_{x,p}^{-1})$$

for $u \in W$, $F \in \Lambda^1(\pi^*(TM); V)_u$ and $p = \pi(u)$. We also have that

$$\tau_{\widetilde{z}}' \circ \tau_z'^{-1}(u, G) = (u, D_2(\widetilde{z}_2 \circ z^{-1})(z(u)) \circ G \circ D(x \circ \widetilde{x}^{-1})(\widetilde{z}_1(u))) \qquad (2)$$

for $u \in W \cap \widetilde{W}$ and $G \in L(\mathbf{R}^n, \mathbf{R}^m)$. Using this, we can define an operation of the vector space $\Lambda^1(\pi^*(TM); V)_u$ on the fibre J^1E_u by

$$\tau_{z,u}(j_p^1\phi + F) = \tau_{z,u}(j_p^1\phi) + \tau_{z,u}'(F) . \qquad (3)$$

In fact, we have that $j_p^1\phi + F = j_p^1\psi$ for a section $\psi \in \Gamma_p(E)$ given by

$$\psi = z^{-1} \circ \{z \circ \phi + t_{z,u} \circ F \circ t_{x,p}^{-1} \circ T_{x(p)}^{-1} \circ x\} ,$$

where $T_{x(p)} : \mathbf{R}^n \to \mathbf{R}^n$ is translation by $x(p)$, since this implies that

$$z_2 \circ \psi \circ x^{-1} = z_2 \circ \phi \circ x^{-1} + pr_2 \circ t_{z,u} \circ F \circ t_{x,p}^{-1} \circ T_{x(p)}^{-1} .$$

It follows from (1) with $G = \tau_{z,u}(j_p^1\phi + F)$ and $G = \tau_{z,u}(j_p^1\phi)$, (2) with $G = \tau_{z,u}'(F)$ and (3) that we also have

$$\tau_{\widetilde{z},u}(j_p^1\phi + F) = \tau_{\widetilde{z},u}(j_p^1\phi) + \tau_{\widetilde{z},u}'(F) ,$$

showing that the operation on J^1E_u given by (3) is well defined and does not depend on the choice of local chart (z, W). □

12.37 Remark From each local chart (z, W) on E with coordinate functions x^i and y^α defined in Remark 12.4, we obtain an induced local chart $((z \times \beta) \circ \tau_z, \pi_{1,0}^{-1}(W))$ on J^1E with coordinate functions x^i, y^α and y_i^α defined by

$$x^i(j_p^1\phi) = x^i(p) \ , \quad y^\alpha(j_p^1\phi) = y^\alpha \circ \phi(p) \quad \text{and} \quad y_i^\alpha(j_p^1\phi) = \frac{\partial(y^\alpha \circ \phi)}{\partial x^i}(p)$$

for $j_p^1\phi \in \pi_{1,0}^{-1}(W)$, $i = 1, 2, \ldots, n$ and $\alpha = 1, 2, \ldots, m$. It will be clear from the context whether x^i and y^α denote coordinate functions on J^1E, on E or on M and F respectively. The functions y_i^α are called *derivative coordinates*.

12.38 Proposition Let $\pi : E \to M$ be a bundle over M^n with fibre F^m. Then we have a canonical embedding $\lambda : J^1E \to \Lambda^1(\pi^*(TM);TE)$ given by

$$\lambda(j_p^1\phi) = \phi_{*p}$$

for $p \in M$ and $\phi \in \Gamma_p(E)$, which is an affine bundle map over E. Its linear part is the map $\alpha : \Lambda^1(\pi^*(TM);V) \to \Lambda^1(\pi^*(TM);TE)$ given by

$$\alpha(F) = i_u \circ F$$

for $u \in E$ and $F \in \Lambda^1(\pi^*(TM);V)_u$, where $i : V \to TE$ is the inclusion map.

PROOF: Let (z,W) be a local chart around a point u_0 in E as described in Proposition 12.3, and let $\tau_z : \pi_{1,0}^{-1}(W) \to W \times L(\mathbf{R}^n,\mathbf{R}^m)$ be the equivalence over W defined in the proof of Proposition 12.36. Furthemore, let $\tau'_z : \pi'^{-1}(W) \to W \times L(\mathbf{R}^n,\mathbf{R}^m)$ and $\tau''_z : \pi''^{-1}(W) \to W \times L(\mathbf{R}^n,\mathbf{R}^{n+m})$ be the equivalences over W in the vector bundles $\pi' : \Lambda^1(\pi^*(TM);V) \to E$ and $\pi'' : \Lambda^1(\pi^*(TM);TE) \to E$ defined in Proposition 5.28. Then we have that

$$\tau''_{z,u} \circ \lambda_u \circ \tau_{z,u}^{-1}(G)(a) = (a, G(a))$$

and

$$\tau''_{z,u} \circ \alpha_u \circ \tau'^{-1}_{z,u}(G)(a) = (0, G(a))$$

for $u \in W$, $G \in L(\mathbf{R}^n,\mathbf{R}^m)$ and $a \in \mathbf{R}^n$, showing that

$$\lambda_u \circ \tau_{z,u}^{-1}(G+G') = \lambda_u \circ \tau_{z,u}^{-1}(G) + \alpha_u \circ \tau'^{-1}_{z,u}(G')$$

for $G, G' \in L(\mathbf{R}^n,\mathbf{R}^m)$, which completes the proof of the proposition. □

12.39 Proposition If $\pi : E \to M$ is a trivial bundle over M with fibre F, then $\pi_{1,0} : J^1E \to E$ is a vector bundle over E.

PROOF: Let $t : E \to M \times F$ be a trivialization over M, and let $pr_1 : M \times F \to M$ and $pr_2 : M \times F \to F$ be the projections on the first and second factor. Then we have a section $s : E \to J^1E$ of $\pi_{1,0}$ given by $s(u) = j_p^1\phi_u$ for $u \in E$, where $p = \pi(u)$ and $\phi_u : M \to E$ is the section of π defined by

$$\phi_u(q) = t^{-1}(q, pr_2 \circ t(u))$$

for $q \in M$. Note that $\phi_u(p) = u$ when $p \in M$ and $u \in E_p$ since $p = \pi(u) = pr_1 \circ t(u)$, which shows that $\pi_{1,0} \circ s = id_E$.

In order to prove that s is smooth, consider a local chart (z,W) around a point u_0 on E as described in Proposition 12.3. Using the same notation as in the proof of Proposition 12.36, we have that

$$z_2 \circ \phi_u(q) = z_2(u)$$

for $u \in W$ and $q \in M$, which implies that

$$\tau_z \circ s(u) = (u, 0)$$

for $u \in W$. The result now follows from Proposition 12.18. □

12.40 Proposition Let $\pi : E \to M$ be a bundle over M^n with fibre F^m. Then $\pi_1 : J^1E \to M$ is also a bundle.

PROOF: Let $(t, \pi^{-1}(U))$ be a local trivialization in $\pi : E \to M$, where U is the coordinate neighbourhood of a local chart (x, U) on M, and consider the trivial bundle $\pi' : \mathbf{R}^n \times F \to \mathbf{R}^n$. Since $\pi'_1 = \pi' \circ \pi'_{1,0}$ is a submersion, we have that $J^1_0(\mathbf{R}^n \times F) = \pi'^{-1}_1(0)$ is a closed submanifold of $J^1(\mathbf{R}^n \times F)$.

Now let $t' : \pi_1^{-1}(U) \to U \times J^1_0(\mathbf{R}^n \times F)$ be the map given by

$$t'(j^1_p\phi) = (p, j^1_0\psi)$$

for $p \in U$ and $\phi \in \Gamma_p(E)$, where $\psi \in \Gamma_0(\mathbf{R}^n \times F)$ is the section defined by

$$\psi = (T_{x(p)}^{-1} \times id_F) \circ (x \times id_F) \circ t \circ \phi \circ x^{-1} \circ T_{x(p)} ,$$

using the translation $T_{x(p)} : \mathbf{R}^n \to \mathbf{R}^n$ by $x(p)$. We want to show that t' is a diffeomorphism. Consider the local chart (z, W) in E obtained from $(t, \pi^{-1}(U))$, (x, U) and a local chart (y, Q) on F as described in Proposition 12.3, so that $W = t^{-1}(U \times Q)$ and $z = (x \times y) \circ t$. We also obtain the local chart $(\widetilde{z}, \mathbf{R}^n \times Q)$ in $\mathbf{R}^n \times F$, where $\widetilde{z} = id_{\mathbf{R}^n} \times y$. With the same notation as in the proof of Proposition 12.36, we have that

$$\widetilde{z}_2 \circ \psi = z_2 \circ \phi \circ x^{-1} \circ T_{x(p)} ,$$

which shows that

$$\widetilde{z}_2 \circ \psi(0) = z_2 \circ \phi(p)$$

and

$$D(\widetilde{z}_2 \circ \psi)(0) = D(z_2 \circ \phi \circ x^{-1})(x(p)) .$$

Using the local charts $((z \times \beta) \circ \tau_z, \pi_{1,0}^{-1}(W))$ and (η, O) in J^1E and $J^1_0(\mathbf{R}^n \times F)$, respectively, where $O = \pi'^{-1}_{1,0}(\{0\} \times Q)$ and $\eta = (\widetilde{z}_2 \times \beta) \circ \tau_{\widetilde{z}}|_O$, we now have that

$$\begin{aligned}(x \times \eta) \circ t'(j^1_p\phi) &= (x(p), \widetilde{z}_2 \circ \psi(0), \beta \circ D(\widetilde{z}_2 \circ \psi)(0)) \\ &= (z \circ \phi(p), \beta \circ D(z_2 \circ \phi \circ x^{-1})(x(p))) = (z \times \beta) \circ \tau_z(j^1_p\phi) .\end{aligned}$$

This completes the proof that $\pi_1 : J^1E \to M$ is a bundle, with $(t', \pi_1^{-1}(U))$ as a local trivialization for each local trivialization $(t, \pi^{-1}(U))$ in $\pi : E \to M$ of the above form. □

12.41 Corollary The 1-jet bundle $\pi_1 : J^1(\mathbf{R}^n \times F) \to \mathbf{R}^n$ of the trivial bundle $\pi : \mathbf{R}^n \times F \to \mathbf{R}^n$ is trivial, having a trivialization $t : J^1(\mathbf{R}^n \times F) \to \mathbf{R}^n \times J^1_0(\mathbf{R}^n \times F)$ over $\mathbf{R}^n$ given by

$$t(j^1_a\phi) = (a, j^1_0\psi)$$

for $a \in \mathbf{R}^n$ and $\phi \in \Gamma_a(\mathbf{R}^n \times F)$, where $\psi \in \Gamma_0(\mathbf{R}^n \times F)$ is the section defined by

$$\psi = (T_a^{-1} \times id_F) \circ \phi \circ T_a .$$

PROOF : Follows from the proof of Proposition 12.40, using the identity as a local trivialization over $\mathbf{R}^n$ in $\pi : \mathbf{R}^n \times F \to \mathbf{R}^n$, and as a coordinate map on $\mathbf{R}^n$. □

12.42 Proposition Let $(\tilde{f}, f)$ be a bundle map between the bundles $\pi : E \to M$ and $\widetilde{\pi} : \widetilde{E} \to \widetilde{M}$, where $f : M \to \widetilde{M}$ is a diffeomorphism. Then we have a smooth map $j^1\tilde{f} : J^1E \to J^1\widetilde{E}$, called the 1-*jet prolongation of* $\tilde{f}$, defined by

$$j^1\tilde{f}\,(j^1_p\phi) = j^1_{f(p)}(\tilde{f} \circ \phi \circ f^{-1})$$

for $p \in M$ and $\phi \in \Gamma_p(E)$. We have that $(j^1\tilde{f}, \tilde{f})$ is an affine bundle map between the affine bundles $\pi_{1,0} : J^1E \to E$ and $\widetilde{\pi}_{1,0} : J^1\widetilde{E} \to \widetilde{E}$. It is modelled on the bundle map $(\lambda, \tilde{f})$ between the vector bundles $\pi''' : \Lambda^1(\pi^*(TM); V) \to E$ and $\widetilde{\pi}''' : \Lambda^1(\pi^*(T\widetilde{M}); \widetilde{V}) \to \widetilde{E}$, where $\lambda(F) = \tilde{f}_{*u} \circ F \circ f_{*p}^{-1}$ for $u \in E$, $F \in \Lambda^1(\pi^*(TM); V)_u$ and $p = \pi(u)$. We also have that $(j^1\tilde{f}, f)$ is a bundle map between the bundles $\pi_1 : J^1E \to M$ and $\widetilde{\pi}_1 : J^1\widetilde{E} \to \widetilde{M}$.

PROOF : Let u_0 be a point on E, and consider local charts (z, W) and $(\widetilde{z}, \widetilde{W})$ around u_0 and $\tilde{f}(u_0)$, respectively, of the form described in Proposition 12.3. Using the same notation as in the proof of Proposition 12.36, we have that

$$\begin{aligned} \tau_{\tilde{z}} \circ j^1\tilde{f} \circ \tau_z^{-1}(u, G) = (\tilde{f}(u), \{D_1(\widetilde{z}_2 \circ \tilde{f} \circ z^{-1})(z(u)) \\ + D_2(\widetilde{z}_2 \circ \tilde{f} \circ z^{-1})(z(u)) \circ G\} \circ D(x \circ f^{-1} \circ \widetilde{x}^{-1})(\widetilde{z}_1 \circ \tilde{f}(u))) \end{aligned} \tag{1}$$

and

$$\begin{aligned} \tau'_{\tilde{z}} \circ \lambda \circ \tau_z'^{-1}(u, G) = \\ (\tilde{f}(u), D_2(\widetilde{z}_2 \circ \tilde{f} \circ z^{-1})(z(u)) \circ G \circ D(x \circ f^{-1} \circ \widetilde{x}^{-1})(\widetilde{z}_1 \circ \tilde{f}(u))) \end{aligned} \tag{2}$$

for $u \in W \cap \tilde{f}^{-1}(\widetilde{W})$ and $G \in L(\mathbf{R}^n, \mathbf{R}^m)$. It now follows from (1) with $G = \tau_{z,u}(j^1_p\phi + F)$ and $G = \tau_{z,u}(j^1_p\phi)$, and (2) with $G = \tau'_{z,u}(F)$ that we have

$$j^1\tilde{f}\,(j^1_p\phi + F) = j^1\tilde{f}\,(j^1_p\phi) + \lambda(F)$$

for $u \in W \cap \tilde{f}^{-1}(\widetilde{W})$, $j^1_p\phi \in J^1E_u$ and $F \in \Lambda^1(\pi^*(TM); V)_u$, showing that $j^1\tilde{f}$ is an affine map with linear part λ on each fibre. We see that $(j^1\tilde{f}, \tilde{f})$ is a bundle map since

$$\begin{aligned} \widetilde{\pi}_{1,0} \circ j^1\tilde{f}\,(j^1_p\phi) &= \widetilde{\pi}_{1,0}(j^1_{f(p)}(\tilde{f} \circ \phi \circ f^{-1})) = \{\tilde{f} \circ \phi \circ f^{-1}\} \circ f(p) \\ &= \tilde{f} \circ \phi(p) = \tilde{f} \circ \pi_{1,0}(j^1_p\phi) \end{aligned}$$

for $p \in M$ and $\phi \in \Gamma_p(E)$, which shows that

$$\widetilde{\pi}_{1,0} \circ j^1\tilde{f} = \tilde{f} \circ \pi_{1,0}\,.$$

From this it also follows that $(j^1\tilde{f}, f)$ is a bundle map since

$$\widetilde{\pi}_1 \circ j^1\tilde{f} = \widetilde{\pi} \circ \widetilde{\pi}_{1,0} \circ j^1\tilde{f} = \widetilde{\pi} \circ \tilde{f} \circ \pi_{1,0} = f \circ \pi \circ \pi_{1,0} = f \circ \pi_1\,.$$ □

12.43 Proposition

(1) Suppose that $(\tilde{f}, f)$ is a bundle map from the bundle $\pi : E \to M$ to the bundle $\pi' : E' \to M'$, and that $(\tilde{g}, g)$ is a bundle map from the bundle $\pi' : E' \to M'$ to the bundle $\pi'' : E'' \to M''$, where $f : M \to M'$ and $g : M' \to M''$ are diffeomorphisms. Then we have that

$$j^1(\tilde{g} \circ \tilde{f}) = j^1\tilde{g} \circ j^1\tilde{f} .$$

(2) Let (id_E, id_M) be the bundle map from the bundle $\pi : E \to M$ to itself, where $id_M : M \to M$ and $id_E : E \to E$ are the identity maps on M and E, respectively. Then we have that

$$j^1(id_E) = id_{J^1E} .$$

(3) If $(\tilde{f}, f)$ is an equivalence from the bundle $\pi : E \to M$ to the bundle $\pi' : E' \to M'$, then $j^1\tilde{f} : J^1E \to J^1E'$ is a diffeomorphism, and we have that

$$(j^1\tilde{f})^{-1} = j^1(\tilde{f}^{-1}) .$$

PROOF: (1) We have that

$$\begin{aligned} j^1(\tilde{g} \circ \tilde{f})\,(j^1_p\phi) &= j^1_{g(f(p))}(\tilde{g} \circ \tilde{f} \circ \phi \circ f^{-1} \circ g^{-1}) \\ &= j^1\tilde{g}\,(j^1_{f(p)}(\tilde{f} \circ \phi \circ f^{-1})) = j^1\tilde{g} \circ j^1\tilde{f}\,(j^1_p\phi) \end{aligned}$$

for $p \in M$ and $\phi \in \Gamma_p(E)$.

(2) We have that

$$j^1(id_E)\,(j^1_p\phi) = j^1_{id_M(p)}(id_E \circ \phi \circ id_M^{-1}) = j^1_p\phi$$

for $p \in M$ and $\phi \in \Gamma_p(E)$.

(3) By (1) and (2) we have that

$$j^1\tilde{f} \circ j^1(\tilde{f}^{-1}) = j^1(\tilde{f} \circ \tilde{f}^{-1}) = j^1(id_E) = id_{J^1E}$$

and

$$j^1(\tilde{f}^{-1}) \circ j^1\tilde{f} = j^1(\tilde{f}^{-1} \circ \tilde{f}) = j^1(id_E) = id_{J^1E} .$$

□

12.44 Remark A section $\phi : M \to E$ in the bundle $\pi : E \to M$ is a bundle map over M between the bundles $id_M : M \to M$ and $\pi : E \to M$. Identifying J^1M with M, the 1-jet prolongation $j^1\phi : M \to J^1E$ of ϕ defined in Proposition 12.42 is given by

$$j^1\phi\,(p) = j^1_p\phi$$

for $p \in M$.

Let $(\tilde{f}, f)$ be a bundle map between the bundles $\pi : E \to M$ and $\widetilde{\pi} : \widetilde{E} \to \widetilde{M}$, where $f : M \to \widetilde{M}$ is a diffeomorphism. If $\phi : M \to E$ is a section of π, then the section $\tilde{f} \circ \phi \circ f^{-1} : \widetilde{M} \to \widetilde{E}$ of $\tilde{\pi}$ is called the *push-forward* of ϕ by $(\tilde{f}, f)$, and it is denoted by $\tilde{f}_*(\phi)$. By Proposition 12.42 we have that

$$j^1\tilde{f} \circ j^1\phi = j^1(\tilde{f}_*(\phi)) \circ f$$

which implies that

$$(j^1\tilde{f})_*(j^1\phi) = j^1(\tilde{f}_*(\phi)) \,.$$

HOLONOMIC TANGENT VECTORS

12.45 Lemma Given a section $\phi \in \Gamma_p(E)$ with $\phi(p) = u$ in the bundle $\pi : E \to M$ over M with fibre F, the tangent space T_uE can be written as a direct sum

$$T_uE = H_u \oplus V_u$$

where $H_u = \phi_{*p}(T_pM)$ and $V_u = \ker \pi_{*u}$. The projections $h_u : T_uE \to T_uE$ and $v_u : T_uE \to T_uE$ on the subspaces H_u and V_u are given by $h_u = \phi_{*p} \circ \pi_{*u}$ and $v_u = id_{T_uE} - h_u$.

PROOF : For each $w \in T_uE$ we have that

$$w = \phi_{*p} \circ \pi_{*u}(w) + \{w - \phi_{*p} \circ \pi_{*u}(w)\}$$

where $\phi_{*p} \circ \pi_{*u}(w) \in H_u$ and $w - \phi_{*p} \circ \pi_{*u}(w) \in V_u$.

Suppose that $w \in H_u \cap V_u$. Since $w \in H_u$, we have that $w = \phi_{*p}(v)$ for a vector $v \in T_pM$. Now using that $w \in V_u$, it follows that $v = \pi_{*u}(w) = 0$ which shows that $w = 0$. □

12.46 Proposition Let $\pi : E \to M$ be a bundle over M^n with fibre F^m, and let $\tilde{\pi}' : \pi_{1,0}^*(TE) \to J^1E$ be the pullback of the tangent bundle $\pi' : TE \to E$ by the target projection $\pi_{1,0} : J^1E \to E$. Then we have two subbundles $\tilde{\pi}'' : \widetilde{H} \to J^1E$ and $\tilde{\pi}''' : \widetilde{V} \to J^1E$ of $\tilde{\pi}'$ so that

$$\pi_{1,0}^*(TE)_\sigma = \widetilde{H}_\sigma \oplus \widetilde{V}_\sigma$$

for $\sigma \in J^1E$, where $\widetilde{H}_\sigma = \{\sigma\} \times \phi_*(T_pM)$ and $\widetilde{V}_\sigma = \{\sigma\} \times V_u$ when $\sigma = j^1_p\phi$ and $\phi(p) = u$. They are called the bundles of *holonomic* and *vertical* tangent vectors, respectively. The maps $\tilde{h} : \pi_{1,0}^*(TE) \to \pi_{1,0}^*(TE)$ and $\tilde{v} : \pi_{1,0}^*(TE) \to \pi_{1,0}^*(TE)$ which are projections on $\widetilde{H}_\sigma$ and $\widetilde{V}_\sigma$ in each fibre $\pi_{1,0}^*(TE)_\sigma$, are bundle maps over J^1E called the *holonomic* and *vertical projection.*

PROOF : Using the same notation as in the proof of Proposition 12.36, we have that

$$pr_1 \circ t_{z,u} = t_{x,p} \circ \pi_{*u}$$

where $pr_1 : \mathbf{R}^n \times \mathbf{R}^m \to \mathbf{R}^n$ is the projection on the first factor and $u = \pi(p)$.

We have a local trivialization $\widetilde{\tau}_z : \widetilde{\pi}'^{-1}(\widetilde{W}) \to \widetilde{W} \times \mathbf{R}^n \times \mathbf{R}^m$ in the vector bundle $\widetilde{\pi}' : \pi_{1,0}^*(TE) \to J^1E$ associated with the local chart (z, W) on E, where $\widetilde{W} = \pi_{1,0}^{-1}(W)$ and

$$\begin{aligned}\widetilde{\tau}_z(j_p^1\phi, w) &= (j_p^1\phi, t_{x,p} \circ \pi_{*u}(w), pr_2 \circ t_{z,u}(w - \phi_{*p} \circ \pi_{*u}(w))) \\ &= (j_p^1\phi, pr_1 \circ t_{z,u}(w), \{pr_2 - \tau_{z,u}(j_p^1\phi) \circ pr_1\} \circ t_{z,u}(w))\end{aligned}$$

for $p \in O$, $\phi \in \Gamma_p(E)$ with $\phi(p) = u \in W$ and $w \in T_uE$, since

$$(\tau_{z,u} \times id_{\mathbf{R}^n \times \mathbf{R}^m}) \circ \widetilde{\tau}_z \circ (\tau_{z,u} \times t_{z,u})^{-1}(G, a, b) = (G, a, b - G(a))$$

for $G \in L(\mathbf{R}^n, \mathbf{R}^m)$ and $(a, b) \in \mathbf{R}^n \times \mathbf{R}^m$. Using that $w - \phi_{*p} \circ \pi_{*u}(w) \in V_u$ so that

$$pr_1 \circ t_{z,u}(w - \phi_{*p} \circ \pi_{*u}(w)) = 0\,,$$

we now have that

$$\widetilde{\tau}_z(\widetilde{\pi}'^{-1}(\widetilde{W}) \cap \widetilde{H}) = \widetilde{W} \times \mathbf{R}^n \times \{0\}$$

and

$$\widetilde{\tau}_z(\widetilde{\pi}'^{-1}(\widetilde{W}) \cap \widetilde{V}) = \widetilde{W} \times \{0\} \times \mathbf{R}^m\,,$$

completing the proof that $\widetilde{\pi}'' : \widetilde{H} \to J^1E$ and $\widetilde{\pi}''' : \widetilde{V} \to J^1E$ are subbundles of $\widetilde{\pi}' : \pi_{1,0}^*(TE) \to J^1E$ satisfying the assertions in the proposition. The last part of the proposition now follows from Proposition 2.64. □

12.47 Remark Suppose that $(\sigma, w) \in \pi_{1,0}^*(TE)$ where $\sigma = j_p^1\phi$, $\phi(p) = u$ and $w \in T_uE$ is the tangent vector given by

$$w = \sum_{i=1}^n a^i \frac{\partial}{\partial x^i}\bigg|_u + \sum_{\alpha=1}^m b^\alpha \frac{\partial}{\partial y^\alpha}\bigg|_u$$

in the local coordinates described in Remark 12.4. Then we have that

$$\begin{aligned}\widetilde{h}(\sigma, w) = (\sigma, \phi_{*p} \circ \pi_{*u}(w)) &= \left(\sigma, \sum_{i=1}^n a^i\, \phi_{*p}\left(\frac{\partial}{\partial x^i}\bigg|_p\right)\right) \\ &= \left(\sigma, \sum_{i=1}^n a^i \left\{\frac{\partial}{\partial x^i}\bigg|_u + \sum_{\alpha=1}^m y_i^\alpha(\sigma)\, \frac{\partial}{\partial y^\alpha}\bigg|_u\right\}\right)\end{aligned}$$

and

$$\widetilde{v}(\sigma,w) = (\sigma, w - \phi_{*p}\circ\pi_{*u}(w)) = \left(\sigma, \sum_{\alpha=1}^{m}\left\{ b^{\alpha} - \sum_{i=1}^{n} a^{i}\, y_{i}^{\alpha}(\sigma)\right\}\frac{\partial}{\partial y^{\alpha}}\bigg|_{u}\right).$$

12.48 Definition Let $\pi : E \to M$ be a bundle over M^n with fibre F^m. A vector field $X : J^1E \to TE$ along $\pi_{1,0}$ is called a *total derivative* if the section $\widetilde{X} : J^1E \to \pi_{1,0}^*(TE)$ determined by X belongs to $\widetilde{H}$, i.e., if

$$X(j_p^1\phi) \in \phi_*(T_pM)$$

for $p \in M$ and $\phi \in \Gamma_p(E)$. Let (z,W) be a local chart on E with coordinate functions described in Remark 12.4. The vector field $X_i : \pi_{1,0}^{-1}(W) \to TE$ along $\pi_{1,0}$ defined by

$$X_i(\sigma) = \phi_*\left(\frac{\partial}{\partial x^i}\bigg|_p\right) = \frac{\partial}{\partial x^i}\bigg|_u + \sum_{\alpha=1}^{m} y_i^{\alpha}(\sigma)\,\frac{\partial}{\partial y^{\alpha}}\bigg|_u$$

for $\sigma \in \pi_{1,0}^{-1}(W)$ where $\sigma = j_p^1\phi$ and $\phi(p) = u$, is called the i-th *coordinate total derivative* and is denoted by $\frac{d}{dx^i}$ for $i = 1, \ldots, n$. The tangent vector $X_i(\sigma)$ in T_uE is denoted by $\frac{d}{dx^i}\Big|_{\sigma}$, and we have that

$$\frac{df}{dx^i}(\sigma) = \frac{d}{dx^i}\bigg|_{\sigma}(f) = \phi_*\left(\frac{\partial}{\partial x^i}\bigg|_p\right)(f) = \frac{\partial}{\partial x^i}\bigg|_p(f\circ\phi) = \frac{\partial(f\circ\phi)}{\partial x^i}(p)$$

when $\sigma = j_p^1\phi$, $\phi(p) = u$ and $f \in \mathscr{F}_u(E)$, so that

$$\frac{df}{dx^i}\circ j^1\phi = \frac{\partial(f\circ\phi)}{\partial x^i}.$$

12.49 Let $\pi : E \to M$ be a bundle over M^n with fibre F^m, and let $\pi' : TE \to E$ and $\pi_1' : TJ^1E \to J^1E$ be the tangent bundles of E and J^1E. By 12.13 the map $(\pi_1', \pi_{1,0\,*}) : TJ^1E \to J^1E \times TE$ induces a bundle map $\rho_1 : TJ^1E \to \pi_{1,0}^*(TE)$ over J^1E, as we see from the commutative diagram

$$\begin{array}{ccc} TJ^1E & \xrightarrow{\ \pi_{1,0\,*}\ } & TE \\ {\scriptstyle \pi_1'}\big\downarrow & & \big\downarrow{\scriptstyle \pi'} \\ J^1E & \xrightarrow{\ \pi_{1,0}\ } & E \end{array}$$

Using the local charts (z,W) and $((z\times\beta)\circ\tau_z,\pi_{1,0}^{-1}(W))$ on E and J^1E with coordinate functions described in Remarks 12.4 and 12.37, we have that

$$\rho_1\left(\left.\frac{\partial}{\partial x^i}\right|_\sigma\right)=\left(\sigma,\left.\frac{\partial}{\partial x^i}\right|_u\right),\quad \rho_1\left(\left.\frac{\partial}{\partial y^\alpha}\right|_\sigma\right)=\left(\sigma,\left.\frac{\partial}{\partial y^\alpha}\right|_u\right)$$

$$\text{and}\qquad \rho_1\left(\left.\frac{\partial}{\partial y_i^\alpha}\right|_\sigma\right)=(\sigma,0)$$

for $\sigma\in\pi_{1,0}^{-1}(W)$ where $\pi_{1,0}(\sigma)=u$.

Let $\widetilde{\omega}\in\Omega^1(J^1E;\pi_{1,0}^*(TE))$ be the bundle-valued 1-form determined by the composed bundle map $\widehat{h}=\widetilde{h}\circ\rho_1:TJ^1E\to\pi_{1,0}^*(TE)$ over J^1E as described in Proposition 5.30. Then it follows from Proposition 5.44 and Remark 12.47 that

$$\phi_1\circ\widetilde{\omega}\,|_{\pi_{1,0}^{-1}(W)}=\sum_{i=1}^n dx^i\otimes\frac{d}{dx^i}\tag{1}$$

where $\phi_1:\Lambda^1(TJ^1E;\,\pi_{1,0}^*(TE))\to\mathscr{T}_1^1(TJ^1E,\pi_{1,0}^*(TE))$ is the equivalence over J^1E defined as in 5.25.

Let $\pi_{1,0}':V_{1,0}\to J^1E$ be the vertical tangent bundle of the bundle $\pi_{1,0}:J^1E\to E$. The composed map $\widehat{v}=\widetilde{v}\circ\rho_1:TJ^1E\to\pi_{1,0}^*(TE)$ is a bundle map over J^1E with $\dim\ker(\widehat{v}_\sigma)=n(1+m)$ for every $\sigma\in J^1E$. By Proposition 2.65 we therefore have that $C=\ker(\widehat{v})$ is a distribution on J^1E, called the *Cartan distribution*, which is given by $C_\sigma=\rho_1^{-1}(\widetilde{H}_\sigma)=(j^1\phi)_*(T_pM)\oplus V_{1,0\,\sigma}$ for $\sigma\in J^1E$ where $\sigma=j_p^1\phi$.

By Proposition 12.36 we know that $\pi_{1,0}:J^1E\to E$ is an affine bundle modelled on the vector bundle $\pi''':\Lambda^1(\pi^*(TM);V)\to E$ over E, where $\pi'':V\to E$ is the vertical tangent bundle of $\pi:E\to M$. According to Proposition 12.19 we therefore have an equivalence $\eta:V_{1,0}\to\pi_{1,0}^*(\Lambda^1(\pi^*(TM);V))$ over J^1E from the vertical tangent bundle $\pi_{1,0}':V_{1,0}\to J^1E$ of $\pi_{1,0}:J^1E\to E$ to the vector bundle $\widetilde{\pi}''':\pi_{1,0}^*(\Lambda^1(\pi^*(TM);V))\to J^1E$ induced from $\pi''':\Lambda^1(\pi^*(TM);V)\to E$ by $\pi_{1,0}$. If $\phi_2:\Lambda^1(\pi_1^*(TM);\pi_{1,0}^*(V))\to\mathscr{T}_1^1(\pi_1^*(TM),\pi_{1,0}^*(V))$ is the equivalence over J^1E defined as in 5.25, we have that

$$\phi_2\circ\eta\left(\left.\frac{\partial}{\partial y_i^\alpha}\right|_\sigma\right)=(\sigma,dx^i(p))\otimes\left(\sigma,\left.\frac{\partial}{\partial y^\alpha}\right|_u\right)\tag{2}$$

for $i=1,2,\ldots,n$, $\alpha=1,2,\ldots,m$ and $\sigma\in\pi_{1,0}^{-1}(W)$, where $\sigma=j_p^1\phi$ and $\phi(p)=u$. Indeed we have that

$$\widetilde{y}_i^\alpha=\beta^{\,b(i,\alpha)}\circ\tau_{z,u}$$

so that

$$t_{\beta\,\circ\,\tau_{z,u},\,\sigma}\left(\left.\frac{\partial}{\partial\widetilde{y}_i^\alpha}\right|_\sigma\right)=e_{b(i,\alpha)}$$

by Remark 2.82, where $\widetilde{y}_i^\alpha = y_i^\alpha|_{\pi_{1,0}^{-1}(u)}$, $\beta : L(\mathbf{R}^n, \mathbf{R}^m) \to \mathbf{R}^{nm}$ is the linear isomorphism defined in Corollary 5.29, $b : I_n \times I_m \to I_{nm}$ is the bijection defined in Remark 4.27, and $\mathscr{E} = \{e_1, ..., e_{nm}\}$ is the standard basis for $\mathbf{R}^{nm}$. Hence it follows that

$$\eta\left(\left.\frac{\partial}{\partial y_i^\alpha}\right|_\sigma\right) = (\sigma, F_i^\alpha),$$

where $F_i^\alpha : T_pM \to V_u$ is the linear map given by

$$F_i^\alpha\left(\left.\frac{\partial}{\partial x^j}\right|_p\right) = \delta_{ij}\left.\frac{\partial}{\partial y^\alpha}\right|_u$$

for $j = 1, ... , n$, which implies formula (2).

Let $\omega \in \Omega^1(M)$ be a 1-form on M. Then we have a bundle-valued 1-form $S_\omega \in \Omega^1(J^1E; V_{1,0})$ on J^1E given by

$$S_\omega(\sigma)(\xi) = \{\phi_2 \circ \eta\}^{-1}\{(\sigma, \omega(p)) \otimes \widehat{v}(\xi)\}$$

for $\sigma \in J^1E$ and $\xi \in T_\sigma J^1E$ where $\pi_1(\sigma) = p$. If

$$\omega|_U = \sum_{i=1}^n a_i\, dx^i,$$

then it follows from Remark 12.47 that

$$\begin{aligned} dy_i^\alpha(\sigma) \circ S_\omega(\sigma)\left(\left.\frac{\partial}{\partial x^j}\right|_\sigma\right) &= \omega(p)\left(\left.\frac{\partial}{\partial x^i}\right|_p\right)\ (\sigma, dy^\alpha(u)) \circ \widehat{v}\left(\left.\frac{\partial}{\partial x^j}\right|_\sigma\right) \\ &= -a_i(p)\, y_j^\alpha(\sigma) \end{aligned}$$

and

$$\begin{aligned} dy_i^\alpha(\sigma) \circ S_\omega(\sigma)\left(\left.\frac{\partial}{\partial y^\beta}\right|_\sigma\right) &= \omega(p)\left(\left.\frac{\partial}{\partial x^i}\right|_p\right)\ (\sigma, dy^\alpha(u)) \circ \widehat{v}\left(\left.\frac{\partial}{\partial y^\beta}\right|_\sigma\right) \\ &= a_i(p)\, \delta_{\alpha\beta}, \end{aligned}$$

showing that

$$\phi_3 \circ S_\omega|_{\pi_{1,0}^{-1}(W)} = \sum_{i,\alpha}(a_i \circ \pi_1)\left\{dy^\alpha - \sum_{j=1}^n y_j^\alpha\, dx^j\right\} \otimes \frac{\partial}{\partial y_i^\alpha} \tag{3}$$

where $\phi_3 : \Lambda^1(TJ^1E; V_{1,0}) \to \mathscr{T}_1^1(TJ^1E, V_{1,0})$ is the equivalence over J^1E defined as in 5.25.

Now let $\theta \in \Omega^1(J^1E)$ be a 1-form on J^1E, and consider the map $D_\theta : \Omega^1(M) \to \Omega^1(J^1E)$ given by

$$D_\theta(\omega) = i_{S_\omega}\theta$$

for $\omega \in \Omega^1(M)$. We see that it is linear over $\mathscr{F}(M)$ since

$$D_\theta(f\omega) = i_{\pi_1^*(f)\,S_\omega}\theta = \pi_1^*(f)\,D_\theta(\omega)$$

for $f \in \mathscr{F}(M)$. By Proposition 12.29 there is therefore a unique bundle-valued 1-form $S_\theta \in \Omega^1(J^1E;\pi_1^*(TM))$ along π_1 so that

$$D_\theta(\omega) = i_{S_\theta}\omega$$

for $\omega \in \Omega^1(M)$. It is given by

$$S_\theta(\sigma)(\xi) = \sum_{i=1}^n D_\theta(\omega^i)(\sigma)(\xi)\left.\frac{\partial}{\partial x^i}\right|_p = \sum_{i=1}^n \theta(\sigma)\circ S_{\omega^i}(\sigma)(\xi)\left.\frac{\partial}{\partial x^i}\right|_p$$

for $\sigma \in J^1E$ and $\xi \in T_\sigma J^1E$ where $\pi_1(\sigma) = p$, and where $\omega^1,\ldots,\omega^n$ are 1-forms on M with $\omega^i(p) = dx^i(p)$ for $i = 1,\ldots,n$. If

$$\theta\,|_{\pi_{1,0}^{-1}(W)} = \sum_i a_i\,dx^i + \sum_\alpha b_\alpha\,dy^\alpha + \sum_{i,\alpha} c_\alpha^i\,dy_i^\alpha\,,$$

then

$$\phi_4\circ S_\theta\,|_{\pi_{1,0}^{-1}(W)} = \sum_{i=1}^n\left\{\sum_{\alpha=1}^m c_\alpha^i\left\{dy^\alpha - \sum_{j=1}^n y_j^\alpha\,dx^j\right\}\right\}\otimes\left\{\frac{\partial}{\partial x^i}\circ\pi_1\right\} \tag{4}$$

where $\phi_4 : \Lambda^1(TJ^1E;\pi_1^*(TM)) \to \mathscr{T}_1^1(TJ^1E,\pi_1^*(TM))$ is the equivalence over J^1E defined as in 5.25.

Finally, suppose that we have an n-form $\varepsilon \in \Omega^n(M)$ on M, and consider the map $D_\varepsilon : \Omega^1(J^1E) \to \Omega^n(J^1E)$ given by

$$D_\varepsilon(\theta) = i_{S_\theta}\varepsilon$$

for $\theta \in \Omega^1(J^1E)$. We see that it is linear over $\mathscr{F}(J^1E)$ since

$$D_\varepsilon(f\theta) = i_{fS_\theta}\varepsilon = f\,D_\varepsilon(\theta)$$

for $f \in \mathscr{F}(J^1E)$, and that $D_\varepsilon(\theta) = 0$ when θ belongs to $A(V_{1,0})$. By Proposition 12.29 there is therefore a unique bundle-valued n-form $S_\varepsilon \in \Omega^n(J^1E;V_{1,0})$ on J^1E so that

$$D_\varepsilon(\theta) = i_{S_\varepsilon}\theta$$

for $\theta \in \Omega^1(J^1E)$. It is given by

$$S_\varepsilon(\sigma)(\xi_1,\ldots,\xi_n) = \sum_{i,\alpha} D_\varepsilon(\theta_i^\alpha)(\sigma)(\xi_1,\ldots,\xi_n)\left.\frac{\partial}{\partial y_i^\alpha}\right|_\sigma$$

$$= \sum_{i,\alpha} i_{S_{\theta_i^\alpha}}\varepsilon(\sigma)(\xi_1,\ldots,\xi_n)\left.\frac{\partial}{\partial y_i^\alpha}\right|_\sigma$$

for $\sigma \in J^1E$ and $\xi_1, \dots, \xi_n \in T_\sigma J^1E$, where θ_i^α are 1-forms on J^1E with $\theta_i^\alpha(\sigma) = dy_i^\alpha(\sigma)$ for $i = 1, \dots, n$ and $\alpha = 1, \dots, m$. If

$$\varepsilon\,|_U = a\, dx^1 \wedge \dots \wedge dx^n\,,$$

then it follows from Proposition 12.23 that

$$i_{S_{\theta_i^\alpha}}\varepsilon\,|_{\pi_{1,0}^{-1}(W)} = \sum_{k=1}^{n} (-1)^{k+1}\, i_{S_{\theta_i^\alpha}}(dx^k) \wedge \pi_1^*(a\, dx^1 \wedge \cdots \wedge dx^{k-1} \wedge dx^{k+1} \wedge \cdots \wedge dx^n)$$

$$= (a \circ \pi_1)\, dx^1 \wedge \cdots \wedge dx^{i-1} \wedge \left\{ dy^\alpha - \sum_{j=1}^{n} y_j^\alpha\, dx^j \right\} \wedge dx^{i+1} \wedge \cdots \wedge dx^n$$

so that

$$\phi_5 \circ S_\varepsilon\,|_{\pi_{1,0}^{-1}(W)} = \tag{5}$$

$$\sum_{i,\alpha} \left\{ (a \circ \pi_1)\, dx^1 \wedge \cdots \wedge dx^{i-1} \wedge \left\{ dy^\alpha - \sum_{j=1}^{n} y_j^\alpha\, dx^j \right\} \wedge dx^{i+1} \wedge \cdots \wedge dx^n \right\} \otimes \frac{\partial}{\partial y_i^\alpha}$$

where $\phi_5 : \Lambda^n(TJ^1E; V_{1,0}) \to \Lambda_1^n(TJ^1E, V_{1,0})$ is the equivalence over J^1E defined as in 5.25. The bundle-valued n-form S_ε will be important in our study of the calculus of variations in 12.90.

CONTACT COTANGENT VECTORS

12.50 Lemma Given a section $\phi \in \Gamma_p(E)$ with $\phi(p) = u$ in the bundle $\pi : E \to M$ over M with fibre F, the cotangent space T_u^*E can be written as a direct sum

$$T_u^*E = A(H_u) \oplus A(V_u)$$

where $A(H_u) = \ker(\phi_{*p})^*$ and $A(V_u) = \alpha(\pi^*(T^*M)_u)$.

PROOF: We have that $A(H_u) = \ker(\phi_{*p})^*$ since

$$\lambda(H_u) = (\phi_{*p})^*(\lambda)(T_pM)$$

for $\lambda \in T_u^*E$. The proposition therefore follows from Proposition 4.20 and Remark 12.15. □

12.51 Proposition Let $\pi : E \to M$ be a bundle over M^n with fibre F^m, and let $\widetilde{\pi}' : \pi_{1,0}^*(T^*E) \to J^1E$ be the pullback of the cotangent bundle $\pi' : T^*E \to E$ by the target projection $\pi_{1,0} : J^1E \to E$. Then we have two subbundles $\widetilde{\pi}'' : A(\widetilde{H}) \to J^1E$ and $\widetilde{\pi}''' : A(\widetilde{V}) \to J^1E$ of $\widetilde{\pi}'$ so that

$$\pi_{1,0}^*(T^*E)_\sigma = A(\widetilde{H})_\sigma \oplus A(\widetilde{V})_\sigma$$

for $\sigma \in J^1E$, where $A(\widetilde{H})_\sigma = \{\sigma\} \times \ker(\phi_{*p})^*$ and $A(\widetilde{V})_\sigma = \{\sigma\} \times \alpha(\pi^*(T^*M)_u)$ when $\sigma = j_p^1\phi$ and $\phi(p) = u$. They are called the bundles of *contact* and *horizontal* cotangent vectors, respectively. The maps $\widetilde{v}^* : \pi_{1,0}^*(T^*E) \to \pi_{1,0}^*(T^*E)$ and $\widetilde{h}^* : \pi_{1,0}^*(T^*E) \to \pi_{1,0}^*(T^*E)$ which are projections on $A(\widetilde{H})_\sigma$ and $A(\widetilde{V})_\sigma$ in each fibre $\pi_{1,0}^*(T^*E)_\sigma$, are bundle maps over J^1E called the *contact* and *horizontal projection*.

PROOF : Follows from Propositions 2.64, 4.22, 12.46 and 12.50. □

12.52 Remark Suppose that $(\sigma,\lambda) \in \pi_{1,0}^*(T^*E)$ where $\sigma = j_p^1\phi$, $\phi(p) = u$ and $\lambda \in T_u^*E$ is the cotangent vector given by

$$\lambda = \sum_{i=1}^{n} a_i\, dx^i(u) + \sum_{\alpha=1}^{m} b_\alpha\, dy^\alpha(u)$$

in the local coordinates described in Remark 12.4. Then it follows from Remark 12.47 that

$$\begin{aligned}
\widetilde{v}^*(\sigma,\lambda) &= (\sigma, \lambda \circ v_u) \\
&= \left(\sigma, \sum_{\alpha=1}^{m} \lambda \circ v_u\left(\left.\frac{\partial}{\partial y^\alpha}\right|_u\right) dy^\alpha(u) + \sum_{i=1}^{n} \lambda \circ v_u\left(\left.\frac{\partial}{\partial x^i}\right|_u\right) dx^i(u)\right) \\
&= \left(\sigma, \sum_{\alpha=1}^{m} \lambda\left(\left.\frac{\partial}{\partial y^\alpha}\right|_u\right) dy^\alpha(u) - \sum_{i=1}^{n} \lambda\left(\sum_{\alpha=1}^{m} y_i^\alpha(\sigma) \left.\frac{\partial}{\partial y^\alpha}\right|_u\right) dx^i(u)\right) \\
&= \left(\sigma, \sum_{\alpha=1}^{m} b_\alpha \left\{dy^\alpha(u) - \sum_{i=1}^{n} y_i^\alpha(\sigma)\, dx^i(u)\right\}\right)
\end{aligned}$$

and

$$\widetilde{h}^*(\sigma,\lambda) = (\sigma, \lambda \circ h_u)$$

$$= \left(\sigma, \sum_{i=1}^{n} \lambda \circ h_u \left(\left.\frac{\partial}{\partial x^i}\right|_u \right) dx^i(u) + \sum_{\alpha=1}^{m} \lambda \circ h_u \left(\left.\frac{\partial}{\partial y^\alpha}\right|_u \right) dy^\alpha(u) \right)$$

$$= \left(\sigma, \sum_{i=1}^{n} \lambda \left(\left.\frac{\partial}{\partial x^i}\right|_u + \sum_{\alpha=1}^{m} y_i^\alpha(\sigma) \left.\frac{\partial}{\partial y^\alpha}\right|_u \right) dx^i(u) \right)$$

$$= \left(\sigma, \sum_{i=1}^{n} \left\{ a_i + \sum_{\alpha=1}^{m} b_\alpha y_i^\alpha(\sigma) \right\} dx^i(u) \right).$$

12.53 Remark By Proposition 12.14 the vector bundle $\widetilde{\pi}' : \pi_{1,0}^*(T^*E) \to J^1E$ can be identified with a subbundle of the cotangent bundle $\pi' : T^*J^1E \to J^1E$ by means of the bundle map $\alpha_1 : \pi_{1,0}^*(T^*E) \to T^*J^1E$ over J^1E defined by

$$\alpha_1(\sigma,\lambda) = \lambda \circ \pi_{1,0*\sigma}$$

for $\sigma \in J^1E$ and $\lambda \in T_u^*E$ where $u = \pi_{1,0}(\sigma)$, which is an embedding. We have that

$$\alpha_1(\sigma, dx^i(u)) = dx^i(\sigma) \quad \text{and} \quad \alpha_1(\sigma, dy^\alpha(u)) = dy^\alpha(\sigma).$$

By Remark 12.52 we see that

$$\left\{ dy^1 - \sum_{i=1}^{n} y_i^1\, dx^i, \ldots, dy^m - \sum_{i=1}^{n} y_i^m\, dx^i \right\} \quad \text{and} \quad \{dx^1, \ldots, dx^n\}$$

are local bases for the subbundles $\pi'' : \alpha_1(A(\widetilde{H})) \to J^1E$ and $\pi''' : \alpha_1(A(\widetilde{V})) \to J^1E$, respectively.

JET FIELDS AND CONNECTIONS

12.54 Definition A *connection* in a bundle $\pi : E \to M$ is a distribution H on E such that

$$T_uE = H_u \oplus V_u$$

for every $u \in E$. The subspace H_u of T_uE is called the *horizontal subspace* at u, and the vectors in H_u are called *horizontal tangent vectors* at u.

By Proposition 2.64 the maps $h : TE \to TE$ and $v : TE \to TE$ which are projections on H_u and V_u in each fibre T_uE, are bundle maps over E called the *horizontal* and *vertical projection* determined by H. The bundle-valued 1-form $\omega \in \Omega^1(E;TE)$

determined by the horizontal projection h as described in Proposition 5.30, is called the *connection form* of H.

12.55 Let $\pi : E \to M^n$ be a bundle with fibre F^m, and let $\Gamma : E \to J^1E$ be a jet field on E. Then we have a commutative diagram

$$\begin{array}{ccccccc}
TE & \xrightarrow{\widetilde{\Gamma}} & \pi_{1,0}^*(TE) & \xrightarrow{\widetilde{v}} & \pi_{1,0}^*(TE) & \xrightarrow{\widetilde{\pi}_{1,0}} & TE \\
\pi' \downarrow & & \widetilde{\pi}' \downarrow & & \widetilde{\pi}' \downarrow & & \pi' \downarrow \\
E & \xrightarrow{\Gamma} & J^1E & \xrightarrow{id_{J^1E}} & J^1E & \xrightarrow{\pi_{1,0}} & E
\end{array}$$

where $\widetilde{\Gamma} = (\Gamma \circ \pi', id_{TE})$ is the map defined by $\widetilde{\Gamma}(w) = (\Gamma(u), w)$ for $u \in E$ and $w \in T_uE$. We see that $v = \widetilde{\pi}_{1,0} \circ \widetilde{v} \circ \widetilde{\Gamma} : TE \to TE$ is a bundle map over E with $\dim \ker(v_u) = n$ for every $u \in E$. By Proposition 2.65 we thereforc have that $H = \ker(v)$ is a connection in E, called the *connection determined by the jet field* Γ, which is given by $H_u = \phi_*(T_pM)$ for $u \in E$ where $\Gamma(u) = j_p^1\phi$.

Every connection H in E is determined by a unique jet field Γ. To define $\Gamma(u)$ for a point $u \in E$, let (z, W) be a local chart around u as described in Proposition 12.3, and let $h_z : W \times \mathbf{R}^{n+m} \to W \times \mathbf{R}^{n+m}$ be the bundle map over W given by $h_z = t_z \circ h \circ t_z^{-1}$, where h is the horizontal projection determined by H. Now we define $\Gamma(u) = j_p^1\phi$ for the section $\phi \in \Gamma_p(E)$ given by

$$\phi = z^{-1} \circ T_{z(u)} \circ h_{z,u} \circ i_1 \circ T_{x(p)}^{-1} \circ x$$

where $\pi(u) = p$, $T_{x(p)}$ and $T_{z(u)}$ are translations by $x(p)$ and $z(u)$ in $\mathbf{R}^n$ and $\mathbf{R}^{n+m}$, respectively, and $i_1 : \mathbf{R}^n \to \mathbf{R}^n \times \mathbf{R}^m$ is the injection given by $i_1(a) = (a, 0)$ for $a \in \mathbf{R}^n$. If $\tau_z : \pi_{1,0}^{-1}(W) \to W \times L(\mathbf{R}^n, \mathbf{R}^m)$ is the equivalence over W defined in the proof of Proposition 12.36, we have that

$$\tau_z \circ \Gamma(u) = (\phi(p), D(z_2 \circ \phi \circ x^{-1})(x(p)) = (u, pr_2 \circ h_{z,u} \circ i_1)$$

where $pr_2 : \mathbf{R}^n \times \mathbf{R}^m \to \mathbf{R}^m$ is the projection on the second factor, showing that Γ is a jet field on E. We also have that

$$t_{z,u} \circ \phi_{*p} \circ t_{x,p}^{-1} = D(z \circ \phi \circ x^{-1})(x(p)) = h_{z,u} \circ i_1 \ ,$$

which is equivalent to

$$\phi_*(T_pM) = t_{z,u}^{-1} \circ h_{z,u}(\mathbf{R}^n \times \{0\}) = h_u \circ t_{z,u}^{-1}(\mathbf{R}^n \times \{0\}) = H_u$$

since

$$T_uE = t_{z,u}^{-1}(\mathbf{R}^n \times \{0\}) \oplus t_{z,u}^{-1}(\{0\} \times \mathbf{R}^m) = t_{z,u}^{-1}(\mathbf{R}^n \times \{0\}) \oplus V_u \,.$$

This shows that H is the connection determined by the jet field Γ, and that Γ is uniquely determined by H.

12.56 Remark If ω is the connection form of H, then it follows from Proposition 5.44 and Remark 12.47 that

$$\phi_1 \circ \omega|_W = \sum_{i=1}^{n} dx^i \otimes \left\{ \frac{\partial}{\partial x^i} + \sum_{\alpha=1}^{m} (y_i^\alpha \circ \Gamma) \frac{\partial}{\partial y^\alpha} \right\}$$

in the local coordinates described in Remarks 12.4 and 12.37, where $\phi_1 : \Lambda^1(TE;TE) \to T_1^1(E)$ is the equivalence over E defined as in 5.25.

12.57 Definition Let H be a connection in a bundle $\pi : E \to M$, and let $\omega \in \Omega^1(E;TE)$ be the connection form of H. Then the bundle-valued 2-form $\Omega \in \Omega^2(E;TE)$ given by

$$\Omega = -\tfrac{1}{2}\,[\omega,\omega]$$

is called the *curvature form* of H. The connection H is said to be *flat* if $\Omega = 0$.

12.58 Proposition Let H be a connection with curvature form Ω in a bundle $\pi : E \to M$, and let h and v be the horizontal and vertical projection determined by H. Then we have that

$$\Omega(X,Y) = -v \circ [h \circ X, h \circ Y]$$

for every pair of vector fields X and Y on E.

PROOF: Since V is an integrable distribution on E, we have that

$$\Omega(X, v \circ Y) = \omega([\omega(X), v \circ Y]) - \omega([X, v \circ Y]) = -\omega([v \circ X, v \circ Y]) = 0$$

by Corollary 12.34. This implies that

$$\begin{aligned}\Omega(X,Y) &= \Omega(h \circ X, h \circ Y) = \omega([h \circ X, h \circ Y]) - [h \circ X, h \circ Y] \\ &= -v \circ [h \circ X, h \circ Y]\,.\end{aligned}$$

□

EQUIVARIANT JET FIELDS

12.59 Definition By an operation or an action of a Lie group G on a bundle $\pi : E \to M$ on the right we mean a pair $(\widetilde{\mu}, \mu)$ of operations $\widetilde{\mu} : E \times G \to E$ and $\mu : M \times G \to M$ of G on E and M, respectively, such that the corresponding pair $(\widetilde{R}_g, R_g)$ of right translations forms a bundle map for each $g \in G$. This implies that the projection $\pi : E \to M$ is equivariant with respect to the actions $\widetilde{\mu}$ and μ of G on E and M.

12.60 Proposition Let $\pi : P \to M$ be a principal G-bundle over M^n, where $\mu : P \times G \to P$ is the action of G^m on P on the right. Then we have an operation $(\widetilde{\mu},\mu)$ of G on the jet bundle $\pi_{1,0} : J^1P \to P$ such that the right translations R_g and $\widetilde{R}_g$ corresponding to μ and $\widetilde{\mu}$, respectively, are related by

$$\widetilde{R}_g = j^1 R_g$$

for $g \in G$.

PROOF: By Proposition 12.42 we know that $(\widetilde{R}_g, R_g)$ is a bundle map for each $g \in G$, and it only remains to prove that $\widetilde{\mu}$ is smooth at every point $(\sigma,h) \in J^1P \times G$. Suppose that $(t,\pi^{-1}(U))$ is a local trivialization around the point $u' = \pi_{1,0}(\sigma)$ in P with $t(u') = (q,h')$. Let (x,O) be a local chart around q on M with $O \subset U$, and let (y'',Q''), (y',Q') and (y,Q) be local charts around $h'h$, h' and h on G with $Q'Q \subset Q''$. Then (z',W') and (z,W), where $W' = t^{-1}(O \times Q'')$, $W = t^{-1}(O \times Q')$, $z' = (x \times y'') \circ t$ and $z = (x \times y') \circ t$, are local charts around $u'h$ and u' on P. The map $f : z(W) \times y(Q) \to \mathbf{R}^{n+m}$ defined by $f = z' \circ \mu \circ (z \times y)^{-1}$ is smooth, and we have that

$$f(a,b) = z' \circ R_{y^{-1}(b)} \circ z^{-1}(a)$$

for $a \in z(W)$ and $b \in y(Q)$.

If $\tau_{z'} : \pi_{1,0}^{-1}(W') \to W' \times L(\mathbf{R}^n,\mathbf{R}^m)$ and $\tau_z : \pi_{1,0}^{-1}(W) \to W \times L(\mathbf{R}^n,\mathbf{R}^m)$ are the equivalences over W' and W associated with the local charts (z',W') and (z,W), respectively, we have that

$$\begin{aligned}
&\tau_{z'} \circ \widetilde{\mu} \circ (\tau_z \times id_Q)^{-1}(u,G,g) = \tau_{z'} \circ \widetilde{R}_g(j_p^1\phi) = \tau_{z'}(j_p^1(R_g \circ \phi)) \\
&= (R_g \circ \phi(p), D(z_2' \circ R_g \circ z^{-1})(z(u)) \circ D(z \circ \phi \circ x^{-1})(x(p))) \\
&= (R_g(u), D(z_2' \circ R_g \circ z^{-1})(z(u)) \circ (id_{\mathbf{R}^n}, G))
\end{aligned}$$

for $g \in Q$ and $(u,G) \in W \times L(\mathbf{R}^n,\mathbf{R}^m)$ where $\tau_z^{-1}(u,G) = j_p^1\phi$. This shows that

$$\begin{aligned}
&(z' \times id) \circ \tau_{z'} \circ \widetilde{\mu} \circ (\tau_z \times id_Q)^{-1} \circ (z \times id \times y)^{-1}(a,G,b) \\
&= (z' \circ R_{y^{-1}(b)} \circ z^{-1}(a), D(z_2' \circ R_{y^{-1}(b)} \circ z^{-1})(a) \circ (id_{\mathbf{R}^n}, G)) \\
&= (f(a,b), D_1 f(a,b) \circ (id_{\mathbf{R}^n}, G))
\end{aligned}$$

for $(a,G,b) \in z(W) \times L(\mathbf{R}^n,\mathbf{R}^m) \times y(Q)$, which completes the proof of the proposition. □

12.61 Proposition Let $\pi : P \to M$ be a principal G-bundle over a smooth manifold M^n, and let H be the connection determined by a jet field $\Gamma : P \to J^1P$ on P. Then Γ

is equivariant with respect to the actions of G on P and J^1P described in Proposition 12.60 if and only if

$$H_{ug} = (R_g)_* H_u$$

for every $u \in P$ and $g \in G$. A connection H satisfying this condition is called a *principal connection*.

PROOF : Let $u \in P$ and $g \in G$, and suppose that $\Gamma(u) = j_p^1\phi$ and $\Gamma(ug) = j_p^1\psi$ so that $H_u = \phi_*(T_pM)$ and $H_{ug} = \psi_*(T_pM)$. Then the condition in the proposition is equivalent to the assertion that

$$\psi_{*p} = (R_g \circ \phi)_{*p} ,$$

which is also equivalent to

$$\Gamma \circ R_g(u) = j_p^1\psi = j_p^1(R_g \circ \phi) = j^1R_g \circ \Gamma(u) .$$

As $u \in P$ and $g \in G$ was arbitrary, this completes the proof of the proposition. □

12.62 Proposition Let $\pi : P \to M$ be a principal G-bundle, and let $\pi' : E \to M$ be the fibre bundle associated with P with fibre F obtained from a Lie group homomorphism $\psi : G \to G'$ into a Lie group G' which acts on the smooth manifold F effectively on the left. If Γ is an equivariant jet field on P, then there is a unique jet field Γ' on E so that the diagram

$$\begin{array}{ccc} J^1P & \xrightarrow{\;j^1\nu_v\;} & J^1E \\ \Gamma \big\uparrow & & \big\uparrow \Gamma' \\ P & \xrightarrow{\;\nu_v\;} & E \end{array}$$

is commutative for every $v \in F$, where $\nu_v : P \to E$ is the smooth map defined in 10.28. The connections H and H' determined by Γ and Γ' are related by

$$H'_w = \nu_{v*}(H_u)$$

for $u \in P$ and $v \in F$, where $w = \nu_v(u)$.

PROOF : Let $w \in E_p$ with $p \in M$. For each $u \in P_p$ there is a unique $v \in F$ so that $\nu_v(u) = w$. If $\nu_{v'}(u') = w$ with $u' \in P_p$ and $v' \in F$, then $u' = u\,g$ and $v' = \psi(g^{-1})\,v$ for a $g \in G$. Using that the jet field Γ is equivariant, we have that

$$j^1\nu_{v'} \circ \Gamma(u') = j^1\nu_{v'} \circ \Gamma \circ R_g(u) = j^1\nu_{v'} \circ j^1R_g \circ \Gamma(u) = j^1\nu_v \circ \Gamma(u) ,$$

showing that $\Gamma'(w)$ is well defined for every $w \in E$. We also have that

$$\pi'_{1,0} \circ \Gamma' \circ \nu_v = \pi'_{1,0} \circ j^1\nu_v \circ \Gamma = \nu_v \circ \pi_{1,0} \circ \Gamma = \nu_v \,,$$

so it only remains to show that Γ' is smooth.

Let $(t, \pi^{-1}(U))$ be a local trivialization in P, where U is the coordinate neighbourhood of a local chart (x,U) around a point $p_0 \in M$, and let $(t', \pi'^{-1}(U))$ be the local trivialization in E associated with $(t, \pi^{-1}(U))$ as described in 10.28. We then have that

$$t' \circ \nu_v \circ t^{-1}(p,g) = (p, \mu(g,v))$$

for $p \in U$, $g \in G$ and $v \in F$, where $\mu : G \times F \to F$ is the group action of G on F defined by $\mu(g,v) = \psi(g)\, v$ for $g \in G$ and $v \in F$. Given a point $v_0 \in F$, we choose local charts (y,Q) and (y',Q') around e and v_0 on G and F, respectively, obtaining local charts (z,W) and (z',W') on P and E where $W = t^{-1}(U \times Q)$, $W' = t'^{-1}(U \times Q')$, $z = (x \times y) \circ t$ and $z' = (x \times y') \circ t'$. Using the same notation as in the proof of Proposition 12.42, the above formula implies that

$$\begin{aligned}
&\tau_{z',2} \circ \Gamma' \circ t'^{-1}(p,v) = \tau_{z',2} \circ \Gamma' \circ \nu_v \circ t^{-1}(p,e) = \tau_{z',2} \circ j^1\nu_v \circ \Gamma \circ t^{-1}(p,e) \\
&= \{\tau_{z',2} \circ j^1\nu_v \circ \tau_z^{-1}\} \circ \{\tau_z \circ \Gamma \circ t^{-1}\}(p,e) = D_1(z'_2 \circ \nu_v \circ z^{-1})(x(p),y(e)) \\
&+ D_2(z'_2 \circ \nu_v \circ z^{-1})(x(p),y(e)) \circ (\tau_{z,2} \circ \Gamma \circ t^{-1}(p,e)) \\
&= \ D_1(y' \circ \mu \circ (y \times y')^{-1})(y(e),y'(v)) \circ (\tau_{z,2} \circ \Gamma \circ t^{-1}(p,e))
\end{aligned}$$

for $p \in U$ and $v \in Q'$, showing that Γ' is a jet field on E.

Now suppose that $w = \nu_v(u)$ where $u \in P$ and $v \in F$, and that $\Gamma(u) = j^1_p\phi$. Then it follows that

$$\Gamma'(w) = \Gamma' \circ \nu_v(u) = j^1\nu_v \circ \Gamma(u) = j^1_p(\nu_v \circ \phi)$$

so that

$$H'_w = (\nu_v \circ \phi)_*(T_pM) = \nu_{v*}(H_u) \,,$$

thus proving that last assertion in the proposition. □

12.63 Theorem Let $\pi : P \to M$ be a principal G-bundle, and let $\pi' : E \to M$ be the associated vector bundle obtained from a representation $\psi : G \to Aut\,(W)$ of G on a finite dimensional vector space W. Let H' be the connection in E obtained from a principal connection H in P as described in Proposition 12.62, and let $\nu' : TE \to TE$ be the vertical projection determined by H'. Using the canonical bundle map $(\widetilde{\eta}, \pi')$ from the vertical tangent bundle $\pi'' : V \to E$ to the vector bundle $\pi' : E \to M$ described in Proposition 12.19, we obtain a bundle map (K, π') from the tangent bundle $\pi'' : TE \to E$ to $\pi' : E \to M$ where

$$K = \widetilde{\eta} \circ \nu' \,.$$

The covariant derivative of a section $s \in \Gamma(M;E)$ of π' is given by

$$\nabla s(p)(v) = K \circ s_*(v)$$

for $p \in M$ and $v \in T_pM$.

PROOF : We use the same notation as in 10.28 and Definition 10.77. Let $(t, \pi^{-1}(U))$ be a local trivialization around p in P, and let $(t', \pi'^{-1}(U))$ be the local trivialization in E associated with $(t, \pi^{-1}(U))$. Now let $u = t^{-1}(p,e) \in P$ so that $\pi(u) = p$. Then there is a unique $w \in H_u$ with $\pi_*(w) = v$, and a unique $\xi \in W$ with $v_\xi(u) = s(p)$. Since

$$\pi' \circ s \circ \pi = \pi = \pi' \circ v_\xi\,,$$

it follows from Proposition 12.62 that

$$s_*(v) = v_{\xi\,*}(w) + \{s_* \circ \pi_*(w) - v_{\xi\,*}(w)\}$$

where $v_{\xi\,*}(w) \in H'_{s(p)}$ and $s_* \circ \pi_*(w) - v_{\xi\,*}(w) \in V_{s(p)}$, so that

$$v' \circ s_*(v) = s_* \circ \pi_*(w) - v_{\xi\,*}(w)\,.$$

From (1) in 10.28 it follows that

$$t' \circ s \circ \pi \circ t^{-1}(p',g') = (p', \psi(g')\, \mu_{u'}^{-1} \circ s \circ \pi(u')) = (p', \psi(g')\, \bar{s} \circ t^{-1}(p',g'))$$

and

$$t' \circ v_\xi \circ t^{-1}(p',g') = t'([u',\xi]) = (p', \psi(g')\, \xi)$$

for $(p',g') \in U \times G$, where $u' = t^{-1}(p',g')$.

Now let $i_e : U \to U \times G$, $i_p : G \to U \times G$, $i'_\xi : U \to U \times W$ and $i'_p : W \to U \times W$ be the embeddings defined as in Proposition 2.74. Then there is a unique $X \in \mathfrak{g}$ with $t_*(w) = i_{e\,*}(v) + i_{p\,*}(X)$. Hence we have that

$$t'_* \circ s_* \circ \pi_*(w) = i'_{\xi\,*}(v) + i'_{p\,*} \circ \bar{s}_*(w) + i'_{p\,*} \circ \omega_\xi^{-1} \circ \psi_*(X)\, \xi$$

and

$$t'_* \circ v_{\xi\,*}(w) = i'_{\xi\,*}(v) + i'_{p\,*} \circ \omega_\xi^{-1} \circ \psi_*(X)\, \xi\,,$$

which implies that

$$t'_{p\,*} \circ v' \circ s_*(v) = \bar{s}_*(w)\,.$$

Using the commutative diagram

$$\begin{array}{ccc}
T_{s(p)}E_p & \xrightarrow{\;t'_{p\,*\,s(p)}\;} & T_\xi W \\
\omega_{s(p)} \Big\downarrow & & \Big\downarrow \omega_\xi \\
E_p & \xrightarrow{\;t'_p\;} & W
\end{array}$$

we conclude that

$$K \circ s_*(v) = \omega_{s(p)} \circ v' \circ s_*(v) = t_p'^{-1} \circ \omega_\xi \circ \bar{s}_*(w) = \mu_u \circ d\bar{s}(u)(w) = \nabla s(p)(v) .$$

□

SECOND ORDER JET BUNDLES

12.64 Proposition Let $\pi : E \to M$ be a bundle over M with fibre F. Then $\pi_* : TE \to TM$ is a bundle with fibre TF.

PROOF: Suppose that $(t, \pi^{-1}(U))$ is a local trivialization in E, and let $pr_1 : U \times F \to U$ and $pr_2 : U \times F \to F$ be the projections on the first and second factor. Then $(((pr_1 \circ t)_*, (pr_2 \circ t)_*), \pi_*^{-1}(TM|_U))$ is a local trivialization in TE. □

12.65 Remark We say that $(t^1, \pi_*^{-1}(TM|_U))$, where $t^1 = ((pr_1 \circ t)_*, (pr_2 \circ t)_*) : \pi_*^{-1}(TM|_U) \to TM|_U \times TF$, is the local trivialization in the bundle $\pi_* : TE \to TM$ associated with the local trivialization $(t, \pi^{-1}(U))$ in $\pi : E \to M$. Here $pr_1 : U \times F \to U$ and $pr_2 : U \times F \to F$ are the projections on the open set U in the base manifold M and the fibre F.

Let (x, O) and (y, Q) be local charts on M^n and F^m, respectively, with $O \subset U$. Then (z, W), where $W = t^{-1}(O \times Q)$ and $z = (x \times y) \circ t$, is a local chart on E, and we let $z_1 : W \to \mathbf{R}^n$ and $z_2 : W \to \mathbf{R}^m$ be the component maps of $z : W \to \mathbf{R}^n \times \mathbf{R}^m$.

We also obtain local charts $(x^1, TM|_O)$ and $(y^1, TF|_Q)$ on TM and TF, where $x^1 = (x \times id_{\mathbf{R}^n}) \circ t_x$ and $y^1 = (y \times id_{\mathbf{R}^m}) \circ t_y$. In the same way as above, we have a local chart $(z^1, TE|_W)$ on TE, where $z^1 = (x^1 \times y^1) \circ t^1$, and we let $z_1^1 : TE|_W \to \mathbf{R}^{2n}$ and $z_2^1 : TE|_W \to \mathbf{R}^{2m}$ be the component maps of $z^1 : TE|_W \to \mathbf{R}^{2n} \times \mathbf{R}^{2m}$.

If $\phi : \widetilde{O} \to E$ is a section in $\pi : E \to M$, we have that

$$t_y \circ (pr_2 \circ t \circ \phi)_* \circ t_x^{-1}(p, \xi) = (pr_2 \circ t \circ \phi(p), D(z_2 \circ \phi \circ x^{-1})(x(p))\xi)$$

for $p \in \widetilde{O} \cap O$ and $\xi \in \mathbf{R}^n$, which implies that

$$z_2^1 \circ \phi_* \circ {x^1}^{-1}(a, \xi) = (z_2 \circ \phi \circ x^{-1}(a), D(z_2 \circ \phi \circ x^{-1})(a)\xi)$$

for $a \in x(\widetilde{O} \cap O)$ and $\xi \in \mathbf{R}^n$.

Denoting the component maps of $z_2^1 : TE|_W \to \mathbf{R}^m \times \mathbf{R}^m$ by $z_{2,1}^1 : TE|_W \to \mathbf{R}^m$ and $z_{2,2}^1 : TE|_W \to \mathbf{R}^m$, it follows that

$$D_1(z_{2,2}^1 \circ \phi_* \circ {x^1}^{-1})(a, \xi)(\zeta) = D^2(z_2 \circ \phi \circ x^{-1})(a)(\xi, \zeta)$$

for $a \in x(\widetilde{O} \cap O)$ and $\xi, \zeta \in \mathbf{R}^n$.

12.66 Proposition Let $\pi : E \to M$ be a bundle over M^n with fibre F^m, and let $\widetilde{\pi}' : \pi_1^*(TM) \to J^1E$ be the pullback of the tangent bundle $\pi' : TM \to M$ by π_1, and $\widetilde{\pi}'' : \pi_{1,0}^*(V) \to J^1E$ be the pullback of the vertical bundle $\pi'' : V \to E$ by $\pi_{1,0}$. Then $\pi_{2,1} : J^2E \to J^1E$ is an affine bundle over J^1E modelled on the vector bundle $\pi''' : \mathscr{T}_s^{\,2}(\pi_1^*(TM);\pi_{1,0}^*(V)) \to J^1E$.

PROOF: Consider a local chart (z,W) around a point u_0 on E as described in Remark 12.65, and let $W_1 = \pi_{1,0}^{-1}(W)$. Then we have a bijection $\tau_z^2 : \pi_{2,1}^{-1}(W_1) \to W_1 \times \mathscr{T}_s^{\,2}(\mathbf{R}^n;\mathbf{R}^m)$ which is given by

$$\tau_z^2(j_p^2\phi) = (j_p^1\phi\,, D^2(z_2 \circ \phi \circ x^{-1})(x(p)))$$

for $p \in O$ and $\phi \in \Gamma_p(E)$ with $\phi(p) = u \in W$. It follows from Remark 12.65 that τ_z^2 is well defined since

$$D(z_2^1 \circ \phi_* \circ {x^1}^{-1})(x^1(v)) = pr_2^1 \circ t_{z^1,w} \circ (\phi_*)_{*v} \circ t_{x^1,v}^{-1}$$

for $v \in T_pM$, where $\phi_*(v) = w$ and $pr_2^1 : \mathbf{R}^{2n} \times \mathbf{R}^{2m} \to \mathbf{R}^{2m}$ is the projection on the second factor.

Suppose that $(\widetilde{z},\widetilde{W})$ is another local chart on E obtained in the same way from a local trivialization $(\widetilde{t},\pi^{-1}(\widetilde{U}))$ in E and local charts $(\widetilde{x},\widetilde{O})$ and $(\widetilde{y},\widetilde{Q})$ on M and F, respectively, with $\widetilde{O} \subset \widetilde{U}$. Then we have that

$$\begin{aligned}\tau_{\widetilde{z}}^2 \circ {\tau_z^2}^{-1}(\sigma,S) &= (\sigma,\{D_1(\widetilde{z}_2 \circ z^{-1})(z(u)) + D_2(\widetilde{z}_2 \circ z^{-1})(z(u)) \circ G\} \\ &\circ D^2(x \circ \widetilde{x}^{-1})(\widetilde{z}_1(u)) + \{D_{11}(\widetilde{z}_2 \circ z^{-1})(z(u)) \\ &+ D_{12}(\widetilde{z}_2 \circ z^{-1})(z(u)) \circ (id_{\mathbf{R}^n} \times G) + D_{21}(\widetilde{z}_2 \circ z^{-1})(z(u)) \circ (G \times id_{\mathbf{R}^n}) \\ &+ D_{22}(\widetilde{z}_2 \circ z^{-1})(z(u)) \circ (G \times G) + D_2(\widetilde{z}_2 \circ z^{-1})(z(u)) \circ S\} \\ &\circ (D(x \circ \widetilde{x}^{-1})(\widetilde{z}_1(u)) \times D(x \circ \widetilde{x}^{-1})(\widetilde{z}_1(u))))\end{aligned} \tag{1}$$

for $\sigma \in W_1 \cap \widetilde{W}_1$ and $S \in \mathscr{T}_s^{\,2}(\mathbf{R}^n;\mathbf{R}^m)$, where $\tau_z(\sigma) = (u,G)$. Hence it follows from Proposition 12.6 that $\pi_{2,1} : J^2E \to J^1E$ is a bundle over J^1E with fibre $\mathbf{R}^{n(n+1)m/2}$, with $((id_{W_1} \times \beta_s) \circ \tau_z^2, \pi_{2,1}^{-1}(W_1))$ as a local trivialization for each local chart (z,W) on E of the above form, using a fixed linear isomorphism $\beta_s : \mathscr{T}_s^{\,2}(\mathbf{R}^n;\mathbf{R}^m) \to \mathbf{R}^{n(n+1)m/2}$ defined in the same way as β in the proof of Proposition 5.31.

By Proposition 5.28 we have a local trivialization $((id_{W_1} \times \beta_s) \circ \widetilde{\tau}_z^2, {\pi'''}^{-1}(W_1))$ in the vector bundle $\pi''' : \mathscr{T}_s^{\,2}(\pi_1^*(TM);\pi_{1,0}^*(V)) \to J^1E$ associated with the local chart (z,W) on E, where $\widetilde{\tau}_z^2 : {\pi'''}^{-1}(W_1) \to W_1 \times \mathscr{T}_s^{\,2}(\mathbf{R}^n;\mathbf{R}^m)$ is the equivalence over W_1 defined by

$$\widetilde{\tau}_z^2(T) = (\sigma, pr_2 \circ t_{z,u} \circ T \circ (t_{x,p}^{-1} \times t_{x,p}^{-1}))$$

for $\sigma \in W_1$ and $T \in \mathscr{T}_s^{\,2}(\pi_1^*(TM);\pi_{1,0}^*(V))_\sigma$, where $u = \pi_{1,0}(\sigma)$ and $p = \pi(u)$. We also have that

$$\begin{aligned}\widetilde{\tau}_{\widetilde{z}}^2 \circ {\widetilde{\tau}_z^2}^{-1}(\sigma,S) &= (\sigma, D_2(\widetilde{z}_2 \circ z^{-1})(z(u)) \circ S \\ &\circ (D(x \circ \widetilde{x}^{-1})(\widetilde{z}_1(u)) \times D(x \circ \widetilde{x}^{-1})(\widetilde{z}_1(u))))\end{aligned} \tag{2}$$

for $\sigma \in W_1 \cap \widetilde{W}_1$ and $S \in \mathscr{T}_s^2(\mathbf{R}^n; \mathbf{R}^m)$. Using this, we can define an operation of the vector space $\mathscr{T}_s^2(\pi_1^*(TM); \pi_{1,0}^*(V))_\sigma$ on the fibre $J^2 E_\sigma$ by

$$\tau_{z,\sigma}^2(j_p^2\phi + T) = \tau_{z,\sigma}^2(j_p^2\phi) + \widetilde{\tau}_{z,\sigma}^2(T). \tag{3}$$

It follows from (1) with $S = \tau_{z,\sigma}^2(j_p^2\phi + T)$ and $S = \tau_{z,\sigma}^2(j_p^2\phi)$, (2) with $S = \widetilde{\tau}_{z,\sigma}^2(T)$ and (3) that we also have

$$\tau_{\widetilde{z},\sigma}^2(j_p^2\phi + T) = \tau_{\widetilde{z},\sigma}^2(j_p^2\phi) + \widetilde{\tau}_{\widetilde{z},\sigma}^2(T),$$

showing that the operation on $J^2 E_\sigma$ given by (3) is well defined and does not depend on the choice of local chart (z, W). $\qquad\square$

12.67 Remark From each local chart (z, W) on E with coordinate functions x^i and y^α defined in Remark 12.4, we obtain an induced local chart $((((z \times \beta) \circ \tau_z) \times \beta_s) \circ \tau_z^2, \pi_{2,0}^{-1}(W))$ on $J^2 E$ with coordinate functions x^i, y^α, y_i^α and y_{ij}^α defined by

$$x^i(j_p^2\phi) = x^i(p), \quad y^\alpha(j_p^2\phi) = y^\alpha \circ \phi(p), \quad y_i^\alpha(j_p^2\phi) = \frac{\partial(y^\alpha \circ \phi)}{\partial x^i}(p)$$

$$\text{and} \quad y_{ij}^\alpha(j_p^2\phi) = \frac{\partial(y^\alpha \circ \phi)}{\partial x^j \partial x^i}(p)$$

for $j_p^2\phi \in \pi_{2,0}^{-1}(W)$, $i, j = 1, 2, \dots, n$ and $\alpha = 1, 2, \dots, m$. It will be clear from the context whether x^i and y^α denote coordinate functions on $J^2 E$, on $J^1 E$, on E or on M and F, respectively, and whether y_i^α denote coordinate functions on $J^2 E$ or on $J^1 E$. The functions y_i^α and y_{ij}^α are called *derivative coordinates*.

12.68 Proposition Let $\pi : E \to M$ be a bundle over M^n with fibre F^m. Then $\pi_2 : J^2 E \to M$ is also a bundle.

PROOF: Let $(t, \pi^{-1}(U))$ be a local trivialization in $\pi : E \to M$, where U is the coordinate neighbourhood of a local chart (x, U) on M, and consider the trivial bundle $\pi' : \mathbf{R}^n \times F \to \mathbf{R}^n$. Since $\pi_2' = \pi_1' \circ \pi_{2,1}'$ is a submersion, we have that $J_0^2(\mathbf{R}^n \times F) = \pi_2'^{-1}(0)$ is a closed submanifold of $J^2(\mathbf{R}^n \times F)$.

Now let $t'' : \pi_2^{-1}(U) \to U \times J_0^2(\mathbf{R}^n \times F)$ be the map given by

$$t''(j_p^2\phi) = (p, j_0^2\psi)$$

for $p \in U$ and $\phi \in \Gamma_p(E)$, where $\psi \in \Gamma_0(\mathbf{R}^n \times F)$ is the section defined by

$$\psi = (T_{x(p)}^{-1} \times id_F) \circ (x \times id_F) \circ t \circ \phi \circ x^{-1} \circ T_{x(p)},$$

using the translation $T_{x(p)} : \mathbf{R}^n \to \mathbf{R}^n$ by $x(p)$. We want to show that t'' is a diffeomorphism. Consider the local chart (z, W) in E obtained from $(t, \pi^{-1}(U))$, (x, U) and a local chart (y, Q) on F as described in Proposition 12.3, so that $W = t^{-1}(U \times Q)$

and $z = (x \times y) \circ t$. We also obtain the local chart $(\widetilde{z}, \mathbf{R}^n \times Q)$ in $\mathbf{R}^n \times F$, where $\widetilde{z} = id_{\mathbf{R}^n} \times y$. With the same notation as in the proof of Proposition 12.66, we have that

$$\widetilde{z}_2 \circ \psi = z_2 \circ \phi \circ x^{-1} \circ T_{x(p)} \,,$$

which shows that

$$D^2(\widetilde{z}_2 \circ \psi)(0) = D^2(z_2 \circ \phi \circ x^{-1})(x(p)) \,.$$

Using the diffeomorphism $t' : \pi_1^{-1}(U) \to U \times J_0^1(\mathbf{R}^n \times F)$ defined in the proof of Proposition 12.40 and the diffeomorphism $\eta : {\pi'_{2,0}}^{-1}(\{0\} \times Q) \to {\pi'_{1,0}}^{-1}(\{0\} \times Q) \times \mathscr{T}_s^{\,2}(\mathbf{R}^n; \mathbf{R}^m)$ induced by $\tau_{\widetilde{z}}^2$, we now have that

$$\begin{aligned}(id_M \times \eta) \circ t''(j_p^2\phi) &= (p, j_0^1\psi, D^2(\widetilde{z}_2 \circ \psi)(0)) \\ &= (t'(j_p^1\phi), D^2(z_2 \circ \phi \circ x^{-1})(x(p))) = (t' \times id) \circ \tau_z^2(j_p^2\phi)\,.\end{aligned}$$

This completes the proof that $\pi_2 : J^2E \to M$ is a bundle, with $(t'', \pi_2^{-1}(U))$ as a local trivialization for each local trivialization $(t, \pi^{-1}(U))$ in $\pi : E \to M$ of the above form. □

12.69 Proposition Let $(\tilde{f}, f)$ be a bundle map between the bundles $\pi : E \to M$ and $\widetilde{\pi} : \widetilde{E} \to \widetilde{M}$, where $f : M \to \widetilde{M}$ is a diffeomorphism. Then we have a smooth map $j^2\tilde{f} : J^2E \to J^2\widetilde{E}$, called the *2-jet prolongation of* $\tilde{f}$, defined by

$$j^2\tilde{f}\,(j_p^2\phi) = j_{f(p)}^2(\tilde{f} \circ \phi \circ f^{-1})$$

for $p \in M$ and $\phi \in \Gamma_p(E)$. We have that $(j^2\tilde{f}, j^1\tilde{f})$ is an affine bundle map between the affine bundles $\pi_{2,1} : J^2E \to J^1E$ and $\widetilde{\pi}_{2,1} : J^2\widetilde{E} \to J^1\widetilde{E}$. It is modelled on the bundle map $(\lambda\,, j^1\tilde{f})$ between the vector bundles $\pi''' : \mathscr{T}_s^{\,2}(\pi_1^*(TM); \pi_{1,0}^*(V)) \to J^1E$ and $\widetilde{\pi}''' : \mathscr{T}_s^{\,2}(\pi_1^*(T\widetilde{M}); \pi_{1,0}^*(\widetilde{V})) \to J^1\widetilde{E}$, where $\lambda\,(T) = \tilde{f}_{*u} \circ T \circ (f_{*p}^{-1} \times f_{*p}^{-1})$ for $\sigma \in J^1E$, $T \in \mathscr{T}_s^{\,2}(\pi_1^*(TM); \pi_{1,0}^*(V))_\sigma$, $u = \pi_{1,0}(\sigma)$ and $p = \pi_1(\sigma)$.

We also have that $(j^2\tilde{f}, f)$ is a bundle map between the bundles $\pi_2 : J^2E \to M$ and $\widetilde{\pi}_2 : J^2\widetilde{E} \to \widetilde{M}$.

PROOF: Let u_0 be a point on E, and consider local charts (z, W) and $(\widetilde{z}, \widetilde{W})$ around u_0 and $\tilde{f}(u_0)$, respectively, of the form described in Proposition 12.3. Using the same notation as in the proof of Proposition 12.66, we have that

$$\begin{aligned}\tau_{\widetilde{z}}^2 \circ j^2\tilde{f} \circ {\tau_z^2}^{-1}(\sigma, S) &= (j^1\tilde{f}(\sigma), \{D_1(\widetilde{z}_2 \circ \tilde{f} \circ z^{-1})(z(u)) \\ &+ D_2(\widetilde{z}_2 \circ \tilde{f} \circ z^{-1})(z(u)) \circ G\} \circ D^2(x \circ f^{-1} \circ \widetilde{x}^{-1})(\widetilde{z}_1 \circ \tilde{f}(u)) \\ &+ \{D_{11}(\widetilde{z}_2 \circ \tilde{f} \circ z^{-1})(z(u)) + D_{12}(\widetilde{z}_2 \circ \tilde{f} \circ z^{-1})(z(u)) \circ (id_{\mathbf{R}^n} \times G) \\ &+ D_{21}(\widetilde{z}_2 \circ \tilde{f} \circ z^{-1})(z(u)) \circ (G \times id_{\mathbf{R}^n}) \\ &+ D_{22}(\widetilde{z}_2 \circ \tilde{f} \circ z^{-1})(z(u)) \circ (G \times G) + D_2(\widetilde{z}_2 \circ \tilde{f} \circ z^{-1})(z(u)) \circ S\} \\ &\circ (D(x \circ f^{-1} \circ \widetilde{x}^{-1})(\widetilde{z}_1 \circ \tilde{f}(u)) \times D(x \circ f^{-1} \circ \widetilde{x}^{-1})(\widetilde{z}_1 \circ \tilde{f}(u))))\end{aligned} \tag{1}$$

and

$$\widetilde{\tau}_{\tilde{z}}^2 \circ \lambda \circ \widetilde{\tau}_z^{2\,-1}(\sigma, S) = (j^1\tilde{f}(\sigma), D_2(\widetilde{z}_2 \circ \tilde{f} \circ z^{-1})(z(u)) \circ S \\ \circ (D(x \circ f^{-1} \circ \widetilde{x}^{-1})(\widetilde{z}_1 \circ \tilde{f}(u)) \times D(x \circ f^{-1} \circ \widetilde{x}^{-1})(\widetilde{z}_1 \circ \tilde{f}(u)))) \tag{2}$$

for $\sigma \in \pi_{1,0}^{-1}(W) \cap \pi_{1,0}^{-1}(\tilde{f}^{-1}(\widetilde{W}))$ and $S \in \mathscr{T}_s^2(\mathbf{R}^n; \mathbf{R}^m)$, where $\tau_z(\sigma) = (u, G)$. It now follows from (1) with $S = \tau_{z,\sigma}^2(j_p^2\phi + T)$ and $S = \tau_{z,\sigma}^2(j_p^2\phi)$, and (2) with $S = \widetilde{\tau}_{z,\sigma}^2(T)$ that we have

$$j^2\tilde{f}\,(j_p^2\phi + T) = j^2\tilde{f}\,(j_p^2\phi) + \lambda\,(T)$$

for $\sigma \in \pi_{1,0}^{-1}(W) \cap \pi_{1,0}^{-1}(\tilde{f}^{-1}(\widetilde{W}))$, $j_p^2\phi \in J^2E_\sigma$ and $T \in \mathscr{T}_s^2(\pi_1^*(TM); \pi_{1,0}^*(V))_\sigma$, showing that $j^2\tilde{f}$ is an affine map with linear part λ on each fibre. We see that $(j^2\tilde{f}, j^1\tilde{f})$ is a bundle map since

$$\widetilde{\pi}_{2,1} \circ j^2\tilde{f}\,(j_p^2\phi) = \widetilde{\pi}_{2,1}(\,j_{f(p)}^2(\tilde{f} \circ \phi \circ f^{-1})) = j_{f(p)}^1(\tilde{f} \circ \phi \circ f^{-1}) \\ = j^1\tilde{f}\,(j_p^1\phi) = j^1\tilde{f} \circ \pi_{2,1}(\,j_p^2\phi)$$

for $p \in M$ and $\phi \in \Gamma_p(E)$, which shows that

$$\widetilde{\pi}_{2,1} \circ j^2\tilde{f} = j^1\tilde{f} \circ \pi_{2,1}\,.$$

From this it also follows that $(j^2\tilde{f}, f)$ is a bundle map since

$$\widetilde{\pi}_2 \circ j^2\tilde{f} = \widetilde{\pi}_1 \circ \widetilde{\pi}_{2,1} \circ j^2\tilde{f} = \widetilde{\pi}_1 \circ j^1\tilde{f} \circ \pi_{2,1} = f \circ \pi_1 \circ \pi_{2,1} = f \circ \pi_2\,. \qquad \square$$

12.70 Remark A section $\phi : M \to E$ in the bundle $\pi : E \to M$ is a bundle map over M between the bundles $id_M : M \to M$ and $\pi : E \to M$. Identifying J^2M with M, the 2-jet prolongation $j^2\phi : M \to J^2E$ of ϕ defined in Proposition 12.69 is given by

$$j^2\phi\,(p) = j_p^2\phi$$

for $p \in M$.

Let $(\tilde{f}, f)$ be a bundle map between the bundles $\pi : E \to M$ and $\widetilde{\pi} : \widetilde{E} \to \widetilde{M}$, where $f : M \to \widetilde{M}$ is a diffeomorphism. As in Remarks 12.44 we then have that

$$j^2\tilde{f} \circ j^2\phi = j^2(\tilde{f}_*(\phi)) \circ f$$

which implies that

$$(j^2\tilde{f})_*(j^2\phi) = j^2(\tilde{f}_*(\phi))$$

for every section $\phi : M \to E$ of π.

12.71 Proposition Let $\pi : E \to M$ be a bundle over M^n with fibre F^m. Then there is a bundle map $i_{1,1} : J^2E \to J^1(J^1E)$ over J^1E between the bundles $\pi_{2,1} : J^2E \to J^1E$ and $(\pi_1)_{1,0} : J^1(J^1E) \to J^1E$, given by

$$i_{1,1}(j_p^2\phi) = j_p^1(j^1\phi)$$

for $p \in M$ and $\phi \in \Gamma_p(E)$, which is an embedding.

PROOF : Consider a local chart (z,W) around a point u_0 on E as described in Proposition 12.3, and let $(\tilde{z},W_1)$ be the induced local chart on J^1E where $W_1 = \pi_{1,0}^{-1}(W)$ and $\tilde{z} = (z \times \beta) \circ \tau_z$. Using the equivalences $\tau_z^2 : \pi_{2,1}^{-1}(W_1) \to W_1 \times \mathscr{T}_s^{\,2}(\mathbf{R}^n;\mathbf{R}^m)$ and $\tau_{\tilde{z}} : (\pi_1)_{1,0}^{-1}(W_1) \to W_1 \times L(\mathbf{R}^n, \mathbf{R}^m \times \mathbf{R}^{nm})$ over W_1, we have that

$$(\tau_z \times id) \circ \tau_z^2(j_p^2\phi) = (\phi(p), D(z_2 \circ \phi \circ x^{-1})(a), D^2(z_2 \circ \phi \circ x^{-1})(a))$$

and

$$(\tau_z \times id) \circ \tau_{\tilde{z}} \circ i_{1,1}(j_p^2\phi) = (\phi(p), D(z_2 \circ \phi \circ x^{-1})(a), D(\tilde{z}_2 \circ j^1\phi \circ x^{-1})(a))$$

where

$$\tilde{z}_2 \circ j^1\phi \circ x^{-1}(a) = (z_2 \circ \phi \circ x^{-1}(a), \beta \circ D(z_2 \circ \phi \circ x^{-1})(a))$$

for $a = x(p) \in x(O)$ and $\phi \in \Gamma_p(E)$ with $\phi(p) \in W$. Let $\widetilde{\beta} : \mathscr{T}_s^{\,2}(\mathbf{R}^n;\mathbf{R}^m) \to L(\mathbf{R}^n, \mathbf{R}^{nm})$ be the linear injection defined by

$$\widetilde{\beta}(T)(a) = \beta(T_a)$$

for $T \in \mathscr{T}_s^{\,2}(\mathbf{R}^n;\mathbf{R}^m)$ and $a \in \mathbf{R}^n$, where $T_a \in L(\mathbf{R}^n, \mathbf{R}^m)$ is the linear map given by $T_a(b) = T(a,b)$ for $b \in \mathbf{R}^n$. Then we have that

$$(\tau_z \times id) \circ \tau_{\tilde{z}} \circ i_{1,1} \circ (\tau_z^2)^{-1} \circ (\tau_z \times id)^{-1}(u,G,T) = (u,G,G,\widetilde{\beta}(T))$$

for $u \in E$, $G \in L(\mathbf{R}^n, \mathbf{R}^m)$ and $T \in \mathscr{T}_s^{\,2}(\mathbf{R}^n;\mathbf{R}^m)$, which shows that $i_{1,1}$ is a well-defined embedding and a bundle map over J^1E. □

12.72 Remark Using the coordinate functions x^i, y^α and y_i^α on J^1E defined in Remark 12.37, we denote the coordinate functions on $J^1(J^1E)$ by x^i, y^α, y_i^α, $y_{;j}^\alpha$ and $y_{i;j}^\alpha$. Note that the coordinate functions y_j^α and $y_{;j}^\alpha$ are different on $J^1(J^1E)$, but that they coincide on $i_{1,1}(J^2E)$ since

$$y_{;j}^\alpha(i_{1,1}(j_p^2\phi)) = y_{;j}^\alpha(j_p^1(j^1\phi)) = \frac{\partial(y^\alpha \circ j^1\phi)}{\partial x^j}(p) = \frac{\partial(y^\alpha \circ \phi)}{\partial x^j}(p) = y_j^\alpha(j_p^2\phi)\,.$$

The same is true for the coordinate functions $y_{i;j}^\alpha$ and $y_{j;i}^\alpha$ when $i \neq j$, since

$$y_{i;j}^\alpha(i_{1,1}(j_p^2\phi)) = y_{i;j}^\alpha(j_p^1(j^1\phi)) = \frac{\partial(y_i^\alpha \circ j^1\phi)}{\partial x^j}(p) = \frac{\partial(y^\alpha \circ \phi)}{\partial x^j\,\partial x^i}(p) = y_{ij}^\alpha(j_p^2\phi)\,.$$

The elements of $J^1(J^1E)$ are called *repeated jets*, and those belonging to $i_{1,1}(J^2E)$ are called *holonomic*.

12.73 Proposition Let $\pi : E \to M$ be a bundle over M^n with fibre F^m. Then there is a bundle map $(D_1, \pi_{1,0})$, called the *Spencer map*, between the bundles $(\pi_1)_{1,0} : J^1(J^1E) \to J^1E$ and $\pi' : \Lambda^1(\pi^*(TM);V) \to E$, where

$$D_1 = j^1(\pi_{1,0}) - (\pi_1)_{1,0}\,.$$

We also have a bundle $\widehat{\pi}_{2,1} : \widehat{J}^2E \to J^1E$, called the *semi-holonomic 2-jet bundle*, where $\widehat{J}^2E$ is the submanifold of $J^1(J^1E)$ given by $\widehat{J}^2E_\sigma = \ker D_{1\,\sigma}$ for $\sigma \in J^1E$, and $\widehat{\pi}_{2,1}$ is the restriction of $(\pi_1)_{1,0}$ to $\widehat{J}^2E$.

PROOF : By Proposition 12.36 we know that $\pi_{1,0} : J^1E \to E$ is an affine bundle over E modelled on the vector bundle $\pi' : \Lambda^1(\pi^*(TM);V) \to E$. To show that D_1 is well defined, we need to prove that $j^1(\pi_{1,0})(\sigma)$ and $(\pi_1)_{1,0}(\sigma)$ are in the same fibre of J^1E over E for each $\sigma \in J^1(J^1E)$. Let $\sigma = j^1_p\psi$ where $\psi \in \Gamma_p(J^1E)$. Then we have that

$$\pi_{1,0} \circ j^1(\pi_{1,0})(j^1_p\psi) = \pi_{1,0}(j^1_p(\pi_{1,0} \circ \psi)) = \pi_{1,0} \circ \psi(p)$$

and

$$\pi_{1,0} \circ (\pi_1)_{1,0}(j^1_p\psi) = \pi_{1,0}(\psi(p)) = \pi_{1,0} \circ \psi(p) \,,$$

showing that $D_1(j^1_p\psi)$ is a well-defined element in the fibre of $\Lambda^1(\pi^*(TM);V)$ over $\pi_{1,0} \circ \psi(p)$.

Using the same notation as in the proof of Proposition 12.71, the diffeomorphism $(\tau_z \times id) \circ \tau_{\tilde{z}} : (\pi_1)^{-1}_{1,0}(W_1) \to W \times L(\mathbf{R}^n,\mathbf{R}^m) \times L(\mathbf{R}^n,\mathbf{R}^m) \times L(\mathbf{R}^n,\mathbf{R}^{nm})$ is given by

$$(\tau_z \times id) \circ \tau_{\tilde{z}} = (\pi_{1,0} \circ (\pi_1)_{1,0},\, pr_2 \circ \tau_z \circ (\pi_1)_{1,0},\, pr_2 \circ \tau_z \circ j^1(\pi_{1,0}),\, pr_3 \circ \tau_{\tilde{z}}) \,,$$

where $pr_2 : W \times L(\mathbf{R}^n,\mathbf{R}^m) \to L(\mathbf{R}^n,\mathbf{R}^m)$ and $pr_3 : W_1 \times L(\mathbf{R}^n,\mathbf{R}^m) \times L(\mathbf{R}^n,\mathbf{R}^{nm}) \to L(\mathbf{R}^n,\mathbf{R}^{nm})$ are the projections on the last factor. If $\tau'_z : \pi'^{-1}(W) \to W \times L(\mathbf{R}^n,\mathbf{R}^m)$ is the equivalence over W described in the proof of Proposition 12.36, we have that

$$\tau'_z \circ D_1 \circ \tau_{\tilde{z}}^{-1} \circ (\tau_z \times id)^{-1}(u,F,G,H) = (u, G-F)$$

for $u \in E$, $F, G \in L(\mathbf{R}^n,\mathbf{R}^m)$ and $H \in L(\mathbf{R}^n,\mathbf{R}^{nm})$, showing that D_1 is a bundle map.

We also obtain an equivalence $\Lambda \circ (\tau_z \times id) \circ \tau_{\tilde{z}}|_{\widehat{\pi}_{2,1}^{-1}(W_1)}$ over W_1, where $\Lambda : W \times L(\mathbf{R}^n,\mathbf{R}^m) \times L(\mathbf{R}^n,\mathbf{R}^m) \times L(\mathbf{R}^n,\mathbf{R}^{nm}) \to W \times L(\mathbf{R}^n,\mathbf{R}^m) \times L(\mathbf{R}^n,\mathbf{R}^{nm})$ is the projection given by

$$\Lambda(u,F,G,H) = (u,F,H) \,,$$

completing the proof that $\widehat{\pi}_{2,1} : \widehat{J}^2E \to J^1E$ is a bundle. □

12.74 Remark Using the coordinate functions on $J^1(J^1E)$ described in Remark 12.72, we see that $\widehat{J}^2E$ is the submanifold of $J^1(J^1E)$ where $y^\alpha_{\tilde{j}} = y^\alpha_{;j}$ for $j = 1, \ldots, n$. From this it follows that

$$i_{1,1}(J^2E) \subset \widehat{J}^2E \subset J^1(J^1E) \,.$$

We use x^i, y^α, $y^\alpha_{\tilde{i}}$ and $y^\alpha_{\tilde{i};j}$ as coordinate functions on $\widehat{J}^2E$.

12.75 Proposition Let $\pi : E \to M$ be a bundle over M^n with fibre F^m, and let $\widetilde{\pi}' : \pi^*_{2,1}(TJ^1E) \to J^2E$ be the pullback of the tangent bundle $\pi' : TJ^1E \to J^1E$ by the 1-jet projection $\pi_{2,1} : J^2E \to J^1E$. Then we have two subbundles $\widetilde{\pi}'' : \widetilde{H} \to J^2E$ and $\widetilde{\pi}''' : \widetilde{V} \to J^2E$ of $\widetilde{\pi}'$ so that

$$\pi_{2,1}^*(TJ^1E)_\sigma = \widetilde{H}_\sigma \oplus \widetilde{V}_\sigma$$

for $\sigma \in J^2E$, where $\widetilde{H}_\sigma = \{\sigma\} \times (j^1\phi)_*(T_pM)$ and $\widetilde{V}_\sigma = \{\sigma\} \times V_{1\,v}$ when $\sigma = j_p^2\phi$ and $\pi_{2,1}(\sigma) = v$. The maps $\widetilde{h} : \pi_{2,1}^*(TJ^1E) \to \pi_{2,1}^*(TJ^1E)$ and $\widetilde{v} : \pi_{1,0}^*(TJ^1E) \to \pi_{1,0}^*(TJ^1E)$ which are projections on $\widetilde{H}_\sigma$ and $\widetilde{V}_\sigma$ in each fibre $\pi_{2,1}^*(TJ^1E)_\sigma$, are bundle maps over J^2E.

PROOF: Apply Proposition 12.46 to the bundle $\pi_1 : J^1E \to M$, and let $\widetilde{\pi}'' : \widetilde{H} \to J^2E$ and $\widetilde{\pi}''' : \widetilde{V} \to J^2E$ be the bundles induced from the subbundles of holonomic and vertical tangent vectors in $\widetilde{\pi}_1' : (\pi_1)_{1,0}^*(TJ^1E) \to J^1(J^1E)$ by the embedding $i_{1,1} : J^2E \to J^1(J^1E)$, using that $(\pi_1)_{1,0} \circ i_{1,1} = \pi_{2,1}$. □

12.76 Remark Suppose that $(\sigma, w) \in \pi_{2,1}^*(TJ^1E)$ where $\sigma = j_p^2\phi$, $\pi_{2,1}(\sigma) = v$ and $w \in T_vJ^1E$ is the tangent vector given by

$$w = \sum_i a^i \frac{\partial}{\partial x^i}\bigg|_v + \sum_\alpha b^\alpha \frac{\partial}{\partial y^\alpha}\bigg|_v + \sum_{i,\alpha} c_i^\alpha \frac{\partial}{\partial y_i^\alpha}\bigg|_v$$

in the local coordinates described in Remark 12.37. Then we have that

$$\widetilde{h}(\sigma, w) = (\sigma, (j^1\phi)_{*p} \circ \pi_{1*v}(w)) = \left(\sigma, \sum_{i=1}^n a^i\, (j^1\phi)_{*p}\left(\frac{\partial}{\partial x^i}\bigg|_p\right)\right)$$

$$= \left(\sigma, \sum_{i=1}^n a^i \left\{ \frac{\partial}{\partial x^i}\bigg|_v + \sum_\alpha y_i^\alpha(\sigma) \frac{\partial}{\partial y^\alpha}\bigg|_v + \sum_{i,\alpha} y_{ij}^\alpha(\sigma)\, \frac{\partial}{\partial y_j^\alpha}\bigg|_v \right\}\right)$$

and

$$\widetilde{v}(\sigma, w) = (\sigma, w - (j^1\phi)_{*p} \circ \pi_{1*v}(w))$$

$$= \left(\sigma, \sum_\alpha \left\{ b^\alpha - \sum_{i=1}^n a^i\, y_i^\alpha(\sigma) \right\} \frac{\partial}{\partial y^\alpha}\bigg|_v + \sum_{i,\alpha} \left\{ c_i^\alpha - \sum_{i=1}^n a^i\, y_{ij}^\alpha(\sigma) \right\} \frac{\partial}{\partial y_i^\alpha}\bigg|_v \right).$$

12.77 Definition Let $\pi : E \to M$ be a bundle over M^n with fibre F^m. A vector field $X : J^2E \to TJ^1E$ along $\pi_{2,1}$ is called a *total derivative* if the section $\widetilde{X} : J^2E \to \pi_{2,1}^*(TJ^1E)$ determined by X belongs to $\widetilde{H}$, i.e., if

$$X(j_p^2\phi) \in (j^1\phi)_*(T_pM)$$

for $p \in M$ and $\phi \in \Gamma_p(E)$. Let (z, W) and $((z \times \beta) \circ \tau_z, \pi_{1,0}^{-1}(W))$ be local charts on E and J^1E with coordinate functions described in Remarks 12.4 and 12.37. The vector field $X_i : \pi_{2,0}^{-1}(W) \to TJ^1E$ along $\pi_{2,1}$ defined by

$$X_i(\sigma) = (j^1\phi)_*\left(\frac{\partial}{\partial x^i}\bigg|_p\right) = \frac{\partial}{\partial x^i}\bigg|_v + \sum_\alpha y_i^\alpha(\sigma)\frac{\partial}{\partial y^\alpha}\bigg|_v + \sum_{j,\alpha} y_{ij}^\alpha(\sigma)\frac{\partial}{\partial y_j^\alpha}\bigg|_v$$

for $\sigma \in \pi_{2,0}^{-1}(W)$ where $\sigma = j_p^2\phi$ and $\pi_{2,1}(\sigma) = v$, is called the i-th *coordinate total derivative* and is denoted by $\frac{d}{dx^i}$ for $i = 1, \ldots, n$. The tangent vector $X_i(\sigma)$ in T_vJ^1E is denoted by $\frac{d}{dx^i}\bigg|_\sigma$, and we have that

$$\frac{dg}{dx^i}(\sigma) = \frac{d}{dx^i}\bigg|_\sigma(g) = (j^1\phi)_*\left(\frac{\partial}{\partial x^i}\bigg|_p\right)(g) = \frac{\partial}{\partial x^i}\bigg|_p(g \circ j^1\phi) = \frac{\partial(g \circ j^1\phi)}{\partial x^i}(p)$$

when $\sigma = j_p^2\phi$, $\pi_{2,1}(\sigma) = v$ and $g \in \mathscr{F}_v(J^1E)$. In particular, it follows by Definition 12.48 that

$$\frac{d}{dx^i}\left(\frac{df}{dx^j}\right)(\sigma) = \frac{\partial^2(f \circ \phi)}{\partial x^i \partial x^j}(p)$$

when $\sigma = j_p^2\phi$, $\phi(p) = u$ and $f \in \mathscr{F}_u(E)$.

12.78 Let $\pi : E \to M$ be a bundle over M^n with fibre F^m, and let $\pi_1' : TJ^1E \to J^1E$ and $\pi_2' : TJ^2E \to J^2E$ be the tangent bundles of J^1E and J^2E. By 12.13 the map $(\pi_2', \pi_{2,1\,*}) : TJ^2E \to J^2E \times TJ^1E$ induces a bundle map $\rho_2 : TJ^2E \to \pi_{2,1}^*(TJ^1E)$ over J^2E, as we see from the commutative diagram

$$\begin{array}{ccc} TJ^2E & \xrightarrow{\pi_{2,1\,*}} & TJ^1E \\ {\scriptstyle \pi_2'}\downarrow & & \downarrow{\scriptstyle \pi_1'} \\ J^2E & \xrightarrow{\pi_{2,1}} & J^1E \end{array}$$

Using the local charts (z, W), $((z \times \beta) \circ \tau_z, \pi_{1,0}^{-1}(W))$ and $((((z \times \beta) \circ \tau_z) \times \beta_s) \circ \tau_z^2, \pi_{2,0}^{-1}(W))$ on E, J^1E and J^2E with coordinate functions described in Remarks 12.37 and 12.67, we have that

$$\rho_2\left(\frac{\partial}{\partial x^i}\bigg|_\sigma\right)=\left(\sigma,\frac{\partial}{\partial x^i}\bigg|_v\right),\ \rho_2\left(\frac{\partial}{\partial y^\alpha}\bigg|_\sigma\right)=\left(\sigma,\frac{\partial}{\partial y^\alpha}\bigg|_v\right),$$

$$\rho_2\left(\frac{\partial}{\partial y^\alpha_i}\bigg|_\sigma\right)=\left(\sigma,\frac{\partial}{\partial y^\alpha_i}\bigg|_v\right)\ \text{ and }\ \rho_2\left(\frac{\partial}{\partial y^\alpha_{ij}}\bigg|_\sigma\right)=(\sigma,0)$$

for $\sigma\in\pi_{2,0}^{-1}(W)$ where $\pi_{2,1}(\sigma)=v$.

Let $\widetilde{\omega}\in\Omega^1(J^2E;\pi_{2,1}^*(TJ^1E))$ be the bundle-valued 1-form determined by the composed bundle map $\widehat{h}=\widetilde{h}\circ\rho_2:TJ^2E\to\pi_{2,1}^*(TJ^1E)$ over J^2E as described in Proposition 5.30. Then it follows from Proposition 5.44 and Remark 12.76 that

$$\phi_1\circ\widetilde{\omega}\,|_{\pi_{2,0}^{-1}(W)}=\sum_{i=1}^n dx^i\otimes\frac{d}{dx^i}\tag{1}$$

where $\phi_1:\Lambda^1(TJ^2E;\,\pi_{2,1}^*(TJ^1E))\to\mathscr{T}^1_1(TJ^2E,\pi_{2,1}^*(TJ^1E))$ is the equivalence over J^2E defined as in 5.25.

Now let $d_{\widetilde{\omega}}:\Omega(J^1E)\to\Omega(J^2E)$ be the derivation along $\pi_{2,1}$ of type d_* and degree 1 obtained from $\widetilde{\omega}$ as described in Proposition 12.30. If $\sigma=j^2_p\phi$, we have that

$$(i_{\widetilde{\omega}}\,dy^\alpha)(\sigma)=dy^\alpha(j^1_p\phi)\circ\widetilde{\omega}(\sigma)=dy^\alpha(j^1_p\phi)\circ(j^1\phi)_{*p}\circ\pi_{2*\sigma}$$

$$=d(y^\alpha\circ j^1\phi)(p)\circ\pi_{2*\sigma}=\sum_{i=1}^n\frac{\partial(y^\alpha\circ\phi)}{\partial x^i}(p)\;dx^i(p)\circ\pi_{2*\sigma}=\sum_{i=1}^n y^\alpha_i(\sigma)\;dx^i(\sigma)$$

so that

$$d_{\widetilde{\omega}}\,dy^\alpha=-\,d\circ i_{\widetilde{\omega}}\,dy^\alpha=\sum_{i=1}^n dx^i\wedge dy^\alpha_i\,.$$

In the same way we see that

$$d_{\widetilde{\omega}}\,dy^\alpha_i=\sum_{j=1}^n dx^j\wedge dy^\alpha_{ij}\ \text{ and }\ d_{\widetilde{\omega}}\,dx^i=0\,.$$

If $f\in\mathscr{F}(J^1E)$, then

$$d_{\widetilde{\omega}}f=\sum_{i=1}^n\frac{df}{dx^i}\,dx^i$$

since

$$(d_{\widetilde{\omega}}f)(\sigma)=(i_{\widetilde{\omega}}\,df)(\sigma)=\sum_{i=1}^n\frac{\partial(f\circ j^1\phi)}{\partial x^i}(p)\;dx^i(p)\circ\pi_{2*\sigma}=\sum_{i=1}^n\frac{df}{dx^i}(\sigma)\;dx^i(\sigma)\,.$$

Consider the commutative diagram

$$\begin{array}{ccc} M & \xrightarrow{j^2\phi} & J^2E \\ {\scriptstyle id_M}\downarrow & & \downarrow{\scriptstyle \pi_{2,1}} \\ M & \xrightarrow[j^1\phi]{} & J^1E \end{array}$$

and let $\varepsilon \in \Omega^1(M;TM)$ be the bundle-valued 1-form defined in Example 5.45 corresponding to the Kronecker delta tensor on M. If $\sigma = j_p^2\phi$, we have that

$$\widetilde{\omega}(\sigma) \circ (j^2\phi)_{*p} = (j^1\phi)_{*p} \circ \pi_{2*\sigma} \circ (j^2\phi)_{*p} = (j^1\phi)_{*p} \circ \varepsilon(p)\,,$$

showing that the bundle-valued 1-forms ε and $\widetilde{\omega}$ along id_M and $\pi_{2,1}$ are $(j^2\phi, j^1\phi)$-related as described in 12.32. Hence it follows that

$$d \circ (j^1\phi)^* = (j^2\phi)^* \circ d_{\widetilde{\omega}}. \qquad (2)$$

The derivation $d_{\widetilde{\omega}}$ will be important in our study of the calculus of variations in 12.90.

PROLONGATION OF VECTOR FIELDS

12.79 Let $\pi : E \to M$ be a bundle over M^n with fibre F^m, and let $\pi'' : V \to E$ and $\pi_1'' : V_1 \to J^1E$ be the vertical tangent bundles of the bundles $\pi : E \to M$ and $\pi_1 : J^1E \to M$. We will construct a map $i_1 : J^1V \to V_1$ which is an equivalence over M between the 1-jet bundle $\pi_1' : J^1V \to M$ of the bundle $\pi' : V \to M$ where $\pi' = \pi \circ \pi''$, and the bundle $\pi_1''' : V_1 \to M$ where $\pi_1''' = \pi_1 \circ \pi_1''$.

Let $(t, \pi^{-1}(U))$ be a local trivialization around a point u on E with $t(u) = (p,v)$, where U is the coordinate neighbourhood of a local chart (x,U) around p on M with $x(p) = a$, and consider the local chart (z,W) on E obtained from $(t,\pi^{-1}(U))$, (x,U) and a local chart (y,Q) around v on F as described in Proposition 12.3, so that $W = t^{-1}(U \times Q)$ and $z = (x \times y) \circ t$.

The bundle $\pi' : V \to M$ has fibre TF, and we have a local trivialization $(t', \pi'^{-1}(U))$ on V with $t' = (\pi_U', (pr_2' \circ t)_* \circ i)$, where $pr_2' : U \times F \to F$ is the projection on the second factor and $i : V \to TE$ is the inclusion map. From this we also obtain a local chart (z', W') on V with $W' = t'^{-1}(U \times TF|_Q)$ and $z' = (x \times y') \circ t'$, using the local chart $(y', TF|_Q)$ on TF where $y' = (y \times id) \circ t_y$. We let $\beta : L(\mathbf{R}^n, \mathbf{R}^m) \to \mathbf{R}^{nm}$ and $\beta' : L(\mathbf{R}^n, \mathbf{R}^{2m}) \to \mathbf{R}^{2nm}$ be linear isomorphisms defined as in Corollary 5.29.

To define the map $i_1 : J^1V \to V_1$, we consider a smooth map $\alpha : I \times O \to E$ satisfying $\pi \circ \alpha = pr_2$, where $pr_2 : I \times O \to M$ is the projection on the second factor, and where O is an open neighbourhood of p on M and I is an open interval containing 0 so that $\alpha(I \times O) \subset W$. Let $\widetilde{\alpha} : I \times x(O) \to \mathbf{R}^m$ be the smooth map given by $\widetilde{\alpha} = z_2 \circ \alpha \circ (id \times x)^{-1}$. Now let $\alpha_q : I \to E$ and $\alpha_t : O \to E$ be the smooth maps defined by $\alpha_q(t) = \alpha_t(q) = \alpha(t,q)$ for $t \in I$ and $q \in O$, and consider the section $\psi : O \to V$ of π' given by $\psi(q) = \alpha_q'(0)$ for $q \in O$. Using the local charts defined above, we have that

$$\begin{aligned} &z_2' \circ \psi \circ x^{-1}(b) = (y \times id) \circ t_y \circ (pr_2' \circ t)_* \circ \psi(q) \\ &= (y \times id) \circ t_y \circ (pr_2' \circ t \circ \alpha_q)'(0) = (z_2 \circ \alpha_q(0), (z_2 \circ \alpha_q)'(0)) \\ &= (\widetilde{\alpha}\,(0,b), D_1\widetilde{\alpha}\,(0,b)) \end{aligned}$$

for $b = x(q) \in x(O)$. As in the proof of Proposition 12.40 it follows that

$$\begin{aligned} &(z_2' \times \beta') \circ \tau_{z'}(j_p^1\psi) = (z_2' \circ \psi \circ x^{-1}(a), \beta' \circ D(z_2' \circ \psi \circ x^{-1})(a)) \\ &= (\widetilde{\alpha}\,(0,a), D_1\widetilde{\alpha}\,(0,a), \beta'(D_2\widetilde{\alpha}\,(0,a), D_2D_1\widetilde{\alpha}\,(0,a)))\,. \end{aligned}$$

We now let $c : I \to J^1E$ be the curve given by $c(t) = j_p^1\alpha_t$ for $t \in I$, and define $i_1(j_p^1\psi) = c'(0)$. Using the local chart $(\widetilde{z}, \widetilde{W})$ on J_p^1E with $\widetilde{W} = J_p^1E \cap \pi_{1,0}^{-1}(W)$ and $\widetilde{z} = (z_2 \times \beta) \circ \tau_z|_{\widetilde{W}}$, we have that

$$\widetilde{z} \circ c(t) = (z_2 \circ \alpha_t \circ x^{-1}(a), \beta \circ D(z_2 \circ \alpha_t \circ x^{-1})(a))$$

for $t \in I$, so that

$$\begin{aligned} &(\widetilde{z} \times id) \circ t_{\widetilde{z}} \circ i_1(j_p^1\psi) = (\widetilde{z} \circ c(0), (\widetilde{z} \circ c)'(0)) \\ &= (\widetilde{\alpha}\,(0,a), \beta \circ D_2\widetilde{\alpha}\,(0,a), D_1\widetilde{\alpha}\,(0,a), \beta \circ D_1D_2\widetilde{\alpha}\,(0,a))\,. \end{aligned}$$

Hence it follows that

$$j_1 \circ (\widetilde{z} \times id) \circ t_{\widetilde{z}} \circ i_1 = (z_2' \times \beta') \circ \tau_{z'}|_{\widetilde{W}'}, \tag{1}$$

where $\widetilde{W}' = J_p^1V \cap \pi_{1,0}'^{-1}(W')$ and $j_1 : \mathbf{R}^m \times \mathbf{R}^{nm} \times \mathbf{R}^m \times \mathbf{R}^{nm} \to \mathbf{R}^m \times \mathbf{R}^m \times \mathbf{R}^{2nm}$ is the linear isomorphism given by

$$j_1(a,b,c,d) = (a, c, \beta' \circ (\beta \times \beta)^{-1}(b,d))$$

for $a, c \in \mathbf{R}^m$ and $b, d \in \mathbf{R}^{nm}$, showing that the map $i_1 : J^1V \to V_1$ is a well-defined equivalence over M. It satisfies the relation

$$\widetilde{\pi}_{1,0} \circ i_1 = \pi_{1,0}' \tag{2}$$

where $\widetilde{\pi}_{1,0} : V_1 \to V$ is the smooth map induced by $\pi_{1,0\,*} : TJ^1E \to TE$, since

$$\pi_{1,0\,*} \circ i_1(j_p^1\psi) = \pi_{1,0\,*} \circ c'(0) = (\pi_{1,0} \circ c)'(0) = \alpha_p'(0) = \psi(p) = \pi_{1,0}'(j_p^1\psi)\,.$$

We also have that

$$(id \times \beta)^{-1} \circ t_\beta \circ (pr_2'' \circ \tau_z)_* \circ i_1 = pr_2''' \circ \tau_{z'} \tag{3}$$

where $pr_2'' : W \times L(\mathbf{R}^n, \mathbf{R}^m) \to L(\mathbf{R}^n, \mathbf{R}^m)$ and $pr_2''' : W' \times L(\mathbf{R}^n, \mathbf{R}^{2m}) \to L(\mathbf{R}^n, \mathbf{R}^{2m})$ are the projections on the second factor, since

$$pr_2'' \circ \tau_z \circ c(t) = pr_2'' \circ \tau_z(j_p^1 \alpha_t) = D(z_2 \circ \alpha_t \circ x^{-1})(a) = D_2 \widetilde{\alpha}(t,a)$$

for $t \in I$ so that

$$(id \times \beta)^{-1} \circ t_\beta \circ (pr_2'' \circ \tau_z)_* \circ i_1(j_p^1 \psi) = (id \times \beta)^{-1} \circ t_\beta \circ (pr_2'' \circ \tau_z \circ c)'(0)$$

$$= (D_2 \widetilde{\alpha}(0,a), D_1 D_2 \widetilde{\alpha}(0,a)) = D(z_2' \circ \psi \circ x^{-1})(a) = pr_2''' \circ \tau_{z'}(j_p^1 \psi) \,.$$

Finally, we have that

$$\pi_1'' \circ i_1 \circ j^1 \psi = j^1(\pi'' \circ \psi) \tag{4}$$

since

$$\pi_1'' \circ i_1(j_p^1 \psi) = \pi_1'' \circ c'(0) = c(0) = j_p^1 \alpha_0 = j_p^1(\pi'' \circ \psi) \,.$$

12.80 Remark Using the local coordinates described in Remarks 12.4, 12.5 and 12.37, we have coordinate functions x^i, y^α and $\dot{y}^\alpha$ on V which give rise to the coordinate functions x^i, y^α, $\dot{y}^\alpha$, y_i^α and $\dot{y}_i^\alpha$ on J^1V for $i = 1,2, \dots, n$ and $\alpha = 1,2, \dots, m$. We also have coordinate functions x^i, y^α and y_i^α on J^1E from which we obtain the coordinate functions x^i, y^α, y_i^α, $\dot{y}^\alpha$ and $\dot{y}_i^\alpha$ on V_1. By formula (1) in 12.79 we have that

$$x^i \circ i_1 = x^i \;,\quad y^\alpha \circ i_1 = y^\alpha \;,\quad y_i^\alpha \circ i_1 = y_i^\alpha \;,\quad \dot{y}^\alpha \circ i_1 = \dot{y}^\alpha \quad \text{and} \quad \dot{y}_i^\alpha \circ i_1 = \dot{y}_i^\alpha$$

for $i = 1,2, \dots, n$ and $\alpha = 1,2, \dots, m$.

12.81 Let X be a vertical vector field on E, i.e., a section on E in the vertical tangent bundle $\pi'' : V \to E$, using the same notation as in 12.79. Regarding X as a bundle map over M between the bundles $\pi : E \to M$ and $\pi' : V \to M$, we obtain a vertical vector field X^1 on J^1E, called the 1-*jet prolongation of* X, defined by

$$X^1 = i_1 \circ j^1 X \,.$$

Indeed, it follows from formula (4) in 12.79 that

$$\pi_1'' \circ X^1(j_p^1 \phi) = \pi_1'' \circ i_1(j_p^1(X \circ \phi)) = j_p^1 \phi$$

for $p \in M$ and $\phi \in \Gamma_p(E)$, showing that X^1 is a section on J^1E in the vertical tangent bundle $\pi_1'' : V_1 \to J^1E$. Using formula (2) in 12.79 we also have that

$$\widetilde{\pi}_{1,0} \circ X^1 = \widetilde{\pi}_{1,0} \circ i_1 \circ j^1 X = \pi_{1,0}' \circ j^1 X = X \circ \pi_{1,0} \,.$$

12.82 Proposition Let $\pi : E \to M$ be a bundle over M^n with fibre F^m, and let $\pi'' : TE \to E$ be the tangent bundle of E and $\pi' : TE \to M$ be the bundle where $\pi' = \pi \circ \pi''$. Then the map $r_1 : J^1TE \to TJ^1E$ defined by

$$r_1(j_p^1\psi) = i_1(j_p^1(\psi - \phi_* \circ \pi_* \circ \psi)) + (j^1\phi)_* \circ \pi_* \circ \psi(p)$$

for $p \in M$ and $\psi \in \Gamma_p(TE)$ where $\phi = \pi'' \circ \psi$, is a bundle map over TE from the 1-jet bundle $\pi'_{1,0} : J^1TE \to TE$ of π' to the bundle $\pi_{1,0\,*} : TJ^1E \to TE$. We have that

$$\pi_1'' \circ r_1 \circ j^1\psi = j^1\phi$$

where $\pi_1'' : TJ^1E \to J^1E$ is the projection in the tangent bundle of J^1E.

PROOF: We first note that the term $i_1(j_p^1(\psi - \phi_* \circ \pi_* \circ \psi))$ is well defined since

$$\pi'' \circ \phi_* \circ \pi_* \circ \psi = \phi \circ \pi \circ \pi'' \circ \psi = \phi = \pi'' \circ \psi$$

and

$$\pi_* \circ (\psi - \phi_* \circ \pi_* \circ \psi) = 0\,.$$

From this and formula (4) in 12.79 it also follows that

$$\pi_1'' \circ i_1(j_p^1(\psi - \phi_* \circ \pi_* \circ \psi)) = j_p^1(\pi'' \circ \{\psi - \phi_* \circ \pi_* \circ \psi\}) = j_p^1\phi$$

and

$$\pi_1'' \circ (j^1\phi)_* \circ \pi_* \circ \psi(p) = j^1\phi \circ \pi \circ \pi'' \circ \psi(p) = j_p^1\phi\,,$$

showing that $r_1(j_p^1\psi)$ is well defined. It also shows the last part of the proposition. Using formula (2) in 12.79 we now have that

$$\pi_{1,0\,*} \circ r_1 = \pi'_{1,0}$$

since

$$\begin{aligned}\pi_{1,0\,*} \circ r_1(j_p^1\psi) &= \pi'_{1,0}\,(j_p^1(\psi - \phi_* \circ \pi_* \circ \psi)) + (\pi_{1,0} \circ j^1\phi)_* \circ \pi_* \circ \psi(p) \\ &= (\psi - \phi_* \circ \pi_* \circ \psi)(p) + \phi_* \circ \pi_* \circ \psi(p) = \psi(p) = \pi'_{1,0}\,(j_p^1\psi)\,.\end{aligned}$$

Formula (3) in 12.79 implies that

$$\begin{aligned}&(id \times \beta)^{-1} \circ t_\beta \circ (pr_2'' \circ \tau_z)_* \circ r_1(j_p^1\psi) \\ &= pr_2''' \circ \tau_{z'}\,(j_p^1(\psi - \phi_* \circ \pi_* \circ \psi)) \\ &+ (id \times \beta)^{-1} \circ t_\beta \circ (pr_2'' \circ \tau_z \circ j^1\phi \circ \pi)_* \circ \psi(p) \\ &= D(z_2' \circ \{\psi - \phi_* \circ \pi_* \circ \psi\} \circ x^{-1})(a) \\ &+ (id \times \beta)^{-1} \circ t_\beta \circ \{D(z_2 \circ \phi \circ x^{-1}) \circ x \circ \pi\}_* \circ \psi(p)\,.\end{aligned}$$

Let O be an open neighbourhood of p on M so that $\phi(O) \subset W$, and let $pr_2 : W \times \mathbf{R}^{n+m} \to \mathbf{R}^{n+m}$ be the projection on the second factor with component maps $pr_{21} : W \times \mathbf{R}^{n+m} \to \mathbf{R}^n$ and $pr_{22} : W \times \mathbf{R}^{n+m} \to \mathbf{R}^m$. Then we have that

$$z_2' \circ (\psi - \phi_* \circ \pi_* \circ \psi)(q) = (y \times id) \circ t_y \circ (pr_2' \circ t)_* \circ (\psi - \phi_* \circ \pi_* \circ \psi)(q)$$
$$= (y \circ pr_2' \circ t \circ \phi(q), \{D(y \circ pr_2' \circ t \circ z^{-1})(z \circ \phi(q))$$
$$- D(y \circ pr_2' \circ t \circ \phi \circ x^{-1})(x(q)) \circ D(x \circ \pi \circ z^{-1})(z \circ \phi(q))\} \circ pr_2 \circ t_z \circ \psi(q))$$
$$= (z_2 \circ \phi(q), \{pr_{22} - D(z_2 \circ \phi \circ x^{-1})(x(q)) \circ pr_{21}\} \circ t_z \circ \psi(q))$$

for $q \in O$, which implies that

$$D(z_2' \circ \{\psi - \phi_* \circ \pi_* \circ \psi\} \circ x^{-1})(a)$$
$$= (D(z_2 \circ \phi \circ x^{-1})(a), D(pr_{22} \circ t_z \circ \psi \circ x^{-1})(a)$$
$$- D(z_2 \circ \phi \circ x^{-1})(a) \circ D(pr_{21} \circ t_z \circ \psi \circ x^{-1})(a)$$
$$- i_{pr_{21} \circ t_z \circ \psi \circ x^{-1}(a)} D^2(z_2 \circ \phi \circ x^{-1})(a)) .$$

We also have that

$$(id \times \beta)^{-1} \circ t_\beta \circ \{D(z_2 \circ \phi \circ x^{-1}) \circ x \circ \pi\}_* \circ \psi(p) = (D(z_2 \circ \phi \circ x^{-1})(a),$$
$$\beta^{-1} \circ D(\beta \circ D(z_2 \circ \phi \circ x^{-1}) \circ x \circ \pi \circ z^{-1})(z \circ \phi(p)) \circ pr_2 \circ t_z \circ \psi(p))$$
$$= (D(z_2 \circ \phi \circ x^{-1})(a), i_{pr_{21} \circ t_z \circ \psi \circ x^{-1}(a)} D^2(z_2 \circ \phi \circ x^{-1})(a)) .$$

Combining this, we see that the second derivatives cancel, and we obtain that

$$(id \times \beta)^{-1} \circ t_\beta \circ (pr_2'' \circ \tau_z)_* \circ r_1(j_p^1 \psi)$$
$$= (D(z_2 \circ \phi \circ x^{-1})(a), D(pr_{22} \circ t_z \circ \psi \circ x^{-1})(a)$$
$$- D(z_2 \circ \phi \circ x^{-1})(a) \circ D(pr_{21} \circ t_z \circ \psi \circ x^{-1})(a)) .$$

The bundle $\pi' : TE \to M$ has fibre $\mathbf{R}^n \times TF$, and we have a local trivialization $(t'', \pi'^{-1}(U))$ on TE with $t'' = (\pi_U', pr_2'''' \circ t_x \circ \pi_*, (pr_2' \circ t)_*)$, where $pr_2'''' : \mathbf{R}^n \times \mathbf{R}^n \to \mathbf{R}^n$ is the projection on the second factor. From this we also obtain a local chart (z'', W'') on TE with $W'' = t''^{-1}(U \times \mathbf{R}^n \times TF|_Q)$ and $z'' = (x \times id \times y') \circ t''$, using the local chart $(y', TF|_Q)$ on TF where $y' = (y \times id) \circ t_y$. We now have that

$$\begin{aligned}
\tau_{z''}(j^1_p\psi) &= (\psi(p), D(z''_2 \circ \psi \circ x^{-1})(a)) \\
&= (\psi(p), D(pr''''_2 \circ t_x \circ \pi_* \circ \psi \circ x^{-1})(a), \\
&\quad D(y' \circ (pr'_2 \circ t)_* \circ \psi \circ x^{-1})(a)) \\
&= (\psi(p), D(pr_{21} \circ t_z \circ \psi \circ x^{-1})(a), \\
&\quad D(z_2 \circ \phi \circ x^{-1})(a), D(pr_{22} \circ t_z \circ \psi \circ x^{-1})(a))
\end{aligned}$$

which shows that

$$(id \times \beta)^{-1} \circ t_\beta \circ (pr''_2 \circ \tau_z)_* \circ r_1 \circ \tau_{z''}^{-1}(u, G_1, G_2, G_3) = (G_2, G_3 - G_2 \circ G_1)$$

for $u \in W''$, $G_1 \in L(\mathbf{R}^n, \mathbf{R}^n)$ and $G_2, G_3 \in L(\mathbf{R}^n, \mathbf{R}^m)$. □

12.83 Remark Let $\pi : E \to M$ be a bundle over M^n with fibre F^m, and consider the tangent bundle $\pi'' : TE \to E$ of E and the bundle $\pi' : TE \to M$ where $\pi' = \pi \circ \pi''$. Let $r_1 : J^1TE \to TJ^1E$ be the bundle map over TE defined in Proposition 12.82, and suppose that $\psi \in \Gamma_U(TE)$ is a section of π' on an open subset U of M, given in the local coordinates described in Remark 12.4 by

$$\psi(p) = \sum_{i=1}^{n} a^i(p) \left.\frac{\partial}{\partial x^i}\right|_{\phi(p)} + \sum_{\alpha=1}^{m} b^\alpha(p) \left.\frac{\partial}{\partial y^\alpha}\right|_{\phi(p)}$$

for $p \in U$ where $\phi = \pi'' \circ \psi$. Using also the local coordinates described in Remarks 12.5 and 12.37, we have that

$$\dot{x}^i \circ r_1(j^1_p\psi) = \dot{x}^i \circ \pi_{1,0\,*} \circ r_1(j^1_p\psi) = \dot{x}^i \circ \pi'_{1,0}(j^1_p\psi) = \dot{x}^i \circ \psi(p) = a^i(p)$$

and

$$\dot{y}^\alpha \circ r_1(j^1_p\psi) = \dot{y}^\alpha \circ \pi_{1,0\,*} \circ r_1(j^1_p\psi) = \dot{y}^\alpha \circ \pi'_{1,0}(j^1_p\psi) = \dot{y}^\alpha \circ \psi(p) = b^\alpha(p)$$

for $p \in U$. We also have that

$$\begin{aligned}
&\dot{y}^\alpha_i \circ r_1(j^1_p\psi) = \dot{y}^\alpha_i \circ i_1(j^1_p(\psi - \phi_* \circ \pi_* \circ \psi)) + \dot{y}^\alpha_i \circ (j^1\phi)_* \circ \pi_* \circ \psi(p) \\
&= \left.\frac{\partial}{\partial x^i}\right|_p \{\dot{y}^\alpha \circ \psi - (\pi_* \circ \psi)(y^\alpha \circ \phi)\} + (\pi_* \circ \psi)(p)(y^\alpha_i \circ j^1\phi) \\
&= \frac{\partial b^\alpha}{\partial x^i}(p) - \left.\frac{\partial}{\partial x^i}\right|_p \left(\sum_{j=1}^{n} a^j \frac{\partial(y^\alpha \circ \phi)}{\partial x^j} \right) + \sum_{j=1}^{n} a^j(p) \left.\frac{\partial}{\partial x^i}\right|_p \left(\frac{\partial(y^\alpha \circ \phi)}{\partial x^j} \right) \\
&= \frac{\partial b^\alpha}{\partial x^i}(p) - \sum_{j=1}^{n} \frac{\partial a^j}{\partial x^i}(p) \frac{\partial(y^\alpha \circ \psi)}{\partial x^j}(p) = \left(\dot{y}^\alpha_i - \sum_{j=1}^{n} \dot{x}^j_i \, y^\alpha_j \right)(j^1_p\psi)
\end{aligned}$$

which implies that

$$r_1(j_p^1\psi) = \sum_i a^i(p)\,\frac{\partial}{\partial x^i}\bigg|_{j_p^1\phi} + \sum_\alpha b^\alpha(p)\,\frac{\partial}{\partial y^\alpha}\bigg|_{j_p^1\phi} + \sum_{i,\alpha}\left(\dot{y}_i^\alpha - \sum_{j=1}^n \dot{x}_i^j\, y_j^\alpha\right)(j_p^1\psi)\,\frac{\partial}{\partial y_i^\alpha}\bigg|_{j_p^1\phi}$$

for $p \in U$.

12.84 Let X be a vector field on E, i.e., a section on E in the tangent bundle $\pi'' : TE \to E$, using the same notation as in Proposition 12.82. Regarding X as a bundle map over M between the bundles $\pi : E \to M$ and $\pi' : TE \to M$, we obtain a vector field X^1 on J^1E, called the 1*-jet prolongation of* X, defined by

$$X^1 = r_1 \circ j^1X\,.$$

Indeed, it follows from Proposition 12.82 that

$$\pi_1'' \circ X^1(j_p^1\phi) = \pi_1'' \circ r_1(j_p^1(X\circ\phi)) = j_p^1\phi$$

for $p \in M$ and $\phi \in \Gamma_p(E)$, showing that X^1 is a section on J^1E in the tangent bundle $\pi_1'' : TJ^1E \to J^1E$. We also have that

$$\pi_{1,0\,*} \circ X^1 = \pi_{1,0\,*} \circ r_1 \circ j^1X = \pi_{1,0}' \circ j^1X = X \circ \pi_{1,0}\,,$$

which shows that the vector fields X^1 and X are $\pi_{1,0}$ - related.

12.85 Remark Suppose that the vector field X on E is given by

$$X\,|_W = \sum_{i=1}^n a^i\,\frac{\partial}{\partial x^i} + \sum_{\alpha=1}^m b^\alpha\,\frac{\partial}{\partial y^\alpha}$$

on an open subset W of E, using the local coordinates described in Remark 12.4. By Definition 12.48 and Remark 12.83 we have that

$$\dot{y}_i^\alpha \circ X^1(j_p^1\phi) = \dot{y}_i^\alpha \circ r_1(j_p^1(X\circ\phi)) = \left(\dot{y}_i^\alpha - \sum_{j=1}^n \dot{x}_i^j\, y_j^\alpha\right)(j_p^1(X\circ\phi))$$

$$= \frac{\partial(b^\alpha\circ\phi)}{\partial x^i}(p) - \sum_{j=1}^n \frac{\partial(a^j\circ\phi)}{\partial x^i}(p)\;\frac{\partial(y^\alpha\circ\phi)}{\partial x^j}(p) = \left(\frac{db^\alpha}{dx^i} - \sum_{j=1}^n y_j^\alpha\,\frac{da^j}{dx^i}\right)(j_p^1\phi)$$

which shows that

$$X^1|_{\pi_{1,0}^{-1}(W)} = \sum_i (a^i\circ\pi_{1,0})\,\frac{\partial}{\partial x^i} + \sum_\alpha (b^\alpha\circ\pi_{1,0})\,\frac{\partial}{\partial y^\alpha} + \sum_{i,\alpha}\left(\frac{db^\alpha}{dx^i} - \sum_{j=1}^n y_j^\alpha\,\frac{da^j}{dx^i}\right)\frac{\partial}{\partial y_i^\alpha}\,.$$

12.86 Let $\pi : E \to M$ be a bundle over M^n with fibre F^m, and let $\pi'' : V \to E$ and $\pi_2'' : V_2 \to J^2E$ be the vertical tangent bundles of the bundles $\pi : E \to M$ and $\pi_2 : J^2E \to M$. We will construct a map $i_2 : J^2V \to V_2$ which is an equivalence over M between the 2-jet bundle $\pi_2' : J^2V \to M$ of the bundle $\pi' : V \to M$ where $\pi' = \pi \circ \pi''$, and the bundle $\pi_2''' : V_2 \to M$ where $\pi_2''' = \pi_2 \circ \pi_2''$.

Let $(t, \pi^{-1}(U))$ be a local trivialization around a point u on E with $t(u) = (p, v)$, where U is the coordinate neighbourhood of a local chart (x, U) around p on M with $x(p) = a$, and consider the local chart (z, W) on E obtained from $(t, \pi^{-1}(U))$, (x, U) and a local chart (y, Q) around v on F as described in Proposition 12.3, so that $W = t^{-1}(U \times Q)$ and $z = (x \times y) \circ t$.

The bundle $\pi' : V \to M$ has fibre TF, and we have a local trivialization $(t', \pi'^{-1}(U))$ on V with $t' = (\pi_U', (pr_2' \circ t)_* \circ i)$, where $pr_2' : U \times F \to F$ is the projection on the second factor and $i : V \to TE$ is the inclusion map. From this we also obtain a local chart (z', W') on V with $W' = t'^{-1}(U \times TF|_Q)$ and $z' = (x \times y') \circ t'$, using the local chart $(y', TF|_Q)$ on TF where $y' = (y \times id) \circ t_y$. We let $\beta : L(\mathbf{R}^n, \mathbf{R}^m) \to \mathbf{R}^{nm}$, $\beta' : L(\mathbf{R}^n, \mathbf{R}^{2m}) \to \mathbf{R}^{2nm}$, $\beta_s : \mathscr{T}_s^{\,2}(\mathbf{R}^n; \mathbf{R}^m) \to \mathbf{R}^{n(n+1)m/2}$ and $\beta_s' : \mathscr{T}_s^{\,2}(\mathbf{R}^n; \mathbf{R}^{2m}) \to \mathbf{R}^{n(n+1)m}$ be linear isomorphisms defined as in the proof of Proposition 5.31.

To define the map $i_2 : J^2V \to V_2$, we consider a smooth map $\alpha : I \times O \to E$ satisfying $\pi \circ \alpha = pr_2$, where $pr_2 : I \times O \to M$ is the projection on the second factor, and where O is an open neighbourhood of p on M and I is an open interval containing 0 so that $\alpha(I \times O) \subset W$. Let $\widetilde{\alpha} : I \times x(O) \to \mathbf{R}^m$ be the smooth map given by $\widetilde{\alpha} = z_2 \circ \alpha \circ (id \times x)^{-1}$. Now let $\alpha_q : I \to E$ and $\alpha_t : O \to E$ be the smooth maps defined by $\alpha_q(t) = \alpha_t(q) = \alpha(t, q)$ for $t \in I$ and $q \in O$, and consider the section $\psi : O \to V$ of π' given by $\psi(q) = \alpha_q'(0)$ for $q \in O$. Using the local charts defined above, we have that

$$
\begin{aligned}
& z_2' \circ \psi \circ x^{-1}(b) = (y \times id) \circ t_y \circ (pr_2' \circ t)_* \circ \psi(q) \\
&= (y \times id) \circ t_y \circ (pr_2' \circ t \circ \alpha_q)'(0) = (z_2 \circ \alpha_q(0), (z_2 \circ \alpha_q)'(0)) \\
&= (\widetilde{\alpha}(0, b), D_1 \widetilde{\alpha}(0, b))
\end{aligned}
$$

for $b = x(q) \in x(O)$. As in the proof of Proposition 12.66 it follows that

$$
\begin{aligned}
& (((z_2' \times \beta') \circ \tau_{z'}) \times \beta_s') \circ \tau_{z'}^2(j_p^2 \psi) \\
&= (z_2' \circ \psi \circ x^{-1}(a), \beta' \circ D(z_2' \circ \psi \circ x^{-1})(a), \beta_s' \circ D^2(z_2' \circ \psi \circ x^{-1})(a)) \\
&= (\widetilde{\alpha}(0, a), D_1 \widetilde{\alpha}(0, a), \beta'(D_2 \widetilde{\alpha}(0, a), D_2 D_1 \widetilde{\alpha}(0, a)), \beta_s'(D_2^2 \widetilde{\alpha}(0, a), D_2^2 D_1 \widetilde{\alpha}(0, a)))\,.
\end{aligned}
$$

We now let $c : I \to J^2E$ be the curve given by $c(t) = j_p^2 \alpha_t$ for $t \in I$, and define $i_2(j_p^2 \psi) = c'(0)$. Using the local chart $(\widetilde{z}, \widetilde{W})$ on J_p^2E with $\widetilde{W} = J_p^2E \cap \pi_{2,0}^{-1}(W)$ and $\widetilde{z} = (((z_2 \times \beta) \circ \tau_z) \times \beta_s) \circ \tau_z^2|_{\widetilde{W}}$, we have that

$$
\widetilde{z} \circ c(t) = (z_2 \circ \alpha_t \circ x^{-1}(a), \beta \circ D(z_2 \circ \alpha_t \circ x^{-1})(a), \beta_s \circ D^2(z_2 \circ \alpha_t \circ x^{-1})(a))
$$

for $t \in I$, so that

$$(\widetilde{z} \times id) \circ t_{\widetilde{z}} \circ i_1(j_p^2\psi) = (\widetilde{z} \circ c(0), (\widetilde{z} \circ c)'(0))$$
$$= (\widetilde{\alpha}(0,a), \beta \circ D_2\widetilde{\alpha}(0,a), \beta_s \circ D_2^2\widetilde{\alpha}(0,a),$$
$$D_1\widetilde{\alpha}(0,a), \beta \circ D_1D_2\widetilde{\alpha}(0,a), \beta_s \circ D_1D_2^2\widetilde{\alpha}(0,a))\ .$$

Hence it follows that

$$j_2 \circ (\widetilde{z} \times id) \circ t_{\widetilde{z}} \circ i_2 = (((z_2' \times \beta') \circ \tau_{z'}) \times \beta_s') \circ \tau_{z'}^2|_{\widetilde{W}'}, \tag{1}$$

where $\widetilde{W}' = J_p^2 V \cap \pi_{2,0}'^{-1}(W')$ and $j_2 : \mathbf{R}^m \times \mathbf{R}^{nm} \times \mathbf{R}^{n(n+1)m/2} \times \mathbf{R}^m \times \mathbf{R}^{nm} \times \mathbf{R}^{n(n+1)m/2} \to \mathbf{R}^m \times \mathbf{R}^m \times \mathbf{R}^{2nm} \times \mathbf{R}^{n(n+1)m}$ is the linear isomorphism given by

$$j_2(a,b,c,d,e,f) = (a, d, \beta' \circ (\beta \times \beta)^{-1}(b,e), \beta_s' \circ (\beta_s \times \beta_s)^{-1}(c,f))$$

for $a, d \in \mathbf{R}^m$, $b, e \in \mathbf{R}^{nm}$ and $c, f \in \mathbf{R}^{n(n+1)m/2}$, showing that the map $i_2 : J^2V \to V_2$ is a well-defined equivalence over M. It satisfies the relation

$$\widetilde{\pi}_{2,1} \circ i_2 = i_1 \circ \pi_{2,1}' \tag{2}$$

where $i_1 : J^1V \to V_1$ is the equivalence over M defined in 12.79 and $\widetilde{\pi}_{2,1} : V_2 \to V_1$ is the smooth map induced by $\pi_{2,1\,*} : TJ^2E \to TJ^1E$, since

$$\pi_{2,1\,*} \circ i_2(j_p^2\psi) = \pi_{2,1\,*} \circ c'(0) = (\pi_{2,1} \circ c)'(0) = i_1(j_p^1\psi) = i_1 \circ \pi_{2,1}'(j_p^2\psi)\ .$$

We also have that

$$(id \times \beta_s)^{-1} \circ t_{\beta_s} \circ (pr_2'' \circ \tau_z^2)_* \circ i_2 = pr_2''' \circ \tau_{z'}^2 \tag{3}$$

where $pr_2'' : W \times \mathscr{T}_s^2(\mathbf{R}^n;\mathbf{R}^m) \to \mathscr{T}_s^2(\mathbf{R}^n;\mathbf{R}^m)$ and $pr_2''' : W' \times \mathscr{T}_s^2(\mathbf{R}^n;\mathbf{R}^{2m}) \to \mathscr{T}_s^2(\mathbf{R}^n;\mathbf{R}^{2m})$ are the projections on the second factor, since

$$pr_2'' \circ \tau_z^2 \circ c(t) = pr_2'' \circ \tau_z^2(j_p^2\alpha_t) = D^2(z_2 \circ \alpha_t \circ x^{-1})(a) = D_2^2\widetilde{\alpha}(t,a)$$

for $t \in I$ so that

$$(id \times \beta_s)^{-1} \circ t_{\beta_s} \circ (pr_2'' \circ \tau_z^2)_* \circ i_2(j_p^2\psi) = (id \times \beta_s)^{-1} \circ t_{\beta_s} \circ (pr_2'' \circ \tau_z^2 \circ c)'(0)$$
$$= (D_2^2\widetilde{\alpha}(0,a), D_1D_2^2\widetilde{\alpha}(0,a)) = D^2(z_2' \circ \psi \circ x^{-1})(a) = pr_2''' \circ \tau_{z'}^2(j_p^2\psi)\ .$$

Finally, we have that

$$\pi_2'' \circ i_2 \circ j^2\psi = j^2(\pi'' \circ \psi) \tag{4}$$

since

$$\pi_2'' \circ i_2(j_p^2\psi) = \pi_2'' \circ c'(0) = c(0) = j_p^2\alpha_0 = j_p^2(\pi'' \circ \psi)\ .$$

12.87 Remark Using the local coordinates described in Remarks 12.4, 12.5 and

12.67, we have coordinate functions x^i, y^α and $\dot{y}^\alpha$ on V which give rise to the coordinate functions x^i, y^α, $\dot{y}^\alpha$, y_i^α, $\dot{y}_i^\alpha$, y_{ij}^α and $\dot{y}_{ij}^\alpha$ on J^2V for $i,j = 1,2,\ldots,n$ and $\alpha = 1,2,\ldots,m$. We also have coordinate functions x^i, y^α, y_i^α and y_{ij}^α on J^2E from which we obtain the coordinate functions x^i, y^α, y_i^α, y_{ij}^α, $\dot{y}^\alpha$, $\dot{y}_i^\alpha$ and $\dot{y}_{ij}^\alpha$ on V_2. By formula (1) in 12.86 we have that

$$x^i \circ i_2 = x^i \ , \quad y^\alpha \circ i_2 = y^\alpha \ , \quad y_i^\alpha \circ i_2 = y_i^\alpha \ , \quad y_{ij}^\alpha \circ i_2 = y_{ij}^\alpha \ ,$$
$$\dot{y}^\alpha \circ i_2 = \dot{y}^\alpha \ , \quad \dot{y}_i^\alpha \circ i_2 = \dot{y}_i^\alpha \quad \text{and} \quad \dot{y}_{ij}^\alpha \circ i_2 = \dot{y}_{ij}^\alpha$$

for $i,j = 1,2,\ldots,n$ and $\alpha = 1,2,\ldots,m$.

12.88 Let X be a vertical vector field on E, i.e., a section on E in the vertical tangent bundle $\pi'' : V \to E$, using the same notation as in 12.86. Regarding X as a bundle map over M between the bundles $\pi : E \to M$ and $\pi' : V \to M$, we obtain a vertical vector field X^2 on J^2E, called the *2-jet prolongation of* X , defined by

$$X^2 = i_2 \circ j^2X \ .$$

Indeed, it follows from formula (4) in 12.86 that

$$\pi_2'' \circ X^2(j_p^2\phi) = \pi_2'' \circ i_2(j_p^2(X \circ \phi)) = j_p^2\phi$$

for $p \in M$ and $\phi \in \Gamma_p(E)$, showing that X^2 is a section on J^2E in the vertical tangent bundle $\pi_2'' : V_2 \to J^2E$. Using formula (2) in 12.86 we have that

$$\widetilde{\pi}_{2,1} \circ X^2 = \widetilde{\pi}_{2,1} \circ i_2 \circ j^2X = i_1 \circ \pi_{2,1}' \circ j^2X = i_1 \circ j^1X \circ \pi_{2,1} = X^1 \circ \pi_{2,1} \ .$$

The vector fields X^2 and X^1 can also be considered as bundle-valued 0-forms along id_{J^2E} and id_{J^1E} which are $(\pi_{2,1}, \pi_{2,1})$-related as described in 12.32. Hence it follows that

$$i_{X^2} \circ \pi_{2,1}^* = \pi_{2,1}^* \circ i_{X^1} \ .$$

12.89 Remark Suppose that the vertical vector field X on E is given by

$$X|_W = \sum_{\alpha=1}^{m} b^\alpha \frac{\partial}{\partial y^\alpha}$$

on an open subset W of E, using the local coordinates described in Remarks 12.4. By Definition 12.48 and Remark 12.87 we have that

$$\dot{y}^\alpha \circ X^2(j_p^2\phi) = \dot{y}^\alpha(j_p^2(X \circ \phi)) = b^\alpha(\phi(p)) \ ,$$

$$\dot{y}_i^\alpha \circ X^2(j_p^2\phi) = \dot{y}_i^\alpha(j_p^2(X \circ \phi)) = \frac{\partial(b^\alpha \circ \phi)}{\partial x^i}(p) = \frac{db^\alpha}{dx^i}(j_p^1\phi)$$

and

$$\dot{y}^\alpha_{ij} \circ X^2(j^2_p\phi) = \dot{y}^\alpha_{ij}(j^2_p(X \circ \phi)) = \frac{\partial^2(b^\alpha \circ \phi)}{\partial x^j \partial x^i}(p) = \frac{d}{dx^j}\left(\frac{db^\alpha}{dx^i}\right)(j^2_p\phi)\,,$$

which shows that

$$X^2|_{\pi_{2,0}^{-1}(W)} = \sum_\alpha (b^\alpha \circ \pi_{2,0})\,\frac{\partial}{\partial y^\alpha} + \sum_{i,\alpha}\left(\frac{db^\alpha}{dx^i} \circ \pi_{2,1}\right)\frac{\partial}{\partial y^\alpha_i} + \sum_{ij,\alpha}\frac{d}{dx^j}\left(\frac{db^\alpha}{dx^i}\right)\frac{\partial}{\partial y^\alpha_{ij}}\,.$$

CALCULUS OF VARIATIONS

12.90 Let M^n be a smooth manifold with a volume element ε, and let $\pi : E \to M$ be a bundle over M with fibre F^m. Then a smooth function $L : J^1E \to \mathbf{R}$ is called a *Lagrangian function* for π, and the horizontal n-form $L\,\pi_1^*\varepsilon$ on J^1E is the corresponding *Lagrangian*. Let (x,U), (z,W) and $((z \times \beta) \circ \tau_z, \pi_{1,0}^{-1}(W))$ be local charts on M, E and J^1E with coordinate functions described in Remarks 12.4 and 12.37. If

$$\varepsilon\,|_U = a\,dx^1 \wedge \ldots \wedge dx^n\,,$$

the function $\mathscr{L} = (a \circ \pi_1)\,L|_{\pi_{1,0}^{-1}(W)}$ on $\pi_{1,0}^{-1}(W)$ is called a *Lagrangian density*. The *Cartan form* of L is the n-form on J^1E given by

$$\Theta_L = L\,\pi_1^*\varepsilon + d_{S_\varepsilon}L\,,$$

where $S_\varepsilon \in \Omega^n(J^1E; V_{1,0})$ is the bundle-valued n-form on J^1E obtained from ε as described in 12.49. Since $d_{S_\varepsilon}L = i_{S_\varepsilon}\,dL$, we have that

$$\begin{aligned}\Theta_L\,|_{\pi_{1,0}^{-1}(W)} &= L\,\pi_1^*\varepsilon\,|_{\pi_{1,0}^{-1}(W)} \\ &+ \sum_{i,\alpha}\frac{\partial\mathscr{L}}{\partial y^\alpha_i}\,dx^1\wedge\cdots\wedge dx^{i-1}\wedge\left\{dy^\alpha - \sum_{j=1}^n y^\alpha_j\,dx^j\right\}\wedge dx^{i+1}\wedge\cdots\wedge dx^n.\end{aligned} \tag{1}$$

The *Euler–Lagrange form* of L is the $(n+1)$-form on J^2E defined by

$$\mathscr{E}_L = \pi_{2,1}^*(dL \wedge \pi_1^*\varepsilon) + d_{\widetilde{\omega}}\,\Theta_L\,,$$

where $d_{\widetilde{\omega}} : \Omega(J^1E) \to \Omega(J^2E)$ is the derivation along $\pi_{2,1}$ of type d_* and degree 1 described in 12.78. Since

$$\begin{aligned}d_{\widetilde{\omega}}\,\Theta_L\,|_{\pi_{2,0}^{-1}(W)} = &-\sum_{i,\alpha}\frac{d}{dx^i}\left(\frac{\partial\mathscr{L}}{\partial y^\alpha_i}\right)\,dy^\alpha\wedge dx^1\wedge\cdots\wedge dx^n \\ &-\sum_{i,\alpha}\left(\frac{\partial\mathscr{L}}{\partial y^\alpha_i}\circ\pi_{2,1}\right)\,dy^\alpha_i\wedge dx^1\wedge\cdots\wedge dx^n\,,\end{aligned}$$

it follows that

$$\mathscr{E}_L\,|_{\pi_{2,0}^{-1}(W)} = \sum_{\alpha=1}^{m}\left\{\left(\frac{\partial\mathscr{L}}{\partial y^\alpha}\circ\pi_{2,1}\right) - \sum_{i=1}^{n}\frac{d}{dx^i}\left(\frac{\partial\mathscr{L}}{\partial y_i^\alpha}\right)\right\} dy^\alpha\wedge dx^1\wedge\cdots\wedge dx^n. \tag{2}$$

A section $\phi\in\Gamma_U(E)$ of π on the open subset U of M is called an *extremal* of L if it satisfies the following assertion :

For each compact submanifold N^n of U with boundary, and each vertical vector field X on E vanishing on $\pi^{-1}(\partial N)$ with global flow $\gamma:\mathscr{D}(X)\to E$, we have that

$$\frac{d}{dt}\bigg|_0 \int_N (j^1(\gamma_t\circ\phi))^*L\,\varepsilon = 0. \tag{3}$$

The map $\alpha=\gamma\circ(id\times\phi):\mathscr{B}\to E$, defined on the open subset $\mathscr{B}=(id\times\phi)^{-1}(\mathscr{D}(X))$ of $\mathbf{R}\times U$, is called the *variation of ϕ induced by* X. Since

$$(X\circ\phi)(p)=\alpha_p'(0)$$

for $p\in U$, it follows from 12.79 and 12.81 that

$$X^1(j_p^1\phi) = i_1\circ j^1X(j_p^1\phi) = i_1(j_p^1(X\circ\phi)) = \frac{d}{dt}\bigg|_0 j_p^1\alpha_t = \frac{d}{dt}\bigg|_0 j_p^1(\gamma_t\circ\phi)$$

so that

$$(j^1\phi)^*(d_{X^1}L)(p) = (i_{X^1}\,dL)(j_p^1\phi) = dL(X^1)(j_p^1\phi) = \frac{d}{dt}\bigg|_0 (j^1(\gamma_t\circ\phi))^*L(p)$$

for $p\in U$ by Proposition 4.14. Condition (3) is therefore equivalent to the condition

$$\int_N (j^1\phi)^*(d_{X^1}L)\;\varepsilon = 0. \tag{4}$$

12.91 Theorem Let M^n be a smooth manifold with a volume element ε, and let $\pi:E\to M$ be a bundle over M with fibre F^m. Let $L\in C^\infty(J^1E)$ be a Lagrangian function for π, and let (x,U), (z,W) and $((z\times\beta)\circ\tau_z,\pi_{1,0}^{-1}(W))$ be local charts on M, E and J^1E with coordinate functions described in Remarks 12.4 and 12.37. Then a section $\phi\in\Gamma_U(E)$ of π with $\phi(U)\subset W$ is an extremal of L if it satisfies the *Euler–Lagrange equations*

$$(j^2\phi)^*\left\{\left(\frac{\partial\mathscr{L}}{\partial y^\alpha}\circ\pi_{2,1}\right) - \sum_{i=1}^{n}\frac{d}{dx^i}\left(\frac{\partial\mathscr{L}}{\partial y_i^\alpha}\right)\right\} = 0$$

on U for $\alpha=1,\dots,m$, where $\mathscr{L}$ is the Lagrangian density on $\pi_{1,0}^{-1}(W)$.

PROOF : Let N^n be a compact submanifold of U with boundary, and let X be a vertical vector field on E vanishing on $\pi^{-1}(\partial N)$. If

$$X|_W = \sum_{\alpha=1}^{m} b^\alpha \frac{\partial}{\partial y^\alpha},$$

we have that

$$i_{X^1}\Theta_L|_{\pi_{1,0}^{-1}(W)} = \sum_{i,\alpha} (-1)^{i+1} (b^\alpha \circ \pi_{1,0}) \frac{\partial \mathscr{L}}{\partial y_i^\alpha} dx^1 \wedge \cdots \wedge dx^{i-1} \wedge dx^{i+1} \wedge \cdots \wedge dx^n$$

so that

$$d_{\tilde{\omega}}\, i_{X^1}\Theta_L|_{\pi_{2,0}^{-1}(W)} = \sum_{i,\alpha} \left\{ \left(\frac{db^\alpha}{dx^i} \circ \pi_{2,1}\right)\left(\frac{\partial \mathscr{L}}{\partial y_i^\alpha} \circ \pi_{2,1}\right) + (b^\alpha \circ \pi_{2,0}) \frac{d}{dx^i}\left(\frac{\partial \mathscr{L}}{\partial y_i^\alpha}\right)\right\}$$

$$dx^1 \wedge \cdots \wedge dx^n = -\, i_{X^2}\, d_{\tilde{\omega}}\, \Theta_L|_{\pi_{2,0}^{-1}(W)}\,.$$

As X vanishes on $\pi^{-1}(\partial N)$, it follows from Stoke's theorem that

$$\begin{aligned} 0 &= \int_{\partial N} (j^1\phi \circ i_N)^*\, i_{X^1}\Theta_L = \int_N d\,(j^1\phi)^*\, i_{X^1}\Theta_L \\ &= \int_N (j^2\phi)^*\, d_{\tilde{\omega}}\, i_{X^1}\Theta_L = -\int_N (j^2\phi)^*\, i_{X^2}\, d_{\tilde{\omega}}\, \Theta_L\,. \end{aligned}$$

where $i_N : \partial N \to N$ is the inclusion map. Now using that

$$\int_N (j^1\phi)^*(d_{X^1}L)\,\varepsilon = \int_N (j^1\phi)^*\, i_{X^1}(dL \wedge \pi_1^*\varepsilon) = \int_N (j^2\phi)^*\, i_{X^2}\, \pi_{2,1}^*(dL \wedge \pi_1^*\varepsilon)\,,$$

we see that condition (4) in 12.90 is equivalent to the condition

$$\int_N (j^2\phi)^*\, i_{X^2}\, \mathscr{E}_L = 0\,.$$

Since

$$i_{X^2}\,\mathscr{E}_L|_{\pi_{2,0}^{-1}(W)} = \sum_{\alpha=1}^{m} (b^\alpha \circ \pi_{2,0}) \left\{\left(\frac{\partial \mathscr{L}}{\partial y^\alpha} \circ \pi_{2,1}\right) - \sum_{i=1}^{n} \frac{d}{dx^i}\left(\frac{\partial \mathscr{L}}{\partial y_i^\alpha}\right)\right\} dx^1 \wedge \cdots \wedge dx^n,$$

this completes the proof of the theorem. □

Appendix A

PRELIMINARIES

MAPS

13.1 Let $f : A \to B$ be a map, and let C and D be sets with $C \subset A$ and $f(C) \subset D \subset B$. Then a map $g : C \to D$ is said to be *induced by f* if $g(x) = f(x)$ for all $x \in C$, which means that we have a commutative diagram

$$\begin{array}{ccc} A & \xrightarrow{f} & B \\ {\scriptstyle i_1}\uparrow & & \uparrow{\scriptstyle i_2} \\ C & \xrightarrow{g} & D \end{array}$$

where $i_1 : C \to A$ and $i_2 : D \to B$ are the inclusion maps. If $D = B$, g is called the *restriction* of f to C and is denoted by $f|_C$. We denote g by f_D when $C = f^{-1}(D)$.

13.2 Given two maps $f : A \to B$ and $g : C \to D$, we define their *composition* $g \circ f : f^{-1}(B \cap C) \to D$ by $g \circ f(a) = g(f(a))$ for $a \in f^{-1}(B \cap C)$. We do not exclude the case where $f^{-1}(B \cap C) = \emptyset$. If the map $f : A \to B$ is injective, we denote by $f^{-1} : f(A) \to A$ the inverse of the bijective map $f_{f(A)}$ defined by $f^{-1}(f(a)) = a$ for $a \in A$.

THE PERMUTATION GROUP

13.3 Definition If $I_n = \{1, ..., n\}$, a bijective map $\sigma : I_n \to I_n$ is called a *permutation* of I_n. This set of permutations is a group S_n called the *permutation group* of n elements, where the group product is composition of mappings. We often write $\sigma\tau$ instead of $\sigma \circ \tau$ for two permutations σ and τ.

13.4 Remark We have an injective homomorphism $\phi : S_{n-1} \to S_n$ given by $\phi(\sigma)(k) = \sigma(k)$ for $k \in I_{n-1}$ and $\phi(\sigma)(n) = n$, so that S_{n-1} may be identified with the subgroup of S_n consisting of all permutations leaving n fixed.

13.5 Definition A *transposition* τ in S_n is a permutation which interchanges two numbers in I_n and leaves all the other numbers fixed, i.e., there are two different numbers $i, j \in I_n$ such that $\tau(i) = j$, $\tau(j) = i$ and $\tau(k) = k$ for $k \in I_n - \{i, j\}$.

13.6 Remark We have that $\tau^{-1} = \tau$ and $\tau^2 = id$ for every transposition τ.

13.7 Proposition Every permutation in S_n can be expressed as a product of transpositions.

PROOF: We prove the proposition by induction on n. We have that S_1 only consists of the identity map which can be considered as a product of transpositions where the number of factors is zero.

Assume next the the assertion in the proposition is true for S_{n-1}, and let σ be a permutation in S_n. If $\sigma(n) = k$ is different from n, we let τ be the transposition in S_n which interchanges n and k, otherwise we let τ be the identity. Then we have that $\tau\sigma(n) = n$, so that $\tau\sigma = \phi(\sigma')$ for a permutation of $\sigma' \in S_{n-1}$, where ϕ is the injective homomorphism defined in Remark 13.4. By the induction hypothesis there exist transpositions $\tau_1', \ldots, \tau_k'$ in S_{n-1} such that $\sigma' = \tau_1' \cdots \tau_k'$. Then we have that $\sigma = \tau^{-1}\tau_1 \cdots \tau_k$ where $\tau_i = \phi(\tau_i')$ is a transposition in S_n for $i = 1, \ldots, k$, and this completes the proof of the proposition. □

13.8 Let $A \times \cdots \times A = A^n$ be the set of n-tuples of elements in a set A, and let σ be a permutation in S_n. Then we have a map $\pi_\sigma : A^n \to A^n$ defined by

$$\pi_\sigma(x_1, \ldots, x_n) = (x_{\sigma(1)}, \ldots, x_{\sigma(n)}). \tag{1}$$

The n-tuple $(x_1, \ldots, x_n)$ is a map $x : I_n \to A$ with $x(i) = x_i$, and we have that $\pi_\sigma(x) = x \circ \sigma$. From this it follows that

$$(\pi_\sigma \circ \pi_\tau)(x) = (x \circ \tau) \circ \sigma = x \circ (\tau\sigma) = \pi_{\tau\sigma}(x)$$

so that

$$\pi_\sigma \circ \pi_\tau = \pi_{\tau\sigma}$$

for all $\tau, \sigma \in S_n$.

If $h : C \to A$ is a map from a set C into A, we define $h^n : C^n \to A^n$ by

$$h^n(x_1, \ldots, x_n) = (h(x_1), \ldots, h(x_n)),$$

and we let $\pi_\sigma' : C^n \to C^n$ be the map on C^n corresponding to π_σ and which is also defined by (1). If the n-tuple $(x_1,...,x_n)$ is again interpreted as a map $x : I_n \to C$ with $x(i) = x_i$, we have that $h^n(x) = h \circ x$. From this it follows that

$$(\pi_\sigma \circ h^n)(x) = (h \circ x) \circ \sigma = h \circ (x \circ \sigma) = (h^n \circ \pi'_\sigma)(x)$$

so that

$$\pi_\sigma \circ h^n = h^n \circ \pi'_\sigma$$

for every $\sigma \in S_n$.

13.9 Given a map $f : A^n \to B$ and a permutation σ in S_n, we define the map $f^\sigma : A^n \to B$ by $f^\sigma = f \circ \pi_\sigma$ so that

$$f^\sigma(x_1,...,x_n) = f(x_{\sigma(1)},...,x_{\sigma(n)}) .$$

If τ and σ are two permutations in S_n, we have that

$$f^{\sigma\tau} = f \circ \pi_{\sigma\tau} = f \circ \pi_\tau \circ \pi_\sigma = (f^\tau)^\sigma .$$

Given a map $h : C \to A$ as in 13.8, we have that

$$f^\sigma \circ h^n = f \circ \pi_\sigma \circ h^n = f \circ h^n \circ \pi'_\sigma = (f \circ h^n)^\sigma .$$

The map $f \circ h^n$ is called the *pull-back* of f by h and is usually denoted by $h^*(f)$. The last relation may then be written as

$$h^*(f^\sigma) = h^*(f)^\sigma .$$

If B is a ring, we may define the sum and product of two function $f : A^n \to B$ and $g : A^n \to B$ as usual. We then have that

$$(f+g)^\sigma = f^\sigma + g^\sigma \quad \text{and} \quad (fg)^\sigma = f^\sigma g^\sigma$$

for any permutation σ in S_n. We also have that $(-f)^\sigma = -f^\sigma$.

13.10 Theorem There is a unique homomorphism $\varepsilon : S_n \to \{1,-1\}$ such that $\varepsilon(\tau) = -1$ for every transposition τ.

$\varepsilon(\sigma)$ is called the *sign* of the permutation σ. We say that σ is *even* if $\varepsilon(\sigma) = 1$ and *odd* if $\varepsilon(\sigma) = -1$. The kernel A_n of ε, consisting of the even permutations, is called the *alternating group* of n elements.

PROOF: If σ is a permutation in S_n, we know by Proposition 13.7 that it may be expressed as a product of transpositions

$$\sigma = \tau_1 \cdots \tau_m .$$

From the conditions in the theorem it then follows that $\varepsilon(\sigma) = (-1)^m$, and this proves the uniqueness of ε.

To prove existence, we must show that the sign $(-1)^m$ is independent of the expression of σ as a product of transpositions, i.e., if $\sigma = \tau_1' \cdots \tau_k'$ is another such expression, then $(-1)^k = (-1)^m$.

Let $S = \{(i,j) \in I_n^2 \mid i < j\}$, and consider the polynomials $\Delta_{ij} : \mathbf{Z}^n \to \mathbf{Z}$ defined by $\Delta_{ij}(x_1, \ldots, x_n) = x_j - x_i$ for $(i,j) \in S$. Let

$$\Delta = \prod_{(i,j)\in S} \Delta_{ij} .$$

If τ is a transposition in S_n interchanging the integers r and s in I_n, where $r < s$, we have that

$$\Delta^\tau = \prod_{(i,j)\in S} \Delta_{ij}^\tau$$

where $\Delta_{rs}^\tau = -\Delta_{rs}$. We see that the factors Δ_{ij} which do not have r or s as an index remain unchanged when we apply τ. The other factors may be grouped in pairs of the following three types

$$1)\ \Delta_{rk}\Delta_{sk} \text{ where } s < k ,$$

$$2)\ \Delta_{rk}\Delta_{ks} \text{ where } r < k < s ,$$

$$3)\ \Delta_{kr}\Delta_{ks} \text{ where } k < r ,$$

which are also all unchanged when we apply τ.

Hence we have that $\Delta^\tau = -\Delta$ for every transposition τ. Using the two expressions for σ given above, we see that

$$\Delta^\sigma = (-1)^k \Delta = (-1)^m \Delta$$

which shows that $(-1)^k = (-1)^m$ and completes the proof of the theorem. □

13.11 Remark We see that if a permutation is expressed as a product of transpositions, then the number of factors is either always odd or always even. In the first case it is an odd permutation with sign -1, and in the second case it is an even permutation with sign 1.

GROUP ACTIONS

13.12 Definition Let S be a set and G a group with identity element e. By an *operation* or an *action of G on S on the left* we mean a map $\mu : G \times S \to S$, where $\mu(g,s)$ is denoted by gs for $g \in G$ and $s \in S$, satisfying

$$(gh)s = g(hs) \quad \text{and} \quad es = s$$

for every $g, h \in G$ and $s \in S$. If there is such an operation μ, we say that G *operates* or *acts on* S *on the left*, or that S is a G-set.

For each $g \in G$, we let $L_g : S \to S$ be the map defined by $L_g(s) = gs$ for $s \in S$. It is a bijective map also called a *permutation* of S, and we obtain a homomorphism $g \mapsto L_g$ from G into the group of permutations of S called a *representation* of G.

If this representation is injective, it is also said to be *faithful*, and G is said to act *effectively* on S. This means that the only element $g \in G$ with $gs = s$ for all $s \in S$, is the identity element e.

If we have the stronger condition that $gs = s$ for some $s \in S$ implies $g = e$, then G is said to act *without fixed point* or *freely* on S.

We say that G acts *transitively* on S if there for any pair of elements $s, t \in S$ always exists a group element $g \in G$ with $gs = t$.

If S and S' are two G-sets, then a map $f : S \to S'$ is called a *G-map*, or it is said to be *equivariant* with respect to the two actions of G, if

$$f(gs) = gf(s)$$

for every $g \in G$ and $s \in S$.

13.13 Remark Similarly, we define an *operation of* G *on* S *on the right* to be a map $\nu : S \times G \to S$, where $\nu(s, g)$ is denoted by sg for $g \in G$ and $s \in S$, satisfying

$$s(gh) = (sg)h \quad \text{and} \quad se = s$$

for every $g, h \in G$ and $s \in S$. For each $g \in G$, we let $R_g : S \to S$ be the bijective map defined by $R_g(s) = sg$ for $s \in S$.

13.14 Example A group G operates on itself on the left by left translation. If $g \in G$, we define the left translation $L_g : G \to G$ by $L_g(h) = gh$ for $h \in G$, and we have an operation $\mu : G \times G \to G$ given by $\mu(g, h) = gh = L_g(h)$.

Similarly, if H is a subgroup of G, then G operates by left translation on the set $G/H = \{gH | g \in G\}$ of left cosets. Indeed, if $g_1 \in G$, then $L_{g_1}(g_2H) = (g_1g_2)H$. The canonical projection $\pi : G \to G/H$ defined by $\pi(g) = gH$ is a G-map.

13.15 Definition Let G be a group which operates on a set S on the left, and let $s \in S$. Then $Gs = \{gs | g \in G\}$ is a subset of S called the *orbit* of s under G, and $G_s = \{g \in G | gs = s\}$ is a subgroup of G called the *isotropy group* of s in G.

13.16 Proposition If G is a group which operates on a set S on the left, then two orbits under G are either disjoint or equal. The orbits therefore form a partition of the set S.

PROOF: If two orbits Gs_1 and Gs_2 have an element s in common, then $s = g_1s_1 = g_2s_2$ for group elements $g_1, g_2 \in G$. Hence we have that $Gs = Gg_is_i = Gs_i$ for $i = 1, 2$. □

13.17 Proposition If G is a group which operates on a set S on the left, then $G_{gs} = g\, G_s\, g^{-1}$ for every $g \in G$ and $s \in S$.

PROOF: If $h \in G_s$, then $(ghg^{-1})(gs) = (gh)s = g(hs) = gs$, which implies that $g\, G_s\, g^{-1} \subset G_{gs}$. By replacing g and s with g^{-1} and gs, respectively, in the last inclusion, we have that $g^{-1}\, G_{gs}\, g \subset G_s$ so that $G_{gs} \subset g\, G_s\, g^{-1}$. □

13.18 Proposition Let G be a group which operates on a set S on the left, and let $s \in S$. If $H = G_s$ is the isotropy group of s in G, and $\pi : G \to G/H$ is the canonical projection, then the map $\phi : G \to S$ given by $\phi(g) = gs$ induces an injective G-map $\overline{\phi} : G/H \to S$ with $\overline{\phi} \circ \pi = \phi$. The image of $\overline{\phi}$ is the orbit Gs.

PROOF: We have that $\phi(g_1) = \phi(g_2) \Leftrightarrow g_1 s = g_2 s \Leftrightarrow (g_2^{-1} g_1) s = s \Leftrightarrow g_2^{-1} g_1 \in H \Leftrightarrow g_1 H = g_2 H$. Hence we have a well-defined injective map $\overline{\phi} : G/H \to S$ given by

$$\overline{\phi}(gH) = gs$$

for $g \in G$, which satisfies $\overline{\phi} \circ \pi = \phi$.

Since both π and ϕ are G-maps and π is surjective, it follows that $\overline{\phi}$ is also a G-map. Indeed, we have that $\overline{\phi}(g_1 \pi(g_2)) = \overline{\phi}(\pi(g_1 g_2)) = g_1 \overline{\phi}(\pi(g_2))$ for every $g_1, g_2 \in G$. □

CATEGORIES AND FUNCTORS

13.19 Definition By a *category* $\mathscr{C}$ we mean a class of objects $\mathrm{Ob}(\mathscr{C})$ such that for each ordered pair of objects $X, Y \in \mathrm{Ob}(\mathscr{C})$ we have a set $\mathrm{Mor}(X,Y)$, called the set of *morphisms* from X to Y, and for each ordered triple of objects $X, Y, Z \in \mathrm{Ob}(\mathscr{C})$ a map

$$\mathrm{Mor}(Y,Z) \times \mathrm{Mor}(X,Y) \to \mathrm{Mor}(X,Z)\,,$$

called the *law of composition*, sending a pair of morphisms $f \in \mathrm{Mor}(X,Y)$ and $g \in \mathrm{Mor}(Y,Z)$ to a morphism $g \circ f \in \mathrm{Mor}(X,Z)$, satisfying the following three axioms:

(i) The sets $\mathrm{Mor}(X,Y)$ and $\mathrm{Mor}(X',Y')$ are disjoint unless $X = X'$ and $Y = Y'$.

(ii) The law of composition is *associative*, i.e., if $f \in \mathrm{Mor}(X,Y)$, $g \in \mathrm{Mor}(Y,Z)$ and $h \in \mathrm{Mor}(Z,W)$, then $(h \circ g) \circ f = h \circ (g \circ f)$.

(iii) For every object Y there is a morphism $id_Y \in \mathrm{Mor}(Y,Y)$, called the *identity morphism* of Y, such that $id_Y \circ f = f$ for every $f \in \mathrm{Mor}(X,Y)$ and $g \circ id_Y = g$ for every $g \in \mathrm{Mor}(Y,Z)$.

A morphism $f \in \mathrm{Mor}(X,Y)$ is also denoted by $f : X \to Y$.

13.20 Definition A morphism $f : X \to Y$ is called an *isomorphism* if there is a morphism $g : Y \to X$ such that $g \circ f = id_X$ and $f \circ g = id_Y$. The morphism g is called the *inverse* of f and is denoted by f^{-1}. If $X = Y$, f is also called an *automorphism.*

13.21 Definition By a *covariant functor* F from a category $\mathscr{C}$ to a category $\mathscr{D}$ we mean a rule which to each object X in $\mathscr{C}$ associates an object $F(X)$ in $\mathscr{D}$, and to each morphism $f : X \to Y$ in $\mathscr{C}$ associates a morphism $F(f) : F(X) \to F(Y)$ in $\mathscr{D}$, satisfying the following two conditions:

(i) If $f : X \to Y$ and $g : Y \to Z$ are two morphisms in $\mathscr{C}$, then $F(g \circ f) = F(g) \circ F(f)$.

(ii) For each object X in $\mathscr{C}$ we have that $F(id_X) = id_{F(X)}$.

If we with $f : X \to Y$ associate a morphism $F(f) : F(Y) \to F(X)$ in $\mathscr{D}$ going in the opposite direction, and replace the formula for composition in (i) by $F(g \circ f) = F(f) \circ F(g)$, then F is called a *contravariant functor* from $\mathscr{C}$ to $\mathscr{D}$.

A functor F from $\mathscr{C}$ to $\mathscr{D}$ is often denoted by $F : \mathscr{C} \to \mathscr{D}$ or by the transformation rule $X \mapsto F(X)$ and $f \mapsto F(f)$ for each object X and morphism f in $\mathscr{C}$.

13.22 Remark A functor $F : \mathscr{C} \to \mathscr{D}$ transforms isomorphisms in $\mathscr{C}$ to isomorphisms in $\mathscr{D}$.

13.23 Definition If $F : \mathscr{C} \to \mathscr{D}$ and $G : \mathscr{D} \to \mathscr{E}$ are two functors, we define their *composed functor* $G \circ F : \mathscr{C} \to \mathscr{E}$ by $G \circ F(X) = G(F(X))$ and $G \circ F(f) = G(F(f))$ for each object X and morphism f in $\mathscr{C}$. The composition $G \circ F$ is covariant if F and G are of the same variance, otherwise it is contravariant.

13.24 Definition A category $\mathscr{C}$ is called a *subcategory* of the category $\mathscr{D}$ if there is a covariant functor $I : \mathscr{C} \to \mathscr{D}$, called the *inclusion functor*, given by $I(X) = X$ and $I(f) = f$ for each object X and morphism f in $\mathscr{C}$.

This means that $\mathrm{Ob}(\mathscr{C}) \subset \mathrm{Ob}(\mathscr{D})$, and that $\mathrm{Mor}_{\mathscr{C}}(X,Y) \subset \mathrm{Mor}_{\mathscr{D}}(X,Y)$ for every pair $X,Y \in \mathrm{Ob}(\mathscr{C})$. Moreover, for each pair of morphisms $f : X \to Y$ and $g : Y \to Z$ in $\mathscr{C}$, their composition in $\mathscr{C}$ coincides with their composition in $\mathscr{D}$, and for each object X in $\mathscr{C}$, the identity morphism in $\mathrm{Mor}_{\mathscr{C}}(X,X)$ coincides with the identity morphism in $\mathrm{Mor}_{\mathscr{D}}(X,X)$.

If in addition $\mathrm{Mor}_{\mathscr{C}}(X,Y) = \mathrm{Mor}_{\mathscr{D}}(X,Y)$ for every pair $X,Y \in \mathrm{Ob}(\mathscr{C})$, $\mathscr{C}$ is called a *full subcategory* of $\mathscr{D}$.

CONNECTIVITY

13.25 Definition A topological space X is said to be *connected* if it is not a disjoint union of two nonempty open subsets. A subset A of X is called connected if it is connected considered as a subspace of X.

A topological space X is said to be *locally connected* if every point in X has a basis of connected neighbourhoods.

13.26 Proposition A topological space X is connected if and only if one of the following equivalent conditions is satisfied:

(1) X it is not a disjoint union of two nonempty closed subsets.

(2) X and $\emptyset$ are the only subsets of X which are both open and closed.

PROOF: If two complimentary sets in X are open, then they are also closed, and vice versa, so we may replace open by closed in Definition 13.25. This gives the equivalent condition (1).

The existence of nonempty open complementary sets U and U^c in X is equivalent to the existence of a set U in X different from X and $\emptyset$ which is both open and closed. This gives the condition (2). □

13.27 Example The closed interval $[a,b]$ in $\mathbf{R}$ is connected. To prove this, suppose that $[a,b]$ is the disjoint union of the nonempty closed subsets A and B. Without loss of generality, we may assume that $b \in B$. If $c = \sup A$, then $c \in A$ since A is closed. Hence we have that $c < b$ and $\langle c,b\,] \subset B$, and this implies that $c \in B$ since B is closed, contradicting the fact that A and B are disjoint.

13.28 Proposition If $f : X \to Y$ is a continuous map from a connected topological space X, then the image $f(X)$ is also connected.

PROOF: If $f(X)$ is the disjoint union of the nonempty subsets U and V which are open relative to $f(X)$, then $X = f^{-1}(U) \cup f^{-1}(V)$ is also a disjoint union of nonempty open sets, contradicting the fact that X is connected. □

13.29 Proposition If $\{A_\lambda \mid \lambda \in \Lambda\}$ is a family of connected subsets of X whose intersection is nonempty, then $\bigcup A_\lambda$ is connected.

PROOF: Let x_0 be a point in $\bigcap A_\lambda$, and let $A = \bigcup A_\lambda$. Suppose that A is the disjoint union of two subsets U and V which are open relative to A, and that U is the one containing x_0. Then $A_\lambda \cap U$ and $A_\lambda \cap V$ are open in A_λ, and since $x_0 \in A_\lambda \cap U$ and A_λ is connected, it follows that $A_\lambda \cap V = \emptyset$ for every $\lambda \in \Lambda$. Hence we have that $V = \emptyset$ which shows that A is connected. □

13.30 Corollary A topological space X is connected if and only if each pair of points in X belongs to a connected subset of X.

PROOF: If X is connected, then the last assertion is obviously true if we let the connected subset be X itself.

Conversely, suppose that the last assertion in the corollary holds, and fix a point x_0 in X. For each point $x \in X$ there is then a connected subset A_x of X containing x_0 and x. Since X is the union of the family $\{A_x | x \in X\}$ whose intersection contains x_0, it follows from Proposition 13.29 that X is connected. □

13.31 Example A set $A \subset \mathbf{R}^n$ is said to be *convex* if $x, y \in A$ implies that $x + t(y-x) \in A$ for $0 \leq t \leq 1$. By Corollary 13.30 every convex set A in $\mathbf{R}^n$ is connected since the line segment $\{x + t(y-x) | 0 \leq t \leq 1\}$ is the image of the continuous map $\sigma : [0,1] \to A$ from the connected set $[0,1]$, defined by $\sigma(t) = tx + (1-t)x$ for $t \in [0,1]$. In particular, the open and closed balls $\mathbf{D}^n = \{y \in \mathbf{R}^n | \, \|y\| < 1\}$ and $\mathbf{E}^n = \{y \in \mathbf{R}^n | \, \|y\| \leq 1\}$ in $\mathbf{R}^n$ are connected.

13.32 Example The n-sphere $\mathbf{S}^n = \{x \in \mathbf{R}^{n+1} | \, \|x\| = 1\}$ is connected when $n \geq 1$ by Proposition 13.29, since it is the union of the closed half spheres $S_1 = \{x \in \mathbf{S}^n | x^{n+1} \geq 0\}$ and $S_2 = \{x \in \mathbf{S}^n | x^{n+1} \leq 0\}$ with nonempty intersection $S_1 \cap S_2 = \{x \in \mathbf{S}^n | x^{n+1} = 0\}$. These half spheres are connected as they are the images of the continuous maps $f_1 : \mathbf{E}^n \to \mathbf{R}^{n+1}$ and $f_2 : \mathbf{E}^n \to \mathbf{R}^{n+1}$ from the closed ball $\mathbf{E}^n = \{y \in \mathbf{R}^n | \, \|y\| \leq 1\}$, given by

$$f_1(y) = (y, (1 - \|y\|^2)^{\frac{1}{2}}) \quad \text{and} \quad f_2(y) = (y, -(1 - \|y\|^2)^{\frac{1}{2}})$$

for $y \in \mathbf{E}^n$.

13.33 Proposition If X and Y are two connected topological spaces, then their product $X \times Y$ is also connected.

PROOF: Let (x_1, y_1) and (x_2, y_2) be two arbitrary points in $X \times Y$. Then the sets $X \times \{y_1\}$ and $\{x_2\} \times Y$ are connected, since they are the images of the continuous maps $f : X \to X \times Y$ and $g : Y \to X \times Y$ given by $f(x) = (x, y_1)$ and $g(y) = (x_2, y)$, respectively. These two sets have the point (x_2, y_1) in common, so their union is connected by Proposition 13.29. Hence the points (x_1, y_1) and (x_2, y_2) belong to the connected subset $(X \times \{y_1\}) \cup (\{x_2\} \times Y)$ of $X \times Y$, and the proposition follows from Corollary 13.30. □

13.34 Proposition Let $\{X_\lambda | \lambda \in \Lambda\}$ be a family of connected topological spaces. Then the product

$$X = \prod_{\lambda \in \Lambda} X_\lambda$$

is connected.

PROOF : If the indexing set Λ is finite, the proposition follows by induction from Proposition 13.33. In the general case, suppose that X is a disjoint union of two nonempty open sets U and V, and let $x \in U$ and $y \in V$. Choose an open neighbourhood

$$W = \prod_{\lambda \in \Lambda} W_\lambda$$

of y contained in V, where each W_λ is an open neighbourhood of y_λ, and where $W_\lambda = X_\lambda$ for all indices λ except those contained in a finite subset Λ_0 of Λ. Let z be the point in X defined by

$$z_\lambda = \begin{cases} y_\lambda & \text{for } \lambda \in \Lambda_0 \\ x_\lambda & \text{for } \lambda \in \Lambda - \Lambda_0 \end{cases} .$$

By the first part of the proof, we know that the space

$$X_0 = \prod_{\lambda \in \Lambda_0} X_\lambda$$

is connected, and hence the same is true for the subspace

$$Z = \prod_{\lambda \in \Lambda} Z_\lambda$$

of X defined by

$$Z_\lambda = \begin{cases} X_\lambda & \text{for } \lambda \in \Lambda_0 \\ \{x_\lambda\} & \text{for } \lambda \in \Lambda - \Lambda_0 \end{cases} ,$$

as Z is the image of the continuous map $f : X_0 \to X$ given by

$$f(u)_\lambda = \begin{cases} u_\lambda & \text{for } \lambda \in \Lambda_0 \\ x_\lambda & \text{for } \lambda \in \Lambda - \Lambda_0 \end{cases} .$$

This however contradicts the fact that Z is the disjoint union of the sets $Z \cap U$ and $Z \cap V$ which are open in Z and nonempty since $x \in Z \cap U$ and $z \in Z \cap V$. Hence we conclude that X must be connected. □

13.35 Proposition If A is a connected subset of a topological space X, and if $A \subset B \subset \overline{A}$, then B is also connected.

PROOF : Suppose that B is the disjoint union of the subsets U_1 and U_2 which are open relative to B. Then the sets $V_i = A \cap U_i$ for $i = 1, 2$ are open in A. If U_i contains a point x, then U_i is a neighbourhood of x in B and must contain a point from A, as A is dense in B. Hence $U_i \neq \emptyset$ implies that $V_i \neq \emptyset$. Since A is connected, it therefore follows that either U_1 or U_2 must be empty, showing that B is also connected. □

13.36 Let x be a point in a topological space X. Then the set $\{x\}$ is connected, and the union of every connected subsets of X containing x is connected by Proposition 13.29. This is the largest connected subset containing x, and it is called the *connected* *component* or just the *component* of x in X. By Proposition 13.35 it follows that it is closed.

The connected components of two points x and y in X are either disjoint or equal by Proposition 13.29. Hence the connected components form a partition of X into closed connected subsets.

13.37 Proposition If A is a nonempty subset of a topological space X which is both open and closed, then A is a union of connected components in X. If A is also connected, then it is a connected component.

PROOF : Let A be a nonempty subset of X which is both open and closed. If C is a connected component in X containing a point $x \in A$, then $C \cap A$ is a nonempty subset of C which is both open and closed relative to C. Since C is connected, it follows that $C \cap A = C$ so that $C \subset A$, and this completes the proof of the first assertion in the proposition.

If A is connected, then we also have $A \subset C$, which shows that A coincides with C and therefore is a connected component. □

13.38 Proposition A topological space X is locally connected if and only if the connected components of open subsets of X are open. If X is locally connected, then every point in X has a basis of open connected neighbourhoods.

PROOF : Suppose that X is locally connected, and let C be a connected component of an open set U in X. Then each point $x \in C$ has a connected neighbourhood V contained in U. As C is the largest connected subset of U containing x, we have that $V \subset C$, showing that C is a neighborhood of x and therefore is open.

Conversely, suppose that the connected components of open subsets of X are open, and let U be an open neighbourhood of a point $x \in X$. Then the connected component C of U containing x is an open connected neighbourhood of x with $C \subset U$. This shows that X is locally connected, and it also shows the last part of the proposition. □

13.39 Corollary In a locally connected topological space X, the connected components are open as well as closed.

PROOF : Follows from 13.36 and Proposition 13.38. □

HOMOTOPY THEORY

13.40 In this chapter we let I denote the closed unit interval $[0,1]$ and ∂I denote its boundary $\{0,1\}$.

13.41 Definition A pair (X,A) consisting of a topological space X and a subspace A is called a *topological pair*. By a *continuous map* $f:(X,A)\to(Y,B)$ between two topological pairs we mean a continuous map $f:X\to Y$ such that $f(A)\subset B$. The set of all continuous maps from (X,A) to (Y,B) is denoted by $C(X,A;Y,B)$. We thus obtain a category $\mathscr{T}^2$ of topological pairs and continuous maps.

The topological pair $(X,\emptyset)$ is identified with X, and the category of topological spaces and continuous maps is a full subcategory of $\mathscr{T}^2$ denoted by $\mathscr{T}$.

If $A=\{x_0\}$ for a point $x_0\in X$, then the topological pair (X,A) is called a *pointed topological space* with *base point* x_0, and it is usually denoted simply by (X,x_0). The category of pointed topological spaces and continuous maps is a full subcategory of $\mathscr{T}^2$ denoted by $\mathscr{T}_0$.

The *cartesian product* between a topological pair (X,A) and the unit interval I is defined by $(X,A)\times I=(X\times I,A\times I)$.

13.42 Definition Let $f,g:(X,A)\to(Y,B)$ be two continuous maps between the topological pairs (X,A) and (Y,B), and let X' be a subset of X. Then we say that f is *homotopic to g relative to* X', and we write $f\simeq g$ rel X', if there is a continuous map $F:(X,A)\times I\to(Y,B)$ such that

$$\begin{aligned} F(x,0)&=f(x) && \text{for } x\in X,\\ F(x,1)&=g(x) && \text{for } x\in X,\\ F(x,t)&=f(x)=g(x) && \text{for } x\in X',\, t\in I. \end{aligned}$$

Such a map F is called a *homotopy from f to g relative to* X', and we write $F:f\simeq g$ rel X'. In particular, it is necessary that $f|_{X'}=g|_{X'}$. If $X'=\emptyset$, we omit the phrase "relative to $\emptyset$", and write simply $f\simeq g$ and $F:f\simeq g$.

13.43 Proposition Let $\{A_k\}_{k=1}^n$ be a finite closed covering of a topological space X. Then a map $f:X\to Y$ into a topological space Y is continuous if and only if $f|_{A_k}$ is continuous for every k.

PROOF: The only if part follows since $f|_{A_k}=f\circ i_k$, where $i_k:A_k\to X$ is the inclusion map which is continuous for every k.

Conversely, suppose that $f|_{A_k}$ is continuous for every k, and let V be a closed subset of Y. Then $f^{-1}(V)\cap A_k=(f|_{A_k})^{-1}(V)$ is closed in A_k, and therefore also closed in X as A_k is closed. Since

$$f^{-1}(V)=\bigcup_{k=1}^{n} f^{-1}(V)\cap A_k\,,$$

it follows that $f^{-1}(V)$ is closed in X, thus showing that f is continuous. □

13.44 Proposition Let (X,A) and (Y,B) be topological pairs, and let $X'\subset X$. Then homotopy relative to X' is an equivalence relation in $C(X,A;Y,B)$.

PROOF: If $f \in C(X,A;Y,B)$, we define the *constant homotopy* $F : f \simeq f \text{ rel } X'$ by $F(x,t) = f(x)$ for $x \in X$ and $t \in I$. This shows that homotopy relative to X' is *reflexive*.

To prove that it is *symmetric*, let $F : f \simeq g \text{ rel } X'$ be any homotopy relative to X'. We then have an *inverse homotopy* $G : g \simeq f \text{ rel } X'$ defined by $G(x,t) = F(x,1-t)$ for $x \in X$ and $t \in I$.

Finally, we show that homotopy relative to X' is *transitive*. If $F : f \simeq g \text{ rel } X'$ and $G : g \simeq h \text{ rel } X'$ are two homotopies relative to X', we define their *combined homotopy* $H : f \simeq h \text{ rel } X'$ by

$$H(x,t) = \begin{cases} F(x,2t) & \text{for } x \in X,\, t \in [0,1/2] \\ G(x,2t-1) & \text{for } x \in X,\, t \in [1/2,1] \end{cases}$$

Since $F(x,1) = g(x) = G(x,0)$ for every $x \in X$, H is well defined, and it follows from Proposition 13.43 that H is continuous. □

13.45 Remark The equivalence class of a continuous map $f \in C(X,A;Y,B)$ with respect to the above equivalence relation is called the *homotopy class* of f relative to X' and is denoted by $[f]_{X'}$. If $X' = \emptyset$, we omit the reference to X'.

13.46 Proposition Let $f_0, f_1 : (X,A) \to (Y,B)$ and $g_0, g_1 : (Y,B) \to (Z,C)$ be continuous maps, and let X' and Y' be subsets of X and Y, respectively, with $f_0(X') \subset Y'$. If $f_0 \simeq f_1 \text{ rel } X'$ and $g_0 \simeq g_1 \text{ rel } Y'$, then $g_0 \circ f_0 \simeq g_1 \circ f_1 \text{ rel } X'$.

PROOF: Let $F : f_0 \simeq f_1 \text{ rel } X'$ and $G : g_0 \simeq g_1 \text{ rel } Y'$ be homotopies. From these we obtain the homotopies

$$g_0 \circ F : g_0 \circ f_0 \simeq g_0 \circ f_1 \text{ rel } X'$$

and

$$G \circ (f_1 \times id) : g_0 \circ f_1 \simeq g_1 \circ f_1 \text{ rel } X',$$

so that $g_0 \circ f_0 \simeq g_1 \circ f_1 \text{ rel } X'$ by Proposition 13.44. □

13.47 Definition If $f \in C(X,A;Y,B)$ and $g \in C(Y,B;Z,C)$, we define the composition of the hopotopy classes $[f]$ and $[g]$ by $[g] \circ [f] = [g \circ f]$. This is well defined by Proposition 13.46, and we thus obtain a category $\mathfrak{h}^2$ of topological pairs and homotopy classes of continuous maps called the *homotopy category of topological pairs*. We have a canonical functor $F : \mathscr{T}^2 \to \mathfrak{h}^2$ given by $F((X,A)) = (X,A)$ and $F(f) = [f]$.

As in Definition 13.41, we have the full subcategories $\mathfrak{h}$ and $\mathfrak{h}_0$ of $\mathfrak{h}^2$, where the objects are topological spaces and pointed topological spaces, respectively.

13.48 Definition A continuous map $f : (X,A) \to (Y,B)$ is called a *homotopy equivalence* if the homotopy class $[f]$ is an isomorphism in $\mathfrak{h}^2$, i.e., if there is a continuous map $g : (Y,B) \to (X,A)$ such that $g \circ f \simeq id_X$ and $f \circ g \simeq id_Y$. The map g is called a *homotopy inverse* of f.

Two topological pairs (X,A) and (Y,B) are said to be *homotopically equivalent* or of the same *homotopy type* if there exists a homotopy equivalence $f:(X,A)\to(Y,B)$.

13.49 Definition A continuous map $f:(X,A)\to(Y,B)$ is said to be *null homotopic* if it is homotopic to a constant map. In particular, if the identity map $id_X:(X,A)\to(X,A)$ is null homotopic, then (X,A) is said to be *contractible*. If $c:(X,A)\to(X,A)$ is the constant map mapping X to a point $x_0\in A$, then a homotopy $F:id_X\simeq c$ is called a *contraction* of (X,A) to x_0.

13.50 Example A set $V\subset \mathbf{R}^n$ is said to be *star-shaped* with respect to a point $x_0\in V$ if $x\in V$ implies that $x_0+t\,(x-x_0)\in V$ for $0\le t\le 1$. Every star-shaped set V in $\mathbf{R}^n$ is contractible since the map $F:V\times I\to V$ defined by $F(x,t)=t\,x_0+(1-t)\,x$ is a contraction of V to x_0. In particular, the Euclidean space $\mathbf{R}^n$ itself and every convex subset of $\mathbf{R}^n$ is star-shaped and therefore contractible.

More generally, we have that every topological pair (V,A), where V and A are star-shaped with respect to a point $x_0\in A$, is contractible since F is a contraction of (V,A) to x_0.

13.51 Proposition A topological space X is contractible if and only if it has the same homotopy type as a one-point space.

PROOF: Suppose first that X is contractible, and let F be a contraction of X to a point $x_0\in X$. If $f:X\to\{x_0\}$ is the canonical map and $g:\{x_0\}\to X$ the inclusion map, we have that $F:id_X\simeq g\circ f$ and $f\circ g=id_{\{x_0\}}$, showing that f is a homotopy equivalence from X to $\{x_0\}$.

Conversely, assume that X is of the same homotopy type as a one-point space $\{x_0\}$, and let $f:X\to\{x_0\}$ be a homotopy equivalence with a homotopy inverse $g:\{x_0\}\to X$. Then $g\circ f:X\to X$ is a constant map, and since $id_X\simeq g\circ f$, it follows that X is contractible. □

13.52 Definition Let X be a topological space, and let $x_0,x_1\in X$. By a *path* in X from x_0 to x_1 we mean a continuous map $\sigma:I\to X$ with $\sigma(0)=x_0$ and $\sigma(1)=x_1$. The points x_0 and x_1 are called the *initial* and *final* point of σ, respectively. If $x_0=x_1$, σ is called a *closed path* or a *loop* at x_0.

If σ maps I to a single point $x_0\in X$, σ is called the *constant* path at x_0 and is denoted by $<x_0>$. Given a path σ from x_0 to x_1, we define the *inverse* path σ^{-1} from x_1 to x_0 by $\sigma^{-1}(t)=\sigma(1-t)$. For two paths σ and τ with $\sigma(1)=\tau(0)$, we define the *combined* path $\sigma*\tau$ by

$$\sigma*\tau(t)=\begin{cases}\sigma(2t) & \text{for } t\in[0,1/2]\\ \tau(2t-1) & \text{for } t\in[1/2,1]\end{cases}$$

13.53 We say that two points x_0 and x_1 in a topological space X are equivalent, and write $x_0\sim x_1$, if there is a path in X from x_0 to x_1. It follows from Definition

13.52 that $\sim$ is an equivalence relation, and the equivalence class of a point $x \in X$ is called the *path component* of x and consists of all points which can be connected to x by a path in X. The set of all path components in X is denoted by $\pi_0(X)$.

If $f : X \to Y$ is a continuous map and σ is a path in X from x_0 to x_1, then $f \circ \sigma$ is a path in Y from $f(x_0)$ to $f(x_1)$. Hence $x_0 \sim x_1$ implies that $f(x_0) \sim f(x_1)$, and we obtain an induced map $f_* : \pi_0(X) \to \pi_0(Y)$ mapping the path component of a point x in X to the path component of $f(x)$ in Y.

If $f, g : X \to Y$ are homotopic, then $f_* = g_*$. Indeed, if $F : f \simeq g$ is a homotopy, we have for each point $x \in X$ a path τ in Y from $f(x)$ to $g(x)$ given by $\tau(t) = F(x,t)$ for $t \in I$, showing that $f(x)$ and $g(x)$ belong to the same path component in Y.

We thus obtain a covariant functor $\pi_0 : \mathfrak{h} \to \mathscr{S}$ from the homotopy category $\mathfrak{h}$ of topological spaces to the category $\mathscr{S}$ of sets, assigning to a topological space X the set $\pi_0(X)$ of path components in X, and to a homotopy class $[f]$ the map f_*.

13.54 Definition A topological space X is said to be *pathwise connected* if every pair of points in X can be connected with a path. A subset A of X is called pathwise connected if it is pathwise connected considered as a subspace of X, i.e., if every pair of points in A can be connected with a path lying entirely in A.

A topological space X is said to be *locally pathwise connected* if every point in X has a basis of pathwise connected neighbourhoods.

13.55 Proposition The path component of a point x in a topological space X is the largest pathwise connected subset of X containing x.

PROOF : Let C be the path component of x. Then we have that $A \subset C$ for every pathwise connected subset A of X containing x, since any point $y \in A$ can be connected to x with a path in A which is certainly also a path in X.

It only remains to show that C itself is pathwise connected. If $y \in C$, there is a path σ in X from x to y, and we contend that σ actually is a path in C. In fact, for every $t \in I$, we have a path σ_t in X from x to $\sigma(t)$ given by $\sigma_t(s) = \sigma(st)$ for $s \in I$, showing that $\sigma(t) \in C$. This proves that every pair of points in C can be connected with a path lying entirely in C, thereby completing the proof that C is pathwise connected. □

13.56 Proposition A topological space X is locally pathwise connected if and only if the path components of open subsets of X are open. If X is locally pathwise connected, then every point in X has a basis of open pathwise connected neighbourhoods.

PROOF : Suppose that X is locally pathwise connected, and let C be a path component of an open set U in X. Then each point $x \in C$ has a pathwise connected neighbourhood V contained in U. As C is the largest pathwise connected subset of U containing x, we have that $V \subset C$, showing that C is a neighborhood of x and therefore is open.

Conversely, suppose that the path components of open subsets of X are open, and let U be an open neighbourhood of a point $x \in X$. Then the path component C of U containing x is an open pathwise connected neighbourhood of x with $C \subset U$. This shows that X is locally pathwise connected, and it also shows the last part of the proposition. □

13.57 Corollary In a locally pathwise connected space X, the path components are open as well as closed.

PROOF: By 13.53 and Proposition 13.56 we know that the path components are open sets forming a partition of X. Hence they are also closed in X. □

13.58 Proposition A pathwise connected space X is connected.

PROOF: If X is pathwise connected, then every pair of points x_0 and x_1 in X can be connected by a path $\sigma : I \to X$. Using Proposition 13.28 we have that $\sigma(I)$ is a connected subset of X containing x_0 and x_1. Hence the result follows from Corollary 13.30. □

13.59 Corollary A locally pathwise connected space X is locally connected.

13.60 Corollary Each connected component in a topological space X is a union of path components in X.

PROOF: Let C be a connected component in X. Then it follows from Propositions 13.55 and 13.58 that the path component in X of a point $x \in C$ is connected, and therefore is contained in C. □

13.61 Proposition In a locally pathwise connected space X the path components coincide with the connected components.

PROOF: By Proposition 13.55 and 13.58 and Corollary 13.57 we know that each path component in X is a nonempty connected set which is both open and closed, and therefore is a connected component in X by Proposition 13.37. □

13.62 Corollary A locally pathwise connected space X is pathwise connected if and only if it is connected.

13.63 Proposition Topological spaces of the same homotopy type always have the same number of path components. In particular, every contractible space is pathwise connected.

PROOF: If $f : X \to Y$ is a homotopy equivalence, then the induced map $f_* : \pi_0(X) \to \pi_0(Y)$ is a bijection by 13.53 and Remark 13.22. □

13.64 Proposition If $f : X \to Y$ is continuous and X is pathwise connected, then $f(X)$ is also pathwise connected.

PROOF : Follows from 13.53. □

13.65 Definition We say that two paths σ and τ in a topological space X are *homotopic*, and write $\sigma \simeq \tau$, if they are homotopic relative to ∂I. In particular, we must have that $\sigma(0) = \tau(0)$ and $\sigma(1) = \tau(1)$. A homotopy F from σ to τ relative to ∂I is called simply a homotopy from σ to τ, and it is denoted by $F : \sigma \simeq \tau$.

The homotopy class of σ relative to ∂I is denoted by $[\sigma]$ and is called the *path class* of σ. The points $\sigma(0)$ and $\sigma(1)$ are called the *initial* and *final* point of $[\sigma]$, respectively.

13.66 Proposition Let σ_0 and σ_1 be homotopic paths in X from x_0 to x_1, and τ_0 and τ_1 be homotopic paths in X from x_1 to x_2. Then the combined paths $\sigma_0 * \tau_0$ and $\sigma_1 * \tau_1$ are homotopic.

PROOF : If $F : \sigma_0 \simeq \sigma_1$ and $G : \tau_0 \simeq \tau_1$ are homotopies, we define a homotopy $H : \sigma_0 * \tau_0 \simeq \sigma_1 * \tau_1$ by

$$H(s,t) = \begin{cases} F(2s,t) & \text{for } s \in [0,1/2] \\ G(2s-1,t) & \text{for } s \in [1/2,1] \end{cases}$$

Since $F(1,t) = x_1 = G(0,t)$ for every $t \in I$, H is well defined, and it follows from Proposition 13.43 that H is continuous. □

13.67 Definition If $[\sigma]$ is a path class from x_0 to x_1 and $[\tau]$ a path class from x_1 to x_2, we define the *combined path class* $[\sigma] * [\tau]$ from x_0 to x_2 by $[\sigma] * [\tau] = [\sigma * \tau]$.

13.68 Proposition For each topological space X we have a category $\mathscr{P}(X)$, called the *path category* of X, where the objects are the points in X, the morphisms from x_1 to x_0 are the path classes from x_0 to x_1, and the law of composition is the combination of path classes as defined above. The identity morphism of x_0 is the path class of the constant path at x_0.

PROOF : We first show that the law of composition is associative. If ρ, σ and τ are paths in X with $\rho(1) = \sigma(0)$ and $\sigma(1) = \tau(0)$, we have a homotopy $F : (\rho * \sigma) * \tau \simeq \rho * (\sigma * \tau)$ defined by

$$F(s,t) = \begin{cases} \rho\left(\frac{4s}{1+t}\right) & \text{for } 0 \le s \le \frac{1+t}{4} \\ \sigma(4s-t-1) & \text{for } \frac{1+t}{4} \le s \le \frac{2+t}{4} \\ \tau\left(\frac{4s-t-2}{2-t}\right) & \text{for } \frac{2+t}{4} \le s \le 1 \end{cases}$$

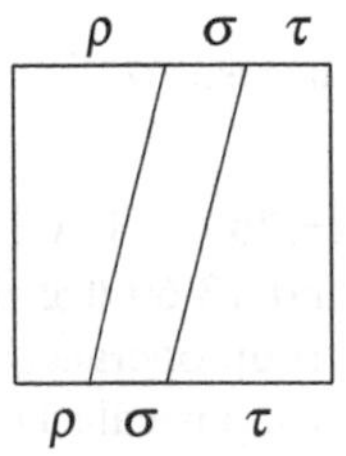

showing that $([\rho] * [\sigma]) * [\tau] = [\rho] * ([\sigma] * [\tau])$.

Furthermore, for any path σ in X from x_0 to x_1, we have the homotopies $G: <x_0> * \sigma \simeq \sigma$ and $H : \sigma * <x_1> \simeq \sigma$ defined by

$$G(s,t) = \begin{cases} \sigma(0) & \text{for } 0 \le s \le \frac{1-t}{2} \\ \sigma\left(\frac{2s+t-1}{1+t}\right) & \text{for } \frac{1-t}{2} \le s \le 1 \end{cases}$$

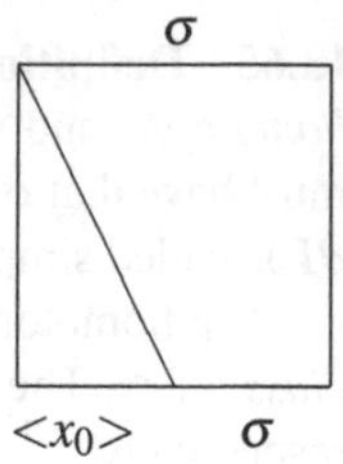

and

$$H(s,t) = \begin{cases} \sigma\left(\frac{2s}{1+t}\right) & \text{for } 0 \le s \le \frac{1+t}{2} \\ \sigma(1) & \text{for } \frac{1+t}{2} \le s \le 1 \end{cases}$$

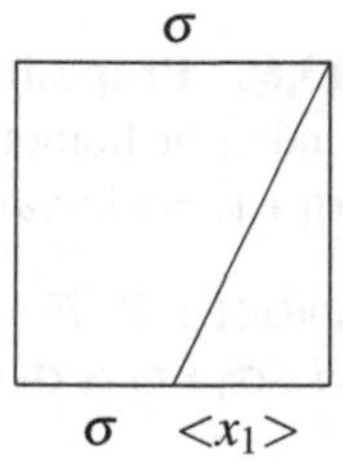

which shows that $[<x_0>] * [\sigma] = [\sigma]$ and $[\sigma] * [<x_1>] = [\sigma]$ and completes the proof of the proposition. □

13.69 Proposition Every morphism in $\mathscr{P}(X)$ is an isomorphism.

PROOF: If σ is a path in X from x_0 to x_1, we have a homotopy $F : \sigma * \sigma^{-1} \simeq <x_0>$ defined by

$$F(s,t) = \begin{cases} \sigma(0) & \text{for } 0 \le s \le \frac{t}{2} \\ \sigma(2s-t) & \text{for } \frac{t}{2} \le s \le \frac{1}{2} \\ \sigma(2-2s-t) & \text{for } \frac{1}{2} \le s \le 1-\frac{t}{2} \\ \sigma(0) & \text{for } 1-\frac{t}{2} \le s \le 1 \end{cases}$$

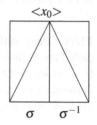

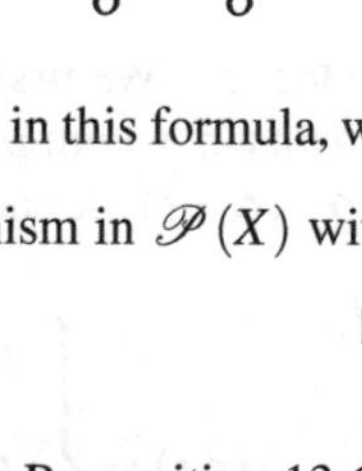

which shows that $[\sigma] * [\sigma^{-1}] = [<x_0>]$. By replacing σ with σ^{-1} in this formula, we also have that $[\sigma^{-1}] * [\sigma] = [<x_1>]$. Hence $[\sigma]$ is an isomorphism in $\mathscr{P}(X)$ with inverse $[\sigma]^{-1} = [\sigma^{-1}]$. □

13.70 If X is a topological space and $x_0 \in X$, it follows from Proposition 13.68 and 13.69 that the path classes of the closed paths at x_0 form a group, where the group operation is combination of path classes as defined in Definition 13.67. This group is called the *fundamental group of* X *at* x_0, and it is denoted by $\pi_1(X, x_0)$.

If $f : (X, x_0) \to (Y, y_0)$ is a continuous map, we have a homomorphism $f_* : \pi_1(X, x_0) \to \pi_1(Y, y_0)$ defined by $f_*([\sigma]) = [f \circ \sigma]$. It follows from Proposition 13.46

that f_* is well defined, and that $f_* = g_*$ for two homotopic maps $f,g : (X,x_0) \to (Y,y_0)$.

We thus obtain a covariant functor $\pi_1 : \mathfrak{h}_0 \to \mathfrak{g}$, called the *fundamental group functor*, from the homotopy category $\mathfrak{h}_0$ of pointed topological spaces to the category $\mathfrak{g}$ of groups, assigning to a pointed topological space (X,x_0) its fundamental group $\pi_1(X,x_0)$, and to a homotopy class $[f]$ the homomorphism f_*.

In particular, every contractible pointed topological space has a trivial fundamental group.

13.71 To describe how the fundamental group $\pi_1(X,x_0)$ depends on the base point x_0, we consider the covariant functor $F : \mathscr{P}(X) \to \mathfrak{g}$ from the path category $\mathscr{P}(X)$ to the category $\mathfrak{g}$ of groups, which assigns to each point x_0 the fundamental group $\pi_1(X,x_0)$ and to to each morphism $[\sigma] : x_0 \to x_1$ the homomorphism

$$h_{[\sigma]} : \pi_1(X,x_0) \to \pi_1(X,x_1)$$

defined by $h_{[\sigma]}([\omega]) = [\sigma] * [\omega] * [\sigma]^{-1}$. From Proposition 13.69 it follows that every $h_{[\sigma]}$ is an isomorphism so that $\pi_1(X,x_0) \cong \pi_1(X,x_1)$ whenever x_0 and x_1 are in the same path component in X.

This isomorphism is determined up to an inner automorphism of the fundamental group. More precisely, if $[\sigma],[\tau] : x_0 \to x_1$ are two morphisms from x_0 to x_1, and if $[\theta] = [\tau] * [\sigma]^{-1} \in \pi_1(X,x_1)$, we have that

$$h_{[\tau]}([\omega]) = [\theta] * h_{[\sigma]}([\omega]) * [\theta]^{-1}$$

which shows that $h_{[\sigma]}$ and $h_{[\tau]}$ are conjugate by the inner automorphism of $\pi_1(X,x_1)$ defined by $[\theta]$.

13.72 Definition A topological space X is said to be *simply connected* if it is pathwise connected and has a trivial fundamental group at every point.

13.73 Example Every star-shaped subset V of $\mathbf{R}^n$ is simply connected. In fact, V is pathwise connected, and if V is star-shaped with respect to x_0, then (V,x_0) is a contractible pointed topological space as described in Example 13.50. The result therefore follows from 13.70 and 13.71.

13.74 Proposition A topological space X is simply connected if and only if the following assertion is satisfied:

For each pair of points x_0 and x_1 in X, there is a unique path class from x_0 to x_1.

PROOF: Suppose first that X is simply connected, and let $x_0,x_1 \in X$. As X is pathwise connected, there is a path σ from x_0 to x_1. If τ is another such path, then $[\sigma] * [\tau]^{-1} \in \pi_1(X,x_0)$ so that $[\sigma] * [\tau]^{-1} = [<x_0>]$ since $\pi_1(X,x_0)$ is trivial. Hence it follows that $[\sigma] = [\tau]$.

Conversely, suppose that the above assertion is satisfied. Then X is pathwise connected, and $\pi_1(X,x_0)$ is trivial for every $x_0 \in X$, thus showing that X is simply connected. □

13.75 Proposition If $f,g : X \to Y$ are homotopic maps with $f(x_0) = y_0$ and $g(x_0) = y_1$, then there is a path σ in Y from y_0 to y_1 such that

$$f_* = h_{[\sigma]} \circ g_* : \pi_1(X,x_0) \to \pi_1(Y,y_0) .$$

PROOF : Let $F : f \simeq g$ be a homotopy, and let σ be the path from y_0 to y_1 in Y defined by $\sigma(t) = F(x_0,t)$. We want to show that

$$[f \circ \omega] = [\sigma] * [g \circ \omega] * [\sigma]^{-1}$$

for every closed path ω at x_0. Let $H : I \times I \to Y$ be the map defined by $H = F \circ (\omega \times id)$, and consider the paths α_0, α_1, β_0 and β_1 in $I \times I$ given by $\alpha_i(s) = (s,i)$ and $\beta_i(t) = (i,t)$ for $i = 0,1$. Since $I \times I$ is star-shaped and therefore simply connected, it follows from Proposition 13.74 that $\alpha_0 \simeq \beta_0 * \alpha_1 * \beta_1^{-1}$ so that

$$f \circ \omega = H \circ \alpha_0 \simeq (H \circ \beta_0) * (H \circ \alpha_1) * (H \circ \beta_1^{-1}) = \sigma * (g \circ \omega) * \sigma^{-1},$$

which completes the proof of the proposition. □

13.76 Proposition If $f : X \to Y$ is a homotopy equivalence with $f(x_0) = y_0$, then the induced map $f_* : \pi_1(X,x_0) \to \pi_1(Y,y_0)$ is an isomorphism.

PROOF : Let $g : Y \to X$ be a homotopy inverse of f, and let $g(y_0) = x_1$ and $f(x_1) = y_1$. Then f induces the maps $f_*^i : \pi_1(X,x_i) \to \pi_1(Y,y_i)$ for $i = 0,1$, where $f_*^0 = f_*$ is the induced map referred to in the proposition.

Since $g \circ f \simeq id_X$ and $f \circ g \simeq id_Y$, we have by Proposition 13.75 a path σ in X from x_1 to x_0 with $g_* \circ f_*^0 = h_{[\sigma]}$, and a path τ in Y from y_1 to y_0 with $f_*^1 \circ g_* = h_{[\tau]}$, so that the diagram

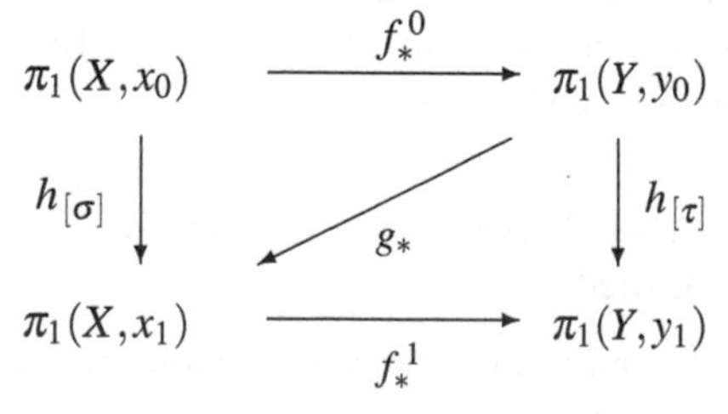

is commutative. As $h_{[\sigma]}$ and $h_{[\tau]}$ are both isomorphisms, it follows that g_* is both surjective and injective. Hence we conclude that g_*, and therefore also f_*, is an isomorphism. □

13.77 Corollary Every contractible space X is simply connected.

PROOF: Follows from Propositions 13.63 and 13.76. □

13.78 Proposition Let $\{X_\lambda | \lambda \in \Lambda\}$ be a family of topological spaces, and let

$$X = \prod_{\lambda \in \Lambda} X_\lambda$$

be the product space. We choose a base point x_λ in X_λ for each $\lambda \in \Lambda$, and use $x = \{x_\lambda | \lambda \in \Lambda\}$ as a base point in X. Then the projections $p_\lambda : X \to X_\lambda$ induce a group isomorphism

$$p_* : \pi_1(X,x) \to \prod_{\lambda \in \Lambda} \pi_1(X_\lambda, x_\lambda)$$

given by $p_{*\lambda} = p_{\lambda *}$ for $\lambda \in \Lambda$.

PROOF: From 13.70 it follows that p_* is a homomorphism, and it is also surjective since

$$p_*([\{\sigma_\lambda | \lambda \in \Lambda\}]) = \{[\sigma_\lambda] | \lambda \in \Lambda\}$$

for every closed path σ_λ in X_λ at x_λ for $\lambda \in \Lambda$.

To show that it is injective, let $\sigma = \{\sigma_\lambda | \lambda \in \Lambda\}$ and $\tau = \{\tau_\lambda | \lambda \in \Lambda\}$ be two closed paths in X at x with $p_*([\sigma]) = p_*([\tau])$. Then we have a homotopy $F_\lambda : \sigma_\lambda \simeq \tau_\lambda$ for each $\lambda \in \Lambda$, from which we obtain a homotopy $F : \sigma \simeq \tau$ given by $F = \{F_\lambda | \lambda \in \Lambda\}$. This shows that $[\sigma] = [\tau]$ and completes the proof that p_* is injective. □

COVERINGS

13.79 Definition Let $p : \widetilde{X} \to X$ be a continuous map. Then an open subset U of X is said to be *evenly covered* by p if $p^{-1}(U)$ is a disjoint union of open subsets $\widetilde{U}_\alpha$ of $\widetilde{X}$, each of which is mapped homeomorphically onto U by p. The sets $\widetilde{U}_\alpha$ are called *sheets* over U.

A continuous map $p : \widetilde{X} \to X$ is called a *covering* of X if every point $x \in X$ has an open neighbourhood which is evenly covered by p. The spaces $\widetilde{X}$ and X are called the *covering space* and the *base space* of the covering p, respectively, and the map p is also called the *covering projection*. For each point $x \in X$, the set $p^{-1}(x)$ is called the *fibre* over x.

13.80 Remark It follows immediately from the definition that a covering projection is surjective, and that each fibre is a discrete subspace of the covering space.

Indeed, if $p : \widetilde{X} \to X$ is a covering projection and U is an evenly covered neighbourhood of $x \in X$, then each sheet $\widetilde{U}_\alpha$ over U is an open subset of $\widetilde{X}$ containing exactly one point from the fibre $p^{-1}(x)$, and every point in the fibre is contained in a sheet.

13.81 Proposition If U is an open set in X which is evenly covered by a continuous map $p : \widetilde{X} \to X$, then every open subset of U is also evenly covered by p.

PROOF: Suppose that

$$p^{-1}(U) = \bigcup_{\alpha \in A} \widetilde{U}_\alpha$$

is a disjoint union of open subsets $\widetilde{U}_\alpha$ of $\widetilde{X}$ which are mapped homeomorphically onto U by p. If V is an open subset of U, we let

$$\widetilde{V}_\alpha = p^{-1}(V) \cap \widetilde{U}_\alpha = (p|_{\widetilde{U}_\alpha})^{-1}(V)$$

for every $\alpha \in A$. Then each $\widetilde{V}_\alpha$ is an open subset of $\widetilde{X}$ which is mapped homeomorphically onto V by p, and we have a disjoint union

$$p^{-1}(V) = \bigcup_{\alpha \in A} \widetilde{V}_\alpha$$

which shows that V is evenly covered by p. □

13.82 Proposition Let U be a nonempty connected open subset of X which is evenly covered by a continuous map $p : \widetilde{X} \to X$. Then the sheets over U are the connected components of $p^{-1}(U)$.

PROOF: Since the sheets over U are open sets forming a partition of $p^{-1}(U)$, they are also closed in $p^{-1}(U)$. Furthermore, each sheet is nonempty and connected since it is homeomorphic to U. The result therefore follows from Proposition 13.37. □

13.83 Proposition Let $p : \widetilde{X} \to X$ be a covering projection. If the base space X is Hausdorff, then so is the covering space $\widetilde{X}$.

PROOF: Let $\tilde{x}_1$ and $\tilde{x}_2$ be two different points in $\widetilde{X}$. If $\tilde{x}_1$ and $\tilde{x}_2$ lie in different fibres they can be separated by the neighbouhoods $p^{-1}(U_1)$ and $p^{-1}(U_2)$, where U_1 and U_2 are disjoint neighbourhoods of $p(\tilde{x}_1)$ and $p(\tilde{x}_2)$.

If $\tilde{x}_1$ and $\tilde{x}_2$ lie in the same fibre over a point $x \in X$, and U is an evenly covered neighbourhood of x, then $\tilde{x}_1$ and $\tilde{x}_2$ lie in different sheets over U which are disjoint open neighbourhoods of $\tilde{x}_1$ and $\tilde{x}_2$ in $\widetilde{X}$. □

13.84 Definition A continuous map $f: X \to Y$ is called a *local homeomorphism* if each point $x \in X$ has an open neighbourhood which is mapped homeomorphically by f onto an open neighbourhood of $f(x)$.

13.85 Proposition A covering projection is a local homeomorphism.

PROOF: Let $p: \widetilde{X} \to X$ be a covering projection and $\tilde{x} \in \widetilde{X}$, and choose an evenly covered neighbourhood U of $p(\tilde{x})$. Then the sheet $\widetilde{U}_\alpha$ over U containing $\tilde{x}$ is an open neighbourhood of $\tilde{x}$ which is mapped homeomorphically onto U by p, thus showing that p is a local homeomorphism. □

13.86 Proposition A local homeomorphism is an open map.

PROOF: If $f: X \to Y$ is mapping an open neighbourhood U of a point $x \in X$ homeomorphically onto an open neighbourhood $f(U)$ of $f(x)$, then this is so also for all open neighbourhoods of x contained in U. The proposition now follows since the collection of these neighbourhoods for all $x \in X$ is a basis for the topology on X. □

13.87 Definition Let Y and B be topological spaces, and let $p: Y \to B$ be a continuous map. Given a continuous map $f: X \to B$ from a topological space X, a continuous map $g: X \to Y$ is called a *lifting* of f with respect to p if $p \circ g = f$. If it is clear from the situation, we omit the reference to p.

13.88 The Unique Lifting Theorem Let $p: \widetilde{X} \to X$ be a covering projection, and let $f: Y \to X$ be a continuous map from a connected space Y. If $\tilde{f}_1, \tilde{f}_2: Y \to \widetilde{X}$ are two liftings of f such that $\tilde{f}_1(y_0) = \tilde{f}_2(y_0)$ for a point $y_0 \in Y$, then $\tilde{f}_1 = \tilde{f}_2$.

PROOF: We will prove that the sets $Y_1 = \{y | \tilde{f}_1(y) = \tilde{f}_2(y)\}$ and $Y_2 = \{y | \tilde{f}_1(y) \neq \tilde{f}_2(y)\}$ are both open in Y. Since Y is connected and $Y = Y_1 \cup Y_2$ with $y_0 \in Y_1$, it then follows that $Y_1 = Y$ so that $\tilde{f}_1 = \tilde{f}_2$.

First let $y \in Y_1$, and choose an evenly covered neighbourhood U of $f(y)$. Let $\widetilde{U}_\alpha$ be the sheet over U containing $\tilde{f}_1(y) = \tilde{f}_2(y)$, and let $s_\alpha: U \to \widetilde{U}_\alpha$ be the inverse of the homeomorphism induced by p and $i_\alpha: \widetilde{U}_\alpha \to \widetilde{X}$ be the inclusion map. Then $V = \tilde{f}_1^{-1}(\widetilde{U}_\alpha) \cap \tilde{f}_2^{-1}(\widetilde{U}_\alpha)$ is an open neighbourhood of y, and we have that $\tilde{f}_1|_V = i_\alpha \circ s_\alpha \circ f|_V = \tilde{f}_2|_V$ so that $V \subset Y_1$, thus showing that Y_1 is open.

Next let $y \in Y_2$, and again choose an evenly covered neighbourhood U of $f(y)$. Let $\widetilde{U}_\alpha$ and $\widetilde{U}_\beta$ be the disjoint sheets over U containing $\tilde{f}_1(y)$ and $\tilde{f}_2(y)$, respectively. Then $V = \tilde{f}_1^{-1}(\widetilde{U}_\alpha) \cap \tilde{f}_2^{-1}(\widetilde{U}_\beta)$ is an open neighbourhood of y, and we have that $\tilde{f}_1(V) \cap \tilde{f}_2(V) = \emptyset$ so that $V \subset Y_2$. Hence we have shown that Y_2 also is open. □

13.89 The Homotopy Lifting Theorem Let $p : \widetilde{X} \to X$ be a covering projection, and let $\tilde{f} : Y \to \widetilde{X}$ be a lifting of a continuous map $f : Y \to X$. Then a homotopy $F : Y \times I \to X$ from f can be lifted to a homotopy $\widetilde{F} : Y \times I \to \widetilde{X}$ from $\tilde{f}$. If F is a homotopy relative to a subset Y' of Y, then $\widetilde{F}$ is also a homotopy relative to Y'.

PROOF : We first fix a point $y \in Y$. For each $t \in I$ there are open neighbourhoods U_t of y in Y and V_t of t in I such that $F(U_t \times V_t)$ is contained in an evenly covered neighbourhood of $F(y,t)$ in X. Since I is compact, there is a finite number of points $t_0, \dots, t_n \in I$ such that $I = \bigcup_{i=0}^n V_{t_i}$, and $N_y = \bigcap_{i=0}^n U_{t_i}$ is an open neighbourhood of y.

Let $0 = k_0 < k_1 < \dots < k_r = 1$ be a partition of I such that each interval $[k_i, k_{i+1}]$ is contained in at least one of the sets $V_{t_0}, \dots, V_{t_n}$. By induction on i we will define continuous maps $\widetilde{F}_i : N_y \times [0, k_i] \to \widetilde{X}$ for $i = 0, \dots, r$ so that $p \circ \widetilde{F}_i = F|_{N_y \times [0,k_i]}$ and $\widetilde{F}_i(u,0) = \tilde{f}(u)$ for every $u \in N_y$.

The map $\widetilde{F}_0$ is uniquely defined by the last equation. Supposing next that $\widetilde{F}_i$ is already defined, we want to define $\widetilde{F}_{i+1}$. By the first part of the proof we know that $F(U_t \times [k_i, k_{i+1}])$ is contained in an evenly covered neighbourhood U in X. For each sheet $\widetilde{U}_\alpha$ over U we let $s_\alpha : U \to \widetilde{U}_\alpha$ be the inverse of the homeomorphism induced by p, and we let $Q_\alpha = \{u \in N_y | \widetilde{F}_i(u, k_i) \in \widetilde{U}_\alpha\}$. The sets Q_α form a disjoint open cover of N_y. We now define

$$\widetilde{F}_{i+1}(u,t) = \begin{cases} \widetilde{F}_i(u,t) & \text{for} \quad u \in N_y \ , \ t \in [0, k_i] \\ s_\alpha \circ F(u,t) & \text{for} \quad u \in Q_\alpha \ , \ t \in [k_i, k_{i+1}] \end{cases}$$

Since $\widetilde{F}_i(u, k_i) = s_\alpha \circ F(u, k_i)$ for every $u \in Q_\alpha$, the map $\widetilde{F}_{i+1}$ is well defined, and it follows from Proposition 13.43 that it is continuous. Furthermore, we have that $p \circ \widetilde{F}_{i+1} = F|_{N_y \times [0, k_{i+1}]}$ and $\widetilde{F}_{i+1}(u,0) = \tilde{f}(u)$ for every $u \in N_y$, which completes the induction step.

For every $y \in Y$ we now have an open neighbourhood N_y of y and a continuous map $\widetilde{F}_y : N_y \times I \to \widetilde{X}$ so that $p \circ \widetilde{F}_y = F|_{N_y \times I}$ and $\widetilde{F}_y(u,0) = \tilde{f}(u)$ for every $u \in N_y$. If $u \in N_y \cap N_z$, then $\widetilde{F}_y|_{\{u\} \times I}$ and $\widetilde{F}_z|_{\{u\} \times I}$ are both liftings of $F|_{\{u\} \times I}$, and we have that $\widetilde{F}_y(u,0) = \tilde{f}(u) = \widetilde{F}_z(u,0)$. Since $\{u\} \times I$ is connected, it follows from the unique lifting theorem that $\widetilde{F}_y|_{\{u\} \times I} = \widetilde{F}_z|_{\{u\} \times I}$. This shows that $\widetilde{F}_y$ and $\widetilde{F}_z$ coincide on $(N_y \cap N_z) \times I$, and we thus obtain a continuous map $\widetilde{F} : Y \times I \to \widetilde{X}$ so that $\widetilde{F}|_{N_y \times I} = \widetilde{F}_y$ for every $y \in Y$. We have that $p \circ \widetilde{F} = F$ and $\widetilde{F}(y,0) = \tilde{f}(y)$ for every $y \in Y$, showing that $\widetilde{F}$ is a homotopy from $\tilde{f}$ which is a lifting of F.

If F is a homotopy relative to Y' and $y \in Y'$, then $\widetilde{F}(\{y\} \times I)$ is a connected subset of the fibre over $f(y)$ containing $\widetilde{F}(y,0) = \tilde{f}(y)$. Since the fibre is a discrete topological space, it follows that $\widetilde{F}(y,t) = \tilde{f}(y)$ for all $t \in I$, which shows that $\widetilde{F}$ is also a homotopy relative to Y'. □

13.90 Remark Let $p : \widetilde{X} \to X$ be a covering projection, and let $f,g : Y \to X$ be two continuous maps which are homotopic. Then it follows from the homotopy lifting theorem that f has a lifting if and only if the same is true for g. Hence the existence of liftings with respect to a covering projection is a problem in the homotopy category.

13.91 Corollary Let $p : (\widetilde{X},\tilde{x}_0) \to (X,x_0)$ be a covering projection. Then a path σ in X from x_0 can be lifted to a unique path $\tilde{\sigma}$ in $\widetilde{X}$ from $\tilde{x}_0$.

PROOF : Let $Y = \{y_0\}$ be a topological space consisting of a single point y_0, and let $\alpha : I \to Y \times I$ and $\beta : Y \times I \to I$ be the canonical homeomorphisms defined by $\alpha(t) = (y_0,t)$ and $\beta(y_0,t) = t$ for $t \in I$. Furthermore, let $f : Y \to X$ and $\tilde{f} : Y \to \widetilde{X}$ be the continuous maps given by $f(y_0) = x_0$ and $\tilde{f}(y_0) = \tilde{x}_0$.

Then $\tilde{f}$ is a lifting of f, and $F = \sigma \circ \beta$ is a homotopy from f which can be lifted to a homotopy $\widetilde{F}$ from $\tilde{f}$ by the homotopy lifting theorem. We now have that $\tilde{\sigma} = \widetilde{F} \circ \alpha$ is a path in $\widetilde{X}$ from $\tilde{x}_0$ which is a lifting of σ, and this completes the existence part of the corollary. The uniqueness of $\tilde{\sigma}$ follows from the unique lifting theorem. □

13.92 Proposition Let $p : (\widetilde{X},\tilde{x}_0) \to (X,x_0)$ be a covering projection, and let $\widetilde{\sigma}$ and $\widetilde{\tau}$ be paths in $\widetilde{X}$ from $\tilde{x}_0$. Then we have that $\widetilde{\sigma} \simeq \widetilde{\tau}$ if and only if $p \circ \widetilde{\sigma} \simeq p \circ \widetilde{\tau}$.

PROOF : If $\widetilde{\sigma} \simeq \widetilde{\tau}$, it follows from Proposition 13.46 that $p \circ \widetilde{\sigma} \simeq p \circ \widetilde{\tau}$.

Conversely, suppose that $p \circ \widetilde{\sigma} \simeq p \circ \widetilde{\tau}$. By the homotopy lifting theorem we know that a homotopy $F : p \circ \widetilde{\sigma} \simeq p \circ \widetilde{\tau}$ can be lifted to a homotopy $\widetilde{F}$ from $\widetilde{\sigma}$ to a path $\widetilde{\omega}$. Since $\widetilde{\omega}$ and $\widetilde{\tau}$ are both paths from $\tilde{x}_0$ which are liftings of $p \circ \widetilde{\tau}$, it follows from Corollary 13.91 that $\widetilde{\omega} = \widetilde{\tau}$, showing that $\widetilde{\sigma} \simeq \widetilde{\tau}$. □

13.93 Corollary If $p : (\widetilde{X},\tilde{x}_0) \to (X,x_0)$ is a covering projection, then the induced homomorphism $p_* : \pi_1(\widetilde{X},\tilde{x}_0) \to \pi_1(X,x_0)$ is injective.

13.94 Let $p : \widetilde{X} \to X$ be a covering projection, and let σ be a path in X from x_0 to x_1. For each point $\tilde{x}_0 \in p^{-1}(x_0)$, there is by Corollary 13.91 a unique lifting of σ

to a path $\widetilde{\sigma}_{\tilde{x}_0}$ in $\widetilde{X}$ from $\tilde{x}_0$, and it follows from Proposition 13.92 that $\widetilde{\sigma}_{\tilde{x}_0}(1)$ only depends on the path class of σ. Hence we have a map

$$f_{[\sigma]} : p^{-1}(x_0) \to p^{-1}(x_1)$$

defined by $f_{[\sigma]}(\tilde{x}_0) = \widetilde{\sigma}_{\tilde{x}_0}(1)$, which is called the *translation* over σ.

We thus obtain a contravariant functor $F : \mathscr{P}(X) \to \mathscr{S}$ from the path category $\mathscr{P}(X)$ to the category $\mathscr{S}$ of sets, which assigns to each point x_0 the fibre $p^{-1}(x_0)$ and to to each morphism $[\sigma] : x_1 \to x_0$ the translation $f_{[\sigma]} : p^{-1}(x_0) \to p^{-1}(x_1)$.

From Proposition 13.69 it follows that every $f_{[\sigma]}$ is a bijection. So if X is pathwise connected, every fibre has the same number of points, called the *multiplicity* of the covering p. A covering is also said to be n*-fold* if it has multiplicity n.

13.95 Proposition Let $p : (\widetilde{X}, \tilde{x}_0) \to (X, x_0)$ be a covering projection, and let $\widetilde{\sigma}$ and $\widetilde{\tau}$ be paths in $\widetilde{X}$ from $\tilde{x}_0$ which are liftings of the paths σ and τ in X from x_0, respectively. Then we have that $\widetilde{\sigma}(1) = \widetilde{\tau}(1)$ if and only if $\sigma(1) = \tau(1)$ and

$$[\sigma * \tau^{-1}] \in p_* \pi_1(\widetilde{X}, \tilde{x}_0). \tag{1}$$

PROOF: If $\widetilde{\sigma}(1) = \widetilde{\tau}(1)$, then $\sigma(1) = \tau(1)$, and $\widetilde{\sigma} * \widetilde{\tau}^{-1}$ is a well-defined closed path at $\tilde{x}_0$ so that $[\sigma * \tau^{-1}] = p_*[\widetilde{\sigma} * \widetilde{\tau}^{-1}] \in p_* \pi_1(\widetilde{X}, \tilde{x}_0)$.

Conversely, suppose that $\sigma(1) = \tau(1)$ and that (1) is satisfied. Then there exists a path class $[\widetilde{\omega}] \in \pi_1(\widetilde{X}, \tilde{x}_0)$ with

$$[\sigma] * [\tau]^{-1} = [p \circ \widetilde{\omega}]$$

from which it follows that

$$[p \circ \widetilde{\sigma}] = [\sigma] = [p \circ \widetilde{\omega}] * [\tau] = [p \circ \widetilde{\omega}] * [p \circ \widetilde{\tau}] = [p \circ (\widetilde{\omega} * \widetilde{\tau})] .$$

This shows that $\widetilde{\sigma}$ and $\widetilde{\omega} * \widetilde{\tau}$ are paths in $\widetilde{X}$ from $\tilde{x}_0$ with $p \circ \widetilde{\sigma} \simeq p \circ (\widetilde{\omega} * \widetilde{\tau})$, so that $\widetilde{\sigma} \simeq \widetilde{\omega} * \widetilde{\tau}$ by Proposition 13.92. In particular, we have that $\widetilde{\sigma}(1) = \widetilde{\tau}(1)$, which completes the proof of the proposition. □

13.96 Definition Two coverings $p_1 : (\widetilde{X}_1, \tilde{x}_1) \to (X, x)$ and $p_2 : (\widetilde{X}_2, \tilde{x}_2) \to (X, x)$ of the pointed topological space (X, x) are said to be *equivalent* if there is a homeomorphism $\phi : (\widetilde{X}_1, \tilde{x}_1) \to (\widetilde{X}_2, \tilde{x}_2)$ with $p_2 \circ \phi = p_1$.

13.97 Remark Let X be a topological space which is pathwise connected and locally pathwise connected, and let $p : (\widetilde{X}, \tilde{x}_0) \to (X, x_0)$ be a covering with a pathwise connected covering space $\widetilde{X}$. If Σ is the set of all paths in X from x_0 and σ_0 is the constant path at x_0, then we have a surjective map $\phi : (\Sigma, \sigma_0) \to (\widetilde{X}, \tilde{x}_0)$ defined by $\phi(\sigma) = \widetilde{\sigma}(1)$, where $\widetilde{\sigma}$ is the unique path in $\widetilde{X}$ from $\tilde{x}_0$ which is a lifting of σ.

By Proposition 13.95 it follows that $\phi(\sigma) = \phi(\tau)$ if and only if $\sigma(1) = \tau(1)$ and $[\sigma * \tau^{-1}] \in p_* \pi_1(\widetilde{X}, \tilde{x}_0)$, in which case we write $\sigma \sim \tau$. Then $\sim$ is clearly an equivalence relation in Σ, and if $\pi : \Sigma \to \Sigma/\sim$ is the canonical projection, we have an induced bijection $\overline{\phi} : \Sigma/\sim \to \widetilde{X}$ so that $\overline{\phi} \circ \pi = \phi$. If $<\sigma>$ denotes the equivalence class of $\sigma \in \Sigma$ and $p' = p \circ \overline{\phi}$, we have that $p'(<\sigma>) = \sigma(1)$.

There is a unique topology τ on $\Sigma/\sim$ such that $\overline{\phi}$ is a homeomorphism. If $\sigma \in \Sigma$, there is a basis for neighbourhoods of $\widetilde{\sigma}(1)$ in $\widetilde{X}$ consisting of open pathwise connected neighbourhoods $\widetilde{U}$ which are mapped homeomorphically by p onto open pathwise connected neighbourhoods U of $\sigma(1)$ in X, since p is a local homeomorphism. Then $\overline{\phi}^{-1}(\widetilde{U})$ consists of all equivalence classes in $\Sigma/\sim$ having a representative of the form $\sigma * \sigma_1$, where σ_1 is a path in U from $\sigma(1)$. The sets $\overline{\phi}^{-1}(\widetilde{U})$, also denoted by $<\sigma, U>$, form a basis for neighbourhoods of $<\sigma>$ in the topology τ. Using this topology, we have that $p' : (\Sigma/\sim, <\sigma_0>) \to (X, x_0)$ is a covering of (X, x_0) which is equivalent to $p : (\widetilde{X}, \tilde{x}_0) \to (X, x_0)$.

This shows that the pointed topological space $(\widetilde{X}, \tilde{x}_0)$ and the projection p can be described up to equivalence by means of (X, x_0) and the subgroup $H = p_* \pi_1(\widetilde{X}, \tilde{x}_0)$ of $\pi_1(X, x_0)$. We also say that the covering $p : (\widetilde{X}, \tilde{x}_0) \to (X, x_0)$ *belongs to H*.

13.98 Let $p : (\widetilde{X}, \tilde{x}_0) \to (X, x_0)$ be a covering projection, and consider the operation of the fundamental group $\pi_1(X, x_0)$ on the fibre $p^{-1}(x_0)$ given by

$$[\sigma]\tilde{x} = f_{[\sigma]^{-1}}(\tilde{x})$$

for $[\sigma] \in \pi_1(X, x_0)$ and $\tilde{x} \in p^{-1}(x_0)$, where $f_{[\sigma]^{-1}}$ is translation over the path σ^{-1} as defined in 13.94.

The isotropy group of $\tilde{x}_0$ is $p_* \pi_1(\widetilde{X}, \tilde{x}_0)$, and if the covering space $\widetilde{X}$ is pathwise connected, the orbit of $\tilde{x}_0$ is the entire fibre $p^{-1}(x_0)$. Then it follows from Proposition 13.18 that the map $\phi : \pi_1(X, x_0) \to p^{-1}(x_0)$ given by $\phi([\sigma]) = [\sigma]\tilde{x}_0$ induces a bijection $\overline{\phi} : \pi_1(X, x_0) / p_* \pi_1(\widetilde{X}, \tilde{x}_0) \to p^{-1}(x_0)$.

13.99 Proposition A covering projection $p : \widetilde{X} \to X$ with a pathwise connected covering space $\widetilde{X}$ is a homeomorphism if and only if $p_* \pi_1(\widetilde{X}, \tilde{x}_0) = \pi_1(X, p(\tilde{x}_0))$ for some point $\tilde{x}_0 \in \widetilde{X}$.

PROOF : If p satisfies the last assertion of the proposition, then it follows from 13.94

and 13.98 that p is injective. Since every covering projection is surjective, continuous and open by Remark 13.80, Definition 13.79 and Propositions 13.85 and 13.86, it follows that p is a homeomorphism.

Assuming conversely that p is a homeomorphism, then the last assertion of the proposition follows from 13.70 for any point $\tilde{x}_0 \in \widetilde{X}$. □

13.100 The Lifting Theorem Let $p\colon (\widetilde{X},\tilde{x}_0) \to (X,x_0)$ be a covering projection, and let Y be a topological space which is pathwise connected and locally pathwise connected. Then a continuous map $f : (Y,y_0) \to (X,x_0)$ can be lifted to a map $\tilde{f} : (Y,y_0) \to (\widetilde{X},\tilde{x}_0)$ if and only if $f_* \pi_1(Y,y_0) \subset p_* \pi_1(\widetilde{X},\tilde{x}_0)$.

PROOF: If there is such a base point preserving lifting $\tilde{f}$ with $p \circ \tilde{f} = f$, then we obtain $p_* \circ \tilde{f}_* = f_*$ by using the canonical functor into the homotopy category of pointed topological spaces followed by the fundamental group functor, and this implies that $f_* \pi_1(Y,y_0) \subset p_* \pi_1(\widetilde{X},\tilde{x}_0)$.

Conversely, supposing that the last inclusion is satisfied, we must show the existence of a lifting $\tilde{f}$ of f preserving base points. Let $y \in Y$, and choose a path σ in Y from y_0 to y, using the fact that Y is pathwise connected. Then $f \circ \sigma$ is a path in X from x_0 which by Corollary 13.91 can be lifted to a unique path $\widetilde{\sigma}$ in $\widetilde{X}$ from $\tilde{x}_0$, and we define $\tilde{f}(y) = \widetilde{\sigma}(1)$.

For $\tilde{f}$ to be well defined, we must show that $\tilde{f}(y)$ does not depend on the choice of the path σ. So let τ be another path in Y from y_0 to y, and let $\widetilde{\tau}$ be the unique path in $\widetilde{X}$ from $\tilde{x}_0$ which is a lifting of $f \circ \tau$. Then $f \circ \sigma(1) = f(y) = f \circ \tau(1)$ and

$$[(f \circ \sigma) * (f \circ \tau)^{-1}] = f_*([\sigma * \tau^{-1}]) \in f_* \pi_1(Y,y_0) \subset p_* \pi_1(\widetilde{X},\tilde{x}_0)\,,$$

so that $\widetilde{\sigma}(1) = \widetilde{\tau}(1)$ by Proposition 13.95.

It follows immediately from the definition that $p \circ \tilde{f} = f$ and $\tilde{f}(y_0) = \tilde{x}_0$, so it only remains to show that $\tilde{f}$ is continuous. Let $\widetilde{U}$ be an arbitrary neighbourhood of $\tilde{f}(y)$ in $\widetilde{X}$. Since p is a local homeomorphism, $\widetilde{U}$ contains an open neighbourhood $\widetilde{V}$ of $\tilde{f}(y)$ which is mapped homeomorphically by p onto an open neighborhood V of $f(y)$. We let $s : V \to \widetilde{V}$ be the inverse homeomorphism and $i : \widetilde{V} \to \widetilde{X}$ be the inclusion map. Since Y is locally pathwise connected, there is a pathwise connected neighbourhood W of y with $f(W) \subset V$, and we will show that $\tilde{f}(W) \subset \widetilde{V}$.

Let $y' \in W$, and choose a path ρ in W from y to y'. Then $\widetilde{\rho} = i \circ s \circ f \circ \rho$ is the unique path in $\widetilde{X}$ from $\tilde{f}(y)$ which is a lifting of $f \circ \rho$. Now $\sigma * \rho$ is a path in Y from y_0 to y', and $\widetilde{\sigma} * \widetilde{\rho}$ is the unique path in $\widetilde{X}$ from $\tilde{x}_0$ which is a lifting of $f \circ (\sigma * \rho)$.

Hence we have that $\tilde{f}(y') = \tilde{\sigma} * \tilde{\rho}(1) = \tilde{\rho}(1) \in \tilde{V}$, which shows that $\tilde{f}(W) \subset \tilde{V} \subset \tilde{U}$ and completes the proof that $\tilde{f}$ is continuous. □

13.101 Corollary Let X be a topological space which is pathwise connected and locally pathwise connected. Then two coverings $p_1 : (\tilde{X}_1, \tilde{x}_1) \to (X, x)$ and $p_2 : (\tilde{X}_2, \tilde{x}_2) \to (X, x)$ with pathwise connected covering spaces $\tilde{X}_1$ and $\tilde{X}_2$ are equivalent if $p_{1*}\, \pi_1(\tilde{X}_1, \tilde{x}_1) = p_{2*}\, \pi_1(\tilde{X}_2, \tilde{x}_2)$.

13.102 Definition A topological space X is said to be *semi-locally simply connected* if every point $x_0 \in X$ has a neighbourhood V such that $i_* : \pi_1(V, x_0) \to \pi_1(X, x_0)$ is trivial, where $i : (V, x_0) \to (X, x_0)$ is the inclusion map. This means that every closed path σ in V at x_0 is null homotopic in X. A neighbourhood V of x_0 having this property and which in addition is open and pathwise connected, is called a *pathwise simple* neighbourhood of x_0.

13.103 Proposition If X is a topological space which is locally pathwise connected and semi-locally simply connected, then every point in X has a basis of pathwise simple neighbourhoods.

PROOF: Let $U \subset V$ be two neighbourhoods of x_0, and suppose that every closed path in V at x_0 is null homotopic in X. Then this holds in particular for every closed path in U at x_0. The proposition hence follows from Proposition 13.56. □

13.104 Theorem Let X be a topological space which is pathwise connected, locally pathwise connected and semi-locally simply connected. If $x_0 \in X$ and H is a subgroup of $\pi_1(X, x_0)$, then there is a covering $p : (\tilde{X}, \tilde{x}_0) \to (X, x_0)$ with a pathwise connected covering space $\tilde{X}$ such that $p_*\, \pi_1(\tilde{X}, \tilde{x}_0) = H$.

PROOF: Let Σ be the set of paths in X from x_0, and let σ_0 be the constant path at x_0. We say that two paths σ and τ in Σ are equivalent, and write $\sigma \sim \tau$, if $\sigma(1) = \tau(1)$ and $[\sigma * \tau^{-1}] \in H$. Then $\sim$ is an equivalence relation in Σ, and we denote the equivalence class of σ by $<\sigma>$. Let $\tilde{X} = \Sigma/\sim$ be the quotient set and $\tilde{x}_0 = <\sigma_0>$ be the equivalence class of the constant path at x_0. We define the map $p : (\tilde{X}, \tilde{x}_0) \to (X, x_0)$ by $p(<\sigma>) = \sigma(1)$.

If $\sigma \in \Sigma$ and U is an open neighbourhood of $\sigma(1)$, we let $<\sigma, U>$ be the set of all equivalence classes in $\tilde{X}$ having a representative of the form $\sigma * \sigma_1$, where σ_1 is a path in U from $\sigma(1)$. We will show that the sets $<\sigma, U>$ form a basis for a topology τ on $\tilde{X}$.

If $<\tau> \in <\sigma, U>$, then $\tau = \sigma * \sigma_1$ for a path σ_1 in U from $\sigma(1)$. For every path τ_1 in U from $\tau(1)$ we then have that $\tau * \tau_1 = \sigma * (\sigma_1 * \tau_1)$, showing that

$<\tau,U> \subset <\sigma,U>$. Since $\sigma = \tau * \sigma_1{}^{-1}$, it also follows that $<\sigma> \in <\tau,U>$ so that $<\sigma,U> \subset <\tau,U>$, thus showing that $<\tau,U> = <\sigma,U>$.

Therefore, if $<\omega> \in <\sigma,U> \cap <\tau,V>$, we have that $<\omega,U\cap V> \subset <\sigma,U> \cap <\tau,V>$, which shows that the sets $<\sigma,U>$ form a basis for a topology on $\widetilde{X}$.

Giving $\widetilde{X}$ this topology τ, the map p is continuous since $<\sigma,U>$ is an open neighbourhood of $<\sigma>$ with $p\,(<\sigma,U>) \subset U$ for each open neighbourhood U of $p\,(<\sigma>)$. The map p is also open since $p\,(<\sigma,U>)$ is the path component of U containing $p\,(<\sigma>)$, which is open since X is locally pathwise connected. In particular, we have that $p\,(<\sigma,U>) = U$ for each open pathwise connected neighbourhood U of $p\,(<\sigma>)$.

We next show that p maps $<\sigma,U>$ homeomorphically onto U when U is a pathwise simple neighbourhood of $p\,(<\sigma>)$. We already know that p is continuous and open and that p maps $<\sigma,U>$ onto U, so it only remains to prove that the restriction of p to $<\sigma,U>$ is injective. Let σ_1 and σ_2 be two paths in U from $\sigma(1)$ such that $p\,(<\sigma*\sigma_1>) = p\,(<\sigma*\sigma_2>)$. Then $\sigma_1(1) = \sigma_2(1)$ so that $\sigma_1 * \sigma_2{}^{-1}$ is a closed path in U at $\sigma(1)$ which must be null homotopic since the neighbourhood U is semi-locally simply connected. Hence it follows that

$$[(\sigma*\sigma_1)*(\sigma*\sigma_2)^{-1}] = [\sigma]*[\sigma_1*\sigma_2{}^{-1}]*[\sigma^{-1}] = [\sigma_0] \in H$$

which implies that $<\sigma*\sigma_1> = <\sigma*\sigma_2>$.

Using this, we can now show that every pathwise simple neighbourhood U of a point $x \in X$ is evenly covered by p. We know that $p^{-1}(U)$ is the union of the open sets $<\sigma,U>$ where $<\sigma> \in p^{-1}(U)$, and these sets are either equal or disjoint. Indeed, if $<\sigma,U>$ and $<\tau,U>$ are not disjoint and contain a common element $<\omega>$, we have that $<\omega,U> = <\sigma,U>$ and $<\omega,U> = <\tau,U>$ showing that they are equal. Hence U is evenly covered by p, and it follows from Proposition 13.103 that p is a covering projection.

Finally, we need to verify that $p_*\pi_1(\widetilde{X},\tilde{x}_0) = H$ and that $\widetilde{X}$ is pathwise connected. If $\sigma \in \Sigma$, we let $\widetilde{\sigma} : I \to \widetilde{X}$ be the map defined by $\widetilde{\sigma}(t) = <\sigma_t>$, where $\sigma_t \in \Sigma$ is the path given by $\sigma_t(s) = \sigma(st)$ for $s \in I$. We want to show that $\widetilde{\sigma}$ is the unique path in $\widetilde{X}$ from $\tilde{x}_0$ which is a lifting of σ, and that $\widetilde{\sigma}(1) = <\sigma>$. The last assertion follows immediately from the definition of $\widetilde{\sigma}$. Since $\widetilde{\sigma}(0) = <\sigma_0> = \tilde{x}_0$ and $p \circ \widetilde{\sigma}(t) = \sigma_t(1) = \sigma(t)$ for $t \in I$, it only remains to show that $\widetilde{\sigma}$ is continuous, as the uniqueness follows from Corollary 13.91.

If $\widetilde{\sigma}(t_0) \in <\tau,U>$, then $<\sigma_{t_0},U> = <\tau,U>$ and $\sigma(t_0) = p \circ \widetilde{\sigma}(t_0) \in U$. Choosing a convex open subset J of I containing t_0 with $\sigma(J) \subset U$, we contend that $\widetilde{\sigma}(J) \subset <\tau,U>$. If fact, if $t \in J$ and ω is the path in U defined by $\omega(s) = \sigma((1-s)t_0 + st)$ for $s \in I$, we have a homotopy $H : \sigma_t \simeq \sigma_{t_0} * \omega$ given by

$$H(s,u) = \begin{cases} \sigma(st + us(2t_0 - t)) & \text{for } s \in [0,1/2] \\ \sigma(st + u(1-s)(2t_0 - t)) & \text{for } s \in [1/2,1] \end{cases}$$

so that

$$\widetilde{\sigma}(t) = <\sigma_t> = <\sigma_{t_0} * \omega> \in <\sigma_{t_0}, U> = <\tau, U>,$$

thus completing the proof that $\widetilde{\sigma}$ is continuous.

From this it now follows that $p_* \pi_1(\widetilde{X}, \tilde{x}_0) = H$, since $[\sigma] \in H \Leftrightarrow \sigma \sim \sigma_0 \Leftrightarrow <\sigma> = \tilde{x}_0 \Leftrightarrow \widetilde{\sigma}$ is a closed path in $\widetilde{X}$ at $\tilde{x}_0 \Leftrightarrow [\widetilde{\sigma}] \in \pi_1(\widetilde{X}, \tilde{x}_0)$ $\Leftrightarrow [\sigma] = p_*([\widetilde{\sigma}]) \in p_* \pi_1(\widetilde{X}, \tilde{x}_0)$, where we have used Corollary 13.93 to obtain the last equivalence. It also follows that $\widetilde{X}$ is pathwise connected, since there is a path $\widetilde{\sigma}$ in $\widetilde{X}$ connecting $\tilde{x}_0$ to $<\sigma>$ for every element $<\sigma>$ in $\widetilde{X}$. □

TOPOLOGICAL GROUPS

13.105 Definition A *topological group* G is a group which is at the same time a topological space such that the maps $\mu : G \times G \to G$ and $\nu : G \to G$ given by the group operations $\mu(g,h) = gh$ and $\nu(g) = g^{-1}$ are continuous. The unit element of G is denoted by e.

13.106 Remark Continuity of μ and ν is equivalent to assuming that the map $\rho : G \times G \to G$ given by $\rho(g,h) = gh^{-1}$ is continuous.

13.107 Definition If G is a topological group, we let $\Omega(G,e)$ denote the set of closed paths in G at e. We define the *product* of two paths $\omega, \theta \in \Omega(G,e)$ to be the path $\omega\,\theta$ given by $\omega\,\theta\,(t) = \omega(t)\,\theta(t)$ for $t \in I$. This defines a group structure on $\Omega(G,e)$. The unit element is the constant path $<e>$ at e.

13.108 Proposition Let G be a topological group, and let $\omega_0, \omega_1, \theta_0, \theta_1 \in \Omega(G,e)$. If $\omega_0 \simeq \omega_1$ and $\theta_0 \simeq \theta_1$, then $\omega_0\,\theta_0 \simeq \omega_1\,\theta_1$.

PROOF: If $F : \omega_0 \simeq \omega_1$ and $G : \theta_0 \simeq \theta_1$ are homotopies, we have a homotopy $H : \omega_0\,\theta_0 \simeq \omega_1\,\theta_1$ given by $H(s,t) = F(s,t)\,G(s,t)$ for $s,t \in I$. □

13.109 Proposition If G is a topological group, the canonical projection $\phi : \Omega(G,e) \to \pi_1(G,e)$ given by $\phi(\omega) = [\omega]$ is a group homomorphism.

PROOF: If $\omega, \theta \in \Omega(G,e)$, we have that $\omega \simeq \omega * <e>$ and $\theta \simeq <e> * \theta$. Hence it follows from Proposition 13.108 that

$$\omega\,\theta \simeq (\omega * <e>)(<e> * \theta) = \omega * \theta$$

which shows that $[\omega\,\theta] = [\omega] * [\theta]$. □

13.110 Proposition If G is a topological group, then $\pi_1(G,e)$ is abelian.

PROOF : If $\omega, \theta \in \Omega(G,e)$, we have that $\omega * <e> \simeq \omega \simeq <e> * \omega$ and $<e> * \theta \simeq \theta \simeq \theta * <e>$. Hence it follows from Proposition 13.108 that

$$\omega * \theta = (\omega * <e>)(<e> * \theta) \simeq (<e> * \omega)(\theta * <e>) = \theta * \omega .$$

□

13.111 Proposition Let $\phi : G \to H$ be a continuous homomorphism between the topological groups G and H. Then ϕ is a covering projection if and only if it is an open surjection with a discrete kernel.

PROOF : If ϕ is a covering projection, then it follows from Remark 13.80 and Propositions 13.85 and 13.86 that ϕ is an open surjection with a discrete kernel.

Conversely, suppose that ϕ satisfies these properties, and let $K = ker\,\phi$ and W be an open neighbourhood of e in G with $W \cap K = \{e\}$. Since the map $\rho : G \times G \to G$ given by $\rho(g,h) = g^{-1}h$ is continuous, there is an open neighbourhood V of e in G such that $V^{-1}V \subset W$. Then $U = \phi(V)$ is an open neighbourhood of e in H which we claim is evenly covered by ϕ.

We first show that

$$\phi^{-1}(U) = \bigcup_{k \in K} Vk ,$$

and that ϕ maps Vk onto U for each $k \in K$. The last assertion as well as the inclusion $\supset$ follows from the fact that

$$\phi(Vk) = \{\phi(vk) | v \in V\} = \{\phi(v) | v \in V\} = \phi(V) = U$$

for every $k \in K$. To prove the inclusion $\subset$, let $g \in \phi^{-1}(U)$, and choose a $v \in V$ such that $\phi(v) = \phi(g)$. Then $k = v^{-1}g \in K$ and $g = vk \in Vk$.

We next show that $\phi|_{Vk}$ is injective for each $k \in K$. Suppose that $\phi(v_1 k) = \phi(v_2 k)$ where $v_1, v_2 \in V$. Then $\phi(v_1) = \phi(v_2)$ so that

$$v_2^{-1} v_1 \in V^{-1}V \cap K \subset W \cap K = \{e\} ,$$

which implies that $v_1 = v_2$.

Finally, we show that the sets Vk are disjoint for $k \in K$. If $g \in Vk_1 \cap Vk_2$ where $k_1, k_2 \in K$, then $g = v_1 k_1 = v_2 k_2$ for $v_1, v_2 \in V$ so that

$$v_2^{-1} v_1 = k_2 k_1^{-1} \in V^{-1}V \cap K \subset W \cap K = \{e\} ,$$

which implies that $k_1 = k_2$. This shows that $Vk_1 \cap Vk_2 = \emptyset$ when $k_1 \neq k_2$ and completes the proof that U is evenly covered by ϕ.

Now let $h \in H$. As ϕ is a surjection, there is a $g \in G$ with $\phi(g) = h$. Since the left translations L_g and L_h in G and H, respectively, are homeomorphisms such that $L_h \circ \phi \circ L_g^{-1} = \phi$, it follows that $h\,U$ is an open neighbourhood of h which is evenly covered by ϕ. Indeed, the inverse image $\phi^{-1}(h\,U)$ is the disjoint union of the open sets $g\,Vk = L_g(Vk)$ which are mapped homeomorphically onto $h\,U = L_h(U)$ by ϕ for $k \in K$. This completes the proof that ϕ is a covering projection. □

13.112 Proposition Let U be an open neighbourhood of the unit element e in a connected topological group G. Then U generates G, i.e., we have that

$$G = \bigcup_{n=1}^{\infty} U^n \,,$$

where U^n denotes the set of all n-fold products of elements in U.

PROOF: Let $V = U \cap U^{-1}$, where $U^{-1} = \{g^{-1} | g \in U\}$. Then V is an open neighbourhood of e contained in U with $V = V^{-1}$, and the set

$$H = \bigcup_{n=1}^{\infty} V^n \subset \bigcup_{n=1}^{\infty} U^n$$

is a subgroup of G which is open, since $hV \subset H$ for every $h \in H$. From this it follows that each left coset of H in G is open. As G is connected, this implies that $H = G$, which completes the proof of the proposition. □

TOPOLOGICAL VECTOR SPACES

13.113 Definition A *topological vector space* V is a vector space which is at the same time a topological space such that the maps $\mu : V \times V \to V$ and $\rho : \mathbf{R} \times V \to V$ given by the vector space operations $\mu(u,v) = u + v$ and $\rho(k,v) = kv$ are continuous.

13.114 Definition A subset C of a vector space V is said to be *convex* if $x, y \in C$ implies that $x + t(y - x) \in C$ for $0 \leq t \leq 1$. A set $B \subset V$ is said to be *balanced* if $x \in B$ and $|t| \leq 1$ implies that $tx \in B$.

13.115 Proposition A topological vector space V has a basis of balanced open neighbourhoods of 0.

PROOF: Let U be a neighbourhood of 0 in V. By the continuity of scalar multiplication there is a real number $r > 0$ and an open neighbourhood W of 0 such that $(-r,r)\,W \subset U$. We have that $(-r,r)\,W$ is a balanced open neighbourhood of 0 since

$$(-r,r)\,W = \bigcup_{0 \,<\, |t| \,<\, r} tW \,.$$

□

13.116 Definition If M is a Banach space, we let $B_r(x_0)$ and $\overline{B}_r(x_0)$ denote, respectively, the open and closed ball of radius r centered at x_0, i.e.,

$$B_r(x_0) = \{x \in M | \, \|x - x_0\| < r\} \quad \text{and} \quad \overline{B}_r(x_0) = \{x \in M | \, \|x - x_0\| \leq r\} .$$

We let $S_r(x_0)$ denote the sphere of radius r centered at x_0, i.e.,

$$S_r(x_0) = \{x \in M | \, \|x - x_0\| = r\} .$$

13.117 Proposition A finite dimensional vector space V has a unique Hausdorff topology compatible with the vector space operations. With this topology, every linear isomorphism $x : V \to \mathbf{R}^n$, where $n = \dim(V)$, is a homeomorphism.

PROOF : If $x : V \to \mathbf{R}^n$ is a linear isomorphism, then there is a unique topology $\mathscr{T}$ on V so that x is a homeomorphism, consisting of the subsets O of V such that $x(O)$ is open in $\mathbf{R}^n$. This is clearly a Hausdorff topology on V compatible with the vector space operations. Is does not depend on the choice of x, for if $y : V \to \mathbf{R}^n$ is another linear isomorphism, then $y = (y \circ x^{-1}) \circ x$ is the composition of the homeomorphisms x and $y \circ x^{-1}$. This completes the proof of the existence part of the proposition.

To show the uniqueness part, suppose that V is given a Hausdorff topology $\mathscr{T}'$ compatible with the vector space operations, and let $x : V \to \mathbf{R}^n$ be any linear isomorphism. Let $\mathscr{E} = \{e_1, ..., e_n\}$ be the standard basis for $\mathbf{R}^n$ and $\mathscr{B} = \{v_1, ..., v_n\}$ be a basis for V so that $x(v_i) = e_i$ for $i = 1, ..., n$. Then we have that

$$x^{-1}(a) = \sum_{j=1}^{n} a^j v_j$$

for $a \in \mathbf{R}^n$, showing that $x^{-1} : \mathbf{R}^n \to V$ is continuous.

We can also prove that x is continuous. Let $\varepsilon > 0$. As the sphere $S_\varepsilon(0)$ in $\mathbf{R}^n$ is compact, it follows that $x^{-1}(S_\varepsilon(0))$ is compact and therefore closed in V since V is Hausdorff. By Proposition 13.115 there is a balanced open neighbourhood U of 0 in V with $U \cap x^{-1}(S_\varepsilon(0)) = \emptyset$. Then $x(U)$ is a balanced open neighbourhood of 0 in $\mathbf{R}^n$ such that $x(U) \cap S_\varepsilon(0) = \emptyset$, which shows that $x(U) \subset B_\varepsilon(0)$. From this it now follows that x is continuous, since $a - b \in U$ implies that $x(a) - x(b) = x(a - b) \in B_\varepsilon(0)$, thus showing that $\mathscr{T}' = \mathscr{T}$ and completing the proof of the uniqueness part of the proposition. □

13.118 Definition A linear map $F : X \to Y$ between the normed spaces X and Y is said to be *bounded* if there is a real number $M \geq 0$ such that

$$\|F(v)\| \leq M \|v\| \tag{1}$$

for every $v \in X$.

13.119 Proposition Let $F : X \to Y$ be a linear map between the normed spaces X and Y. Then the following assertions are equivalent

(1) F is continuous at 0.

(2) F is uniformly continuous.

(3) F is bounded.

PROOF : The equivalence of (1) and (2) follows since $\|F(v)-F(w)\| = \|F(v-w)\|$ by the linearity of F.

Suppose that F is bounded, and choose a real number $M > 0$ so that formula (1) in Definition 13.118 is satisfied for every $v \in X$. Then F is continuous at 0 since $\|F(v)\| < \varepsilon$ whenever $\|v\| < \varepsilon/M$.

Suppose conversely that F is continuous at 0, and choose a $\delta > 0$ so that $\|F(v)\| < 1$ when $\|v\| < \delta$. Then formula (1) in Definition 13.118 is satisfied with any $M > 1/\delta$, since $\|F(v/M\|v\|)\| < 1$ for all $v \neq 0$. The formula is clearly also satisfied for $v = 0$, thus showing that F is bounded. □

13.120 Proposition Let X and Y be normed spaces. Then the set $L(X,Y)$ of bounded linear maps from X to Y is a normed space with the norm defined by

$$\|F\| = \sup\{\|F(v)\| \mid \|v\| \leq 1\} \tag{1}$$

for $F \in L(X,Y)$. We have that

$$\|F(v)\| \leq \|F\|\,\|v\| \tag{2}$$

for every $F \in L(X,Y)$ and $v \in X$, so that $\|F\|$ is the least real number $M \geq 0$ satisfying inequality (1) in Definition 13.118 for every $v \in X$. If Y is a Banach space, so is $L(X,Y)$.

PROOF : If $F, G \in L(X,Y)$, then $F+G \in L(X,Y)$ and

$$\|F+G\| \leq \|F\| + \|G\|$$

by the triangle inequality in Y which implies that

$$\|(F+G)(v)\| = \|F(v)+G(v)\| \leq \|F(v)\| + \|G(v)\| \leq \|F\| + \|G\|$$

for every $v \in X$ with $\|v\| \leq 1$. Furthermore, if $F \in L(X,Y)$ and a is a scalar, we have that $aF \in L(X,Y)$ and

$$\|aF\| = |a|\|F\|$$

since

$$\|(aF)(v)\| = \|aF(v)\| = |a|\|F(v)\|$$

for every $v \in X$ with $\|v\| \leq 1$. Finally, if $F \neq 0$, then $F(v) \neq 0$ for some $v \in X$ with $\|v\| \leq 1$, thus showing that $\|F\| > 0$. This completes the proof that $L(X,Y)$ is a vector space, and that (1) defines a norm in $L(X,Y)$.

Inequality (2) is clearly true when $v = 0$, and it follows from

$$\frac{1}{\|v\|}\|F(v)\| = \left\| F\left(\frac{v}{\|v\|}\right) \right\| \leq \|F\|$$

when $v \neq 0$.

Now suppose that Y is a Banach space, and let $\{F_i\}_{i=1}^{\infty}$ be a Cauchy sequence in $L(X,Y)$. Since

$$\|F_i(v) - F_j(v)\| \leq \|F_i - F_j\| \, \|v\|$$

for every positive integer i and j, it follows that $\{F_i(v)\}_{i=1}^{\infty}$ is a Cauchy sequence in Y which must converge to a vector $F(v) \in Y$ for every $v \in X$. As each F_i is linear, the same is true for F.

For each $\varepsilon > 0$ there is an N so that $\|F_i - F_j\| < \varepsilon$ for $i, j \geq N$. Hence we have that

$$\|F_i(v) - F_j(v)\| \leq \varepsilon \, \|v\|$$

for $v \in X$ and $i, j \geq N$, and letting $j \to \infty$ we obtain

$$\|F_i(v) - F(v)\| \leq \varepsilon \, \|v\|$$

for $v \in X$ and $i \geq N$. This shows that $F_i - F \in L(X,Y)$ so that $F \in L(X,Y)$, and $\|F_i - F\| \leq \varepsilon$ for $i \geq N$ which shows that $\{F_i\}_{i=1}^{\infty}$ converges to F in the norm of $L(X,Y)$. □

13.121 Proposition Let X,Y and Z be normed spaces. Then the composition of the bounded linear maps $F \in L(X,Y)$ and $G \in L(Y,Z)$ is a bounded linear map $G \circ F \in L(X,Z)$ with

$$\|G \circ F\| \leq \|G\| \|F\| .$$

PROOF : Using inequality (2) in Proposition 13.120 we have that

$$\|G \circ F(v)\| \leq \|G\| \|F(v)\| \leq \|G\| \|F\| \|v\|$$

for every $v \in X$. □

13.122 Corollary Let X be a Banach space. Then $L(X,X)$ is a Banach algebra.

PROOF : Follows from Propositions 13.120 and 13.121. □

13.123 Proposition Let $\psi : X \to Y$ be an isometric isomorphism between the normed spaces X and Y. Then the map $\phi : L(X,X) \to L(Y,Y)$ given by $\phi(F) = \psi \circ F \circ \psi^{-1}$ for $F \in L(X,X)$, is an isometric isomorphism.

PROOF : We have that

$$\begin{aligned}
\|\phi(F)\| &= \sup\{\|\psi \circ F \circ \psi^{-1}(w)\| \mid w \in Y \text{ and } \|w\| \leq 1\} \\
&= \sup\{\|F \circ \psi^{-1}(w)\| \mid w \in Y \text{ and } \|\psi^{-1}(w)\| \leq 1\} \\
&= \sup\{\|F(v)\| \mid v \in X \text{ and } \|v\| \leq 1\} = \|F\|
\end{aligned}$$

for every $F \in L(X,X)$. □

13.124 Lemma Let A be an element in a Banach algebra X with $\|A\| < 1$. Then $I - A$ is invertible.

PROOF: The elements $S_n = \sum_{k=0}^{n} A^k$ form a Cauchy sequence in X since

$$\|S_m - S_n\| = \left\| \sum_{k=n+1}^{m} A^k \right\| \le \sum_{k=n+1}^{m} \|A\|^k$$

when $n < m$, and $\sum_{k=0}^{\infty} \|A\|^k$ is a convergent geometric series. As X is complete, the sequence $\{S_n\}_{n=0}^{\infty}$ must therefore converge to an element $B \in X$, and we have that

$$S_n (I - A) = (I - A) S_n = I - A^{n+1}$$

for each n. Letting $n \to \infty$ and using that $\|A^{n+1}\| \le \|A\|^{n+1}$, we thus obtain

$$B (I - A) = (I - A) B = I,$$

showing that $I - A$ is invertible with inverse B. □

13.125 Proposition The set U of invertible elements in a Banach algebra X is open.

PROOF: If $A \in U$, then the map $L_A : X \to X$ given by $L_A(B) = AB$ for $B \in X$, is a homeomorphism with inverse $L_{A^{-1}}$, mapping U onto U. Hence the open ball $B_1(I) = \{C \in X \,|\, \|C - I\| < 1\}$ is mapped by L_A onto an open neighbourhood of A which is contained in U by Lemma 13.124. □

13.126 Proposition Let X be a Banach algebra, and let U be the set of invertible elements in X. Then we have a continuous map $exp : X \to U$ given by

$$exp\,(A) = \sum_{k=0}^{\infty} \frac{A^k}{k!} \tag{1}$$

for $A \in X$. If $A, B \in X$ with $AB = BA$, then

$$exp\,(A + B) = exp\,(A)\; exp\,(B), \tag{2}$$

and we have that $exp\,(0) = I$ and $exp\,(-A) = exp\,(A)^{-1}$ for $A \in X$. If $A \in X$ and $B \in U$, then

$$B\; exp\,(A)\; B^{-1} = exp\,(BAB^{-1}). \tag{3}$$

PROOF: We see that the series in (1) converges since

$$\left\| \sum_{k=n+1}^{m} \frac{A^k}{k!} \right\| \le \sum_{k=n+1}^{m} \frac{\|A\|^k}{k!}$$

when $n < m$, using the fact that

$$e^{\|A\|} = \sum_{k=0}^{\infty} \frac{\|A\|^k}{k!}$$

is a convergent series for all $A \in X$. The map exp is also continuous, since

$$\left\| exp\,(A) - \sum_{k=0}^{n} \frac{A^k}{k!} \right\| = \left\| \sum_{k=n+1}^{\infty} \frac{A^k}{k!} \right\| \leq \sum_{k=n+1}^{\infty} \frac{a^k}{k!}$$

for every n when $\|A\| \leq a$, so that the series in (1) is uniformly convergent on every bounded subset of X.

If $A, B \in X$ with $AB = BA$, we have the usual Binomial formula

$$(A+B)^k = \sum_{r=0}^{k} \binom{k}{r} A^{k-r} B^r$$

which implies that

$$\frac{(A+B)^k}{k!} = \sum_{r=0}^{k} \frac{A^{k-r}}{(k-r)!} \frac{B^r}{r!}$$

so that

$$\left\| \sum_{k=0}^{2n} \frac{(A+B)^k}{k!} - \left(\sum_{k=0}^{n} \frac{A^k}{k!} \right) \left(\sum_{k=0}^{n} \frac{B^k}{k!} \right) \right\|$$
$$\leq \left(\sum_{k=n+1}^{2n} \frac{\|A\|^k}{k!} \right) \left(\sum_{k=0}^{2n} \frac{\|B\|^k}{k!} \right) + \left(\sum_{k=0}^{2n} \frac{\|A\|^k}{k!} \right) \left(\sum_{k=n+1}^{2n} \frac{\|B\|^k}{k!} \right)$$

for every n. As the last expression converges to 0 when $n \to \infty$, this completes the proof of formula (2).

Finally, formula (3) follows from the relation

$$B \left(\sum_{k=0}^{n} \frac{A^k}{k!} \right) B^{-1} = \sum_{k=0}^{n} \frac{(BAB^{-1})^k}{k!}$$

when $n \to \infty$. □

Bibliography

[1] Abraham, R. and Marsden, J. *Foundations of Mechanics*, Second edition. Benjamin-Cummings, New York. (1978)

[2] Apostol, T.M. *Mathematical Analysis*, Second edition. Addison Wesley, Reading, MA. (1974)

[3] Bishop, R.L. and Crittenden, R.J. *Geometry of Manifolds*. Chelsea Publishing, Providence, RI. (2001)

[4] Chevalley, C. *Theory of Lie Groups*. Princeton University Press, Princeton, NJ. (1946)

[5] Curtis, W.D. and Miller, F.R. *Differential Manifolds and Theoretical Physics*. Academic Press, Orlando, FL. (1985)

[6] Frankel, T. *Gravitational Curvature*. Freeman, San Francisco, CA. (1979)

[7] Goldstein, H. *Classical Mechanics*, Second edition. Addison Wesley, Reading, MA. (1980)

[8] Grøn, Ø. and Hervik, S. *Einstein's General Theory of Relativity*. Springer, New York. (2007)

[9] Kobayashi, S. and Nomizu, K. *Foundations of Differential Geometry*, Volume I. Interscience, New York. (1963)

[10] Kobayashi, S. and Nomizu, K. *Foundations of Differential Geometry*, Volume II. Interscience, New York. (1969)

[11] Lang, S. *Algebra*. Addison Wesley, Reading, MA. (1965)

[12] Lang, S. *Differential Manifolds*. Springer-Verlag, New York. (1985)

[13] Lang, S. *Linear Algebra*. Addison Wesley, Reading, MA. (1966)

[14] Lang, S. *Real Analysis*. Addison Wesley, Reading, MA. (1969)

[15] Mangiarotti, L. and Sardanashvily, G. *Connections in Classical and Quantum Field Theory*. World Scientific, Singapore. (2000)

[16] Matsushima, Y. *Smooth Manifolds*. Marcel Dekker, New York. (1972)

[17] Miller, W. *Symmetry Groups and Their Applications*. Academic Press, New York. (1972)

[18] Misner, C.W., Thorn, K.S. and Wheeler, J.A. *Gravitation*. Freeman, San Francisco, CA. (1973)

[19] O'Neill, B. *Semi-Riemannian Geometry*. Academic Press, San Diego, CA. (1983)

[20] Rudin, W. *Principles of Mathematical Analysis*, Second edition. McGraw-Hill, New York. (1964)

[21] Sauders, D.J. *The Geometry of Jet Bundles*. Cambridge University Press, Cambridge. (1989)

[22] Schubert, H. *Topology*. MacDonald & Co, London. (1968)

[23] Spanier, E.H. *Algebraic Topology*. McGraw-Hill, New York. (1966)

[24] Spivak, M. *Differential Geometry*, Volume I, Second edition. Publish or Perish, Houston, TX. (1979)

[25] Spivak, M. *Differential Geometry*, Volume II, Second edition. Publish or Perish, Houston, TX. (1979)

[26] Spivak, M. *Differential Geometry*, Volume III, Second edition. Publish or Perish, Houston, TX. (1979)

[27] Steenrod, N. *The Topology of Fibre Bundles*. Princeton University Press, Princeton, NJ. (1951)

[28] Varadarajan, V.S. *Lie Groups, Lie Algebras and Their Representations*. Prentice-Hall, Englewood Cliffs, NJ. (1974)

[29] Warner, F.W. *Foundations of Smooth Manifolds and Lie Groups*. Scott, Foresman and Company, Glenview, IL. (1971)

[30] Westenholz, C. von *Differential Forms in Mathematicel Physics*. North-Holland Publishing Company, Amsterdam. (1981)

Index

Milton Keynes UK
Ingram Content Group UK Ltd.
UKHW021933071024
449327UK00022B/1792

9 780367 658250